Brief Contents

This textbook is available in two versions:

Biology: Science for Life, Second Edition (0-13-148969-0) consists of **Chapters 1-15,** which provide non-majors biology students with a thorough overview of the foundations of modern biological science: cell biology, genetics, evolution, and ecology.

Biology: Science for Life with Physiology, Second Edition (0-13-225770-X), an expanded version of the text, contains **Chapters 1-23:** the fifteen chapters noted above, plus eight chapters on animal and plant anatomy and physiology.

BIOLOGY
Science for Life with Physiology
SECOND EDITION

Colleen Belk
University of Minnesota–Duluth

Virginia Borden
University of Minnesota–Duluth

PEARSON
Prentice Hall

Upper Saddle River, NJ 07458

Library of Congress Cataloging-in-Publication Data

Belk, Colleen M.
  Biology: science for life/Colleen Belk, Virginia Borden. — 2nd ed.
    p. cm.
  Includes bibliographical references.
  ISBN 0-13-225770-X (alk. paper)
  1. Biology. I. Borden, Virginia. II. Title.

QH307.2.B43 2007
570 — dc22

2005030024

Executive Editor: Teresa Ryu Chung
Development Editor: Elaine Page
Production Editor: Donna King
Senior Media Editor: Patrick Shriner
Assistant Editor: Andrew Sobel
Editor in Chief: Daniel Kaveney
Editor in Chief of Development: Carol Trueheart
Assistant Managing Editor: Beth Sweeten
Executive Managing Editor: Kathleen Schiaparelli
Art Development Editors: Kim Quillin, Jay McElroy, Jennifer Henke
Art Director: Jonathan Boylan
AV Production Manager: Ronda Whitson
AV Production Editor: Rhonda Aversa
Manufacturing Buyer: Alan Fischer
Director of Marketing, Science: Patrick Lynch
Managing Editor, Science Media: Nicole M. Jackson
Manufacturing Manager: Alexis Heydt-Long
Composition: Progressive Information Technologies
Assistant Managing Editor, Science Supplements: Karen Bosch
Director of Creative Services: Paul Belfanti
Creative Director: Juan López
Interior Designers: Joseph Sengotta, Jonathan Boylan
Senior Managing Editor, Audio and Visual Assets: Patricia Burns

Manager, Production Technologies: Matthew Haas
Managing Editor, Art Management: Abigail Bass
Art Studio: J. B. Woolsey
Director, Image Resource Center: Melinda Reo
Manager, Rights and Permissions: Zina Arabia
Interior Image Specialist: Beth Brenzel
Cover Image Specialist: Karen Senatar
Image Permission Coordinator: Debbie Latronica
Photo Researcher: Elaine Soares
Editorial Assistant: Gina Kayed
Cover Designer: Jonathan Boylan
Cover Photos: Rubberball Productions/Getty Images, Inc. (Single Tree); Blickwinkel/Alamy (Ermine Stoat); Jerry Young/Dorling Kindersley Media Library (Butterfly); Jerry Young/Dorling Kindersley Media Library (Finch and Gouldiae); Don W. Fawcett/Photo Researchers, Inc. (Human Sperm); Dennis Degnan/Corbis Bettmann (Twin Sisters); Colin Keates/Dorling Kindersley Media Library (Fossil); Photo 24/Brand X Pictures/Getty Images, Inc (Humming Bird); Photo 24/Brand X Pictures/Getty Images, inc (Peacock Feathers); Mike Hill/Taxi/Getty Images, Inc (Sea Nettles); Stevie Grand/Photo Researchers, Inc (MRI); Addenbrookes Hospital/Photo Researchers, Inc. (Chromosomes)

© 2007, 2004 Pearson Education, Inc.
Pearson Prentice Hall
Pearson Education, Inc.
Upper Saddle River, NJ 07458

Printed in the United States of America

10 9 8 7 6 5 4 3 2

ISBN 0-13-225770-X

Pearson Education Ltd., *London*
Pearson Education Australia Pty. Ltd., *Sydney*
Pearson Education Singapore, Pte. Ltd.
Pearson Education North Asia Ltd., *Hong Kong*
Pearson Education Canada, Inc., *Toronto*
Pearson Educación de Mexico, S.A. de C.V.
Pearson Education—Japan, *Tokyo*
Pearson Education Malaysia, Pte. Ltd.

Preface

To the Student

As you worked your way through high school, or otherwise worked to prepare yourself for college, you were probably unaware that an information explosion was taking place in the field of biology. This explosion, brought on by advances in biotechnology and communicated by faster, more powerful computers, has allowed scientists to gather data more quickly and disseminate data to colleagues in the global scientific community with the click of a mouse. Every discipline of biology has benefited from these advances, and today's scientists collectively know more than any individual could ever hope to understand.

Paradoxically, as it becomes more and more difficult to synthesize huge amounts of information from disparate disciplines within the broad field of biology, it becomes more vital that we do so. The very same technologies that led to the information boom, coupled with expanding human populations, present us with complex ethical questions. These questions include whether or not it is acceptable to clone humans, when human life begins and ends, who owns living organisms, what our responsibilities toward endangered species are, and many more. No amount of knowledge alone will provide satisfactory answers to these questions. Addressing these kinds of questions requires the development of a scientific literacy that surpasses the rote memorization of facts. To make decisions that are individually, socially, and ecologically responsible, you must not only understand some fundamental principles of biology but also be able to use this knowledge as a tool to help you analyze ethical and moral issues involving biology.

To help you understand biology and apply your knowledge to an ever-expanding suite of issues, we have structured each chapter of *Biology: Science for Life* around a compelling story in which biology plays an integral role. Through the story you will not only learn the relevant biological principles, but you will also see how science can be used to help answer complex questions. As you learn to apply the strategies modeled by the text, you will begin developing your critical thinking skills.

By the time you finish this book, you should have a clear understanding of many important biological principles. You will also be able to critically evaluate which information is most reliable instead of simply accepting all the information you hear or read about. Even though you may not be planning to be a practicing biologist, well-developed critical thinking skills will enable you to make decisions that affect your own life, such as whether or not to take nutritional supplements, and decisions that affect the lives of others, such as whether or not to believe the DNA evidence presented to you as a juror in a criminal case.

It is our sincere hope that understanding how biology applies to important personal, social, and ecological issues will convince you to stay informed about such issues. On the job, in your community, at the doctor's office, in the voting booth, and at home reading the paper or surfing the web, your knowledge of the basic biology underlying so many of the challenges that we as individuals and as a society face will enable you to make well-informed decisions for your home, your nation, and your world.

To the Instructor

Virginia Borden and **Colleen Belk** have collaborated on teaching the non-majors biology course at the University of Minnesota–Duluth for over a decade. This collaboration has been enhanced by their differing but complementary areas of expertise. In addition to the nonmajors course, Colleen Belk teaches General Biology for majors, Genetics, Cell Biology, and Molecular Biology courses. Virginia Borden teaches General Biology for majors, Evolutionary Biology, Plant Biology, Ecology, and Conservation Biology courses.

After several somewhat painful attempts at teaching all of biology in a single semester, the two authors came to the conclusion that they needed to find a better way. They realized that their students were more engaged when they understood how biology directly affected their lives. Colleen and Virginia began to structure their lectures around stories they knew would interest students. When they began letting the story drive the science, they immediately noticed a difference in student interest, energy, and willingness to work harder at learning biology. Not only has this approach increased student understanding, it has increased the authors' enjoyment in teaching the course—presenting students with fascinating stories infused with biological concepts is simply a lot more fun. This approach served to invigorate their teaching. Knowing that their students are learning the biology that they will need now and in the future gives the authors a deep and abiding satisfaction.

By now you are probably all too aware that teaching nonmajor students is very different from teaching biology majors. You know that most of these students will never take another formal biology course; therefore your course may be the last chance for these students to see the relevance of science in their everyday lives and the last chance to appreciate how biology is woven throughout the fabric of their lives. You recognize the importance of engaging these students because you know that these students will one day be voting on issues of scientific importance, holding positions of power in the community, serving on juries, and making healthcare decisions for themselves and their families. You know that your students' lives will be enhanced if they have a thorough grounding in basic biological principles and scientific literacy.

Themes Throughout *Biology: Science for Life*

Helping nonmajors to appreciate the importance of learning biology can be a tremendously rewarding job—and a challenging one as well. We sometimes struggled to actively engage students in lectures and to raise their scientific literacy and critical thinking skills, and we knew we were not alone. In fact, when we asked instructors from around the country what challenges they faced while teaching the nonmajors introductory biology course, these dedicated teachers echoed our concerns. This book was written to help biology instructors meet these challenges.

The Story Drives the Science.   We have found that students are much more likely to be engaged in the learning process when the textbook and lectures capitalize on their natural curiosity. This text accomplishes this by using a story to drive the science in every chapter. Students get caught up in the story and become interested in learning the biology so they can see how the story is resolved. This approach allows us to cover the key areas of biology, including basic chemistry, the unity and diversity of life, cell structure and function, classical and molecular genetics, evolution, and ecology, in a manner that makes students want to learn. Not only do students want to learn, this approach allows students to both connect the science to their everyday lives and integrate the principles and concepts for later application to other situations. This approach will give you flexibility in teaching and will support you in developing students' critical thinking skills.

The Process of Science.   This book also uses another novel approach in the way that the process of science is modeled. The first chapter is dedicated to the scientific method and hypothesis testing, and each subsequent chapter weaves the scientific method and hypothesis testing throughout the story. The development of students' critical thinking skills is thus reinforced for the duration of the course. Students will see that the application of the scientific method is often the best way to answer questions raised in the story. This practice not only allows students to develop their critical thinking skills but, as they begin to think like scientists, helps them understand why and how scientists do what they do.

Integration of Evolution.   Another aspect of *Biology: Science for Life* that sets it apart from many other texts is the manner in which evolutionary principles are integrated throughout the text. The role of evolutionary processes is highlighted in every chapter, even when the chapter is not specifically focused on an evolutionary question. For example, when discussing infectious diseases, the evolution of antibiotic-resistant strains of bacteria is addressed. The physiology unit includes an essay on evolution in each chapter. These essays illustrate the importance of natural selection in the development of various organs and organ systems across a wide range of organisms. With evolution serving as an overarching theme, students are better able to see that all of life is connected through this process.

Pedagogical Elements

Open the book and flip through a few pages and you will see some of the most inviting, lively, and informative illustrations you have ever seen in a biology text. The illustrations are inviting because they have a warm, hand-drawn quality that is clean and uncluttered. Most importantly, the illustrations are informative, not only because they were carefully crafted to enhance concepts in the text but also because they employ techniques like the "pointer" that help draw the students' attention to the important part of the figure (see page 50). Likewise, tables are more than just tools for organizing information; they are illustrated to provide attractive, easy references for the student. We hope that the welcoming nature of the art and tables in this text will encourage nonmajors to explore ideas and concepts instead of being overwhelmed before they even get started.

In addition to lively illustrations, this text also strives to engage the nonmajor student through the use of analogies. For example, the process of translation is likened to baking a cake, and the heterozygote advantage is likened to the advantage conferred by having more than one type of jacket (see p. 144 and pp. 201–202). These clever illustrations are peppered throughout the text.

Students can reinforce and assess what they are learning in the classroom by reading the chapter, studying the figures, and answering the end-of-chapter questions. We have written these questions in every format likely to be used by an instructor during an exam so that students have practice in answering many different types of questions. We have also included "Connecting the Science" questions that would be appropriate for essay exams, class discussions, or use as topics for term papers.

Improvements in the Second Edition

Bringing the first edition of *Biology: Science for Life* to instructors around the country was extremely gratifying to us. The success of the first edition and the positive feedback that it garnered assured us that this approach works for students and instructors alike. We also received feedback from users and reviewers on how to enhance the second edition to most closely meet the needs of biology instructors in diverse settings.

- To make your transition to teaching from this text as simple as possible, we have reorganized the sequence of topics to more closely resemble the traditional topical order found in most biology courses.
- To better meet the needs of instructors teaching a wide diversity of nonmajors biology courses who employed the first edition of *Biology: Science for Life*, we have created two versions of the book. The shorter version provides a thorough overview of the four pillars of modern biological science: cell biology, genetics, evolution, and ecology. We have increased coverage of basic chemistry and metabolism so that students develop a more solid understanding of the basic process that support and constrain much of life on Earth. The longer version includes an expanded animal physiology section, including coverage of all major animal organ systems, as well as additional chapters on plant structure and function.
- We have increased coverage of non-human biology, often providing examples of how the biology of other organisms differs from that of humans. For example, while the storyline of the natural selection chapter is the evolution and control of the HIV virus in an infected person, the chapter is peppered with non-human examples of the effects of natural selection—from the evolution of bill size in Galapagos finches to the adaptations for wind pollination in grasses. Likewise, we have included additional photos to provide more vivid illustrations of the wonderful variety of life on Earth, from amoebas reproducing asexually (Chapter 5) to peacocks demonstrating the effects of sexual selection (Chapter 11).
- The overall goal of the text remains providing a thorough overview of the essentials of biological science while trying to avoid overloading students with information. We worked closely with instructors using the first edition, as well as other reviewers, to pinpoint essential content to include in the second edition. When necessary, we judiciously added more detailed content where the students' understanding of the material would be enhanced by additional information. Chapter 2:

Are We Alone in the Universe? and Chapter 4: Is the Earth Warming? provide examples of where coverage was expanded (in this case, of chemistry and metabolism), while staying true to the book's philosophy of teaching essential information in a story format.

The development of the second edition has truly been a collaborative process among ourselves, the students and instructors who used the first edition, and the many thoughtful reviewers of these chapters. We look forward to learning about your experience with *Biology: Science for Life, 2e*.

Supplements

A group of talented and dedicated biology educators teamed up with us to build a set of resources that equip nonmajors with the tools to achieve scientific literacy that will allow them to make informed decisions about the biological issues that affect them daily. The student resources offer several ways of reviewing and reinforcing the concepts and facts covered in this textbook. The instructor resources provide a thoroughly revised collection of test questions, an updated and expanded suite of lecture presentation materials, and a valuable source of ideas for educators to enrich their instruction efforts. Available in print and media formats, the *Biology: Science for Life* resources are easy to navigate and support a variety of learning and teaching styles.

We believe you will find that the design and format of this text and its supplements will help you meet the challenge of helping students both succeed in your course and develop science skills—for life.

Acknowledgments

Reviewers

Each chapter of this book was thoroughly reviewed several times as it moved through the development process. Reviewers were chosen on the basis of their demonstrated talent and dedication in the classroom. Many of these reviewers were already trying various approaches to actively engage students in lectures, and to raise the scientific literacy and critical thinking skills among their students. Their passion for teaching and commitment to their students was evident throughout this process. These devoted individuals scrupulously checked each chapter for scientific accuracy, readability, and coverage level. In addition to general reviewers, we also had a team of expert reviewers evaluate individual chapters to ensure that the content was accurate and that all the necessary concepts were included.

All of these reviewers provided thoughtful, insightful feedback, which improved the text significantly. Their efforts reflect their deep commitment to teaching nonmajors and improving the scientific literacy of all students. We are very thankful for their contributions.

We express sincere gratitude to the expert reviewers who worked so carefully with the author in reviewing manuscript to ensure the scientific accuracy of the text and art.

Paul Beardsley
Idaho State University

Carl T. Bergstrom
University of Washington

Robert S. Boyd
Auburn University

Warren Burggren
University of North Texas

William F. Collins III
Stony Brook University

Susan Dunford
University of Cincinnati

Craig M. Hart
Louisiana State University

Peter Heywood
Brown University

Loren Knapp
University of South Carolina

Charles Mallery
University of Miami

T. D. Maze
Lander University

James E. Mickle
North Carolina State University

Shawn Nordell
St. Louis University

Marilee Benore Parsons
University of Michigan, Dearborn

William E. Rogers
Texas A&M University

Gray Scrimgeour
University of Toronto

Robert R. Wise
University of Wisconsin, Oshkosh

Donna Young
University of Winnipeg

Second Edition Reviewers

Daryl Adams
Minnesota State University, Mankato

James S. Backer
Daytona Beach Community College

Gail F. Baker
LaGuardia Community College

Ellen Baker
Santa Monica College

Neil R. Baker
The Ohio State University

Tamatha R. Barbeau
Francis Marion University

Andrew Baldwin
Mesa Community College

Sarah Barlow
Middle Tennessee State University

Andrew M. Barton
University of Maine, Farmington

Vernon Bauer
Francis Marion University

Donna H. Bivans
Pitt Community College

John Blamire
City University of New York, Brooklyn College

Barbara Blonder
Flagler College

Bruno Borsari
Winona State University

Eric Brenner
New York University

Carol Britson
University of Mississippi

Carole Browne
Wake Forest University

Neil Buckley
State University of New York, Plattsburgh

Stephanie Burdett
Brigham Young University

Nancy Butler
Kutztown University

Wilbert Butler
Tallahassee Community College

David Byres
Florida Community College, Jacksonville

Tom Campbell
Pierce College, Los Angeles

Deborah Cato
Wheaton College

Bruce Chase
University of Nebraska, Omaha

Thomas F. Chubb
Villanova University

Gregory Clark
University of Texas, Austin

Kimberly Cline-Brown
University of Northern Iowa

William H. Coleman
University of Hartford

Jerry L. Cook
Sam Houston State University

Erica Corbett
*Southeastern Oklahoma State
    University*

James B. Courtright
Marquette University

Judith D'Aleo
Plymouth State University

Juville Dario-Becker
Central Virginia Community College

Garry Davies
University of Alaska, Anchorage

Edward A. DeGrauw
Portland Community College

Heather DeHart
Western Kentucky University

Lisa Delissio
Salem State College

Elizabeth Desy
Southwest Minnesota State University

Gregg Dieringer
Northwest Missouri State

Diane Dixon
*Southeastern Oklahoma State
    University*

Christopher Dobson
Grand Valley State University

Cecile Dolan
*New Hampshire Community Technical
    College, Manchester*

Matthew Douglas
Grand Rapids Community College

Lee C. Drickamer
Northern Arizona University

Patrick J. Enderle
East Carolina University

William Epperly
Robert Morris College

Dan Eshel
*City University of New York,
    Brooklyn College*

Lynn Firestone
Brigham Young University

Brandon L. Foster
Wake Technical Community College

Richard A. Fralick
Plymouth State University

Barbara Frank
Idaho State University

Lori Frear
Wake Technical Community College

Suzanne Frucht
Northwest Missouri State University

Edward Gabriel
Lycoming College

Patrick Galliart
North Iowa Area Community College

Alexandros Georgakilas
East Carolina University

Tammy Gillespie
Eastern Arizona College

Mac F. Given
Neumann College

Andrew Goliszek
*North Carolina Agricultural and Technical
    State University*

Beatriz Gonzalez
Sante Fe Community College

Lara Gossage
Hutchinson Community College

Tamar Goulet
University of Mississippi

Becky Graham
University of West Alabama

Robert S. Greene
Niagara University

Tony J. Greenfield
Southwest Minnesota State University

Stanley Guffey
University of Tennessee

Robert Harms
St. Louis Community College

Bethany Henderson-Dean
University of Findlay

Julie Hens
*University of Maryland University
    College*

Julia Hinton
McNeese State University

Phyllis C. Hirsh
East Los Angeles College

Elizabeth Hodgson
York College of Pennsylvania

Margaret Horton
University of North Carolina, Greensboro

Michael E. S. Hudspeth
Northern Illinois University

Pamela D. Huggins
Fairmont State University

Sue Hum-Musser
Western Illinois University

James Hutcheon
Georgia Southern University

Carl Johansson
Fresno City College

Thomas Jordan
Pima Community College

Mary K. Kananen
Penn State University, Altoona

Arnold Karpoff
University of Louisville

Judy Kaufman
Monroe Community College

Judith Kelly
Henry Ford Community College

Andrew Keth
Clarion University

Stacey Kiser
Lane Community College

Dennis Kitz
Southern Illinois University, Edwardsville

Carl Kloock
California State, Bakersfield

Michael A. Kotarski
Niagara University

Michelle LaBonte
Framingham State College

Dale Lambert
Tarrant County College

Brenda Leady
University of Toledo

Mary Lehman
Longwood University

Abigail Littlefield
Landmark College

Andrew D. Lloyd
Delaware State University

Judy Lonsdale
Boise State University

Kate Lormand
Arapahoe Community College

Kimberly Lyle-Ippolito
Anderson University

Stephen E. MacAvoy
American University

Molly MacLean
University of Maine

Mary McNamara
*Albuquerque Technical Vocational
Institute*

Susan T. Meiers
Western Illinois University

Diane Melroy
*University of North Carolina,
Wilmington*

Joseph Mendelson
Utah State University

Paige A. Mettler-Cherry
Lindenwood University

Debra Meuler
Cardinal Stritch University

Craig Milgrim
Grossmont College

Ali Mohamed
Virginia State University

James Mone
Millersville University

Daniela Monk
Washington State University

David Mork
Yakima Valley Community College

John J. Natalini
Quincy University

Alissa A. Neill
University of Rhode Island

Dawn Nelson
Community College of Southern Nevada

David L.G. Noakes
University of Guelph

Tonye E. Numbere
University of Missouri, Rolla

Igor Oksov
Union County College

Arnas Palaima
University of Mississippi

Anthony Palombella
Longwood University

Steven L. Peck
Brigham Young University

John Peters
College of Charleston

Polly Phillips
Florida International University

Francis J. Pitocchelli
Saint Anselm College

Roberta L. Pohlman
Wright State University

Robert Pozos
San Diego State University

Marion Preest
The Claremont Colleges

Gregory Pryor
Francis Marion University

Rongsun Pu
Kean University

Narayanan Rajendran
Kentucky State University

Anne E. Reilly
Florida Atlantic University

Michael H. Renfroe
James Madison University

Gwynne S. Rife
University of Findlay

Todd Rimkus
Marymount University

Wilma Robertson
Boise State University

Troy Rohn
Boise State University

Christel Rowe
Hibbing Community College

Joanne Russell
Manchester Community College

Michael Rutledge
Middle Tennessee State University

Kim Sadler
Middle Tennessee State University

Brian Sailer
Albuquerque Technical Vocational Institute

Louis Scala
Kutztown University

Daniel C. Scheirer
Northeastern University

Beverly Schieltz
Wright State University

Nancy Schmidt
Pima Community College

Julie Schroer
Bismarck State College

Fayla Schwartz
Everett Community College

Steven Scott
Merritt College

Roger Seeber
West Liberty State College

Allison Shearer
Grossmont College

Cara Shillington
Eastern Michigan University

Beatrice Sirakaya
Pennsylvania State University

Cynthia Sirna
Gadsden State Community College

Brian Smith
Black Hills State University

Douglas Smith
Clarion University of Pennsylvania

Mark Smith
Chaffey College

Anna Bess Sorin
University of Memphis

Gregory Smutzer
Temple University

Carol St. Angelo
Hofstra University

Susan L. Steen
Idaho State University

Bradley J. Swanson
Central Michigan University

Joyce Tamashiro
University of Puget Sound

Jeffrey Taylor
Slippery Rock University

Tania Thalkar
Clarion University of Pennsylvania

Jeffrey Thomas
*University of California, Los
Angeles*

Janis Thompson
*Lorain County Community
College*

Martin Tracey
Florida International University

Jeffrey Travis
*State University of New York,
Albany*

Michael Tveten
*Pima Community College, Northwest
Campus*

Brandi Van Roo
Framingham State College

Mark Venable
Appalachian State University

Janet Vigna
Grand Valley State University

Sean Walker
*California State University,
Fullerton*

Tracy Ware
Salem State College

Jennifer Warner
*University of North Carolina,
Charlotte*

Marcia Wendeln
Wright State University

Shauna Weyrauch
Ohio State University, Newark

Howard Whiteman
Murray State University

Gerald Wilcox
Potomac State College

Peter J. Wilkin
Purdue University North Central

Kenneth Wunch
Sam Houston State University

Michelle L. Zjhra
Georgia Southern University

Michelle Zurawski
Moraine Valley Community College

First Edition Reviewers

Karen Aguirre
Clarkson University

Susan Aronica
Canisius College

Mary Ashley
University of Chicago

Thomas Balgooyen
San Jose State University

Donna Becker
Northern Michigan University

Steve Berg
Winona State University

Lesley Blair
Oregon State University

Susan Bornstein-Forst
Marian College

James Botsford
New Mexico State University

Bryan Brendley
Gannon University

Peggy Brickman
University of Georgia

Carole Browne
Wake Forest University

Neil Buckley
*State University of New York,
    Plattsburgh*

Suzanne Butler
Miami-Dade Community College

David Byres
Florida Community College, Jacksonville

Peter Chabora
Queens College

Mary Colavito
Santa Monica College

Walter Conley
State University of New York, Potsdam

Melanie Cook
Tyler Junior College

Scott Cooper
University of Wisconsin, La Crosse

George Cornwall
University of Colorado

Angela Cunningham,
Baylor University

Garry Davies
University of Alaska, Anchorage

Miriam del Campo
Miami-Dade Community College

Veronique Delesalle
Gettysburg College

Beth De Stasio
Lawrence University

Donald Deters
Bowling Green State University

Douglas Eder
Southern Illinois University, Edwardsville

Deborah Fahey
Wheaton College

Richard Firenze
Broome Community College

David Froelich
Austin Community College

Anne Galbraith
University of Wisconsin, La Crosse

Wendy Garrison
University of Mississippi

Robert George
*University of North Carolina,
    Wilmington*

Sharon Gilman
Coastal Carolina University

John Green
Nicholls State University

Robert Greene
Niagara University

Bruce Goldman
University of Connecticut, Storrs

Eugene Goodman
University of Wisconsin, Parkside

Tamar Goulet
University of Mississippi

Mark Grobner
California State University, Stanislaus

Stan Guffey
University of Tennessee, Knoxville

Mark Hammer
Wayne State University

Blanche Haning
North Carolina State University

Patricia Hauslein
St. Cloud State University

Stephen Hedman
University of Minnesota, Duluth

Julie Hens
Yale University

Leland Holland
Pasco-Hernando Community College

Jane Horlings
Saddleback Community College

David Howard
University of Wisconsin, La Crosse

Michael Hudecki
State University of New York, Buffalo

Laura Huenneke
New Mexico State University

Carol Hurney
James Madison University

Jann Joseph
Grand Valley State University

Michael Keas
Oklahoma Baptist University

Karen Kendall-Fite
Columbia State Community College

David Kirby
American University

Dennis Kitz
*Southern Illinois University,
    Edwardsville*

Jennifer Knapp
Nashville State Technical Community College

Loren Knapp
University of South Carolina

Phyllis Laine
Xavier University

Tom Langen
Clarkson University

Lynn Larsen
Portland Community College

Mark Lavery
Oregon State University

Mary Lehman
Longwood University

Doug Levey
University of Florida

Jayson Lloyd
College of Southern Idaho

Paul Lurquin
Washington State University

Douglas Lyng
Indiana University/Purdue University

Michelle Mabry
Davis and Elkins College

Ken Marr
Green River Community College

Kathleen Marrs
Indiana University/Purdue University

Steve McCommas
Southern Illinois University, Edwardsville

Colleen McNamara
Albuquerque Technical Vocational Institute

John McWilliams
Oklahoma Baptist University

Diane Melroy
University of North Carolina, Wilmington

Joseph Mendelson
Utah State University

Hugh Miller
East Tennessee State University

Jennifer Miskowski
University of Wisconsin, La Crosse

Stephen Molnar
Washington University

Bertram Murray
Rutgers University

Ken Nadler
Michigan State University

Joseph Newhouse
California University of Pennsylvania

Jeffrey Newman
Lycoming College

Kevin Padian
University of California, Berkeley

Javier Penalosa
Buffalo State College

Rhoda Perozzi
Virginia Commonwealth University

John Peters
College of Charleston

Patricia Phelps
Austin Community College

Calvin Porter
Xavier University

Linda Potts
University of North Carolina Wilmington

Gregory Pryor
University of Florida

Laura Rhoads
State University of New York, Potsdam

Laurel Roberts
University of Pittsburgh

Deborah Ross
Indiana University/Purdue University

Michael Rutledge
Middle Tennessee State University

Wendy Ryan
Kutztown University

Christopher Sacchi
Kutztown University

Jasmine Saros
University of Wisconsin, La Crosse

Ken Saville
Albion College

Robert Schoch
Boston University

Robert Shetlar
Georgia Southern University

Thomas Sluss
Fort Lewis College

Douglas Smith
Clarion University of Pennsylvania

Sally Sommers Smith
Boston University

Amanda Starnes
Emory University

Timothy Stewart
Longwood College

Shawn Stover
Davis and Elkins College

Bradley Swanson
Central Michigan University

Martha Taylor
Cornell University

Alice Templet
Nicholls State University

Nina Thumser
California University of Pennsylvania

Alana Tibbets
Southern Illinois University, Edwardsville

Jeffrey Travis
State University of New York, Albany

Robert Turgeon
Cornell University

James Urban
Kansas State University

John Vaughan
St. Petersburg Junior College

Martin Vaughan
Indiana State University

Paul Verrell
Washington State University

Tanya Vickers
University of Utah

Janet Vigna
Grand Valley State University

Don Waller
University of Wisconsin, Madison

Jennifer Warner
University of North Carolina, Charlotte

Lisa Weasel
Portland State University

Carol Weaver
Union University

Frances Weaver
Widener University

Elizabeth Welnhofer
Canisius College

Wayne Whaley
Utah Valley State College

Vernon Wiersema
Houston Community College

Michelle Withers
Louisiana State University

Art Woods
University of Texas, Austin

Elton Woodward
Daytona Beach Community College

Student Focus Group Participants

California State University, Fullerton:
Danielle Bruening
Leslie Buena
Andrés Carrillo
Victor Galvan
Jessica Ginger
Sarah Harpst
Robin Keber
Ryan Roberts

Melissa Romero
Erin Seale
Nathan Tran
Tracy Valentovich
Sean Vogt

Fullerton College:
Michael Baker
Mahetzi Hernandez

Heidi McMorris
Daniel Minn
Sam Myers
David Omut
Jonathan Pistorino
James W. Pura
Samantha Ramirez
Tiffany Speed
Tristan Terry

Supplements Authors

Print and media supplements were prepared by a very creative, energetic, and fun team of nonmajors biology instructors from colleges and universities across the country. As a result, students will see dynamic animations of many complex processes and will have the opportunity to practice newly learned skills. The work of these instructors helped ensure that the supplements were reinforcing the chapter content. We cannot thank them enough.

Supplements Contributors

Deborah Cato	*Wheaton College*
William H. Coleman	*University of Hartford*
Cynthia Ghent	*Towson University*
Diane Melroy	*University of North Carolina, Wilmington*
Anthony Palombella	*Longwood University*
Gregory S. Pryor	*Francis Marion University*
Troy Rohn	*Boise State University*
Carol St. Angelo	*Hofstra University*

Supplements Reviewers

James S. Backer	*Daytona Beach Community College*
David Belt	*Penn Valley Community College*
Barbara Blonder	*Flagler College*
Kimberly Cline-Brown	*University of Northern Iowa*
Judy Dacus	*Cedar Valley College*
Lisa Delissio	*Salem State College*
Chris Farrell	*Trevecca Nazarene University*
Stewart Frankel	*University of Hartford*
Anthony Ippolito	*DePaul University*
Michael A. Kotarski	*Niagara University*
Michelle Mabry	*Davis and Elkins College*
Matthew J. Maurer	*University of Virginia's College at Wise*
Wilma Robertson	*Boise State University*
Joanne Russell	*Manchester Community College*
Carol St. Angelo	*Hofstra University*
Joyce Tamashiro	*University of Puget Sound*
Michael Troyan	*Pennsylvania State University*

The Book Team

When we set out to write this book, we would not have predicted that we would so thoroughly enjoy the experience. Our enjoyment stems directly from the enthusiasm and talent of the Prentice Hall team. It has been an honor to work with all of these talented, dedicated people.

The book team came together due to the efforts of our executive editor Teresa R. Chung. Teresa is a woman of tremendous vision, insight, integrity, humor, energy, and style. She has guided every aspect of this project from its inception to its delivery. It was heartening to be in such capable hands and to be able to thoroughly trust your editor's judgment. It was also a pleasure to work with someone who is so cheerful and upbeat. For keeping us on track and inspiring us to do our best work, we sincerely thank her.

Another important book team member was Elaine Page, who served as our Development Editor. Elaine read every word we wrote from a student's perspective and helped us effectively address issues raised by users and reviewers. She was essential in helping us develop the "two version approach" we applied for this edition. Her keen insights, hard work, and patience are very much appreciated. Becky Strehlow, the Development Editor for the first edition, also deserves our continued thanks for helping to refine the vision of this text.

Art Development Editor Jay McElroy is responsible for taking our ideas about art and transferring them to the page. In addition to being a talented artist, he has been a pleasure to work with. Our Illustration Designer for the first edition, Kim Quillin, provided the essential outline for the illustration program, which is maintained in this edition. We owe much to her for helping to create the unique look of *Biology: Science for Life*.

Senior Media Editor Patrick Shriner and Assistant Editor Andrew Sobel were the wizards behind our print and media supplements and have brought so much creativity to the entire package. We are grateful for their careful work. Editorial Assistants Gina Kayed and Lisa Tarabokjia were always ready to cheerfully and competently solve any problems that arose along the way.

At the very early stages of production, this text and its illustrations and images were in the hands of four very capable people. Art Director Jonathan Boylan guided the book design with much talent and creativity. Art Editors Connie Long and Rhonda Aversa skillfully managed the production of the illustration program. Copyeditor Chris Thillen did an excellent job of working the text into its final form, making sure no mistakes crept in. Photo Researcher Elaine Soares has located most of the striking images in the text. She did an excellent job of translating our photo wishes into beautiful images. Donna King was the Production Editor for this text. She managed to seamlessly coordinate the work of the copyeditor, photo researcher, illustrators, and authors under a tight schedule.

This book is dedicated to our families, friends, and colleagues who have supported us over the years. Having loving families, great friends, and a supportive work environment enabled us to make this heartfelt contribution to nonmajors biology education.

Colleen Belk and Virginia Borden

University of Minnesota-Duluth

"*Because science,*

told as a story, can intrigue and inform the

non-scientific minds among us, it has the

potential to bridge the two cultures into

which civilization is split—the sciences

and the humanities. For educators, stories

are an exciting way to draw young minds

into the scientific culture."

E. O. Wilson

Support Your Biology Lectures with the Best Animated Resources...

69 Topics   Cells   Plants

Chemistry   Genetics   Animals

Energy   Evolution   Ecology

Teach biological concepts the way you want with Prentice Hall BLAST!!

- Designed to support individual teaching approaches and class needs

- Interactive variable controls and stunning 3-D graphics engage students and enliven lectures

- Available only with Prentice Hall textbooks

All the Support You Need to Teach Your Course the Way You Want:

- **Belk / Borden Instructor Resource Center on CD/DVD** — 3 CD/1 DVD set plus TestGen® Test Item File CD – a full suite of JPEG and PowerPoint® options, 100+ Instructor Animations, and additional lecture and course support items (0-13-188801-3)

- **Test Item File** — printed version (0-13-219743-X)

- **Instructor Resource Guide** — lecture outlines and activities (0-13-219746-4)

- **Transparency Pack** — 300 transparencies (0-13-196318-X)

—PLUS—

- **Prentice Hall BLAST!!** — 3 CD/1 DVD Biology Lecture Animation and Simulation Tool set – *see above* (0-13-237034-4)

...and the **Most Complete JPEG and PowerPoint® Options Available**

All found on the Belk & Borden Instructor Resource Center CD/DVD

JPEG Files | PowerPoint® Resources

- Every **figure** and **table** from the textbook
- All illustrations **optimized** for clear viewing
- All figures offered in **labeled, unlabeled,** and **editable-label** versions

- All text figure and table JPEG files in an **Image Gallery** file for each chapter
- Two suggested **Lectures** for each chapter
- 100+ **Instructor Animations** pre-loaded into PowerPoint® in PC and Mac versions
- **Classroom Response System** files for each chapter

Label sizes increased and colors enhanced for clarity when projected

Figure "splits" made with teaching needs in mind

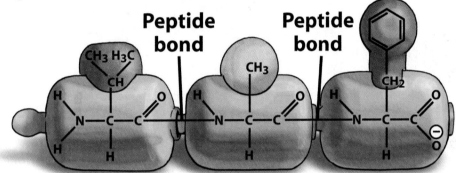

Peptide bond formation

Student Support Items

- **Companion Website with Grade Tracker** — Online Study Guide, 46 Web Tutorials, key term tool, and printable PDF files of all line drawings appearing in the text: www.prenhall.com/belk
- **Student CD** — Interactive quizzes and 23 Web Tutorials (packaged with textbook)
- **Audio Study Guide** — 20+ minutes of study tools on MP3 files for each chapter: www.prenhall.com/belk
- **Student Study Companion** — Printed study guide (0-13-188802-1)
- **Student Lecture Notebook** — All the line drawings from the text in print in full color (0-13-219744-8)
- **SafariX Textbook Online** — eTextbook (0-13-227673-9)

Laboratory Manual

- Contains 75 Lab Exercises under 17 Topics, all written by the textbook authors and tied directly to the textbook (0-13-188804-8)
- Annotated *Instructor's Edition* includes answers, helpful suggestions, lists of materials, and pre-lab quizzes. (0-13-173373-7)

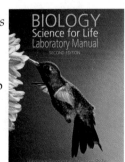

Course Management Solutions

Visit http://cms.prenhall.com for details.

For more information on the variety of student study options with Belk/Borden, see the inside front cover of the textbook

Contents

Chapter 4

Is the Earth Warming?   The Greenhouse Effect, Cellular
Respiration, and Photosynthesis   **70**

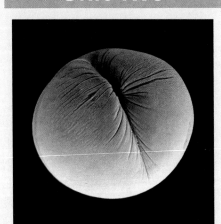

Chapter 7

DNA Detective   Extensions of Mendelism, Sex Linkage, Pedigree Analysis, and DNA Fingerprinting   **166**

Chapter 8

Genetic Engineering   Transcription, Translation, and Genetically Modified Organisms   **192**

Unit Three

Evolution

Chapter 9

Where Did We Come From? The Evidence for Evolution 224

Chapter 10

An Evolving Enemy Natural Selection 256

Chapter 11

Who Am I?   Species and Races   **282**

Chapter 12

Prospecting for Biological Gold   Biodiversity and Classification   **316**

Unit Four

Ecology

Chapter 13

Is the Human Population Too Large?
Population Ecology **344**

Chapter 14

Is Earth Experiencing a Biodiversity Crisis?
Community Ecology, Ecosystem Ecology, and
Conservation Biology **358**

Chapter 15

Where Do You Live?   Climate and Biomes   **390**

Animal Structure and Function   Unit Five

Chapter 16

Organ Donation   Tissues, Organs, and Organ Systems   **420**

Chapter 17

Clearing the Air   Respiratory, Circulatory, and Excretory Systems   442

Chapter 18

Will Mad Cow Disease Become an Epidemic?

Immune System, Bacteria, Viruses, and Other Pathogens   **468**

Chapter 19

Sex Differences and Athleticism   Endocrine, Skeletal, and Muscular Systems   **494**

Chapter 20

Is There Something in the Water?   Reproductive and Developmental Biology   **514**

Chapter 21

Attention Deficit Disorder   Brain Structure and Function   546

Unit Six   Plant Biology

Chapter 22

Feeding the World   Plant Structure and Growth   570

Chapter 23

Growing a Green Thumb   Plant Physiology   602

CHAPTER 1

Can Science Cure the Common Cold?

Introduction to the Scientific Method

Another cold! What can I do?

Take massive doses of Vitamin C?

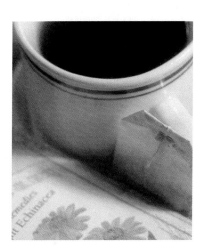

Drink echinacea tea?

We have all been there—you just recover from one bad head cold, and on a morning soon after you notice that scratchy feeling in your throat that signals a new one is about to begin. It is always at the worst time, too, when you have a big exam coming up, a term paper due, and a packed social calendar. Why are you sick yet again? What can you do about it?

If you ask your friends and relatives, you will hear the usual advice on how to prevent and treat colds: Take massive doses of vitamin C, suck on zinc lozenges, drink cups of echinacea tea, spend time meditating, get plenty of exercise, wear your hat and gloves outside, get more rest, take two aspirin and call the doctor in the morning. What should you do? Most people follow the advice that makes the most sense to them, and if they find that they still feel terrible, they try another remedy. This approach to a problem is the kind of science that most of us perform daily. We see a problem, think of several possible causes, and try to solve the problem by addressing what we feel is the most likely cause. If our solution fails to work, we move to another possible solution that addresses other possible causes.

This brand of science may eventually give us an answer to a question about how to prevent and treat colds. But we cannot know if it is the best answer unless we try out all the potential treatments. We also will not know if

How would a scientist determine which advice is best?

1

any treatment that seemed to be effective was simply one that we tried at the right time—when our cold was getting better anyway. Luckily for us, legions of professional scientists spend their time trying to answer questions such as these. Scientists use the same basic process of testing ideas about how the world works and discarding (or modifying) ideas that are inadequate.

There are, however, some key differences between the ways that scientists approach questions and the daily scientific investigations illustrated by our own quests for cold relief. In this chapter, we introduce you to the process of science as it is practiced in the research setting and help you understand how to evaluate scientific claims by evaluating techniques that are often used to prevent or treat the common cold.

1.1   The Process of Science

The statements our friends and family make about what actions will help us remain healthy (for example, the advice to wear a hat) are in some part based on the advice giver's understanding of how our bodies resist colds. Ideas about "how things work" are called **hypotheses**. Or, more formally, a hypothesis is a proposed explanation for one or more **observations**. All of us generate hypotheses about the causes of some phenomenon based on our understanding of the world. Hypotheses in biology are generated as a result of knowledge about how the body and other biological systems work, experiences in similar situations, our understanding of other scientific research, and logical reasoning; but they are also shaped by our creative mind (Figure 1.1). When mom tells you to dress warmly in order to avoid colds, she is basing her advice on her belief in the accuracy of the following hypothesis: Becoming chilled makes an individual more susceptible to illness.

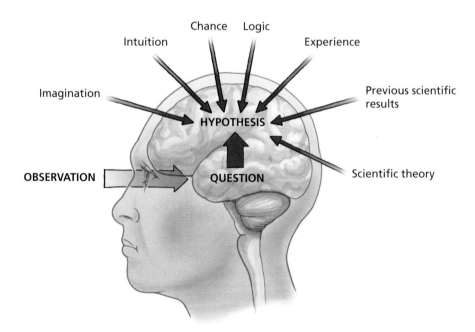

Figure 1.1   Hypothesis generation. All of us generate hypotheses. Many different factors, both logical and creative, influence the development of a hypothesis. Scientific hypotheses are both testable and falsifiable.

The hallmark of science is that hypotheses are subject to rigorous testing. Therefore, scientific hypotheses must be **testable**—it must be possible to evaluate a hypothesis through observations of the measurable universe. Not all hypotheses are testable. For instance, the statement that "colds are generated by disturbances in psychic energy" is not a scientific hypothesis because psychic energy cannot be seen or measured—it does not have a material nature and therefore cannot be put to a test. In addition, hypotheses that require the intervention of a supernatural force cannot be tested scientifically. If something is supernatural, it is not constrained by the laws of nature, and its behavior cannot be predicted using our current understanding of the natural world.

Scientific hypotheses must also be **falsifiable**; that is, a hypothesis must be stated in such a way that you can imagine an observation or set of observations that would prove it false. The hypothesis that exposure to cold temperatures increases your susceptibility to colds is falsifiable because we can imagine an observation that would cause us to reject this hypothesis (for instance, the observation that people exposed to cold temperatures do *not* catch more colds than people protected from chills). Of course, not all hypotheses are proved false, but it is essential to scientific progress that incorrect ideas can be discarded. And that can occur only if it is possible to prove those ideas false. This is why statements that are judgments, such as "It is wrong to cheat on an exam," are not scientific because different people have different ideas about right and wrong. It is impossible to falsify these types of statements. To find answers to questions of morality, ethics, or justice, we turn to other methods of gaining understanding—such as philosphy and religion.

The Logic of Hypothesis Testing

One very common hypothesis about cold prevention is that taking vitamin C supplements keeps you healthy. This hypothesis is very appealing, especially given the following generally known facts:

1. Fruits and vegetables contain a lot of vitamin C.
2. People with diets rich in fruits and vegetables are generally healthier than people who skimp on these food items.
3. Vitamin C is known to be an anti-inflammatory agent, reducing throat and nose irritation.

Given these facts, we can state the following falsifiable hypothesis:

Consuming vitamin C decreases the risk of catching a cold.

This hypothesis makes sense, given the statements just listed and the experiences of the many people who insist that vitamin C keeps them healthy. The process used to construct this hypothesis is called **inductive reasoning**—combining a series of specific observations (statements 1–3) to discern a general principle. The set of observations certainly seems like enough information on which to base a decision about how to proceed—start taking vitamin C supplements if you want to avoid future colds. However, a word of caution is in order: just because an inductively deduced hypothesis makes sense does not mean that it is true. The following example clearly demonstrates the point that inductively reasoned hypotheses may later be proven false.

Consider the ancient hypothesis, asserted by Aristotle in approximately 350 B.C., that the sun revolves around Earth. This hypothesis was sensible, based on the observation that the sun appeared on the eastern horizon every day at sunrise and disappeared behind the western horizon at sunset. For 2000 years, this hypothesis was considered to be a "fact" by nearly all of Western society. To most people, the hypothesis made perfect sense, especially since the common religious belief in western Europe was that Earth had been created and then surrounded by the vault of heaven. It was not until the early seventeenth century that this hypothesis was falsified as the result of Galileo Galilei's observations of

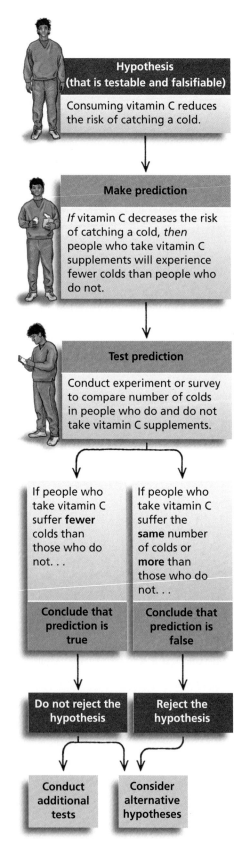

Figure 1.2 The scientific method. Tests of hypotheses follow a logical path. This flowchart illustrates the process.

the movements of Venus. Galileo's work helped to confirm the more modern hypothesis, proposed by Nicolai Copernicus, that Earth revolves around the sun.

So even though the hypothesis about vitamin C is perfectly sensible, it needs to be tested. Hypothesis testing is based on a process called **deductive reasoning** or deduction. Deduction involves using a general principle to predict an expected observation. The predicted observation, or **prediction**, concerns the outcome of an action, test, or systematic investigation. In other words, the prediction is the result we would expect from a particular test of the hypothesis.

Deductive reasoning takes the form of "if/then" statements. That is, *if* our general principle is correct, *then* we would expect to observe a specific set of outcomes. A prediction based on the vitamin C hypothesis could be:

If vitamin C decreases the risk of catching a cold, *then* people who take vitamin C supplements with their regular diets will experience fewer colds than will people who do not take supplements.

Deductive reasoning, with its resulting predictions, is a powerful method for testing hypotheses. However, the structure of such a statement means that hypotheses can be clearly rejected if untrue, but impossible to prove if true (Figure 1.2). This shortcoming is illustrated using the if/then statement above.

Consider the possible outcomes of a comparison between people who supplement with vitamin C and those who do not. People who take vitamin C supplements may suffer through more colds than will people who do not; they may have the same number of colds as the people who do not supplement, or supplementers may in fact experience fewer colds. What do these results tell us about the hypothesis?

If people who take vitamin C have more colds or the same number of colds as those who do not supplement, the hypothesis that vitamin C alone provides protection against colds can be rejected. But what if people who supplement with vitamin C *do* experience fewer colds? If this is the case, should we be out proclaiming the news, "Vitamin C—A Wonder Drug That Prevents the Common Cold"? No, we should not. We need to be much more cautious than that; we can only say that the hypothesis has been supported and not disproven.

Why is it impossible to say that the hypothesis that vitamin C prevents colds is true? Primarily because there could be other factors (that is, there are **alternative hypotheses**) that explain why people with different vitamin-taking habits are different in their cold susceptibility. In other words, demonstrating the truth of the *then* portion of a deductive statement does not guarantee that the *if* portion is true.

Consider the alternative hypothesis that frequent exercise reduces susceptibility to catching a cold. Perhaps people who take vitamin C supplements are more likely to engage in regular exercise than those who do not supplement. What if the alternative hypothesis were true? If so, the prediction that people who take vitamin C supplements experience fewer colds than people who do not supplement would be true, but not because the original hypothesis (vitamin C reduces the risk of colds) is true. Instead, people who take vitamin C supplements experience fewer colds than people who do not supplement because they are more likely to exercise, and it is exercise that reduces cold susceptibility.

A hypothesis that seems to be true because it has not been rejected by an initial test may be rejected later based on the results of a different test. As a matter of fact, this is the case for the hypothesis that vitamin C consumption reduces susceptibility to colds. The argument for the power of vitamin C was popularized in 1970 by Nobel Prize–winning chemist Linus Pauling in his book, *Vitamin C and the Common Cold*. Pauling based his assertion—that large doses of vitamin C reduce the incidence of colds by as much as 45%—on the results of a few studies that had been published since the 1930s. However, repeated, careful tests of this hypothesis have since failed to support it. In many of the studies Pauling cited, it appears that one or more alternative hypotheses may explain the difference in cold frequency between vitamin C supplementers and non-supplementers. Today, most researchers studying the common cold agree that the hypothesis that vitamin C prevents colds has been convincingly falsified.

The example of the vitamin C hypothesis also illustrates one of the challenges of communicating scientific information to the general public. You can see why the belief that vitamin C prevents colds is so widespread. For readers unfamiliar with the notion that scientific knowledge progresses by the rejection of incorrect ideas, a book by a Noble Prize–winning scientist may have seemed like the last word on the role of vitamin C in cold prevention. It required many years of careful research to show that this "last word" was, in fact, wrong.

The Experimental Method

The previous discussion may seem discouraging—how can scientists determine the truth of any hypothesis when there is always a chance that the hypothesis could be falsified? Even if one of the hypotheses about cold prevention is supported, does the difficulty of eliminating alternative hypotheses mean that we will never know which approach is truly best? The answer is yes—and no. Hypotheses cannot be proven absolutely true; it is always possible that the true cause of a particular phenomenon may be found in a hypothesis that has not yet been evaluated. However, in a practical sense, a hypothesis can be proven beyond a reasonable doubt. That is, when most of the reasonable alternative hypotheses have been eliminated and one hypothesis has not been disproven through repeated testing, scientists accept that this hypothesis is, in a practical sense, true. One of the most effective ways to test many hypotheses is through rigorous scientific experiments.

Experiments are contrived situations designed to test specific hypotheses. Generally, an experiment allows a scientist to control the conditions under which a given phenomenon occurs. Being able to manipulate the environment allows a scientist to minimize the number of alternative hypotheses that may explain the result. This process is equivalent to what a mechanic does when he diagnoses a car problem. If the engine does not turn over, the problem could be the battery; if charging or replacing the battery fails to solve the problem, then the next step might be to replace the starter motor. If the mechanic had started by replacing both, he would not know which component had caused the problem, and he would have an unhappy customer who was charged for the replacement of both parts. Likewise, a scientist who changes many variables and sees a different experimental result cannot determine which of the changes she made was the one that made the difference. The information collected by scientists during hypothesis testing is known as **data**. Data collected from well-designed experiments should allow researchers to either reject or support a hypothesis.

Not all scientific hypotheses can be tested through experimentation. For instance, hypotheses about historical events, such as those concerning how life on Earth originated or the cause of dinosaur extinction, are usually not testable in this way. These hypotheses must instead be tested via careful observation of the natural world; for instance, the examination of fossils and other evidence preserved in rocks allows scientists to test hypotheses regarding the extinction of the dinosaurs. In addition, not all testable hypotheses are subjected to experimentation; the science that is actually performed is a reflection of the priorities of the decision makers in our society. But hypotheses about the origin and prevention of colds can be, and are, tested experimentally.

Experimentation has enabled scientists to prove beyond a reasonable doubt that the common cold is caused by a virus. A virus has a very simple structure—it typically contains a short strand of genetic material and a few proteins encased in a relatively tough outer shell composed of more proteins and sometimes a fatty membrane. A virus must enter, or infect, a cell in order to reproduce. Of the over 200 types of viruses that are known to cause varieties of the common cold, most infect the cells in our noses and throats. The sneezing, coughing, congestion, and

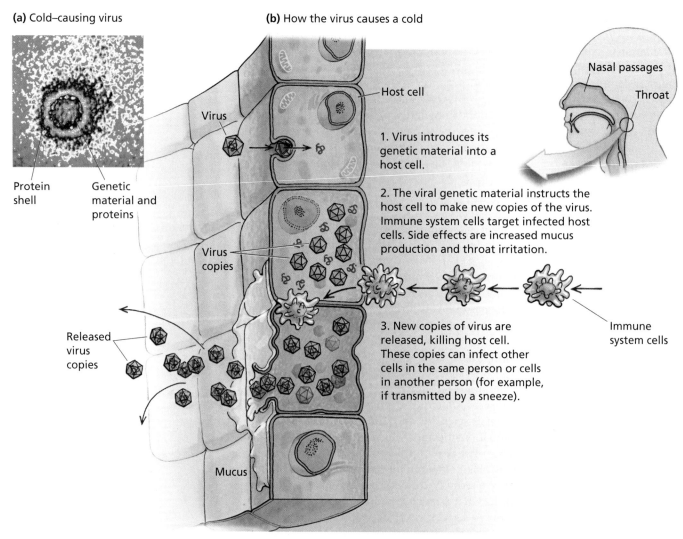

(a) Cold–causing virus

Protein shell

Genetic material and proteins

(b) How the virus causes a cold

Virus

Host cell

1. Virus introduces its genetic material into a host cell.

2. The viral genetic material instructs the host cell to make new copies of the virus. Immune system cells target infected host cells. Side effects are increased mucus production and throat irritation.

Virus copies

Released virus copies

3. New copies of virus are released, killing host cell. These copies can infect other cells in the same person or cells in another person (for example, if transmitted by a sneeze).

Mucus

Nasal passages

Throat

Immune system cells

Figure 1.3 A cold-causing virus. (a) An image from an electron microscope of a typical rhinovirus, one of the many viruses that cause the common cold. (b) A rhinovirus causes illness by invading cells in the lining of the nose and throat and using those cells as "factories" to make virus copies. Cold symptoms result when your immune system attempts to control and eliminate this invader.

sore throat characteristic of infection by most cold viruses appear to be the result of the body's immune response to a viral invasion (Figure 1.3). Because there are literally millions of varieties of cold viruses, many of which will be attacked by the immune system, we are susceptible to "catching cold" throughout our lives; we are not as susceptible to coming down with chicken pox, which is caused by a single virus, more than once. The enormous capacity for viruses to change over time is discussed in Chapter 10.

The statement, "viruses cause the common cold," is generally accepted as a fact for three reasons. Here are the first two: (1) All reasonable alternative hypotheses about the causes of colds (for instance, simply exposure to cold air) have been rejected in numerous experimental tests; and (2) the hypothesis has not been rejected after carefully designed experiments measured cold incidence in people exposed to purified virus samples. "Truth" in science can therefore be defined as *what we know and understand based on all available information.* If a hypothesis appears to explain all instances of a particular phenomenon and has been repeatedly tested and supported, it may eventually be accepted as accurate.

The third reason for considering the relationship between colds and viruses as a fact is that it conforms to a well-accepted scientific principle, the germ theory of disease. A **scientific theory** is an explanation of a set of related observations based on well-supported hypotheses from several different, independent lines of research. The germ theory, like many other scientific theories, arose as a result of inductive reasoning, whereby many observations and the results of hypothesis tests are pulled together to point to a logical conclusion. The **germ theory** arose in the early 20th century from the accumulated observations of biologists such as Louis Pasteur and Robert Koch, who first noted the relationship between particular diseases and particular microorganisms (that is, organisms too small to be seen with the naked eye). The basic premise of the germ theory is that microorganisms are the cause of some or all human diseases. Pasteur inductively reasoned from his observation that bacteria cause milk to become sour and that these same types of organisms can injure humans. Koch's demonstration of the link between anthrax bacteria and a specific set of fatal symptoms in mice provided additional evidence for this theory. Germ theory is further supported by the observation that treating some ill patients with drugs that are known to be deadly to certain microorganisms can relieve the patients of their illness.

In everyday usage, the word *theory* is synonymous with "untested hypothesis" or "induction based on limited information." For example, a father may have a theory about why his young daughter has awakened at 3 a.m. every night for the past week, and a golfer may have a theory about why she is missing so many putts lately. In contrast, scientists use the term *theory* when referring to explanatory models about how the natural world "works" that are typically well-supported by observation and experiment. The germ theory of disease joins other well-accepted theories that form the basis of biology, including the cornerstone—evolutionary theory. The **theory of evolution** is that all organisms derive from a single common ancestor and have changed and diverged through time. Charles Darwin formulated this theory over 150 years ago by drawing together observations of the distribution of species across Earth, the distribution and appearance of fossils, similarities in form among living organisms, and observations of change in traits of a species in response to human-caused selection, among other pieces of information. We discuss the theory of evolution in detail in Chapter 9, but because it is the foundational theory in biology, it is addressed in every chapter of this book.

The supporting foundation of all scientific theories consists of multiple hypothesis tests. Germ theory has been supported by numerous tests of hypotheses suggested by the theory—for instance, observations that infectious human diseases ranging from chicken pox to malaria to tuberculosis to yeast infections have all been clearly associated with infection with viruses, protozoans, bacteria, and fungi. The most unambiguous support for a theory comes in the form of hypothesis tests called controlled experiments.

Controlled Experiments.   **Control** has a very specific meaning in science. A control for an experiment is a subject who is similar to an experimental subject, except that the control is not exposed to the experimental treatment. Measurements of the control group are used as baseline values for comparison to measurements of the experimental group; if the control and experimental groups differ and the experiment is well designed, then the difference is likely due to the experimental treatment.

An extract of *Echinacea purpurea* (a common North American prairie plant) in the form of echinacea tea has been touted as a treatment to reduce the likelihood as well as the severity and duration of colds (Figure 1.4). A recent scientific experiment on the effectiveness of echinacea tea involved asking individuals in both the control and the experimental group to rate their perception of the effectiveness of the cold treatment they were using. In this study, people who used echinacea tea felt that it was 33% more effective at reducing symptoms. The "33% more effective" is in comparison to the opinions of people about the effectiveness of a tea that did not contain *Echinacea* extract—that is, the results from the control group.

A good controlled experiment eliminates as many alternative hypotheses that could explain the observed result as possible. The first step is to select a pool

Figure 1.4 *Echinacea purpurea*, an **American coneflower.** Extracts from the leaves and roots of this plant are among the most popular herbal remedies sold in the United States.

(a)

Control group	Experimental group
Experiencing early cold symptoms	Experiencing early cold symptoms
Sought treatment from clinic	Sought treatment from clinic
Received **placebo** tea	Received **echinacea** tea

(b)

Echinacea tea 33% more effective

Control group   Experimental group

y-axis: Effectiveness of tea at relieving cold symptoms — Not effective ↕ Effective (0–5)

Figure 1.5   A controlled experiment.
(a) In the case of a controlled experiment testing the efficacy of echinacea tea as a treatment for colds, individuals in both groups were treated identically except for the type of tea they consumed.
(b) The results of the experiment indicated that echinacea tea was 33% more effective than the placebo tea.

of subjects in such a way as to eliminate differences in participants' ages, diets, stress levels, and likelihood of visiting a health-care provider. One effective way of eliminating differences between groups is the **random assignment** of individuals to experimental and control groups. For example, a researcher might put all of the volunteers' names in a hat, draw out half, and designate these people as the experimental group and the remainder as the control group. As a result, there is unlikely to be a systematic difference between the experimental and control groups—each group should be a rough cross section of the population in the study. In the echinacea tea experiment just described, members of both the experimental and control groups were female employees of a nursing home who sought relief from their colds at their employer's clinic. Imagine what would happen if the colds experienced in the nursing home changed over the course of the experiment—that is, if one cold virus affected a number of individuals for a few weeks, and then a different cold virus affected other individuals in the next few weeks. If the researchers had simply assigned the first 25 visitors to the clinic to the control group and the next 25 to the experimental group, they would run the risk of the two groups actually being infected with different cold viruses as well as drinking different teas. Then, if the women in the experimental group felt that they recovered more quickly than those in the control group, the researchers would not know if this occurred because echinacea tea is effective or because most of the individuals in the experimental group had a milder cold than did the women in the control group. To avoid this kind of problem, the volunteers were randomly assigned into either the experimental or control group as they came into the clinic.

The second step in designing a good control is to attempt to treat control subjects and experimental subjects identically during the course of the experiment. In this study, all participants received the same information about the purported benefits of echinacea tea, and during the course of the experiment, all participants were given tea with instructions to consume 5 to 6 cups daily until their symptoms subsided. However, individuals in the control group received "sham tea" that did not contain *Echinacea* extract. This sham tea would be equivalent to "sugar pills" that are given to control subjects when testing a particular drug. Like other intentionally ineffective medical treatments, sham tea and sugar pills are called **placebos**. Employing a placebo generates only one consistent difference between individuals in the two groups—in this case, the type of tea they consumed.

Good controls are the basis of **strong inference**. In the echinacea tea study, the data indicated that cold severity was lower in the experimental group compared to those who received the placebo. Because their study utilized controls, the researchers can have high confidence that the two groups differed because *Echinacea* extract relieves cold symptoms. Since their treatment of the control group had greatly reduced the likelihood that alternative hypotheses could explain their results, the researchers could strongly infer that they were measuring a real, positive effect of echinacea tea on colds (Figure 1.5).

The study described here supports the hypothesis that echinacea tea reduces the severity of colds. However, it is extremely rare that a single experiment will cause the scientific community to accept a hypothesis beyond a reasonable doubt. Dozens of studies, each using different experimental designs and many using extracts from different parts of the plant, have investigated the effect of *Echinacea* on common colds and other illnesses. Some of these studies have shown a positive effect, but others have shown none. In the medical community as a whole, the jury is still out regarding the effectiveness of this popular herb as a cold treatment. Only through continued controlled tests of the hypothesis, using a variety of extracts, ways to deliver the extract (e.g., pill or tea form), and common cold varieties, will an accurate answer to the question "Is *Echinacea* an effective cold treatment?" be discerned.

Our confidence in the results of any single experiment can be increased if efforts are made to remove bias from the generation and collection of experimental results.

Minimizing Bias in Experimental Design. Scientists and human research subjects may have strong opinions about the veracity of a particular hypothesis even before it is tested. These opinions may cause participants to unfairly influence, or **bias**, the results of an experiment.

One potential source of bias is subject expectation, which is sometimes called the onstage effect. Individual experimental subjects may consciously or unconsciously model the behavior they feel the researcher expects from them. For example, an individual who knew she was receiving echinacea tea may have felt confident that she would recover more quickly. This might cause her to underreport her cold symptoms. This potential problem is avoided by designing a **blind experiment**, wherein individual subjects are not aware of exactly what they are predicted to experience. In experiments on drug treatments, this means not telling participants whether they are receiving the drug or a placebo.

Another source of bias arises when a researcher makes consistent errors in the measurement and evaluation of results. This phenomenon is called observer bias. In the echinacea tea experiment, observer bias could take various forms. Expecting a particular outcome might lead a scientist to give slightly different instructions about what symptoms constituted a cold to subjects who received echinacea tea. Or, if the researcher expected people who drank echinacea tea to experience fewer colds, she might make small errors in the measurement of cold severity that influenced the final result. To avoid the problem of experimenter bias, the data collectors themselves should be "blind." Ideally, the scientist, doctor, or technician applying the treatment does not know which group (experimental or control) any given subject is part of until after all data have been collected and analyzed (Figure 1.6). Blinding the data collector ensures that the data are **objective**, or in other words, without bias.

We call experiments **double-blind** when *both* the research subjects and the technicians performing the measurements are unaware of either the hypothesis or whether a subject is in the control or experimental group. Double-blind experiments nearly eliminate the effects of human bias on results. When both researcher and subject have few expectations about the hypothesized outcome of a particular experimental treatment, the results obtained from the experiment should be considered more credible.

What they know

Technician "blind"
• **Limited knowledge** of experimental hypothesis
• **No knowledge** of which group participants belong to

Subject "blind"
• **Limited knowledge** of experimental hypothesis
• **No knowledge** of which group he or she belongs to

How they behave

• **No difference** in instructions to participants
• **No difference** in treatment of participants
• **No difference** in data collection

• **Unbiased** reporting of symptoms or effects of treatment

Figure 1.6 Double-blind experiments. Double-blind experiments result in more objective data.

(a) Rat

(b) Dog

(c) Chimpanzee

Figure 1.7   Model organisms in science.
Testing hypotheses on human health and
disease often occurs using nonhuman
organisms that share many aspects of
biology with us, including (a) the classic
"lab rat," which is easy to raise and care for
and has little genetic diversity to confound
the results of an experiment; (b) dogs,
whose cardiovascular systems are very similar to those of humans and easier to work
with than the systems in rats; and (c) chimpanzees, which are humans' closest
biological relatives.

Using Correlation to Test Hypotheses

Well-controlled experiments can be difficult to perform when humans are
the experimental subjects. As you can see from the echinacea tea study, the
requirement that both experimental and control groups be treated nearly
identically means that some people receive no treatment. In the case of cold
sufferers who have limited means of reducing cold duration and severity,
the placebo treatment does not substantially hurt those who receive it.
However, placebo treatments are impossible or unethical in many cases. For
instance, imagine testing the effectiveness of a birth control drug by giving
one group of women the drug and comparing their rate of pregnancies to
that for another group of women who thought they were using effective
birth control but were actually getting a placebo!

One method scientists may use to test hypotheses on human biology using
experiments that could not be performed on humans is to use **model organisms**
as stand-ins for people. In the case of research on human health and disease,
model organisms are typically other mammals. Mammals are especially useful
as model organisms in medical research because they are closely related to us
evolutionarily—like us, they have hair and give birth to live young, and thus
they also share with us similarities in anatomy and physiology. The vast majority of animals used in biomedical research are rodents such as rats, mice, and
guinea pigs, although some areas of research require animals that are more similar to humans in size, such as dogs or pigs, or share a closer evolutionary
relationship, such as chimpanzees (Figure 1.7). The use of model organisms
allows experimental testing on potential drugs and other therapies before these
methods are employed on people. Research on model organisms has contributed to a better understanding of nearly every serious human health threat,
including cancer, heart disease, Alzheimer's disease, and AIDS. However, ethical concerns about the use of animals in research persist and can complicate
such studies. Additionally, the results of animal studies are not always directly
applicable to humans because despite a shared evolutionary history, animals
still can have important differences from humans in physiology. Testing
hypotheses about human health in human beings still provides the clearest
answer to these questions.

When controlled experiments on humans are difficult or impossible to perform, scientists will test hypotheses using **correlations**. A correlation is a relationship between two variables. Suggestions about using meditation to
reduce susceptibility to colds are based on a correlation between high levels of
psychological stress and increased susceptibility to cold-virus infections
(Figure 1.8). This correlation was generated by researchers who collected data
on a number of individuals' psychological stress levels before giving them
nasal drops that contained a cold virus. Doctors later reported on the incidence and severity of colds among participants in the study. Note that while
the cold virus was applied to each participant in the study, the researchers had
no influence on the stress level of the study participants—in other words, this
is not a controlled experiment because people were not randomly assigned to
different "treatments" of low or high stress.

Let's examine the results of this study, presented in Figure 1.8. The horizontal axis of the graph, or x-axis, contains a scale of stress level—from low
stress on the left edge of the scale to high stress on the right. The vertical axis
of the graph, the y-axis, indicates the percentage of study participants who
developed "clinical colds," that is, colds reported by their doctors. Each point
on the graph represents a group of individuals and tells us what percentage of
people in each stress category had clinical colds. The line connecting the
5 points on the graph illustrates a correlation—the relationship between
stress level and susceptibility to cold-virus infections. Because the line rises to
the right, these data tell us that people who have higher stress levels were
more likely to come down with colds. In fact, it appears from the data in the
graph that individuals experiencing high levels of stress are more than twice

as likely to become ill than those with low stress. But does this relationship mean that high stress causes increased cold susceptibility?

To conclude that stress causes illness, we need the same assurances that are given by a controlled experiment. In other words, we must assume that the individuals measured for the correlation are similar in every way, except for their stress levels. Is this a good assumption? Not necessarily. Most correlations cannot control for alternative hypotheses. People who feel more stressed may have poorer diets because they feel time limited and rely on fast food more often. Additionally, people who feel highly stressed may be in situations where they are exposed to more cold viruses. These differences among people who differ in stress level may also influence their cold susceptibility (Figure 1.9). Therefore, even with a strong correlational relationship between the two factors, we cannot strongly infer that stress *causes* decreased resistance to colds.

Researchers who use correlational studies do their best to ensure that their subjects are similar in many characteristics. For example, this study on stress and cold susceptibility evaluated whether individuals in the different stress categories were different in age, weight, sex, education, and their exposure to infected individuals. None of these other factors differed among low-stress and high-stress groups. Eliminating some of the alternative hypotheses that could explain this correlation increases the strength of the inference that high stress levels truly do increase susceptibility to colds. However, people with high-stress lifestyles still may be fundamentally different from those with low-stress lifestyles, and it is possible that one of those important differences is the real cause of disparities in cold frequency.

You may see from this discussion that it is difficult to demonstrate a cause-and-effect relationship between two factors simply by showing a correlation between them. In other words, *correlation does not equal causation*. For example, a commonly understood correlation exists between exposure to cold air and epidemics of the common cold. It is true that as outdoor temperatures drop, the incidence of colds increases. But numerous controlled experiments indicate that chilling does not increase susceptibility to colds. Instead, cold outdoor temperatures mean increased close contact with other people (and their viruses). Despite the correlation, cold air does not cause colds—exposure to viruses does.

Figure 1.8 Correlation between stress level and illness. This graph summarizes the results of an experiment that compared rates of virus infection in groups of individuals with different self-reported stress levels. The graph indicates that people experiencing higher levels of stress became infected by a virus more often than did people experiencing low levels of stress.

(a) Does high stress cause high cold frequency?

(b) Or does one of the causes of high stress cause high cold frequency?

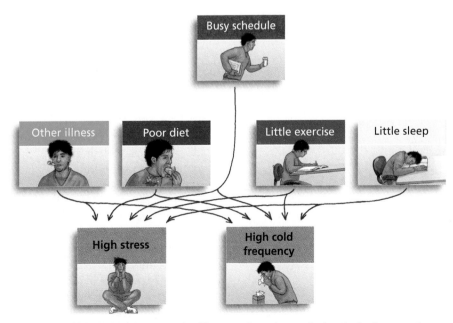

Figure 1.9 Correlation does not signify causation. A correlation typically cannot eliminate all alternative hypotheses.

Understanding Statistics

During a review of scientific literature on cold prevention and treatment, you may come across statements about the "significance" of the effects of different cold-reducing measures. For instance, one report may state that factor A reduced cold severity but that the results of the study were "not significant." Another study may state that factor B caused a "significant reduction" in illness. We might then assume that this statement means factor B will help us feel better, whereas factor A will have little effect. This is an incorrect assumption, because in scientific studies, *significance* is defined a bit differently from its usual sense. To evaluate the scientific use of the term *significance*, we need a basic understanding of statistics.

What Statistical Tests Can Tell Us.   **Statistics** is a specialized branch of mathematics used in the evaluation of experimental data. An experimental test utilizes a small subgroup, or **sample**, of a population. Statistics can be employed to summarize data from the sample—for instance, we can describe the average length of colds experienced by experimental and control groups. **Statistical tests** can then be used to extend the results from a sample to the entire population. When scientists conduct an experiment, they hypothesize that there is a true, underlying effect of their experimental treatment on the entire population. An experiment on a sample of a population can only estimate this true effect because a sample is always an imperfect "snapshot" of an entire population.

Imagine trying to determine the average hair length of all women students in a class of 400 by taking a snapshot of a random subset of 10 women. If hairstyles were identical among the women in the snapshot, you could reasonably assume that the average hair length in the class is approximately equal to the average hair length in the snapshot. However, you are much more likely to see that women in the snapshot have a variety of hairstyles. When you calculate the average hair length in the snapshot, it is unlikely to be identical to the average hair length you would calculate if you measured everyone in the entire class. If the women in the snapshot exhibited a wide variety of hairstyles, from crew cuts to long braids, it would be more difficult to determine whether the average hair length in the snapshot is at all close to the average for the class. Your snapshot could, by chance, contain the one woman in the class with extremely long hair, causing the average length in this sample to be much longer than the average length for the class. Statistical tests help scientists evaluate, given the number of individuals sampled and the variance among individuals in how they responded to treatment, how likely it is that their snapshot provides an accurate picture of the whole population.

In the experiment with the echinacea tea, statistical tests tell us the likelihood that the 33% reduction in cold severity observed by the researchers is an accurate measure of how well echinacea tea works, versus the likelihood that the 33% reduction resulted from chance differences between the experimental and control groups.

Statistical Significance.   We can explore the role that statistical tests play by evaluating a study on another proposed treatment to reduce the severity of colds—lozenges containing zinc. Some forms of zinc can block certain common cold viruses from entering the cells that line the nasal cavity. This observation led scientists to hypothesize that consuming zinc at the beginning of a cold decreases the number of cells that become infected, which in turn decreases the length and severity of cold symptoms. To test this hypothesis, a group of researchers at the Cleveland Clinic performed a study using a sample of 100 of their employees who enrolled in the study within 24 hours of developing cold symptoms. The researchers randomly assigned subjects to control or experimental groups. Members of the experimental group received lozenges containing zinc, while members of the control group received placebo lozenges. Members of both groups received the same instructions for using the lozenges and were asked to rate their symptoms until they had recovered. The experiment was double-blind.

When the data from the experiment were summarized, researchers observed that the mean length of time to recovery was more than 3 days shorter in the zinc group than in the placebo group (Figure 1.10). Superficially, this result appears to support the hypothesis. However, recall the example of the snapshot of women's hair length. A statistical test is necessary because even with well-designed experiments, if a population is variable, chance will always result in some difference between the control and experimental groups. The effect of chance on experimental results is known as **sampling error**—more specifically, sampling error is the difference between a sample of a population and the population as a whole. In the snapshot described earlier, sampling error is the difference between the women in the snapshot and the women in the population as a whole in average hair length. Similarly, in any experiment, the group of individuals assigned to the experimental treatment and the group assigned to the control treatment will differ from each other in random ways. Even if there is *no* true effect of an experimental treatment, the results observed in the experimental and control groups will never be exactly the same.

For example, we know that people differ in their ability to recover from a cold infection. If we give zinc lozenges to one volunteer and placebo lozenges to another, it is likely that the two volunteers will have colds of different lengths. But even if the zinc-taker had a shorter cold than the placebo-taker, you would probably say that the test did not tell us much about our hypothesis—the zinc-taker might just have had a less severe cold for other reasons. Now imagine that we had 5 volunteers in each group and saw a difference, or that the difference was only 1 day instead of 3 days. How would we determine if the lozenges had an effect? Statistical tests allow researchers to look at their data and determine how likely it is that the result is due to sampling error. In the case of the zinc lozenge experiment, the researchers used a statistical test to determine if the large reduction in cold duration in the experimental group was likely to be due to the effectiveness of zinc as a cold treatment or due to a chance difference between the experimental and control groups. In the case of the experiment with zinc lozenges, the statistical test indicated that there was a low probability, less than 1 in 10,000 (0.01%), that the experimental and control groups were so different simply by chance. In other words, the result is **statistically significant**.

Figure 1.10   Zinc lozenges reduce the duration of colds. This graph illustrates the results of an experiment testing the effectiveness of zinc lozenges on decreasing cold duration. Individuals in the experimental group had colds lasting about 4½ days as opposed to approximately 7½ days for the placebo group.

Factors Influencing Statistical Significance.   One characteristic of experiments influencing the power of statistical tests is sample size—the number of individuals in the experimental and control groups. If a treatment has no effect, a small sample size might mislead researchers into thinking that it does; this was the case with the vitamin C hypothesis described earlier. Subsequent tests with larger sample sizes, encompassing a wider variety of individuals with different underlying susceptibilities to colds, allowed scientists to finally reject the hypothesis that vitamin C prevents colds. Conversely, if the effect of a treatment is real, but the sample size of the experiment is small, a single experiment may not allow researchers to determine convincingly that their hypothesis has support. Several of the experiments that test the efficacy of *Echinacea* extract demonstrate this phenomenon. For example, one study performed at a clinic in Wisconsin with 48 participants indicated that members of the echinacea-consuming experimental group were 30% less likely to experience any cold symptoms after being exposed to a virus as compared to individuals who received a placebo; however, the small sample size of the study meant that this result was not statistically significant.

The more participants there are in a study, the more likely it is that researchers will see a true effect of an experimental treatment, if one exists. If the sample size is large, any difference between an experimental and control group, no matter how small, is more likely to be statistically significant. For example, a study of over 21,000 male smokers over 6 years in Finland demonstrated that men who took vitamin E supplements had 5% fewer colds than did the men who did not take these supplements. In this case, the large sample size allowed researchers to see that vitamin E has a real, but relatively tiny, effect on cold incidence.

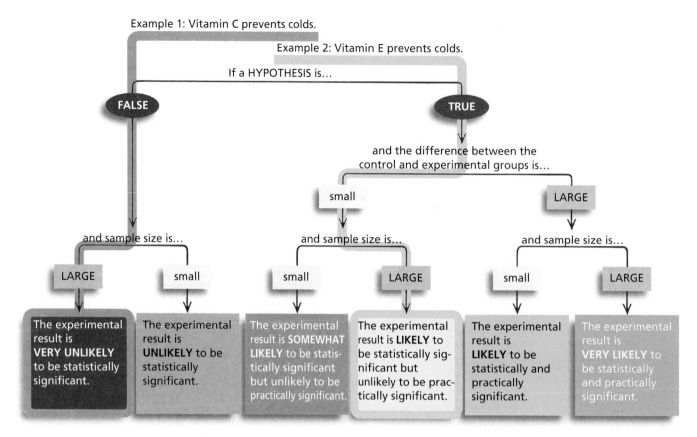

Figure 1.11 Factors that influence statistical significance. This flowchart summarizes the relationship between the true effect of a treatment and the sample size of an experiment on the likelihood of obtaining statistical significance.

The relationship between hypotheses, experimental tests, sample size, and statistical and practical significance is summarized in Figure 1.11. However, there is one final caveat to this discussion. A statistically significant result is typically defined as one that has a 5% probability or less of being due to chance alone. If all scientific research uses this same standard, as many as one in every 20 statistically significant results (that is, 5% of the total) is actually reporting an effect that is *not real*. In other words, some statistically significant results simply represent a surprisingly large difference between the experimental and control group that occurred only as a result of sampling error. An experiment with a statistically significant result will still be considered to support the hypothesis it was meant to test. However, the small but important probability that the results are due to chance explains why one supportive experiment is usually not enough to convince all scientists that a particular hypothesis is accurate.

Even with a statistical test indicating that the result had a likelihood of less than 0.01% of occurring by chance, we should begin to feel assured that taking zinc lozenges will reduce the duration of colds only after locating additional tests of this hypothesis that give similar results. In fact, scientists continue to test this hypothesis, and there is still no consensus among them about the effectiveness of zinc as a cold treatment.

What Statistical Tests Cannot Tell Us. Statistical significance by itself is not a sufficient measure of the accuracy of an experiment, and all statistical tests operate with the assumption that the experiment was *designed and carried out correctly*. In other words, a statistical test evaluates the chance of sampling error, not observer error, and a statistically significant result should never be taken as the last word on an experimentally tested hypothesis. An examination of the experiment itself is required. In the test of the effectiveness of zinc lozenges, the experimental design minimized the likelihood that alternative hypotheses could explain the results by randomly assigning subjects to treatment groups, using an effective placebo, and blinding both the data collectors and the subjects. Given such a well-designed experiment, this statistically significant result allows researchers to strongly infer that consuming zinc lozenges reduces the duration of colds.

Statistical significance is also not equivalent to *significance* as we define the term more practically, that is, as "meaningful or important." A very large sample size can reveal a very small effect, as described earlier in the study reporting the positive effect of vitamin E on cold prevention. Unfortunately, experimental results reported in the news often use the term *significant* without clarifying this important distinction. Understanding that problem, as well as other sometimes misleading aspects of how science is popularly presented, will enable you to better use scientific information.

1.2 Evaluating Scientific Information

Given the challenges inherent in establishing scientific "truth"—the rigorous requirements for using controls to eliminate alternative hypotheses and the problem of sampling error—we can see why definitive scientific answers to our questions are slow in coming. A well-designed experiment can certainly allow us to approach the truth. Looking critically at reports of experiments can help us make well-informed decisions about actions to take. Most of the research on cold prevention and treatment is first published as **primary sources**, written by the researchers themselves and reviewed within the scientific community (Figure 1.12). The process of **peer review** helps increase confidence in scientific information because other scientists critique the results and conclusions of an experiment before it is published in a professional journal. Research articles in journals such as *Science, Nature,* the *Journal of the American Medical Association,* and hundreds of others represent the first and most reliable sources of current scientific knowledge.

However, evaluating the hundreds of scientific papers that are published weekly is a task no one of us can perform. Even if we focused only on a particular field of interest, the technical jargon used in many scientific papers may be a significant barrier to those of us who are not members of that

Scientists write a paper and submit it to a journal.

The journal sends the paper to other scientists for peer review.

Paper is rejected because of poor experimental design, or the findings are not appropriate for the journal.

Paper is returned to authors for revision (sometimes more than once).

OR

Primary sources

The journal publishes the revised paper.

Secondary sources

Media reports appear in radio, newspaper, magazines, and/or TV.

Comments from other scientists are published in subsequent volumes of the journal.

Figure 1.12 Primary sources: publishing scientific results. After an experiment is complete, the researchers write a scientific paper for publication in a journal. Both before and after publication, the paper is reviewed by other scientists who evaluate the research presented in the paper and the researchers' conclusions. Peer review provides checks and balances that help maintain the integrity of the science presented.

particular scientific subfield. Instead of reading the primary literature, most of us receive our scientific information from **secondary sources** such as books, news reports, and advertisements. How can we evaluate information in this context?

Information from Anecdotes

Information about dietary supplements such as echinacea tea and zinc lozenges is often in the form of **anecdotal evidence**—meaning that the advice is based on one individual's personal experience. A friend's enthusiastic plug for vitamin C, because she felt it helped her, is an example of a testimonial—a common form of anecdote. Advertisements that use a celebrity to pitch a product "because it worked for them" are classic forms of testimonials. You should be very cautious about basing decisions on anecdotal evidence, which is not in any way equivalent to well-designed scientific research. For example, countless hours of research have established that there is a clear link between cigarette smoking and lung cancer. Although many of us have heard anecdotes along the lines of the grandpa who was a pack-a-day smoker and lived to the age of 94, the risk of premature death due to smoking is very well established by careful scientific research. Although anecdotes may indicate that a product or treatment has merit, only well-designed tests of the hypothesis can help determine if it is likely to be safe and effective for most people.

Science in the News

Popular news sources provide a steady stream of health information. However, stories about research results in the general media often do not contain information about the adequacy of controls, the number of subjects, the experimental design, or the source of the scientist's funding. How can anyone evaluate the quality of research that supports statements like these? "Supplement Helps Melt Fat and Build Muscle," or "Curry Spice Might Prevent Bowel Cancer."

First, you must consider the source of media reports. Certainly news organizations will be more reliable reporters of fact than will entertainment tabloids, and news organizations with science writers should be considered better reporters of the substance of a study than those without. Television talk shows, which need to fill air time, regularly have guests who promote a particular health claim. Too often these guests may be presenting information that is based on anecdotes or an incomplete survey of the primary literature as well as work that has not been subjected to peer review.

Paid advertisements are a legitimate means of disseminating information. However, claims in advertising should be very carefully evaluated. Our pursuit of health fuels a multibillion-dollar industry—companies that succeed need to be very effective at getting the attention of consumers. Advertisements of over-the-counter and prescription drugs must conform to rigorous government standards regarding the truth of their claims, but advertisements for herbal supplements, many health food products, and diet plans have lower standards. Be sure to examine the fine print, because advertisers often are required to clarify the statements made in bold type in their ads.

Another commonly used source for health information is the Internet. As you know, anyone can post information on the Internet. Typing in "common cold prevention" on a standard web search engine will return thousands of web pages—from highly respected academic and government sources to small companies trying to sell their products, or individuals who have strong, sometimes completely unsupported, ideas about cures. Often it can be difficult to determine the reliability of a well-designed website. Here are some things to consider when using the web as a resource for health information:

1. Choose sites maintained by reputable medical establishments, such as the National Institute of Health (NIH) or the Mayo Clinic.

2. It costs money to maintain a website. Consider whether the website seems to be promoting a product or agenda. Advertisements for a specific product should alert you to a website's bias.

3. Check the date when the website was last updated, and see whether the page has been updated since its original posting. Science and medicine are disciplines that must frequently incorporate new data into hypotheses. A reliable website will be updated often.

4. Determine whether unsubstantiated claims are being made. Look for references, and be suspicious of any studies that are not from peer-reviewed journals.

Understanding Science from Secondary Sources

Once you are satisfied that a media source is relatively reliable, you can examine the scientific claim that it presents. Use your understanding of the process of science and of experimental design to evaluate the story and the science. Does the story about the claim present the results of a scientific study, or is it built around an untested hypothesis? Is the story confusing correlation with causation? Does it seem that the information is applicable to non-laboratory situations, or is it based on results from preliminary or animal studies? Look for clues about how well the reporters did their homework. Scientists usually discuss the limitations of their research in their papers; are these cautions noted in an article or television piece? If not, the reporter may be overemphasizing the applicability of the results.

Then, note if the scientific discovery itself is controversial. That is, does it reject a hypothesis that has long been supported? Does it concern a subject that is controversial in human society (like the origin of racial differences or homosexuality)? Might it lead to a change in social policy? In these cases, be extremely cautious. New and unexpected research results must be evaluated in light of other scientific evidence and understanding. Reports that lack comments from other experts in related fields may omit important problems with a study or fail to place the study in context with other research. Table 1.1 provides a checklist of questions to ask and answer as you evaluate a news report about a recent scientific development.

Finally, realize that even among the most credible organizations, the news media generally highlights only those stories about experiments that editors and producers find newsworthy. As we have seen, scientific understanding accumulates relatively slowly, with many tests of the same hypothesis finally leading to an accurate understanding of a natural phenomenon. News organizations are also more likely to report a study that supports a hypothesis rather than one that gives less supportive results, even if both types of studies exist. In addition, even the most respected media sources may not be as thorough as readers would like. For example, a recent review published in the *New England Journal of Medicine* evaluated the news media's coverage of new medications. Of 207 randomly selected news stories, only 40% that cited experts who had financial ties to a drug disclosed this relationship. This potential conflict of interest may influence the expert's credibility. Another 40% of the news stories did not provide basic statistics about the drugs' benefits. Most of the news reports also failed to distinguish between absolute benefits (how many people were helped by the drug) and relative benefits (how many people were helped by the drug relative to other therapies for the condition). The *Journal*'s review is a vivid reminder that we need to be cautious when reading or viewing news reports on scientific topics.

Even after following all of these guidelines, you may still find situations where reports on several scientific studies seem to give conflicting and confusing results. This could mean one of two things: Either the reporter is not giving you enough information, in which case you may want to read the researchers' papers yourself, or the researchers themselves are just as confused as you are. Such

Table 1.1   A guide for evaluating science in the news. For each question, check the appropriate box.

Questions	Possible answers			
	Preferred answer ☑		**Raises a red flag** ☑	
1. What is the basis for the story?	Hypothesis test	☐	Untested assertion *No data to support claims in the article.*	☐
2. What is the affiliation of the scientist?	Independent (university or government agency)	☐	Employed by an industry or advocacy group *Data and conclusions could be biased.*	☐
3. What is the funding source for the study?	Government or nonpartisan foundation (without bias)	☐	Industry group or other partisan source (with bias) *Data and conclusions could be biased.*	☐
4. **If the hypothesis test is a correlation:** Did the researchers attempt to eliminate reasonable alternative hypotheses?	Yes	☐	No *Correlation does not equal causation. One hypothesis test provides poor support if alternatives are not examined.*	☐
If the hypothesis test is an experiment: Is the experimental treatment the only difference between the control group and the experimental group?	Yes	☐	No *An experiment provides poor support if alternatives are not examined.*	☐
5. Was the sample of individuals in the experiment a good cross section of the population?	Yes	☐	No *Results may not be applicable to the entire population.*	☐
6. Was the data collected from a relatively large number of people?	Yes	☐	No *Study is prone to sampling error.*	☐
7. Were participants blind to the group they belonged to and/or to the "expected outcome" of the study?	Yes	☐	No *Subject expectation can influence results.*	☐
8. Were data collectors and/or analysts blinded to the group membership of participants in the study?	Yes	☐	No *Observer bias can influence results.*	☐
9. Did the news reporter put the study in the context of other research on the same subject?	Yes	☐	No *Cannot determine if these results are unusual or fit into a broader pattern of results.*	☐
10. Did the news story contain commentary from other independent scientists?	Yes	☐	No *Cannot determine if these results are unusual or if the study is considered questionable by others in the field.*	☐
11. Did the reporter list the limitations of the study or studies he or she is reporting on?	Yes	☐	No *Reporter may not be reading study critically and could be overstating the applicability of the results.*	☐

Note the number of red flags raised by the time you have reached this point. Some of the issues raised may be more serious than others; but in general, the fewer the red flags, the more thorough the report and the more reliable the scientific study.

confusion is the nature of the scientific process—early in our search for understanding a phenomenon, many hypotheses are proposed and discussed; some are tested and rejected immediately, and some are supported by one experiment but later rejected by more thorough experiments. It is only by clearly understanding the process and pitfalls of scientific research that you can distinguish "what we know" from "what we don't know."

1.3 Is There a Cure for the Common Cold?

So where does our discussion leave us? Will we ever find the best way to prevent a cold or reduce its effects? In the United States, over 1 billion cases of the common cold are reported per year, costing billions of dollars in medical visits, treatment, and lost work days. Consequently, there is an enormous effort to find effective protection from the different viruses that cause colds. A search of the leading medical publication database indicates that every year, nearly 100 scholarly articles regarding the biology, treatment, and consequences of common cold infection are published. This research has led to several important discoveries about the structure and biochemistry of common cold viruses, how they enter cells, and how the body reacts to these infections.

Despite all of the research and the emergence of some promising possibilities, the best prevention method known for common colds is still the old standby—keep your hands clean. Numerous studies have indicated that rates of common-cold infection are 20% to 30% lower in populations who employ effective hand-washing procedures. Cold viruses can survive on surfaces for many hours; if you pick them up from a surface on your hands and transfer them to your mouth, eyes, or nose, you may inoculate yourself with a 7-day sniffle.

Of course, not everyone gets sick when exposed to a cold virus. The reason one person has more colds than another might not be due to a difference in personal hygiene. The correlation that showed a relationship between stress and cold susceptibility appears to have some merit. Research indicates that among people exposed to viruses, the likelihood of ending up with an infection increases with high levels of psychological stress—something that many college students clearly experience. Research also indicates that vitamin C intake, diet quality, exposure to cold temperatures, and exercise frequency appear to have *no effect* on cold susceptibility; although, along with echinacea tea and zinc lozenges, there is some evidence that vitamin C may reduce cold symptoms after infection. Surprisingly, even though medical research has led to the elimination of killer viruses such as smallpox and polio, scientists are still a long way from "curing" the common cold.

CHAPTER REVIEW

Summary

1.1 The Process of Science

- Science is a process of testing hypotheses—statements about how the natural world works. Scientific hypotheses must be testable and falsifiable (pp. 2–3).

- Hypotheses are often generated as a result of inductive reasoning, whereby scientists infer a general principle through observing a number of specific events (p. 3).

- Hypotheses are tested via the process of deductive reasoning, which allows researchers to make specific predictions about expected observations (p. 4).

- Absolutely proving hypotheses is impossible. However, well-designed scientific experiments can allow researchers to strongly infer that their hypothesis is correct (p. 4).

- A scientific theory is an explanation of a set of related observations based on well-supported hypotheses from several different, independent lines of research (pp. 6–7).

- Controlled experiments test hypotheses about the effect of experimental treatments by comparing a randomly assigned experimental group with a control group. Controls are individuals who are treated identically to the experimental group except for application of the treatment (pp. 7–8).

- Bias in scientific results can be minimized with double-blind experiments that keep subjects and data collectors unaware of which individuals belong in the control or experimental group (p. 9).

- In situations where performing controlled experiments on humans is considered unethical, scientists sometimes employ model organisms, such as other mammals (p. 10).

- Some hypotheses can be tested using a correlational approach, which looks for associations between two factors. A correlation can show a relationship between two factors, but it does not eliminate all alternative hypotheses (pp. 10–11).

- Statistics help scientists evaluate the results of their experiments by determining if results appear to reflect the true effect of an experimental treatment on a sample of a population (p. 12).

- A statistical test indicates the role that chance plays in the experimental results; this is called sampling error. A statistically significant result is one that is very unlikely to be due to chance differences between the experimental and control group. (pp. 12–13).

- Even when an experimental result is highly significant, hypotheses are tested multiple times before scientists come to consensus on the true effect of a treatment (p. 14).

Web Tutorial 1.1 **Science as a Process: Arriving at Scientific Insights**

1.2 Evaluating Scientific Information

- Primary sources of information are experimental results published in professional journals and peer reviewed by other scientists before publication (p. 15).

- Most people get their scientific information from secondary sources such as the news media. Being able to evaluate science from these sources is an important skill (pp. 15–16).

- Anecdotal evidence is an unreliable means of evaluating information, and media sources are of variable quality; distinguishing between news stories and advertisements is important when evaluating the reliability of information. The Internet is a rich source of information, but users should look for clues to a particular website's credibility (p. 16).

- Stories about science should be carefully evaluated for information on the actual study performed, the universality of the claims made by the researchers, and other studies on the same subject. Sometimes confusing stories about scientific information are a reflection of controversy within the scientific field itself (p. 19).

Learning the Basics

1. How does a double-blind experiment decrease the amount of bias introduced into experimental results?

2. What does statistical significance mean?

3. What are the advantages and disadvantages of using correlations to test hypotheses?

4. A scientific hypothesis is _____.

   A. an opinion; **B.** a proposed explanation for an observation; **C.** a fact; **D.** easily proved true; **E.** an idea proposed by a scientist

5. Which of the following is a prediction of the hypothesis: Eating chicken noodle soup is an effective treatment for colds?

   A. People who eat chicken noodle soup have shorter colds than do people who do not eat chicken noodle soup.; **B.** People who do not eat chicken noodle soup experience unusually long and severe colds.; **C.** Cold viruses cannot live in chicken noodle soup.; **D.** People who eat chicken noodle soup feel healthier than do people who do not eat chicken noodle soup.; **E.** Consuming chicken noodle soup causes people to sneeze.

6. Control subjects in an experiment _____.

   A. should be similar in most ways to the experimental subjects; **B.** should not know whether they are in the control or experimental group; **C.** should have essentially the same interactions with the researchers as the experimental subjects; **D.** help eliminate alternative hypotheses that could explain experimental results; **E.** all of the above

7. A relationship between two factors, for instance between outside temperature and the number of people with active colds in a population, is known as a(n) _____.

   A. significant result; **B.** correlation; **C.** hypothesis; **D.** alternative hypothesis; **E.** experimental test

8. If the results of an experiment are exactly what was predicted by the hypothesis, then _____.

   A. the hypothesis is proved; **B.** the alternative hypotheses are falsified; **C.** the hypothesis is supported; **D.** the hypothesis was scientific; **E.** none of the above

9. Statistical tests tell us _____.

   A. if an experimental treatment showed more of an effect than would be predicted by chance; **B.** if a hypothesis is true; **C.** whether an experiment was well designed; **D.** if the experiment suffered from any bias; **E.** how similar the sample was to the population from which it was drawn

10. A primary source of scientific results is _____.

    A. the news media; **B.** anecdotes from others; **C.** articles in peer-reviewed journals; **D.** the Internet; **E.** all of the above

Analyzing and Applying the Basics

1. There is a strong correlation between obesity and the occurrence of a disease known as type 2 diabetes—that is, obese individuals have a higher instance of diabetes than non-obese individuals do. Does this mean that obesity causes diabetes? Explain.

2. In an experiment on the effect of vitamin C on reducing the severity of cold symptoms, college students visiting their campus health service with early cold symptoms either received vitamin C or were treated with over-the-counter drugs. Students then reported on the length and severity of their colds. The timing of dosages and the type of pill were very different; thus, both the students and the clinic health providers knew which treatment students were receiving. This study reported that vitamin C significantly reduced the length and severity of colds experienced in this population. Why might this result be questionable, given the experimental design?

3. Samuel George Morton, who published data in the 1840s, reported differences in brain size among human races. His research indicated that Europeans had larger brains than did Native Americans and Africans. His measures of brain size were based on skull volume, calculated by packing individual skulls with mustard seed and then measuring the volume of the seeds they contained. When the biologist Stephen Jay Gould reexamined Morton's data in the 1970s, he found that Morton had systematically erred in his measurements—consistently underestimating the size of the African and Native American skulls. According to Gould, Morton appeared not to realize that he was affecting his own results to support his hypothesis that Europeans had larger brains than the other groups. How could Morton have designed his experiment to minimize the effect of this bias on his results?

Connecting the Science

1. Much of the research on common cold prevention and treatment is performed by scientists employed or funded by drug companies. Often these companies do not allow scientists to publish the results of their research, for fear that competitors at other drug companies will use this research to develop a new drug before they do. Should our society allow scientific research to be owned and controlled by private companies?

2. Should society restrict the kinds of research performed by government-funded scientists? For example, many people believe that research performed on tissues from human fetuses should be restricted; these people believe that such research would justify abortion. If most Americans feel this way, should the government avoid funding this research? Are there any risks associated with *not* funding research with public money?

CHAPTER 2

Are We Alone in the Universe?

Water, Biochemistry, and Cells

Does life exist anywhere besides Earth?

Popular images aside, we know of no other intelligent life in our solar system.

But there is intriguing evidence that life once existed on Mars.

In the summer of 1996, Dr. David McKay announced to the world that he and his colleagues had found persuasive, if not conclusive, evidence of life on Mars. People around the world were astounded. How could this cold, dry, harsh planet harbor life?

The evidence of life found by Dr. McKay's team did not in any way resemble the cartoon images often used to depict Martians. Instead, what these scientists found was evidence of life in a 3.6-billion-year-old, potato-sized rock. They believe that the rock had been ejected from the surface of Mars around 15 million years ago and had traveled through space for nearly that entire time. Ultimately the rock crashed to Earth, landing in Antarctica about 13,000 years ago, and remained there until discovered by scientists in 1984. This meteorite, drably named ALH84001, appeared to contain the same features that scientists use to demonstrate the existence of life in 3.6-billion-year-old Earth rocks—there were fossils, various minerals that are characteristic of life, and evidence of complex chemicals typically produced by living organisms. The announcement of this discovery injected new energy into Mars exploration. Since then, multiple robotic rovers and mapping satellites have been sent to the planet, and in January 2004, President George W. Bush announced an initiative to send astronauts to the red planet by the 2020s. While there are many reasons to explore Mars, the question that remains most

Humans are willing to expend enormous resources to look for life outside Earth.

intriguing—and is a significant focus of several of these missions—is whether life ever existed there.

The fascination about potential Martian life speaks to a fundamental question that many humans share: Are the creatures on Earth the only living organisms in the universe? Our galaxy is filled with countless stars and planets, and the universe teems with galaxies. Even if we find no convincing evidence of life on Mars, there is a seemingly infinite number of places to look for other living beings. In this chapter, we discuss the characteristics and requirements of life and examine techniques that scientists use to search the universe for other living creatures.

2.1   What Does Life Require?

Because the galaxy likely contains billions of planets, scientists looking for life elsewhere seek to identify the range of conditions under which they would expect life to arise. What is it that scientists look for when identifying a planet (or moon) as a candidate for hosting life?

A Definition of Life

In science-fiction movies, alien life-forms are often obviously alive, and even somewhat familiar looking. But in reality, living organisms may be truly alien; that is, they may look nothing like organisms we are familiar with on Earth. So how would we determine whether an entity found on another planet was actually *alive*?

Surprisingly, biologists do not have a simple definition for a "living organism." A list of the attributes found in most earthly life-forms includes growth, **metabolism** (all of the chemical processes that occur in cells, including the breakdown of substances to produce energy, the synthesis of substances necessary for life, and the excretion of wastes generated by these processes), movement, reproduction, and response to external environmental stimuli. However, this definition could apply to things that no one considers to be living. For example, fire can grow, consume energy, give off waste, move, reproduce by sending off sparks, and change in response to environmental conditions. And some organisms that are clearly living do not conform to this definition. For instance, male mules grow, metabolize, move, and respond to stimuli, but they are sterile (unable to reproduce). In practice, most biologists do not attempt to define the characteristics of living organisms; they are content to apply the same standard that U.S. Supreme Court Justice Potter Stewart did when struggling to define "obscene material" in 1964: "I know it when I see it." Unfortunately, when people on Earth are faced with a truly alien life-form, it is not clear that this approach will be sufficient. In fact, it may not be safe to assume that living things on other planets would have the same characteristics as living things on Earth.

If we examine more closely the characteristics of living organisms on Earth, we will see that all organisms contain a common set of biological molecules, are composed of cells, and can maintain **homeostasis**, that is, a roughly constant internal environment despite an ever-changing external environment (Figure 2.1). In addition, populations of living organisms can evolve, that is, change in average physical characteristics over time. If we search the universe for planets that could support life similar to that found on Earth—and thus organisms that we would clearly identify as

Figure 2.1 Homeostasis. Black-capped chickadees can maintain a core body temperature of 108°F during the day, even when the air temperature is well below zero. This constant internal temperature requires a complex feedback system between multiple sensory and physiological systems that is possible only in living organisms.

"living"—the list of planetary requirements becomes more stringent. In particular, an Earth-like planet should have abundant liquid **water** available.

The Properties of Water

Water is a requirement for life. Although Mars does not currently appear to have any liquid water, frozen water is found at its poles, and features of its surface indicate that it once contained salty seas and flowing water (Figure 2.2). The presence of liquid water on Mars fulfills an essential prerequisite for the appearance of life. But why is water such an important feature?

One characteristic of water that makes it a suitable medium for life is that many different substances will dissolve in it. A substance that dissolves in a liquid is called the **solute**. The liquid in which another substance is dissolved is called the **solvent**. Many different molecules can be dissolved in water. Once dissolved, a particular molecule can pass freely throughout the water and may come into contact with another solute with which it can break existing chemical bonds and form new ones. Such changes in the chemical composition of substances are called chemical reactions. Chemical reactions occur when the starting materials or **reactants** are converted into the ending materials or **products**. Water facilitates chemical interactions between molecules. Part of the reason water functions as such a powerful solvent is due to its structure.

Water (H_2O) consists of hydrogen and oxygen atoms. **Atoms** are the smallest units into which a substance can be broken down; they are composed of **protons**, **neutrons**, and **electrons**. The positively charged protons (H^+) and electrically neutral neutrons make up the **nucleus**. The negatively charged electrons are found outside the nucleus in an "electron cloud." Electrons are attracted to the positively charged nucleus. A water molecule is composed of 2 hydrogen atoms and 1 oxygen atom that are bonded together by shared electrons. In general, a **molecule** consists of two or more atoms held together by chemical bonds, and chemical bonds occur between two atoms in a molecule that share electrons—often the shared electrons orbit around the nucleus of both atoms. The structure of chemical bonds within molecules is described in detail on page 27 of this chapter.

Oxygen is more electron-pulling, or electronegative, than most other atoms, including hydrogen. This means that the electrons in a water molecule spend more time near the nucleus of the oxygen atom than near the nuclei of the hydrogen atoms. When shared electrons are closer to one atom than another, the atom to which they are the closest will have a partial negative charge, symbolized by the Greek delta, δ^-. The atom from which the electrons are pulled away will have a partial positive charge, symbolized by δ^+ (Figure 2.3). This unequal sharing of electrons makes water a **polar** molecule, since different regions, or poles, of the molecule have different charges. When atoms of a molecule carry no charge, and thus do not have differing poles, they are said to be **nonpolar**.

Water molecules tend to orient themselves so that the hydrogen atom (with its partial positive charge) of one molecule is near the oxygen atom (with its

Figure 2.2   Polar ice. This image from the Mars rover indicates that frozen water exists on Mars.

Electrons are pulled toward oxygen

Figure 2.3   Polarity in a water molecule. Water is a polar molecule. Its atoms do not share electrons equally.

(a) Bonds between two water molecules   **(b)** Bonds between many water molecules

Figure 2.4   Hydrogen bonding between water molecules. The weak attraction between the hydrogen and oxygen atoms of different molecules is an example of hydrogen bonding.

partial negative charge) of another molecule (Figure 2.4a). The weak attraction between the hydrogen atom and the oxygen atom is called a **hydrogen bond**. Hydrogen bonding is a type of weak chemical bond that forms between hydrogen and another atom and is based on the attraction of partial charges for each other. Figure 2.4b shows hydrogen bonding that occurs among water molecules.

Water molecules stick together as a result of this hydrogen bonding. This tendency of molecules to stick together is called **cohesion**. Cohesion is much stronger in water than in most liquids and is an important property of many biological systems. For instance, many plants depend on cohesion to help transport water from the roots to the leaves.

The flowing water that was once found on Mars (Figure 2.5a) is now only in the form of ice. Until scientists can land on Mars and collect ice samples for analysis, its actual composition is a matter of conjecture. However, images taken by a NASA rover have led scientists to believe that some of the rocks on Mars were probably produced from deposits at the bottom of a body of salt

Figure 2.5   Water as a solvent. (a) This photograph, taken by the European Mars express orbiter, shows a channel on Mars that may have been formed by running water. (b) Salty water, such as seawater on Earth and the water that may have existed on Mars, is rich in sodium chloride. Each molecule of sodium chloride is composed of one sodium (Na^+) and one chloride (Cl^-). In this polar molecule, the shared electrons associate more closely to chlorine, giving chlorine a negative charge and sodium a positive charge. When salt is placed in water, the negatively charged regions of the water molecules surround the positively charged sodium, and the positively charged regions of the water molecules surround the negatively charged chlorine, breaking the bond holding sodium and chloride together and dissolving the salt.

Figure 2.6   The pH scale. The pH scale is a measure of hydrogen ion concentration. The more acidic a solution is, the higher the H^+ concentration is relative to the OH^- ions. Basic solutions have fewer H^+ ions relative to OH^- ions. The scale ranges from 0 (most acidic) to 14 (most basic). Each pH unit actually represents a tenfold ($10\times$) difference in the concentration of H^+ ions. For example, a substance with a pH of 5 has 100 times more H^+ ions than a substance with a pH of 7 does. Water has a pH of 7 and is therefore neutral; that is, it has as many H^+ ions as OH^- ions. The pH of most cells is very close to 7.

water. Salt water on Mars is likely to be the same salt as water on Earth, a solution of the salt sodium chloride. We know from surveying Earth's oceans that salt water is hospitable to millions of different life-forms. In fact, most hypotheses about the origin of life on Earth presume that our ancestors first arose in the salty oceans.

The ability of water to dissolve salts such as sodium chloride is illustrated in Figure 2.5b and is a direct result of its polarity. In the case of sodium chloride, the negative pole of water molecules will be attracted to a positively charged sodium and separate it from a negatively charged chloride. Water can also dissolve other polar molecules, such as alcohol, in a similar manner. Polar molecules are called **hydrophilic** ("water loving") because of their ability to dissolve in water.

Salts are produced by the reaction of an **acid** (a substance that donates H^+ ions to a solution) with a **base** (a substance that accepts H^+ ions). Water can break apart or dissociate into H^+ and OH^- ions. The **pH** scale is a measure of the relative amounts of these ions in a solution (Figure 2.6). These ions can react with other charged molecules and help to bring them into the water solution. At any given time, a small percentage of water molecules in a pure solution will be dissociated. There are equal numbers of these ions in pure water, and so it is neutral, which on the pH scale is 7.

Nonpolar molecules, such as oil, do not contain charged atoms and are referred to as **hydrophobic** ("water hating") because they do not easily mix with water (Figure 2.7). In early 2005, a European Space Agency probe landed on Titan, one of Saturn's moons and a place where the chemical composition of the atmosphere may be similar to that found on early Earth. Photos transmitted by the probe indicate that liquid was present on the surface of this bitterly cold place. At atmospheric temperatures of approximately $-292°F$ the liquid is obviously not water; instead, it is most likely a mixture of ethane and methane, both nonpolar molecules. As a result, oceans on Titan are much poorer solvents than are oceans on Earth, and conditions in these oceans are probably not suitable for the evolution of life.

Organic Chemistry

The Martian meteorite ALH84001 had one characteristic that provided some evidence that the rock once contained living organisms—the presence of complex molecules containing the element carbon. **Elements** are simple chemical substances composed of atoms that cannot be broken down by normal chemical means. Elements differ in the number of subatomic particles in their atoms. All atoms of a particular element have the same number of protons, giving the element its **atomic number**. An atom's **mass number** is the sum of the numbers of protons and neutrons in its nucleus.

All life on Earth is based on the chemistry of the element carbon. In fact, the branch of chemistry that is concerned with complex carbon-containing molecules is called **organic chemistry**, implying that it is the chemistry of life.

Carbon is only one of many essential elements that living organisms require, but it makes up most of their mass. Carbon is an ideal element as a foundation for organic chemistry because of its ability to make bonds with up to four other elements. Like a Tinkertoy™ connector, carbon has multiple

Figure 2.7   Oil and water. Oil is nonpolar and will not mix with water.

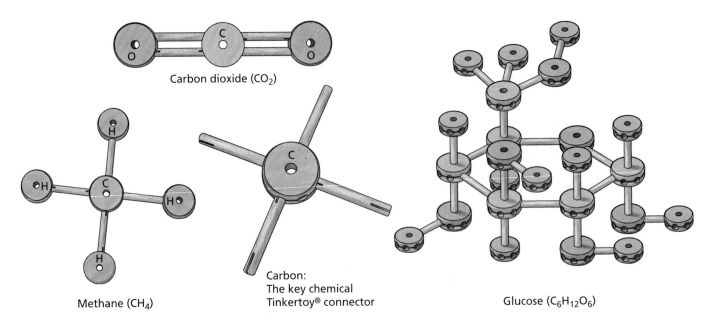

Figure 2.8 Carbon, the chemical Tinkertoy™ connector. Because carbon can connect with up to four other elements, carbon-containing compounds can be very diverse in shape.

sites for connections that allow carbon-containing molecules to take an almost infinite variety of shapes (Figure 2.8).

The ability of carbon, or any element, to make chemical bonds depends on its electron configuration. The electrons in the electron cloud that surrounds the nucleus have different energy levels based on their distance from the nucleus. The first energy level, or **energy shell**, is closest to the nucleus, and the electrons located there have the lowest energy. The second energy level is a little farther away, and the electrons located in the second shell have a little more energy. The third energy level is even farther away, and its electrons have even more energy, and so on. Figure 2.9a shows the electron configuration of carbon.

Each energy level can hold a specific maximum number of electrons. The first shell holds 2 electrons, and the second and third shells each hold a maximum of 8. Electrons fill the lowest energy shell before advancing to fill a higher energy-level shell. Atoms with the same number of electrons in their outermost energy shell, called the **valence shell**, exhibit similar chemical behaviors. When the valence shell is full of electrons, the atom will not normally form chemical bonds with other atoms.

Atoms that have space in their valence shell will combine to form compounds. A **compound** is a substance formed from two or more elements with a fixed ratio determining the composition. For example, water is a compound composed of two hydrogen atoms for every oxygen atom. Atoms with only 1 or 2 electrons in their valence shell tend to lose electrons and therefore become positively charged ions, while atoms with 6 or 7 electrons in the valence shell tend to gain electrons and thus become negatively charged ions. These two types of atoms will often form chemical compounds that are made up of 2 ions; the electrical attraction between a positive ion and a negative ion keeps them together loosely in an **ionic bond**. In aqueous solutions, these bonds are easily disrupted, for instance, when the compound is mixed with a polar solvent such as water.

Atoms with 4 or 5 electrons in the valence shell tend to share electrons to complete their valence shells. When atoms share electrons, a type of bond called a **covalent bond** is formed. Covalent bonds are stronger than ionic bonds and will not break apart in water. Figure 2.9b shows carbon covalently bonded to 4 hydrogens to produce methane, an organic compound that is common in the atmosphere of Titan.

(a)

(b)

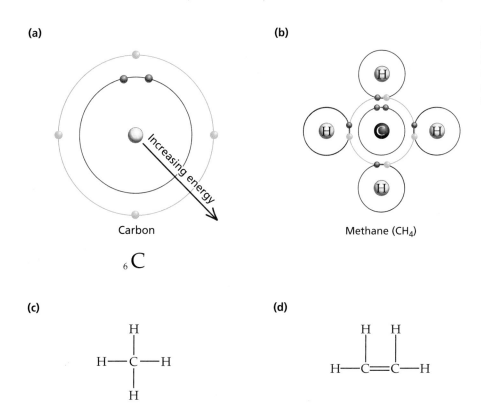

Carbon

$_6$C

Methane (CH$_4$)

(c)

H
|
H—C—H
|
H

(d)

H   H
| |
H—C═C—H

Carbon atoms are often involved in covalent bonding. Covalent bonds are symbolized by a short line indicating a shared pair of electrons (Figure 2.9c). When an element such as carbon enters into bonds involving two *pairs* of shared electrons, this is called a double bond. A carbon-to-carbon double bond is symbolized by two horizontal lines (Figure 2.9d).

The simple organic molecules found in the Martian meteorite that appear to have formed on Mars are carbonates, molecules containing carbon and oxygen, and **hydrocarbons**, made up of chains and rings of carbon and hydrogen. Carbonates and hydrocarbons can form under certain natural conditions even without the presence of life. However, what the meteorite lacked was convincing evidence of organic molecules that are known to be produced only by living organisms called **macromolecules**.

Structure and Function of Macromolecules

The macromolecules present in living organisms are carbohydrates, proteins, lipids, and nucleic acids. To date, every living organism on Earth has been found to contain these same macromolecules, whether bacteria, plant, or animal.

Carbohydrates.   Sugars, or **carbohydrates**, are found in every living organism on Earth and provide the major source of energy for daily activities. Carbohydrates also play important structural roles in cells. The simplest carbohydrates are composed of carbon, hydrogen, and oxygen in the ratio (CH$_2$O). For example, the carbohydrate sugar glucose is symbolized as 6(CH$_2$O) or C$_6$H$_{12}$O$_6$. Glucose is a simple sugar, or monosaccharide, which consists of a single ring-shaped structure. Disaccharides are two rings joined together. Table sugar, called sucrose, is a disaccharide composed of glucose and fructose, a sugar found in fruits.

Joining many individual subunits, or monomers, together produces polymers (*poly* means "many"). Polymers of sugar monomers are called **polysaccharides** (Figure 2.10). Plants use tough polysaccharides in their cell walls as a sort of structural skeleton. The polysaccharide cellulose, found in plant cell walls, is the most abundant carbohydrate on Earth. The external skeletons of insects, spiders, and

Figure 2.10 Carbohydrates. Monosaccharides, such as the glucose molecule shown here, are individual sugar molecules, while disaccharides are 2 monosaccharides joined together, and polysaccharides are long chains of sugars joined together. The monosaccharide glucose and the disaccharide sucrose are important sources of energy; and cellulose plays a structural role in plant cell walls.

Glucose monomer

Sucrose—a disaccharide

Glucose   Fructose

Cellulose—a polysaccharide

lobsters are composed of polysaccharides, and the cell walls that surround bacterial cells are rich in structural polysaccharides.

According to David McKay and his colleagues, the particular set of hydrocarbons found on the Martian meteorite is identical to the set formed when carbohydrates in certain bacteria on Earth break down. These trace remains of possible Martian carbohydrates are an important piece of evidence that scientists use to argue that Mars once harbored Earth-like life. Evidence of the presence of proteins on the meteorite is less convincing.

Proteins. Living organisms require **proteins** for a wide variety of processes. Proteins are important structural components of cells; in fact, they make up half the dry weight of most cells. Some cells, such as animal muscle cells, are largely composed of proteins. Proteins called **enzymes** accelerate and help regulate all the chemical reactions that build up and break down molecules inside cells. The catalytic power of enzymes (their ability to drastically increase reaction rates) allows metabolism to occur under normal cellular conditions. Proteins can also serve as channels through which substances are brought into cells, and they can function as hormones that send chemical messages throughout an organism's body.

Proteins are large molecules made of monomer subunits called **amino acids**. There are 20 commonly occurring amino acids. Like carbohydrates, amino acids are made of carbons, hydrogens, and oxygens; but in addition, they have nitrogen as part of an amino ($-NH_2^+$) group along with various side groups. Side groups are chemical groups that give amino acids different chemical properties (Figure 2.11a).

(a) General formula for amino acid

Amino group   Side group   Carboxyl group

(b) Peptide bond formation

Peptide bond   Peptide bond

Valine   Alanine   Phenylalanine

(c) Protein

Figure 2.11 Amino acids. (a) All amino acids have the same backbone but different side groups. (b) Amino acids are joined together by chemical bonds called peptide bonds. Long chains of these are called polypeptides. (c) Polypeptide chains fold upon themselves to produce proteins, and different combinations of amino acids produce distinct proteins.

Polymers of amino acids are sometimes called polypeptides. The chemical bond joining adjacent amino acids is a **peptide bond** (Figure 2.11b). Amino acids are joined together in various orders to produce different proteins in much the same manner that children can use differently shaped beads to produce a wide variety of structures (Figure 2.11c). Each amino acid side group has unique chemical properties, including being polar or nonpolar. Since each protein is composed of a particular sequence of amino acids, each protein has a unique shape and therefore specialized chemical properties.

Scientists have found no evidence of proteins on the Martian meteorite, although one group of investigators did report the presence of tiny amounts of three amino acids within the rock. However, it may be the case that these amino acids are contaminants; that is, they are present in the meteor because the meteor has been on protein-rich Earth for several thousand years. In addition, some amino acids are known to form under conditions where life is not present, so the presence of amino acids is not necessarily evidence of life.

Lipids.   One type of organic molecule, abundant in living organisms, that has not been found in the Martian meteorite is lipids. **Lipids** are partially or entirely hydrophobic organic molecules made primarily of hydrocarbons. Important lipids include fats, steroids, and phospholipids.

(a) Fat

Glycerol   Hydrocarbons (fatty acid tails)

(b) Cholesterol

Cholesterol

(c) Phospholipid

Hydrocarbons (fatty acid tails)

Glycerol

Figure 2.12   Three types of lipids. (a) Fats are composed of a glycerol molecule with three hydrocarbon-rich fatty acid tails attached. (b) Cholesterol is a steroid common in animal cell membranes. (c) Phospholipids are composed of a glycerol backbone with 2 fatty acids attached and 1 phosphate head group. They have hydrophilic heads and hydrophobic tails. In the cartoon drawing to the right, the phosphate head group is symbolized by a circle and the fatty acid tails are red.

Fat.   The structure of a **fat** is that of a 3-carbon glycerol molecule with up to three long, hydrocarbon chains attached to it (Figure 2.12a). Like the hydrocarbons present in gasoline, these can be burned to produce energy. The long hydrocarbon chains are called **fatty acid tails** of the fat molecule. Fats are hydrophobic and function in energy storage within living organisms.

Steroids.   **Steroids** are composed of 4 fused carbon-containing rings. Cholesterol (Figure 2.12b) is one steroid that you are probably familiar with; its primary function in animal cells (plant cells do not contain cholesterol) is to help maintain the fluidity of membranes. Other steroids include the sex hormones testosterone, estrogen, and progesterone, which are produced by the sex organs and have effects throughout the body.

Phospholipids.   **Phospholipids** are similar to fats except that each glycerol molecule is attached to 2 fatty acid tails (not 3, as you would find in a dietary fat). The third bond in a phospholipid is to a phosphate head group. The phosphate head group is hydrophilic, and the two tails are hydrophobic (Figure 2.12c). Phospholipids often have an additional head group, attached to the phosphate, which also confers unique chemical properties on the individual phospholipid. Phospholipids are important constituents of the membranes that surround cells and that designate compartments within cells.

Even if the Martian meteorite contained unambiguous traces of carbohydrates, proteins, and lipids, the source of these molecules would not clearly be living organisms without a mechanism for passing information about their traits to the next generation. The hereditary, or genetic, information common to all life on Earth is in the form of nucleic acids.

Nucleic Acids.   Nucleic acids are composed of long strings of monomers called **nucleotides.** A nucleotide is made up of a sugar, a phosphate, and a nitrogen-containing base. The nucleic acid that serves as the primary storage of genetic information in nearly all living organisms is **deoxyribonucleic acid (DNA).** Figure 2.13 shows the three-dimensional structure of a DNA molecule and

(a) DNA double helix is made of two strands.   **(b)** Each strand is a chain of of antiparallel nucleotides.

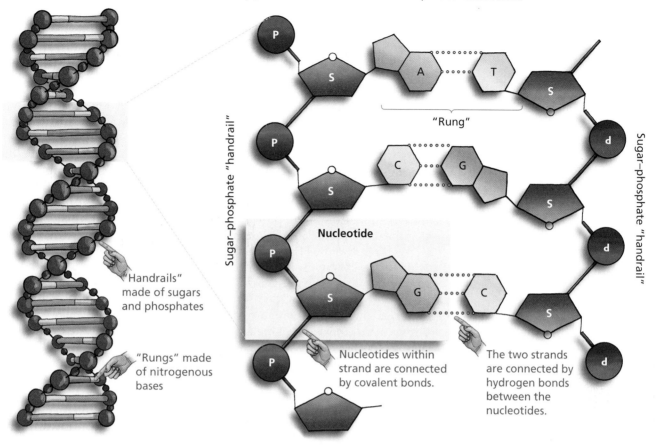

(c) Each nucleotide is composed of a phosphate, a sugar, and a nitrogenous base.

Figure 2.13 DNA structure. (a) DNA is a double-helical structure composed of sugars, phosphates, and nitrogenous bases. (b) Each strand of the helix is composed of repeating units of sugars and phosphates, making the sugar-phosphate backbone, and of nitrogenous bases. (c) A phosphate, a sugar, and a nitrogenous base comprise the structure of a nucleotide. Adenine and guanine are purines, which have a double-ring structure; cytosine and thymine are pyrimidines, which have a single-ring structure.

zooms inward to the chemical structure. You can see that DNA is composed of two curving strands that wind around each other to form a double helix. The sugar in DNA is the 5-carbon sugar deoxyribose. The nitrogen-containing bases, or **nitrogenous bases**, of DNA have one of four different chemical structures, each with a different name: **adenine (A), guanine (G), thymine (T),** and **cytosine (C)**. Nucleotides are joined to each other along the length of the helix by covalent bonds.

Nitrogenous bases form hydrogen bonds with each other across the width of the helix. On a DNA molecule, an adenine (A) on one strand always pairs with a thymine (T) on the opposite strand. Likewise, guanine (G) always pairs with cytosine (C). The term **complementary** is used to describe these pairings. For example, A is complementary to T, and C is complementary to G. Therefore, the order of nucleotides on one strand of the DNA helix predicts the order of nucleotides on the other strand. Thus, if one strand of the DNA molecule is composed of nucleotides AACGATCCG, then we know that the order of nucleotides on the other strand is TTGCTAGGC.

As a result of this **base-pairing rule** (A pairs with T; G pairs with C), the width of the DNA helix is uniform. There are no bulges or dimples in the structure of the DNA helix, because A and G, called **purines**, are structures composed of two rings; and C and T are single-ring structures called **pyrimidines**. A purine always pairs with a pyrimidine and vice versa, so there are always 3 rings across the width of the helix. A to T base pairs have 2 hydrogen bonds holding them together. G to C pairs have 3 hydrogen bonds holding them together.

Each strand of the helix thus consists of a series of sugars and phosphates alternating along the length of the helix, the **sugar-phosphate backbone**. The strands of the helix align so that the nucleotides face "up" on one side of the helix and "down" on the other side of the helix. For this reason, the two strands of the helix are said to be antiparallel.

The overall structure of a DNA molecule can be likened to a rope ladder that is twisted, with the sides of the ladder composed of sugars and phosphates (the sugar-phosphate backbone), and the rungs of the ladder composed of the nitrogenous-base sequences A, C, G, and T. The structure of DNA was determined by a group of scientists in the 1950s, most notably James Watson and Francis Crick (Figure 2.14).

How Might Macromolecules on Other Planets Differ?   Many scientists argue that the fundamental constituents described here—carbohydrates, proteins, lipids, and nucleic acids—will be essentially similar wherever life is found. They will readily admit that the finer details are very likely to differ, however. For example, all proteins known on Earth contain only 20 different amino acids, despite an infinite number of possibilities. Presumably, proteins on other planets could contain completely different amino acids and many more than 20.

Not all scientists agree with this position, which they call "carbon chauvinism." Carbon is not the only chemical Tinkertoy™ connector; other elements, including silicon, can also make connections with four other atoms. Silicon is also relatively abundant in the universe and could theoretically form the backbone of an alternative organic chemistry. The basic constituents of silicon-based life may be very different from the chemical building blocks of life on Earth.

Even if all life in the universe is based on carbon chemistry, it is very unlikely that the suite of organisms found on another planet will look much like life on our planet. However, understanding the history of life on Earth also provides insight into the possible nature of life elsewhere in the universe.

Figure 2.14 American James Watson (left) and Englishman Francis Crick are shown with the 3-dimensional model of DNA they devised while working at the University of Cambridge in England.

2.2 Life on Earth

One of the most dramatic features of the Martian meteorite is the presence of fossils that look remarkably like the tiniest living organisms known from Earth. The largest of these fossils is less than 1/100th of the diameter of a human hair, and most are about 1/1000th of the diameter of a human hair—small enough that it would take about 1000 laid end to end to span the dot at the end of this sentence. Some are egg shaped, while others are tubular. These fossils appear similar to the simplest and most ancient of known organisms and are the strongest piece of evidence supporting the hypothesis that Mars once was home to living organisms.

Prokaryotic and Eukaryotic Cells

David McKay and his colleagues argue that the fossil structures in the Martian meteorite are the remains of tiny cells. A **cell** is the fundamental structural unit of life on Earth, separated from its environment by a membrane and often an external wall. Bacteria are composed of single cells, which perform all of the activities required for life. Other organisms such as humans and oak trees are composed of trillions of cells working together and do not have any cells that could survive and reproduce independently.

All cells can be placed into one of two categories, prokaryotic or eukaryotic, based on the presence or absence of certain cellular structures. Bacteria are **prokaryotic** cells. Prokaryotes do not have a nucleus, a separate membrane-bound compartment that contains primarily genetic material in the form of DNA. Nor do they contain any membrane-bound internal structures. Prokaryotic cells are much smaller than eukaryotic cells (Figure 2.15), and according to the fossil record, they predate eukaryotic cells. The fossils in the Martian meteorite resemble modern prokaryotic cells known as nanobacteria.

Eukaryotic cells have a nucleus and other internal structures with specialized functions, called **organelles**, that are surrounded by membranes. Eukaryotic organisms include single-celled organisms such as amoebas and yeast as well as multicellular plants, fungi, and animals. As you will learn in Chapter 12, scientists believe that the first prokaryotic cells appeared on Earth over 3.5 billion years ago and that the first eukaryotes appeared about 1.7 billion years later.

Many scientists dispute David McKay's interpretation of the tubular structures in ALH84001. In fact, similar structures can be formed in the absence of life by certain minerals under extremes of heat and pressure. If the Martian fossils are indeed cells, they likely contained features found inside earthly cells, some of which should be visible in the fossils. Each living cell can be considered a veritable factory working to break down nutrients and to recycle its components. We will start from the outside of the cell and examine the structure and function of various cell components as we work our way in.

Cell Structure

Plasma Membrane. All cells are enclosed by a structure called a **plasma membrane** (Table 2.1). The plasma membrane defines the outer boundary of each cell, isolates the cell's contents from the environment, and serves as a semi-permeable barrier that determines which nutrients are allowed into and out of the cell. Membranes that enclose structures inside the cell are usually referred to as cell membranes, while the outer boundary is the plasma membrane.

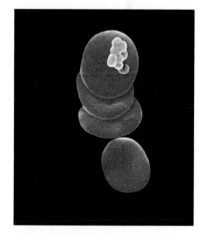

Figure 2.15 Bacterial and eukaryotic cells. Prokaryotic cells, like the blue-tinted bacteria above, are many times smaller than eukaryotic cells. The eukaryotic cell in this image is a red blood cell.

Table 2.1 Cell components. Illustrations and descriptions of cell components and their functions.

Component	Function
Plasma Membrane — Sugar chains, Cholesterol, Phospholipid bilayer, Head, Tail, Protein	All cells are surrounded by a plasma membrane. It is composed of a bilayer of phospholipids (tails toward the center), perforated by proteins. Proteins in the bilayer help transport substances across the hydrophobic core of the membrane. Cholesterol in the membranes of animal cells helps maintain the fluidity of the membrane. The sugar chains function as identification tags, marking cells as a particular cell type (liver cell, heart cell, etc.)
Nucleus — Nuclear pore, Nuclear envelope, Nucleolus, Chromatin	Eukaryotic cells contain a nucleus. The nucleus is a spherical structure surrounded by two membranes, together called the nuclear envelope. The nuclear envelope is studded with nuclear pores that regulate traffic into and out of the nucleus. Inside the nucleus is chromatin, composed of DNA and proteins. The nucleolus is where ribosomes are produced.
Mitochondrion — Outer membrane, Intermembrane space, Inner membrane, Matrix	Eukaryotic cells contain mitochondria. Mitochondria are energy-producing organelles surrounded by two membranes. The inner and outer mitochondrial membranes are separated by the intermembrane space. The highly convoluted inner membrane carries many of the proteins involved in producing ATP. The matrix of the mitochondrion is the location of many of the reactions of cellular respiration.
Chloroplast — Outer membrane, Inner membrane, Stroma, Thylakoids, Granum	An important organelle present in plant cells, the chloroplast uses the sun's energy to convert carbon dioxide and water into sugars. Each chloroplast has an outer membrane, an inner membrane, a liquid material called the stroma, and a network of flattened membranes called thylakoids that stack on one another to form structures called grana (singular: granum). Chloroplasts also contain pigment molecules that give green parts of plants their color.

Component	Function
Ribosomes Ribosome	Ribosomes are found in eukaryotic and prokaryotic cells. Ribosomes are built in the nucleus and shipped out through nuclear pores to the cytoplasm, where they are used as work benches for protein synthesis. They can be found floating in the cytoplasm or tethered to the ER.
Endoplasmic Reticulum (ER) Nuclear envelope, Rough endoplasmic reticulum, Ribosomes, Vesicle, Smooth endoplasmic reticulum	The ER is a large network of membranes that begins at the nuclear envelope and extends into the cytoplasm. ER with ribosomes attached is called rough ER. Proteins synthesized on rough ER will be secreted from the cell or will become part of the plasma membrane. ER without ribosomes attached is called smooth ER. The function of the smooth ER depends on cell type but includes tasks such as detoxifying harmful substances and synthesizing lipids. Vesicles are pinched-off pieces of membrane that transport substances to the Golgi apparatus or plasma membrane.
Golgi Apparatus Vesicle from ER arriving at Golgi apparatus, Vesicle departing Golgi apparatus	The Golgi apparatus is a stack of membranous sacs. Vesicles from the ER fuse with the Golgi apparatus and empty their protein contents. The proteins are then modified, sorted, and sent to the correct destination in new transport vesicles that bud off from the sacs.
Lysosome Membrane, Digestive enzymes and digested material	A lysosome is a membrane-enclosed sac of digestive enzymes that degrade proteins, carbohydrates, and fats. Lysosomes roam around the cell, and engulf targeted molecules and organelles for recycling.
Centrioles Microtubules	Centrioles are barrel-shaped rings composed of nine microtubule triplets. Microtubules help move chromosomes around when a cell divides. Centrioles are involved in microtubule formation during cell division and the formation of cilia and flagella.
Cytoskeletal Elements Microfilaments, Intermediate filaments, Microtubules	Cytoskeletal elements are protein fibers in the cytoplasm that give shape to a cell, hold and move organelles (including transport vesicles), and are typically involved in cell movement.

(*table continued on next page*)

(Table 2.1   continued)

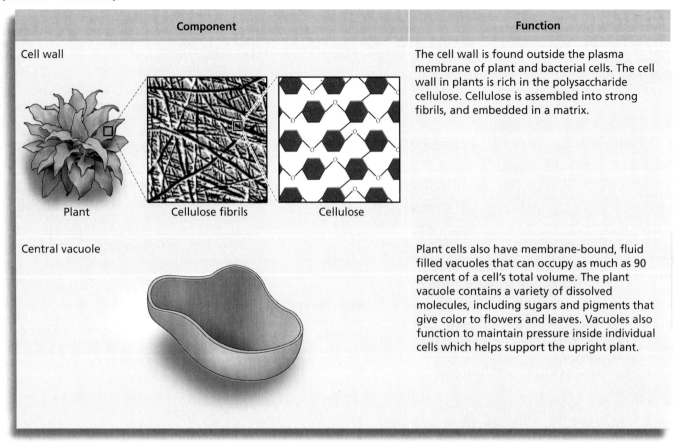

Component			Function
Cell wall			The cell wall is found outside the plasma membrane of plant and bacterial cells. The cell wall in plants is rich in the polysaccharide cellulose. Cellulose is assembled into strong fibrils, and embedded in a matrix.
Plant	Cellulose fibrils	Cellulose	
Central vacuole			Plant cells also have membrane-bound, fluid filled vacuoles that can occupy as much as 90 percent of a cell's total volume. The plant vacuole contains a variety of dissolved molecules, including sugars and pigments that give color to flowers and leaves. Vacuoles also function to maintain pressure inside individual cells which helps support the upright plant.

Internal and external cell membranes are composed, in part, of phospholipids. The chemical properties of these lipids make membranes flexible and self-sealing. When phospholipid molecules are placed in a watery solution, such as in a cell, they orient themselves so that their hydrophilic heads are exposed to the water and their hydrophobic tails are away from the water. They cluster into a form called a **phospholipid bilayer**, in which the tails of the phospholipids interact with themselves and exclude water, while the heads maximize their exposure to the surrounding water both inside and outside of the membrane. The bilayer of phospholipids is stuffed with proteins that carry out enzymatic functions and help transport substances.

A Fluid Mosaic of Lipids and Proteins.   All of the lipids and proteins in the plasma membrane are free to bob about, sliding from one location in the membrane to another. However, they cannot readily cross from one layer of the bilayer to another, because this would require the hydrophilic portions of a phospholipid or protein to traverse the hydrophobic core of the bilayer. Because lipids and proteins can move about laterally within the membrane, the membrane is a **fluid mosaic** of lipids and proteins. The membrane is fluid since the composition of any one location on the membrane can change. In the same manner that a patchwork quilt is a mosaic (different fabrics making up the whole quilt), so too is the membrane a mosaic with different regions of membrane being composed of different types of phospholipids and proteins.

Cell membranes are **semipermeable** in the sense that they allow some substances to cross and prevent others from crossing. This characteristic allows cells to maintain a degree of independence from the surrounding solution.

The permeability of the plasma membrane to water presents a problem for many organisms. In environments where the water contains low levels of solutes, water from outside the cell will flow into the cell, causing it to burst. Bacterial cells have a **cell wall** that helps them maintain their shape in these conditions. The fossils in the Martian meteorite show no evidence of similar walls, calling into question their supposed biological origin.

Nucleus.   In addition to being surrounded by a plasma membrane, all eukaryotic cells contain a nucleus, which houses the DNA.

Cytosol.   Between the nucleus and the plasma membrane lies the cytosol, a watery matrix containing water, salts, and many of the enzymes required for cellular reactions. The cytosol houses the subcellular structures called organelles. The term **cytoplasm** includes the cytosol and organelles.

Organelles.   Organelles are to cells as organs are to the body. Each organelle performs a specific job required by the cell, and all organelles work together to keep an individual cell healthy and to produce the raw materials that the cell needs to survive. Some organelles are involved in metabolism. For example, organelles called **mitochondria** help the cells convert food energy into a form usable by cells, called ATP, while **chloroplasts** in plant cells use energy from sunlight to make sugars. **Lysosomes** help break down food that is ingested before it is sent to the mitochondria. Other organelles are involved in producing proteins. The **ribosomes** are workbenches where proteins are assembled. Many proteins are assembled on the membranes of the **endoplasmic reticulum** and modified and sorted in a membranous structure called the **Golgi apparatus**. Some subcellular structures help cells divide and maintain their shape. **Centrioles** are involved in moving genetic material around when a cell divides, and many subcellular fibers help maintain the cell shape. Some subcellular structures are found in certain cell types only. For instance, in addition to having a cell wall, the plant cell also has a **vacuole** to store sugars and pigments. Table 2.1 describes the structures and functions of most cellular organelles in greater detail. Figure 2.16 shows an animal cell and a plant cell complete with their complement of organelles.

(a) Animal cell   **(b)** Plant cell

Figure 2.16   Animal and plant cells. (a) Animal cells contain lysosomes and centrioles, and plant cells do not. (b) Plant cells have a cell wall, vacuole, and chloroplasts, and animal cells do not.

The Tree of Life and Evolutionary Theory

Biologists disagree about the total number of different **species**, or types of living organisms, that are present on Earth today. This uncertainty stems from lack of knowledge. Although scientists likely have identified most of the larger organisms—such as land plants, mammals, birds, reptiles, and fish—millions of species of insects, fungi, bacteria, and other microscopic organisms remain unknown to science. Amazingly, credible estimates of the number of species on Earth range from 5 million to 100 million; given the uncertainty, most biologists think that the likeliest number is near 10 million.

Theory of Evolution.   While the diversity of living organisms is tremendous, there exist remarkable similarities among all known species. All have the same basic biochemistry, including carbohydrates, lipids, proteins, and nucleic acids. All consist of cells surrounded by a plasma membrane. All eukaryotic organisms (including fungi, animals, and plants) contain nearly the same suite of cellular organelles. The best explanation for the shared characteristics of all species, what biologists refer to as "the unity of life," is that all living organisms share a common ancestor that arose on Earth nearly 4 billion years ago. The divergence and differences among modern species arose as a result of changes in the characteristics of populations, both in response to environmental change (a process called natural selection) and due to chance. These ideas underlie the entire science of biology and are known as the **theory of evolution**.

The common ancestor can be thought of as the starting place for life on Earth, and the continual divergence among species and groups of species can be thought of as life's branching. Modern organisms can therefore be arranged on a tree of life that reflects their basic unity and relationships. According to current understanding, living organisms can be grouped into three large groups: two that are prokaryotic and one containing all eukaryotes. Eukaryotes can be further grouped into several categories made up primarily of free-living, single-celled organisms (such as amoebas and algae) and the three major multicellular groups—plants, fungi, and animals (Figure 2.17). Chapter 12 provides a deeper exploration of the diversity of life on Earth.

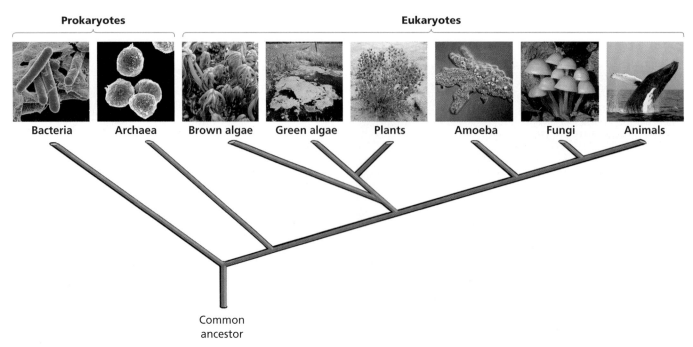

Figure 2.17   Tree of life. All life on Earth shares basic characteristics and can be arranged into a tree of life based on more specific similarities. In this illustration, many groups are omitted for simplicity.

Because evolutionary change results from chance events and environmental changes (including the appearance of other species), the group of species present on Earth today represents only one set of an infinite number of possibilities. In other words, life on other planets need not look identical to life on Earth. For example, instead of the common body form found in animals, called bilateral symmetry, where bodies can be visually divided into two mirror-image halves, life on other planets could be primarily radially symmetric and thus look very different (Figure 2.18). In fact, it is possible that life on other planets might not even be based on carbon. Scientists have no examples of what living organisms would look like on a planet where the organic molecules were based on silicon or bathed in liquid methane.

Life in the Universe.   Do other living organisms exist in the universe? Given the universe's sheer size and complexity, most scientists who study this question think that the existence of life on other planets is nearly certain. While the evidence of life in the Martian meteorite is unconvincing to many scientists, we may find out in our lifetimes that life exists, or once existed, on our planetary neighbor.

What about the existence of intelligent life that could communicate with us? Some scientists argue that as a result of natural selection, the evolution of intelligence is inevitable wherever life arises. Others point to the history of life on Earth—consisting of at least 2.5 billion years, during which all life was made up of single-celled organisms—to argue that most life in the universe must be "simple and dumb." It is clear from our explorations of the solar system that none of the sun's other planets host intelligent life. The nearest sun-like stars that could host an Earth-like planet, Alpha Centauri A and B, are over 4 light years away—nearly 40 trillion miles. With current technologies, it would take nearly 50,000 years to reach the Alpha Centauri stars, and there is certainly no guarantee that intelligent life would be found on any planets that circle them. For all practical purposes, at this time in human history, we are still unique and alone in the universe.

Figure 2.18   Diversity of body form. Not all animals are bilaterally symmetric like us, with two eyes, two ears, two arms, two legs, a clear head and "tail," and one central axis. A sea star is radially symmetric—it can be divided into two equal halves in any direction. If most animals on another planet were radially symmetric, then that world would look very different from Earth.

CHAPTER REVIEW

Summary

2.1 What Does Life Require?

- Living organisms must be able to grow, metabolize substances, reproduce, and respond to external stimuli (p. 24).

- Living organisms contain a common set of biological molecules, are composed of cells, and can maintain homeostasis and evolve (p. 24).

- A water molecule is composed of 2 hydrogen atoms and 1 oxygen atom, bonded together by shared electrons (p. 25).

- Water is a good solvent, in part because the weak attraction between the hydrogen atom and the oxygen atom forms a hydrogen bond (p. 26).

- Hydrogen bonding is also responsible for cohesion among water molecules (p. 26).

- The polarity of water also facilitates the dissolving of salts. Salts are produced by the reaction of an acid with a base (pp. 26–27).

- The pH scale is a measure of the relative percentages of these H^+ and OH^- ions in a solution and ranges from 0 (acidic or rich in H^+ ions) to 14 (basic or rich in OH^- ions) (p. 27).

- Life on Earth is based on the chemistry of the element carbon, which can make bonds with up to four other elements (pp. 27–28).

- Chemical bonding depends on an element's electron configuration. Electrons closer to the nucleus have less energy than those that are farther away from the nucleus. Each energy level can hold a specific maximum number of electrons. Atoms that have space in their valence shell will combine to form compounds (p. 28).

- Ionic bonds form between positively and negatively charged ions. These tend to be weak bonds (p. 28).

- Covalent bonds form when atoms share electrons. These tend to be strong bonds (p. 28).

- Carbohydrates function in energy storage and play structural roles. They can be single-unit monosaccharides or multiple-unit polysaccharides with sugar monomers arranged in different orders (p. 29).

- Proteins play structural, enzymatic, and transport roles in cells. They are composed of amino acid monomers arranged in different orders (p. 30).

- Lipids are partially or entirely hydrophobic and come in three different forms. Fats are composed of glycerol and three fatty acids. Fats store energy. Phospholipids are composed of glycerol, two fatty acids, and a phosphate group. They are important structural components of cell membranes. Steroids are composed of four fused rings. Cholesterol is a steroid found in some animal-cell membranes and helps maintain fluidity. Other steroids function as hormones (pp. 31–32).

- Nucleic acids are polymers of nucleotides, each of which is composed of a sugar, a phosphate, and a nitrogen-containing base (pp. 32–34).

Web Tutorial 2.1 **Chemistry and Water**
Web Tutorial 2.2 **Nucleic Acids**

2.2 Life on Earth

- There are two main categories of cells: Those with nuclei and membrane-bound organelles are eukaryotes; those lacking a nucleus and membrane-bound organelles are prokaryotes (p. 35).

- The plasma membrane that surrounds cells is a semipermeable boundary composed of a phospholipid bilayer that has embedded proteins and cholesterol (pp. 35–38).

- Lipids and proteins can move about the membrane. This fluidity of the membrane allows changes in the protein and lipid composition (p. 38).

- Some organisms, such as plants and bacterial cells, have a cell wall outside the plasma membrane that helps protect these cells and maintain their shape (p. 39).

- Subcellular organelles and structures perform many different functions within the cell. Mitochondria and chloroplasts are involved in energy conversions. Lysosomes are involved in breakdown of macromolecules. Ribosomes serve as sites for protein synthesis. Membranous endoplasmic reticulum to help localize protein synthesis. The Golgi apparatus sorts proteins and sends them to their cellular destination. Centrioles help cells divide. The plant cell vacuole stores water and other substances (p. 39).

- The number of living species on Earth is unknown, but there may be nearly 10 million unique life-forms. Despite all of this diversity, all life on Earth shares the same organic chemistry, genetic material, and basic cellular structures (p. 40).

- The similarities among living organisms on Earth provide support for the theory of evolution, which states that all life on Earth derives from a common ancestor. The process of evolutionary change since the origin of that ancestor led to the modern relationships among organisms, described as the "tree of life." (pp. 40–41)

Web Tutorial 2.3 **A Comparison of Prokaryotic and Eukaryotic Cells**

Learning the Basics

1. What are the building-block molecules of a carbohydrate, a protein, and a fat?

2. List the roles of the organelles discussed in this chapter.

3. Describe the structure of the plasma membrane.

4. Water _____.

   A. is a good solute; **B.** dissociates into H^+ and OH^- ions; **C.** serves as an enzyme; **D.** makes strong covalent bonds with other molecules; **E.** has an acidic pH

5. Electrons _____.

   A. are negatively charged; **B.** along with neutrons comprise the nucleus; **C.** are attracted to the negatively charged nucleus; **D.** located closest to the nucleus have the most energy; **E.** all of the above are true

6. Which of the following terms is least like the others?
   A. monosaccharide; **B.** phospholipid; **C.** fat; **D.** steroid; **E.** lipid

7. Different proteins are composed of different sequences of _____.
   A. sugars; **B.** glycerols; **C.** fats; **D.** amino acids

8. Proteins may function as _____.
   A. the genetic material; **B.** cholesterol molecules; **C.** fat reserves; **D.** enzymes; **E.** all of the above

9. A fat molecule consists of _____.
   A. carbohydrates and proteins; **B.** complex carbohydrates only; **C.** saturated oxygen atoms; **D.** glycerol and fatty acids

10. Eukaryotic cells differ from prokaryotic cells in that _____.
   A. only eukaryotic cells contain DNA; **B.** only eukaryotic cells have a plasma membrane; **C.** only eukaryotic cells are considered to be alive; **D.** only eukaryotic cells have a nucleus; **E.** only eukaryotic cells are found on Earth

Analyzing and Applying the Basics

1. A virus is made up of a protein coat surrounding a small segment of genetic material (either DNA or RNA) and a few proteins. Some viruses are also enveloped in membranes derived from the virus's host cell. Viruses cannot reproduce without taking over the genetic "machinery" of their host cell. Based on this description and biologists' definition of life, should a virus be considered a living organism?

2. Any molecule containing oxygen can be polar. The structure of methanol (CH_3OH) is drawn in Figure 2.19. Which part of this molecule will have a partial negative charge, and which will have a partial positive charge?

3. Some scientists have argued that silicon (Si) could also be an appropriate basis for organic chemistry because it is abundant and can form bonds with many other atoms. Carbon contains 6 electrons, and silicon contains 14. Recalling that the lowest electron shell contains 2 electrons, and the next 2 shells can contain a maximum of 8, how many "spaces" does silicon have in its valence shell? How does this compare to carbon?

Figure 2.19   Methanol.

Connecting the Science

1. Water's characteristic as an excellent solvent means that many human-created chemicals (including some that are quite toxic) can be found in water bodies around the globe. How would our use and manufacture of toxic chemicals be different if most of these chemicals could not be dissolved and diluted in water, but instead accumulated where they were produced and used?

2. Do you believe that humans should expend considerable energy and resources looking for life, even intelligent life, elsewhere in the universe? Why or why not?

CHAPTER 3

The Only Diet You Will Ever Need

Cells and Metabolism

A high premium is placed on thinness in U.S. culture.

Most college students are, for the first time, making all their own choices about food.

Making unhealthy choices now can lead to future health problems associated with obesity . . .

The average college student spends a lot of time thinking about his or her body and ways to make it more attractive. While people come in all shapes and sizes, attractiveness tends to be more narrowly defined by the images of men and women we see in the popular media. Nearly all media images equate attractiveness and desirability with a very limited range of body types.

For men, the ideal includes a tall, broad-shouldered, muscular physique with so little body fat that every muscle is visible. The standards for female beauty are equally unforgiving and include small hips; long, thin limbs; large breasts; and no body fat.

Into this milieu steps the average college student—worried about appearance, trying to find time to study, exercise, and socialize—and now making all of his or her own decisions about food, often on a limited budget.

Making choices that are good for long-term health is not easy. Typical meal-plan choices at college dining centers are often greasy and fat laden—and available in unlimited portions. Making healthful choices is further complicated by the presence of campus snack shops, vending machines, and conveniently located fast-food restaurants that offer time-pressed students easily accessible, inexpensive foods containing little nutrition.

. . . or anorexia.

45

Coupling tremendous pressure to be thin with a glut of readily available unhealthful foods can lead students to establish poor eating habits that persist far beyond college life. In many cases, these conflicting pressures may cause students to develop eating habits that result in lifelong battles with obesity or starvation.

Learning about the kinds and amounts of foods you should be eating, and understanding how much body fat is right for you, will help you make good decisions about eating that will set you up for a lifetime of good health.

3.1   Nutrients

The food that organisms ingest provides building-block molecules that can be broken down and used as raw materials for growth, for maintenance and repair, and as a source of energy. Another name for the substances in food that provide structural materials or energy is **nutrients**. Nutrients that are required in large amounts are called **macronutrients**. These include water, carbohydrates, proteins, and fats. Nutrients that are essential in minute amounts, such as vitamins and minerals, are called **micronutrients**.

Macronutrients

Macronutrients include water, carbohydrates, proteins, and fats.

Water and Nutrition.   Most animals can survive for several weeks with no nutrition other than water. However, survival without water is limited to just a few days. Besides helping the body disperse other nutrients, water helps dissolve and eliminate the waste products of digestion. Water helps to maintain blood pressure and is involved in virtually all cellular activities.

A decrease in the body's required water level, called **dehydration**, can lead to muscle cramps, fatigue, headaches, dizziness, nausea, confusion, or increased heart rate. Large water deficits can result in hallucinations, heat stroke, and fatality. These symptoms occur because water in the circulatory system helps deliver oxygen and other nutrients to all parts of the body, including the brain. When water is low, this delivery becomes less effective. In addition, evaporation of water from the skin (sweating) helps maintain body temperature. When water is low and sweating decreases, the body temperature can rise to a harmful level.

Every day, humans lose about 3 liters of water as sweat, in urine, and in feces. To avoid the negative health impacts of dehydration, we must replace this water. We can replace some of it by consuming food that contains water. A typical adult obtains about 1.5 liters of water per day from food consumption, leaving a deficit of about 1.5 liters that he or she must drink to replenish.

In addition to obtaining a healthy dose of water every day, people must consume foods that contain carbohydrates, proteins, and fats. In Chapter 2, we explored the structure of these macromolecules, and now we will focus on how they function in the body.

Carbohydrates as Nutrients.   Foods such as bread, cereal, rice, and pasta, as well as fruits and vegetables, are rich in sugars called carbohydrates. Carbohydrates are the major source of energy for cells. Energy is stored in the chemical bonds between the carbons, hydrogens, and oxygens that comprise carbohydrate molecules. Carbohydrates can exist as single-unit monomers or can be bonded to each other to produce longer-chain polysaccharide polymers.

Plants, such as potatoes, store their excess carbohydrates as polymers of starch. Animals store their excess carbohydrates as glycogen in muscles and the liver. Both starch and glycogen are polymers of glucose (Figure 3.1).

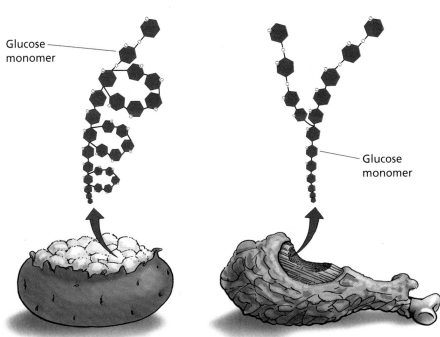

Glucose monomer

Glucose monomer

Potatoes contain **starch.**   Animal muscle contains **glycogen.**

Figure 3.1   Stored carbohydrates. All cells contain carbohydrate, protein, and fat, but some have more than others. When carbohydrate is in excess, plants store the excess as starch, and animals store the excess carbohydrates as glycogen.

When multi-subunit sugars are composed of many different branching chains of sugar monomers, they are called **complex carbohydrates**. Complex carbohydrates, such as those found in fruits and vegetables, are often involved in storing energy for later use. The body digests complex carbohydrates more slowly than it does simpler sugars, because complex carbohydrates have more chemical bonds to break. Endurance athletes will load up on complex carbohydrates for several days before a race to increase the amount of easily accessible energy they can draw on during competition.

Nutritionists agree that most of the carbohydrates in a healthful diet should be in the form of complex carbohydrates and that we should consume only minimal amounts of refined and processed sugars. When you consume complex carbohydrates in fruits, vegetables, and grains, you are also consuming many vitamins and minerals as well as fiber.

Dietary fiber, also called roughage, is composed mainly of complex carbohydrates that humans cannot digest into component monosaccharides. For this reason, dietary fiber is passed into the large intestine; some fiber is digested by bacteria living there, and the remainder gives bulk to the feces.

Although fiber is not a nutrient, because it is not absorbed by the body, it is still an important part of a healthful diet. Fiber helps lower the amount of the membrane lipid, **cholesterol**, without changing the level of high-density lipoprotein (HDL, or "good" cholesterol); at the same time, it helps decrease low-density lipoprotein (LDL, the "bad" form of cholesterol-carrying molecules). Fiber may also decrease your risk of various cancers. Fruits and vegetables tend to be rich in dietary fiber.

Proteins as Nutrients.   Protein-rich foods include beef, poultry, fish, beans, eggs, nuts, and dairy products such as milk, yogurt, and cheese.

Chapter 2 acquaints you with the idea that proteins are composed of amino acids and that amino acids differ from each other based on their side groups. Amino acids are bonded to each other in an infinite variety of combinations in order to produce a diverse array of proteins with many different functions.

Your body is able to synthesize many of the commonly occurring amino acids. Those your body cannot synthesize are called **essential amino acids** and must be supplied by the foods you eat. **Complete proteins** contain all the

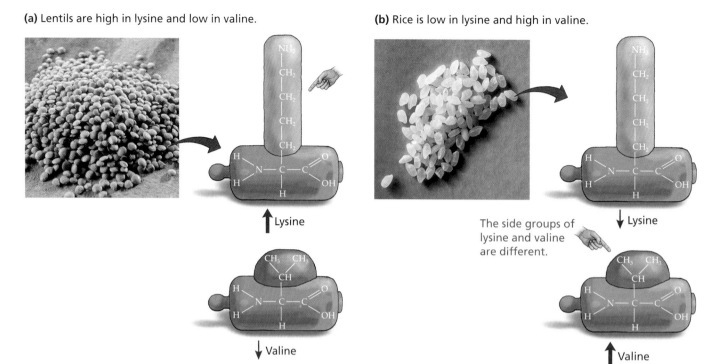

(a) Lentils are high in lysine and low in valine.

(b) Rice is low in lysine and high in valine.

↑ Lysine

↓ Lysine

The side groups of lysine and valine are different.

↓ Valine

↑ Valine

Figure 3.2   Essential amino acids. Essential amino acids, such as lysine and valine, cannot be synthesized by the body and must be obtained from the diet. Some proteins are high in one amino acid but low in another. Eating a wide variety of proteins thus ensures that all the necessary amino acids are available for growth, development, and maintenance.

(a) Fat within muscle

(b) Fat on surface of muscle

Figure 3.3   Fat storage. Fat can be intertwined with muscle tissue, as seen in this marbled piece of beef (a), or it can lie on the surface, as seen on this chicken breast (b).

essential amino acids your body needs. Proteins obtained by eating meat are more likely to be complete than are those obtained by eating plants; plant proteins can often be missing one or more essential amino acids (Figure 3.2).

In the past, some nutritionists believed that vegetarians might be at risk for deficiencies in certain amino acids. However, scientific studies have shown that there is little cause for concern. If a vegetarian's diet is rich in a wide variety of plant-based foods, the body will have little trouble obtaining all the amino acids it needs to build proteins.

Even though you can obtain all the proteins you need by eating a variety of plants, Americans tend to eat a lot of meat as well. In addition to being rich in proteins, meat tends to be rich in fat.

Fats as Nutrients.   The body uses fat, a type of lipid, as a source of energy. Gram for gram, fat contains a little more than twice as much energy as carbohydrates or protein. This energy is stored in the carbon, oxygen, and hydrogen bonds of the fat molecule.

Foods that are rich in fat include meat, milk, cheese, vegetable oils, and nuts. Muscle is often surrounded by stored fat; but some animals store fat throughout muscle, leading to the marbled appearance of some red meat. Other animals—chickens, for example—store fat on the surface of the muscle, making it easy to remove for cooking (Figure 3.3). Most mammals store fat just below the skin, to help cushion and protect vital organs, insulate the body from cold weather and to store energy in case of famine. Some scientists believe that prehistoric humans often faced times of famine and may have evolved to crave fat.

Fats consist of a glycerol molecule with hydrogen and carbon-rich fatty acid tails attached. Your body can synthesize most of the fatty acids it requires. Those that cannot be synthesized are called **essential fatty acids**. Like essential amino acids, essential fatty acids must be obtained from the diet.

The fatty acid tails of a fat molecule can differ in the number and placement of double bonds (Figure 3.4a). When the carbons of a fatty acid are bound to as many hydrogens as possible, the fat is said to be a **saturated fat** (saturated in hydrogens). When there are carbon-to-carbon double bonds, the fat is not saturated in hydrogens, and it is therefore an **unsaturated fat** (Figure 3.4b). The more double bonds, the higher the degree of unsaturation. When it contains many unsaturated carbons, the fat is referred to as **polyunsaturated**. The double bonds in unsaturated fats make the structures kink instead of lying flat. This form prevents the adjacent fat molecules from packing tightly together, so unsaturated fat tends to be liquid at room temperature. Cooking oil is an example of an unsaturated fat. Unsaturated fats are more likely to come from plant sources. Saturated fats, with their absence of carbon-to-carbon double bonds, pack tightly together to make a solid structure. This is why saturated fats, such as butter, are solid at room temperature (Figure 3.4c).

Commercial food manufacturers sometimes add hydrogen atoms to unsaturated fats by combining hydrogen gas with vegetable oils under pressure. This process, called **hydrogenation**, increases the fat's level of saturation; it retards spoilage and solidifies liquid oils, thereby making food seem less greasy. Margarine is vegetable oil that has undergone hydrogenation.

When hydrogen atoms are on the same side of the carbon-to-carbon double bond, they are said to be in the cis configuration—naturally occurring unsaturated fats have their hydrogen atoms in cis configuration.

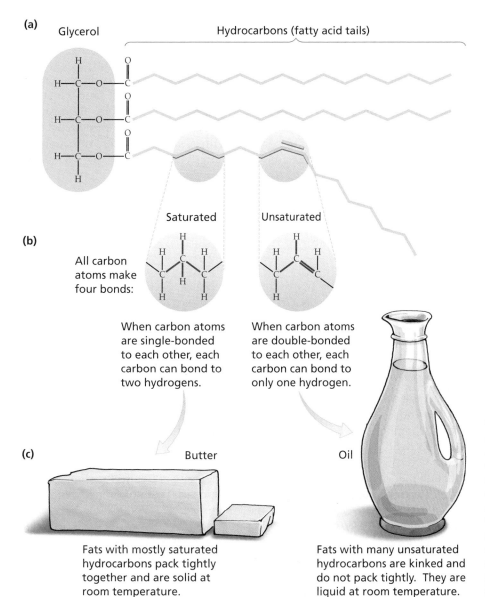

Figure 3.4 Saturated and unsaturated fats. (a) Fats are long chains of hydrogen and carbon (fatty acids) attached to a 3-carbon glycerol skeleton. Fatty acids differ in length and the placement of double bonds. (b) Carbon can make chemical bonds with up to 4 other atoms. The carbon atoms in a saturated fat are bonded to 4 other atoms. The carbon atoms in an unsaturated fat are double-bonded to other carbon atoms. (c) Saturated fats are solid at room temperature. Unsaturated fats are liquids at room temperature.

Figure 3.5 Hydrogenation. Adding hydrogen gas to vegetable oil forces the addition of hydrogen atoms to the fatty acid chains, some of which are incorporated in the unnatural trans configuration. The addition of hydrogens means that there are fewer double bonds; hence liquids can be solidified by this process.

When hydrogen atoms are on opposite sides of the double bond, they are said to be in the trans configuration (Figure 3.5). This form of fatty acid is not found in nature. During hydrogenation, there is no way to control whether hydrogen atoms bond in the cis or trans configuration.

While definitive studies are currently under way, the potential health risks of consuming foods rich in trans-fatty acids, common in fast foods, include increased risk of clogged arteries, heart disease, and diabetes. Because fat contains more stored energy per gram than carbohydrate and protein do, and because excess fat intake is associated with several diseases, nutritionists recommend that you limit the amount of fat in your diet.

Micronutrients

Micronutrients are required in very small amounts and include vitamins and minerals. They are neither destroyed by the body during use nor burned for energy.

Vitamins. Vitamins are organic substances (*organic* means "carbon containing"), most of which the body cannot synthesize. Most vitamins function as **coenzymes**, or molecules that help enzymes, and thus speed up the body's chemical reactions. When a vitamin is not present in sufficient quantities, deficiencies can affect every cell in the body because many different enzymes, all requiring the same vitamin, are involved in many different bodily functions. Vitamins also help with the absorption of other nutrients; for example, vitamin C increases the absorption of iron from the intestine. Some vitamins may even help protect the body against cancer and heart disease, and slow the aging process.

Vitamin D, also called calcitrol, is the only vitamin your cells can synthesize. Because sunlight is required for synthesis, people living in cold climates can develop deficiencies in vitamin D. All other vitamins must be supplied by the foods you eat. Many vitamins are water soluble, so boiling causes them to leach out into the water—this is why fresh vegetables are more nutritious than cooked ones. Steaming vegetables or using the vitamin-rich broth of canned vegetables when making soup helps preserve the vitamin content. Because the body does not store them, water-soluble vitamins are more likely than fat-soluble vitamins to be the source of dietary deficiencies. Vitamins A, D, E, and K are fat soluble and build up in stored fat; allowing an excess of these vitamins to accumulate in the body can be toxic. Table 3.1 lists some vitamins and their roles in the body.

Table 3.1 Vitamins. Water-soluble and fat soluble vitamins.

Water-soluble vitamins

- Small organic molecules (containing carbon)
- Will dissolve in water
- Cannot be synthesized by body
- Supplements packaged as pressed tablets
- Excesses usually not a problem because water-soluble vitamins are excreted in urine, not stored

Vitamin	Sources	Functions	Effects of Deficiency
Thiamin (B₁)	Pork, whole grains, leafy green vegetables	Required component of many enzymes	Water retention and heart failure
Riboflavin (B₂)	Milk, whole grains, leafy green vegetables	Required component of many enzymes	Skin lesions
Folic acid	Dark green vegetables, nuts, legumes (dried beans, peas, and lentils), whole grains	Required component of many enzymes	Neural-tube defects, anemia, and gastrointestinal problems
B₁₂	Chicken, fish, red meat, dairy	Required component of many enzymes	Anemia and impaired nerve function
B₆	Red meat, poultry, fish, spinach, potatoes, and tomatoes	Required component of many enzymes	Anemia, nerve disorders, and muscular disorders
Pantothenic acid	Meat, vegetables, grains	Required component of many enzymes	Fatigue, numbness, headaches, and nausea
Biotin	Legumes, egg yolk	Required component of many enzymes	Dermatitis, sore tongue, and anemia
C	Citrus fruits, strawberries, tomatoes, broccoli, cabbage, green pepper	Collagen synthesis; improves iron absorption	Scurvy and poor wound healing
Niacin (B₃)	Nuts, leafy green vegetables, potatoes	Required component of many enzymes	Skin and nervous system damage

Fat-soluble vitamins

- Small organic molecules (containing carbon)
- Will not dissolve in water
- Cannot be synthesized by body (except vitamin D)
- Supplements packaged as oily gel caps
- Excesses can cause problems since fat-soluble vitamins are not excreted readily

Vitamin	Sources	Functions	Effects of Deficiency	Effects of Excess
A	Leafy green and yellow vegetables, liver, egg yolk	Component of eye pigment	Night blindness, scaly skin, skin sores, and blindness	Drowsiness, headache, hair loss, abdominal pain, and bone pain
D	Milk, egg yolk	Helps calcium be absorbed and increases bone growth	Bone deformities	Kidney damage, diarrhea, and vomiting
E	Dark green vegetables, nuts, legumes, whole grains	Required component of many enzymes	Neural-tube defects, anemia, and gastrointestinal problems	Fatigue, weakness, nausea, headache, blurred vision, and diarrhea
K	Leafy green vegetables, cabbage, cauliflower	Helps blood clot	Bruising, abnormal clotting, and severe bleeding	Liver damage and anemia

Minerals. Minerals are substances that do not contain carbon but are essential for many cell functions. Because they lack carbon, minerals are said to be inorganic. They are important for proper fluid balance, in muscle contraction and conduction of nerve impulses, and for building bones and teeth. Calcium, chloride, magnesium, phosphorus, potassium, sodium, and sulfur are all minerals. Like some vitamins, minerals are water soluble and can leach out into the water during boiling. Also like vitamins, minerals are not synthesized in the body and must be supplied through your diet. Table 3.2 lists the various functions of minerals that your body requires and discribes what happens when there is a deficiency or an excess in certain minerals.

Processed Versus Whole Foods

Food that has undergone extensive processing has been stripped of much of its nutritive value. For example, refined flour has had the nutrient-rich, outer parts of the grain (called bran) along with the nutrient-rich, inner germ portion removed during processing, resulting in the loss of many vitamins and minerals and much of the fiber. It is best to limit your consumption of processed foods in general. Likewise, sweets (highly processed, sugar-rich foods) should occupy a

Table 3.2   Minerals. The minerals we require and their roles in the body.

Minerals

- Will dissolve in water
- Inorganic elements (do not contain carbon)
- Cannot be synthesized by body
- Supplements packaged as pressed tablets

Mineral	Sources	Functions	Effects of Deficiency	Effects of Excess
Calcium	Milk, cheese, dark green vegetables, legumes	Bone strength, blood clotting	Stunted growth, osteoporosis	Kidney stones
Chloride	Table salt, processed foods	Formation of HCl acid in stomach	Muscle cramps, reduced appetite, poor growth	High blood pressure
Magnesium	Whole grains, leafy green vegetables, legumes, dairy, nuts	Required component of many enzymes	Muscle cramps	Neurological disturbances
Phosphorus	Dairy, red meat, poultry, grains	Bone and tooth formation	Weakness, bone damage	Impaired ability to absorb nutrients
Potassium	Meats, fruits, vegetables, whole grains	Water balance, muscle function	Muscle weakness	Muscle weakness, paralysis, and heart failure
Sodium	Table salt, processed foods	Water balance, nerve function	Muscle cramps, reduced appetite	High blood pressure
Sulfur	Meat, legumes, milk, eggs	Components of many proteins	None known	None known

very small portion of your diet because they provide no real nutrition, just calories. This is why sweets are often referred to as "empty" calories.

Foods that have not been stripped of their nutrition by processing are called **whole foods**. Eating a wide variety of whole foods such as fruits, vegetables, and grains gives you a much better chance of achieving a healthful diet than eating highly refined, fatty foods that are low in complex carbohydrates and vitamins—also known as junk food.

In addition to containing vitamins and minerals, whole foods contain substances called **antioxidants** that are thought to play a role in the prevention of many diseases, including cancer. Biologists are currently investigating antioxidants to see whether these substances can slow the aging process. Antioxidants protect cells from damage caused by highly reactive molecules that are generated by normal cell processes. These highly reactive molecules, called **free radicals**, have an incomplete electron shell, which makes them more chemically reactive than molecules with complete electron shells. Antioxidants are abundant in fruits and vegetables, nuts, grains, and some meats. Table 3.3 describes food sources of common antioxidants.

Table 3.3   Antioxidants. Antioxidants are being investigated for their disease-preventing abilities.

Antioxidants

- Present in whole foods
- Protect cells from damage caused by free radicals
- Thought to have a role in disease prevention

Antioxidant	Source
Beta-carotene	Foods rich in beta-carotene are orange in color; they include carrots, cantaloupe, squash, mangoes, pumpkin, and apricots. Beta-carotene is also found in some leafy green vegetables such as collard greens, kale, and spinach.
Lutein	Lutein, which is known to help keep eyes healthy, is also found in leafy green vegetables such as collard greens, kale, and spinach.
Lycopene	Lycopene is a powerful antioxidant found in watermelon, papaya, apricots, guava, and tomatoes.
Selenium	Selenium is a mineral (not an antioxidant) that serves as a cofactor for many antioxidant enzymes, thereby increasing their effectiveness. Rice, wheat, meats, bread, and Brazil nuts are major sources of dietary selenium.
Vitamin A	Foods rich in vitamin A include sweet potatoes, liver, milk, carrots, egg yolks, and mozzarella cheese.
Vitamin C	Foods rich in vitamin C include most fruits, vegetables, and meats.
Vitamin E	Vitamin E is found in almonds, many cooking oils, mangoes, broccoli, and nuts.

3.2   Enzymes and Metabolism

In addition to eating a well-balanced diet rich in unprocessed foods, it is important for people to eat the right amount of food. All food, whether carbohydrate, protein, or fat, can be turned into fat when too much is consumed. In this manner, energy stored in the chemical bonds of food is converted into fat and stored for later use.

The amount of fat that a given individual will store depends partly on how quickly or slowly he or she breaks down food molecules into their component parts. **Metabolism** is a general term used to describe all of the chemical reactions occurring in the body.

Enzymes

All metabolic reactions are regulated by proteins called **enzymes** that speed up, or **catalyze**, the rate of reactions. Enzymes help your body break down the foods you ingest and liberate the energy stored in their chemical bonds.

To break chemical bonds, molecules must absorb energy from their surroundings, often by absorbing heat. This is why heating a chemical reaction will speed up the reaction. Heating cells to an excessively high temperature can damage or kill them. Enzymes help catalyze the body's chemical reactions without requiring heat for the reactants to break their chemical bonds. Therefore, by decreasing the energy required to start the reaction, enzymes allow chemical reactions to occur more quickly.

Activation Energy.   The energy required to start the metabolic reaction serves as a barrier to catalysis and is called the **activation energy**. If not for the activation energy barrier, all of the chemical reactions in cells would occur relentlessly, whether the products of the reactions were needed or not. Because most metabolic reactions need to surpass the activation energy barrier before proceeding, they can be regulated by the presence or absence of enzymes. In other words, a given chemical reaction will occur only if the proper enzyme is available. How do enzymes decrease the activation energy barrier?

Induced Fit.   The chemicals that are metabolized by an enzyme-catalyzed reaction are called the enzyme's **substrate**. Enzymes decrease activation energy by binding to their substrate and placing stress on its chemical bonds, decreasing the amount of initial energy required to break the bonds. The region of the enzyme where the substrate binds is called the enzyme's **active site**. Each active site has its own shape and chemistry. When the substrate binds to the active site, the enzyme changes shape slightly in order to envelop the substrate. This shape change by the enzyme in response to substrate binding is called **induced fit**, because the substrate induces the enzyme to change shape to conform to the substrate's contours. When the enzyme changes shape, it places stress on the chemical bonds of the substrate, making them easier to break. In this manner, the enzyme helps convert the substrate to a reaction product and then resumes its original shape so that it can perform the reaction again (Figure 3.6).

Different enzymes catalyze different reactions by a property called **specificity**. Enzymes are usually named for the reaction they catalyze and end in the suffix *-ase*. For example, sucrase is the enzyme that breaks down table sugar (sucrose). The specificity of an enzyme is the result of its shape and the shape of its active site. Different enzymes have unique

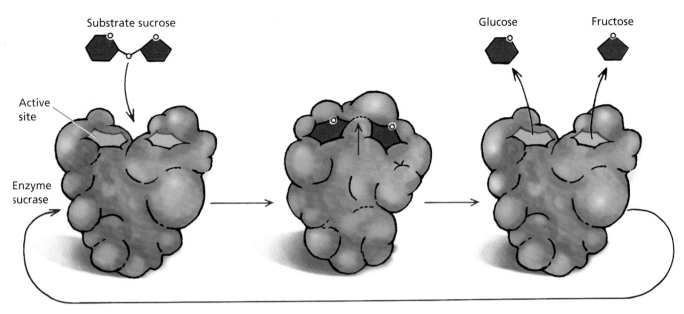

1. The shape of the substrate matches the shape of the enzyme's active site.

2. When the substrate binds to the active site, the enzyme changes shape, and the bond between the sugars is stressed.

3. The shape change splits the substrate and releases the two subunits. The enzyme is able to perform the reaction again.

Figure 3.6   Enzymes. The enzyme sucrase is cleaving (splitting) the disaccharide sucrose into its monosaccharide subunits, fructose and glucose. The enzyme can then be recycled to perform the same reaction again and again.

shapes because they are composed of amino acids in varying sequences. The 20 amino acids, each with its own unique side group, are arranged in distinct orders for each enzyme, producing enzymes of all shapes and sizes, each with an active site that can bind with its particular substrate. Although an infinite variety of enzymes could be produced, it is quite often the case that different organisms will utilize the same enzyme, likely due to their evolution from a common ancestor.

Lactose intolerance is a common dietary problem caused by an enzyme deficiency. People with lactose intolerance are unable to digest large amounts of lactose (the most common sugar in milk). This condition results from a shortage of the enzyme **lactase**, typically produced by the cells of the small intestine. Lactase breaks down lactose into simpler forms that can then be absorbed into the bloodstream. When there is not enough lactase available to digest lactose, bacteria in the intestine metabolize the sugar, producing lactic acid. The buildup of lactic acid causes bloating, gas, cramps, and diarrhea.

Most babies can digest milk sugars efficiently, but as they age, their ability to produce this enzyme declines. Scientists hypothesize that in the evolutionary past, the production of this enzyme after weaning was unnecessary, and most adults were probably lactose intolerant. However, with the addition of dairy products to the human diet, selective pressure now favors continuing to produce this enzyme into adulthood. Lactose intolerance can be treated by taking dietary supplements that contain the lactase enzyme.

Enzymes do more than just the break down milk sugars, they mediate all of the metabolic reactions occurring in an organism's cells. Enzymes even affect the rate of an individual's metabolism. This means that two similarly sized individuals might need to consume different amounts of food in order to meet their daily energy requirements.

Figure 3.7   Energy in food. The food you eat is broken down by the digestive system and transported into cells, where nutrients are used to make energy (ATP).

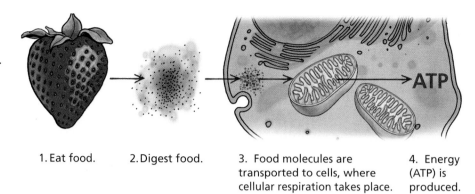

1. Eat food.   2. Digest food.   3. Food molecules are transported to cells, where cellular respiration takes place.   4. Energy (ATP) is produced.

Calories and Metabolic Rate

Energy is measured in units called calories. A **calorie** is the amount of energy required to raise the temperature of 1 gram of water by 1 degree Celsius. In scientific literature, energy is usually reported in kilocalories, and 1 kilocalorie equals 1000 calories of energy. However, in physiology—and on nutritional labels—the prefix *kilo–* is dropped, and a kilocalorie is referred to as a **Calorie** (with a capital C). Calories are consumed to supply the body with energy to do work, which includes maintaining body temperature.

Balancing energy intake versus energy output means eating the correct amount of food to maintain health. When foods are eaten, they are broken down into their component subunits. The energy stored in the chemical bonds of food can be used to make a form of energy that the cell can use (Figure 3.7). Cells power their activities by using the chemical **adenosine triphosphate** or **ATP** as their energy currency. ATP, a nucleotide, is discussed in detail in Chapter 4 along with the process of cellular respiration, which uses the chemical energy stored in food to produce ATP. When the supply of Calories is greater than the demand, the excess Calories can be stored by the body as fat.

The speed and efficiency of many different enzymes will lead to an overall increase or decrease in the rate at which a person can metabolize food. Thus, when you say that your metabolism is slow or fast, you are actually referring to the speed at which enzymes catalyze chemical reactions in your body.

A person's **metabolic rate** is a measure of his or her energy use. This rate changes according to the person's activity level. For example, people require less energy when asleep than we do when exercising. The **basal metabolic rate** represents the resting energy use of an awake, alert person. The average basal metabolic rate is 70 Calories per hour, or 1680 Calories per day. However, this rate varies widely among individuals because many factors influence each person's basal metabolic rate: exercise habits, body weight, nutritional status, sex, age, and genetics. Overall nutritional status can also affect metabolism, because an enzyme that is missing its vitamin coenzyme may not be able to perform at its optimal rate.

Exercise requires energy, which allows you to consume more Calories without having to store them. As for body weight, a heavy person utilizes more Calories during exercise than a thin person does. Figure 3.8 shows the number of Calories used per hour for various activities based on body weight. Males require more Calories per day than females do because testosterone, a hormone produced in larger quantities by males, increases the rate at which fat breaks down. Men also have more muscle than women, which requires more energy to maintain than fat does.

Age and genetics also play a role in metabolic rate. Two people of the same size and sex, who consume the same number of Calories and exercise the

To calculate the number of calories you are burning per hour, multiply your weight by these numbers.

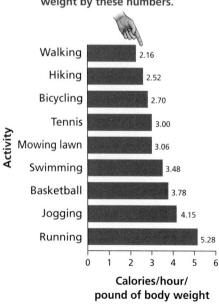

Activity	Calories/hour/pound of body weight
Walking	2.16
Hiking	2.52
Bicycling	2.70
Tennis	3.00
Mowing lawn	3.06
Swimming	3.48
Basketball	3.78
Jogging	4.15
Running	5.28

Figure 3.8   Energy expenditures for various activities. This bar graph can help you determine how many Calories you burn during certain activities.

same amount, will not necessarily store the same amount of fat. The rate at which the foods you eat are metabolized slows as you age, and some people are simply born with lower basal metabolic rates.

The properties of metabolic enzymes, like those of all proteins, are determined by the genes that encode them. Genes that influence a person's rate of fat storage and utilization are passed from parents to children. All of these variables help explain why some people seem to eat and eat and never gain an ounce, while others struggle with their weight for their entire lives.

To obtain a rough measure of how many Calories you should consume per day, multiply the weight you wish to maintain by 11 and add the number of Calories you burn during exercise (see Figure 3.8). If you are trying to lose weight, decrease your caloric intake and increase your exercise level. Losing 1 pound of fat requires you to burn 3500 more Calories than you consume. For example, reducing your intake by 300 Calories per day and increasing your exercise level to burn off an additional 200 Calories should result in a weight loss of 1 pound per week. To gain weight, increase the amount of Calories consumed.

To be fully metabolized, food is broken down by the digestive system and then transported to individual cells via the bloodstream (Figure 3.9). Once nutrients arrive at cells, they must traverse the membrane that surrounds cells, called the **plasma membrane**.

Figure 3.9   Nutrients move from the bloodstream to cells. Substances absorbed from the small intestine into the bloodstream make their way into individual cells. To do so, they must first cross the plasma membrane; once inside a cell, the food can be broken down further to release the energy stored in its chemical bonds.

3.3   Transport Across Membranes

Building-block molecules must cross the plasma membrane to gain access to the inside of the cell, where they can be used to synthesize cell components or be metabolized to provide energy for the cell. The chemistry of the membrane facilitates the transport of some substances and prevents the transport of others. The plasma membrane that surrounds cells is composed of two layers of phospholipids, called a **phospholipid bilayer**. The interior of the bilayer is water hating, or hydrophobic. Therefore, hydrophobic substances can pass through the membrane more easily than hydrophilic (water loving) ones. In this sense, the membrane of the cell is differentially permeable to the transport of molecules, allowing some to pass and blocking others from passing.

Passive Transport: Diffusion, Facilitated Diffusion, and Osmosis

Imagine opening a bottle of your favorite carbonated beverage. The hissing sound you hear when removing the bottle cap is the sound of carbon dioxide gas (CO_2) trickling out of the beverage and into the surrounding air. In fact, if you leave the top off for long enough, all the CO_2 will escape. So, why does the carbon dioxide leave the beverage? Because all molecules contain energy that makes them vibrate and bounce against each other, scattering around like billiard balls during a game of pool. In fact, molecules will bounce against each other until they are spread out over all the available area. In other words, molecules will move from their own high concentration (in the bottle) to their own low concentration (in the surrounding air). This movement of molecules from where they are in high concentration to where they are in low concentration is called **diffusion**. During diffusion, the net movement of molecules is *down* a concentration gradient. This movement does not require an input of outside energy; it is spontaneous.

Diffusion also occurs in living organisms. When substances diffuse across the plasma membrane, we call the movement **passive transport**. Passive transport has the name *passive* because it does not require an input of energy. The structure of the phospholipid bilayer that comprises the plasma membrane prevents many substances from diffusing across it. Only very small, hydrophobic molecules are able to cross the membrane by diffusion. In effect, these molecules dissolve in the membrane and slip from one side to the other (Figure 3.10a). The carbon dioxide present in carbonated beverages are molecules that can cross the membrane unaided, as can oxygen.

Hydrophilic molecules and charged molecules such as ions are unable to simply diffuse across the water-hating, hydrophobic core of the membrane. For example, when you have a meal of chicken, rich in charged amino acids, and a green salad, rich in hydrophilic carbohydrates and ions such as calcium (Ca^+), these amino acids, sugars, and ions cannot gain access to the inside of the cell on their own. Instead, these molecules are transported across membranes by proteins embedded in the lipid bilayer. This type of transport does not require an input of energy and is called **facilitated diffusion**. As with passive transport, facilitated diffusion moves substances with, or down, their concentration gradient. Facilitated diffusion is so named because the specific membrane proteins are helping or "facilitating" the diffusion of substances across the plasma membrane (Figure 3.10b).

The movement of water across a membrane is a type of passive transport called **osmosis**. Like other substances, water moves from its own high

(a) Diffusion

(b) Facilitated diffusion

(c) Osmosis in animal cell

(d) Osmosis in plant cell

Higher concentration of solute outside the cell

Shriveled cell

Higher concentration of solute outside the cell

Wilted cell

Figure 3.10   Passive transport. (a) Diffusion of molecules across the plasma membrane occurs with the concentration gradient and does not require energy. Small hydrophobic molecules, carbon dioxide, and oxygen can diffuse across the membrane. (b) Facilitated diffusion is the diffusion of molecules assisted by substrate specific proteins. Molecules move with their concentration gradient, which does not require energy. (c) Osmosis is the movement of water in response to a concentration gradient. Water moves toward a region that has more dissolved solute. When water leaves an animal cell, it shrinks. (d) When water leaves a plant cell, the plant wilts instead of shrinks due to the presence of the cell wall.

concentration to its own low concentration. Water can move through proteins in the membrane, called **aquaporins**, but even without these, water can still cross the membrane. When an animal cell is placed in a solution of salt water, water leaves the cell, causing the cell to shrivel (Figure 3.10c). Likewise, plants that are overfertilized wilt because water leaves the cells to equilibrate the concentration of water on either side of the plasma membrane (Figure 3.10d).

Active Transport: Pumping Substances Across the Membrane

In some situations, a cell will need to maintain a concentration gradient. For example, nerve cells require a high concentration of certain ions inside the cell in order to transmit nerve impulses. To maintain this difference in concentration across the membrane requires the input of energy. Think of a hill with a steep incline or grade. Riding your bike down the hill requires no energy, but riding your bike up the grade requires energy in the form of ATP. **Active transport** is transport that uses proteins, powered by the energy currency ATP, to move substances up a concentration gradient (Figure 3.11).

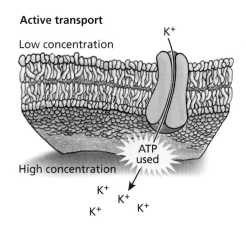

Figure 3.11   Active transport. Active transport moves substances against their concentration gradient and requires energy (ATP) to do so.

Figure 3.12   Movement of large substances. (a) Exocytosis is the movement of substances out of the cell. (b) Endocytosis is the movement of substances into the cell.

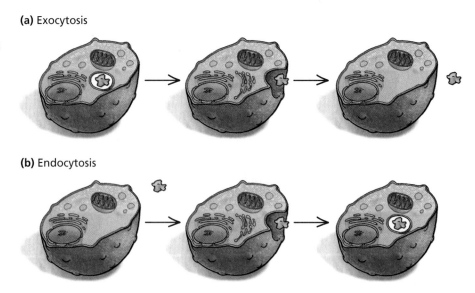

(a) Exocytosis

(b) Endocytosis

Exocytosis and Endocytosis: Movement of Large Molecules Across the Membrane

Larger molecules are often too big to diffuse across the membrane or to be transported through a protein, regardless of whether they are hydrophobic or hydrophilic. Instead, they must be moved around inside membrane-bound vesicles that can fuse with membranes. **Exocytosis** (Figure 3.12a) occurs when a membrane-bound vesicle, carrying some substance, fuses with the plasma membrane and releases its contents into the exterior of the cell. **Endocytosis** (Figure 3.12b) occurs when a substance is brought into the cell by a vesicle pinching inward, bringing the substance with it.

All the nutrients you consume and dismantle must find some way into your cells, so that they can be used for energy and to build cellular components. Knowing all this should give heightened meaning to the phrase, "you are what you eat" (Figure 3.13). Once inside your cells, nutrients will be used to build structural components and provide energy, or they can be stored for later use as fat.

3.4   Body Fat and Health

A clear understanding of how much body fat is healthful is hard to come by, because cultural and biological definitions of the term *overweight* differ markedly. Cultural definitions of overweight have changed over the years. Men and women who were considered to be of normal weight in the past might not be seen as meeting today's standards. In the United States, the evolution of this trend has been paralleled by changes in children's action figures and dolls and by Miss America pageant winners over the last several decades. You need only compare the physiques of action figures from the 1960s and 1970s to the physiques seen on today's action figures to see how the standards have changed for males (Figure 3.14a). Standards for women have also changed—the average weight of Miss America contestants has decreased significantly over the last few decades (Figure 3.14b). The unrealistic nature of these standards is illustrated by the fact that the average woman in the United States weighs 140 pounds and wears a size 12. The average model weighs 103 pounds and wears a size 4.

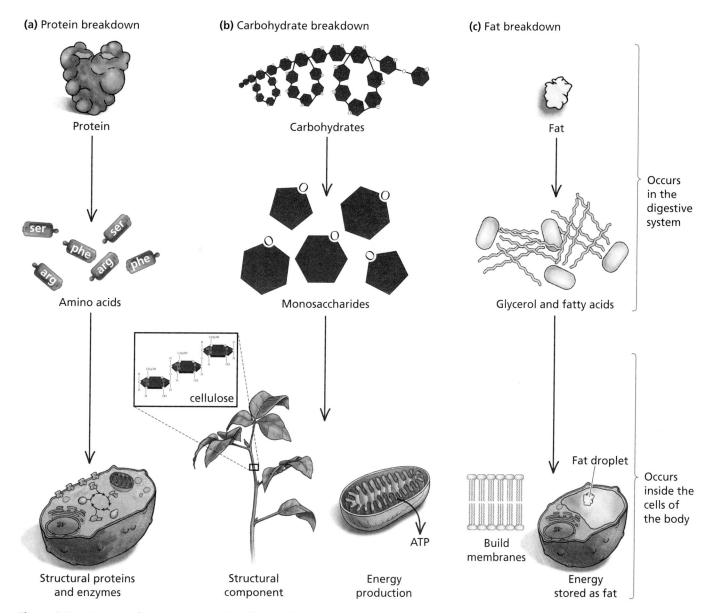

(a) Protein breakdown

Protein

ser
ser
phe
arg
arg
phe

Amino acids

Structural proteins
and enzymes

(b) Carbohydrate breakdown

Carbohydrates

O
O
O
O
O

Monosaccharides

cellulose

Structural
component

ATP

Energy
production

(c) Fat breakdown

Fat

Glycerol and fatty acids

Occurs
in the
digestive
system

Fat droplet

Build
membranes

Energy
stored as fat

Occurs
inside the
cells of
the body

Figure 3.13   You are what you eat. Food is digested into component molecules that are used to build cellular structures and generate ATP.

(a) GI Joe has become more muscular over time.

(b) Miss America has become thinner.

Figure 3.14   The perception of beauty.
(a) GI Joe in 1964 and in 2005. (b) Miss America 1964, Donna Axum, and Miss America 2005, Deidre Downs.

The next time you read the newspaper, take note of advertisements for diets featuring men and women of healthful weights promoting diet products. It is often the case that "before" pictures show individuals of healthful weights, and "after" pictures show people who are too thin. With these distorted messages about body fat, it is difficult for average people to know how much body fat is right for them.

Evaluating How Much Body Fat Is Healthful

A person's sex, along with other factors, determines his or her ideal amount of body fat. Women need more body fat than men do to maintain their fertility. On average, healthy women have 22% body fat, and healthy men have 14%. To maintain essential body functions, women need at least 12% body fat but not more than 32%; for men, the range is between 3% and 29%. This difference between females and males, a so-called sex difference, exists because women store more fat on their breasts, hips, and thighs than men do. A difference in muscle mass leads to increased energy use by males because muscles use more energy than does fat.

Women also have an 8% thicker layer of tissue called the dermis under the outer epidermal layer of the skin as compared to men. This means that in a woman and a man of similar strength and body fat, the woman's muscles would look smoother and less defined than the man's muscles would.

A person's frame size also influences body fat—larger-boned people carry more fat. In addition, body fat tends to increase with age.

Determining Ideal Weight. Unfortunately, it is a bit tricky to determine what any individual's ideal body weight should be. In the past, you simply weighed yourself and compared your weight to a chart showing a range of acceptable weights for given heights. The weight ranges on these tables were associated with the weights of a group of people who bought life insurance in the 1950s and whose health was monitored until they died. The problem with using these tables is that the subjects may not have been representative of the whole population. As you learned in Chapter 1, generalizing results seen in one group to another group can lead to erroneous conclusions. People who had the money to buy life insurance tended to have the other benefits of money as well, including easier access to health care, better nutrition, *and* lower body weight. Their longer lives may have had more to do with better health care and nutrition than with their weight.

To deal with some of the ambiguities associated with the insurance company's weight tables, a new measure of weight and health risk, the **body mass index (BMI)**, has been developed. BMI is a calculation that uses both height and weight to determine a value that correlates an estimate of body fat with the risk of illness and death (Table 3.4).

Although the BMI measurement is a better approximation of ideal weight than are the insurance charts of the past, it is not perfect; BMI still does not account for differences in frame size, gender, or muscle mass. In fact, studies show that as many as one in four people may be misclassified by BMI tables, because this measurement provides no means to distinguish between lean muscle mass and body fat. For example, an athlete with a lot of muscle will weigh more than a similarly sized person with a lot of fat, because muscle is heavier than fat.

If your BMI falls within the healthy range (BMI of 20–25), you probably have no reason to worry about health risks from excess weight. If your BMI is high, you may be at increased risk for diseases associated with obesity.

Obesity

One in four Americans has a BMI of 30 or greater and is therefore considered to be obese. This crisis in **obesity** is the result you would expect when constant access to cheap, high-fat, energy-dense, unhealthful food is combined with

Table 3.4 Body Mass Index (BMI). A chart based on height and weight correlations.

Height	Weight											
4'10" →	91	96	100	105	110	115	119	124	129	134	138	143
4'11" →	94	99	104	109	114	119	124	128	133	138	143	148
5'0" →	97	102	107	112	118	123	128	133	138	143	148	153
5'1" →	100	106	111	116	122	127	132	137	143	148	153	158
5'2" →	103	109	115	120	126	131	136	142	148	153	158	164
5'3" →	107	113	118	124	130	135	141	146	152	158	163	169
5'4" →	110	116	122	128	134	140	145	151	157	163	169	174
5'5" →	114	120	126	132	138	144	150	156	162	168	174	180
5'6" →	117	124	130	136	142	148	155	161	167	173	179	186
5'7"	121	127	134	140	146	153	159	→166	172	178	185	191
5'8" →	125	131	138	144	151	158	164	171	177	184	190	197
5'9" →	129	135	142	149	155	162	169	176	183	189	196	203
5'10" →	132	139	146	153	160	167	174	181	188	195	202	207
5'11" →	136	143	150	157	165	172	179	186	193	200	208	215
6'0" →	140	147	154	162	169	177	184	191	198	206	213	221
6'1" →	144	151	159	166	174	182	189	197	205	212	219	227
6'2" →	148	155	163	171	179	186	194	202	210	218	225	233
6'3" →	151	160	168	176	184	192	200	208	216	224	232	240
6'4" →	156	164	172	180	189	197	205	213	221	230	238	246
BMI	19	20	21	22	23	24	25	26	27	28	29	30

16 — Anorexic
20 — Underweight and possibly anorexic
— Healthy
25 — Overweight
30 — Obese

lack of exercise. This relationship is clearly illustrated by the case of the Pima Indians.

Several hundred years ago, the ancestral population of Pima Indians split into two tribes. One branch moved to Arizona and adopted the American diet and lifestyle; the typical Pima of Arizona gets as much exercise as the average American and, like most Americans, eats a high-fat, low-fiber diet. Unfortunately, the health consequences for these people are more severe than they are for most other Americans—close to 60% of the Arizona Pima are obese and diabetic. In contrast, the Pima of Mexico maintained their ancestral farming life; their diet is rich in fruits, vegetables, and fiber. The Pima of Mexico also engage in physical labor for close to 22 hours per week and are on average 60 pounds lighter than their Arizona relatives. Consequently, diabetes is virtually unheard of in this group.

This example illustrates the impact of lifestyle on health: the Pima of Arizona share many genes with their Mexican relatives but have far less healthful lives due to their diet and lack of exercise. The example also shows that genes influence body weight because the Pima of Arizona have higher rates of obesity and diabetes than those of other Americans whose lifestyle they share.

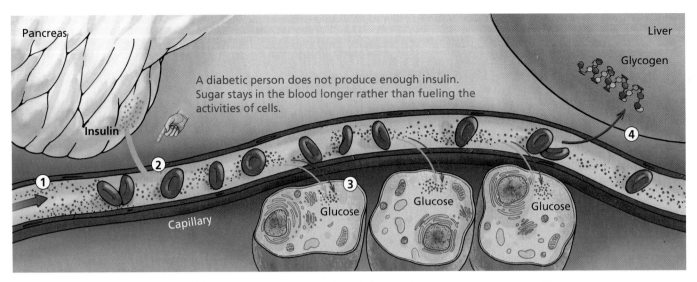

① Blood sugar is higher following a meal.

② When blood sugar is high, a healthy pancreas secretes insulin into the bloodstream.

③ Insulin triggers all the cells of the body to take up glucose.

④ Excess glucose is stored in the liver as glycogen.

Figure 3.15   Diabetes. The cells of the pancreas secrete insulin, which helps glucose move into body cells. Because diabetics produce less insulin, sugar stays in the blood longer.

Whether obesity is the result of genetics, diet, or lack of exercise, the health risks associated with obesity are the same. As your weight increases, so do your risks of diabetes, hypertension, heart disease, stroke, and joint problems.

Diabetes.   **Diabetes** is a disorder of carbohydrate metabolism characterized by the impaired ability of the body to produce or respond to insulin. **Insulin** is a hormone secreted by beta cells, which are located within clusters of cells in the pancreas. Insulin's role in the body is to trigger cells to take up glucose so that they can convert the sugar into energy. People with diabetes are unable to metabolize glucose; as a result, the level of glucose in the blood rises (Figure 3.15).

There are two forms of the disease. Type I, **insulin-dependent diabetes mellitus** (IDDM, formerly referred to as juvenile-onset diabetes), usually arises in childhood. People with IDDM cannot produce insulin, because their immune systems mistakenly destroy their own beta cells. When the body is no longer able to produce insulin, daily injections of the hormone are required. Type I diabetes is *not* correlated with obesity.

Type II, **non-insulin-dependent diabetes mellitus** (NIDDM, sometimes called adult-onset diabetes), usually occurs after 40 years of age and is more common in the obese. NIDDM arises either from decreased pancreatic secretion of insulin or from reduced responsiveness to secreted insulin in target cells. People with NIDDM are able to control blood-glucose levels through diet and exercise and, if necessary, by insulin injections.

Hypertension.   **Hypertension**, or high blood pressure, places increased stress on the circulatory system and causes the heart to work too hard. Compared to a person with normal blood pressure, a hypertensive person is six times more likely to have a heart attack.

Blood pressure is the force, originated by the pumping action of the heart, exerted by the blood against the walls of the blood vessels. Blood vessels expand and contract in response to this force. Blood pressure is reported as two numbers: the higher number, called the **systolic blood pressure**, represents the pressure exerted by the blood against the walls of the blood vessels; the lower number, called the **diastolic blood pressure**, is the pressure that exists between contractions of the heart when the heart is relaxing. Normal blood pressure is around 120 over 80 (symbolized as 120/80). Blood pressure is considered to be high when it is persistently above 140/90.

Problematic weight gain is typically the result of increases in the amount of fatty tissue versus increases in muscle mass. Fat, like all tissues, relies on oxygen and other nutrients from food to produce energy. As the amount of fat on your body increases, so does the demand for these substances. Therefore, the amount of blood required to carry oxygen and nutrients also increases. Increased blood volume means that the heart has to work harder to keep the blood moving through the vessels, thus placing more pressure on blood-vessel walls and leading to increased heart rate and blood pressure.

Heart Attack and Stroke.   A **heart attack** occurs when there is a sudden interruption in the supply of blood to the heart caused by the blockage of a vessel supplying the heart. A **stroke** is a sudden loss of brain function that results when blood vessels supplying the brain are blocked or ruptured. Heart attack and stroke are more likely in obese people because the elevated blood pressure caused by obesity also damages the lining of blood vessels and increases the likelihood that cholesterol will be deposited there. Cholesterol-lined vessels are said to be atherosclerotic.

Because lipids like cholesterol are not soluble in aqueous (water-based) solutions, cholesterol is carried throughout the body, attached to proteins in structures called lipoproteins. **Low-density lipoproteins (LDL)** have a high proportion of cholesterol (in other words, they are low in protein). LDLs distribute both the cholesterol synthesized by the liver and the cholesterol derived from diet throughout the body. LDLs are also important for carrying cholesterol to cells, where it is used to help make plasma membranes and hormones. **High-density lipoproteins (HDL)** contain more protein than cholesterol. HDLs scavenge excess cholesterol from the body and return it to the liver, where it is used to make bile. The cholesterol-rich bile is then released into the small intestine, and from there much of it exits the body in the feces. The LDL/HDL ratio is an index of the rate at which cholesterol is leaving body cells and returning to the liver.

Your physician can measure your cholesterol level by determining the amounts of LDL and HDL in your blood. If your total cholesterol level is over 200 or your LDL level is above 100 or so, then your physician may recommend that you decrease the amount of cholesterol and saturated fat in your diet. This may mean eating more plant-based foods and less meat, since plants do not have cholesterol, as well as reducing the amount of saturated fats in your diet. Saturated fat is thought to raise cholesterol levels by stimulating the liver to step up its production of LDLs and slowing the rate at which LDLs are cleared from the blood.

Cholesterol is not all bad; in fact, some cholesterol is necessary—it is present in cell membranes to help maintain their fluidity, and it is the building block for steroid hormones such as estrogen and testosterone. You do, however, synthesize enough cholesterol so that you do not need to obtain much from your diet.

For some people, those with a genetic predisposition to high cholesterol, controlling cholesterol levels through diet is difficult because dietary cholesterol makes up only a fraction of the body's total cholesterol. People with high cholesterol who do not respond to dietary changes may have inherited genes that increase the liver's production of cholesterol. These people may require prescription medications to control their cholesterol levels.

Cholesterol-laden, atherosclerotic vessels increase your risk of heart disease and stroke. Fat deposits narrow your heart's arteries, so less blood can flow to your heart. Diminished blood flow to your heart can cause chest pain, or angina. A complete blockage can lead to a heart attack. Lack of blood flow to the heart during a heart attack can cause the oxygen-starved heart tissue to die, leading to irreversible heart damage.

The same buildup of fatty deposits also occurs in the arteries of the brain. If a blood clot forms in a narrowed artery in the brain, it can completely block blood flow to an area of the brain, resulting in a stroke. If oxygen-starved brain tissue dies, permanent brain damage can result.

Anorexia and Bulimia

Eating disorders that make you underweight cause health problems that are as severe as those caused by too much weight (Table 3.5). **Anorexia**, or self- starvation, is rampant on college campuses. Estimates suggest that 1 in 5 college women and 1 in 20 college men restrict their intake of Calories so severely that they are essentially starving themselves to death. Others allow themselves to eat—sometimes very large amounts of food (called binge eating)—but prevent the nutrients from being turned into fat by purging themselves, often by vomiting. Binge eating followed by purging is called **bulimia**.

Anorexia has serious long-term health consequences. Anorexia can starve heart muscles to the point that altered rhythms develop. Blood flow is reduced, and blood pressure drops so much that the little nourishment present cannot get to the cells. The lack of fat accompanying anorexia can also lead to the cessation of menstruation, a condition known as amenorrhea. **Amenorrhea** occurs when a protein called **leptin**, which is secreted by fat cells, signals the brain that there is not enough body fat to support a pregnancy. Hormones (such as estrogen) that regulate menstruation are blocked, and menstruation ceases. Lack of menstruation can be permanent and causes sterility in a substantial percentage of anorexics.

The damage done by the lack of estrogen is not limited to the reproductive system—bones are affected as well. Estrogen secreted by the ovaries during the menstrual cycle acts on bone cells to help them maintain their strength and size. Anorexics reduce the development of dense bone and put themselves at a much higher risk of breaking their weakened bones, in a condition called **osteoporosis**.

Besides experencing the same health problems that anorexics face, bulimics can rupture their stomachs through forced vomiting. They often have dental and gum problems caused by stomach acid being forced into their mouths during vomiting, and they can become fatally dehydrated.

As you have seen, the health problems associated with obesity, anorexia, and bulimia are severe. You can avoid all of these problems—and improve your overall health—by focusing more on fitness and less on body weight.

Health problems resulting from **OBESITY**	Health problems resulting from **ANOREXIA and BULIMIA**
• Adult-onset diabetes	• Altered heart rhythms
• Hypertension (high blood pressure)	• Amenorrhea (cessation of menstruation)
• Heart attack	• Osteoporosis (weakened bones)
• Stroke	• Dental/gum problems
• Joint problems	• Ruptured stomach
	• Dehydration

Table 3.5   Obesity and anorexia or bulimia. Health problems result from being either overweight or underweight.

Figure 3.16 USDA Food Guide Pyramid.
The newly designed pyramid stresses the importance of physical activity. The size of each triangle represents the relative proportion of your diet that should be composed of each food group.

As the newly designed USDA Food Guide Pyramid attempts to illustrate (Figure 3.16), working slowly toward being fit and eating healthfully rather than trying the latest fad diet are more realistic and attainable ways to achieve the positive health outcomes that we all desire. In fact, fitness may be more important than body weight in terms of health. Studies show that fit but overweight people have better health outcomes than unfit slender people. In other words, lack of fitness is associated with higher health risks than excess body weight. Therefore, it makes more sense to focus on eating right and exercising than it does to focus on the number on the scale.

CHAPTER REVIEW

Summary

3.1 Nutrients

- Nutrients provide structural units and energy for cells. Although not technically a nutrient, water is an important dietary constituent that helps dissolve and eliminate the waste and maintain blood pressure and body temperature (p. 46).

- Macronutrients are required in large amounts for proper growth and development. Macronutrients include carbohydrates, proteins, and fats. All of these molecules are composed of subunits that can be broken down for use by the cell (p. 46).

- Micronutrients are dietary substances required in minute amounts for proper growth and development; they include vitamins and minerals (p. 50).

- Vitamins are organic substances, most of which the body cannot synthesize. Many vitamins serve as coenzymes to help enzymes function properly (p. 50).

- Minerals are inorganic substances essential for many cell functions (p. 52).

- Processing foods decreases their nutritive value (p. 52).

3.2 Enzymes and Metabolism

- The chemical reactions that occur in cells to build up or break down macromolecules are called metabolic reactions (p. 54).

- Metabolism is governed by enzymes. Enzymes are proteins that catalyze specific cellular reactions, first by binding the substrate to the enzyme's active site. This binding causes the enzyme to change shape (induced fit), placing stress on the bonds of the substrate and thereby lowering the activation energy (p. 54).

- Energy is measured in units called Calories (p. 56).

- The energy stored in the chemical bonds of food can be released by metabolic reactions and stored in the bonds of ATP. Cells use ATP to power energy-requiring processes (p. 56).

- An individual's metabolic rate is affected by many factors, including age, sex, exercise level, body weight, and genetics (p. 56).

Web Tutorial 3.1 Enzymes

3.3 Transport Across Membranes

- To gain access to cells, nutrients move across the plasma membrane, which functions as a semipermeable barrier that allows some substances to pass and prevents others from crossing (p. 58).

- The plasma membrane is composed of two layers of phospholipids, in which are embedded proteins and cholesterol (p. 58).

- Passive transport mechanisms include simple diffusion and facilitated diffusion (diffusion through proteins).

- Passive transport always moves substances with their concentration gradient and does not require energy (p. 58).

- Osmosis, the diffusion of water across a membrane, can involve the movement of water through protein pores in the membrane (p. 58).

- Active transport is an energy requiring process which requires proteins in cell membranes to move substances against their concentration gradients (p. 59).

- Larger molecules move into and out of cells enclosed in membrane-bound vesicles (p. 60).

Web Tutorial 3.2 Transport Across Membranes
Web Tutorial 3.3 Exocytosis and Endocytosis

3.4 Body Fat and Health

- Determining ideal body weight is difficult with conventional methods (p. 62).

- Obesity is associated with many health problems, including hypertension, heart attack and stroke, diabetes, and joint problems (pp. 62–65).

- Anorexia and bulimia are very common on college campuses and result in serious long-term health problems (p. 66).

Learning the Basics

1. What is metabolism, and what factors affect an individual's metabolic rate?

2. What types of substances can pass through cell membranes unaided? What types require help to pass through membranes?

3. Macronutrients _____.
   A. include carbohydrates and vitamins; **B.** should comprise a small percentage of a healthful diet; **C.** are essential in minute amounts to help enzymes function; **D.** include carbohydrates, fats, and proteins; **E.** are synthesized by cells and not necessary to obtain from the diet

4. The function of low-density lipoprotein (LDL) is to _____.
   A. break down proteins; **B.** digest starch; **C.** transport cholesterol from the liver; **D.** carry carbohydrates into the urine

5. Which of the following is a *false* statement regarding enzymes?
   A. Enzymes are proteins that speed up metabolic reactions.; **B.** Enzymes have specific substrates.; **C.** Enzymes supply ATP to their substrates.; **D.** An enzyme may be used many times over.

6. Enzymes speed up chemical reactions by _____.
   A. heating cells; **B.** binding to substrates and placing stress on their bonds; **C.** changing the shape of the cell; **D.** supplying energy to the substrate

7. A substance moving across a membrane against a concentration gradient is moving by _____.
   A. passive transport; **B.** osmosis; **C.** facilitated diffusion; **D.** active transport; **E.** diffusion

8. A cell that is placed in salty seawater will _____.
   A. take sodium and chloride ions in by diffusion; **B.** move water out of the cell by active transport; **C.** use facilitated diffusion to break apart the sodium and chloride ions; **D.** lose water to the outside of the cell via osmosis

9. Which of the following forms of membrane transport require specific membrane proteins?
   A. diffusion; **B.** exocytosis; **C.** facilitated diffusion; **D.** active transport; **E.** facilitated diffusion and active transport

10. Water crosses cell membranes _____.
    A. by active transport; **B.** through protein channels called aquaporins; **C.** against its concentration gradient; **D.** in plant cells but not in animal cells

Analyzing and Applying the Basics

1. A friend of yours does not want to eat meat, so instead she consumes protein powders that she buys at a nutrition store. What would be the disadvantages of this practice?

2. Two people with very similar diets and similar exercise levels have very different amounts of body fat. Why might this be the case?

3. A friend has his cholesterol level checked and tells you that he is really relieved because his cholesterol is normal—that is, under 200. Should your friend actually have no health concerns about his cholesterol level, or does he need to consider other factors?

Connecting the Science

1. Why do you think that anorexia and bulimia are more common among women than men?

2. What would you say to a friend who believes that he is fat, even though his BMI places him in the "normal" range? How about a friend who qualifies as obese on a BMI chart but who exercises regularly and eats a well-balanced diet?

Is the Earth Warming?

The Greenhouse Effect, Cellular Respiration, and Photosynthesis

Inhabitants of the Nation of Tuvalu . . .

will have to leave their tiny island communities.

High tides are eroding and flooding these Pacific islands.

A nine-island chain in the South Pacific, located between Australia and Hawaii, comprises the nation of Tuvalu. Each of the tropical islands that is part of the chain is an atoll—a circular column of coral rising up from the sea floor and extending above sea level. Many of the islands are covered with coconut trees and sandy beaches; together, they have close to 10 square miles of land above sea level but not very far above sea level—13 feet at the highest point.

The 10,000 or so people inhabiting these islands farm and fish in order to feed themselves and their families. Crops produced on the islands include coconuts, taro, and bananas. Each of the inhabited islands has a cooperative market where goods are sold or traded.

Tuvaluans live peaceful lives in a nation where crime is virtually unheard of; in fact, most residents sleep with their doors open to allow in the cooling night breezes. The islands have no television service, and the lone bank closes at 1:00 p.m. each day; no one takes credit cards. Soccer is the most popular sport, but Tuvaluans consider it bad sportsmanship to tackle an opponent aggressively enough to cause him or her to fall down. Transportation is largely via bicycle or motor scooter.

But there is one big problem in this seemingly idyllic island paradise—the islands of Tuvalu are disappearing. The storm surges and high tides that have

Many scientists agree that increased sea levels are caused by global warming.

71

become more common in recent years are eroding protective offshore barriers and beaches, destroying roads, and flooding homes and plantations. Recently, a record high tide submerged much of the country, causing week-long telephone service outages and flooding Tuvalu's only airport. Some roads have been moved inland as the Pacific eats away asphalt closer to the shore. The flooding of plantations with salty seawater kills the crops, forcing citizens to grow crops in metal containers filled with compost. Many people have simply left the island nation altogether. Nearly 3,000 native Tuvaluans have relocated, and a government program is moving its citizens at the rate of 75 per year.

What is causing the sudden rise in sea levels of the Tuvaluan islands? The prime minister of Tuvalu believes that global warming, which is causing the polar ice caps to melt and deposit water into the seas, is at fault. Concern that the tides will soon be high enough to submerge the nation and force the exodus of its remaining residents has led the prime minister to attempt to sue those he believes to be responsible—namely, the United States. Per capita, this country is producing more of the gasses associated with global warming than any other country.

4.1   The Greenhouse Effect

Global warming is the progressive increase of Earth's average temperature that has been occurring over the past century. The prime minister of Tuvalu is not alone in his belief that global warming is caused by increased emissions of certain gases. Most scientists agree that global warming is caused by recent increases in the concentrations of particular atmospheric gases including methane, nitrous oxide, water vapor, and carbon dioxide. Because increases in carbon dioxide seem to be the major source of problems related to global warming, we will focus mainly on that gas for the rest of this discussion.

The presence of carbon dioxide in the atmosphere leads to a phenomenon called the **greenhouse effect**. The greenhouse effect works like this: Warmth from the sun heats Earth's surface, which then radiates the heat energy absorbed outward. Most of this heat is radiated back into space, but some of the heat is retained in the atmosphere. The retention of heat is facilitated by carbon dioxide molecules, which act like a blanket to trap the heat radiated by Earth's surface (Figure 4.1). When you sleep under a blanket at night, your body heat is trapped and helps keep you warm. When the levels of greenhouse gases in the atmosphere increase, the effect is similar to sleeping under too many blankets—the temperature increases. The trapping of this warmth radiating from Earth is known as the greenhouse effect.

This is not exactly how panes of glass in a greenhouse function, that is, by allowing radiation from the sun to penetrate into the greenhouse and then trapping the heat that radiates from the warmed-up surfaces inside the greenhouse. But the overall effect is the same—the air temperature increases.

The greenhouse effect is not in itself a dangerous or unhealthy phenomenon. If Earth's atmosphere did not have some greenhouse gases, too much heat would be lost to space, and Earth would be too cold to support life. It is the excess warming due to more and more carbon dioxide accumulating in the atmosphere as a result of coal, oil, and natural gas burning that is causing problems.

In the absence of excess greenhouse gases, water vapor and carbon dioxide work together to keep temperatures on Earth hospitable for life.

Figure 4.1   The greenhouse effect. Heat from the sun is trapped in the atmosphere by water vapor, carbon dioxide, and other greenhouse gases. Increased levels of carbon dioxide contribute to the greenhouse effect.

Water, Heat, and Temperature

Bodies of water absorb heat and help maintain stable temperatures on Earth. Heat and temperature are measures of energy. **Heat** is the total amount of energy associated with the movement of atoms and molecules in a substance. **Temperature** is a measure of the intensity of heat—for example, how fast the molecules in the substance are moving.

Figure 4.2 Hydrogen bonding in water. Hydrogen bonds break as they absorb heat and reform as water releases heat. Water remains in the liquid form because not all the hydrogen bonds are broken at any one time.

Water molecules are attracted to each other, resulting in the formation of weak chemical bonds, called hydrogen bonds, between neighboring molecules. When water is heated, the heat energy disrupts the hydrogen bonds. Only after the hydrogen bonds have been broken can heat cause individual water molecules to move faster, thus increasing the temperature. In other words, the initial input of heat used to break hydrogen bonds between water molecules does not immediately raise the temperature of water; instead, it breaks hydrogen bonds. Therefore, water can absorb and store a large amount of heat while warming up only a few degrees in temperature. When water cools, hydrogen bonds re-form between adjacent molecules, releasing heat into the atmosphere. Water can release a large amount of heat into the surroundings while not decreasing the temperature of the body of water very much (Figure 4.2).

Water's high heat-absorbing capacity has important effects on Earth's climate. The vast amount of water contained in Earth's oceans and lakes moderates temperatures by storing huge amounts of heat radiated by the sun and giving off heat that warms the air during cooler times. Therefore, the balance between releasing and maintaining heat energy is vital to the maintenance of climate conditions on Earth. This balance can be disrupted when increasing levels of carbon dioxide cause more heat to be trapped.

Carbon Dioxide

Many of the atoms found in complex molecules of living organisms are broken down into simpler molecules and recycled for use in different capacities. Carbon dioxide (CO_2) is no different. The carbon dioxide you exhale is released into the atmosphere, where it can absorb heat, diffuse into the oceans, or be absorbed by forests and soil. Volcanic eruptions return carbon dioxide trapped within Earth's surface to the atmosphere. As you can see in Figure 4.3, carbon dioxide naturally flows back and forth between living organisms, the atmosphere, bodies of water, and soil.

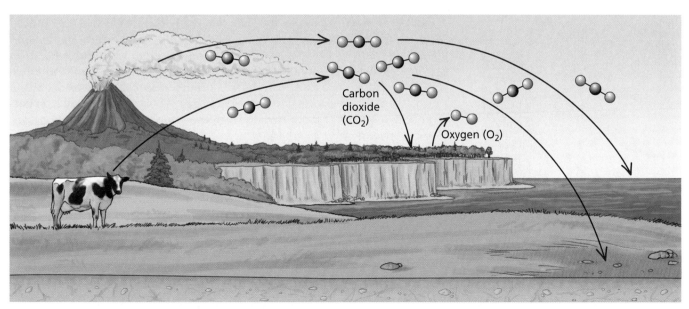

Figure 4.3 The flow of carbon. All living organisms and volcanoes produce CO_2. Forests, oceans, and soil absorb CO_2 from the air.

Figure 4.4   The flow of chemicals and energy. Energy enters biological systems in the form of sunlight, which is used to convert carbon dioxide and water into sugars during photosynthesis. The products of photosynthesis are broken down during cellular respiration to produce carbon dioxide and water and release energy.

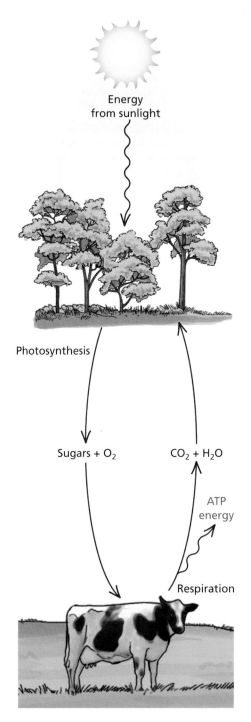

Energy from sunlight

Photosynthesis

Sugars + O₂   CO₂ + H₂O

ATP energy

Respiration

It is not just carbon that flows between plants and other organisms—energy does also. Plants use energy from the sun to produce sugars and other organic molecules that other organisms consume. The energy is stored in the bonds of these organic molecules and can be used to produce energy for the cell. When plants use energy from sunlight to produce organic molecules, by a process called **photosynthesis**, they also release oxygen (O_2) into the atmosphere. The metabolism of organic molecules by **cellular respiration** produces not only energy but also carbon dioxide and water (Figure 4.4).

The carbon dioxide produced by respiration is taken up by plants during the process of photosynthesis, is absorbed by both chemicals and organisms in the ocean, or accumulates in the atmosphere. The ocean has served as Earth's largest carbon dioxide and heat reservoir, but oceanic and atmospheric scientists are very concerned about the ocean's ability to absorb carbon dioxide at the rate that it is being emitted into the atmosphere. This is because human activities have rapidly increased the rate of carbon dioxide release into the atmosphere, largely by burning fossil fuels (Figure 4.5).

Fossil fuels are the buried remains of ancient plants and microorganisms that have been transformed by heat and pressure into coal, oil, and natural gas. These fuels are rich in carbon because plants remove carbon from the atmosphere during photosynthesis; consequently, plant structures are rich in organic carbon. Dead plant materials that are buried before they decompose, and thus before their carbon is released in the form of carbon dioxide, can produce fossil fuels. Humans combust this stored organic carbon to produce energy. The plants that made up the majority of fossil fuels lived from 362 to 290 million years ago, during a geological period called the Carboniferous period.

Burning these fossil fuels to generate electricity, power our cars, and heat our homes releases carbon dioxide into the atmosphere. Increases in carbon dioxide are well documented by direct measurements of the atmosphere over the past 50 years (Figure 4.6).

Figure 4.5   Burning fossil fuels. The burning of fossil fuels by industrial plants and automobiles adds more carbon dioxide to the environment.

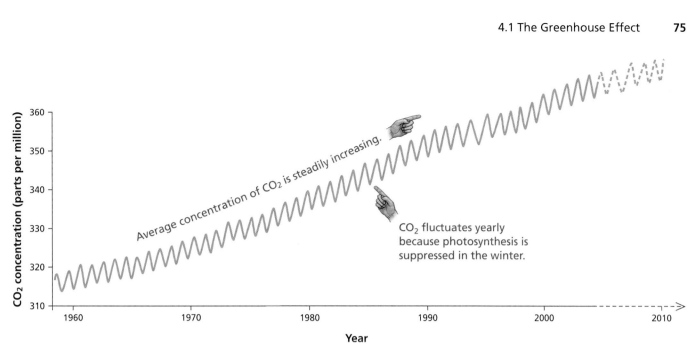

Figure 4.6   Increases in atmospheric carbon dioxide. Carbon dioxide levels have increased over the years.

Scientists can also directly measure the amount of carbon dioxide that was present in the atmosphere in the past by examining cores of ice sheets that have existed for thousands of years. This is because snow near the surface of ice traps air. As more snow accumulates, the underlying snow is compressed into ice that contains air bubbles. Cores can be removed from long-lived ice sheets and analyzed to determine the concentration of carbon dioxide trapped in air bubbles. These bubbles are actual samples of the atmosphere from up to hundreds of thousands of years ago (Figure 4.7). In addition, certain characteristics of gases trapped in the bubbles of ice cores can provide indirect information about temperatures at the time the bubbles formed. Ice core data from Antarctica, shown in Figure 4.8 (shown on page 76), indicate that the concentration of carbon dioxide in the atmosphere is much higher now than at any time in the past 400,000 years and that increased levels of carbon dioxide are correlated with increased temperatures.

Although Earth has gone through temperature cycles many times in the past, the concerns regarding current warming trends are that human activities are inflating the rate of increase and that these increases may persist for thousands of years. Many scientists believe that the effects of increased temperatures will be far reaching. Even now, the Tuvaluans and their Pacific islands are not the only organisms and environments being affected.

The Greenhouse Effect, Organisms, and Their Environments

Several million tourists visit Glacier National Park, located in the northwest corner of Montana, every year. With each passing year, the glaciers in this park decrease in size and number. As the glaciers shrink, they take with them natural habitat set aside for protection in this national park. Some of the park's glaciers have already shrunk to half their original size, and the total number of glaciers has decreased from approximately 150 in 1850 to around 35 today.

Figure 4.7   Ice core. By analyzing ice cores, scientists can measure the concentration of carbon dioxide that was present in early atmospheres.

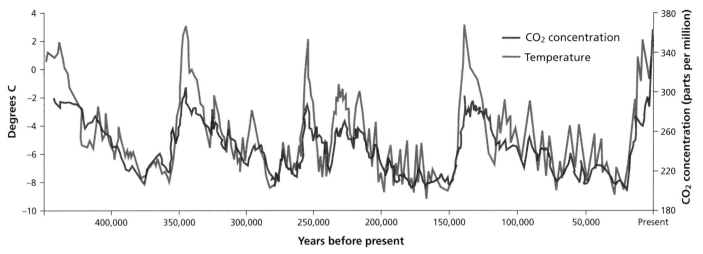

Figure 4.8   Records of temperature and atmospheric carbon dioxide concentration from Antarctic ice cores. These data indicate that increases in carbon dioxide levels are correlated with higher temperatures.

Like ice masses all over the world, these glaciers are slowly succumbing to warmer temperatures. According to the U.S. National Climate Data Center, the entire planet has warmed by 0.25°C (0.5°F) during the twentieth century. If this trend continues, scientists predict that by the year 2030, not a single glacier will be left in the park.

Melting glaciers are not confined to this park; in fact, they are melting worldwide. Mountain glaciers are receding from their peaks as far away as Tanzania, where Mount Kilimanjaro has lost 82% of its ice cap since 1912. The Greenland ice sheet is becoming thinner at its margins every year. Alpine glaciers contain half as much ice as they did in the mid-nineteenth century, when climbers first hiked to their peaks. In Antarctica, rising temperatures have led to the collapse of massive ice shelves. In the past 3 years, two massive chunks of ice, each about the size of Rhode Island, have fallen into the ocean.

The loss of ice has been a problem for the polar bear population in Hudson Bay. Seals, the bears' main food source, live on the ice of Hudson Bay, but this ice is breaking up earlier and earlier. The amount of time ice exists on western Hudson Bay has decreased by 3 weeks over the last 20 years. Rising temperatures thin the ice pack, making it too fragile to support seals and the bears that hunt them and driving the bears to shore in poor condition for hibernation. The average weight of polar bears in this region is declining, and fewer cubs are being born.

Sea levels have risen by 10 to 20 cm (4–8 inches) in the twentieth century. Increased ocean volumes due to the addition of water from melted ice can also lead to changes in climate. Worldwide rain and snowfall over land has increased by about 1%, and rain storms, as seen in Tuvalu, are expected to become more frequent and more severe. In addition to its impact on the humans who live there, flooding of tropical oceanic islands disturbs some of the most unique and diverse habitats on the planet.

A review published in the journal *Nature* in March 2002 described various species that have been affected by climate change. Many of these species are temperature sensitive, and they must move closer to the poles or to higher elevations to find regions with the proper climate. Arctic foxes are retreating northward and being replaced by the less cold-hardy red fox. Edith's checkerspot butterfly is now found 124 meters higher in elevation and 92 kilometers north of its range in 1900, and a wide variety of corals have experienced a dramatic increase in the frequency and extent of damage resulting from increased ocean temperatures.

It is not just animals that need to migrate along with changing temperatures. Plant species with specific temperature requirements will have to move as well. Those that cannot migrate quickly enough will likely become extinct. One example of a plant that will need to undergo this forced migration is the sugar maple, the source of maple syrup.

New England risks losing its profitable maple syrup industry along with its leaf-watching tourists as the cool-weather-adapted sugar maple population declines in a warming climate. Turning the maples' sugar into syrup requires nighttime temperatures that are below freezing and daytime temperatures in the mid-forties. Warmer temperatures overall have led to tapping seasons that start earlier, end sooner, and produce syrup of a lesser quality. A report by the U.S. Office of Science and Technology Policy indicates that the ideal range of the sugar maple is now close to 300 miles north of New England. The effects of warming temperatures on species of less commercial importance are not as well documented, and these species are less likely to receive human aid in making the transition.

The cost of global warming to Tuvaluans is even more dramatic. While migrating to drier climes would not mean extinction of the Tuvaluan people, it might well mean the extinction of their culture, since its members would likely disperse to many different countries. Reducing the biological, economic, and social losses caused by global warming will require not only slowing the rate of warming but also mediating the effects of increasing temperatures that are inevitable given current atmospheric carbon dioxide levels. Before they can effectively mediate these effects, scientists need to understand how warming temperatures affect not only climate factors such as average temperature, rainfall, and storm intensity and frequency but also biological processes. For the remainder of the chapter, we will focus on the effects of increased carbon dioxide and increased temperatures on the physiology of living organisms.

4.2   Cellular Respiration

Increasing temperatures can change an organism's energy needs and can affect how rapidly it grows, develops, and reproduces. For some organisms, increasing energy needs associated with higher temperatures can cause them to be outcompeted for resources by other organisms; this is ultimately what requires many of the species described earlier to move toward the poles or higher in elevation. For other organisms, higher temperatures allow them to go through their life cycles more rapidly, leading to increased populations. In both cases, the process that causes these effects is cellular respiration.

The main function of cellular respiration is to convert the energy stored in chemical bonds of food into energy that cells can use. Energy is stored in the electrons of chemical bonds, and when bonds are broken, electrons can be moved from one molecule to another. Cells use a chemical called **adenosine triphosphate**, or **ATP**, as their energy source. ATP can supply energy to cells because it stores energy obtained from the movement of electrons that originated in food into its own bonds.

Structure and Function of ATP

As described in Chapter 2, nucleic acids are one of the four main categories of biological molecules required by cells (the other three are carbohydrates, lipids, and proteins). Nucleic acids are polymers of nucleotides. Nucleotides consist of a nitrogenous, or nitrogen-containing, base—adenine (A), guanine (G), cytosine (C), or thymine (T)—plus a sugar and a phosphate group (made up of the elements phosphorus and oxygen). ATP is a nucleotide *tri*phosphate. It contains the nitrogenous base adenine, a sugar, and not one but three phosphates (Figure 4.9). Each phosphate in the series of three is negatively charged.

Figure 4.9   The structure of ATP. ATP is a nucleotide (sugar + phosphate + nitrogenous base) with a total of 3 phosphates. Notice that the phosphates are all negatively charged.

Figure 4.10 Stored energy. (a) A slingshot uses energy stored in the rubber band, supplied by arm muscles, to perform the work of propelling an object such as a rock. (b) ATP uses energy stored in its bonds to perform cellular work.

(a) Released energy can be used to perform work.

Stored energy

Released energy

(b) Releasing a phosphate group from ATP generates released energy.

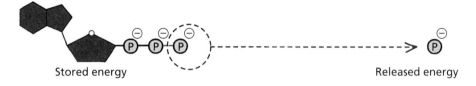

Stored energy

Released energy

These negative charges repel each other, which contributes to the stored energy in this molecule. ATP behaves in a manner similar to a stretched rubber band. Think about using a slingshot. Pulling back the rubber band on the slingshot requires energy from your arm muscles, and much of the energy you use will be stored in the stretched band (Figure 4.10a). When you release the rubber band, the energy is released and is used to perform some work—in this case, sending a projectile through the air. Likewise, releasing a phosphate group from ATP liberates stored energy that can be used by cells to perform work (Figure 4.10b). After the removal of a phosphate group, ATP is converted into **adenosine diphosphate (ADP)**, which has two phosphates (hence *di*phosphate instead of *tri*phosphate).

The phosphate group that is removed from ATP can be transferred to another molecule. Thus, one way for ATP to energize other compounds is through **phosphorylation**, which means that it adds a phosphate. You can think of the donated phosphate as a little bag of energy. When a molecule, say an enzyme, needs energy, the phosphate group is transferred from ATP to the enzyme, and the enzyme now has the energy it needs to perform its job (Figure 4.11). The energy released by the removal of the outermost phosphate of ATP can be used to help cells perform many different kinds of work. ATP helps power *mechanical work* such as the movement of proteins in muscles,

ATP

Enzyme

Phosphorylated (energized) enzyme

ADP

Figure 4.11 Phosphorylation. The terminal phosphate group of an ATP molecule can be transferred to another molecule, in this case an enzyme, to energize it. When ATP loses a phosphate, it becomes ADP. The enzyme that gained the phosphate group becomes energized.

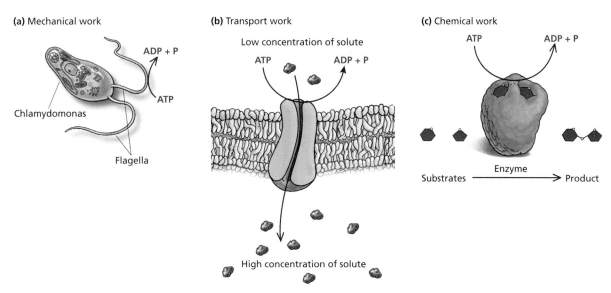

(a) Mechanical work

ADP + P

ATP

Chlamydomonas

Flagella

(b) Transport work

Low concentration of solute

ATP   ADP + P

High concentration of solute

(c) Chemical work

ATP   ADP + P

Substrates —— Enzyme ——→ Product

Figure 4.12 ATP and cellular work. ATP powers (a) mechanical work, such as the moving of the flagella of this single-celled green algae, Chlamydomonas; (b) transport work, such as the active transport of a substance across a membrane from its own low to high concentration; and (c) chemical work, such as the enzymatic conversion of substrates to a product.

transport work such as the movement of substances across membranes during active transport, and *chemical work* such as the making of complex molecules from simpler ones (Figure 4.12).

Cells are continuously using ATP. Exhausting the supply of ATP means that more ATP must be regenerated. ATP is synthesized by adding back a phosphate group to ADP during the process of cellular respiration (Figure 4.13). During this process, cells produce carbon dioxide and use oxygen to produce water. Because some of the steps in cellular respiration require oxygen, they are said to be **aerobic** reactions, and cellular respiration is called **aerobic respiration**. Humans and other animals with lungs breathe in oxygen, which is then delivered to cells. The carbon dioxide that is exhaled during breathing removes this waste product of cellular respiration from your body (Figure 4.14, shown on page 80). Plants and other organisms without lungs can respire and produce carbon dioxide as well.

Most foods can be broken down to produce ATP as they are routed through a complex pathway. Carbohydrate metabolism begins at the beginning of the pathway, while proteins and fats enter at later points.

A General Overview of Cellular Respiration

The equation for carbohydrate breakdown is:

$$C_6H_{12}O_6 + 6\,O_2 \longrightarrow 6\,CO_2 + 6\,H_2O$$
$$\text{glucose} + \text{oxygen} \longrightarrow \text{carbon dioxide} + \text{water}$$

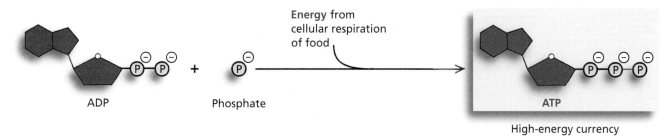

ADP   Phosphate

Energy from cellular respiration of food

ATP

High-energy currency

Figure 4.13 Regenerating ATP. ATP is regenerated from ADP and phosphate during the process of cellular respiration.

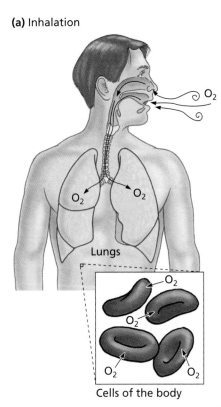

(a) Inhalation

(b) Exhalation

O_2

CO_2

Lungs

O_2 O_2

Lungs

CO_2 CO_2

O_2

O_2

O_2 O_2

Cells of the body

CO_2 CO_2

CO_2

CO_2 CO_2

Cells of the body

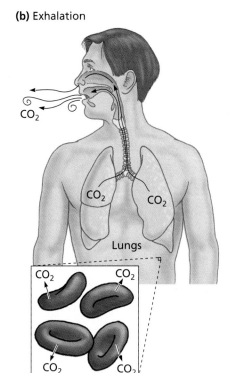

Figure 4.14 Breathing and cellular respiration. (a) When you inhale, you bring oxygen from the atmosphere into your lungs. This oxygen is delivered through the bloodstream to tissues that use it to drive cellular respiration. (b) The carbon dioxide produced by cellular respiration is released from cells and diffuses into the blood and to the lungs. Carbon dioxide is released from the lungs when you exhale.

Glucose is an energy-rich sugar, but the products of its digestion—carbon dioxide and water—are energy poor. The energy released during the conversion of glucose to carbon dioxide and water is used to synthesize ATP. Many of the chemical reactions in this process occur in sausage-shaped organelles called mitochondria through a series of complex reactions that break apart the glucose molecule. In doing so, the carbons and oxygens that make up the original glucose molecule are released from the cell as carbon dioxide. The hydrogens present in the original glucose molecule combine with oxygen to produce water (Figure 4.15). Gaining an appreciation for *how* this happens requires a more in-depth look.

Glycolysis, the Krebs Cycle, and Electron Transport

To harvest energy from glucose, the 6-carbon glucose molecule is first broken down into two 3-carbon **pyruvic acid** molecules. This part of the process of cellular respiration actually occurs outside of any organelle in the fluid cytosol and is called **glycolysis** (Figure 4.16). Glycolysis does not require oxygen but does produce a small amount of ATP. Bacteria on early Earth, which lacked oxygen, may have obtained their energy by glycolysis. Even today, many bacteria and organisms that live in the absence of oxygen, or **anaerobic** environments, rely on glycolysis for energy generation (Essay 4.1 on page 82). After glycolysis, the pyruvic acid is decarboxylated (loses a carbon dioxide molecule), and the 2-carbon fragment that is left is further metabolized inside the mitochondria.

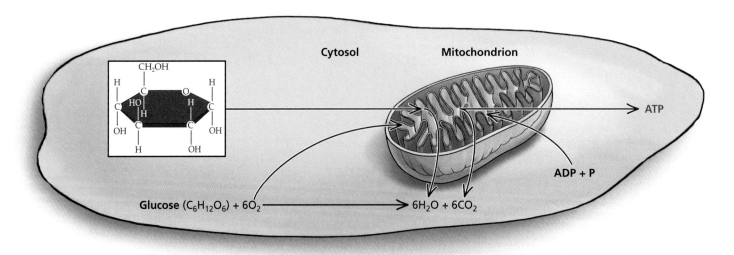

Cytosol   **Mitochondrion**

CH_2OH

ATP

ADP + P

Glucose ($C_6H_{12}O_6$) + $6O_2$ ⟶ $6H_2O + 6CO_2$

Figure 4.15 Overview of cellular respiration. The breakdown of glucose by cellular respiration requires oxygen and ADP plus phosphate. The energy stored in the bonds of glucose is harvested to produce ATP (from ADP and P), releasing carbon dioxide and water.

Figure 4.16 Glycolysis. Glycolysis occurs in the cytosol and does not require oxygen. Glycolysis is the enzymatic conversion of glucose into two pyruvic acid molecules. The pyruvic acid molecules are further broken down in the mitochondrion. A very small amount of ATP is made during glycolysis.

   Mitochondria (singular: mitochondrion) are organelles found in both plant and animal cells. These organelles are surrounded by an inner and an outer membrane (Figure 4.17). The space between the two membranes is called the intermembrane space. The semifluid medium inside the mitochondrion is called the matrix. Once inside the mitochondrion, the energy stored in the bonds of pyruvic acid is converted into the energy stored in the bonds of ATP. The first step of this conversion is called the Krebs cycle.

Krebs Cycle.   The Krebs cycle is a series of reactions catalyzed by eight different enzymes, located in the matrix of each mitochondrion. The Krebs cycle breaks down the remains of a carbohydrate, harvesting its

(a) Cross section of a mitochondrion   **(b)** Mitochondrial features

Figure 4.17   Mitochondria. (a) Mitochondria are microscopic organelles. (b) Mitochondria have an inner and an outer membrane. The area enclosed by these two membranes is called the intermembrane space. The fluid enclosed by the inner membrane is called the matrix. Different cell types have different numbers of mitochondria, and energy-requiring cells have more mitochondria than less active cells.

Essay 4.1 Metabolism Without Oxygen: Anaerobic Respiration and Fermentation

Aerobic respiration is one way for organisms to generate energy. It is also possible for cells to generate energy in the absence of oxygen, by a process called **anaerobic respiration**. Anaerobic organisms do not use oxygen as their final electron acceptor to pull electrons down the electron transport chain; a different molecule must be used. Some bacteria, called nitrate reducers, can transfer electrons to nitrate (NO_3^-), reducing it to nitrite (NO_2^-) (Figure E4.1a). Anaerobic respiration is an ATP-generating process.

Muscle cells normally produce ATP by aerobic respiration. However, oxygen supplies diminish with intense exercise. When muscle cells run low on oxygen, they must get most of their ATP from glycolysis, which does not require oxygen. When glycolysis happens without aerobic respiration, the cells run low on NAD^+, which is converted into NADH by glycolysis. The cells use a process called **fermentation** to regenerate NAD^+. No usable energy is produced by fermentation; fermentation simply recycles NAD^+. Fermentation cannot, however, be used for very long because one of the by-products of this reaction leads to the buildup of a compound called lactic acid. Lactic acid is produced by the actions of the electron acceptor NADH, which has no place to dump its electrons during fermentation since there is no electron transport chain and no oxygen to accept the electrons. Instead, NADH deposits its electrons by giving them to the pyruvic acid produced by glycolysis (Figure E4.1b). Adding electrons to pyruvic acid produces lactic acid

that accumulates and causes the muscle burn or cramping you feel after a vigorous workout. Lactic acid is transported to the liver, where liver cells use oxygen to convert it back to pyruvic acid. This requirement for oxygen, to convert lactic acid to pyruvic acid, explains why you continue to breathe heavily even after you have stopped working out. Your body needs to supply oxygen to your liver for this conversion, sometimes referred to as "paying back your oxygen debt." The accumulation of lactic acid also explains the phenomenon called "hitting the wall." Anyone who has ever felt as though their legs were turning to wood while running or biking knows this feeling. When your muscles are producing lactic acid by fermentation for a long time, the oxygen debt becomes too large, and muscles shut down until the rate of oxygen supply outpaces the rate of oxygen utilization.

Some fungi and bacteria also produce lactic acid during fermentation. Certain microbes placed in an anaerobic environment transform the sugars in milk into yogurt, sour cream, and cheese. It is the lactic acid present in these dairy products that gives them their sharp or sour flavor. Yeast in an anaerobic environment produces ethyl alcohol instead of lactic acid. Ethyl alcohol is formed when carbon dioxide is removed from pyruvic acid (Figure E4.1c). The yeast used to help make beer and wine converts sugars present in grains (beer) or grapes (wine) into ethyl alcohol and carbon dioxide. Carbon dioxide produced by baker's yeast helps bread to rise.

Figure E4.1

(a) Bacterial cell

(b) Human muscle cell

(c) Yeast cell

$$e^-$$
$$e^- \rightarrow ATP \quad \text{Electron transport chain}$$
$$e^-$$
$$NO_3^- + 2H^+ \rightarrow NO_2^- + H_2O$$

Glucose $\rightarrow$ 2 pyruvic acids
NADH $\searrow$
$\rightarrow NAD^+$
2 lactic acids

Glucose $\rightarrow$ 2 pyruvic acids
NADH $\searrow \rightarrow CO_2$
$NAD^+ \swarrow$
2 ethyl alcohols

electrons and releasing carbon dioxide into the atmosphere (Figure 4.18). These reactions are a cycle because every turn of the cycle regenerates the first reactant in the cycle. Therefore, the first reactant in the cycle, a 4-carbon molecule called oxaloacetate (OAA), is always available to react with carbohydrate fragments entering the Krebs cycle.

In addition to removing carbon dioxide, the Krebs cycle also removes electrons for use in producing ATP. These electrons do not simply float around in a cell; they are carried by molecules called electron carriers. One of the electron carriers utilized by cellular respiration is a chemical called **nicotinamide adenine dinucleotide (NAD)** (Figure 4.19 on page 84). **NAD^+** picks up 2 hydrogen atoms and releases 1 positively charged proton (H^+). Each **hydrogen atom** is composed of 1 negatively charged electron and 1 positively charged proton. When NAD^+ picks up 2 hydrogen atoms (each with 1 proton and 1 electron), it utilizes 1 proton and 2 electrons, releasing the remaining proton.

A careful look back at Figure 4.16 on page 81 shows that the conversion of glucose ($C_6H_{12}O_6$) into 2 pyruvic acid molecules ($2C_3H_3O_3{}^-$) results in the loss of 6 hydrogens. This is because glycolysis also results in the production of some NADH from NAD^+.

NADH serves as a sort of taxicab for electrons. The empty taxicab (NAD^+) picks up electrons. The full taxicab (NADH) carries electrons to their destination, where they are dropped off, and the empty taxicab returns

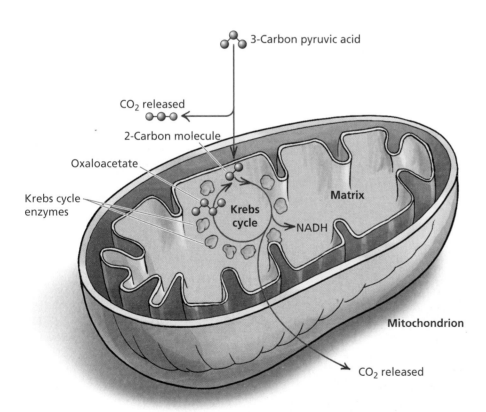

Figure 4.18   The Krebs cycle. The 3-carbon pyruvic acid molecules generated by glycolysis are decarboxylated, leaving a 2-carbon molecule that enters the Krebs cycle within the mitochondrial matrix. The 2-carbon fragment reacts with a 4-carbon OAA molecule and proceeds through a stepwise series of reactions that results in the production of more carbon dioxide and regenerates OAA. NADH is also produced.

Figure 4.19 Nicotinamide adenine dinucleotide (NAD). Nicotinamide adenine dinucleotide (NAD) is a dinucleotide in the sense that it contains 2 sugars, 2 phosphates, and the nitrogenous base adenine. This molecule can pick up a hydrogen atom along with its electrons. Hydrogen atoms are composed of 1 negatively charged electron that circles around the 1 positively charged proton. When NAD$^+$ encounters 2 hydrogen atoms (from food), it utilizes each hydrogen atom's electron and only 1 proton, thus releasing 1 proton.

for more electrons. NADH deposits its electrons at the top of the electron transport chain (Figure 4.20).

The Electron Transport Chain.   This series of proteins embedded in the inner mitochondrial membrane functions as a sort of conveyer belt for electrons, moving them from one protein to another. The electrons are

Figure 4.20 Electron carriers. NADH serves as an electron carrier, bringing electrons removed from the original glucose molecule to the electron transport chain. After dropping off its electrons, the electron carrier can be loaded up again and bring more electrons to the electron transport chain.

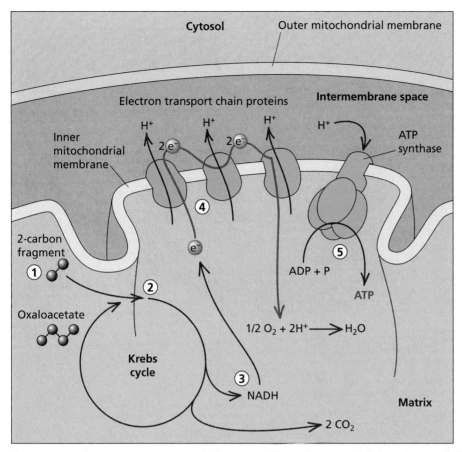

Cytosol   Outer mitochondrial membrane

Electron transport chain proteins   **Intermembrane space**

Inner mitochondrial membrane

H^+ H^+ H^+ H^+

$2\,e^-$ $2\,e^-$

ATP synthase

④

e^-

2-carbon fragment

①

⑤

ADP + P

ATP

②

Oxaloacetate

$1/2\,O_2 + 2H^+ \longrightarrow H_2O$

Krebs cycle

③

NADH

Matrix

2 CO₂

① Glucose is broken down into pyruvic acid by glycolysis. Pyruvic acid is decarboxylated producing a 2-carbon fragment.

② The 2-carbon fragment is fed into the Krebs cycle in the matrix of the mitochondrion.

③ NADH molecules carry electrons from the Krebs cycle to the electron transport chain.

④ As electrons are moved through the proteins of the electron transport chain, hydrogen ions are pumped into the intermembrane space.

⑤ Hydrogen flows back into the matrix through an ATP synthase protein, which converts ADP and P to ATP.

Figure 4.21   The electron transport chain. Energy from electrons added to the top of the electron transport chain is used to produce ATP.

pulled toward the bottom of the electron transport chain by oxygen in the matrix of the mitochondrion. One property of oxygen is that because it is very electronegative, or electron-loving, it pulls electrons toward itself. Each time an electron is picked up by a protein or handed off to another protein, the protein moving it changes shape. This shape change facilitates the movement of protons (H^+) from the matrix of the mitochondrion to the intermembrane space. So, while the proteins in the electron transport chain are moving electrons down the electron transport chain toward oxygen, they are also moving H^+ ions across the inner mitochondrial membrane and into the intermembrane space. This decreases the concentration of H^+ ions in the matrix and increases their concentration within the intermembrane space. As you learned in Chapter 3, whenever a concentration gradient of a molecule exists, molecules will diffuse from an area of high concentration to an area of low concentration. Since charged ions cannot diffuse across the hydrophobic core of the membrane, they escape through a protein channel in the membrane called **ATP synthase**. This enzyme uses the energy generated by the rushing H^+ ions to synthesize ATP from ADP and phosphate in the same manner that water rushing through a mechanical turbine can be used to generate electricity. The electrons that were pulled down the electron transport chain then combine with the oxygen at the bottom of the chain and 2 hydrogens in the matrix to produce water (Figure 4.21).

Overall, the two pyruvic acids produced by the breakdown of glucose during glycolysis are converted into carbon dioxide and water. Carbon dioxide is produced when it is removed from the pyruvic acid molecules during the Krebs cycle, and water is formed when oxygen combines with hydrogens at

Figure 4.22   Summary of cellular respiration. This figure diagrams the inputs and outputs of cellular respiration.

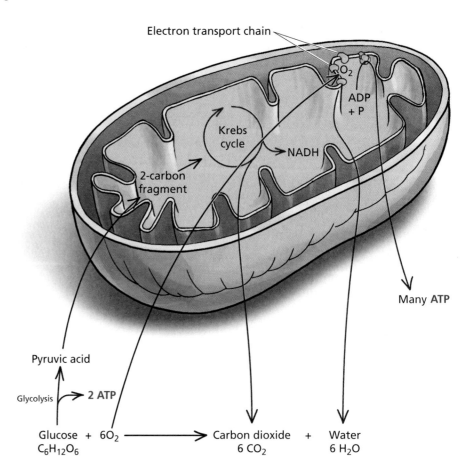

Electron transport chain

O_2

ADP + P

Krebs cycle

NADH

2-carbon fragment

Many **ATP**

Pyruvic acid

Glycolysis → **2 ATP**

Glucose + 6O_2 $C_6H_{12}O_6$ ⟶ Carbon dioxide 6 CO_2 + Water 6 H_2O

the bottom of the electron transport chain. A summary of the process is shown in Figure 4.22.

Metabolism of Other Nutrients.   Proteins and fats are broken down and their subunits merge with the carbohydrate breakdown pathway. Figure 4.23 shows the points of entry for proteins and fats. Protein is broken down into component amino acids, which are then used to synthe-

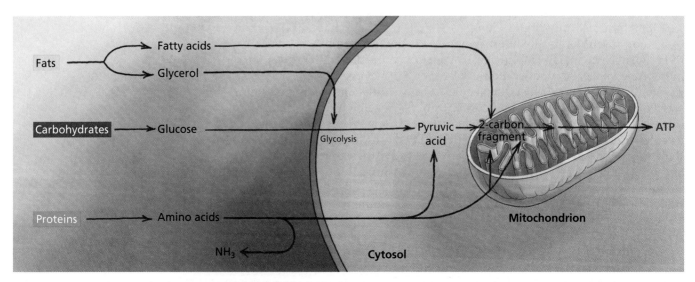

Fatty acids

Fats

Glycerol

Carbohydrates → Glucose

Glycolysis

Pyruvic acid

2-carbon fragment → ATP

Mitochondrion

Proteins → Amino acids

NH_3

Cytosol

Figure 4.23   Metabolism of other macromolecules. Carbohydrates, proteins, and fats can all undergo cellular respiration; they just feed into different parts of the metabolic pathway.

size new proteins. Most organisms can also break down proteins to supply energy. However, this process takes place only when fats or carbohydrates are unavailable. In humans and other animals, the first step in producing energy from the amino acids of a protein is to remove the nitrogen-containing amino group of the amino acid. Amino groups are then converted to a compound called urea, which is excreted in the urine. The carbon, oxygen, and hydrogen remaining after the amino group is removed undergo further breakdown and eventually enter the mitochondria, where they are fed through the Krebs cycle and produce carbon dioxide, water, and ATP. The subunits of fats (glycerol and fatty acids) also go through the Krebs cycle and produce carbon dioxide, water, and ATP. Most cells will break down fat only when carbohydrate supplies are depleted.

Whether carbohydrate, protein, or fat, these nutrients are used to generate energy. It turns out that this energy generation can be affected by rising temperatures, leading to some devastating effects.

Global Warming and Cellular Respiration

Alaska's Kenai Peninsula is experiencing firsthand some effects of global warming on cellular respiration. Increases in temperature have helped to speed up the life cycle of the spruce bark beetle (Figure 4.24a). These beetles, about the size of a grain of rice, attack spruce trees by boring through the outer bark to the phloem, a thin layer directly beneath the outer bark that transports food manufactured by photosynthesis from the foliage down to the roots. Once inside the phloem, the beetle feeds and lays eggs. The resulting damage and blockage of the phloem prevents nutrient transport to the roots, and the tree dies (Figure 4.24b). Over the past decade, the spruce population has suffered huge losses—close to 4 million acres of trees on southeastern Alaska's Kenai Peninsula; nearly all of the spruce trees there have been killed by infestations of these bark beetles.

Populations of spruce bark beetles are normally kept in check by cool summers and bitterly cold winters. Cooler summers help control the number of beetles because they cannot fly when temperatures are below 60°F. This limits the beetles' ability to colonize other trees. Cold winters can kill beetles and their larvae. The warmer temperatures not only fail to kill beetles in the winter but also speed up this insect's rate of reproduction. Typically it takes a spruce bark beetle 2 years to develop from an egg to an adult. But the warmer summers and winters have allowed the beetle to develop into an adult and lay new eggs during just one summer. More beetles mean more destruction to forests.

The accelerated life cycle of spruce bark beetles can be credited, in part, to the speeding up of cellular respiration. The enzymes that catalyze the reactions of cellular respiration, like all enzymes, are affected by temperature. Warmer temperatures typically speed up the rate at which enzymes catalyze reactions. That is, unless the temperature gets too hot, in which case the enzyme loses its characteristic shape and can no longer perform its job, and then a process called **denaturation** takes place. In a sense, warmer temperatures make the Krebs cycle spin faster, producing more energy, which allows the beetles to grow and reproduce more quickly as well as to fly earlier in the year and thus disperse to a greater number of trees.

As the beetles do their damage, trees drop their dried-out dead needles and limbs on the ground, providing fuel for forest fires. Forest fires release even more carbon dioxide into the atmosphere as the carbohydrate that comprises much of the plant tissue, such as cellulose that makes up the cell wall, is burned. Cellular respiration is a controlled burn of carbohydrates whereby the energy released from breaking the bonds of sugars is used to make ATP.

(a) Spruce bark beetle

(b) Spruce tree

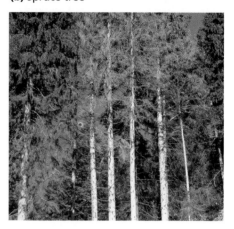

Figure 4.24   Spruce bark infestation. The spruce bark beetle (a) kills spruce trees (b) by blocking water and nutrient flow.

Figure 4.25   Cellular respiration is a controlled burn. Burning carbohydrates releases energy. (a) Plant cells have rigid cell walls composed of cellulose, a polymer of glucose. When carbohydrate burning is uncontrolled, as in a forest fire, energy is released as heat and light. (b) Cellular respiration is a controlled burn. Carbohydrates that are eaten have electrons removed during cellular respiration. As these electrons fall toward oxygen, they release energy that is used to drive the synthesis of ATP.

(a) Forest fire—burning quickly releases heat and light energy.

(b) Cellular respiration—ATP energy release is slow, controlled.

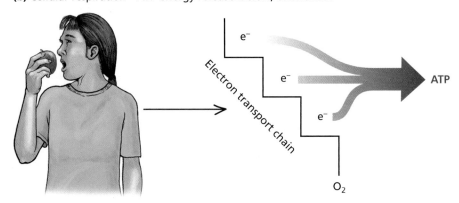

Combustion by fire releases all the stored energy without harvesting any for ATP production (Figure 4.25).

You have learned that carbon dioxide is released by cellular respiration and that increases in carbon dioxide levels are causing global warming. However, it would be a mistake to assume that cellular respiration is causing global warming. It is the increased carbon dioxide production caused by the burning of enormous amounts of fossil fuels that is shifting the balance of carbon dioxide production and uptake. The effects of carbon dioxide production (due to cellular respiration) on global warming is minimal and is mitigated by carbon dioxide uptake that occurs via photosynthesis.

4.3   Photosynthesis

Plants and other photosynthetic organisms remove carbon dioxide from the atmosphere and use it to make sugars and other macromolecules by the process of photosynthesis. The vast majority of the organic carbon found on Earth is a result of photosynthesis.

A General Overview of Photosynthesis

The equation summarizing photosynthesis follows:

$$6\,CO_2 + 6\,H_2O + \text{light energy} \longrightarrow C_6H_{12}O_6 + 6\,O_2$$
$$\text{carbon dioxide} + \text{water} + \text{light energy} \longrightarrow \text{glucose} + \text{oxygen}$$

The sun is the ultimate source of energy for living organisms. Plants transform energy from the sun into chemical energy through the process of photosynthesis by using light energy to rearrange the atoms present in carbon dioxide and water into energy-rich carbohydrates, producing oxygen as a waste product. In land plants, carbon dioxide enters, and oxygen gas is released through adjustable microscopic structures called **stomata** that are located on the surface of the leaf (Figure 4.26).

Plants use the carbohydrates that they produce by photosynthesis to grow and supply energy to their cells. They, along with the organisms that eat them, liberate the energy stored in the chemical bonds of sugars by undergoing the process of cellular respiration. Both plants and animals perform cellular respiration, but animals cannot perform photosynthesis.

The Light Reactions and Calvin Cycle

Green tissues in plants contain specialized organelles that serve as the sites of photosynthesis. Structurally, **chloroplasts** (Figure 4.27) are surrounded by two membranes. The inner and outer membranes together are called the chloroplast envelope. The chloroplast envelope encloses a compartment filled with **stroma**, a thick fluid that houses some of the enzymes of photosynthesis. Suspended in the stroma are disk-like membranous structures called **thylakoids**. When thylakoids are stacked on top of each other, like pancakes, the stacks are called **grana**. The large amount of membrane inside the chloroplast provides more surface area upon which some of the reactions of photosynthesis can occur. On the surface of the thylakoid membrane are millions of pigment molecules, called **chlorophyll**, that absorb energy from the sun.

It is the chlorophyll molecule that gives leaves and other plant structures their green color. Like all pigments, chlorophyll absorbs light. Light is made up of rays with different colors, or levels of energy, and each energy level has a different wavelength—to the human eye, shorter and middle wavelengths appear violet to green, and longer wavelengths appear yellow to red. Different organisms can perceive different wavelengths of light. For example, bees can see ultraviolet light, which is invisible to humans. Differences in wavelength visibility help bees see colors and patterns in floral structures as an aid to direct them to the sexual organs of the plant, like the landing lights at an airport direct jets to the runway.

Chlorophyll looks green to human eyes because it absorbs the shorter (blue) and longer (red) wavelengths of visible light and reflects the middle

Figure 4.26   Stomata. Stomata are adjustable microscopic pores found on the surface of leaves that allow for gas exchange. Carbon dioxide enters the plant, and oxygen leaves through these openings.

(a)   **(b)**

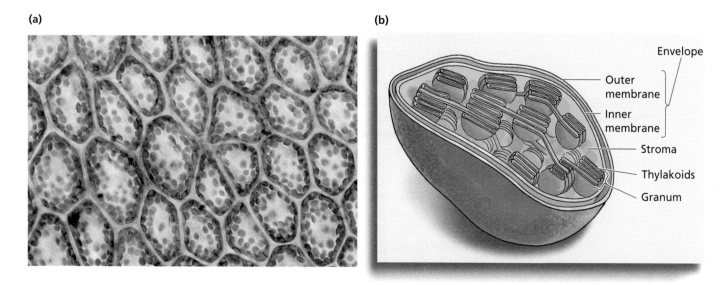

Figure 4.27   Chloroplasts. (a) Chloroplasts are microscopic organelles found in plant cells. (b) The chloroplast is enveloped—it has an inner and an outer membrane. More membranes are found inside the chloroplast housed in the liquid stroma. These membranes are called *thylakoids* when separate and *grana* when stacked.

(green) range of wavelengths. Leaves on deciduous trees change color in the fall because the abundant chlorophyll breaks down, making visible other less-abundant pigments present in the leaf that reflect red, orange, and yellow wavelengths.

Photosynthesis can be broken down into two steps. The first or "photo" step harvests energy from the sun during a series of reactions called the **light reactions**, which occur when there is sunlight. The second or "synthesis" step, called the **Calvin cycle**, uses the harvested energy to synthesize sugars either in the presence or absence of sunlight. For this reason, the Calvin cycle is also sometimes referred to as the dark reactions.

The Light Reactions.   When a pigment such as chlorophyll absorbs sunlight, electrons associated with the pigment become excited or increase their energy level. In effect, the light energy is transferred to the chlorophyll and becomes chemical energy. For most pigments, the molecule remains excited for a very brief amount of time before the chemical energy is lost as heat. This is why a surface that looks black (that is, one composed of a pigment that absorbs all visible light wavelengths) heats up quickly in comparison to a surface that looks white (one that absorbs no visible light wavelengths). Inside a chloroplast, however, the chemical energy of the excited chlorophyll molecules is not allowed to be released as heat; instead, the energy is captured.

When sunlight strikes the chlorophyll molecule and electrons are excited, they move to a higher energy level. The electrons are then transferred to other molecules in an electron transport chain located in the thylakoid membrane. As the electrons are handed down the electron transport chain, some ATP is produced. Some of the proteins in the electron transport chain not only move electrons to a lower energy level, they also pump protons into the interior of the thylakoid, setting up a gradient in protons. The protons then rush through an ATP synthase enzyme located in the thylakoid membrane and produce ATP in the same way that mitochondria make ATP. The newly synthesized ATP is released into the stroma, where it can be used by the enzymes of the Calvin cycle to produce sugars and other organic molecules.

During the light reactions, oxygen is produced when water (H_2O) is "split" into $2\,H^+$ ions and a single oxygen atom (O). Two oxygen atoms combine to produce O_2, which is released from the chloroplast. Since the hydrogen atom usually contains a single proton, around which orbits a single electron, the splitting of water to produce two H^+ ions also releases 2 electrons. These 2 electrons are transferred back to the chlorophyll molecule to replace those passed along the electron transport chain.

At the end of the electron transport chain, electrons are transferred to the electron carrier for plants, which is nicotinamide adenine dinucleotide phosphate, or NADP. Just like the NAD^+ involved in cellular respiration, $NADP^+$ functions as an electron taxicab. The difference between NAD^+ and $NADP^+$ is simply the presence of an extra phosphate group. The $NADP^+$ used during photosynthesis picks up 2 electrons and 1 H^+ ion to become NADPH. NADPH ferries electrons to the stroma, where the enzymes of the Calvin cycle will use the electrons in assembling sugars (Figure 4.28). Thus, the light reactions produce ATP, a source of electrons for the synthesis step, and release oxygen as a by-product.

Calvin Cycle.   The Calvin cycle is a series of enzymes located in the stroma that uses the ATP and NADPH produced during photosynthesis to convert carbon dioxide (CO_2) into sugars (CH_2O). CH_2O is the general formula for sugars. For example, glucose is $C_6H_{12}O_6$ or $6(CH_2O)$. A quick comparison of the composition of these molecules makes it obvious that converting CO_2 into CH_2O requires the incorporation of hydrogen atoms and their associated

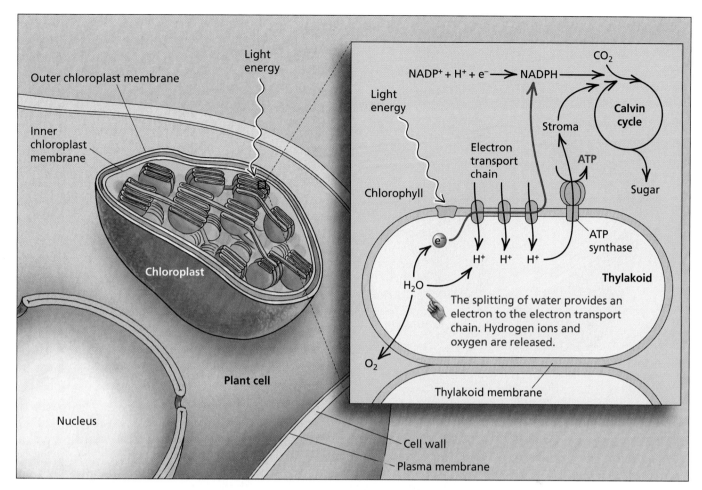

Figure 4.28   The light reactions. During the light reactions of photosynthesis, light strikes chlorophyll molecules located in the thylakoid membrane, exciting electrons that then move to a higher energy level. The energy of the excited electrons is harvested in a stepwise manner as the electrons are handed down an electron transport chain, producing ATP and NADPH that will be used by the Calvin cycle. The proton gradient is generated inside the thylakoid. ATP and NADPH are produced in the stroma where they will be available to the enzymes of the Calvin cycle.

electrons. Hydrogen atoms are removed from NADPH in order to produce sugars, thereby regenerating $NADP^+$.

During the Calvin cycle, carbon dioxide from the environment reacts with a 5-carbon molecule that is generated by the Calvin cycle and called ribulose bisphosphate or RuBP. The enzyme that performs this reaction is ribulose bisphosphate carboxylase oxygenase, or **rubisco**. This reaction produces an unstable 6-carbon molecule that immediately breaks down into two 3-carbon molecules called three-phosphoglyceric acid, or 3-PGA, which is rearranged to produce glyceraldehyde three-phosphate or G3P, a 3-carbon sugar that the cell uses to produce glucose and other compounds. RuBP is regenerated, completing the cycle (Figure 4.29 on page 92).

A careful look at the reactions of photosynthesis and cellular respiration show that the products of photosynthesis are used as reactants in cellular respiration and vice versa. This does not mean that these reactions are simply the reverse of each other. Instead, photosynthesis uses the products of cellular respiration as raw materials for the synthesis of sugars.

Without this reaction in the chloroplasts—when solar energy is transformed into the chemical energy in glucose—cellular respiration could not occur, and glucose would not be used to synthesize ATP. For this reason, *all* living things are dependent on photosynthesis for food, even meat-eating organisms, since they consume organisms that eat plants. In fact, the only two items in the human diet that do not come from plants (either directly or indirectly) are water and salt. Plants and other photosynthetic organisms make all the oxygen in the atmosphere that humans require for respiration.

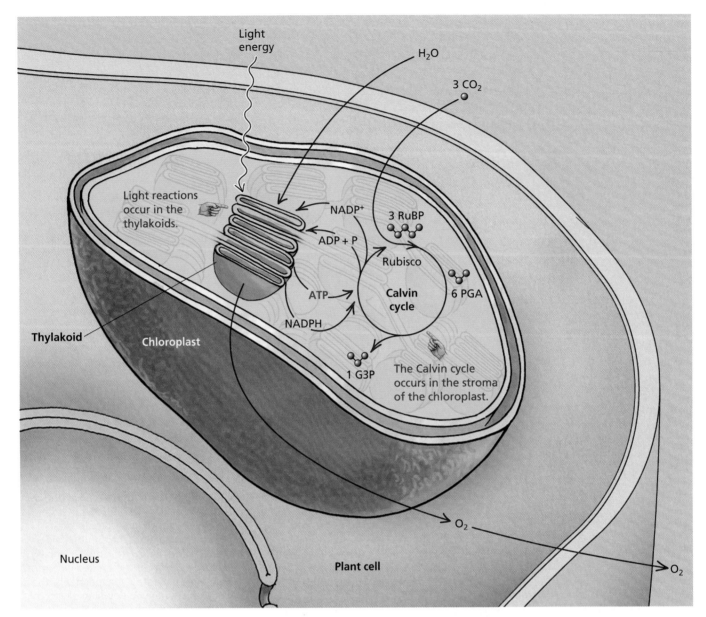

Figure 4.29 The Calvin cycle. Carbon dioxide enters the Calvin cycle. The energy of ATP is used to add hydrogens and electrons from NADPH to produce sugars.

Over time, plants have been able to capture the energy from light to form carbon-rich fossil fuels now buried in the earth. It took over 100 million years to form these nonrenewable resources such as coal, oil, and gas, which are now being consumed at a much faster rate than they were formed. The result is that more carbon dioxide is being released into the atmosphere than can be absorbed via photosynthesis.

Global Warming and Photosynthesis

Global warming is made worse by deforestation and may affect the distribution of plants that undergo different variations of photosynthesis.

Deforestation.   The process that occurs when forests are cleared for logging, farming, and ever-expanding human settlements is called **deforestation**. Deforestation contributes directly to the increase in carbon dioxide within the atmosphere. Current estimates are that up to 25% of the carbon dioxide introduced into the atmosphere originates from the cutting and burning of forests in the tropics. The loss of trees as a result of deforestation also indirectly contributes to rising carbon dioxide levels when these forests are replaced by pasture or cropland. Because net rates of photosynthesis (as measured by grams of carbon dioxide removed from the atmosphere per acre per year) in grasslands or agricultural fields are 30% to 60% less than rates in forests, the loss of forests significantly decreases the removal of carbon dioxide from the atmosphere.

Replanting trees in deforested areas may help increase the rate at which carbon dioxide is removed from the atmosphere because young trees have a faster net photosynthetic rate than older trees. This is because older trees have lots of non-photosynthetic, woody tissues that use the products of photosynthesis and that seedlings have yet to develop. In other words, young trees can put more of their organic carbon into storage as wood, while older trees use more of the carbon simply to maintain themselves. However, when these trees are logged, their roots and branches are left behind to decompose. In addition, most of the wood that is harvested is turned into paper, which will decompose after a few years, or fuel, which will be burned. Decomposition and burning result in the release into the environment of carbon compounds that were once part of the trees. Therefore, even though replanting after deforestation helps remove some of the carbon dioxide from the atmosphere, it does not result in a return to previous levels.

C₃, C₄, and CAM Plants. You learned earlier that plants can bring carbon dioxide into leaves through openings called stomata and that the carbon dioxide brought in is used to produce sugars during the Calvin cycle. Stomatal openings are located on the surface of a leaf and are surrounded by two kidney-bean-shaped cells called **guard cells**. When the guard cells are compressed against each other, the stomata are closed, thus restricting the flow of gases into or out of the plant. When the guard cells change shape to create an opening between them, the stomata are open, and carbon dioxide and oxygen gases can be exchanged. In addition to the exchange of gases, water can move out of the plant through the stomatal opening via a process called **transpiration** (Figure 4.30). The transpired water is replaced when water from the soil is brought into the plant, bringing with it minerals that the plant needs to synthesize many compounds.

In most plant species, the Calvin cycle produces 3-carbon sugars, which are converted into the sugars that are either stored as food for the plant or transported to growing leaves, roots, and reproductive structures. Plants that produce the 3-carbon molecule are called **C₃ plants**. C₃ plants are the most abundant type of plant on Earth and include agriculturally important species such as soybeans, wheat, and rice.

Rising temperatures can reduce the rate of photosynthesis because on hot, dry days, plants close their stomata in order to reduce the rate of water lost to transpiration. Closing stomatal openings prevents carbon dioxide from entering the plant, and the rate of photosynthesis declines.

Closing the stomatal openings to prevent water loss causes another series of reactions to occur, called **photorespiration**. During photorespiration, the first enzyme in the Calvin cycle uses oxygen instead of carbon dioxide as its substrate. While most enzymes display a high degree of specificity for their particular substrate, some enzymes have additional substrates to which they have lesser affinities. The enzyme that catalyzes the first step of the Calvin cycle, rubisco, is one such enzyme. The most common protein on Earth, rubisco is the also most common protein in leafy tissue. Carbon dioxide is its preferred substrate, but when carbon dioxide is low, rubisco will also allow oxygen into its active site. That is, it behaves as an oxygenase. When oxygen is high, such as when photosynthesis is occurring but the stomata are closed, oxygen will be used as the substrate of rubicso, and the plant will undergo photorespiration. During photorespiration, the rubisco enzyme catalyzes the incorporation of oxygen into a compound called glycolate. Glycolate cannot be used for food or for the synthesis of structural components. In fact, it must be destroyed by the plant since high levels of glycolate inhibit photosynthesis. The breakdown of glycolate requires ATP and releases carbon dioxide that had been previously incorporated into sugars during the Calvin cycle.

(a) Open

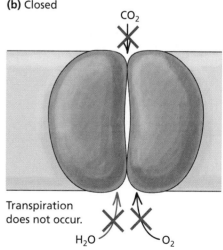

(b) Closed

Figure 4.30 Gas exchange and water loss. (a) When the stomata are open, carbon dioxide and oxygen gases can be exchanged. Water can be lost through a process called transpiration. (b) When the guard cells change shape to block the opening, gas exchange and transpiration do not occur.

You might wonder why it is that plant cells perform this wasteful process. It has to do with the evolution of photosynthesis on early Earth. Photosynthesis evolved when the atmosphere was largely devoid of oxygen. Under these conditions, the enzyme's affinity for oxygen did not present the problems it does in today's oxygen-rich atmosphere. All descendants of the first photosynthesizers inherited the instructions for producing the same rubisco enzyme, and so modern plants are "stuck" with this somewhat inefficient system. In dry environments, natural selection should favor plants that can minimize photorespiration despite having closed stomata much of the time. Two mechanisms for reducing photorespiration are known as C_4 and CAM photosynthesis.

C_4 plants, like all plants, conserve water during hot, sunny periods by closing their stomata. However, these plants carry an additional enzyme (compared to C_3 plants) that allows them to avoid photorespiration and continue to make sugars even though carbon dioxide levels within the plants are low during these periods. This enzyme is present in cells closest to the stomata and has a much higher affinity for carbon dioxide than does rubisco. It is able to procure carbon dioxide even when the stomata are almost closed. The carbon dioxide is combined with a 3-carbon acceptor molecule to produce a 4-carbon compound, hence the name C_4 plants. The 4-carbon compound then migrates to cells deeper within the leaf, where the carbon dioxide is released and produces a locally high concentration of carbon dioxide that enables rubisco to function as a carboxylase in the Calvin cycle. The 3-carbon acceptor molecule returns to the cells nearest the stomata. Corn and sugar cane are C_4 plants that can keep making sugars even though their stomata are almost closed. However, C_4 photosynthesis carries a cost—C_3 plants require 3 ATP molecules to convert 1 molecule of carbon dioxide into sugar, but C_4 plants require 5 molecules of ATP. The enzymes used in C_4 photosynthesis are also more sensitive to cold temperatures than are the enzymes of the Calvin cycle. Thus, C_3 plants have an advantage in certain environmental situations (cool and shady), and C_4 plants have an advantage in others (hot and sunny).

One other photosynthetic adaptation involves **CAM plants**. *CAM* stands for crassulacean acid metabolism, named for the plant family Crassulaceae, in which this mechanism was first discovered. Members of the Crassulaceae include the jade plant and other succulent (water-storing) plants. A CAM plant conserves water by opening its stomata at night only. The carbon dioxide that comes in during the night cannot immediately be used for photosynthesis because that process requires energy from the sun. During the night, the carbon is stored as an acid that is broken down during the day and releases carbon dioxide while the stomata are closed. This carbon dioxide can then be used for photosynthesis when sunlight becomes available. Therefore, carbon dioxide can enter at night, be stored as an acid, and then be used by the Calvin cycle during the subsequent day, even if the stomata are closed during the day to conserve water. Growth is limited in CAM plants because the amount of carbon dioxide stored in acid during the night is limited; the plants use it all up early in the day and cannot perform any more photosynthesis. Table 4.1 summarizes C_3, C_4, and CAM plant strategies.

Scientists are concerned that increasing temperatures may favor plants with these water-saving adaptations and change the relative percentages of C_3, C_4, and CAM plants in existence. This change could negatively affect some agricultural crops and could change the relative percentages of native plant species in a given region since certain adaptations gain relative advantages as the climate changes. Like the sugar maple, those species that cannot migrate to more appropriate environments or cannot compete with other plants in a changing environment may become extinct. As plant communities change and lose species, many of the animals that depend on these communities may suffer. The disruption of biological communities caused by global warming may prove to be the most damaging of all the changes caused by humans' consumption of fossil fuels.

Table 4.1　C_3, C_4, and CAM plant photosynthesis. Plants have evolved adaptations that prevent water loss.

Type of Plant and Example	Stomata Status	Description
C_3 Soybean		The Calvin cycle converts two 3-carbon sugars into glucose.
C_4 Corn		Enzymes scavenge CO_2 to produce 4-carbon sugars, which donate carbon dioxides to the Calvin cycle.
CAM Jade		Water loss is slowed by opening stomata only at night.

The negative consequences of global warming are substantial and may prove to be too much for many species to adapt to, including humans. The changes that are now threatening the Tuvaluans may eventually threaten all of us.

4.4　Decreasing the Effects of Global Warming

The prime minister of Tuvalu is threatening to sue the United States because this country emits immense amounts of carbon dioxide and has not taken official measures to reduce emissions. While Tuvalu is one of the least polluting countries in the world, the United States is the greatest.

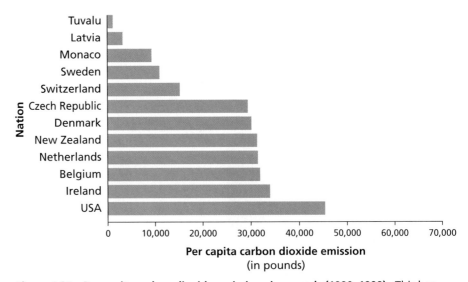

Figure 4.31   Per capita carbon dioxide emissions in pounds (1990–1999). This bar graph shows some of the highest and lowest carbon-dioxide-emitting countries for this time period.

Home to only 4% of the world's population, the U.S. produces close to one-fourth of the carbon dioxide emitted by fossil fuel burning. The per capita emissions rate of carbon dioxide for Americans is twice that of the Japanese or Germans, three times that of the global average, four times that of Swedes, and 20 times that of the average Indian. Figure 4.31 shows average per capita emissions for the highest- and lowest-emitting countries.

The United States has refused to sign the 1997 Kyoto Protocol, an international agreement that sets goals for decreases in emissions of around 5%. To be in compliance with the Kyoto agreement, by the year 2010, the United States would need to decrease its emissions by 24%.

Most of the emissions for an individual country come from industry, followed by transportation and then by commercial, residential, and agricultural emissions. All of us can work to reduce our personal contribution to global warming by decreasing residential and transport emissions. About 2700 pounds of carbon per year come from residential emissions. Most of that is from energy used to heat and cool homes and to power electrical appliances. After residential emissions, transportation is the next highest emitter of carbon dioxide. About 2300 pounds of carbon dioxide per person are released into the atmosphere through personal transportation. Adding to these emissions is the fact that the fuel economy of passenger vehicles has not increased, and the average number of miles traveled has increased. Table 4.2 describes many ways that you can decrease your greenhouse gas emissions and indicates the number of pounds of carbon dioxide that each action would save annually. These reductions may seem trivial in comparison to the scope of the problem, but when they are multiplied by the almost 300 million people in the United States, the savings become significant.

Having an effect on industrial, commercial, and agricultural sectors is difficult for any one individual. Instead, this will take leadership from the policy makers who are committed to reducing emissions. To do so requires that our leaders, and all of us, understand that even though the causes, implications, and solutions of global warming may be open to debate, the fact that it is occurring at an unprecedented rate is not.

Table 4.2   Decreasing your greenhouse gas emissions. Here are some ideas that you can use to help slow the rate of global warming.

Action		Annual decrease in carbon dioxide production
Buy high-efficiency appliances.		400 pounds per appliance
Plant shade trees around your home to decrease energy consumption and to remove carbon dioxide by photosynthesis.		50 pounds
Use a push mower instead of a power mower.		80 pounds
Recycle glass bottles, aluminum cans, plastic, newspapers, and cardboard.		850 pounds
Carpool two days per week.		1590 pounds
Buy food and other products with reusable or recyclable packaging, or reduced packaging, to save the energy required to manufacture new containers.		230 pounds
Drive an energy-efficient vehicle. SUVs average 16 miles per gallon, while smaller cars average 25 miles per gallon.		13,000 pounds
Walk 10 miles per week instead of driving.		590 pounds

CHAPTER REVIEW

Summary

4.1 The Greenhouse Effect

- The planet is warming. This warming will lead to a rise in ocean levels, changes in weather patterns, and the disruption of biological communities (p. 72).

- Greenhouse gases, particularly carbon dioxide, increase the amount of heat retained in Earth's atmosphere, which then leads to increased temperatures (p. 72).

- Water can absorb large amounts of heat without undergoing rapid or drastic changes in temperature because heat must first be used to break hydrogen bonds between adjacent water molecules. A high heat-absorbing capacity is a characteristic of water (p. 73).

- Carbon dioxide cycles between animals, plants, soil, oceans, and the atmosphere (p. 73).

- Carbon dioxide levels in the atmosphere are increasing. This increase is caused by human activities such as the burning of fossil fuels and is leading to global warming (pp. 74–75).

- Glaciers are melting and causing sea levels to rise. This changes habitats for many organisms and forces migrations (pp. 75–76).

4.2 Cellular Respiration

- Cellular respiration converts the energy stored in chemical bonds of food into ATP (p. 77).

- ATP is a nucleotide triphosphate. The nucleotide found in ATP contains a sugar and the nitrogenous base, adenine (p. 77).

- Breaking the terminal phosphate bond of ATP releases energy that can be used to perform cellular work and produces ADP plus a phosphate (p. 78).

- ATP is generated in most organisms by the process of cellular respiration, which consumes carbohydrates and releases water and carbon dioxide as waste products (p. 79).

- Cellular respiration begins in the cytosol, where a 6-carbon sugar is broken down into two 3-carbon pyruvic acid molecules during the anaerobic process of glycolysis (p. 81).

- The pyruvic acid molecules then move across the two mitochondrial membranes where they are decarboxylated. The remaining 2-carbon fragment then moves into the matrix of the mitochondrion, where the Krebs cycle strips them of carbon dioxide and electrons (pp. 81–83).

- The electrons are carried by electron carriers to the inner mitochondrial membrane; there they are added to a series of proteins called the electron transport chain. At the bottom of the electron transport chain, electronegative oxygen pulls the electrons toward itself. As the electrons move down the electron transport chain, the energy that they release is used to drive protons (H^+) into the intermembrane space. Once there, the protons rush through the enzyme ATP synthase and produce ATP from ADP and phosphate (p. 84).

- When electrons reach the oxygen at the bottom of the electron transport chain, they combine with the oxygen and hydrogen ions to produce water (p. 85).

- Increased temperatures lead to an increased rate of cellular respiration, which can then cause increases in the populations of some species and decreases in the populations of others (p. 86).

Web Tutorial 4.1 Glucose Metabolism

4.3 Photosynthesis

- Photosynthesis utilizes carbon dioxide from the atmosphere to make sugars and other substances (p. 87).

- During photosynthesis, energy from sunlight is used to rearrange the atoms of carbon dioxide and water to produce sugars and oxygen (pp. 87–88).

- Photosynthesis occurs in chloroplasts. Sunlight strikes the chlorophyll molecule within chloroplasts, boosting electrons to a higher energy level. These excited electrons are dropped down an electron transport chain located in the thylakoid membrane, and ATP is made (p. 88).

- Electrons are also passed to electron carriers that transport electrons to the Calvin cycle. The electrons that are lost become replaced by electrons acquired during the splitting of water, and oxygen is released. The Calvin cycle utilizes the products of the light reactions (ATP and the electron carrier NADPH) to incorporate carbon dioxide into sugars (pp. 89–90).

- Photosynthesis removes carbon dioxide from the air, potentially reducing the risk of global warming. However, humans are also deforesting Earth's land surface, reducing the global rate of photosynthesis (p. 91).

- Stomata on a plant's surface not only allow in carbon dioxide for photosynthesis but also allow water to escape from the plant. Guard cells surrounding the stomata can change shape to close the stomata and restrict water loss. However, when stomata are closed, carbon dioxide declines in the plant, and the energy-wasting process of photorespiration may occur. C_4 and CAM plants have evolved to perform photosynthesis while reducing the risk of photorespiration in dry conditions (pp. 92–94).

- Global warming may disrupt biological communities as the climate in an area changes and the relative abundance of C_3, C_4, and CAM plants changes (p. 95).

Web Tutorial 4.2 Leaves: The Site of Photosynthesis
Web Tutorial 4.3 Photosynthesis

4.4 Decreasing the Effects of Global Warming

- The effects of global warming will not be as severe if humans can reduce carbon dioxide emissions (pp. 95–97).

Learning the Basics

1. What are the reactants and products of cellular respiration and photosynthesis?

2. Carbon dioxide functions as a greenhouse gas by _____.

    A. interfering with water's ability to absorb heat;
    B. increasing the random molecular motions of oxygen;
    C. allowing radiation from the sun to reach Earth and absorbing the re-radiated heat; **D.** splitting into carbon and oxygen and increasing the rate of cellular respiration

3. Water has a high heat-absorbing capacity because _____.

    A. the sun's rays penetrate to the bottom of bodies of water, mainly heating the bottom surface; **B.** the strong covalent bonds that hold individual water molecules together require large inputs of heat to break; **C.** it has the ability to dissolve many heat-resistant solutes; **D.** initial energy inputs are used to break hydrogen bonds between water molecules and then to raise the temperature; **E.** all of the above are true

4. Cellular respiration involves _____.
    A. the aerobic metabolism of sugars in the mitochondria by a process called glycolysis; **B.** an electron transport chain that releases carbon dioxide; **C.** the synthesis of ATP, which is driven by the rushing of protons through an ATP synthase; **D.** electron carriers that bring electrons to the Krebs cycle; **E.** the production of water during the Krebs cycle

5. The electron transport chain _____.
    A. is located in the matrix of the mitochondrion; **B.** has the electronegative carbon dioxide at its base; **C.** is a series of nucleotides located in the inner mitochondrial membrane; **D.** is a series of enzymes located in the intermembrane space; **E.** moves electrons from protein to protein and moves protons from the matrix into the intermembrane space

6. Which of the following **does not** occur during the light reactions of photosynthesis?

    A. Oxygen is split, releasing water.; **B.** Electrons from chlorophyll are added to an electron transport chain.; **C.** An electron transport chain drives the synthesis of ATP for use by the Calvin cycle.; **D.** NADPH is produced and will carry electrons to the Calvin cycle.; **E.** Oxygen is produced when water is split.

7. Which of the following is a **false** statement about photosynthesis?

    A. During the Calvin cycle, electrons and ATP from the light reactions are combined with atmospheric carbon dioxide to produce sugars.; **B.** The enzymes of the Calvin cycle are located in the chloroplast stroma.; **C.** Oxygen produced during the Calvin cycle is released into the atmosphere.; **D.** Sunlight drives photosynthesis by boosting electrons found in chlorophyll to a higher energy level.; **E.** Electrons released when sunlight strikes chlorophyll are replaced by electrons from water.

8. Hydrogen atoms are composed of _____.
    A. 1 electron; **B.** 1 proton; **C.** 2 electrons; **D.** 2 protons; **E.** 1 proton and 1 electron

9. Select the **true** statement regarding metabolism in plant and animal cells.

    A. Plant and animal cells both perform photosynthesis and aerobic respiration.; **B.** Animal cells perform aerobic respiration only, and plant cells perform photosynthesis only.; **C.** Plant cells perform aerobic respiration only, and animal cells perform photosynthesis only.; **D.** Plant cells perform aerobic respiration and photosynthesis, and animal cells perform aerobic respiration only.

Analyzing and Applying the Basics

1. Are sugars the only macromolecules that can be broken down to produce ATP? If not, how are other nutrients metabolized?

2. How do the different strategies employed by C_3, C_4, and CAM plants help them adapt to their environments?

3. Describe the sites of aerobic respiration and photosynthesis. What organelles are involved? Where in each organelle do the different steps of each process occur?

Connecting the Science

1. What can individuals do to slow the effects of global warming?

2. Do you think it is okay for the individuals of one country to produce more greenhouse gases than do individuals of other countries? Why or why not?

CHAPTER 5

Cancer

DNA Synthesis, Mitosis, and Meiosis

Cancer cells divide
when they should not.

Nicole got sick during her junior year of college.

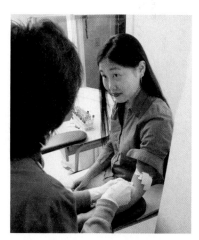

She had to undergo some procedures to see if she had cancer.

Nicole's early college career was similar to that of most students. She enjoyed her independence and the wide variety of courses required for her double major in biology and psychology. She worried about her grades and finding ways to balance her course work with her social life. She also tried to find time for lifting weights in the school's athletic center and snowboarding at a local ski hill. Some weekends, to take a break from school, she would ride the bus home to see her family.

Managing to get schoolwork done, see friends and family, and still have time left to exercise had been difficult, but possible, for Nicole during her first two years at school. That changed drastically during her third year of school.

One morning in October of her junior year, Nicole began having severe pains in her abdomen. The first time this happened, she was just beginning an experiment in her cell biology laboratory. Hunched over and sweating, she barely managed to make it through the two-hour respiration experiment that she and her lab partner were performing. Over the next few days, the pain intensified so much that she was unable to walk from her apartment to her classes without stopping several times to rest.

She wants to understand why she got cancer.

101

Later that week, as she was preparing to leave for class, the pain was so severe that she had to lie down in the hallway of her apartment. When her roommate got home a few minutes later, she took Nicole to the student health center for an emergency visit. The physician at the health center first determined that Nicole's appendix had not burst and then made an appointment for Nicole to see a gynecologist the next day.

After hearing Nicole's symptoms, the gynecologist pressed on her abdomen and felt what he thought was a mass on her right ovary. He used a noninvasive procedure called ultrasound to try to get an image of her ovary. This procedure requires the use of high-frequency sound waves. These waves, which cannot be heard by humans, bounce off tissues and produce a pattern of echoes that can be used to create a picture called a sonogram. Healthy tissues, fluid-filled cysts, and tumors all look different on a sonogram.

Nicole's sonogram convinced her gynecologist that she had a large growth on her ovary. He told her that he suspected this growth was a cyst, or fluid-filled sac. Her gynecologist told her that cysts often go away without treatment, but this one seemed to be quite large and would need to be surgically removed.

Even though the idea of having an operation was scary for Nicole, she was relieved to know that the pain would stop. Her gynecologist also assured her that she had nothing to worry about because cysts are not cancerous. A week after the abdominal pain began, Nicole's gynecologist removed the cyst and her completely engulfed right ovary through an incision just below her navel. He then sent the cystic ovary to a physician who specializes in determining whether tissues are cancerous. This physician, called a pathologist, determined that Nicole's doctor had been right—there was no sign of cancer.

After the operation, Nicole's gynecologist assured her that the remaining ovary would compensate for the missing ovary by ovulating (producing an egg cell) every month. He added that he would have to monitor her remaining ovary carefully to make sure that it did not become cystic, or even worse, cancerous. She could not afford to lose another ovary if she wanted to remain fertile and have children some day.

Monitoring her remaining ovary involved monthly visits to her gynecologist's office, where Nicole had her blood drawn and analyzed. The blood was tested for the level of a protein called CA125, which is produced by ovarian cells. Higher-than-normal CA125 levels usually indicate that the ovarian cells have increased in size or number and are thus associated with the presence of an ovarian tumor.

Nicole went to her scheduled checkups for 5 months after the original surgery. The day after her March checkup, Nicole received a message from her doctor asking that she come to see him the next day. Because she needed to study for an upcoming exam, Nicole tried to push aside her concerns about the appointment. By the time she arrived at her gynecologist's office, she had convinced herself that nothing serious could be wrong. She thought a mistake had probably been made and that he just wanted to perform another blood test.

The minute her gynecologist entered the exam room, Nicole could tell by his demeanor that something was wrong. As he started speaking to her, she began to feel very anxious—when he said that she might have a tumor on her remaining ovary, she could not believe her ears. When she heard the words *cancer* and *biopsy*, Nicole felt as though she was being pulled underwater. She could see that her doctor was still talking, but she could not hear or understand him. She felt too nauseous to think clearly, so she excused herself from the exam room, took the bus home, and immediately called her mother.

After speaking with her mom, Nicole realized that she had many questions to ask her doctor. She did not understand how it was possible for such a young woman to have lost one ovary to a cyst and then possibly to have a tumor on the other ovary. She wondered how this tumor would be treated and what her prognosis would be. Before seeing her gynecologist again, Nicole decided to do some research in order to make a list of questions for her doctor.

5.1   What Is Cancer?

Cancer is a disease that begins when a single cell replicates itself although it should not. **Cell division** is the process a cell undergoes in order to make copies of itself. This process is normally regulated so that a cell divides only when more cells are required and when conditions are favorable for division. A cancerous cell is a rebellious cell that divides without being given the go-ahead.

Unregulated cell division leads to a pileup of cells that form a lump or **tumor**. A tumor is a mass of cells that has no apparent function in the body. Tumors that stay in one place and do not affect surrounding structures are said to be **benign**. Some benign tumors remain harmless; others become cancerous. Tumors that invade surrounding tissues are **malignant** or cancerous. The cells of a malignant tumor can break away and start new cancers at distant locations through a process called **metastasis** (Figure 5.1).

Cancer cells can travel virtually anywhere in the body via the lymphatic and circulatory systems. The **lymphatic system** collects fluids lost from microscopic blood vessels called **capillaries**. **Lymph nodes** are structures that filter the lost fluids, or lymph. When a cancer patient is undergoing surgery, the surgeon will often remove a few lymph nodes to see if any cancer cells are in the nodes. If cancer cells appear in the nodes, then some cells have left the original tumor and are moving through the bloodstream. If this has

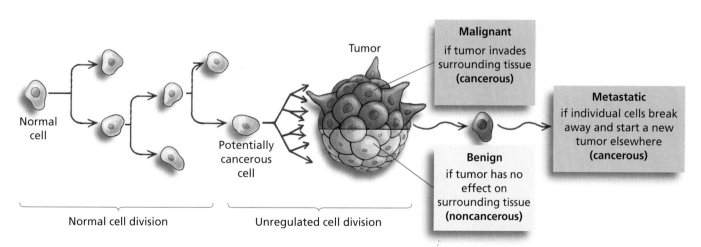

Figure 5.1   What is cancer? A tumor is a clump of cells with no function. Tumors may remain benign, or they can invade surrounding tissues and become malignant. Tumor cells may move, or metastasize, to other locations in the body. Malignant and metastatic tumors are cancerous.

(a) Lymphatic system

(b) Cancer cells travel in lymph and blood.

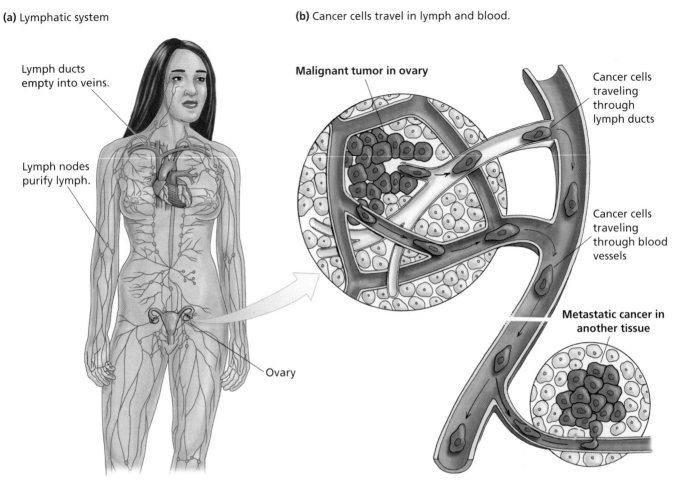

Lymph ducts empty into veins.

Lymph nodes purify lymph.

Ovary

Malignant tumor in ovary

Cancer cells traveling through lymph ducts

Cancer cells traveling through blood vessels

Metastatic cancer in another tissue

Figure 5.2   Metastasis. (a) The lymphatic system contains a series of vessels that remove excess fluids or lymph from tissues. Lymph moves throughout the body as a result of pressure applied by muscle contractions near the vessels. (b) The vessels of the circulatory and lymphatic systems provide a pipeline for cancer cells to move to other locations in the body through a process called metastasis.

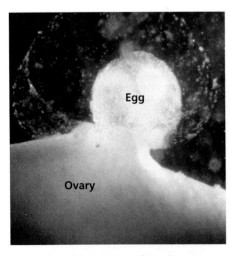

Egg

Ovary

Figure 5.3   Ovulation. When the ovary releases an egg cell, the tissue of the ovary is damaged. Cell division occurs to heal the rupture.

happened, cancerous cells likely have metastasized to other locations in the body (Figure 5.2a).

When cancer cells metastasize, they can gain access not only to the **circulatory system**, which includes blood vessels to transport the blood, but also to the heart, which pumps the blood. Once inside a blood vessel, cancer cells can drift virtually anywhere in the body (Figure 5.2b).

Cancer cells differ from normal cells in three ways: (1) They divide when they should not; (2) they invade surrounding tissues; and (3) they move to other locations in the body. All tissues that undergo cell division are susceptible to becoming cancerous.

Nicole's cancer affected ovarian tissue. When an egg cell is released from the ovary during ovulation, the tissue of the ovary becomes perforated (Figure 5.3). Cells near the perforation site undergo cell division to heal the damaged surface of the ovary. For Nicole, these cell divisions may have become uncontrolled, leading to the growth of a tumor.

5.2 Cell Division Overview

Cell division produces new cells in order to heal wounds, replace damaged cells, and help organisms grow and reproduce themselves. Each of us begins life as a single fertilized egg cell that undergoes millions of rounds of cell division in order to produce all the cells that comprise the tissues and organs of our bodies. Some organisms reproduce by producing exact copies of themselves via cell division. Reproduction of this type, called **asexual reproduction**, does not require genetic input from two parents and results in offspring that are genetically identical to the original parent cell. Single-celled organisms, such as bacteria and amoeba, reproduce in this manner (Figure 5.4a). Some multicellular organisms can reproduce asexually also. For example, most plants can grow from clippings of the stem, leaves, or roots and thereby reproduce asexually (Figure 5.4b). Organisms whose reproduction requires genetic information from two parents undergo **sexual reproduction**. Humans reproduce sexually when sperm and egg cells combine their genetic information at fertilization.

Whether reproducing sexually or asexually, all dividing cells must first make a copy of their genetic material or **DNA (deoxyribonucleic acid)**. The DNA is located within the nucleus and carries the instructions, called **genes**, for building all of the proteins that cells require. The DNA in the nucleus is organized into structures called **chromosomes**. Chromosomes can carry hundreds of genes along their length. Different organisms have different numbers of chromosomes in their cells. For example, dogs have 78 chromosomes in each cell; humans have 46, and dandelions have 24.

Chromosomes are in an uncondensed, string-like form prior to cell division (Figure 5.5a). Before cell division occurs, the DNA in each chromosome is condensed (compressed) into a short, linear form (Figure 5.5b). Condensed linear chromosomes are easier to maneuver during cell division and are less likely to become tangled or broken than the uncondensed and string-like structures are. When a chromosome is replicated, a copy is produced that carries the same genes. The copied chromosomes are called **sister chromatids**, and each sister chromatid is composed of one DNA molecule. Sister chromatids are attached to each other at a region toward the middle of the replicated chromosome, called the **centromere**.

(a) Amoeba

(b) English ivy

Figure 5.4   Asexual reproduction. (a) This single-celled amoeba divides by copying its DNA and producing offspring that are genetically identical to the original, parent amoeba. (b) Some multicellular organisms, such as this English ivy plant, can reproduce asexually from cuttings.

(a) Uncondensed DNA

(b) DNA condensed into chromosomes

Centromere

Sister chromatids

Chromosomes

Figure 5.5   DNA condenses during cell division. (a) DNA in its replicated but uncondensed form prior to cell division. (b) During cell division, each copy of DNA is wrapped neatly around many small proteins, forming a condensed structure called a chromosome. After DNA replication, two identical sister chromatids are produced and joined to each other at a region called the centromere.

(a) DNA double helix is made of two strands.   **(b)** Each strand is a chain of nucleotides.

"Handrails" made of sugars and phosphates

"Rungs" made of nitrogenous bases

Figure 5.6   DNA structure. (a) DNA is a double-helical structure composed of sugars, phosphates, and nitrogenous bases. (b) Each strand is composed of nucleotides. The two strands are anti-parallel.

The DNA molecule itself is double stranded and can be likened to a twisted rope ladder. The backbone or "handrails" of each strand are composed of alternating sugar and phosphate groups (Figure 5.6a). Across the width or "rungs" of the DNA helix are the nitrogenous bases, paired together via hydrogen bonds such that adenine (A) makes a base pair with thymine (T), and guanine (G) makes a base pair with cytosine (C). The strands of the helix align so that the nucleotides face "up" on one side of the helix and "down" on the other side of the helix. For this reason, the two strands of the helix are said to be antiparallel (Figure 5.6b).

Two of the people credited with determining DNA structure are James Watson and Francis Crick. Watson and Crick reported their hypothesis about the structure of the DNA molecule in a 1953 paper for the journal *Nature*. Although they did not go so far as to propose a detailed model for how the DNA molecule was replicated, they did say, "It has not escaped our notice that the specific pairing we have postulated immediately suggests a copying mechanism for the genetic material." The copying mechanism that Watson and Crick referred to is also called DNA replication.

DNA Replication

During the process of **DNA replication** that precedes cell division, the double-stranded DNA molecule is copied, first by splitting the molecule in half up the middle of the helix. New nucleotides (molecules comprised of a sugar, a phosphate, and a nitrogenous base) are added to each side of the original parent molecule, maintaining the A-to-T and G-to-C base pairings. This process results in two daughter DNA molecules, each composed of one strand of parental nucleotides and one newly synthesized strand (Figure 5.7a).

To replicate the DNA, an enzyme that assists in DNA synthesis is required. This enzyme, called **DNA polymerase**, moves along the length of the unwound helix and helps bind incoming nucleotides to each other on the newly forming daughter strand (Figure 5.7b). When free nucleotides floating in the nucleus have an affinity for each other (A for T and G for C), they bind to each other across the width of the helix. Nucleotides that bind to each other are said to be complementary to each other.

The DNA polymerase enzyme triggers or catalyzes the formation of the covalent bond between nucleotides along the length of the helix. The paired

(a) DNA replication   **(b)** The DNA polymerase enzyme facilitates replication.

Figure 5.7   DNA replication. (a) DNA replication results in the production of two identical daughter DNA molecules from one parent molecule. Each daughter DNA molecule contains half of the parental DNA and half of the newly synthesized DNA. (b) The DNA polymerase enzyme moves along the unwound helix, tying together adjacent nucleotides on the newly forming daughter DNA strand. Free nucleotides have three phosphate groups, two of which are removed before the nucleotide is added to the growing chain.

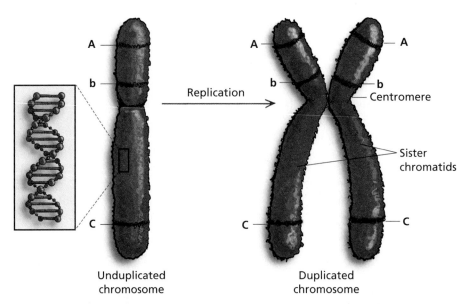

Figure 5.8   Unreplicated and replicated chromosomes. An unreplicated chromosome is composed of one double-stranded DNA molecule. A replicated chromosome is X-shaped and composed of two identical double-stranded DNA molecules. Each DNA molecule of the duplicated chromosome is a copy of the original chromosome and is called a sister chromatid.

nitrogenous bases are joined across the width of the backbone by hydrogen bonding, and the DNA polymerase advances along the parental DNA strand to the next unpaired nucleotide. When an entire chromosome has been replicated, the newly synthesized copies are identical to each other. They are attached at the centromere as sister chromatids (Figure 5.8). After the DNA is replicated in both normal cells and cancer cells, the division of the nucleus separates them into different daughter cells. Mitosis is one such nuclear division.

5.3   The Cell Cycle and Mitosis

Mitosis is an asexual division that produces daughter cells that are exact copies of the parent cell. Mitosis is part of the **cell cycle** or life cycle of the cell.

For cells that divide by mitosis, the cell cycle includes three steps: (1) *interphase*, when the DNA replicates; (2) *mitosis*, when the copied chromosomes move into the daughter nuclei; and (3) *cytokinesis*, when the cytoplasm of the parent cell splits (Figure 5.9a).

Interphase

A normal cell spends most of its time in **interphase** (Figure 5.9b). During this phase of the cell cycle, the cell performs its normal functions and produces the proteins required for the cell to do its particular job. For example, a muscle cell would be producing proteins required for muscle contraction, and a blood cell would be producing proteins required to transport oxygen during interphase. Different cell types spend varying amounts of time in interphase. Cells that frequently divide, like skin cells, spend less time in interphase than do those that seldom divide, such as some nerve cells. A cell that will divide also begins preparations for division during interphase. Interphase can be separated into three phases: G_1, S, and G_2.

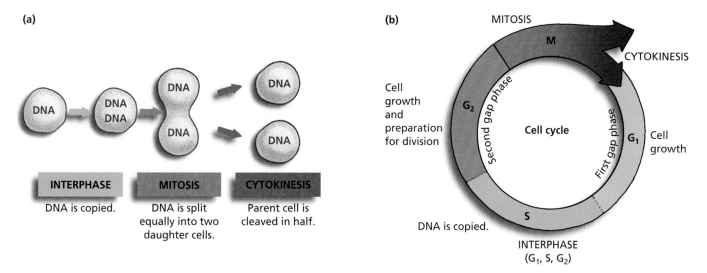

Figure 5.9   The cell cycle. (a) During interphase, the DNA is copied. Separation of the DNA into two daughter cells occurs during mitosis. Cytokinesis is the division of the cytoplasm, creating two cells. (b) During interphase, there are two stages when the cell grows in preparation for cell division, G_1 and G_2. During the S stage of interphase, the DNA replicates.

During the G_1 (first gap or growth) phase, most of the cell's inner machinery, the organelles, duplicate. Consequently, the cell grows larger during this phase. During the S (synthesis) phase, the DNA in the chromosomes replicates. During the G_2 (second gap) phase of the cell cycle, proteins are synthesized that will help drive mitosis to completion. The cell continues to grow and prepare for the division of chromosomes that will take place during mitosis.

Mitosis

The movement of chromosomes into new cells occurs during **mitosis**. Mitosis takes place in all cells with a nucleus, although some of the specifics of cell division differ among kingdoms. Whether these phases occur in a plant or an animal, the outcome of mitosis and the next phase, cytokinesis, is the production of genetically identical daughter cells. To achieve this outcome, the sister chromatids of a replicated chromosome are pulled apart, and one copy of each is placed into each newly forming nucleus. Mitosis is accomplished during 4 stages: prophase, metaphase, anaphase, and telophase.

During **prophase**, the replicated chromosomes condense, allowing them to move around in the cell without becoming entangled. Protein structures called **microtubules** also form and grow, ultimately radiating out from opposite ends, or **poles**, of the dividing cell. The growth of microtubules helps the cell to expand. Microtubules also help to move the chromosomes around during cell division by lengthening and binding to the chromosomes. The membrane that surrounds the nucleus, called the **nuclear envelope**, breaks down so that the microtubules can gain access to the replicated chromosomes. At the poles of each dividing animal cell, structures called **centrioles** physically anchor one end of each forming microtubule. Plant cells do not contain centrioles, but microtubules in these cells do remain anchored at a pole.

During **metaphase**, the replicated chromosomes are aligned across the middle, or equator, of each cell. To do this the microtubules, which are attached to each chromosome at the centromere, line up the chromosomes in single file across the middle of the cell.

During **anaphase**, the centromere splits and the microtubules shorten, pulling each sister chromatid of a chromosome to opposite poles of the cell.

In the last stage of mitosis, **telophase**, the nuclear envelopes reform around the newly produced daughter nuclei, and the chromosomes revert to their uncondensed form. Cytokinesis divides the cytoplasm, and daughter cells are produced. Figure 5.10 summarizes the cell cycle in animal cells. The 4 stages of mitosis are nearly identical in plant cells.

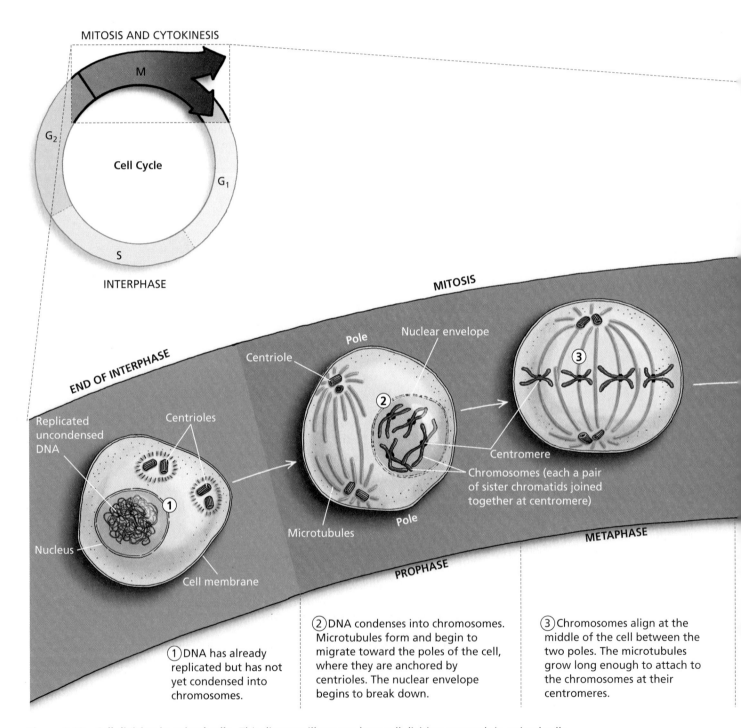

MITOSIS AND CYTOKINESIS

M

G_2

Cell Cycle

G_1

S

INTERPHASE

MITOSIS

END OF INTERPHASE

Pole

Nuclear envelope

Centriole

③

Replicated uncondensed DNA

Centrioles

②

Centromere

Chromosomes (each a pair of sister chromatids joined together at centromere)

Nucleus

Microtubules

Pole

Cell membrane

PROPHASE

METAPHASE

① DNA has already replicated but has not yet condensed into chromosomes.

② DNA condenses into chromosomes. Microtubules form and begin to migrate toward the poles of the cell, where they are anchored by centrioles. The nuclear envelope begins to break down.

③ Chromosomes align at the middle of the cell between the two poles. The microtubules grow long enough to attach to the chromosomes at their centromeres.

Figure 5.10 Cell division in animal cells. This diagram illustrates how cell division proceeds in animal cells.

Cytokinesis

Cytokinesis means "cellular movement." Cytokinesis in plant cells differs from that in animal cells due to the presence of the **cell wall**, an inflexible structure surrounding the plant cells. During telophase of mitosis in a plant cell, membrane-bound vesicles deliver the materials required for building the cell wall to the center of the cell. The materials brought from the cells sorting and packaging the Golgi apparatus include a tough, fibrous carbohydrate

④ The microtubules contract and separate the sister chromatids from each other, pulling them toward the two poles of the cell.

⑤–⑥ During telophase, nuclear envelopes reform and chromosomes decondense. During cytokinesis in an animal cell, a band of filaments contracts around the equator of the cell, causing two cells to form from the original parent cell.

⑦ The two daughter cells enter G₁ phase of interphase.

(a) Cytokinesis in a plant cell.

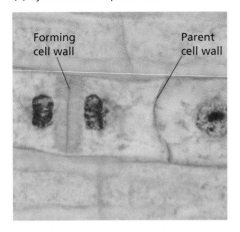

Forming cell wall   Parent cell wall

(b) Cytokinesis in an animal cell.

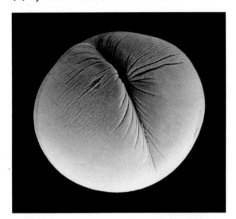

Figure 5.11   Cytokinesis in plant and animal cells. (a) Plant cells undergoing mitosis must do so within the confines of a rigid cell wall. During cytokinesis, plant cells form a cell plate that grows down the middle of the parent cell and eventually forms a new cell wall. (b) Animal cells, such as the frog cell shown here, produce a band of filaments that divide the cell in half.

called **cellulose** as well as some proteins. The membranes surrounding the vesicles gather in the center of the cell to form a structure called a **cell plate**. The cell plate and cell wall grow across the width of the cell and form a barrier that eventually separates the products of mitosis into two daughter cells (Figure 5.11a).

Because animal cells do not have a rigid cell wall, they have evolved a different method for separating the products of mitosis into daughter cells. During cytokinesis in animal cells, a band of proteins encircle the cell at the equator and divide the cytoplasm. This band of proteins contracts to pinch apart the two cells that have formed from the original parent cell. Each daughter cell is genetically identical, having its own nucleus with an exact copy of the parent cells' chromosomes and all the necessary organelles and cytoplasm as well (Figure 5.11b).

After cytokinesis, the cell reenters interphase, and if the conditions are favorable, the cell cycle may repeat itself. Cells that go through the cell cycle even if conditions are unfavorable can give rise to tumors.

5.4   Cell Cycle Control and Mutation

When cell division is working properly, it is a tightly controlled process. Cells are given signals for when and when not to divide. The normal cells in Nicole's ovary and the rest of her body were responding properly to the signals telling them when and how fast to divide. However, the cell that started her tumor was not responding properly to these signals.

Controls in the Cell Cycle

Instead of proceeding in lockstep through the cell cycle, normal cells halt cell division at a series of **checkpoints**. During this stoppage, proteins survey the cell to ensure that conditions for a favorable cellular division have been met. Three checkpoints must be passed before cell division can occur; one takes place during G_1, one during G_2, and the last during metaphase (Figure 5.12).

Proteins at the G_1 checkpoint determine whether it is necessary for a cell to divide. To do this, they survey the cell environment for the presence of other proteins called **growth factors** that stimulate cells to divide. When growth factors are limited in number, cell division does not occur. If enough growth factors are present to trigger cell division, then other proteins check to see if the cell is large enough to divide and if all the nutrients required for cell division are available. At the G_2 checkpoint, other proteins ensure that the DNA has replicated properly and double-check the cell size, again making sure that the cell is large enough to divide. The third and final checkpoint occurs during metaphase. Proteins present at metaphase verify that all the chromosomes have attached themselves to microtubules so that cell division can proceed properly.

If proteins surveying the cell at any of these three checkpoints determine that conditions are not favorable for cell division, the process is halted. When this happens, the cell may die.

Proteins that regulate the cell cycle, like all proteins, are coded for by genes. When these proteins are normal, cell division is properly regulated. When these cycle-regulating proteins are unable to perform their jobs,

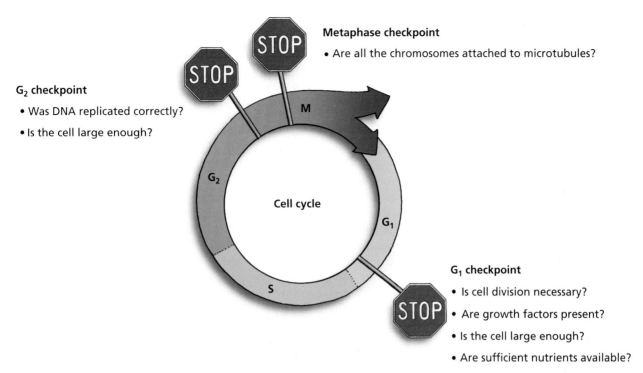

G₂ checkpoint

- Was DNA replicated correctly?
- Is the cell large enough?

Metaphase checkpoint

- Are all the chromosomes attached to microtubules?

G₁ checkpoint

- Is cell division necessary?
- Are growth factors present?
- Is the cell large enough?
- Are sufficient nutrients available?

Figure 5.12 Controls of the cell cycle. Checkpoints at G_1, G_2, and metaphase determine whether a cell will continue to divide.

unregulated cell division leads to large masses of cells called tumors. Mistakes in cell cycle regulation arise when the genes controlling the cell cycle are altered, or mutated, versions of the normal genes. A **mutation** is a change in the sequence of DNA. Changes to DNA can change a gene and in turn can alter the protein that the gene encodes, or provides instructions for. Mutant proteins do not perform their required cell functions in the same way that normal proteins do. If mutations occur to genes that encode the proteins regulating the cell cycle, cells can no longer regulate cell division properly. One or more cells in Nicole's ovary must have accumulated mutations in the cell-cycle control genes, leading to the development of cancer.

How Nicole acquired cancer-causing mutations is unknown. Mutations may be inherited or induced by exposure to substances called **carcinogens** that damage DNA and chromosomes. For a substance to be considered carcinogenic, exposure to the substance must be correlated with an increased risk of cancer. Examples of carcinogens include cigarette smoke, radiation, ultraviolet light, asbestos, and some viruses.

Mutations to Cell-Cycle Control Genes

Those genes that encode the proteins regulating the cell cycle are called **proto-oncogenes** (*proto* means "before," and *onco* means "cancer"). Proto-oncogenes are normal genes located on many different chromosomes that enable organisms to regulate cell division. When they become mutated, these genes are called **oncogenes**. It is when the normal proto-oncogenes undergo mutations and become oncogenes that they become capable of causing cancer. A wide variety of organisms carry proto-oncogenes, from

Figure 5.13   Cancer in nonhuman organisms. Many organisms carry proto-oncogenes, which can undergo mutations. This results in the production of oncogenes and causes the development of tumors or cancer.

mice to trees to dogs to humans, which means that any of these organisms can develop cancer (Figure 5.13).

Many proto-oncogenes provide the cell with instructions for building growth factors. A normal growth factor stimulates cell division only when the cellular environment is favorable and all conditions for division have been met. Oncogenes can overstimulate cell division (Figure 5.14a).

One gene involved in many cases of ovarian cancer is called *HER2*. (Names of genes are italicized, while names of the proteins they produce are not.) The *HER2* gene carries instructions for building a **receptor** protein, which "receives" a growth factor. When the shape of the receptor on the cell's surface is normal, it signals the inside of the cell to allow division to occur. Mutations to the gene that encodes this receptor can result in a receptor protein with a different shape from that of the normal receptor protein. When mutated or misshapen, the receptor protein functions as if many growth factors are present, even when there are actually few to no growth factors.

Receptor function can be compared to the role of a baseball glove in catching the ball and stimulating a ballplayer to make a throw. The normal situation occurs when the glove is open to receive the ball. Once the ball hits the pocket of the glove, the glove closes, triggering the throwing response. When a mutation occurs, the glove closes even without the ball, so the player inappropriately displays the throwing response.

Another class of genes involved in cancer are **tumor suppressors**. These genes, also present in all humans and many other organisms, carry the instructions for producing proteins that suppress or stop cell division if conditions are not favorable. These proteins can also detect and repair damage to the DNA. For this reason, normal tumor suppressors serve as backups in case the proto-oncogenes undergo mutation. If a growth factor overstimulates cell division, the normal tumor suppressor impedes tumor formation by preventing the mutant cell from moving through a checkpoint (Figure 5.14b).

When a tumor-suppressor protein is not functioning properly, it does not force the cell to stop dividing even though conditions are not favorable. Mutated tumor suppressors also allow cells to override cell-cycle checkpoints. One well-studied tumor suppressor, named p53, helps to determine whether cells will repair damaged DNA or commit cellular suicide if the damage is too severe. Mutations to the gene that encodes p53 result in damaged DNA being allowed to proceed through mitosis, thereby passing on even more mutations. Over half of all human cancers involve mutations to the gene that encodes p53.

To use another baseball analogy, a tumor suppressor could be likened to an umpire who, when behaving normally, keeps the game from getting out of

(a) Mutations to proto-oncogenes

Proto-oncogene (such as *HER2*)		Mutated proto-oncogene (oncogene)

DNA ⟶ Mutation ⟶

Protein

Functional protein stimulates cell division only when conditions are right.

Mutated protein can lead to overstimulation of cell division by overriding checkpoint control.

(b) Mutations to tumor-suppressor genes

Tumor suppressor (such as *BRCA2*)		Mutated tumor suppressor

⟶ Mutation ⟶

Mutation

Tumor-suppressor protein stops tumor formation by suppressing cell division.

Mutated tumor-suppressor protein fails to stop tumor growth.

Figure 5.14   Mutations to proto-oncogenes and tumor suppressor genes. (a) Mutations to proto-oncogenes and (b) tumor-suppressor genes can increase the likelihood of cancer developing.

control by warning or ejecting pitchers who are throwing too close to a batter's head. A mutated umpire would ignore such infractions, leading to a bench-clearing brawl.

Mutations to a tumor-suppressor gene are often found in ovarian cells that have become cancerous. Researchers believe that a normal *BRCA2* gene encodes a protein that is involved in helping to repair damaged DNA. The misshapen, mutant version of the protein cannot help to repair damaged DNA. This means that damaged DNA will be allowed to undergo mitosis, thus passing new mutations on to their daughter cells. As more and more mutations occur, the probability that a cell will become cancerous increases. Figure 5.15 summarizes the roles of *HER2* and *BRCA2* in ovarian cancer.

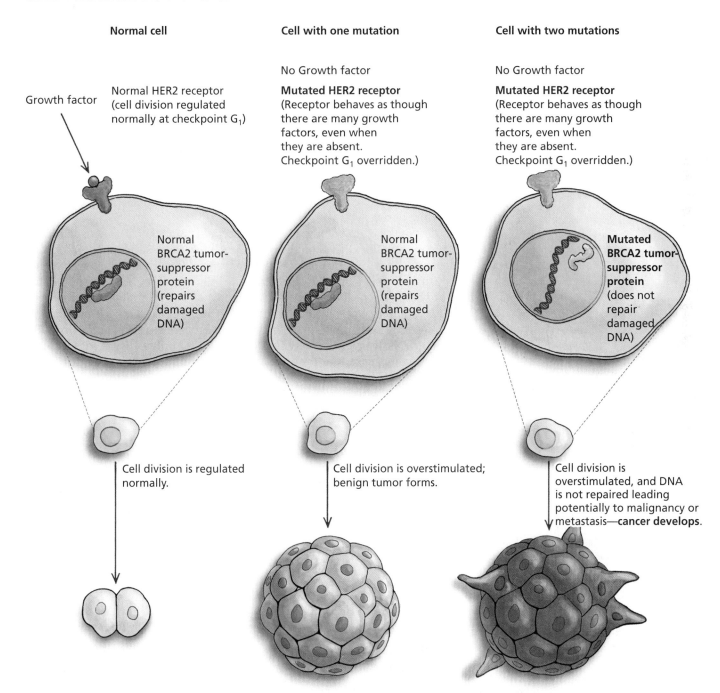

Figure 5.15   The roles of *HER2* and *BRCA2* in ovarian cancer. When a woman's ovarian cells have normal versions of these genes, cell division is properly regulated. If a cell in the woman's ovary has a *HER2* mutation, cell division is overstimulated, and a benign tumor can form. If a cell in the woman's ovary has both the *HER2* and *BRCA2* mutations, a cancerous tumor and metastasis are likely.

From Benign to Malignant.   Some mutations that occur as a result of damaged DNA being allowed to undergo mitosis are responsible for the progression of a tumor from a benign state, to a malignant state, to metastasis. For example, some cancer cells can stimulate the growth of surrounding blood vessels, through a process called **angiogenesis**. These cancer cells secrete a substance that attracts and reroutes blood vessels so that they supply a developing tumor with oxygen (necessary for cellular respiration) and other nutrients. When a tumor has its own blood supply, it can grow at the expense of other, noncancerous cells. Because the growth of rapidly dividing cancer cells occurs more quickly than the growth of normal cells in this process, entire organs can eventually become filled with cancerous cells. When this occurs, an organ can no longer work properly, leading to weakened functioning or organ failure. Damage to organs also explains some of the pain associated with cancer.

Normal cells also display a property called **contact inhibition**, which prevents them from dividing when doing so would require them to pile up on each other. Cancer cells, conversely, continue to divide and form a tumor (Figure 5.16a). In addition, normal cells do need some contact with an underlayer of cells in order to stay in place. This phenomenon is the result of a process called **anchorage dependence** (Figure 5.16b). Cancer cells override this requirement for some contact with other cells because cancer cells are dividing too quickly and do not expend enough energy to secrete adhesion molecules that glue the cells together. Once a cell loses its anchorage dependence, it may leave the original tumor and move to the blood, lymph, or surrounding tissues.

Most cells are programmed to divide a certain number of times—usually 60 to 70 times—and then they stop dividing. This limits most developing tumors to a small mole, cyst, or lump, all of which are benign. Cancer cells, however, do not obey these life-span limits. Instead, they are **immortal**. They achieve immortality by activating a gene that is usually turned off after early development. This gene produces an enzyme called **telomerase** that helps prevent the degradation of chromosomes. As chromosomes degrade with age, a cell loses its ability to divide. In cancer cells, telomerase is reactivated, allowing the cells to divide without limit.

In Nicole's case, the progression from normal ovarian cells to cancerous cells may have occurred as follows: (1) One single cell in her ovary may have acquired a mutation to its *HER2* growth factor receptor gene.

(a) Contact inhibition

Normal cells stop dividing when they come in contact with other cells.

Flask

Growth medium

Cells

Cancer cells continue to divide, piling on top of each other.

(b) Anchorage dependence

Normal cells usually stay anchored to other cells (or to a surface).

Flask

Growth medium

Cells

Cancer cells lose their anchorage dependence and can travel to new locations.

Figure 5.16   Contact inhibition and anchorage dependence. (a) When normal cells are grown on a solid support such as the bottom of a flask, they grow and divide until they cover the bottom of the flask; when they come in contact with other cells, they stop dividing. Cancer cells pile on top of each other and do not stop dividing. (b) Cancer cells lose the ability to adhere to other cells. They break away from the site of the original tumor and travel to new locations, where they set up secondary tumors.

(2) The descendants of this cell would have been able to divide faster than neighboring cells, forming a small, benign tumor. (3) Next, a cell within the tumor may have undergone a mutation to its *BRCA2* tumor suppressor gene, resulting in the inability of the BRCA2 protein to fix damaged DNA in the cancerous cells. (4) Cells produced by the mitosis of these doubly mutant cells would continue to divide even though their DNA is damaged, thereby enlarging the tumor and producing cells with more mutations. (5) Subsequent mutations could result in angiogenesis, lack of contact inhibition, reactivation of the telomerase enzyme, or may override anchorage dependence. If Nicole were very unlucky, the end result of these mutations could be that cells carrying many mutations would break away from the original ovarian tumor and set up a cancer at one or more new locations in her body.

Multiple Hit Model. Because multiple mutations are required for the development and progression of cancer, scientists describe the process of cancer development using the phrase **multiple hit model**. Nicole may have inherited some of these mutations, and/or they may have been induced by environmental exposures. Even though cancer is a disease caused by malfunctioning genes, most cancers are not caused only by the inheritance of mutant genes. In fact, scientists estimate that close to 70% of cancers are caused by mutations that occur during a person's lifetime.

Most of us will inherit few if any mutant cell-cycle control genes. It is the impact of our environment that will determine whether enough mutations will accumulate during our lifetime to cause cancer. Therefore, it is important to be aware of the exposures and behaviors, called **risk factors**, that increase a person's risk of obtaining a disease (see Essay 5.1 on page 118). Risk factors for ovarian cancer include smoking and uninterrupted ovulation. Ovarian cancer risk is thought to decrease for many years if ovulation is prevented for periods of time, as it is when a woman is pregnant, breastfeeding, or taking the birth control pill. When an egg cell is released from the ovary during ovulation, some tissue damage occurs. In the absence of ovulation, there is no damaged ovarian tissue and hence no need for extra cell divisions to repair damaged cells. Because mutations are most likely to occur when DNA is replicating before division, fewer divisions equal a lower likelihood of the appearance of cancer-causing mutations.

Regardless of its origin, once cancer is suspected, detection and treatment follow.

5.5 Cancer Detection and Treatment

Early detection and treatment of cancer increase the odds of survival dramatically. Being on the lookout for warning signs (Figure 5.17) can help alert individuals to developing cancers.

Detection Methods: Biopsy

Different cancers are detected using different methods. Some cancers are detected by the excess production of proteins that are normally produced by a particular cell type. Ovarian cancers often show high levels of CA125 protein in the blood. Nicole's level of CA125 led her gynecologist to think that a tumor might be forming on her remaining ovary. Once he suspected a tumor, Nicole's physician scheduled a biopsy.

A **biopsy** is the surgical removal of cells, tissue, or fluid that will be analyzed to determine whether they are cancerous.

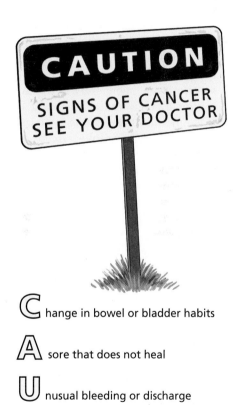

C hange in bowel or bladder habits

A sore that does not heal

U nusual bleeding or discharge

T hickening or lump

I ndigestion or difficulty swallowing

O bvious change in wart or mole

N agging cough or hoarseness

Figure 5.17   Warning signs of cancer. Self-screening for cancer is possible by looking for these signs. If you experience one or more of these warning signs, see your doctor.

Essay 5.1 Cancer Risk Factors

General risk factors for virtually all cancers include tobacco use, a high-fat and low-fiber diet, lack of exercise, obesity, and increasing age. Table E5.1 (on pages 119–120) outlines other risk factors that are linked to particular cancers.

Tobacco Use

The use of tobacco of any type, whether cigarettes, cigars, or chewing tobacco, increases your risk of many cancers. While smoking is the cause of 90% of all lung cancers, it is also the cause of about one-third of all cancer deaths. Cigar smokers have increased rates of lung, larynx, esophagus, and mouth cancers. Chewing tobacco increases the risk of cancers of the mouth, gum, and cheeks. People who do not smoke but are exposed to secondhand smoke have increased lung cancer rates.

Tobacco smoke contains over 20 known carcinogens. The carcinogens that are inhaled during smoking come into contact with live cells deep inside the lungs. Chemicals present in cigarettes and cigarette smoke have been shown to increase cell division, inhibit a cell's ability to repair damaged DNA, and prevent cells from dying when they should. Chemicals in cigarette smoke also disrupt the transport of substances across cell membranes and alter many of the enzyme reactions that occur within cells. They have also been shown to increase the generation of oxygen-free radicals, which remove electrons from other molecules. The removal of electrons from DNA or other molecules causes damage to these molecules—damage that, over time, may lead to cancer. Cigarette smoking provides so many different opportunities for DNA damage and cell damage that tumor formation and metastasis are quite likely for smokers. In fact, people who smoke cigarettes increase their odds of developing virtually every human cancer.

A High-Fat, Low-Fiber Diet

Cancer risk may also be influenced by diet. The American Cancer Society recommends eating at least 5 servings of fruits and vegetables every day as well as 6 servings of food from other plant sources such as breads, cereals, grains, rice, pasta, or beans. This is because plant foods are low in fat and high in fiber. A high-fat (greater than 15% of calories obtained from fat), low-fiber (less than 30 grams per day) diet is associated with increased cancer risk. Fruits and vegetables are also rich in antioxidants that help to neutralize the electrical charge on oxygen-free radicals and thereby prevent the free radicals from taking electrons from other molecules, including DNA. There is some evidence that antioxidants may help prevent certain cancers by minimizing the number of free radicals that may damage the DNA in our cells.

Lack of Exercise and Obesity

Regular exercise decreases the risk of most cancers, partly because exercise keeps the immune system functioning effectively. The immune system helps destroy cancer cells when it can recognize them as foreign to the host body. Unfortunately, since cancer cells are actually your own body's cells run amok, the immune system cannot always differentiate between normal cells and cancer cells. Exercise also helps prevent obesity, which is associated with increased risk for many cancers including cancers of the breast, uterus, ovary, colon, gallbladder, and prostate. The abundance of fatty tissue has been hypothesized to increase the odds of hormone-sensitive cancers such as breast, uterine, ovarian, and prostate cancer because fat is a source of cholesterol. Cholesterol is the building block, or precursor, for estrogen and testosterone. This excess body fat might alter hormonal pathways and increase the risk of developing cancer in hormone-sensitive tissues.

Excess Alcohol Consumption

Drinking alcohol is associated with increased risk of some types of cancer. Men should have no more than 2 alcohol drinks per day, and women 1 or none. People who both drink and smoke increase their odds of cancer in a multiplicative rather than additive manner. In other words, if one type of cancer occurs in 10% of smokers and in 2% of drinkers, someone who smokes and drinks has a chance of obtaining this cancer that is closer to 20% than 12%.

Increasing Age

As you age, your immune system weakens, and its ability to distinguish between cancer cells and normal cells decreases. This is part of the reason many cancers are far more likely in elderly people. Additional factors that help explain the higher cancer risk with increasing age include simple probability. If we are all exposed to carcinogens during our lifetime, then the longer we are alive, the greater the probability that some of those carcinogens will mutate genes involved in regulating the cell cycle. Also, because multiple mutations are necessary for a cancer to develop, it often takes many years to progress from the initial mutation to a tumor and then to full-blown cancer. Scientists estimate that most cancers large enough to be detected have been growing for at least 5 years and are composed of close to 1 billion cells.

Table E5.1 Cancer risk. Risk factors for particular cancers appear below. Cancers are listed in order of occurrence from most to least common.

Cancer Location	Risk Factors	Detection	Comments
Ovary Oviduct Ovary Uterus Vagina	• Smoking • Mutation to *BRCA2* gene • Advanced age • Oral contraceptive use and pregnancy decrease risk.	• Blood test for elevated CA125 level • Gynecological exam	• Fifth leading cause of death among women in the United States
Breast Milk-producing glands Nipple Fatty tissue	• Smoking • Mutation to *BRCA1* gene • High-fat, low-fiber diet • Use of oral contraceptives may slightly increase risk.	• Monthly self exams, look and feel for lumps or changes in contour • Mammogram	• Only 5% of breast cancers are due to *BRCA1* mutations. • Second-highest cause of cancer-related deaths • 1% of breast cancer occurs in males.
Cervix Uterus Cervix Vagina	• Smoking • Exposure to sexually transmitted Human Papilloma Virus (HPV)	• Annual Pap-smear tests for the presence of precancerous cells	• Precancerous cells can be removed by laser surgery or cryotherapy (freezing) before they become cancerous.
Skin Epidermis Dermis	• Smoking • Fair skin • Exposure to ultraviolet light from the sun or tanning beds	• Monthly self-exams, look for growths that change in size or shape	• Skin cancer is the most common of all cancers and is usually curable if caught early.
Lung Trachea Bronchi Lungs	• Smoking • Exposure to secondhand smoke • Asbestos inhalation	• X-ray	• Lung cancer is the most common cause of death from cancer, and the best prevention is to quit, or never start, smoking.

(Continued)

Table E5.1

Cancer Location	Risk Factors	Detection	Comments
Colon and rectum Small intestine Colon Rectum	• Smoking • Polyps in the colon • Advanced age • High-fat, low-fiber diet	• Change in bowel habits • Colonoscopy is an examination of the rectum and colon using a lighted instrument.	• Benign buds called polyps can grow in the colon; removal prevents them from mutating and becoming cancerous.
Prostate Bladder Prostate Rectum	• Smoking • Advanced age • High-fat, low-fiber diet	• Blood test for elevated level of prostate specific antigen (PSA) • Physical exam by physician, via rectum	• More common in African-American men than Asian, white, or Native American men
Testicle Penis Testicle Scrotum	• Abnormal testicular development	• Monthly self exam, inspect for lumps and changes in contour	• Testicular cancer accounts for only 1% of all cancers in men but is the most common form of cancer found in males between the ages of 15 and 35.
Blood (Leukemia) Platelet Red blood cell White blood cell	• Exposure to high-energy radiation such as that produced by atomic bomb explosions in Japan during World War II	• A sample of blood is examined under a microscope.	• Cancerous white blood cells cannot fight infection efficiently: people with leukemia often succumb to infections.

(a) Normal ovarian tissue   **(b)** Benign ovarian tumor   **(c)** Malignant ovarian tumor

Figure 5.18   Biopsy. When stained and viewed under a microscope, (a) normal cells have a different appearance than (b) benign and (c) malignant tumors.

When viewed under a microscope, benign tumors consist of orderly growths of cells that resemble the cells of the tissue from which they were taken. Malignant or cancerous cells do not resemble other cells found in the same tissue; they are dividing so rapidly that they do not have time to produce all the proteins necessary to build normal cells. This leads to the often abnormal appearance of cancer cells as seen under a microscope (Figure 5.18).

A needle biopsy is usually performed if the cancer is located on or close to the surface of the patient's body. For example, breast lumps are often biopsied with a needle to determine whether the lump contains fluid and is a noncancerous cyst, or whether it contains abnormal cells and is a tumor. When a cancer is diagnosed, surgery is often performed to remove as much of the cancerous growth as possible without damaging neighboring organs and tissues.

In Nicole's case, getting at the ovary to find tissue for a biopsy required the use of a surgical instrument called a **laparoscope**. For this operation, the surgeon inserted a small light and a scalpel-like instrument through a tiny incision above Nicole's navel.

Nicole's surgeon preferred to use the laparoscope since he knew Nicole would have a much easier recovery from laparoscopic surgery than she had from the surgery to remove her other, cystic ovary. Laparoscopy had not been possible when removing Nicole's other ovary—the cystic ovary had grown so large that her surgeon had to make a large abdominal incision to remove it.

A laparoscope has a small camera that projects images from the ovary onto a monitor that the surgeon views during surgery. These images showed that Nicole's tumor was a different shape, color, and texture from the rest of her ovary. They also showed that the tumor was not confined to the surface of the ovary; in fact, it appeared to have spread deeply into her ovary. Nicole's surgeon decided to shave off only the affected portion of the ovary and leave as much intact as possible, with the hope that the remaining ovarian tissue might still be able to produce egg cells. He then sent the tissue to a laboratory so that the pathologist could examine it. Unfortunately, when the pathologist looked through the microscope this time, she saw the disorderly appearance characteristic of cancer cells. Nicole's ovary was cancerous, and further treatment would be necessary.

Treatment Methods: Chemotherapy and Radiation

A treatment that works for one woman with ovarian cancer might not work for another ovarian cancer patient because a different suite of mutations may have led to the cancer in each woman's ovary. Luckily for Nicole, her ovarian cancer was diagnosed very early. Regrettably, this is not the case for most women with ovarian cancer, many of whom are diagnosed when symptoms become severe, leading them to see their physician. The symptoms of ovarian cancer tend to be vague and slow to develop. These symptoms include abdominal swelling, pain, bloating, gas, constipation, indigestion, menstrual disorders, and fatigue. Unfortunately, many women simply overlook these discomforts. The difficulty of diagnosis is compounded because no routine screening tests are available. For instance, CA125 levels are checked only when ovarian cancer is suspected because (1) ovaries are not the only tissues that secrete this protein; (2) CA125 levels vary from individual to individual; and (3) these levels depend on the phase of the woman's menstrual cycle. Elevated CA125 levels usually mean that the cancer has been developing for a long time. Consequently, by the time the diagnosis is made, the cancer may have grown quite large and metastasized, making it much more difficult to treat.

Nicole's cancer was caught early. However, her physician was concerned that some of her cancerous ovarian cells may have spread through blood vessels or lymph ducts on or near the ovaries, or into her abdominal cavity, so he started Nicole on chemotherapy after her surgery.

Chemotherapy.   During **chemotherapy**, chemicals are injected into the bloodstream. These chemicals selectively kill dividing cells. A variety of chemotherapeutic agents act in different ways to interrupt cell division. The manner in which a chemotherapeutic agent acts to kill dividing cells is called its mechanism of action.

Chemotherapy involves many drugs since most chemotherapeutic agents affect only one type of cellular activity. Cancer cells are rapidly dividing and do not take the time to repair mistakes in replication that lead to mutations. These cells are allowed to proceed through the G_2 checkpoint with many mutations. Therefore, cancer cells can randomly undergo mutations, a few of which might allow them to evade the actions of a particular chemotherapeutic agent. Cells that are resistant to one drug proliferate when the chemotherapeutic agent clears away the other cells that compete for space and nutrients. Cells with a preexisting resistance to the drugs are selected for and produce more daughter cells with the same resistant characteristics, requiring the use of more than one chemotherapeutic agent.

Scientists estimate that cancer cells become resistant at a rate of approximately 1 cell per million. Because tumors contain about 1 billion cells, the average tumor will have close to 1,000 resistant cells. Therefore, treating a cancer patient with a combination of chemotherapeutic agents aimed at different mechanisms increases the chances of destroying all the cancerous cells in a tumor.

Unfortunately, normal cells that divide rapidly are also affected by chemotherapy treatments. Hair follicles, cells that produce red blood cells, white blood cells, and cells that line the intestines and stomach are often damaged or destroyed. The effects of chemotherapy therefore include temporary hair loss, anemia (dizziness and fatigue due to decreased numbers of red blood cells), and lowered protection from infection due to decreases in the number of white blood cells. Additionally, damage to the cells of the stomach and intestines can lead to nausea, vomiting, and diarrhea.

Several hours after each chemotherapy treatment, Nicole became nauseous; she often had diarrhea and vomited for a day or so after her treatments. Midway through her chemotherapy treatments, Nicole lost most of her hair.

Radiation Therapy.   Cancer patients often undergo radiation treatments as well as chemotherapy. **Radiation therapy** uses high-energy particles to injure or destroy cells by damaging their DNA, making it impossible for these cells to continue to grow and divide. A typical course of radiation involves a series of 10 to 20 treatments performed after the surgical removal of the tumor, although sometimes radiation is used before surgery in order to decrease the size of the tumor. Radiation therapy is typically used only when cancers are located close to the surface of the body because it is difficult to focus a beam of radiation on internal organs such as an ovary. Therefore, Nicole's physician recommended chemotherapy only.

Nicole's treatments consisted of many different chemotherapeutic agents, spread over many months. The treatments took place at the local hospital on Wednesdays and Fridays. She usually had a friend drive her to the hospital very early in the morning and return later in the day to pick her up. The drugs were administered through an intravenous (IV) needle into a vein in her arm (Figure 5.19). During the hour or so that she was undergoing chemotherapy, Nicole usually studied for her classes. She did not mind the actual chemotherapy treatments that much. The hospital personnel were kind to her, and she got some studying done. It was the aftermath of these treatments that she hated. Most days during her chemotherapy regimen, Nicole was so exhausted that she did not get out of bed until late morning, and on the day after her treatments, she often slept until late afternoon. Then she would get up and try to get some work done or make some phone calls before going back to bed early in the evening. After 6 weeks of chemotherapy, Nicole's CA125 levels started to drop. After another 2 months of chemotherapy, her CA125 levels were back down to their normal, precancerous level. If Nicole has normal CA125 levels for 5 years, she will be considered to be in **remission**, or no longer suffering negative impacts from cancer. After 10 years of normal CA125 levels, she will be considered cured of her cancer. Because Nicole's cancer responded to chemotherapy, she was spared from having to undergo any other, more experimental treatments.

Figure 5.19 Chemotherapy. Many chemotherapeutic agents, are administered through an intravenous (IV) needle.

Even though her treatments seemed to be going well, Nicole had other worries. She worried that her remaining ovary would not recover from the surgery and chemotherapy, which meant that she would never be able to have children. Nicole had always assumed that she would have children some day, and although she did not currently have a strong desire to have a child, she wondered if her feelings would change. Even though she was not planning to marry any time soon, she also wondered how her future husband would feel if she were not able to become pregnant.

In addition to her concerns about being able to become pregnant, Nicole also became worried that she might pass on mutated, cancer-causing genes to her children. For Nicole, or anyone, to pass on genes to his or her children, reproductive cells must be produced by another type of nuclear division called meiosis.

5.6 Meiosis

Meiosis is a form of cell division that occurs only in specialized cells within the **gonads** or sex organs. In humans, and in most animals, the male gonads are the testes, and the female gonads are the ovaries. During meiosis, specialized sex cells called **gametes** are produced. In animals, the male gametes are the sperm cells, while the gametes produced by the female are the egg cells.

Meiosis produces gametes that have half as many chromosomes as the parent cell. Both sperm and egg cells contain half as many chromosomes as other non-gamete cells, called **somatic cells**. Animal somatic cells include cells of the skin, muscle, liver, and stomach tissues, etc. Because human somatic cells have 46 chromosomes and meiosis reduces the chromosome number by one-half, the gametes produced during meiosis contain 23 chromosomes each. When an egg cell and a sperm cell combine their 23 chromosomes at fertilization, the developing embryo will then have the required 46 chromosomes.

The placement of chromosomes into gametes is not random; that is, meiosis does not simply place any 23 of the 46 human chromosomes into a gamete. Instead, meiosis apportions chromosomes in a very specific manner. The 46 chromosomes in human body cells are actually 23 different pairs of chromosomes, and meiosis produces cells that contain 1 chromosome of every pair.

It is possible to visualize chromosome pairs by preparing a highly magnified photograph of the chromosomes arranged in pairs called a **karyotype**. A karyotype is usually prepared from chromosomes that have been removed from the nuclei of white blood cells, which have been treated with chemicals to stop mitosis at metaphase. Because these chromosomes are at metaphase of mitosis, they are composed of replicated sister chromatids and are shaped like the letter X. It is possible to photograph chromosomes and then digitally arrange them in pairs. The 46 human chromosomes can be arranged into 22 pairs of non-sex chromosomes, or **autosomes**, and one pair of **sex chromosomes** (the X and Y chromosomes) to make a total of 23 pairs. Human males have an X and a Y chromosome, while females have two X chromosomes. Each chromosome is paired with a mate that is the same size and shape and has its centromere in the same position (Figure 5.20).

The pairs of non-sex chromosomes are called **homologous pairs**. Each member of a homologous pair of chromosomes carries the same genes along its

Figure 5.20 Karyotype. The pairs of chromosomes in this karyotype are arranged in order of decreasing size and numbered from 1 to 22. The X and Y sex chromosomes are the 23rd pair. The sex chromosomes from a female and a male are shown in the insets.

(a) Homologous pair of chromosomes

Two alleles of the
same gene

(b) Nonhomologous pair of chromosomes

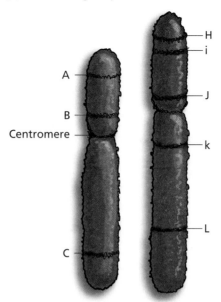

Figure 5.21   Homologous and nonhomologous pairs of chromosomes. (a) Homologous pairs of chromosomes have the same genes (shown here as A, B, and C) but may have different alleles. The dominant allele is represented by an uppercase letter, while the recessive allele is shown with the same letter in lowercase. Note that the chromosomes of this pair each have the same size, shape, and positioning of the centromere. (b) Nonhomologous pairs of chromosomes are different sizes and shapes and carry different genes.

length, although not necessarily the same versions of those genes (Figure 5.21a). Two chromosomes that are not homologous would be different in size, shape, and position of the centromere and would carry genes that encode different traits (Figure 5.21b).

Different versions of the same gene are called **alleles** of a gene. For example, there are normal and mutant versions of the *BRCA2* gene. Note that there is a difference between the same type of information in the sense that both alleles of this gene code for a cell-cycle control protein, but they happen to code for different versions of the same protein. Alleles are alternate forms of a gene, in the same way that chocolate and vanilla are alternate forms of ice cream. A different gene, encoding a different type of information, may also have different alleles. For instance, the sandwich gene might have alleles for peanut butter and bologna.

When a chromosome is replicated, during the S phase of the cell cycle, the DNA is duplicated. Replication faithfully results in two copies, called sister chromatids, that are genetically identical. For this reason, we would find exactly the same information on the sister chromatids that comprise a replicated chromosome (Figure 5.22).

Meiosis separates the members of a homologous pair from each other. Once meiosis is completed, there is 1 copy of each chromosome (1–23) in every gamete. When only 1 member of each homologous pair is present in a cell, we say that the cell is **haploid (*n*)**—both egg cells and sperm cells are haploid. All somatic cells in

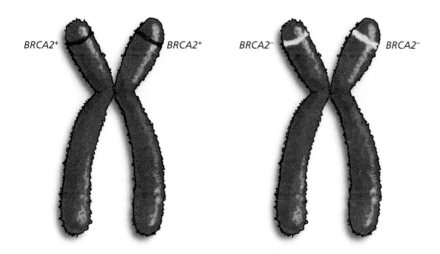

Figure 5.22   Replicated chromosomes. This homologous pair of chromosomes has been replicated. Note that the normal version of the *BRCA2* gene (symbolized *BRCA2*$^+$) is present on both sister chromatids of the chromosome on the left. Its homologue on the right carries the mutant version of this allele (symbolized *BRCA2*$^-$).

(a) Gamete formation in humans

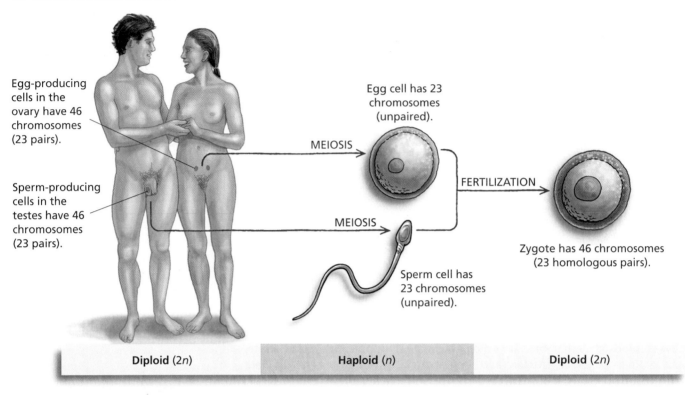

Egg-producing cells in the ovary have 46 chromosomes (23 pairs).

Sperm-producing cells in the testes have 46 chromosomes (23 pairs).

MEIOSIS

MEIOSIS

Egg cell has 23 chromosomes (unpaired).

Sperm cell has 23 chromosomes (unpaired).

FERTILIZATION

Zygote has 46 chromosomes (23 homologous pairs).

Diploid (2n)   **Haploid (n)**   **Diploid (2n)**

(b) Gamete formation in flowering plants

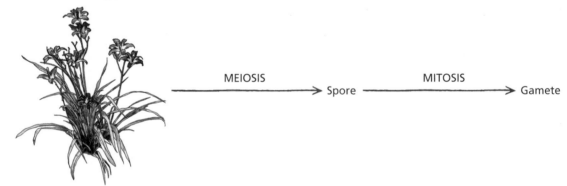

MEIOSIS → Spore   MITOSIS → Gamete

Figure 5.23   Gamete production in humans and flowering plants. (a) In humans, the diploid cells of the ovaries and testes undergo meiosis and produce haploid gametes. At fertilization, the diploid condition is restored. (b) Flowering plants undergo a slightly different sequence of events during gamete formation. Structures called spores are produced by meiosis and then undergo mitosis to produce the actual gametes.

humans contain homologous pairs of chromosomes and are therefore diploid. For a diploid cell in a person's testes or ovary to become a haploid gamete, it must go through meiosis. After the sperm and egg fuse, the fertilized cell, or **zygote**, will contain 2 sets of chromosomes and is said to be **diploid (2n)** (Figure 5.23a). It is not just humans that undergo meiosis to produce gametes; all cells with nuclei and membrane-bound organelles—eukaryotic cells—undergo meiosis. Some eukaryotes, however, produce gametes in a slightly different manner. Flowering plants, for instance, first undergo meiosis to produce spores and then mitosis to produce the actual gametes (Figure 5.23b).

Like mitosis, meiosis is preceded by an interphase stage that includes G_1, S, and G_2. Interphase is followed by two phases of meiosis, called meiosis I and meiosis II, in which divisions of the nucleus take place (Figure 5.24). Meiosis I separates the members of a homologous pair from each other. Meiosis II separates the chromatids from each other. Both meiotic divisions are followed by cytokinesis, during which the cytoplasm is divided between the resulting daughter cells.

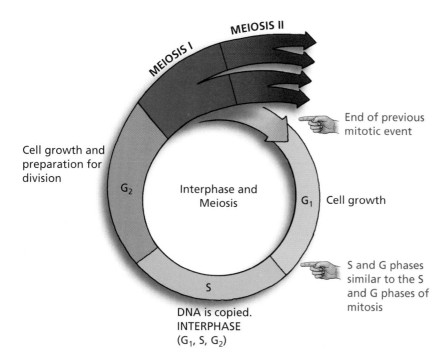

Figure 5.24   Interphase and meiosis. Interphase consists of G_1, S, and G_2 and is followed by two rounds of nuclear division, meiosis I and meiosis II.

Interphase

The interphase that precedes meiosis consists of G_1 S, and G_2. This interphase of meiosis is similar in most respects to the interphase that precedes mitosis. The centrioles from which the microtubules will originate are present. The G phases are times of cell growth and preparation for division. The S phase is when DNA replication occurs. Once the cell's DNA has been replicated, it can enter meiosis I.

Meiosis I

The first meiotic division, meiosis I, consists of prophase I, metaphase I, anaphase I, and telophase I (see Figure 5.25 on pp. 128-129).

During prophase I of meiosis, the nuclear envelope starts to break down, and the microtubules begin to assemble. The previously replicated chromosomes condense so that they can be moved around the cell without becoming entangled. The condensed chromosomes can be seen under a microscope. At this time, the homologous pairs of chromosomes exchange genetic information in a process called crossing over, which we will explain in a moment.

At metaphase I, the homologous pairs line up at the cell's equator, or middle of the cell. Microtubules bind to the metaphase chromosomes near the centromere. Homologous pairs are arranged arbitrarily regarding which member faces which pole. This process is called random alignment. At the end of this section, you will find detailed descriptions of crossing over and random alignment along with their impact on genetic diversity.

At anaphase I, the homologous pairs are separated from each other by the shortening of the microtubules; and at telophase I, nuclear envelopes reform around the chromosomes. DNA is then partitioned into each of the 2 daughter cells by cytokinesis. Because each daughter cell contains only 1 copy of each member of a homologous pair, at this point the cells are haploid. Now both of these daughter cells are ready to undergo meiosis II.

Meiosis II

Meiosis II consists of prophase II, metaphase II, anaphase II, and telophase II. This second meiotic division is virtually identical to mitosis and serves to separate the sister chromatids of the replicated chromosome from each other.

MEIOSIS I AND II

G₂

Interphase and Meiosis

G₁

S

CYTOKINESIS

MEIOSIS I

END OF INTERPHASE

Diploid

Nuclear envelope

Replicated uncondensed DNA

Centrioles

① DNA is replicated during S phase of interphase.

PROPHASE I
② Nuclear envelope starts to break down. Microtubules start to assemble. DNA condenses into chromosomes.

Crossing over may occur.

Microtubules

Nucleus

Cell membrane

METAPHASE I
③ Homologous chromosomes align at middle of cell.

Random alignment

ANAPHASE I
④ Homologous chromosomes are separated by shortening of microtubules.

Figure 5.25 The cell cycle. This diagram illustrates interphase, meiosis I, meiosis II, and cytokinesis in an animal cell.

At prophase II of meiosis, the cell is readying for another round of division, and the microtubules are lengthening again. At metaphase II, the chromosomes align across the equator in much the same way as they do during mitosis—not as pairs, as was the case with metaphase I. At anaphase II, the sister chromatids separate from each other and move to opposite poles of the cell. At telophase II, the separated chromosomes each become enclosed in their own nucleus. In this

MEIOSIS II

CYTOKINESIS

Haploid

ELOPHASE I and CYTOKINESIS	**PROPHASE II**	**METAPHASE II**	**ANAPHASE II**	**TELOPHASE II and CYTOKINESIS**
Cytokinesis sults in two ughter cells. clear envelopes form.	⑥ Microtubules lengthen.	⑦ Chromosomes align at middle of cell.	⑧ Sister chromatids are separated by shortening of microtubules.	⑨ Four haploid daughter cells result. Nuclear envelopes reform.

fashion, half of a person's genes are physically placed into each gamete; thus, children carry one-half of each parent's genes.

Each parent can produce millions of different types of gametes due to two events that occur during meiosis I—crossing over and random alignment. Both of these processes greatly increase the number of different kinds of gametes that an individual can produce and therefore increase the variation in individuals that can be produced when gametes combine.

Crossing Over and Random Alignment

Crossing over occurs during prophase I of meiosis I. It involves the exchange of portions of chromosomes from one member of a homologous pair to the other member. Crossing over is believed to occur several times on each homologous pair during each occurrence of meiosis.

To illustrate crossing over, consider an example using genes involved in the production of flower color and pollen shape in sweet pea plants. These two genes are on the same chromosome and are called **linked genes**. Linked genes move together on the same chromosome to a gamete, and they may or may not undergo crossing over.

If a pea plant has red flowers and long pollen grains, the chromosomes may appear as shown in Figure 5.26. It is possible for this plant to produce 4 different types of gametes with respect to these 2 genes. Two types of gametes would result if no crossing over occurred between these genes—the gamete containing the red flower and long pollen chromosome, and the gamete containing the white flower and short pollen chromosome. Two additional types of gametes could be produced if crossing over did occur—one type containing the red flower and short pollen grain chromosome, and the other type containing the reciprocal white flower and long pollen grain chromosome. Therefore, crossing over increases genetic diversity by increasing the number of distinct combinations of genes that may be present in a gamete.

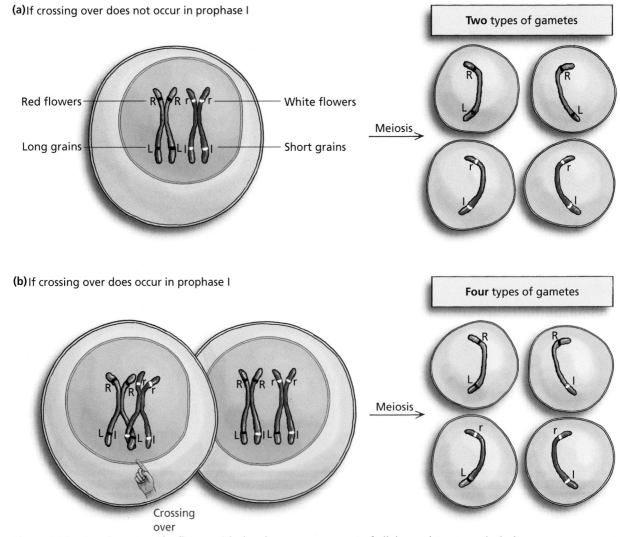

Figure 5.26 Crossing over. If a flower with the above arrangement of alleles undergoes meiosis, it can produce (a) two different types of gametes for these two genes if crossing over does not occur, or (b) four different types of gametes for these two genes if crossing over does occur.

Random alignment of homologous pairs also increases the number of genetically distinct types of gametes that can be produced. Using Nicole's chromosomes as an example (Figure 5.27), let us assume that she did in fact inherit mutant versions of both the *BRCA2* and *HER2* genes and that these genes are located on different chromosomes. The arrangement of homologous pairs of chromosomes at metaphase I determines which chromosomes will end up together in a gamete. If we consider only these 2 homologous pairs of chromosomes, then 2 different alignments are possible, and 4 different gametes can be produced. For example, when Nicole produces egg cells, the 2 chromosomes that she inherits from her dad could move together to the gamete, leaving the 2 chromosomes she inherited from her mom to move to the other gamete. It is equally probable that Nicole could undergo meiosis in which one chromosome from each parent will align randomly together, resulting in two more types of gametes being produced. In Chapters 6 and 7, you will see how random alignment and crossing over affect the inheritance of genes in greater detail.

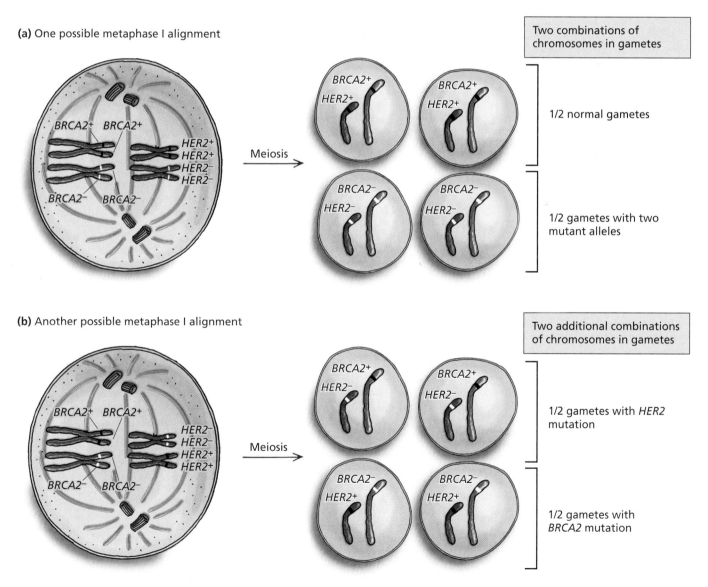

Figure 5.27 Random alignment. Two possible alignments, (a) and (b), can occur when there are two homologous pairs of chromosomes. These different alignments can lead to novel combinations of genes in the gametes.

Figure 5.28 Comparing mitosis and meiosis. Mitosis is a type of cell division that occurs in somatic cells and gives rise to daughter cells that are exact genetic copies of the parent cell. Meiosis occurs in cells that will give rise to gametes and decreases the chromosome number by one half.

From the previous discussions, you have learned that cells undergo mitosis for growth and repair and meiosis to produce gametes. Figure 5.28 compares the significant features of mitosis and meiosis.

It is now possible to revisit the question of whether Nicole will pass on cancer-causing genes to any children she may have. Because Nicole developed cancer at such a young age, it seems likely that she may have inherited at least one mutant cell-cycle control gene; thus, she may or may not pass that gene on. If Nicole has both a normal and a mutant version of a cell-cycle control gene, then she will be able to make gametes with and without the mutant allele. Therefore, she could pass on the mutant allele if a gamete containing that allele is involved in fertilization. We have also seen that it takes many "hits" or mutations for a cancer to develop. Therefore, even if Nicole does pass on one or a few mutant versions of cell-cycle control genes to a child, environmental conditions will dictate whether enough other mutations will accumulate to allow a cancer to develop.

Mutations caused by environmental exposures are not passed from parents to children, unless the mutation happens to occur in a cell of the gonads that will be used to produce a gamete. Nicole's cancer occurred in the ovary, the site of meiosis, but not all cells in the ovary undergo meiosis. Nicole's cancer originated in the outer covering of the ovary, a tissue that does not undergo meiosis. The cells involved in ovulation are located inside the ovary. A skin cancer that develops from exposure to ultraviolet light will not be passed on, nor will most of the mutations that Nicole obtained from environmental exposures. Therefore, for any children that Nicole (or any of us) might have, it is the combined effects of inherited mutant alleles and any mutations induced by environmental exposures that will determine whether cancers will develop.

CHAPTER REVIEW

Summary

5.1 What Is Cancer?

- Unregulated cell division can lead to the formation of a tumor. Benign or noncancerous tumors stay in one place and do not prevent surrounding organs from functioning. Malignant tumors are those that are invasive, or those that metastasize to surrounding tissues, starting new cancers (p. 103).

5.2 Cell Division Overview

- Cell division is a process required for growth and development (p. 105).

- For cell division to occur, the DNA must be copied and passed on to daughter cells (p. 105).

- The DNA molecule is composed of antiparallel strands of nucleotides. Nucleotides are composed of a sugar and a phosphate and a nitrogenous base. Nitrogenous bases include A, C, G, and T. A and T make a base pair with each other, as do C and G (pp. 105–106).

- Once copied, the duplicated chromosome is composed of 2 sister chromatids that are attached to each other at the centromere (p. 105).

- During DNA replication or synthesis, one strand of the double-stranded DNA molecule is used as a template for the synthesis of a new daughter strand of DNA. The newly synthesized DNA strand is complementary to the parent strand. The enzyme DNA polymerase ties together the nucleotides on the daughter strand (pp. 106–107).

Web Tutorial 5.1 The Structure of DNA

5.3 The Cell Cycle and Mitosis

- The cell cycle includes all of the events that occur as one cell gives rise to daughter cells (p. 108).

- Interphase consists of 2 gap phases of the cell cycle (G_1 and G_2), during which the cell grows and prepares to enter mitosis or meiosis. During the S (synthesis) phase, the DNA replicates. The S phase of interphase occurs between G_1 and G_2 (pp. 108–109).

- During mitosis, the sister chromatids are separated from each other into daughter cells. During prophase, the replicated DNA condenses into linear chromosomes. At metaphase, these replicated chromosomes align across the middle of the cell. At anaphase, the sister chromatids separate from each other and align at opposite poles of the cells. At telophase, a nuclear envelope reforms around the chromosomes lying at each pole (pp. 109–110).

- Cytokinesis is the last phase of the cell cycle. During cytokinesis, the cytoplasm is divided into two portions, one for each daughter cell (p. 111).

Web Tutorial 5.2 Mitosis

5.4 Cell Cycle Control and Mutation

- When cell division is working properly, it is a tightly controlled process. Normal cells divide only when conditions are favorable (p. 112).

- Proteins survey the cell and its environment at checkpoints as the cell moves through G_1, G_2, and metaphase and can halt cell division if conditions are not favorable (p. 112).

- Mistakes in regulating the cell cycle arise when genes that control the cell cycle are mutated. Mutated genes can be inherited, or mutations can be caused by exposure to carcinogens (p. 113).

- Proto-oncogenes regulate the cell cycle. Oncogenes are mutated versions of these genes. Many proto-oncogenes encode growth factors (p. 113).

- Tumor suppressors are normal genes that can encode proteins to stop cell division if conditions are not favorable and can repair damage to the DNA. They serve as backups in case the proto-oncogenes undergo mutation (p. 114).

- As a tumor progresses from benign to malignant, it often undergoes angiogenesis, loses contact inhibition and anchorage dependence, and becomes immortal. Thus, many changes or hits to the cancer cell are required for malignancy (p. 116).

Web Tutorial 5.3 The Cell Cycle

5.5 Cancer Detection and Treatment

- Early cancer detection and treatment increase survival odds (p. 117).

- A biopsy is a common method for detecting cancer. It involves removing some cells or tissues suspected of being cancerous and analyzing them (pp. 117, 121).

- Typical cancer treatments include chemotherapy, which involves injecting chemicals that kill rapidly dividing cells, and radiation, which involves killing tumor cells by exposing them to high-energy particles (pp. 121–122).

5.6 Meiosis

- Meiosis is a type of sexual cell division, occurring in cells, that gives rise to gametes. Gametes contain half as many chromosomes as somatic cells do. The reduction division of meiosis begins with diploid cells and ends with haploid cells (p. 123).

- Meiosis is preceded by an interphase stage, in which the DNA is replicated (p. 127).

- During meiosis I, the members of a homologous pair of chromosomes are separated from each other (p. 127).

- During meiosis II, the sister chromatids are separated from each other (p. 127).

- Homologous pairs of chromosomes exchange genetic information during crossing over at prophase I of meiosis, thereby increasing the number of genetically distinct gametes that an individual can produce (p. 130).

- The alignment of members of a homologous pair at metaphase I is random with regard to which member of a pair faces which pole. This random alignment of homologous chromosomes increases the number of different kinds of gametes an individual can produce (p. 131).

Web Tutorial 5.4 Meiosis

Learning the Basics

1. List two types of genes that, when mutated, increase the likelihood of cancer occurring.

2. What property of cancer cells do chemotherapeutic agents attempt to exploit?

3. A cell that begins mitosis with 46 chromosomes produces daughter cells with _____.

   A. 13 chromosomes; **B.** 23 chromosomes; **C.** 26 chromosomes; **D.** 46 chromosomes

4. The centromere is a region at which _____.

   A. sister chromatids are attached to each other; **B.** metaphase chromosomes align; **C.** the tips of chromosomes are found; **D.** the nucleus is located

5. Proto-oncogenes _____.

   A. are mutant genes that a person might inherit; **B.** are normal genes that encode cell-cycle control proteins;

   C. can become oncogenes if a person smokes cigarettes; **D.** are proteins that fail to suppress tumor formation; **E.** B and C are true

6. At metaphase of mitosis, _____.

   A. the chromosomes are condensed and found at the poles; **B.** the chromosomes are composed of 1 sister chromatid; **C.** cytokinesis begins; **D.** the chromosomes are composed of 2 sister chromatids and are lined up along the equator of the cell

7. Sister chromatids _____.

   A. are 2 different chromosomes attached to each other; **B.** are exact copies of 1 chromosome that are attached to each other; **C.** arise from the centrioles; **D.** are broken down by mitosis; **E.** are chromosomes that carry different genes

8. DNA polymerase _____.

   A. attaches sister chromatids at the centromere; **B.** synthesizes daughter DNA molecules from fats and phospholipids; **C.** is the enzyme that facilitates DNA synthesis; **D.** causes cancer cells to stop dividing

9. After telophase I of meiosis, each daughter cell is _____.

   A. diploid, and the chromosomes are composed of 1 double-stranded DNA molecule; **B.** diploid, and the chromosomes are composed of 2 sister chromatids; **C.** haploid, and the chromosomes are composed of 1 double-stranded DNA molecule; **D.** haploid, and the chromosomes are composed of 2 sister chromatids

10. State whether the chromosomes depicted in parts (a)–(d) of Figure 5.29 are haploid or diploid.

Figure 5.29   Haploid or diploid chromosomes?

Analyzing and Applying the Basics

1. Would a skin cell mutation that your father obtained from using tanning beds make you more likely to get cancer? Why or why not?

2. Assume that you have graduated and become a pharmaceutical researcher attempting to treat cancer. If you could design a drug to disrupt any biological process you chose, what would you try to disrupt and why?

3. Why are some cancers treated with radiation therapy while others are treated with chemotherapy?

Connecting the Science

1. Should members of society be forced to pay the medical bills of smokers when the cancer risk from smoking is so evident and publicized? Explain your reasoning.

2. Are there changes you could make in your own life to decrease your odds of getting cancer?

CHAPTER 6

Are You Only As Smart As Your Genes?

Mendelian and Quantitative Genetics

Can a woman create a more perfect child if she chooses the right sperm?

If a woman chooses a donor with the right genes, will her child look like her partner?

If she chooses a donor with the right genes, will her child be a genius?

Or is a child's intelligence more influenced by his environment?

The Fairfax Cryobank is a nondescript brick building located in a quiet, tree-lined suburb of Washington, DC. Stored inside this unremarkable edifice are the hopes and dreams of thousands of women and their partners. The Fairfax Cryobank is a sperm bank; inside its many freezers are vials containing sperm collected from hundreds of men. Women can order these sperm for a procedure called artificial insemination, which may allow them to conceive a child despite the lack of a fertile male partner.

Women who purchase sperm from the Fairfax Cryobank can choose from hundreds of potential donors. The donors are categorized into three classes, and their sperm is priced accordingly. Most women who choose artificial insemination want detailed information about the donor before they purchase a sample; while all Fairfax Cryobank donors submit to comprehensive physical exams and disease testing and provide a detailed family health history, not all provide childhood pictures, audio CDs of their voices, or personal essays. Sperm samples from men who did not provide this additional information are sold at a discount because most women seek a donor who seems compatible in interests and aptitudes. Some women even hope that by choosing a donor who is very similar to their current partner, they will have a child who appears to be the biological son or daughter of this partner.

However, in addition to the information-rich donors, there is also a set of premium sperm donors referred to by Fairfax Cryobank as its "Doctorate" category. These men either are in the process of earning or have completed a doctorate degree. Sperm from this category of donor is 30% more expensive than sperm from the standard donor.

Why would some women be willing to pay significantly more for sperm from a donor who has an advanced degree? Because academic achievement is associated with intelligence. These women want intelligent children, and they are willing to pay more to provide their offspring with "extra smart" genes. But are these women putting their money in the right place? Is intelligence about genes, or is it a function of the environmental conditions in which a baby is raised? In other words, is who we are a result of our "nature" or our "nurture"? As you read this chapter, you will see that the answer to this question is not a simple one—our characteristics come both from our biological inheritance and the environment in which we developed.

6.1 The Inheritance of Traits

Most of us recognize similarities between our birth parents and ourselves. Family members also display resemblances—for instance, all the children of a single set of parents may have dimples. However, it is usually quite easy to tell siblings apart. Each child of a set of parents is unique, and none of us is simply the "average" of our parents' traits. We are each more of a combination—one child may be similar to her mother in eye color and face shape, another similar to mom in height and hair color.

To understand how your parents' traits were passed to you and your siblings, you need a basic understanding of the human life cycle. A **life cycle** is a description of the growth and reproduction of an individual. Figure 6.1 is a diagram of a very simplified human life cycle. Notice that a

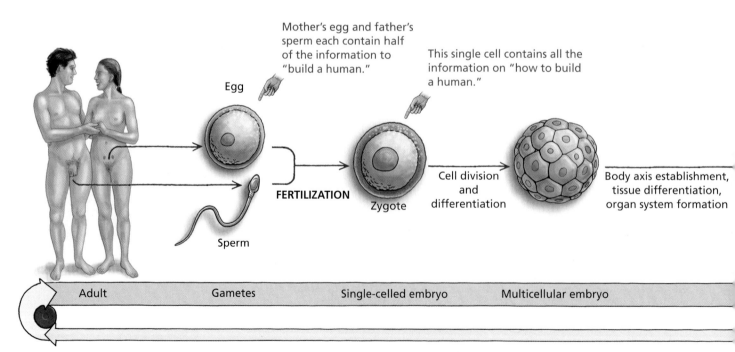

Figure 6.1   The human life cycle. A human baby forms from the fusion of an egg cell from its mother and a sperm cell from its father. The single cell that results from this fusion will grow and divide into trillions of cells, each carrying the same information.

human baby is typically produced from the fusion of a single **sperm** cell produced by the male parent and a single **egg** cell produced by the female parent. After the egg and sperm, or **gametes**, fuse at **fertilization**, the resulting single, fertilized cell (the **zygote**) duplicates all of its contents and undergoes a type of cell division called mitosis (described in Chapter 5) to produce 2 identical daughter cells. Each of these daughter cells divides dozens of times in the same way. The cells in this resulting mass then differentiate into specialized cell types, which continue to divide and organize to produce the various structures of a developing human, called an embryo. Continued division of this same cell and its progeny leads to the production of a full-term infant, and eventually an adult. We are made up of trillions of individual cells, all of them the descendants of that first product of fertilization, and nearly all containing exactly the same information originally found in the zygote.

Genes and Chromosomes

Each normal sperm and egg contains information about "how to build an organism." A large portion of that information is in the form of genes—segments of DNA that contain specific pieces of information about the traits of an organism.

Genes Are Instructions for Making Proteins.   Genes carry instructions about how to make proteins. These proteins may be either structural (like the protein that makes up hair) or functional (like the protein lactase, which breaks down milk sugar). Proteins give cells—and by extension, organs and individuals—nearly all of their characteristics. In Chapters 5, 7, and 8, we explore the physical nature of genes in more detail; here we will use an analogy to describe how genes function in the inheritance of traits.

Imagine genes as being roughly equivalent to the words used in an instruction manual. Words can have one meaning when they are alone (for instance, *saw*) and another meaning when used in combination with other

Birth

Cell division and differentiation

Cell division and differentiation

Cell division and differentiation

Fetus   Baby   Child   Adult

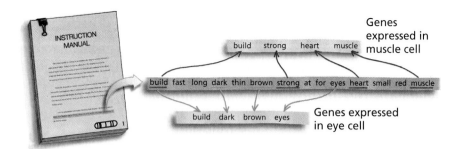

Figure 6.2   Genes as words in an instruction manual. Different words from the manual are used in different parts of the body, and even when the same words are used, they are often used in distinctive combinations. In this way, the manual can provide instructions for making and operating the variety of body parts we possess.

words (*see-saw*). Words can even change meaning in different contexts ("saw the wood" versus "sharpen the saw"). Some words are repeated frequently in any set of directions, but other words are not. It is the presence of certain words and their combination with other words that determines which instruction is given.

Similarly, all cells have the same genes, but it is the timing and combining of these genes that determines the activities of a particular cell. For instance, eye cells and heart cells in mammals both carry instructions for the protein rhodopsin, which helps detect light, but rhodopsin is produced only in eye cells, not in heart cells. Rhodopsin requires assistance from another protein to translate the light it senses into the actions of the eye cell. This other protein may also be produced in heart cells, but there it is combined with a third protein to help coordinate contractions of the heart muscle. Thus, a protein may serve two different functions depending on its context. Because genes, like words, can be used in many combinations, the instruction manual for building a living organism is very flexible (Figure 6.2).

Genes Are on Chromosomes.   In all organisms, genes are carried on chromosomes, structures made up of DNA and proteins. Prokaryotes such as bacteria typically contain a single, circular chromosome that floats freely inside the cell and is passed in its entirety to each offspring. In contrast, eukaryotes, which house their genetic information in a separate cellular compartment called a nucleus, carry their genes on more than one linear chromosome. The number of chromosome pairs in eukaryotes can vary greatly, from a single pair in the jumper ant (*Myrmecia pilosula*) to an incredible 630 in the stalked adder's tongue, a species of fern (*Ophioglossum reticulatum*).

When passing information to their offspring during sexual reproduction, eukaryotic parents each contribute comparable information. Using our earlier analogy, both parents give a complete instruction manual to their offspring. In humans, this information is contained on 23 different chromosomes, and each parent contributes 1 copy of each of these chromosomes. Each set of 2 equivalent chromosomes from both parents is referred to as a **homologous pair**. Thus humans have 23 homologous pairs in each typical body cell—one of each pair is from mom, and its partner is from dad. In our analogy, each chromosome is equivalent to a page in the instruction manual, and two complete instruction manuals, one from each parent, are required to build a normal human baby (Figure 6.3a).

Genetic Variation Is Caused by Mutation.   The information contained on one member of a homologous pair is equivalent to the information on the other member of the pair, but it is not identical. You can imagine how a page of instructions might change over many generations if each parent had to type a copy for each offspring. Typographical errors made by a father would be passed on in the manual he gave to his son. If the son made additional typographical errors, the instruction manual he passed on to his own children would contain the changes made by his parents plus those he made himself. Because these typographical errors are random and not expected to occur in the same words in various individuals, different families should have slightly differing versions of certain words in the manual.

The analogy can be extended to real chromosomes—a sort of "typographical error" in genes, called a **mutation**, can occur when chromosomes are copied during the production of sperm or eggs. Over time, and as a result of many dissimilar mutations passed on in various families, a population of organisms will contain many different versions of a single gene.

(a) Both parents give a complete instruction manual to their offspring.

The 23 pages of each instruction manual are roughly equivalent to the 23 chromosomes in each egg and sperm.

The zygote has 46 pages, equivalent to 46 chromosomes.

(b) Mutations are errors in copying the instructions.

| Normal allele: | grey | strong | nerve |
| Mutant allele: | gray | string | nzrve |

The mutant allele has the same meaning (mutant allele functions the same as the original allele).

The mutant allele has a different meaning (mutant allele functions differently than the original allele).

The mutant allele has no meaning (mutant allele is no longer functional).

Figure 6.3   The formation of different alleles. (a) Each parent provides a complete set of instructions to each offspring. (b) The instructions are first copied, and different alleles for a gene may form as a result of copying errors. Some of these misspellings do not change the meaning of the word, but some may result in altered meanings or have no meaning at all.

These diverse versions are called **alleles**. Various alleles for a single gene may produce the same basic outputs, or their outputs could be very diverse. A mutation often results in a nonfunctional allele, or in our analogy, a nonsensical word. However, some mutations result in alleles with brand-new functions, equivalent to a brand-new word. It is these mutations that lead to individuals with novel characteristics (Figure 6.3b). As we will discuss in Chapter 10, when a novel characteristic increases an individual's chance of survival and reproduction, the mutation contributes to a population's adaptation to its environment. Genetic misspellings are thus the engine that drives evolution itself.

Diversity in Offspring: Segregation, Independent Assortment, Crossing Over, and Random Fertilization

It is the combination of alleles from both parents that helps determine what traits an individual has. The environment in which the alleles are used also plays a role. Just as two people reading the same set of instructions may build slightly different dollhouses, embryos developing in slightly different environments may vary. One part of the reason non-twin offspring are dissimilar is that each twin developed in dissimilar conditions. These conditions could include their mother's nutrition during pregnancy, the presence of toxic compounds in her environment, or even the number of other siblings in the family at the time of birth. Although differences in the environment of development can lead to differences between siblings, the primary reason non-twin siblings are not identical is that their parents did not give all of their offspring exactly the same set of alleles.

Segregation and Independent Assortment Create Gamete Diversity.
Gametes develop from cells that have 2 copies of each chromosome; however, each gamete carries only 1 copy. Meiosis is the type of cell division that reduces the number of chromosomes in daughter cells in preparation for making gametes. In Chapter 5 we described the process of meiosis in detail. However, the instruction manual analogy can help us understand how meiosis caused the egg and sperm that produced one offspring to be different from the egg and sperm that produced its sibling.

Cells that produce eggs and sperm are like other body cells—they carry homologous pairs of each chromosome, or in our analogy, 2 copies of each page of the instruction manual. When a cell undergoes meiosis, it places 1 chromosome (1 copy of every page) into each of 2 daughter cells. The separation of pairs of alleles (instruction manual words) during the production of gametes is called the **law of segregation**. Because the words are carried on pages, we now know that the law of segregation results from the separation of homologous chromosome pairs during meiosis.

Early geneticists recognized that many of the traits they investigated were inherited independently from each other—a fact termed the **law of independent assortment**. We now understand this observation as also resulting from the process of meiosis. Each pair of pages (homologous chromosome pair) is segregated into daughter cells independently of all the other page pairs during the production of gametes. When traits are located on different chromosomes, genes do assort independently. As we discuss in Chapter 7, genes located on the same chromosome do not assort independently.

Due to independent assortment, the instruction manual contained in a man's sperm is made up of a unique combination of pages from the manuals he received from each of his parents. In fact, almost every sperm he makes

Precursor cells to sperm and eggs have 2 copies of each chromosome — that is, 2 full sets of instruction manual pages, 1 from each parent.

Sperm and egg cells have only 1 full set — a random combination of maternal and paternal instruction manual pages.

In humans, over 8 million combinations are possible.

Possible sperm cell 1

Possible sperm cell 2

Possible sperm cell 3

| Page 9 Blood-group gene from **dad** | Page 15 Eye-color genes from **mom** | Page 9 Blood-group gene from **mom** | Page 15 Eye-color genes from **mom** | Page 9 Blood-group gene from **mom** | Page 15 Eye-color genes from **dad** |

Figure 6.4   Each egg and sperm is unique. This figure uses the instruction manual analogy to show how a single man can produce an enormous diversity of sperm. The same process of independent assortment results in enormous diversity in eggs as well. The cells that are the source of a man's sperm carry 2 of each chromosome—that is, 2 full copies of the instruction manual—1 set from his mom, the other from his dad. When a sperm cell is produced, it ends up with only 1 copy of each page. Since each sperm is produced independently, the set of pages in each sperm will be a unique combination of the pages that the man inherited from his mom and dad.

will contain a unique subset of chromosomes—and thus a unique subset of his alleles. In Figure 6.4, you can see that independent assortment causes an allele for an eye-color gene to end up in a sperm cell independently from an allele for the blood-group gene. Since the process of independently placing chromosomes into daughter cells is repeated every time a sperm is produced, the set of alleles that each child receives from a father is different for all of his offspring. The sperm that contributed half of your own genetic information might have carried the eye-color allele from your father's mom and the blood-group allele from his dad, while the sperm that produced your sister might have contained both the allele for eye color and the allele for blood group from your paternal grandmother. Because chromosomes line up independently during every meiotic division, only about 50% of an individual's alleles are identical to those found in another offspring of the same parents—that is, for each gene, you have a 50% chance of being like your sister or brother.

Another analogy may help you visualize how independent assortment generates diversity in offspring. A pair of shoes can be likened to a homologous pair of chromosomes; the two shoes are similar in size, shape,

Figure 6.5   Independent assortment of shoes. A pair of shoes is comparable to a homologous pair of chromosomes. Meiosis separates the members of one pair independently of other pairs—much like what would happen if students created two piles of shoes by each independently separating and sorting their own pair. Here you can see how two different piles of shoes could be created if one student put her right sneaker in the "left shoe" pile, and vice versa.

and style but are not exact duplicates since they fit left and right feet (Figure 6.5). If you ask 23 students to take off their shoes and place them in a row across the front of the classroom, and they arrange their shoes so that the left shoe is on the left, and the right shoe is on the right, the students could then separate all of the left shoes from the right shoes, just as meiosis separates homologous chromosomes. This would produce two different piles—one containing all left shoes, the other containing all right shoes. Each of these piles is equivalent to the set of chromosomes contained within a single human gamete. Different piles of shoes would result if the very first pair of shoes was reversed so that the left shoe and right shoe exchange places, but the other 22 pairs of shoes stayed as they were. When the shoes are separated this time, one pile would have 22 right shoes and 1 left shoe, and the other pile would have 22 left shoes and 1 right shoe. These piles represent two different gametes that could be produced by an individual parent. The students could continue making different combinations of left and right shoes for a very long time because there are 2^{23} (over 8 million) possible ways to line up these 23 pairs of shoes. The same is true of chromosomes; due to independent assortment, each individual human can make at least 8 million different types of egg and sperm.

Crossing Over Leads to Novel Chromosomes.   In addition to independent assortment, there is another event during meiosis that increases the diversity of eggs and sperm. This process is called crossing over, and it is described in detail in Chapter 5. Crossing over occurs when chromosome pairs "swap" information.

In our instruction manual analogy, crossing over is equivalent to tearing a pair of pages in half and reassembling them so that the top part of the page is from one instruction manual and the bottom part is from the other instruction manual (Figure 6.6). After crossing over, the pages represent a novel combination of instructions, consisting of "family misspellings" from both grandparents. In other words, each chromosome that results from crossing over consists of a combination of alleles that, in all likelihood, have never been found together before. The processes of independent assortment and crossing over create almost limitless variation in eggs or sperm from a single parent.

Random Fertilization Results in a Large Variety of Potential Offspring.
Consider that *each* of your parents was able to produce such an enormous diversity of gametes. Theoretically, any sperm produced by your father had an equal chance of fertilizing any egg produced by your mother—a process known as **random fertilization**. In other words, gametes combine without regard to the alleles they carry. Hence, even without considering the diversity introduced by crossing over, the odds of you receiving your particular combination of chromosomes are 1 in 8 million × 1 in 8 million—or 1 in 64 trillion. Remarkably, together your parents could have made over 64 trillion genetically different children, and you are only one of the possibilities. Mutation creates new alleles, and independent assortment, crossing over, and random fertilization result in unique combinations of alleles every generation. These processes produce the

(a)

(b)

Paired chromosomes at the beginning of meisois are analogous to pairs of instruction manual pages.

Chromosomes after crossing over contain information from maternal and paternal chromosomes.

Figure 6.6   Crossing over increases diversity in gametes. When chromosomes pair at the beginning of meiosis, information may be exchanged via the process of crossing over. (a) In our instruction manual analogy, this means that meiosis can result in the formation of completely new chromosomes. (b) A photomicrograph of chromosomes in the process of crossing over during prophase of meiosis I.

diversity of individuals found not only in humans, but in all sexually reproducing biological populations.

Finally, multiply the incredible number of genetically different offspring that can be produced by a single set of parents by the differences in the environment during the development of any single offspring. You can see that the likelihood that two people are exactly alike, or in fact, the likelihood that two individuals of any sexually reproducing species are exactly alike, is virtually zero.

A Special Case—Identical Twins.   Although the process of sexual reproduction can produce two siblings that are very different from each other, an event that occurs after fertilization in some mothers can result in the birth of two children who share 100% of their genes. These children are called identical twins and are the result of a single fertilization event—the fusion of one egg with one sperm.

Nonidentical twins, or fraternal twins, are called **dizygotic** (*di* means "two"), and they occur when two separate eggs fuse with different sperm. The resulting embryos, which develop together, are genetically no more similar than siblings born at different times (Figure 6.7a). However, identical twins are referred to as **monozygotic twins** (*mono* means "one") because they develop from a single egg and sperm. Recall that after fertilization the fertilized egg cell (the zygote) grows and divides, producing an embryo made up of many daughter cells containing the same genetic information. Monozygotic twinning occurs when cells in an embryo separate from each other. If this happens early in development, each cell or clump of cells can develop into a complete individual, yielding twins who carry identical genetic information (Figure 6.7b). In humans, about 1 in every 80 pregnancies produces dizygotic twins, while only approximately 1 of every 285 pregnancies results in identical twins.

Identical twins provide a unique opportunity to study the relative effects of our genes and environment in determining who we are. Because identical twins carry the same genetic information, researchers are able to study how important genes are in determining these twins' health, tastes, intelligence, and personality. We will explore the use of twin studies in the next section.

Figure 6.7   The formation of twins.
(a) Dizygotic twins form when 2 different eggs combine with 2 different sperm cells, resulting in 2 embryos who are only as similar as siblings. (b) Monozygotic twins form when a newly fertilized embryo splits in half resulting in 2 identical embryos.

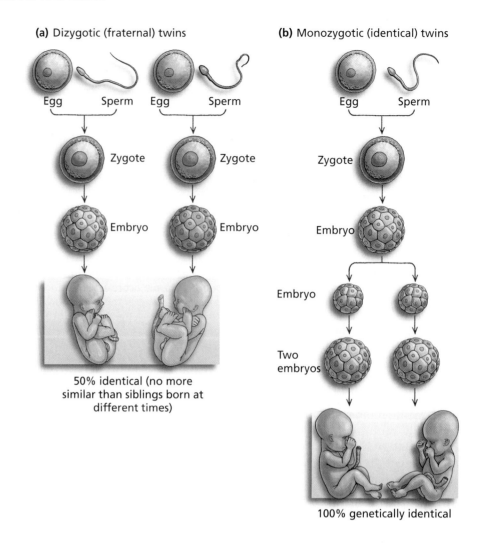

(a) Dizygotic (fraternal) twins

(b) Monozygotic (identical) twins

Egg   Sperm   Egg   Sperm

Egg   Sperm

Zygote   Zygote

Zygote

Embryo   Embryo

Embryo

Embryo

Two embryos

50% identical (no more similar than siblings born at different times)

100% genetically identical

6.2   Mendelian Genetics: When the Role of Genes Is Clear

A few human genetic traits have easily identifiable patterns of inheritance. These traits are said to be "Mendelian" because Gregor Mendel (Essay 6.1) was the first person to accurately describe their inheritance. Mendel developed his understanding of this type of inheritance by studying pea plants, which are easy to grow, are able to produce seeds within only a few months of germinating, and have the potential to make thousands of offspring. Additionally, reproduction in pea plants is easily controlled by selecting which plants fertilize others. Although Mendel himself did not understand the chemical nature of genes, he was able to determine how traits were inherited by carefully analyzing the appearance of parent pea plants and their offspring. The pattern of inheritance Mendel described occurs primarily in traits that are the result of a single gene with a few distinct alleles.

Although Mendel's work forms the basis for genetics, later observations indicated that not all traits affected by a single gene display the straightforward pattern of inheritance he described. Many of the Mendelian traits identified in humans are the result of genes with mutant alleles that result in some type of disease or dysfunction. In our instruction manual analogy, an allele that is dysfunctional is equivalent to a misspelled, nonsensical word on a page.

Essay 6.1 Gregor Mendel

The scientist who first helped to explain how genes are inherited was Johann Gregor Mendel (Figure E6.1a). Due to his groundbreaking experiments with genetic acrosses among pea plants, traits that are inherited, according to the rules elucidated by Mendel's experiments, are said to follow a Mendelian pattern of inheritance.

Mendel was born in Austria in 1822. Because his family was poor, he entered a monastery to obtain an education. After completing his monastic studies, Mendel attended the University of Vienna; he studied math and botany there in addition to other sciences. Mendel attempted to become an accredited teacher but was unable to pass the examinations. After leaving the university, he returned to the monastery and began his experimental studies of inheritance in garden peas.

Mendel studied close to 30,000 pea plants over a 10-year period. His careful experiments consisted of controlled matings between plants with different traits. Mendel was able to control the types of mating that occurred by hand-pollinating the peas' flowers—for example, by applying pollen, which produce sperm, from a tall pea plant to the carpel (or egg-containing structure) of a short pea plant and then growing the seeds resulting from that cross. By doing this, he could evaluate the role of each parent in producing the traits of the offspring (Figure E6.1b). Mendel published the results of his studies in 1865, but his contemporaries did not fully appreciate the significance of his work. This was partly because biologists were still wrestling with the implications of Charles Darwin's theory of evolution, published just 6 years before Mendel's paper.

Prior to Mendel's discoveries, many scientists believed that the traits of a child's parents were blended together, producing a child with characteristics intermediate to those of both parents. In fact, this was Darwin's belief, and it was

Figure E6.1b Pea plants

a major barrier to general acceptance of his ideas on natural selection and evolution. Some scientists, the "spermists," believed that sperm contained a completely formed miniature adult, known as a homunculus, which they imagined they saw when examining sperm under the microscope (Figure E6.1c). Other scientists, the "ovists," believed that the egg cell contained the homunculus. Some even believed that children would resemble the parent who initiated the intercourse! In the climate of these seemingly sensational hypotheses, Mendel's patient, scientifically sound experiments showing that both parents contribute equal amounts of genetic information to their offspring went largely unnoticed.

Mendel eventually gave up his genetic studies and focused his attention on running the monastery until his death in 1884. His work was independently rediscovered by three scientists in 1900; only then did its significance to the new science of genetics become apparent.

Figure E6.1a Gregor Mendel

Figure E6.1c Homunculus

Where two copies of each word are available, the correctly spelled word can usually substitute for its misspelled partner without changing the meaning of the instructions. Likewise, since we have two copies of nearly every gene, carrying one dysfunctional allele for a gene is usually not a serious problem because the functional allele acts as a backup for its mutated partner within a cell.

Genotype and Phenotype

We call the genetic composition of an individual his **genotype** and his physical traits his **phenotype**. An individual who carries two different alleles for a gene has a **heterozygous** genotype (*hetero*, meaning "different," and *zygous*, meaning "origin"). An individual who carries two copies of the same allele has a **homozygous** genotype (*homo*, meaning "same").

The effect of an individual's genotype on her phenotype depends on the nature of the alleles she carries (Figure 6.8). Some alleles are **recessive**, meaning that their effects can be seen only if a copy of a dominant allele (described later) is not also present. Mutations resulting in alleles that cannot produce functional proteins are often recessive. With these types of mutations, a heterozygous individual carrying 2 different alleles—one a normal allele and the other a recessive, nonfunctional allele—has a normal phenotype because the instruction produced by the normal allele substitutes for its nonfunctional partner. For example, the gene that gives a plant root cell instructions for making a surface "hair" to allow for water absorption has a nonfunctional allele that is found in some plant populations. In homozygous individuals that carry 2 copies of this recessive allele, the root hair is not produced, and the plant is a poor absorber of water. In heterozygote plants that carry 1 copy of this allele and 1 copy of the normally functioning allele, the roots have the normal amount of hair and absorb water well.

Sometimes mutations can result in alleles that code for instructions having powerful effects—ones that essentially mask the effects of the nonmutant allele. These types of alleles are termed **dominant** because their effects are seen even when a nondominant allele is present. "American Albino" horses—known for their snow-white coats, tails, and manes, pink skin, and dark eyes—result from a dominant allele that stops a horse's hair-color genes from being expressed during its development. A white coat will be produced if the horse is heterozygous and thus carries only 1 copy of this dominant allele. Even though these horses also possess a nonmutant allele that does not suppress coat-color genes, the suppression still occurs, and the phenotype of the dominant allele is expressed.

For some genes, two identical copies of a dominant allele are required for expression of the full effect of a phenotype. In this case, the phenotype of the heterozygote is intermediate between both homozygotes—a situation called **incomplete dominance**. For example, the alleles that determine flower color in snapdragons are incompletely dominant: one homozygote produces red flowers, the other, presumably carrying two non-functional copies of a color gene, produces white flowers, and the heterozygote, carrying 1 "red" and 1 "white" allele, produces pink flowers.

Other mutations result in a new protein with a different, but not dominant, activity compared to the normal protein. Alternatively, the phenotype of a heterozygote where neither allele is dominant to the other may be a combination of both

Figure 6.8   Genotypes and phenotypes. When 2 alleles for a gene exist in a population, there are 3 possible genotypes and 2 or 3 possible phenotypes for the trait.

fully expressed traits. This situation is known as **codominance.** In cattle, for example, the allele that codes for red hair color and the allele that codes for white hair color are both expressed in a heterozygote. These individuals have patchy coats that consist of an approximately equal mixture of white hairs and red hairs. Incomplete dominance and codominance are further explored in Chapter 7.

Different alleles for a gene are symbolized with letters or number codes that reference a trait the gene affects. Traditionally, for genes with 1 dominant and 1 recessive allele, a capital letter refers to the dominant form, and the same letter in lowercase refers to the recessive allele. For example, in pea plants, height is determined by a gene with at least 2 alleles—a dominant allele that leads to tall plants, and a recessive allele that produces short plants. Thus the symbol *T* signifies the "tall" allele, and the symbol *t* signifies the "short" allele. We use this convention for the remainder of this chapter.

Genetic Diseases in Humans

Scientists have identified genetic diseases in humans that are produced by recessive alleles, dominant alleles, and codominant alleles. We should note, however, that most alleles in humans do not cause disease or dysfunction; they are simply alternative versions of genes. The diversity of alleles in the human population contributes to diversity among us in our appearance, physiology, and behaviors.

Cystic Fibrosis Is a Recessive Condition.   Cystic fibrosis occurs in individuals with 2 copies of an allele that directs the production of a nonfunctional protein—in particular, a protein that when functional helps transport the chloride ion into and out of cells lining the lungs, intestines, and other organs. When this ion transporter is not functional, the balance between sodium and chloride in the cell is disrupted, and it produces a thick, sticky mucus layer instead of the thin, slick mucus produced by cells with the normal allele. Affected individuals suffer from progressive deterioration of their lungs and have difficulties absorbing nutrients across the lining of their intestines. Most children born with cystic fibrosis do not live past their 30s. People who carry 1 copy of the nonfunctional allele and 1 copy of the normal allele are not affected because the nonmutant protein can still function effectively as a chloride transporter. Cystic fibrosis is among the most common genetic diseases in European populations; nearly 1 in 2500 individuals in these populations is affected with the disease, and 1 in 25 is heterozygous for the allele. Sperm banks can test donor sperm for several recessive disorders, including cystic fibrosis. The Fairfax Cryobank tests donor semen for the presence of cystic fibrosis alleles; any men who carry the allele are excluded from their donor program.

Huntington's Disease Is Caused by a Dominant Allele.   Huntington's disease is an example of a fatal genetic condition caused by a dominant allele. Early symptoms of Huntington's disease include restlessness, irritability, and difficulty in walking, thinking, and remembering. The disease is progressive and incurable—the nervous, mental, and muscular symptoms gradually become worse and eventually result in the death of the affected individual. The Huntington's allele causes production of a protein that clumps up inside the nuclei of cells. Nerve cells in certain areas of the brain are especially likely to contain these protein clumps, and these cells gradually die off over the course of the disease. Because this allele produces a protein that behaves abnormally, an individual needs only 1 copy of the allele to be affected by the disease— even heterozygotes have the disease. The symptoms of Huntington's disease usually do not manifest themselves until middle age, so people who carry this allele can unknowingly pass it on to their own children. Only since the mid-1980s has genetic testing allowed people with a family history of Huntington's disease to learn whether they are affected before they show signs of the

Figure 6.9   The effect of the sickle-cell allele. The sickle-cell anemia allele causes red blood cells to look sickled as seen here, on the right. At left, a normal red blood cell appears round and pillow-like.

disease. Although most sperm banks do not test for the presence of the Huntington's allele, the detailed family medical histories required of sperm bank donors enable Fairfax Cryobank to exclude men with a family history of Huntington's disease from their donor list.

The Sickle-cell Allele Is Codominant.   The allele that causes the disease sickle-cell anemia is an example of a codominant mutation that results in an intermediate phenotype in heterozygotes. Sickle-cell anemia is a condition that occurs when an individual carries 2 mutant copies of the gene required to produce normal red blood cells. Red blood cells contain hemoglobin, a protein that carries oxygen in the bloodstream. The mutation that causes sickle-cell anemia results in hemoglobin that functions differently from hemoglobin produced by someone with a normal allele. The abnormal hemoglobin forms rigid structures when red blood cells are low in oxygen (such as during exercise), causing the cells to deform into a sickle shape, in a process called sickling (Figure 6.9). These cells can become lodged in narrow blood vessels, further blocking the flow of oxygen to tissues. Individuals carrying 2 copies of the sickle-cell allele are prone to painful and debilitating sickling attacks and often die at an early age. Individuals who have 1 copy of the sickle-cell allele and 1 copy of the normal allele make both normal and abnormal hemoglobin. Thus, some of their blood cells may sickle, but not to the degree seen in homozygous individuals.

Heterozygotes with 1 copy of the sickle-cell allele are generally much less seriously ill than are homozygotes for the sickle-cell allele. In some situations, heterozygotes are also healthier than individuals who are homozygous for the normal hemoglobin allele. Surprisingly, people with 1 copy of the sickle-cell allele are resistant to malaria. Among populations in sub-Saharan Africa, Northern India, and around the Mediterranean Sea—where malaria is common—heterozygotes for sickle-cell anemia have higher rates of survival than do homozygotes of either type. The malaria parasite is unable to reproduce within cells containing the abnormal hemoglobin; thus, infection with the parasite does not cause severe malarial disease in heterozygotes. Because heterozygotes have higher rates of survival, they contribute more children to the next generation than homozygotes do. Since 50% of their children will carry the sickle-cell allele, the allele has stayed common in these populations. Thus, the advantage of the sickle-cell allele in a high-malaria environment results in a human population that has evolved a high frequency of this allele. The sickle-cell allele is much more common in these populations than we would expect if its only effect was to cause disease. Sperm banks routinely test donor sperm for the presence of the sickle-cell allele and exclude men who carry it from their donor list.

Numerous diseases caused by a single mutation exist in the human population. However, many genetic diseases do not have an easily performed test that would allow sperm banks to exclude men who carry these diseases, and no sperm bank tests donors for all possible mutations—only the most common and devastating ones. Tests performed by the scientists at Fairfax Cryobank greatly reduce the risk that the sperm they sell will carry dangerous genetic mutations, but the tests do not eliminate the possibility that a child produced by these sperm will have a genetic disease. Of course, the same is true of any source of sperm—either from a cryobank or from a male partner.

Using Punnett Squares to Predict Offspring Genotypes

Traits such as cystic fibrosis, Huntington's disease, and sickle-cell anemia are the result of a mutation in a single gene, and the inheritance of these conditions and of other single-gene traits is relatively easy to understand. We can

Figure 6.10   What are the risks of accepting sperm from a carrier of cystic fibrosis? This Punnett square helps determine the likelihood that a woman who carries the cystic fibrosis allele would have a child with cystic fibrosis if her sperm donor was also a carrier. With a 25% chance of producing an affected child, most women would consider sperm from a carrier unacceptable, which is why sperm banks do not accept sperm from carrier males.

follow the inheritance of small numbers of genes by using a tool developed by the British geneticist Reginald Punnett. A **Punnett square** is a table that lists the different kinds of sperm or eggs parents can produce relative to the gene or genes in question and then predicts the possible outcomes of a **cross**, or mating, between these parents (Figure 6.10).

Using a Punnett Square with a Single Gene.   Imagine a couple in which both members are heterozygotes for cystic fibrosis. If we use the letters *F* and *f* to represent the dominant functional and recessive nonfunctional allele, respectively, a heterozygote would have the genotype *Ff*. The genetic cross could then be symbolized:

$$Ff \times Ff$$

We know that the female in this cross can produce eggs that carry either the *F* or *f* allele since the process of egg production will separate the 2 alleles from each other. We place these 2 egg types across the horizontal axis of what will become a Punnett square. The male in this cross is also heterozygous, so he too can make sperm containing either the *F* or *f* allele. We place the kinds of sperm he can produce along the vertical axis. Thus, the letters on the horizontal and vertical axes represent all the possible types of eggs and sperm that the mother and father can produce by meiosis, if we consider only the gene that codes for the chloride transport protein.

Inside the Punnett square are all the genotypes that can be produced from a cross between these two heterozygous individuals. The content of each box is determined by combining the alleles from the egg column and the sperm row. Note that for a single gene with 2 alleles, there are 3 possible offspring types. The chance of this couple having a child affected with cystic fibrosis is 1 in 4, or 25%, because the *ff* combination of alleles occurs once out of the four possible outcomes. The *FF* genotype is represented once out of four times, meaning that the probability of this couple having a homozygous unaffected child is also 25%. The probability of the couple producing a child who is a **carrier** of cystic fibrosis (that is, heterozygous, or *Ff*) is 1 in 2, or 50%, since two of the possible outcomes inside the Punnett square are unaffected heterozygotes—one produced by an *F* sperm and an *f* egg, and the other produced by an *f* sperm and an *F* egg.

Figure 6.11 Calculating the likelihood of genetic traits in children. (a) This Punnett square illustrates the outcome of a cross between a man who carries a single copy of the dominant Huntington's disease allele and an unaffected woman. (b) The outcome of a cross between two individuals with the sickle-cell trait. The two alleles are codominant—allele *S* codes for the normal blood protein, and allele *s* codes for the "sickling" blood protein.

When parents know which alleles they carry for a single-gene trait, they can easily determine the probability that a child they produce will have the disease phenotype (Figure 6.11). You should note that this probability is generated independently for each child—in other words, every offspring of two carriers has a 25% chance of being affected.

Using Punnett Squares with Multiple Genes. Punnett squares can also be employed to predict the likelihood of a particular genotype when considering multiple genes. For example, in Mendel's peas, seed color and seed shape are each determined by a single gene, and each are carried on different chromosomes. The two seed-color gene alleles are *Y*, which is dominant and codes for yellow color, and *y*, the recessive allele, which results in green seeds when homozygous. The two seed-shape alleles are *R*, the dominant allele, which codes for a smooth, round shape; and *r*, which is recessive and codes for a wrinkled shape. Because these genes are on different chromosomes, they are placed in eggs and sperm independently of each other. In other words, a pea plant that is heterozygous for both genes (genotype *Yy Rr*) can make 4 different types of eggs: one carrying dominant alleles for both genes (*Y R*), one carrying recessive alleles for both genes (*y r*), one carrying the dominant allele for seed color and the recessive allele for seed shape (*Y r*), and one carrying the recessive allele for color and the dominant allele for shape (*y R*). A Punnett square for a cross between two individuals who are heterozygous for both genes (called a **dihybrid cross**) would contain 16 boxes; a cross between these two individuals can generate 9 different genotypes (Figure 6.12). As you might imagine, as the number of genes in a Punnett-square type analysis increases, the number of boxes in the square, and the number of possible genotypes, increases. If we follow the inheritance of 3 genes on separate chromosomes, the Punnett square has 64 boxes and 22 different possible genotypes; with 4 genes, the square has 256 boxes; and with 5 genes, there are over 1000 boxes! Predicting the

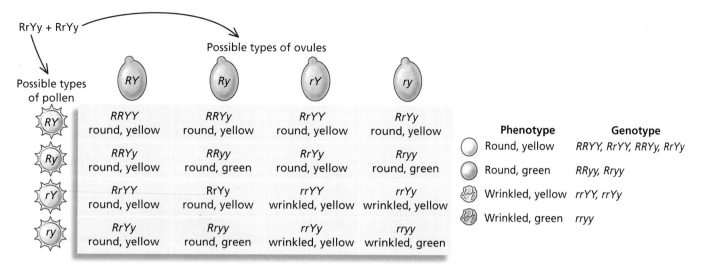

Figure 6.12   A dihybrid cross. Punnett squares can be generated to predict the outcome of a cross between individuals when we know their genotypes for more than one gene as long as those genes are on separate chromosomes. This square shows a cross between two pea plants that are heterozygous for both the seed-color and the seed-shape genes.

outcome of a cross becomes significantly more difficult as the number of genes we are following increases.

Although much of the work in identifying and developing tests for human alleles has so far concentrated on disease alleles, there are many non-disease-causing alleles in the human population which contribute to diversity among us in eye color, hair curliness, and blood type, among other features. As scientists identify more of these genes and alleles, the amount of information about the genes of sperm donors or any potential parent will also increase. Identifying and testing for particular genes in potential parents will allow us to predict the *likelihood* of numerous genotypes in their off-spring. Unfortunately, this increase in genetic testing is not necessarily equaled by an increase in our understanding of how most traits develop, as we shall see in the next section.

6.3   Quantitative Genetics: When Genes and Environment Interact

The single-gene traits discussed in the previous section have a distinct "off or on" character; individuals have either one phenotype (for example, the pea seed is yellow) or the other (the pea seed is green). Traits like this are known as qualitative traits. However, many of the traits that interest women who are choosing a sperm donor do not have this off-or-on character.

Quantitative Traits

Traits such as height, weight, eye color, musical ability, susceptibility to cancer, and intelligence are called **quantitative traits**. Quantitative traits show **continuous variation**; that is, we can see a large range of phenotypes in a population—for instance, from very short people to very tall people. Wide variation in quantitative traits leads to the great diversity we see in the human population.

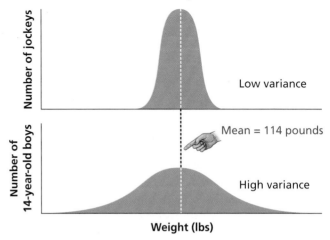

Figure 6.13   A quantitative trait. (a) This graph of the number of men in each category of height is a normal distribution with a center around the mean height of 1.73 m. (b) Fourteen-year-old boys and professional jockeys have the same average weight—approximately 114 pounds. However, to be a jockey, you must be within about 4 pounds of this average. Thus the variance among jockeys in weight is much smaller than the variance among 14-year-olds.

The distribution of phenotypes of a quantitative trait in a population can be displayed on a graph and typically takes the form of a bell-shaped curve called a normal distribution. Figure 6.13a graphs a normal distribution that represents the height of men in an Amherst College class in 1884. Each column on the graph shows the number of men measured at the height indicated along the bottom of the column. The curved line on the figure is an idealized bell-shaped curve that summarizes this data. A trait that is normally distributed in a population may be described in a number of ways. We are used to thinking about the average, or **mean**, value for data. The mean is calculated by adding all of the values for a trait in a population and dividing by the number of individuals in that population. Figure 6.13a shows that the mean height of men in this population is 1.73 meters (about 5 ft 8 in.). However, the average value of a continuously variable trait does not tell you very much about the population. Examine Figure 6.13a closely: Does an average height of 1.73 meters in this particular population imply that most men were this height? Were most men in this population close to the mean, or was there a wide range of heights?

In addition to knowing the mean value for a trait, we must understand how much variability exists in the population for the trait. The amount of variation in a population is described with a mathematical term called **variance.** If a low variance for a trait indicates a small amount of variability in the population; a high variance indicates a large amount of variability (Figure 6.13b). Scientists who study the inheritance of quantitative traits are interested in determining the genetic basis for the variance in human traits.

Why Traits Are Quantitative

One reason we may see a range of phenotypes in a human population is because numerous genotypes exist among the individuals in the population. This happens when a trait is influenced by more than one gene; traits influenced by many genes are called **polygenic traits**. As we saw above, when a single gene with two alleles determines a trait, three possible genotypes are present: *FF, Ff,* and *ff,* for example. However, as you saw in Figure 6.12, when two genes each with two alleles influence a trait, nine genotypes are possible. For example, eye color in humans is a polygenic trait influenced by at least three genes, each with more than one allele. These genes help produce and distribute the pigment melanin to the iris. People with very dark eyes have a lot of melanin in each iris, and brown eyes have a little less melanin. Blue eyes result when there is very little melanin present. When different alleles for the genes for eye-pigment production and distribution interact, a range of eye colors, from dark brown to pale blue, is produced in humans. The continuous variation in eye color among people is a result of several genes, each with several alleles, influencing the phenotype.

Continuous variation also may occur in a quantitative trait due to the influence of environmental factors. In this case, each genotype is capable of producing a range of phenotypes depending on outside influences. Thus, even if all individuals have the same genotype, many different phenotypes can result if they are raised in a variety of environments. For a clear example of the effect of the environment on phenotype, see Figure 6.14. These identical twins share 100% of their genes but are quite different in appearance. This is due to variations in their environment—the twin on the bottom smoked cigarettes and had much greater sun exposure than did the twin on top.

Most traits that show continuous variation are influenced by both genes and the effect of differing environmental factors. Skin color in humans is an example of this type of trait. The shade of an individual's skin is dependent on the amount of melanin present near the skin's surface. As discussed in Chapter 11, differences among human populations in skin color evolved because, in less sunny climates, light-skinned individuals were better able to produce vitamin D; dark-skinned individuals were better able to protect themselves from destruction of another important vitamin, folate, in sunnier climates. A number of genes have an effect on skin-color phenotype—both those that influence melanin production and those that affect the distribution of melanin in the skin. However, the environment, particularly the amount of exposure to the sun during a season or lifetime, also influences the skin color of individuals (Figure 6.15); melanin production increases, and any melanin that is present darkens in sun-exposed skin. In fact, after many years of intensive sun exposure, skin may become permanently darker.

Both genetic factors *and* environmental factors influence most traits that are of interest to women choosing sperm donors. Women choosing Doctorate-category sperm donors from Fairfax Cryobank are presumably interested in having smart, successful children; but intelligence has both a genetic and an

Figure 6.14   The effect of the environment on phenotype. These identical twins have exactly the same genotype, but they are quite different in appearance due to environmental factors.

(a) Genes

(b) Environment

Figure 6.15   Skin color is influenced by genes and environment. (a) The difference in skin color between these two women is due primarily to variations in several alleles that control skin pigment production. (b) The difference in color between the sun-protected and sun-exposed portions of the individual in this picture is entirely due to environmental effects.

environmental component. Intelligence partly depends on brain structure and function, and many alleles that interfere with brain structure and function—and thus intelligence—have been identified. But intelligence also depends on environmental factors. For example, we know that if a developing baby is exposed to high levels of cigarette smoke or alcohol before birth, its brain will develop differently, and it may have delayed or diminished intellectual development.

Using Heritability to Analyze Inheritance

Unlike qualitative traits, where the relationship between genes and traits is very clear, the inheritance of quantitative traits is difficult to understand. For instance, if the variation among individuals could be a result of many different genes, a variety of environmental effects on phenotype, or most likely, the interaction of genes and environment, then how can we predict if the child of a father with a doctorate will also be capable of earning a doctorate? Scientists most often approach this question by attempting to determine the relative importance of genes in determining the variation in phenotype among individuals.

Researchers working with domestic animals and crop plants were the first to develop the scientific model used to measure the importance of genes in determining various quantitative traits. These researchers were trying to find the best way to improve the production of domesticated animals and plants. For example, farmers who wish to increase their dairy herd's milk output can use two basic strategies: (1) Change the herd's environment by changing the way the cows are reared, housed, and fed; or (2) change the herd genetically by choosing only the offspring of the best milk producers for the next-generation herd. The technique of controlling the reproduction of individual organisms to influence the phenotype of the next generation is known as **artificial selection** (Figure 6.16). Artificial selection is similar to

Figure 6.16 Artificial selection increases milk production in cows. Cows that produce exceptional amounts of milk are bred to produce the next generation of dairy cattle. In this example, the female calves of the cow that produces 3.6 gallons of milk daily produce an average of 3.2 gallons of milk per day—23% more than the previous herd.

natural selection, which we describe more thoroughly in Chapter 10 and is a primary cause of evolution in natural populations.

If the quantitative trait of milk production in cows is strongly influenced by genes—in other words, if it has high **heritability**—then artificial selection is an effective way to boost milk output. In fact, heritability of milk production can easily be measured by how well a herd responds to artificial selection. If milk production increases in a herd of cows as a result of artificial selection, it is because alleles for proteins that increase milk production in an individual (for instance, alleles for genes that control the size of the udder or those that influence the activity of milk-producing cells) have become more common in the herd. In other words, the trait must be strongly influenced by genes. If milk production does *not* increase as a result of artificial selection, then the alleles that the high-production cows possess must not be very important in determining milk output; that is, the trait must be more strongly influenced by the environment than by genes. Scientists have calculated an average heritability of milk production in dairy cattle of 0.30. This means that, in a typical dairy herd, about 30% of the variation among cows in milk production is due to differences in their genes; the remainder of their production variation is due to differences among the cows in their environmental conditions. Heritability has been estimated for many traits in several different crops and livestock animals.

Calculating Heritability in Human Populations

Studies that use response to artificial selection to determine the relative influence of genes and environment cannot be performed in human populations. It is ethically and socially unacceptable to design breeding programs to produce people with various traits or to select the men and women who will produce the next generation. Instead, researchers compare the "value" of a trait (for instance, height in centimeters or IQ test scores) among individuals who have varying degrees of genetic similarity. This comparison takes the form of a correlation, whereby researchers estimate how accurately the trait value of an individual can be predicted when the value in a related individual is known.

Correlations between Parents and Children.   Correlations between parents and offspring are often used to determine the heritability of various traits in wild, freely reproducing populations. For example, Figure 6.17 shows a correlation between parents and offspring in a bird population in the strength of their response to injection with the tetanus vaccine. An individual who responded strongly produced a large number of anti-tetanus proteins, called antibodies, while ones that responded weakly produced a lower number of antibodies. As you can see from the graph, parents with weak responses tended to have offspring with weak responses, and parents with strong responses had offspring with strong responses. This strong correlation indicates that the ability to respond to tetanus is highly heritable and that most of the difference between birds in their immune system strength results from differences among them in their genes.

The correlation between the intelligence of human parents and their children helps us determine how important the intelligence of a donor may be to the mental capacity of his children. Intelligence is often measured by performance on an IQ test. Alfred Binet, a French psychologist, developed the intelligence quotient (or IQ), in the early 1900s to more efficiently identify Paris schoolchildren who were in need of remedial help in school. The IQ is not an absolute score on a test; it represents a comparison between an individual and his peers in test performance. The average IQ was arbitrarily set at 100, and IQ tests are designed to produce a normal distribution in scores

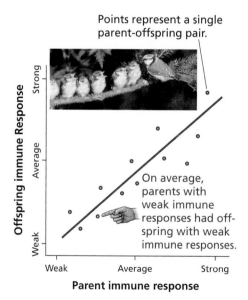

Points represent a single parent-offspring pair.

On average, parents with weak immune responses had offspring with weak immune responses.

Offspring immune Response — Weak · Average · Strong

Parent immune response — Weak · Average · Strong

Figure 6.17 Using correlation between parents and offspring to calculate heritability. In this example, the close correlation of immune system response between parents and offspring of a bird species called the blue tit indicates that much of the variation among birds in the strength of their immune response is due to variation in their genes.

taken from a sample population. This test design strategy has led to one of the most significant criticisms of IQ as a measure of intelligence—because IQ tests are designed for a "standard" population, they may be biased toward the majority culture in that population. Thus, minority subgroups of a population may be expected to perform differently on these tests than does the population as a whole.

Binet's IQ test was not based on any theory of intelligence and was not meant to comprehensively measure mental ability, but the tests remain a commonly used way to measure innate or "natural" intelligence. There is significant controversy over the legitimacy of using IQ tests in this way. Even if IQ tests do not really measure general intelligence, IQ scores have been correlated with academic success—meaning that individuals at higher academic levels usually have higher IQs. So, even without knowing their IQ scores, we can reasonably expect that donors in the Doctorate category have higher IQs than do other available sperm donors. However, the question of whether the high IQ of a prospective sperm donor has a genetic basis still remains.

The average correlation between IQs of parents and their children is 0.42—in other words, according to this analysis, 42% of IQ variation among people is the result of differences among them in their genes. However, parents and children are typically raised in a similar social and economic environment. Since parents and children are similar in genes *and* environment, correlations of IQ between the two groups cannot distinguish the relative importance of genes from the importance of the environment on influencing IQ score. This is the problem found in most arguments about "nature versus nurture"—do children resemble their parents because they are "born that way" or because they are "raised that way"?

The impossibility of using traditional selection studies and the difficulty of separating genetic and environmental influences in most families compels researchers interested in the heritability of human traits to use **natural experiments**. These are situations in which unique circumstances allow a hypothesis test without prior intervention by researchers. Human twins are one source of a natural experiment to test hypotheses about the heritability of quantitative traits in humans.

Correlations between Twins.   By comparing monozygotic twins to dizygotic twins, researchers can begin to separate the effects of shared genes from the effects of shared environments. Twins raised in the same family presumably have similar childhood experiences. Thus, some scientists argue that the only real difference between monozygotic and dizygotic twins is the percentage of genes they share. Recall that monozygotic twins share all of their alleles, while dizygotic twins share, on average, only 50% of their alleles. The average heritability of IQ calculated from a number of different twin studies of this type is about 0.52. According to these studies, 52% of the variability in IQ among humans is due to differences in genotypes. It is somewhat surprising that this value is even higher than the 42% calculated from the correlation between parents and children.

However, other scientists have criticized the heritability value arrived at through the twin studies just described. They argue that monozygotic twins and dizygotic twins *do* differ in more than just genotype. In particular, identical twins are treated as being more alike than nonidentical twins are. This occurs both because of the greater similarity in appearance of monozygotic twins and because of the expectation by parents, relatives, friends, and teachers that identical twins are identical in all respects. If monozygotic twins are *expected* to be more alike than dizygotic twins, their IQ scores may be similar because they are continually encouraged to have the same experiences and to achieve at the same level.

Since this difference in treatment of the types of twins cannot be eliminated, researchers studying the heritability of IQ are especially interested in twins who have been raised apart. By comparing identical twins raised in different environments with nonidentical twins who have also been raised apart, the problem of differential treatment of the two *types* of twins is minimized because no one would know that the individual members of a pair have a

Table 6.1 To what extent is IQ heritable? A summary of various estimates of IQ heritability, their shortcomings, and the problems with using them to understand the role of genes in determining an individual's potential intelligence.

Method of measurement	Result	Warnings when interpreting this result	Warnings that apply to all measurements of heritability
Correlation between parents' IQ and children's IQ in a population	0.42	Since parents and children are similar in genes and environment, a correlation cannot be used to indicate the relative importance of genes and environment in determining IQ.	Heritability values are specific to the populations for which they were measured.
Natural experiment comparing IQ in pairs of identical twins versus nonidentical twins	0.52	Identical twins are treated as more alike than nonidentical twins. Therefore their environment is different from that of nonidentical twins—the heritability value could be an overestimate.	High heritability for a trait does not mean that it is not heavily influenced by environmental conditions; we cannot predict how the trait will respond to a change in the environment.
Natural experiment comparing IQ of individual twins raised apart versus nonidentical twins raised apart	0.72	Small sample size may skew results.	Heritability is a measure of a population, not an individual.

twin. If variation in genes does not explain much of the variation among peoples' IQ scores, then identical twins raised apart should be no more similar than any two unrelated people of the same age and sex.

The frequency of early twin separation is extremely low. Researchers have estimated the heritability of IQ at a remarkable 0.72 in a small sample of twins raised apart. These studies support the hypothesis that differences in our genes explain much of the variation in IQ among people. Table 6.1 summarizes the estimates of IQ heritability and previews the cautions discussed in the next section of this chapter.

6.4 Genes, Environment, and the Individual

Perhaps we can now determine the importance of a sperm-donor father who has earned a doctorate to his child's intellectual development. We know that a sperm donor will definitely influence some of his child's traits—eye and skin color, and perhaps even susceptibility to certain diseases. In addition, according to the twin studies discussed earlier, the donor will probably pass on some intellectual traits to the child. In fact, the high value for heritability of IQ *appears* to indicate that the environment has relatively little influence in determining an individual's intelligence. It seems that it might be a good idea to pay a premium price for Doctorate-category sperm after all.

However, we need to be very careful when applying the results of twin studies to questions about the relative value of sperm donors. The results of twin studies actually give us very limited information about how closely a child will match a sperm donor in intelligence and preferences. To understand why, we will take a closer look at the practical significance of heritability.

The Use and Misuse of Heritability

Remember that heritability is a measure of the relative importance of genes in determining variation in quantitative traits among individuals. For example, with a heritability of 0.30, we can say that only 30% of the variation in dairy cows' milk production is due to variation in genes among these cows. However, the calculated heritability value is unique to the population in which it was measured and to the environment of that population. The specificity of heritability to a particular environment means that we should be very cautious when using heritability to measure the *general* importance of genes to the development of a trait.

A famous misapplication of heritability comes from the book *The Bell Curve* (1994), by Charles Herrnstein and Richard Murray. In this book, the authors report that IQ scores differ among subpopulations in the United States. Among white Americans, IQ averages around 100; among African American populations, IQ averages nearly 15 points lower. Using an estimate of 0.60 for the heritability of intelligence (generally based on the IQ studies described earlier), these authors argued that the IQ differences between whites and blacks are primarily due to a genetic difference in intelligence between these groups. Recall that a major criticism of IQ tests is that they are biased toward the majority culture in a population; this was also a major criticism of the conclusions of *The Bell Curve*. However, if we evaluate the meaning of heritability carefully, we can see that regardless of the legitimacy of IQ tests, Herrnstein and Murray's conclusion is flawed in several ways.

Differences between Groups May Be Environmental. A "thought experiment" can help illustrate this point. Body weight in laboratory mice has a strong genetic component, with a calculated heritability of about 0.90. In a population of mice where weight is variable, bigger mice have bigger offspring, and smaller mice have smaller offspring. Imagine that we randomly divide a population of variable mice into two groups—one group is fed a rich diet, and the other group is fed a poor diet. The mice are treated identically in all other respects. As anticipated, the mice receiving high levels of calories store some excess as fat, while the mice receiving low levels of calories store very little fat. Keeping the mice in the same conditions, imagine that we allow them to reproduce and then weigh their adult offspring. The mice in the rich-diet environment may be twice as heavy, on average, as the mice in the poor-diet environment. If we use Herrnstein and Murray's logic, since the heritability is 0.90 and the average size in the two groups is different, then the groups must be genetically different. However, we know this is not the case; both the heavy and the light mice are offspring of the same original population of parents. These two groups of mice simply live in very different environments (Figure 6.18).

Now extend the same thought experiment to human groups. Say we have two groups of humans, and we have determined that IQ had high heritability. In this case, one group of people was raised in an enriched environment, and their average IQ was higher. The other group was raised in a restricted environment, and their

1. Start with a population of mice that are variable in size.

2. Randomly divide mice into two groups. Feed half a poor diet and the other half a rich diet.

3. Allow the mice in both groups to breed. Measure the weight of adult offspring.

Average weight of the mice in the rich-diet environment is twice the average weight of the population in the poor-diet environment.

Average genetic difference = 0

Figure 6.18   The environment can have powerful effects on highly heritable traits. If genetically similar populations of mice are raised in radically diverse environments, then differences between the populations are entirely due to environment.

average IQ was lower. What conclusions could you draw concerning the genetic differences between these two populations? The answer is none—as with the laboratory mice, these differences could be entirely due to environment. There is definitely a difference in the environment experienced by the average African American child compared to that of the average white child. For example, while 13.5% of white children live in families defined by the U.S. government as impoverished, 31.6% of African American children live at this income level. Given a history of oppression and the current social and economic status of many African Americans in the United States, it is certainly possible that the environment experienced by African Americans is less enriched in the factors that lead to high IQ than is the average environment experienced by a white individual. The high heritability of IQ cannot tell us if two groups with various social environments vary in IQ because of deviations in genes or because of differences in environment.

A Highly Heritable Trait Can Still Respond to Environmental Change.
Sometimes heritability is reported as a "percent" of a trait that is due to genetic factors. That is, a heritability of 0.60 is interpreted as meaning that 60% of your IQ is due to genes, and the remainder (40%) is due to your environment. This seems to imply that highly heritable traits are not strongly influenced by environmental conditions. In *The Bell Curve*, Herrnstein and Murray state that, given IQ's high heritability, policies that increase financial resources to schools will ultimately fail to increase IQ since such a predominantly genetic trait will not respond well to environmental change. However, intelligence in other animals can be demonstrated to be both highly heritable *and* strongly influenced by the environment. The following experiment demonstrates this point.

Rats can be bred for maze-running ability. Using the same sort of selection process that dairy farmers use to produce herds with higher milk production, researchers have produced rats that are "maze-bright" and rats that are "maze-dull." Maze-running ability is highly heritable in the laboratory environment; that is, bright rats have bright offspring, and dull rats have dull offspring. In a normal laboratory environment, maze-dull rats made an average of 165 mistakes every time they attempted to run the maze, while maze-bright rats made only about 115 errors. However, in different environments, the differences between the maze-dull and maze-bright rats were less extreme (Table 6.2). When both rat populations were raised in environments with very little visual variety (that is, with no other rats, no running wheel, and no ability to see activities in the lab), both maze-bright and maze-dull rats made 170 errors per maze. If both types of rats were raised in a high-stimulus, enriched environment (that is, in cages with

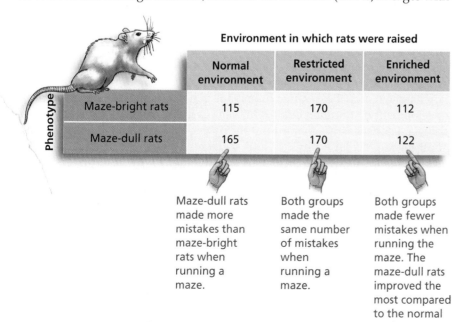

Environment in which rats were raised

Phenotype	Normal environment	Restricted environment	Enriched environment
Maze-bright rats	115	170	112
Maze-dull rats	165	170	122

Maze-dull rats made more mistakes than maze-bright rats when running a maze.

Both groups made the same number of mistakes when running a maze.

Both groups made fewer mistakes when running the maze. The maze-dull rats improved the most compared to the normal environment.

Table 6.2   A highly heritable trait is not identical in all environments. This table describes the average number of mistakes made by rats of two different genotypes in three different environments.

passageways and places to hide), maze-dull rats made only 10 more errors per maze than the maze-bright rats did. All rats did better at maze running in enriched environments, regardless of their genotype. In fact, maze-dull rats improved to a much greater extent than maze-bright rats did when the environment was enriched. What this example demonstrates is that we cannot predict the response of a trait to a change in the environment, even when that trait is highly heritable. Thus, even if IQ has a strong genetic component, environmental factors affecting IQ can have dramatic effects on an individual's intelligence. If the rat experiment represents a typical relationship between genes, environment, and intelligence, then enriching the intellectual environment of all children certainly could increase performance on IQ tests in all groups, and this might minimize currently observed differences between groups.

Heritability Does Not Tell Us Why Two Individuals Differ.   Erroneously stating that heritability is the percentage of a given trait that is determined by genes also seems to imply that, for traits with high heritability, variability *among individuals* is mostly due to differences in genes. However, heritability is a measure of a population. Thus, even if genes explain 90% of the differences among individuals in a particular environment, the reason one individual differs from another may be entirely a function of environment (as an example of this, see the twins in Figure 6.14). Currently, there is no way to determine if a particular child is a poor student because of his genes, because of a poor environment, or a combination of both factors. There is also no way to predict whether a child produced from the sperm of a man with a doctorate will be an accomplished scholar. All we can say is that given our current understanding of the heritability of IQ and the current social environment, the alleles in Doctorate-category sperm *may* increase the probability of having a child with a high IQ.

How Do Genes Matter?

We know that genes can have a strong influence on eye color, risk of genetic diseases such as cystic fibrosis, susceptibility to a heart attack, and even the structure of the brain. But what *really* determines who we are—nature or nurture?

Even with single-gene traits, the outcome of a cross between a woman and a sperm donor is not a certainty; it is only a probability. Couple this with traits being influenced by more than one gene, and independent assortment greatly increases the offspring types possible from a single mating. Knowing the phenotype of potential parents gives you relatively little information about the phenotype of their children. So, even if genes have a strong effect on traits, we cannot "program" the traits of children by selecting the traits of their parents.

In truth, we are really asking ourselves the wrong question when we wonder if nature or nurture has a more powerful influence on who we are. Both our genes and our environment profoundly influence our physical and mental characteristics. Possessing functional genes is imperative to the proper development of a human being. Our cells carry instructions for all the essential characteristics of humanity, but the process of developing from embryo to adult takes place in a physical and social environment that influences how these genes are expressed. Scientists are still a long way from understanding how all of these complex, interacting circumstances result in who we are.

What is the message for women and couples who are searching for a sperm donor from Fairfax Cryobank? Donors in the Doctorate category may indeed have higher IQs than donors in the cryobank's other categories, but there is no real way to predict if a particular child of one of these donors will be smarter than average. According to the current data on the heritability of IQ, sperm from high-IQ donors may "load the dice" and increase the odds of having an offspring with a high IQ but only if parents provide them with a stimulating, healthy, and challenging environment in which to mature. That would be good for children with any alleles.

CHAPTER REVIEW

Summary

6.1 The Inheritance of Traits

- Children resemble their parents in part because they inherit their parents' genes (p. 138).

- Most genes are segments of DNA that contain information about how to make proteins (p. 139).

- Mutations in gene copies can cause slightly different proteins to be produced; different gene versions are called alleles (p. 141).

- Due to independent assortment, parents contribute a unique subset of alleles to each of their nonidentical twin offspring (p. 142).

- Crossing over occurs during meiosis, when portions of chromosomes within homologous pairs are exchanged (p. 144).

- On average, two offspring of the same set of parents share 50% of their alleles, although identical twins share 100% of their alleles (p. 145).

6.2 Mendelian Genetics: When the Role of Genes Is Clear

- The phenotype of a given individual for a particular gene depends on which alleles it carries (its genotype) and whether the alleles are dominant, recessive, incompletely dominant, or codominant (p. 148).

- Diseases such as cystic fibrosis, Huntington's disease, and sickle-cell anemia all have a genetic basis and illustrate the effects of recessive, dominant, and codominant alleles, respectively (pp. 149–150).

- A Punnett square helps us determine the probability that two parents of known genotype will produce a child with a particular genotype (p. 151).

  Web Tutorial 6.1 Mendel's Experiments

6.3 Quantitative Genetics: When Genes and Environment Interact

- Many traits—such as height, IQ, and musical ability—show quantitative variation, which results in a range of values for the trait within a given population (p. 153).

- Quantitative variation in a trait may be generated because the trait is influenced by several genes, because the trait can be influenced by environmental factors, or due to a combination of both factors (pp. 154–155).

- The role of genes in determining the phenotype for a quantitative trait is estimated by calculating the heritability of the trait (p. 156).

- Heritability is calculated by examining the correlation between parents and offspring or by comparing pairs of monozygotic twins to pairs of dizygotic twins (p. 157).

- Twin studies have revealed that the heritability of IQ is relatively high (pp. 158–159).

6.4 Genes, Environment, and the Individual

- Calculated heritability values are unique to a particular population in a particular environment. The environment may cause large differences among individuals, even if a trait has high heritability (p. 159).

- Knowing the heritability of a trait does not tell us why two individuals differ for that trait (p. 160).

- Our current understanding of the relationship between genes and complex traits does not allow us to predict the phenotype of a particular offspring from the phenotype of its parents (p. 162).

Learning the Basics

1. Describe the relationship between a gene, a protein, and a trait.

2. What factors cause quantitative variation in a trait within a population?

3. Can we predict the phenotypes of offspring from the phenotype of their parents? If yes, how? If no, why not?

4. An allele is a _____.

   A. version of a gene; **B.** dysfunctional gene; **C.** protein; **D.** spare copy of a gene; **E.** phenotype

5. Sperm or eggs in humans always _____.

   A. each have 2 copies of every gene; **B.** each have 1 copy of every gene; **C.** each contain either all recessive alleles or all dominant alleles; **D.** are genetically identical to all other sperm or eggs produced by that person; **E.** each contain all of the genetic information from their producer

6. A mistake or "misspelling" that occurs during the copying of a gene and results in a change in a gene is called a(n) _____.

   A. dominant allele; **B.** mutation; **C.** mistakes never occur in gene copying; **D.** dysfunction; **E.** improvement

7. Which of the following genotypes is heterozygous?
   A. *AA*; **B.** *Aa*; **C.** *a*; **D.** *AA BB*; **E.** More than one of the above

8. When the effects of an allele are seen only when an individual carries 2 copies of the allele, the allele is termed _____.

   A. dominant; **B.** incompletely dominant; **C.** recessive;
   D. codominant; **E.** genotypic

9. A quantitative trait _____.

A. may be one that is strongly influenced by the environment; **B.** varies continuously in a population; **C.** may be influenced by many genes; **D.** has more than a few values in a population; **E.** all of the above are correct

10. When a trait is highly heritable, _____.
    A. it is influenced by genes; **B.** it is not influenced by the environment; **C.** the variance of the trait in a population can be explained primarily by variance in genotypes; **D.** a and c are correct; **E.** a, b, and c are correct

Genetics Problems

1. A single gene in pea plants has a strong influence on plant height. The gene has two alleles: tall (*T*), which is dominant, and short (*t*), which is recessive. What are the genotypes and phenotypes of the offspring of a cross between a *TT* and a *tt* plant?

2. What are the genotypes and phenotypes of the offspring of *Tt* × *Tt*?

3. Albinism occurs when individuals carry two recessive alleles (*aa*) that interfere with the production of melanin, the pigment that colors hair, skin, and eyes. If an albino child is born to two individuals with normal pigment, what is the genotype of each parent?

4. Pfeiffer syndrome is a dominant genetic disease that occurs when certain bones in the skull fuse too early in the development of a child, leading to distorted head and face shape. If a man with 1 copy of the allele that causes Pfeiffer syndrome marries a woman who is homozygous for the nonmutant allele, what is the chance that their first child will have this syndrome?

5. Flower color in pea plants is controlled by a gene with codominant alleles: *RR* plants produce red flowers, *rr* plants produce white flowers, and *Rr* plants produce pink flowers.
   a. What percentage of the offspring of a cross between a white-flowered plant and a red-flowered plant is expected to be white?
   b. What percentage of the offspring of a cross between two pink-flowered plants will have white flowers?

6. A cross between a pea plant that produces yellow peas and a pea plant that produces green peas results in 100% yellow-pea offspring.
   a. Which allele is dominant in this situation?
   b. What are the genotypes of the yellow-pea and green-pea plants in the initial cross?

7. A cross between a pea plant that produces yellow peas and a pea plant that produces green peas results in 50% yellow-pea offspring and 50% green-pea offspring. What are the genotypes of the plants in the initial cross?

8. A woman who is a carrier for the sickle-cell allele marries a man who is also a carrier.
   a. What percentage of the woman's eggs will carry the sickle-cell allele?
   b. What percentage of the man's sperm will carry the sickle-cell allele?
   c. The probability that they will have a child who carries 2 copies of the sickle-cell allele is equal to the percentage of eggs that carry the allele times the percentage of sperm that carry the allele. What is this probability?
   d. Is this the same result you would generate when doing a Punnett square of this cross?

9. In cattle, the alleles producing coat color are codominant. *RR* individuals are red, *rr* individuals are white, and *Rr* individuals are "roan"—meaning that their coat is red with a liberal sprinkling of white hairs. In the same animals, the allele for horn production (*H*) is completely dominant over the allele that codes for the lack of horns (*h*). These two genes (for coat color and horn production) are found on different chromosomes. Imagine that a hornless roan bull is mated with a red female who is heterozygous for the horn gene. Describe the different genotypes and associated phenotypes that are possible among their offspring.

10. The allele *BRCA2* was identified in families with unusually high rates of breast and ovarian cancer. Up to 80% of women with 1 copy of the *BRCA2* allele develop one of these cancers in their lifetime.
    a. Is *BRCA2* a dominant or a recessive allele?
    b. How is *BRCA2* different from the typical pattern of Mendelian inheritance?

Analyzing and Applying the Basics

1. Cystic fibrosis occurs when an individual carries 2 mutated copies of a gene that codes for a protein that regulates materials migrating in and out of cells. If two heterozygous individuals already have a child with cystic fibrosis, what is the probability that their next child will be affected? What about a third child? Why is the risk of having another affected child no different after the parents have one affected child?

2. Does a high value of heritability for a trait indicate that the average value of the trait in a population will not change if the environment changes? Explain your answer.

3. The heritability of IQ has been estimated at about 72%. If John's IQ is 120 and Jerry's IQ is 90, does John have stronger "intelligence" genes than Jerry does? Explain your answer.

Connecting the Science

1. If scientists find a gene that is associated with a particular "undesirable" personality trait (for instance, a tendency toward aggressive outbursts), will it mean greater or lesser tolerance toward people with that trait? Will it lead to proposals that those affected by the "disorder" should undergo treatment to be "cured," and that measures should be taken to prevent the birth of other individuals who are also afflicted?

2. Does a genetic basis for differences in IQ between people with Down syndrome and people without this condition mean that we should put fewer resources into education for people with Down syndrome? How does your answer to this question relate to questions about how we should treat individuals with other genetic conditions?

DNA Detective

Extensions of Mendelism, Sex Linkage, Pedigree Analysis, and DNA Fingerprinting

The Romanov family ruled Russia until their overthrow, exile, and 1918 execution.

The fall of communist Soviet Union (symbolized here by people toppling the statue of Lenin) prompted the desire for a proper burial of the Romanov family.

People believed that bones found in a grave in Ekaterinburg were those of the slain Romanovs.

On the night of July 16, 1918, the tsar of Russia, Nicholas II, his wife Alexandra Romanov, their five children, and four family servants were executed in a small room in the basement of the home in which they were exiled. These murders ended three centuries of Romanov rule over imperial Russia.

In February 1917, in the wake of protests throughout Russia, Nicholas II had relinquished his power over the monarchy by abdicating for both himself and on behalf of his only son Alexis, then 13 years old. The tsar had hoped that these abdications would protect his son, the heir to the throne, as well as the rest of the family from harm.

The political climate in Russia at that time was explosive. During the summer of 1914, Russia and other European countries became involved in World War I. This war proved to be a disaster for the imperial government. Russia faced severe food shortages, and the poverty of the common people contrasted starkly with the luxurious lives of their leaders. The Russian people felt deep resentment toward the royal family. This sentiment sparked the first Russian Revolution in February 1917. Following Nicholas's abdication, the Romanovs were kept under guard at one of the family's palaces outside St. Petersburg.

In November 1917, the Bolshevik Revolution brought the Communist regime led by Vladimir Lenin to power. Ridding the country of the last vestige

Scientists made use of DNA evidence to confirm that the bones buried in this shallow grave belonged to the Romanovs.

of Romanov rule became a priority for Lenin and the Bolshevik Party. Lenin believed that doing so would solidify his regime as well as garner support among people who felt that the exiled Romanovs and their opulent lifestyle had come to represent all that was wrong with Imperial Russia. Fearing any attempt by pro-Romanov forces to save the family, Lenin ordered them to the town of Ekaterinburg in Siberia.

Shortly after midnight on July 16, the family was awakened and asked to dress. Nicholas, Alexandra, and their children—Olga, Tatiana, Maria, Anastasia, and Alexis—along with the family physician, cook, maid, and valet, were escorted to a room in the basement of the house in which they had been kept. Believing they were to be moved, the family waited. A soldier entered the room and read a short statement indicating they were to be killed. Armed men stormed into the room, and after a hail of bullets, the royal family and their entourage lay dead.

After the murder, the men loaded the bodies of the Romanovs and their servants into a truck and drove to a remote, wooded area in Ekaterinburg. Historical accounts differ as to whether the bodies were dumped down a mine shaft, later to be removed, or were immediately buried. There is also some disagreement regarding the burial of two of the people who were executed. Some reports indicate that all 11 people were buried together, and two of them either were badly decomposed by acid placed on the ground of the burial site or were burned to ash. Other reports indicate that two members of the family were buried separately. Some people even believe that two victims actually escaped the execution. In any case, the bodies of at least nine people were buried in a shallow grave, where they lay undisturbed until 1991.

The bodies were not all that remained buried. For decades, details of the family's murder were hidden in the Communist Party archives in Moscow. However, after the dissolution of the Soviet Union, postcommunist leaders allowed the bones to be exhumed so that they could be given a proper burial. This exhumation took on intense political meaning because the people of Russia hoped to do more than just give the family a proper burial. The event took on the symbolic significance of laying to rest the brutality of the communist regime that took power after the murders of the Romanov family.

All that remained of these bodies when they were exhumed was a pile of bones, so it was difficult to know whether these were actually the remains of the royal family. However, a great deal of circumstantial evidence pointed to that conclusion. The sizes of the bones seemed to indicate that they belonged to six adults and three children. Investigators electronically superimposed the photographs of the skulls on archived photographs of the family. They compared the skeletons' measurements with clothing known to have belonged to the family. Five of the bodies had gold, porcelain, and platinum dental work, which had been available only to aristocrats. These and other forensic data were consistent with the hypothesis that the bodies could be those of the tsar, the tsarina, three of their five children, and the four servants.

However, at this point, the scientists had shown only that these skeletons *might* be the Romanovs; they had not yet shown with any degree of certainty that these bodies *did* belong to the slain royals. The new Russian leaders did not want to make a mistake when symbolically burying a former regime. Concrete proof was necessary because so much was at stake politically. To begin to answer these questions, scientists turned to the field of genetics.

7.1   Extensions of Mendelian Genetics

Solving one piece of this puzzle would ultimately require that scientists be able to show relatedness between two of the adult-sized skeletons (the tsar and tsarina) and the child-sized skeletons (the Romanov children). In

Chapter 6, we introduced you to the idea that patterns of inheritance can be predicted when genes are inherited in a straightforward manner—for example, a trait that is controlled by one gene with two versions or alleles, one of them dominant and the other recessive. Patterns of inheritance that are a little more complex are called extensions of Mendelian genetics (Table 7.1 on page 170). When the offspring of two different parents has a phenotype that is intermediate to that of either parent, the trait is said to display **incomplete dominance**. For example, when crossing certain red-flowered plants with white-flowered plants, pink-flowered plants are produced. Predicting inheritance is far more difficult for traits that are **polygenic**, or controlled by many genes, especially when the environment can also influence the trait. Because all that was left of the Romanovs was a pile of bones, the scientists could study only a few genetic traits to show the relatedness of the adult-sized skeletons to two of the four child-sized skeletons. Genetic traits that were obvious, such as bone size and structure, are controlled by many genes and affected by environmental components like nutrition and exercise. Therefore, using bone size and structure to predict which of the adult-sized skeletons were related to the child-sized skeletons would have been a matter of guesswork, and the scientists needed to use more sophisticated analyses.

One such analysis that forensic scientists often use to help determine relatedness of people is called blood typing. It involves determining whether certain carbohydrates are located on the surface of red blood cells. Essay 7.1 on page 171 outlines the genetics of the **ABO blood system**, which displays two extensions of Mendelism—**codominance**, whereby both alleles of a gene are expressed in an individual, and **multiple allelism**, which occurs when there are more than two alleles of a gene in the population.

Blood type analysis is an inexpensive, easily performed technique for which results can often be obtained quickly. However, blood type analysis can show only that two people are not related. A parent and child must share at least one allele, but two people who share an allele are not always related. Although blood type analysis is a good forensic tool that has helped answer some important questions, it is not always an option. For the Romanovs, whose remains were very old and contained no blood, blood typing was not possible.

One trait that scientists knew was present in some members of the extended Romanov family displays another extension of Mendelism. The trait, known to be present in Alexis, is a genetic blood-clotting disorder called **hemophilia**. A person with the most common form of hemophilia cannot produce a protein called clotting factor VIII. When this protein is absent, blood does not form clots to stop bleeding from a cut or internal blood vessel damage. Affected individuals bleed excessively, even from small cuts. The effects of this defect in a single gene can lead to multiple effects on an individual's phenotype, by a phenomenon called **pleiotropy**. In addition to the direct effects of excessive bleeding, hemophilia can lead to excessive bruising, pain and swelling in the joints, vision loss from bleeding into the eye, and low red blood cell counts (anemia), resulting in fatigue. Neurological problems may occur if bleeding or blood loss occurs in the brain.

Historical records indicate that Alexis Romanov, heir to the throne, was so ill with hemophilia that his father actually had to carry him to the basement room where he was executed. It appears that Alexis inherited this gene from his mother. We can deduce this pattern of inheritance because we now know that the clotting factor gene is inherited in a sex-specific manner.

Table 7.1 Extensions of Mendelian genetics.

Extension	Example
Incomplete dominance Heterozygote is intermediate to either homozygote.	Flower color in snapdragons Red = *RR* × White = *rr* = Pink = *Rr*
Codominance Both alleles are expressed in a heterozygote.	ABO blood system
Multiple allelism One gene has many alleles.	Coat color in rabbits is controlled by four different alleles. **Genotype Phenotype** *cc* Albino C^hC^h Himalayan $C^{ch}C^{ch}$ Chinchilla C^+C^+ Darkly pigmented rabbit
Polygenic inheritance More than one gene determines the phenotype.	Skin color, eye color, height, IQ
Pleiotropy One gene has many effects.	The singed gene in fruit flies controls both bristle formation and fertility. **Genotype Phenotype** *BB* and *Bb* Normal bristles and fertile *bb* Singed bristles, infertile

Essay 7.1 Blood Group Genetics

Blood typing can be used to help determine whether two people may be related. ABO blood group analysis takes advantage of the fact that various people will have different sugars on the surface of red blood cells. This genetic system displays an extension of Mendelian genetics in which one gene has many different alleles. Genes that have more than two alleles are said to have multiple alleles. Three distinct alleles of one blood-group gene code for the enzymes that synthesize these sugars.

Two of the three alleles in the ABO blood system display another extension of Mendelism that occurs when a heterozygote expresses both different alleles in a condition called codominance. The three alleles of this blood-type gene are I^A, I^B, and i. A given individual will carry only two alleles, one on each member of his or her homologous pairs of chromosomes, but three alleles are being passed on in the entire population. In other words, one person may carry the I^A and I^B alleles, and another might carry the I^A and i alleles. Among these two individuals, there are three different alleles, but each individual carries only two alleles.

The symbols used to represent these alternate forms of the blood-type gene tell us something about their effects. The lowercase i allele is recessive to both the I^A and I^B alleles. Therefore a person with the genotype $I^A i$ has type A blood, and a person with the genotype $I^B i$ has type B blood. A person with both recessive alleles, genotype ii, has type O blood. The uppercase I^A and I^B alleles display codominance in that neither one masks the expression of the other. Both of these alleles are expressed. Thus a person with the genotype $I^A I^B$ has type AB blood. Table E7.1

summarizes the phenotypes and genotypes of the ABO blood system.

Clinicians must take ABO blood groups into account when performing blood transfusions. Persons receiving transfusions from incompatible blood groups will mount an immune response against those sugars that they do not carry on their own red blood cells. The presence of these foreign red blood cell sugars causes a severe reaction in which the donated, incompatible red blood cells form clumps. This clumping can block blood vessels and kill the recipient. Table E7.2 shows compatible recipients and donors for the ABO blood system.

Another molecule on the surface of red blood cells is called the **Rh factor**. Someone who is positive (+) for

Table E7.2   Blood transfusion compatibility and incompatibility.

Recipient	Compatible donors	Incompatible donors
Type O	Type O	Type A Type B Type AB
Type A	Type O Type A	Type B Type AB
Type B	Type O Type B	Type A Type AB
Type AB	Type O Type A Type B Type AB	None

Table E7.1   Red blood cell phenotypes and corresponding genotypes.

Red blood cell phenotype	Red blood cell genotype
Type O	ii
Type A Sugar A	$I^A I^A$ or $I^A i$
Type B Sugar B	$I^B I^B$ or $I^B i$
Type AB Sugars A and B	$I^A I^B$

(Continued)

this trait has the Rh factor on their red blood cells, while someone who is positive (−) does not. This trait, unlike the ABO blood system, is inherited in a straightforward two-allele, completely dominant manner with Rh^+ dominant to Rh^-. Persons who are Rh^+ can have the genotype $Rh^+ Rh^+$ or $Rh^+ Rh^-$. An Rh^- individual has the genotype $Rh^- Rh^-$.

Blood typing is often used to help establish whether a given set of parents could have produced a particular child. For example, a child with type AB blood and parents who are type A and type B could be related, but a child with type O blood could not have two parents who both have type AB blood.

Likewise, if a child has blood type B and the known mother has blood type AB, then the father of that child could have type AB, A, B, or O blood, which does not help to establish parentage. Therefore, if a child has a blood type consistent with alleles that he or she may have inherited from a man who might be his or her father, this does not mean that the man is the father, only that he could be because many other men would also have that blood type. Therefore, blood type analyses can be used only to eliminate people from consideration; it cannot be used to positively identify someone as the father of a particular child.

7.2 Sex Determination and Sex Linkage

The clotting factor VIII gene (the gene that, when mutated, causes hemophilia) is located on the X chromosome. Of the 23 pairs of chromosomes present in the cells of human males, 22 pairs are **autosomes**, or non-sex chromosomes, and one pair, X and Y, are the **sex chromosomes**. Females also have 22 pairs of autosomes, but their sex chromosomes are comprised of two X chromosomes.

Sex Determination

The X and Y chromosomes in humans are involved in determining the sex of an individual. In humans, **sex determination** occurs as follows: When men produce sperm and the chromosome number is divided in half through meiosis, their sperm cells contain one member of each autosome and either an X or a Y chromosome. Females produce gametes with 22 unpaired autosomes and one of their two X chromosomes. Therefore, human egg cells normally contain 1 copy of an X chromosome, but sperm cells can contain either an X or a Y chromosome. It is the sperm cell that determines the sex of the offspring resulting from a particular fertilization. If an X-bearing sperm unites with an egg cell, the resulting offspring will be female (XX). If a sperm bearing a Y chromosome unites with an egg cell, the resulting offspring will be male (XY). Figure 7.1 shows the mechanism of sex determination in humans. The XY system is not the only method of determining sex among living organisms. Table 7.2 outlines some mechanisms of sex determination in nonhumans.

Figure 7.1 Sex determination in humans. In humans, sex is determined by the male since males produce sperm that carry either an X or a Y chromosome (in addition to 22 autosomes). The egg cell always carries an X chromosome along with the 22 autosomes. When an X-bearing sperm fertilizes the egg cell, a female (XX) results. When a Y-bearing sperm fertilizes an egg cell, an XY male results.

Table 7.2 Sex determination in nonhuman organisms.

Type of Organism	Mechanism of Sex Determination
Vertebrates (fish, amphibians, reptiles, birds, and mammals)	In most vertebrates, sex is determined at fertilization by the suite of chromosomes present but differs from human sex determination in that the male may have two of the same chromosomes and the female two different chromosomes. Organisms with two of the same sex chromosomes are called homogametic; organisms with two different sex chromosomes are called heterogametic. Females are the homogametic sex in most mammal and fly species, but the reverse is true in butterflies, fish, and birds. For these species, the female determines the sex of the offspring.
Egg-laying reptiles	In many egg-laying species, two organisms with the same suite of sex chromosomes could become different sexes. In these organisms, sex depends on which genes are activated during embryonic development. For example, the sex of some reptiles is determined by the incubation temperature of the egg. Incubation temperature modifies the number and placement of several enzymes and hormone receptors in the egg. Some researchers have found that applying a drop of the female hormone estrogen to the shell of an incubating egg will produce female offspring in temperature conditions that would normally yield all male hatchlings.
Wasps, ants, bees	In these species, sex is determined by the presence or absence of fertilization. In bees, males (drones) develop from unfertilized eggs. Females (workers or queens) develop from fertilized eggs.
Bony fishes	Some species of bony fishes change their sex after maturation. This mode of sex determination is one in which all individuals will become females unless they are deflected from that pathway due to social signals such as dominance interactions.
Caenorhabditis elegans	The nematode *C. elegans* can either be male or have both male and female reproductive organs. Organisms with both male and female reproductive organs are called hermaphrodites.

Karyotyping is a type of chromosomal analysis that allows scientists to view an individual's chromosomes. Chapter 5 outlines this process and gives an example of a finished karyotype (Figure 5.20 on page 124). This analysis requires that cells be dividing; the chromosomes are best visualized when they are duplicated and condensed, as they would be at metaphase of mitosis. It was not possible to perform karyotyping on the cells of the bones recovered from the Ekaterinburg grave because these cells were no longer dividing. If the scientists had been able to perform karyotyping analysis, they could have determined the sex of the individuals buried in the grave.

Scientists can sometimes determine the sex of an individual based on the structure of the pelvis. Females have evolved to have smaller pelvises with wider pelvic openings to accommodate the passage of a child through the birth canal. Russian scientists had determined that all three of the smaller skeletons and two of the adult-sized skeletons were probably female (and four of the adult skeletons were male). However, the pelvises had decayed, and an unequivocal conclusion was impossible.

Meiosis and Sex Chromosomes

Since the X and Y chromosomes do not carry the same genes, nor are they the same size and shape, they are not considered to be a homologous pair. There is, however, a region at the tip of each chromosome that is similar enough so that they can pair up during meiosis. When meiosis separates the chromosomes during gamete formation, both the autosomes and the sex chromosomes are separated from each other. Sperm and egg cells thus contain one member of each autosome pair and one sex chromosome. Sometimes mistakes occur in meiosis where the members of a homologous pair or the sex chromosomes fail to separate from each other. The failure of chromosomes to separate from each other is called **nondisjunction** and can result in the production of gametes and offspring with too many or too few chromosomes (Essay 7.2).

Essay 7.2 Chromosomal Anomalies

Sometimes mistakes occur during meiosis that result in the production of offspring with too many or too few chromosomes. Too many or too few chromosomes can result when there is a failure of the homologues (or sister chromatids) to separate during meiosis. This failure of chromosomes to separate is termed **nondisjunction** (Figure E7.1). The presence of an extra chromosome is termed **trisomy**. The absence of one chromosome of a homologous pair is termed **monosomy**. Nondisjunction can occur on autosomes or sex chromosomes.

Typically, early embryos with too many or too few chromosomes will die because they have too much or too little genetic information. However, in some situations, such as when the extra or missing chromosome is very small (such as in chromosomes 21, 13, and 18), and/or contains very little genetic information (such as the Y chromosome or an X chromosome that will later be inactivated), the embryo can survive. Table E7.3 lists some chromosomal anomalies in humans that are compatible with life.

Figure E7.1   Nondisjunction. This cell, with two homologous pairs of chromosomes, underwent nondisjunction. Nondisjunction can occur at autosomes or sex chromosomes and results in the production of gametes with too many or too few chromosomes. When a gamete with an extra chromosome is fertilized by a gamete with the normal number of chromosomes, the embryo will be trisomic for the extra chromosome. When a gamete that is short a chromosome is fertilized by a normal gamete, the embryo will be monosomic for the missing chromosome.

Table E7.3   Autosomal and sex-linked chromosomal anomalies.

Conditions Caused by Nondisjunction of Autosomes	Approximate Frequency Among Live Births	Comments
Trisomy 21—Down syndrome	The probability that a woman will have a child with Down syndrome increases with age. In mothers younger than age 35, Down Syndrome occurs in approximately 1 per 1000 births and at age 45, around 4 per 1000 births.	People with Down syndrome tend to be mentally retarded, have abnormal skeletal development, and heart defects.
Trisomy 13—Patau syndrome	1 in 5000	Affected individuals are mentally retarded, deaf, and have a cleft lip and palate.
Trisomy 18—Edwards syndrome	1 in 6000	Babies with Edwards syndrome have malformed organs, ears, mouth, and nose, leading to an elfin appearance. They are mentally retarded and usually die within 6 months of birth.

Conditions Caused by Nondisjunction of Sex Chromosomes	Approximate Frequency Among Live Births	Comments
XO— Turner syndrome	1 in 5000 females	People with only one X chromosome are females with retarded sexual development. They are usually sterile since their ovaries often fail to develop. They can have webbing of the neck, shorter stature, and some hearing impairment. Since they are missing an X chromosome, affected females have 45 chromosomes.
Trisomy X— Meta female	1 in 1000 females	Meta females have three X chromosomes. Two of the X chromosomes are condensed to Barr bodies, and most XXX females develop normally. Since these women have an extra X chromosome, the cells of their bodies have 47 chromosomes.
XXY— Kleinfelter syndrome	1 in 1000 males	Males with the XXY genotype are less fertile than XY males, have small testes, sparse body hair, some breast enlargement, and may have mental retardation. Testosterone injections can reverse some of the anatomical abnormalities but not the mental retardation and lowered fertility.
XYY Condition	1 in 1000 males	Males with two Y chromosomes tend to be taller than average but have a normal male phenotype.

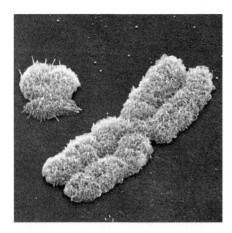

Figure 7.2   The X and Y chromosomes. The Y chromosome (left) is smaller than the X chromosome.

Sex Linkage

Genes located on the X or Y chromosome are called **sex-linked genes** because biological sex is inherited along with, or "linked to," the X or Y chromosome. Sex-linked genes found on the X chromosomes are said to be X-linked, while those on the Y chromosome are Y-linked. The X chromosome is much larger than the Y chromosome, which carries very little genetic information. (Figure 7.2).

X-Linked Genes.   **X-linked genes** are located on the X chromosome. The fact that males have only one X chromosome leads to some peculiarities in inheritance of sex-linked genes. Males always inherit their X chromosome from their mother (they must inherit the Y chromosome from their father in order to be male); thus, males will inherit X-linked genes only from their mothers. Males are more likely to suffer from diseases caused by recessive alleles on the X chromosome because they have only 1 copy of any X-linked gene. Females are less likely to suffer from these diseases, because they carry 2 copies of the X chromosome and thus have a greater likelihood of carrying at least one functional version of each X-linked gene.

A **carrier** of a recessively inherited trait has 1 copy of the recessive allele and 1 copy of the normal allele and will not exhibit symptoms of the disease. Only females can be carriers of X-linked recessive traits because males with a copy of the recessive allele have the trait. Both males and females can be carriers of non-sex-linked, autosomal traits.

Even though female carriers of an X-linked recessive trait will not display the recessive trait, they can pass the trait on to their offspring. For this reason, most women carrying the hemophilia allele will not even realize that they are a carrier until their son becomes ill. Figure 7.3a illustrates that a cross between a male who does not have hemophilia and a female carrier can produce unaffected females, carrier females, unaffected males, and affected males. Figure 7.3b illustrates that all male children produced by a cross between an affected male and an unaffected female would have hemophilia.

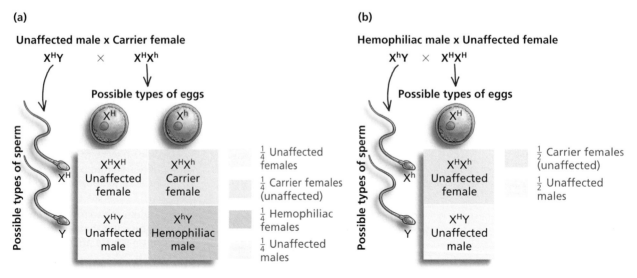

Figure 7.3   Genetic crosses involving the X-linked hemophilia trait. Cross (a) shows the possible outcomes and associated probabilities of a mating between a nonhemophilic male and a female carrier of hemophilia. Cross (b) shows the possible outcomes and associated probabilities of a cross between a hemophilic male and an unaffected female.

Table 7.3   Traits controlled by genes on the X chromosome.

X-Linked Genes	Example
Eye color in _Drosophila melanogaster_ 	Eye color in the fruit fly _Drosophila melanogaster_ is determined by an X-linked gene. Red eyes are the normal eye color, and white eyes are a mutant version of the eye-color gene.
Red-green color blindness 	_Red-green color blindness_ affects approximately 4% of all human males. Red blindness is an inability to see red as a distinct color. Green blindness is an inability to see green as a distinct color. When normal (in this case, the dominant alleles are normal), these genes code for the production of proteins called opsins that help absorb different wavelengths of light. A lack of opsins causes insensitivity to light of red and green wavelengths.
Muscular dystrophy 	_Muscular dystrophy_ is a progressive, fatal disease of muscle wasting that affects approximately 1 in 3,500 males. The onset of muscle wasting occurs between 1 and 6 years of age, and by 12 years of age, affected boys are often confined to a wheelchair. The gene is one that normally codes for the dystrophin protein and is located on the X chromosome. When at least one allele is normal, dystrophin stabilizes cell membranes during muscle contraction. It is thought that the absence of normal dystrophin proteins causes muscle cells to break down and muscle tissue to die.

In the United States, there are over 20,000 hemophiliacs, nearly all of whom are male. Table 7.3 illustrates some additional X-linked traits in humans and other organisms.

Many female organisms actually shut off or inactivate one of their two X chromosomes early in development. This process, called **X inactivation**, is explored fully in Essay 7.3 on page 178.

Y-Linked Genes.   **Y-linked genes** are located on the Y chromosome and are passed from fathers to sons. Although this distinctive pattern of inheritance should make Y-linked genes easy to identify, very few genes have been localized to the Y chromosomes. One gene known to be located exclusively on the Y chromosome is called the _SRY_ gene (for sex-determining region of the Y chromosome). The expression of this gene triggers a series of events leading to development of the testes and some of the specialized cells required for male sexual characteristics. Genes other

Essay 7.3 X Inactivation

Most of the protein products of over 100 genes on the X chromosome have nothing at all to do with the production of biological sex differences. Accordingly, females and males should require equal doses of the products of X-linked genes. How can we account for the fact that females, with their two X chromosomes, could receive two doses of X-linked genes, while males receive only one? The answer comes from a phenomenon called **X inactivation** that occurs in all of the cells of a developing female embryo. This inactivation guarantees that all females actually receive only one dose of the proteins produced by genes on the X chromosomes. Inactivation of the genes on one of the two X chromosomes takes place in the embryo at about the time that the embryo implants in the uterus. One chromosome is inactivated when a string of RNA is wrapped around it (Figure E7.2). This inactivation is random with respect to the parental source—either of the two X chromosomes can be inactivated in a given cell. Inactivation is also irreversible and, as such, is inherited during cell replication. In other words, once a particular X chromosome is inactivated in a cell, all descendants of that cell continue inactivating the same chromosome.

The example of cats with tortoiseshell coat coloring is a good one to illustrate the effects of X inactivation. The coats of tortoiseshell cats are a mixture of black and orange patches. When the background color is white, the pattern of coloration is called calico.

The genes for fur color are located on the X chromosomes. If a male cat with orange fur mates with a female cat with black fur (or vice versa), a female kitten could have one X chromosome with the gene for orange fur, and one X chromosome with the gene for black fur (Figure E7.3a). Early in development, when the embryo consists of about 16 cells, one of the two X chromosomes is randomly inactivated in each cell. Thus, some cells will be expressing the orange fur-color gene, and others will be expressing the black fur-color gene. The pattern of inactivation (the X chromosome that the kitten inherited from its mother or the X chromosome that the kitten inherited from its father) is passed on to the daughter cells of the 16-celled embryo, resulting in the patches of orange and black fur color seen in tortoiseshell cats (Figure E7.3b). Because this pattern of coat coloration requires the expression of both alleles of the

Figure E7.2 X inactivation. One of a female's two X chromosomes is inactivated when a strand of RNA wraps itself around the chromosome.

than *SRY*, on chromosomes other than the Y, code for proteins that are unique to males but are not expressed unless testes develop. One of the other genes on the Y chromosome may be required for the production of healthy sperm. Deletion of this gene results in infertility due to an inability to produce sperm.

7.3 Pedigrees

Because the hemophilia gene is X-linked, Alexis Romanov inherited the disease from his mother, who must have been a carrier of the disease. We can trace the lineage of this disease through the Romanov family by using a chart

(a)

Orange male		Black female		Tortoise shell female
Phenotype	X		=	

Genotype

Allele for orange fur — X^b Y

Allele for black fur — X^B X^B

X^b X^B

(b)

X^b X^B

→ Mitosis →

Inactive X chromosome — X^b X^b

Active X chromosome — X^B

X^B

X^b

X^B

Random X chromosome inactivation

Early embryo

→ Mitosis →

Tortoiseshell cat with patches of orange and black fur.

Figure E7.3 X inactivation in female cats. (a) An orange male cat and a black female cat can produce female offspring with tortoiseshell coats. (b) Random inactivation of one X chromosome early in development leads to patches of coat color.

color gene in different patches of cells, a cat must have two different alleles of this X-linked gene. Therefore, these cats are almost always female. On rare occasions, male cats can have this pattern of coloring. This can happen only if a male cat has two X chromosomes and one Y chromosome—a situation that does occur, though infrequently, via nondisjunction (see Essay 7.2)—and results in sterility.

called a pedigree. A **pedigree** is a family tree that follows the inheritance of a genetic trait for many generations of relatives. Pedigrees are often used in studying human genetics since it is impossible and unethical to set up controlled matings between humans the way one can with fruit flies or plants. Pedigrees allow scientists to study inheritance by analyzing matings that have already occurred. Figure 7.4 identifies some of the symbols used in pedigrees, and Figure 7.5 on page 180 shows how scientists can use pedigrees to determine whether a trait is inherited as an autosomal dominant or recessive, or as a sex-linked recessive.

Information is available about the Romanovs' ancestors because they were royalty and because scientists interested in hemophilia had kept very good records of the inheritance of that trait. Hemophilia was common among European royal families but rare among the rest of the population.

Pedigree analysis symbols

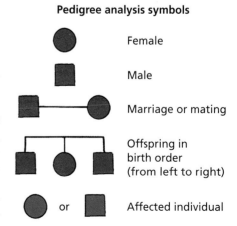

Female

Male

Marriage or mating

Offspring in birth order (from left to right)

or Affected individual

Figure 7.4 Pedigree analysis. Symbols used in pedigrees.

(a) Dominant trait: Polydactyly

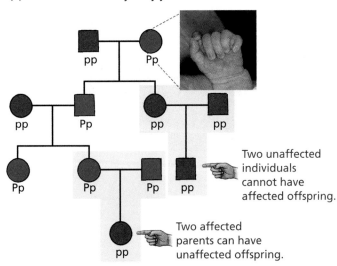

Two unaffected individuals cannot have affected offspring.

Two affected parents can have unaffected offspring.

(b) Recessive trait: Attached earlobes

The recessive trait can skip a generation completely, producing unaffected individuals.

Two affected individuals have affected offspring.

(c) Sex-linked trait: Muscular dystrophy

Affected offspring are most often male.

Carrier

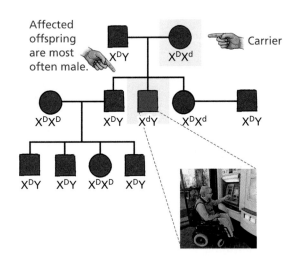

Figure 7.5   Pedigrees showing different modes of inheritance.
(a) Polydactyly is a dominantly inherited trait. People with this condition have extra fingers and/or toes. (b) Having attached ear lobes is a recessively inherited trait. (c) Muscular dystrophy is inherited as an X-linked recessive trait.

This was because members of the royal families intermarried so as to preserve the royal bloodlines. The tsarina must have been a carrier of the hemophilia allele because her son has the trait. Her mother, Alice, must also have been a carrier because the tsarina's brother Fred had the disease, as did two of her sister Irene's sons, Waldemar and Henry. The tsarina's grandmother, Queen Victoria, seems to be the first carrier of this allele in this family because there is no evidence of this disease before her eighth child, Leopold, is affected. Queen Victoria's mother, Princess Victoria, most likely incurred a mutation to the clotting factor VIII gene while the cells of her ovary were undergoing DNA synthesis to produce egg cells. The egg cell that carried the mutant clotting factor VIII gene was passed to her daughter Victoria. The fertilized egg cell that produced Queen Victoria carried this mutation. When the cell divided by mitosis to produce her body cells, the mutation was passed on to each of her cells. When she underwent meiosis to produce gametes, she passed the mutant version on to three of her nine children (Figure 7.6). The extensive pedigree available to the scientists working on the Romanov case, in concert with a powerful technique called DNA fingerprinting, would provide the key data in solving the mystery of the buried bones.

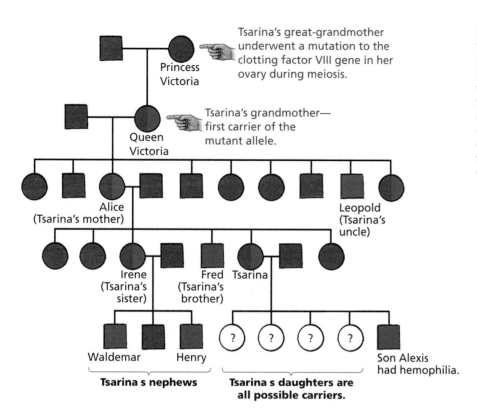

Tsarina's great-grandmother underwent a mutation to the clotting factor VIII gene in her ovary during meiosis.

Tsarina's grandmother— first carrier of the mutant allele.

Figure 7.6   Origin and inheritance of the hemophilia allele. This abbreviated pedigree shows the origin of the hemophilia allele and its inheritance among the tsarina's family. It appears that the tsarina's great-grandmother underwent a mutation that she passed on to her daughter, who then passed it on to three of her nine children—one of whom was the tsarina's mother, Alice. Apparent carriers are half-shaded.

7.4 DNA Fingerprinting

Limits on the power of conventional genetic techniques such as blood typing and karyotyping to identify the bones found in the Ekaterinburg grave necessitated the use of more sophisticated techniques. To do so, scientists took advantage of the fact that any two individuals who are not identical twins have small differences in the sequences of nucleotides that comprise their DNA. To test the hypothesis that the bones buried in the Ekaterinburg grave belonged to the Romanov family, the scientists had to answer the following questions:

1. Which of the bones from the pile are actually different bones from the same individuals?

2. Which of the adult bones could have been from the Romanovs, and which bones could have belonged to their servants?

3. If these are the Romanovs, which two children are missing from the grave?

4. Are these bones actually from the Romanovs, not some other related set of individuals?

All of these questions were answered using **DNA fingerprinting**. This technique allows unambiguous identification of people in the same manner that traditional fingerprinting has been used in the past.

To begin this process, it is necessary to isolate the DNA to be fingerprinted. Scientists can isolate DNA from blood, semen, vaginal fluids, a hair root, skin, and even (as was the case in Ekaterinburg) degraded skeletal remains. When very small amounts of DNA are available, scientists can make make many copies of the DNA by first performing a DNA synthesizing reaction.

(a)

(b)

Figure 7.7 **The polymerase chain reaction (PCR).** (a) PCR is used to amplify, or make copies of, DNA. During a PCR reaction, the DNA to be amplified is added to a small test tube, along with the building-block nucleotides required for synthesizing more DNA and the enzyme *Taq* polymerase. The DNA is heated to separate, or denature, the two strands. The *Taq* polymerase uses the single strands as a template for the synthesis of the complementary strand, producing more double-stranded DNA. The new daughter DNA molecules are then heated again, and the cycle repeats itself. (b) Each round of PCR doubles the number of DNA molecules. This type of exponential growth can yield millions of copies of DNA for scientists to work with.

Polymerase Chain Reaction (PCR)

The **polymerase chain reaction (PCR)** is used to amplify or produce large quantities of DNA (Figure 7.7a). To perform PCR, scientists place the double-stranded DNA to be amplified, along with the individual building-block subunits of DNA—the nucleotides adenine (A), cytosine (C), guanine (G), and thymine (T)—in a small test tube. Next an enzyme called ***Taq*** **polymerase** is added to the tube containing the DNA. This enzyme was given the first part of its name (*Taq*) because it was first isolated from the single-celled organism *Thermus aquaticus*, which lives in hydrothermal vents and can withstand very high temperatures. The second part of the enzyme's name (polymerase) describes its synthesizing activity—it acts as a DNA polymerase. DNA polymerases use one strand of DNA as a template for the synthesis of a daughter strand that carries complementary nucleotides (A:T base pairs are complementary as are G:C base pairs). The main difference between human DNA polymerase and *Taq* polymerase is that the *Taq* polymerase is resistant to extremely high temperatures, temperatures at which human DNA polymerase would be inactivated. The heat-resistant abilities of *Taq* polymerase thus allow PCR reactions to be run at very high temperatures. High temperatures are necessary since the DNA molecule being amplified must first be **denatured**, or split up the middle of the double helix, to produce single strands. After heating, the DNA solution is allowed to cool, and the *Taq* polymerase adds complementary nucleotides to the single strands of the DNA molecule, producing double-stranded DNA molecules.

This cycle of heating and cooling is repeated many times, with each round of PCR doubling the amount of double-stranded DNA present in the test tube (Figure 7.7b).

Once scientists have produced enough DNA by PCR, they can treat the DNA with enzymes that cleave, or cut, the DNA at specific nucleotide sequences. These enzymes are called **restriction enzymes**, and they act like highly specific molecular scissors. As you will see more fully in Chapter 8, individual restriction enzymes cut DNA only at specific nucleotide sequences. Because each individual has distinct nucleotide sequences, cutting different people's DNA with the same enzymes produces fragments of different sizes.

Restriction Fragment Length Polymorphism (RFLP) Analysis

The differently sized fragments produced by restriction digestion of different individuals' DNA are called **restriction fragment length polymorphisms (RFLPs)**. During DNA fingerprinting, restriction fragments are produced in a variety of lengths due to the presence of DNA sequences that vary in number, called **variable number tandem repeats (VNTRs)**. These are nucleotide sequences that all of us carry but in different numbers. For example, one person may have 4 copies of the following sequence (CGATCGA) on one chromosome and 5 copies of the same sequence (CGATCGA) on the other, homologous, chromosome. Another person may have 6 copies of that same repeat sequence in the same location on the same chromosome and 3 copies on the other member of the homologous pair (Figure 7.8). Thus, within a population, there are variable numbers of these known tandem repeat sequences. When restriction enzymes cut DNA around these VNTRs, those segments of DNA that carry more repeats will be longer (and heavier) than those that carry fewer repeats. Since it is impossible to look at a test tube with digested DNA and determine the size of fragments, techniques that allow for the separation and visualization of the DNA are required.

Gel Electrophoresis

The fragments of DNA generated by restriction enzyme cleavage can be separated from each other by allowing the fragments to migrate through a solid support called an **agarose gel**, which resembles a thin slab of gelatin. When an electric current is applied, the gel impedes the progress of the larger DNA fragments more than it does the smaller ones. The current also pulls the negatively charged DNA molecules toward the bottom (positively charged) edge of the gel, further facilitating the size-based separation. The size-based separation of molecules when an electric current is applied to a gel is a technique called **gel electrophoresis**. Segments of DNA with more repeats would be heavier than those with fewer repeats. Heavier DNA segments will not migrate as far in the gel as lighter DNA fragments. Thus, agarose gel electrophoresis separates the DNA fragments on the basis of their size (Figure 7.9a on page 184).

For further analysis, the DNA fragments must be physically removed from the gel. Removing the DNA from the gel is accomplished by drawing the DNA up through the gel and onto a piece of filter paper. When the filter paper is placed on top of the gel, the DNA is wicked up through the gel and adheres to the filter paper. Thus the DNA has been removed from the gel and is now attached to the filter paper. The filter paper is then treated with chemicals that break the hydrogen bonds between the two strands of the DNA helix. The resulting single-stranded DNA now resembles a ladder that has been sliced up the middle, each rung having been cut in half.

Figure 7.8 Variable number tandem repeats (VNTRs). Student 1 has 4 repeat sequences comprising the VNTR on one of his chromosomes and 5 on the other. Student 2 has 6 copies of the same repeat sequence on one of her chromosomes and 3 on the other member of the homologous pair. The repeat sequence is represented as a box. Restriction enzymes (represented here as scissors) cut around the VNTRs, generating fragments of different sizes.

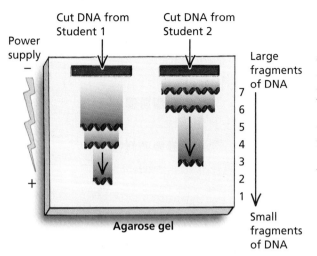

(a) DNA from two different individuals is cut with restriction enzymes and loaded on an agarose gel. When these fragments are subjected to an electric current, shorter fragments migrate through the gel faster than do larger fragments.

Student 1 has DNA sequences that carry 4 and 5 repeat sequences. Student 2 has 3 and 6 repeats. The remaining DNA is DNA that does not carry repeat sequences. Even though the DNA is visible in this figure, DNA is not visible with the unaided eye.

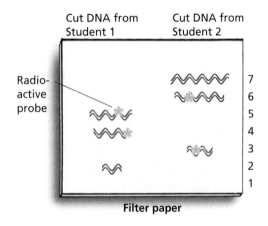

(b) The DNA is transferred from the gel onto filter paper and chemically treated to make it single stranded. The DNA is not visible on the filter paper and must be probed with single-stranded, radioactively labeled DNA that is complementary to the repeat sequence. DNA that does not contain the repeat sequence will not bind to the probe.

(c) When the filter is exposed to X-ray film, radioactive DNA sequences (where the probe is bound) produce a characteristic banding pattern, or DNA fingerprint. Student 1's DNA has 4 and 5 VNTRs, and Student 2's DNA has 3 and 6 VNTRs. The two bands represent the number of VNTR repeats.

Figure 7.9   DNA fingerprinting. Steps a–c show how cut DNA can be electrophoresed and probed to produce a DNA fingerprint.

The single-stranded DNA on the filter paper is then mixed with specially prepared single-stranded DNA that has been radioactively labeled, called a **probe**. The probe is synthesized so that the phosphate molecules in its sugar-phosphate backbone are the radioactive form of phosphorus, ^{32}P. The probe DNA can be designed so that it will base pair with the VNTR sequences present. Such a probe will bind to each repeat sequence on the DNA. For example, DNA sequences with 3 repeats will bind to three separate probes. When the single-stranded probe molecules come into contact with the single-stranded repeat sequences, complementary base pairing ensues, and a double-stranded, radioactive DNA molecule is formed. Note that the only radioactive DNA portion of a DNA molecule would be the segment containing VNTRs. Since not all

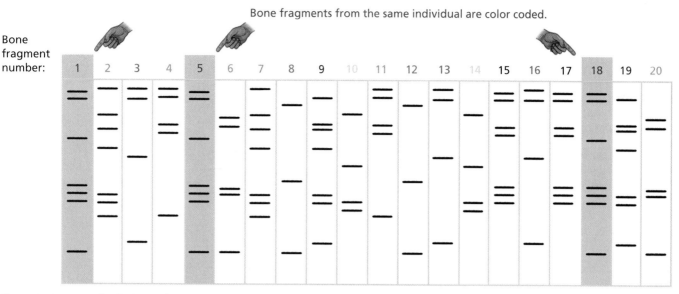

Figure 7.10   Hypothetical DNA fingerprint. Shown are DNA sequences that have been isolated from 20 bone fragments, representing nine different individuals.

of the DNA present in a particular cell will have the repeat sequence, most fragments will be unlabeled (Figure 7.9b).

Once it is bound to the labeled probe, the presence of radioactive DNA can be detected on photographic film, which records the location of DNA when the radiation emitted from radioactive DNA molecules bombards the film. A piece of X-ray film placed over the filter paper shows the locations of the radioactive DNA as a series of bands. Different individuals have different bands since the probe binds to each repeat in a VNTR location. The specific banding pattern that is produced makes up the fingerprint (Figure 7.9c).

In 1992, a team of Russian and English scientists used DNA fingerprinting to determine which of the bones discovered at Ekaterinburg belonged to the same skeleton. Their fingerprinting analysis confirmed that the pile of decomposed bones in the Ekaterinburg grave belonged to nine different individuals. Figure 7.10 shows a hypothetical DNA fingerprint derived from 20 bone fragments.

Since karyotype and pelvic bone analyses were difficult to perform on the decayed bones, scientists also probed the DNA for sequences known to be present only on the Y chromosome. When DNA that was isolated from the children's remains was probed with a Y-specific probe, it became clear that only girls were present in the grave. If these bones did belong to the Romanovs, one of the two missing children was Alexis, the Romanovs' only son.

Once scientists had established that the bones from nine different people were buried in the grave, they tried to determine which bones might belong to the adult Romanovs and which belonged to the servants. For the answer to these questions, scientists took advantage of the fact that Romanov family members would have more DNA sequences in common with each other than they would with the servants. This is because the tsar and tsarina each passed half of their DNA on to each of their children via the process of meiosis.

Meiosis and DNA Fingerprinting

Because the process of meiosis produces gametes containing one member of each homologous pair, each VNTR region that a child carries was inherited from one of his or her parents. Therefore, each band produced by probe binding in a DNA fingerprint of a child must be present in the DNA fingerprint of one of that child's parents (Figure 7.11). By comparing DNA fingerprints made from the smaller skeletons, scientists were able to determine which of the six adult skeletons could

Figure 7.11   DNA fingerprint. This photograph of a DNA fingerprint shows a mother (M), a father (F), and their children (C). Note that every band present in a child must also be present in one of the parents.

Figure 7.12   Hypothetical fingerprint of adult- and child-sized skeletons. Shown is a hypothetical DNA fingerprint made by using many probes to the bone cells of individuals found in the Ekaterinburg grave. From the results of this fingerprint, it is evident that children 1, 2, and 3 are the offspring of adults 1 and 3. Note that each band from each child has a corresponding band in either adult 1 or adult 3. The remaining DNA from adults does not match any of the children, so these adults are not the parents of any of these children.

have been the tsar and tsarina. Figure 7.12 shows a hypothetical DNA fingerprint that illustrates how the banding patterns produced can be used to determine which of the bones belonged to the parents of the smaller skeletons and which bones may have belonged to the unrelated servants.

DNA fingerprint evidence helped scientists determine that the two missing skeletons could have belonged to Alexis and one daughter. Thus DNA evidence put to rest claims made by many pretenders to the throne. Numerous people from all over the world had alleged themselves to be either a Romanov who had escaped execution or their descendant.

The most compelling of these claims was made by a young woman who was rescued from a canal in Berlin, Germany, two years after the murders. This young woman suffered from amnesia and was cared for in a mental hospital, where the staff named her Anna Anderson (Figure 7.13). She later came to believe that she was Anastasia Romanov, a claim she made until her death in 1984. The 1956 Hollywood film *Anastasia*, starring Ingrid Bergman, made Anna Anderson's claim seem plausible; and another, more recent, animated version of the escaped princess story, also titled *Anastasia*, has many young viewers convinced that Anna Anderson was indeed the Romanov heiress.

Since the sex-typing analysis showed only that one daughter was missing from the grave, but not which daughter, scientists again looked to the fingerprinting data. DNA fingerprinting had been done in the early 1990s on intestinal tissue removed during a surgery performed prior to Anna Anderson's death. The analysis showed that Anna was not related to anyone buried in the Ekaterinburg grave; therefore, she could not be Anastasia Romanov.

How can scientists know that Anna Anderson's DNA fingerprint could not have been produced from the DNA of a child of the tsar and tsarina? Again we turn to meiosis to answer this question. Because of meiosis, there is a huge, but limited, number of possible gametes that any person can produce.

The limitation in types of gametes that a person can produce arises because all people carry only two alleles of every gene. Therefore, variation for that particular gene is limited to those two alleles. For example, let us assume that both the tsar and tsarina had attached earlobes. This characteristic is inherited as a recessive trait; so, if the assumption is true, they both would have carried 2 copies of the recessive allele and could only pass the recessive allele to their children. Neither parent could have given their children the dominant allele because they did not carry it.

The tsar and tsarina were able to produce five genetically distinct children (and could have produced millions more) in part due to events occurring during meiosis. Recall from Chapters 5 and 6 that members of a homologous pair of chromosomes align at the cell's equator during meiosis and that their alignment is random with respect to which member of a homologous pair faces which pole. Figure 7.14 is a review of random alignment illustrating the process for two chromosomes that carry some of the genes for hair texture and eye color.

Figure 7.13   Anna Anderson. This woman claimed she was Anastasia Romanov.

(a) One possible Metaphase I alignment

Two types of gametes

Tsarina
Cc Dd

Wavy hair   Dark eyes

Meiosis →

(b) Another possible Metaphase I alignment

Two other types of gametes

Tsarina
Cc Dd

Wavy hair   Dark eyes

Meiosis →

(c) Punnett square for the mating of the Tsar and the Tsarina

Tsar ccDd   x   **Tsarina CcDd**
(straight hair, dark eyes)   (wavy hair, dark eyes)

Possible types of eggs

Possible types of sperm

	CD	cd	Cd	cD
cD	CcDD Wavy hair Dark eyes	ccDd Straight hair Dark eyes	CcDd Wavy hair Dark eyes	ccDD Straight hair Dark eyes
cd	CcDd Wavy hair Dark eyes	ccdd Straight hair Blue eyes	Ccdd Wavy hair Blue eyes	ccDd Straight hair Dark eyes

Figure 7.14   Random alignment. Two possible alignments, (a) and (b), can occur when there are two homologous pairs of chromosomes. (c) Due to random alignment, the tsarina could produce four different types of gametes relative to these two genes. Since the tsar had straight hair and dark eyes (genotype *ccDd*), he could produce *cD* and *cd* gametes. Possible offspring from the mating of the tsar and tsarina are shown inside the Punnett square.

187

The inheritance of hair color is an example of a more complex pattern of inheritance. Curly hair (*CC*) is dominant over wavy (*Cc*) or straight hair (*cc*), and darkly pigmented eyes (*DD* or *Dd*) are dominant over blue eyes (*dd*). Eye color is actually determined by three different genes; but for simplicity, we will follow the inheritance of only one of these genes. Because there are color photos of the Romanovs that allow us to determine their eye color and hair texture, we will use these traits to illustrate random alignment. The tsar had straight hair and dark eyes (*ccDd*), while the tsarina had wavy hair and dark eyes (*CcDd*). We know that the tsar and tsarina were heterozygous for the eye color gene since they had children with blue eyes. Likewise, each member of the royal couple must also have had at least one recessive hair-texture allele, since they had children with straight hair. Because these genes are located on different chromosomes, together the royal couple could produce children with wavy hair and brown eyes (Tatiana and Maria), wavy hair and blue eyes (Anastasia and Olga), straight hair and blue eyes (Alexis), or straight hair and brown eyes.

Since it seems possible that the tsar and tsarina could have produced a child with Anna Anderson's hair and eye color, scientists used VNTRs instead of genes to evaluate whether Anna could have been related to the bones in the grave. Scientists compared the pattern of VNTRs generated by Anna Anderson's

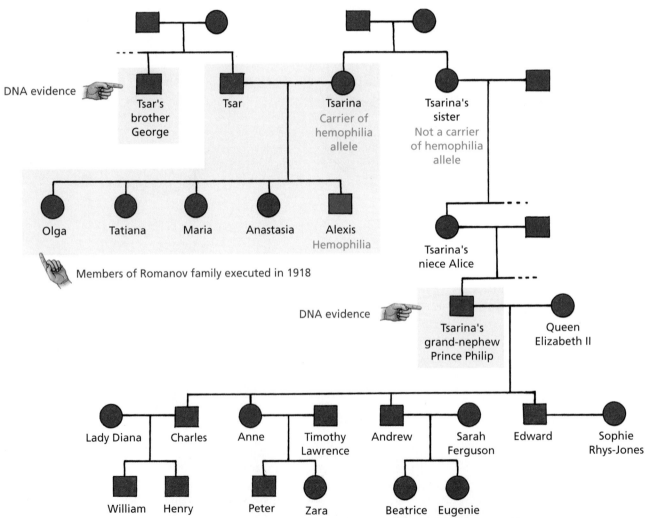

Figure 7.15   Romanov family pedigree. This pedigree shows only the pertinent family members. DNA from the tsar's brother George showed that he was related to the tsar. Note that Prince Philip is the tsarina's grand-nephew. Prince Philip married Queen Elizabeth II. Together they had four children, Charles, Anne, Andrew, and Edward, the current British royal family. The tsarina's sister does not appear to have been a carrier of the hemophilia allele because none of her descendants have been affected by the disease.

Table 7.4 The scientific method. A summary of tests and the conclusions that were drawn from them.

Hypothesis: The bones found in the Ekaterinburg grave belonged to the Romanov family and their servants.	
Test	**Description of Results**
Analyze teeth	Expensive dental work was typically seen only in royalty.
Measure skeletons	The skeletons are those of 6 adults and 3 children.
Sex typing	One male child is missing from grave.
DNA fingerprinting	Children in grave are related to two adults in grave.
DNA fingerprinting	Claims to be one of the missing Romanov children or their descendants are disproved.
DNA fingerprinting	The buried Romanovs are related to known Romanovs.

Conclusion: When you look at each result individually, the evidence is less compelling than when you look at all the evidence together. As a whole, the evidence strongly supports the hypothesis that it was indeed the Romanovs who were buried in the Ekaterinburg grave.

fingerprint with those of the potential Romanovs, generated from bone samples. Since each child can receive only VNTRs that their parents carry, each band of the DNA fingerprint that is found in a child must also be present in one of the parents. This was not the case when Anna Anderson's DNA fingerprint was compared to those of the Romanov parents—Anna Anderson had DNA sequences in her fingerprint that did not match those of either the presumed tsar or tsarina. In other words, if these were the Romanovs, she was unrelated to them.

Thus far, scientists had answered three of the questions posed at the beginning of Section 7.4. They had determined (1) that nine different individuals were buried in the Ekaterinburg grave; (2) that, if these were indeed the Romanovs, the only son, Alexis, was missing from the grave along with one unspecified daughter; and (3) that two of the adult skeletons were the parents of the three children. The last question was still unanswered. How would the scientists show that these were bones from the Romanov family, not just some other set of related individuals?

To answer this question, the scientists turned to a living relative of the Romanovs. DNA testing was performed on England's Prince Philip, who is a grand-nephew of Tsarina Alexandra. In addition, Nicholas II's deceased brother George was exhumed, and his DNA was tested. Figure 7.15 shows the Romanov family pedigree, so that you can see how these individuals are related to each other. The DNA testing performed on these individuals showed that George was genetically related to the adult male skeleton that was related to the child-sized skeletons, and that Prince Philip was genetically related to the adult female skeleton shown to be related to the child-sized skeletons. This evidence strongly supported the hypothesis that the adult skeletons were indeed those of the tsar and tsarina. The process of elimination suggests that the remaining four skeletons had to be the servants.

Having shown that the two adult skeletons belonged to the parents of the three smaller skeletons and that they were genetically related to known Romanov relatives, scientists were convinced that the bones found in the Ekaterinburg grave were those of the tsar, tsarina, three of their five children, and their servants. Table 7.4 summarizes how scientists used the scientific method to test the hypothesis that the remains were indeed those of the Romanovs.

In 1998, 80 years after their execution, the Romanov family finally received a church burial (Figure 7.16). The people of postcommunist Russia have now symbolically laid to rest this part of their country's political history.

Figure 7.16 A church burial for the Romanovs. In 1998, the remains of the Romanov family found in Ekaterinburg were laid to rest.

CHAPTER REVIEW

Summary

7.1 Extensions of Mendelian Genetics

- Some traits are not inherited in the straightforward manner described by Mendel (p. 169).

- Incomplete dominance is an extension of Mendelian genetics whereby the phenotype of the progeny is intermediate to that of both parents (p. 169).

- Polygenic inheritance occurs when many genes control one trait (p. 169).

- Codominance occurs when both alleles of a given gene are expressed (p. 169).

- Genes that have more than two alleles segregating in a population are said to have multiple alleles (p. 169).

- Hemophilia is a genetic blood-clotting disorder that illustrates another extension of Mendelian genetics. It can trigger a phenomenon known as pleiotropy, which occurs when a single gene leads to multiple effects (p. 169).

- The ABO blood system displays both multiple allelism (alleles I^A, I^B, and i) and codominance since both I^A and I^B are expressed in the heterozygote (p. 171).

7.2 Sex Determination and Sex Linkage

- One mechanism of sex determination involves the suite of sex chromosomes present. In humans, males have an X and a Y chromosome and can produce gametes containing either sex chromosome. Females have two X chromosomes and always produce gametes containing an X chromosome. When an X-bearing sperm fertilizes an egg cell, a female baby will result. When a Y-bearing sperm fertilizes an egg cell, a male baby will result (p. 172).

- Meiosis separates both the autosomes and the sex chromosomes from each other. Sperm and egg cells contain one member of each autosome pair and one sex chromosome. When this separation fails, nondisjunction occurs and can result in the production of gametes and offspring with too many or too few chromosomes (p. 174).

- Genes linked to the X and Y chromosomes show characteristic patterns of inheritance. Males need only one recessive X-linked allele to display the associated phenotype. Females can be carriers of an X-linked recessive allele and may pass an X-linked disease on to their sons (p. 176).

- Y-linked genes are passed from fathers to sons (p. 177).

- One of the two X chromosomes in females is inactivated, and the genes residing on it are not expressed. This inactivation is faithfully propagated to all daughter cells (p. 178).

Web Tutorial 7.1 **X-Linked Recessive Traits**

7.3 Pedigrees

- Pedigrees are charts that scientists use to study the transmission of genetic traits among related individuals (p. 179).

7.4 DNA Fingerprinting

- DNA fingerprinting is a technique that is used to show the relatedness of individuals based on similarities in their DNA sequences (p. 181).

- The polymerase chain reaction (PCR) utilizes a special temperature-resistant polymerase called *Taq* polymerase to make millions of copies of a DNA sequence (p. 182).

- Length differences in DNA fragments, generated by restriction digestion of DNA are characteristic of a given individual (p. 183).

- DNA samples in an agarose gel subjected to an electric current will separate according to their size (p. 183).

- Radioactive probes can be used to identify sequences of DNA that have been lifted from a gel (p. 184).

- Related individuals share similar DNA sequences because parents pass DNA on to their children via the process of meiosis (pp. 185–186).

Web Tutorial 7.2 **The Polymerase Chain Reaction (PCR)**

Learning the Basics

1. List and define the extensions of Mendelian genetics that you have learned about in this chapter.

2. Describe the technique of DNA fingerprinting.

3. How is sex determined in humans?

4. If a man with blood type A and a woman with blood type B have a child with type O blood, what are the genotypes of each parent?

5. A man with type A^+ blood whose father had type O^- blood, and a woman with type AB^- blood could produce

children with what phenotypes relative to these blood-type genes?

6. Which of the following is *not* part of the procedure used to make a DNA fingerprint?

   A. DNA is treated with restriction enzymes.; **B.** Cut DNA is placed in a gel to separate the various fragments by size.; **C.** The genes that encode fingerprint patterns are cloned into bacteria.; **D.** DNA from blood, semen, vaginal fluids, or hair-root cells can be used for analysis.; **E.** An electrical current is used to separate DNA fragments.

7. Which of the following statements is consistent with the DNA fingerprint shown in Figure 7.17?

   A. B is the child of A and C.; **B.** C is the child of A and B.; **C.** D is the child of B and C.; **D.** A is the child of B and C.; **E.** A is the child of C and D.

A	B	C	D
–			
	–	–	
–			–
–		–	–
	–	–	
	–		–
–			

Figure 7.17   DNA fingerprint.

8. To bind to DNA, a probe must be _____.

   A. double stranded; **B.** single stranded; **C.** radioactive; **D.** fluorescent

9. The pedigree in Figure 7.18 illustrates the inheritance of hemophilia (sex-linked recessive trait) in the royal family. What is the genotype of individual II-5 (Alexis)?

   A. $X^H X^H$; **B.** $X^H X^h$; **C.** $X^h X^h$; **D.** $X^H Y$; **E.** $X^h Y$

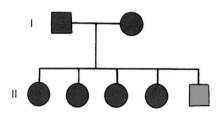

Figure 7.18   Pedigree.

10. A woman is a carrier of the X-linked recessive color-blindness gene. She mates with a man with normal color vision. What genotypes and phenotypes are possible in their offspring?

Analyzing and Applying the Basics

1. Compare and contrast codominance and incomplete dominance.

2. Draw a DNA fingerprint that might be generated by two sisters and their parents.

3. Draw a pedigree of a mating between first cousins.

Connecting the Science

1. Science helped solve the riddle of who was buried in the Ekaterinburg grave, leading to a church burial for some of the Romanov family. In this manner, science played a role in helping the people of Russia come to terms with the brutal communist regime that followed the deaths of the royal family. Can you think of other examples where science has been used to help answer a question with great social implications?

2. Some of the tsar's living relatives have refused to allow their DNA to be tested. Why might they have made that decision?

CHAPTER 8

Genetic Engineering

Transcription, Translation, and Genetically Modified Organisms

Scientists who manipulate genes are often depicted as mad scientists.

Can scientists bring extinct species back to life, as seen in the movie *Jurassic Park*?

Scientists can already clone pets such as Cc the cat.

You hear about them all the time. They are often depicted in cartoons, comic books, movies, and science fiction as mad scientists. These are the scientists who take a gene from one organism and place it into an unrelated organism. These are the scientists who make hormones that farmers inject into the cows that produce the milk we drink. These are the scientists who modify the crops we eat, creating what some people call "Frankenfoods." You may have wondered if it might soon be possible to replace a beloved family member or pet, bring back extinct species through cloning, or even clone yourself. You might worry about a future where parents unwilling to fix their children's "genetic defects" face discrimination.

Who are these scientists? Who pays them? Is anyone regulating their work? Is anyone trying to determine if it is unhealthy to eat these modified foods, whether genetically modified plants will cause environmental problems, or if genetically modified animals are less healthy than their counterparts?

With all kinds of unreliable information coming from so many different sources, it is often hard to separate fact from fiction. To help you sort this out, let us first look at the scientists who are involved in manipulating genes and then learn how they do what they do, as well as about the regulations affecting their work. Finally, we will examine the real prospects and perils of genetic engineering.

Will they one day be able to clone humans as in the movie Multiplicity?

8.1 # Genetic Engineers

Genetic engineers are scientists who manipulate genes. They make their living working at colleges and universities, for the government, and for private companies. Most of them have had extensive training in genetics. The manipulations that genetic engineers perform include changing a gene, changing how a gene is regulated (turned on or off), or moving a gene from one organism to another.

The training for the typical genetic engineer involves many years of schooling. After completing an undergraduate degree, some will obtain a master's degree, which takes two to three years and requires course work as well as a thesis research project. If the student does not continue past the master's level, he or she will probably work in a laboratory under the supervision of a more senior scientist.

Students who want to continue their education can apply to graduate schools with Ph.D. (doctor of philosophy) programs. Scientists holding a Ph.D. have the title of "Doctor" because they have a doctorate in their chosen field (a medical doctor, or M.D., has a doctorate in medicine). A Ph.D. program involves more course work and an expanded research component; the results of this research must also be published in a peer-reviewed journal. Most Ph.D. scientists have gone to school five or more years after earning their undergraduate degree.

In scientific fields, graduate students generally get paid a small salary and have their tuition waived by the university. In exchange for tuition and salary, students work as teaching assistants overseeing undergraduate laboratory courses. If your biology course has a laboratory component, then you may have had experience with a teaching assistant who is a pursuing an advanced degree in biology.

Most colleges and universities, especially the larger ones, expect faculty members to combine teaching with research. In this way, college professors not only pass information to the next generation; they also add to the knowledge base of their field.

The federal government employs many of these biologists—for example, the National Cancer Institute (NCI) employs genetic engineers. In addition, many genetic engineers work in private industry, which tends to focus on for-profit product and drug development.

Genetic engineers in academia, government, and industry are involved in many different research projects. These projects vary from trying to produce a protein in the laboratory, to changing the genetic characteristics of crop plants, to trying to understand how human genes interact. One of the first genetic engineering projects to seize the attention of the public was the genetic engineering of a protein normally produced by cows.

8.2 # Protein Synthesis and Gene Expression

During the early 1980s, genetic engineers at Monsanto® Company began to produce **recombinant bovine growth hormone (rBGH)** in their laboratories. Recombinant (r) bovine growth hormone is a protein that has been made by genetically engineered bacteria. These bacterial cells have had their DNA manipulated so that it carries the instructions for, or encodes, a cow growth hormone that can be produced in the laboratory. Hormones are substances that are secreted from specialized glands and travel through the bloodstream to affect their target organs. Growth hormones act on many different organs to increase the overall size of the body and, in cows, to increase milk production.

Before the advent of genetic technologies, growth hormone was extracted from the pituitary glands of slaughtered cows and then injected into live cows (Figure 8.1). It is also possible to obtain human growth hormone from the pituitary glands of human cadavers. When the human growth hormone is injected into humans who have a condition called **pituitary dwarfism**, their size increases. However, harvesting growth hormone from the pituitary glands of cows and humans is laborious, and many cadavers are necessary to obtain small amounts of the protein.

Genetic engineers at Monsanto realized that genetic technology would allow them to produce large quantities of bovine growth hormone in the laboratory, inject it into dairy cows, and increase their milk production, completely bypassing the less-efficient, surgical isolation of BGH. These scientists understood that if they were successful, Monsanto would stand to make a healthy profit from the dairy farmers who would buy the engineered growth hormone to increase the milk yield of their herds. Let us first examine how cells normally use DNA instructions to produce proteins and then how scientists manipulate this process to have bacteria produce proteins normally made by other organisms.

Figure 8.1   Bovine growth hormone (BGH). Bovine growth hormone is a protein produced by the pituitary gland (in red) of the cow brain.

From Gene to Protein

Protein synthesis involves using the instructions carried by a gene to build a particular protein. Genes do not build proteins directly; instead, they carry the instructions that dictate how a protein should be built. Understanding protein synthesis requires that we review a few basics about DNA, genes, and RNA. First, DNA is a polymer of nucleotides that make chemical bonds with each other based on their complementarity (A to T, and C to G). Second, a gene is a sequence of DNA that encodes a protein. Proteins are large molecules composed of amino acids. Each protein has a unique function that is dictated by its particular structure. The structure of a protein is the result of the order of amino acids that comprise it because the chemical properties of amino acids cause a protein to fold in a particular manner.

Before a protein can be built, the instructions carried by a gene are first copied. When the gene is copied, the copy is comprised not of DNA (deoxyribonucleic acid) but of **RNA (ribonucleic acid)**. Therefore, it is important to understand the differences between DNA and RNA.

RNA is also a polymer of nucleotides. A nucleotide is composed of a sugar, a phosphate group, and a nitrogen-containing base. For DNA nucleotides, the sugar is deoxyribose, and the nitrogenous bases are adenine (A), cytosine (C), guanine (G), and thymine (T). The nucleotides that join together to produce RNA are composed of the sugar ribose, a phosphate group, and the nitrogenous bases A, C, G, and U (uracil). There are no thymines (T) in RNA because uracil (U) replaces them. In addition, RNA is usually single stranded,

(a) DNA is double stranded.

(b) RNA is single stranded.

Figure 8.2   DNA and RNA. (a) DNA is double stranded. Each DNA nucleotide is composed of the sugar deoxyribose, a phosphate group and a nitrogen containing base (A, G, C, or T). (b) RNA is single stranded. RNA nucleotides are composed of the sugar ribose, a phosphate group and a nitrogen containing base (A, G, C, or U). Note the difference in the sugars is shown in the pink boxes.

not double stranded like DNA (Figure 8.2). When a cell requires a particular protein, a strand of RNA is produced by using DNA as a template. RNA nucleotides are able to make base pairs with DNA nucleotides. C and G make a base pair, and U pairs with A.

The RNA copy then serves as a blueprint that tells the cell which amino acids to join together to produce a protein. Thus, the flow of genetic information in a eukaryotic cell is from DNA to RNA to protein (Figure 8.3).

How does this flow of information actually take place in a cell? Going from gene to protein involves two steps. The first step, called **transcription**, involves producing the copy of the required gene. In the same way that a transcript of a speech is a written version of the words spoken by the speak-

Figure 8.3   The flow of genetic information. Genetic information flows from a DNA to an RNA copy of the DNA gene, to the amino acids that are joined together to produce the protein coded for by the gene.

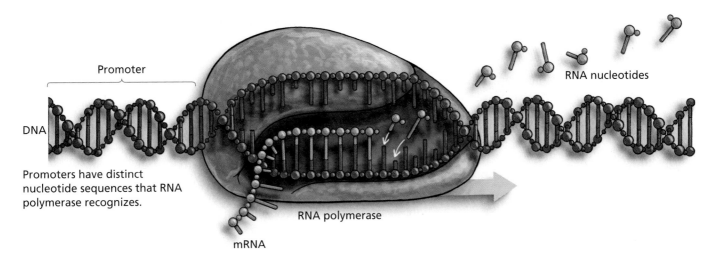

Promoter

DNA

Promoters have distinct
nucleotide sequences that RNA
polymerase recognizes.

RNA nucleotides

RNA polymerase

mRNA

Figure 8.4 Transcription. After locating the start site of the gene, a region called the promoter, the enzyme RNA polymerase ties together nucleotides within the growing RNA strand as they bind to their complementary base on the DNA. Only when a complementary base pair is made between DNA and RNA does the polymerase add an RNA nucleotide to the growing strand. Complementary bases are formed via hydrogen bonding of A with U and G with C. Note that one strand of the DNA is used as a template for the synthesis of the mRNA. When the RNA polymerase reaches the end of the gene, the mRNA transcript is released.

er, transcription inside a cell is a process that produces a copy with the RNA nucleotides substituted for DNA nucleotides. The second step, called **translation**, involves decoding the copied RNA sequence and producing the protein for which it codes. In the same way that a translator helps determine the meaning of words in two different languages, translation in a cell involves moving from the language of nucleotides (DNA and RNA) to the language of amino acids and proteins.

Transcription

Transcription is the copying of a DNA gene into RNA (Figure 8.4). The copy is synthesized by an enzyme called **RNA polymerase**. To begin transcription, the RNA polymerase binds to a nucleotide sequence at the beginning of every gene, called the **promoter**. Once the RNA polymerase has located the beginning of the gene by binding to the promoter, it then rides along the strand of the DNA helix that comprises the gene. As it is traveling along the gene, the RNA polymerase unzips the DNA double helix and ties together RNA nucleotides that are complementary to the DNA strand it is using as a template. This results in the production of a single-stranded RNA molecule that is complementary to the DNA sequence of the gene. This complementary RNA copy of the DNA gene is called **messenger RNA (mRNA)**, since it carries the message of the gene that is to be expressed.

Translation

The second step from gene to protein requires that the mRNA be used to produce the actual protein for which the gene encodes through a process called translation. For translation to occur, a cell needs mRNA, a supply of amino acids to join in the proper order, and some energy in the form of ATP. Translation also requires structures called ribosomes and transfer RNA molecules.

Ribosome: Workbench
for translation

rRNA wrapped
around a
ribosomal
protein

Figure 8.5   Ribosome. Ribosomes are composed of two subunits. Each subunit, in turn, is composed of rRNA and protein. Ribosomes are the site of protein synthesis.

Transfer RNA: The Translator

Amino
acid

Binding site for
amino acid

Region of internal
complementarity
(e.g., A : U or C : G)

tRNA

Binding site for
mRNA codon

Anticodon

mRNA

Codon

Figure 8.6   Transfer RNA (tRNA). Transfer RNA molecules are composed of an RNA strand that has regions of internal complementarity. Each ribosome has a characteristic 3-nucleotide sequence, called the anticodon, that binds to the mRNA codon. Each tRNA also carries the amino acid corresponding to the mRNA codon to which it binds. Transfer RNAs translate the language of nucleotides into the language of amino acids.

Ribosomes.   **Ribosomes** are subcellular, globular structures (Figure 8.5) that are composed of another kind of RNA called **ribosomal RNA (rRNA)**, which is wrapped around many different proteins. Each ribosome is composed of two subunits—one large and one small. When assembled in this fashion, the mRNA can be threaded through the ribosome. In addition, the ribosome can bind to structures called **transfer RNA (tRNA)** that carry amino acids.

Transfer RNA (tRNA).   Transfer RNA (Figure 8.6) is yet another type of RNA found in cells (in addition to mRNA and rRNA). Transfer RNA is single stranded but has regions of internal complementarity, where complementary nucleotides (A and U; G and C) bind to each other, resulting in a structure that is single stranded in some regions and double stranded in others. Even though there are some regions of internal complementarity, transfer RNA as a whole is a single strand of nucleotides that folds on itself in some isolated regions. Individual transfer RNAs carry specific amino acids. As mRNA moves through the ribosome, small sequences of mRNA are exposed. These sequences of mRNA are 3 nucleotides long and encode an amino acid; they are called **codons**. Transfer RNAs bind to codons through interactions between the RNA nucleotides at the base of the tRNA, a region called the **anticodon**, and the mRNA codon. The anticodon on a particular tRNA binds to the complementary mRNA codon. Thus, the codon calls for the incorporation of a specific amino acid. When a tRNA anticodon binds to a mRNA codon, the ribosome adds the amino acid that the tRNA is carrying to the growing chain of amino acids that will eventually constitute the finished protein. Therefore, the transfer RNA functions as a sort of cellular translator, fluent in both the language of nucleotides and the language of amino acids.

In this manner, the sequence of bases in the DNA dictates the sequence of bases in the RNA, which in turn dictates the order of amino acids that will be joined together to produce a protein. Protein synthesis ends when a codon that does not code for an amino acid, called a **stop codon**, moves through the ribosome. When a stop codon is present in the ribosome, no new amino acid can be added, and the growing protein is released. Once released, the protein folds up on itself and moves to where it is required in the cell. A summary of the process of translation is shown in Figure 8.7.

The process of translation allows cells to determine which amino acid sequence a particular gene encodes. Scientists can determine the sequence of amino acids that a gene calls for by looking at a chart called the **genetic code**.

Genetic Code.   The genetic code shows which mRNA codons code for which amino acids (Table 8.1 on page 200). As Table 8.1 shows, there are 64 codons, 61 of which code for amino acids. Three of the codons are stop codons that occur near the end of a mRNA. Since stop codons do not code for an amino acid, protein synthesis ends when a stop codon enters the ribosome. In the table, you can see that the codon AUG functions both as a start codon (and thus is found near the beginning of each mRNA) and as a codon dictating that the amino acid methionine (met) be incorporated into the protein being synthesized. Notice also that there are many examples of situations when the same amino acid can be coded for by more than one codon. For example, the amino acid threonine (thr) is incorporated into a protein in response to the codons ACU, ACC, ACA, and ACG. The fact that more than one codon can code for the same amino acid is referred to as **redundancy** in the genetic code. There is, however, no situation where a given codon can call for more than one amino acid. For example, AGU codes for serine (ser) and nothing else. Therefore, there is no **ambiguity** in the genetic code as to what amino acid any codon will call for. The genetic code is also **universal** in the sense that different organisms typically decode the same gene to produce the same protein.

1. Amino acids and tRNAs float freely in the cytoplasm.

2. Enzymes facilitate the binding of a specific tRNA to its appropriate amino acid.

3. A tRNA will dock if the complementary RNA codon is present on the ribosome.

4. The amino acids link together to form a polypeptide.

Amino acids

tRNA

mRNA

Ribosome

5. The ribosome moves on to the next codon to receive the next tRNA.

6. When the ribosome reaches the stop codon, no tRNA can base-pair with the codon on the mRNA. RNA and the newly synthesized protein are released.

7. The chain of amino acids folds into its globular form, and the protein is ready to perform its job.

Protein (such as BGH)

8. The subunits of the ribosome separate but can reassemble and begin translation of another mRNA.

Figure 8.7 Translation. During translation, mRNA directs the synthesis of a protein. The mRNA codon that is exposed in the ribosome binds to its complementary tRNA molecule, which carries the amino acid coded for by the DNA gene. When many amino acids are joined together, the required protein is produced. When the translation machinery reaches a stop codon, the newly synthesized protein is released into the cytoplasm.

Table 8.1 The genetic code. It is possible to determine which amino acid is coded for by each mRNA codon using a chart called the genetic code. Look at the left-hand side of the chart for the first-base nucleotide in the codon; there are 4 rows, one for each possible RNA nucleotide—A, C, G, or U. By then looking at the intersection of the second-base columns at the top of the chart and the first-base rows, you can narrow your search for the codon to 4 different codons. Finally, the third-base nucleotide in the codon on the right-hand side of the chart determines the amino acid that a given mRNA codon codes for. Note the 3 codons, UAA, UAG, and UGA, that do not code for an amino acid; these are stop codons. The codon AUG is a start codon, found at the beginning of most protein-coding sequences.

<table>
<tr><td></td><td colspan="2" align="center">Second base</td><td></td><td></td></tr>
<tr><td></td><td align="center">U</td><td align="center">C</td><td align="center">A</td><td align="center">G</td><td></td></tr>
<tr>
<td rowspan="4">U</td>
<td>UUU] Phenyl-
UUC] alanine (phe)
UUA] Leucine (leu)
UUG]</td>
<td>UCU]
UCC] Serine (ser)
UCA]
UCG]</td>
<td>UAU] Tyrosine (tyr)
UAC]
UAA   Stop codon
UAG   Stop codon</td>
<td>UGU] Cysteine (cys)
UGC]
UGA   Stop codon
UGG   Tryptophan (trp)</td>
<td>U
C
A
G</td>
</tr>
<tr>
<td rowspan="4">C</td>
<td>CUU]
CUC] Leucine (leu)
CUA]
CUG]</td>
<td>CCU]
CCC] Proline (pro)
CCA]
CCG]</td>
<td>CAU] Histidine (his)
CAC]
CAA] Glutamine (gln)
CAG]</td>
<td>CGU]
CGC] Arginine (arg)
CGA]
CGG]</td>
<td>U
C
A
G</td>
</tr>
<tr>
<td rowspan="4">A</td>
<td>AUU]
AUC] Isoleucine (ile)
AUA]
AUG   Methionine (met)
     Start codon</td>
<td>ACU]
ACC] Threonine (thr)
ACA]
ACG]</td>
<td>AAU] Asparagine (asn)
AAC]
AAA] Lysine (lys)
AAG]</td>
<td>AGU] Serine (ser)
AGC]
AGA] Arginine (arg)
AGG]</td>
<td>U
C
A
G</td>
</tr>
<tr>
<td rowspan="4">G</td>
<td>GUU]
GUC] Valine (val)
GUA]
GUG]</td>
<td>GCU]
GCC] Alanine (ala)
GCA]
GCG]</td>
<td>GAU] Aspartic acid (asp)
GAC]
GAA] Glutamic acid (glu)
GAG]</td>
<td>GGU]
GGC] Glycine (gly)
GGA]
GGG]</td>
<td>U
C
A
G</td>
</tr>
</table>

(Left margin label: First base; Right margin label: Third base)

Mutations

Sometimes changes to the DNA sequence, called **mutations**, can affect the order of amino acids incorporated into a protein during translation. Mutations to a gene can result in the production of different forms, or alleles, of a gene. Different alleles result from changes in the DNA that alter the amino acid order of the encoded protein. Mutations can result in the production of either no functional protein or a protein different from the one previously called for. If this protein does not have the same amino acid composition, it may not be able to perform the same job (Figure 8.8). Chapter 6 indicates that a substitution of a single nucleotide results in the incorporation of a new amino acid in the hemoglobin protein and compromises the ability of cells to carry oxygen, producing sickle-cell disease.

There are also cases in which a mutation has no effect on a protein. They may occur when changes to the DNA result in the production of a mRNA

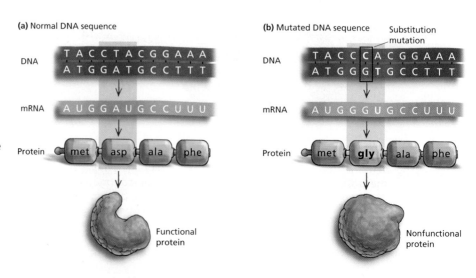

Figure 8.8 Mutation. A single nucleotide change from the normal sequence (a) to the mutated sequence (b) can result in the incorporation of a different amino acid. If the substituted amino acid has chemical properties different from those of the original amino acid, then the protein may assume a different shape and thus lose its ability to perform its job.

codon that codes for the same amino acid as was originally called for. Due to the redundancy of the genetic code, a mutation that changes the mRNA codon from ACU to ACC will have no impact because both of these codons code for the amino acid threonine. This is called a **neutral mutation** (Figure 8.9a). In addition, mutations can result in the substitution of one amino acid for another with similar chemical properties, which may have little or no effect on the protein.

Inserting or deleting a single nucleotide can have a severe impact since the addition (or deletion) of a nucleotide can change the groupings of nucleotides in every codon that follows (Figure 8.9b). Changing the triplet groupings is called altering the **reading frame**. All nucleotides located after an insertion or deletion will be regrouped into different codons, producing a **frameshift mutation**. For example, inserting an extra letter "H" after the fourth letter of the sentence, "The cat ate his dog," could change the reading frame to the nonsensical statement, "The cHa tat ehi sdo g." Inside cells, this often results in the incorporation of a stop codon and the production of a shortened, nonfunctional protein.

To help you understand protein synthesis, let us consider its similarity to an everyday process such as baking a cake. To bake a cake, you would consult a recipe book (genome) for the specific recipe (gene) to make your cake (protein). You may copy the recipe (mRNA) out of the book so that the original recipe (gene) does not become stained or damaged. The original recipe (gene) is left in the book (genome) on a shelf (nucleus), so that you can make another copy when you need it. The original recipe (gene) can be copied again and again. The copy of the recipe (mRNA) is placed on the kitchen counter (ribosome) while you assemble the ingredients (amino acids). The ingredients (amino acids) for your cake (protein) include flour, sugar, butter, milk, and eggs. The ingredients are measured in

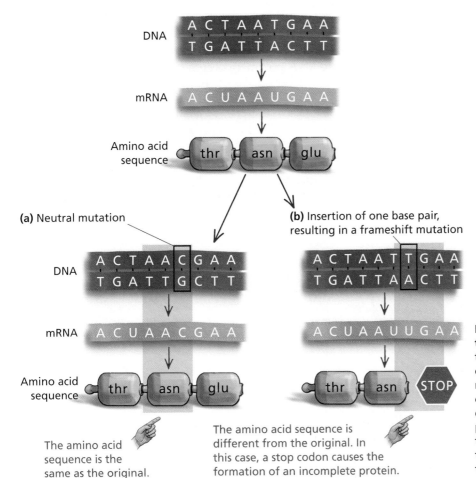

Figure 8.9 Neutral and frameshift mutations. (a) Neutral mutations are changes to the DNA that, due to the redundancy of the genetic code, result in the incorporation of the same amino acid that was originally called for. (b) The insertion (or deletion) of a nucleotide can result in the production of an mRNA that produces the wrong protein or one that terminates translation too soon. This type of mutation is called a frameshift mutation.

measuring spoons and cups (tRNAs) that are dedicated to one specific ingredient. Like the amino acids that are combined in different orders to produce a specific protein, the ingredients in a cake can be used in many ways to produce a variety of foods. The ingredients (amino acids) are always added according to the instructions specified by the original recipe (gene). Changes to the original recipe (mutations) can result in a different or inedible cake being produced.

All cells in all organisms undergo this process of protein synthesis, with different cell types selecting different genes from which to produce proteins. Figure 8.10a shows the coordination of these two processes as they occur in cells with nuclei, that is, eukaryotic cells. In eukaryotic cells, transcription and translation are spatially separate with transcription occurring in the nucleus and translation occurring in the cytoplasm. Cells lacking a membrane-bound nucleus and organelles are called prokaryotic cells. Prokaryotic cells (such as bacterial cells) also undergo protein synthesis, but transcription and translation occur simultaneously in the same location instead of occurring in separate places. As a mRNA is being transcribed, ribosomes attach and begin translating (Figure 8.10b).

Regulating Gene Expression

Different cell types transcribe and translate different genes. Each cell in your body, except sperm or egg cells, has the same complement of genes you inherited from your parents but expresses only a small percentage of those genes. For example, since your liver and pancreas perform a specialized suite of jobs, the cells of your liver turn on or express one suite of genes and the cells of your pancreas, another. Turning a gene on or off, or modulating it more subtly, is called **regulating gene expression**. The expression of a given gene is regulated so that it is turned on and turned off in response to the cell's needs.

Regulation of Transcription.   Gene expression is most commonly regulated by controlling the rate of transcription. Regulation of transcription can occur at the promoter, the sequence of nucleotides adjacent to a gene to which the RNA polymerase binds in order to initiate transcription. When a cell requires a particular protein, the RNA polymerase enzyme binds to the promoter for

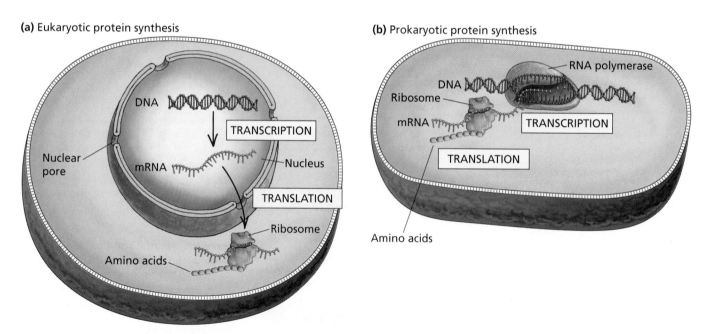

(a) Eukaryotic protein synthesis

(b) Prokaryotic protein synthesis

Figure 8.10   Protein synthesis in eukaryotic and prokaryotic cells. (a) In eukaryotes, transcription occurs in the nucleus and translation in the cytoplasm. (b) In prokaryotic cells, which lack nuclei, transcription and translation occur simultaneously in the same location.

that particular gene and transcribes the gene. Prokaryotic and eukaryotic cells both regulate gene expression by regulating transcription but have different strategies for doing so. Prokaryotic cells typically regulate gene expression by blocking transcription via proteins called **repressors** that bind to the promoter and prevent the RNA polymerase from binding. When the gene needs to be expressed, the repressor will be released from the promoter so that the RNA polymerase can bind (Figure 8.11a). This is the main mechanism by which simple single-celled prokaryotes regulate gene expression.

The more complex eukaryotic cells have evolved more complex mechanisms to control gene expression. To control transcription, eukaryotic cells more commonly enhance gene expression using proteins called **activators** that help the RNA polymerase bind to the promoter, thus facilitating gene expression (Figure 8.11b). The rate at which the polymerase binds to the promoter is also affected by substances that are present in the cell. For example, the presence of alcohol in a liver cell might result in increased transcription of a gene involved in the breakdown of alcohol.

Regulation by Chromosome Condensation.   It is also possible to regulate gene expression by condensing all or part of a chromosome. This prevents RNA polymerase from being able to access genes. Essay 7.3 outlines how the inactivation of an X chromosome turns off the expression of X-linked genes in organisms that have two X chromosomes. Entire chromosomes are also inactivated when they condense during mitosis.

(a) Repression of transcription

(b) Activation of transcription

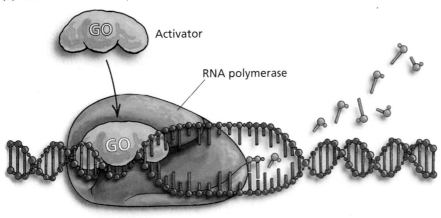

Figure 8.11   Regulation of gene expression. (a) Gene expression can be regulated by (a) repression, during which repressors prevent RNA polymerase from binding the promoter or (b) activation, during which activators help RNA polymerase bind the promoter.

Regulation by mRNA Degradation. Eukaryotic cells can also regulate the expression of a gene by regulating how long a messenger RNA is present in the cytoplasm. Enzymes called **nucleases** roam the cytoplasm, cutting RNA molecules by binding to one end and breaking the bonds between nucleotides. If a particular mRNA has a long "tail," it will survive longer in the cytoplasm and be translated more times. All mRNAs are eventually degraded in this manner; otherwise, once a gene had been transcribed one time, it would be expressed forever.

Regulation of Translation. It is also possible to regulate many of the steps of translation. For example, the binding of the mRNA to the ribosome can be slowed or hastened, as can the movement of the mRNA through the ribosome.

Regulation of Protein Degradation. Once a protein is synthesized, it will persist in the cell for a characteristic amount of time. Like the mRNA that provided the instructions for its synthesis, the life of a protein can be affected by enzymes inside the cell that degrade the protein. Speeding up or slowing down the activities of these enzymes can change the amount of time that a protein is able to be active inside a cell.

The problem of regulating gene expression is easily solved in the case of *r*BGH. Farmers can simply decide how much protein to inject into the bloodstream of a cow.

8.3 Producing Recombinant Proteins

The first step in the production of the *r*BGH protein is to transfer the *BGH* gene from the nucleus of a cow cell into a bacterial cell. Bacteria are single-celled prokaryotes that copy themselves very rapidly. They can thrive in the laboratory if they are allowed to grow in a liquid broth containing the nutrients necessary for survival. Bacteria with the *BGH* gene can serve as factories to produce millions of copies of this gene and its protein product. Making many copies of a gene is called **cloning** the gene.

Cloning a Gene Using Bacteria

The following three steps are involved in moving a *BGH* gene into a bacterial cell (Figure 8.12).

Step 1. Remove the Gene from the Cow Chromosome. The gene is sliced out of the cow chromosome on which it resides by exposing the cow DNA to enzymes that cut DNA. These enzymes, called **restriction enzymes**, act like highly specific molecular scissors. Restriction enzymes cut DNA only at specific sequences, called palindromes, such as:

Note that the bottom middle sequence is the reverse of the top sequence. Many restriction enzymes cut the DNA in a staggered pattern, leaving "sticky ends" such as:

Sticky end

Sticky end

Circular bacterial chromosome

Bacterial cell

Plasmid

Cow cell

DNA

Restriction enzyme

BGH gene

rBGH

Recombinant plasmid

② A plasmid from a bacterium is cut with the same restriction enzymes, creating the same "sticky ends" as the cow gene.

The cut gene and plasmid are placed together in a test tube. Complementary "sticky ends" bind, resulting in a recombinant plasmid.

① BGH gene is cut from the cow chromosome using restriction enzymes that leave "sticky ends" with specific base sequences.

③ The recombinant plasmid is reinserted into a bacterial cell.

The plasmids and the bacterial cells replicate, making millions of copies of the rBGH gene.

The rBGH genes produce large quantities of rBGH proteins that are harvested, purified, and injected into cows to increase milk production.

rBGH protein

Figure 8.12   Cloning genes using bacteria. Bacteria can be used as factories for the production of human or other animal proteins.

The unpaired bases form bonds with any complementary bases with which they come in contact. The enzyme selected by the scientist cuts on both ends of the *BGH* gene but not inside the gene.

| T A T | C | G T A C G | BGH gene | C G T A C G | A A C |
| A T A | G C A T G | C | BGH gene | G C A T G | C | T T G |

Since different individual restriction enzymes cut DNA only at specific points, scientists need some information about the entire suite of genes present in a particular organism, called the **genome**, to determine which restriction enzyme cutting sites surround the gene of interest. Cutting the DNA generates many different fragments, only one of which will carry the gene of interest.

Step 2. Insert the *BGH* Gene into the Bacterial Plasmid.   Once the gene is removed from the cow genome, it is inserted into a bacterial structure called a **plasmid**. A plasmid is a circular piece of DNA that normally exists separate from the bacterial chromosome and can replicate independently of the bacterial chromosome. Think of the plasmid as a ferry that carries the gene into the bacterial cell where it can be replicated. To incorporate the *BGH* gene into the plasmid, the plasmid is also cut with the same restriction enzyme used to cut the gene. Cutting both the plasmid and gene with the same enzyme allows the "sticky ends" that are generated to base-pair with each other (A to T and G to C). When the cut plasmid and the cut gene are placed together in a test tube, they reform into a circular plasmid with the extra gene incorporated.

The bacterial plasmid has now been genetically engineered to carry a cow gene. At this juncture, the *BGH* gene, is referred to as the *rBGH* gene, with the *r* indicating that this product is genetically engineered, or recombinant, because it has been removed from its original location in the cow genome and recombined with the plasmid DNA.

Step 3. Insert the Recombinant Plasmid into a Bacterial Cell.   The recombinant plasmid is now inserted into a bacterial cell. Bacteria can be treated so that their cell membranes become porous. When they are placed into a suspension of plasmids, the bacterial cells allow the plasmids back into the cytoplasm of the cell. Once inside the cell, the plasmids replicate themselves, as does the bacterial cell, making thousands of copies of the *rBGH* gene. Using this procedure, scientists can grow large amounts of bacteria capable of producing BGH.

Once scientists successfully clone the *BGH* gene into bacterial cells, the bacteria produce the protein encoded by the gene. Bacteria can be genetically engineered to produce many proteins of importance to humans. For example, bacteria are now used to produce the clotting protein missing from people with hemophilia as well as human insulin for people with diabetes.

Scientists at Monsanto engineered the bacteria so that they could synthesize the *rBGH* protein by placing the growth hormone gene from cows into a bacterial plasmid. The plasmid was placed back into the bacterial cells, which then transcribed the gene and translated the protein. Then the scientists were able to break open the bacterial cells, isolate the BGH protein, and inject it into cows.

Close to one-third of all dairy cows in the United States now undergo daily injections with recombinant bovine growth hormone. These injections increase the volume of milk that each cow produces by around 20%.

Prior to marketing the recombinant protein to dairy farmers, the Monsanto Company had to demonstrate that its product would not be harmful to cows or to humans who consume the cows' milk. This involved obtaining approval from the U.S. Food and Drug Administration (FDA).

FDA Regulations

The FDA is the governmental organization charged with ensuring the safety of all domestic and imported foods and food ingredients (except for meat and poultry, which are regulated by the United States Department of Agriculture). The manufacturer of any new food that is not **generally recognized as safe (GRAS)** must obtain FDA approval before marketing its product. Adding substances to foods also requires FDA approval, unless the additive is GRAS.

According to both the FDA and Monsanto, there is no detectable difference between milk from treated and untreated cows and no way to distinguish between the two. Even if there were increased levels of rBGH in the milk of treated cows, there should be no effect on the humans consuming the milk because we drink the milk and do not inject it. Drinking the milk ensures that any protein in it will be digested by the body, just like any other protein that is present in food. Therefore, in 1993, the FDA deemed the milk from rBGH-treated cows as safe for human consumption.

In addition, since the milk from treated and untreated cows is indistinguishable, the FDA does not require that milk obtained from rBGH-treated cows be labeled in any manner. Vermont is the only state that requires labeling of rBGH-treated milk. However, many distributors of milk from untreated cows label their milk as "hormone free," even though there is no evidence of the hormone in milk from treated cows.

It is not unusual that most of this work was performed for a corporation (Monsanto), not a university. There are some fundamental differences between the types of research performed by scientists in industry as compared to the work being done at universities and colleges.

Basic Versus Applied Research

Scientists in academia often seek answers to questions for which there is no profit motive or direct commercial application. This type of research, for which there is not necessarily a commercial application, is called **basic research** and is largely funded by taxpayers through government agencies such as the National Institute of Health (NIH) or the National Science Foundation (NSF). The premise behind basic research is that scientists cannot always predict which kinds of scientific understanding will be valuable to society in the future. For instance, scientists might study transcription or translation simply to better understand these processes. Genetic engineers may spend their entire careers trying to understand the conditions under which a particular protein is synthesized.

Funding for basic research is important because no one knows where the next piece of invaluable information will come from. When scientists first began studying the genes in the single-celled eukaryote, *Saccharomyces cerevisiae* (baker's yeast), they probably had no idea that most of the genes present in this yeast were also present in humans. Today, scientists manipulate the environmental conditions of yeast to better understand how genes are regulated in humans. Likewise, scientists interested in studying the diversity of tropical plants (Chapter 12) did not suspect that their work would assist the development of many pharmaceutical agents, including some anticancer agents.

Scientists in industry typically seek to answer questions that will have an immediate and profitable application, like the production of rBGH. This **applied research** is important for scientists in industry because new products and improvements to existing ones increase profitability, which in turn determines the success or failure of the business. One example of applied research that has proven to be very lucrative has been the genetic engineering of crop plants.

Genetic Engineers Can Modify Foods

Whether you realize it or not, you have probably been eating genetically modified foods for your entire life. Some of these modifications have occurred over the last several thousand years due to farmers' use of selective breeding techniques—breeding those cattle that produce the most milk or crossing crop plants that are easiest to harvest. While this artificial selection does not involve moving a gene from one organism to another, it does change the overall frequency of certain alleles for a gene in the population.

The genetic engineering techniques described earlier have allowed scientists to move genes between organisms. Unless you eat only certified organic foods, you have been eating food that has been modified in this way for some time. This may lead you to wonder why and how plants are genetically modified, whether eating them is bad for your health, and whether growing them is bad for the environment.

Why Genetically Modify Crop Plants?

Crop plants are genetically modified to increase their shelf life, yield, and nutritive value. The first genetically engineered fresh produce, tomatoes, became available in American grocery stores in 1994. These tomatoes were engineered to soften and ripen more slowly. The longer ripening time meant that tomatoes would stay on the vine longer, thus making them taste better. The slower ripening also increased the amount of time that tomatoes could be left on grocery store shelves without becoming overripe and mushy. An enzyme called **pectinase** mediates the ripening process in some produce, including tomatoes. This enzyme breaks down pectin, a naturally occurring substance found in plant cells. When the enzyme pectinase is active, it helps break down the pectin, and the produce softens.

To genetically modify tomatoes, genetic engineers inserted a gene that produces a mRNA transcript complementary to the mRNA produced by the transcription of a pectinase gene. In double-stranded DNA, one of the two strands codes for the protein. The mRNA produced by this template strand is called the **sense** RNA. Transcription of the non-template strand produces a version of the mRNA called the **antisense** strand. When the antisense version of the pectinase gene is transcribed, it produces a mRNA that is complementary to the mRNA normally transcribed from the pectinase gene. When the mRNA from the genetically engineered antisense gene is produced, it pairs with its naturally occurring pectinase mRNA complement. Binding the antisense and sense mRNAs leaves less of the sense pectinase mRNA available for translation. Thus, less of the pectinase enzyme is produced, and ripening occurs more slowly (Figure 8.13).

Increasing the economic return on crop plants by improving yield has been the driving force behind the vast majority of genetic engineering projects. Yield can be increased when plants are engineered to be resistant to pesticides, herbicides, drought, and freezing.

Many people believe that improving farmers' yields may help decrease world hunger problems. Others argue that, since there is already enough food being produced to feed the entire population, it might make more sense to use less technological approaches to feeding the hungry. Significant numbers of people around the world are malnourished or starving, not due to a shortage of food but because access to food is tied to money or land. However, as the population increases, it may become imperative to increase the yield of crop plants in order to feed all of the world's people.

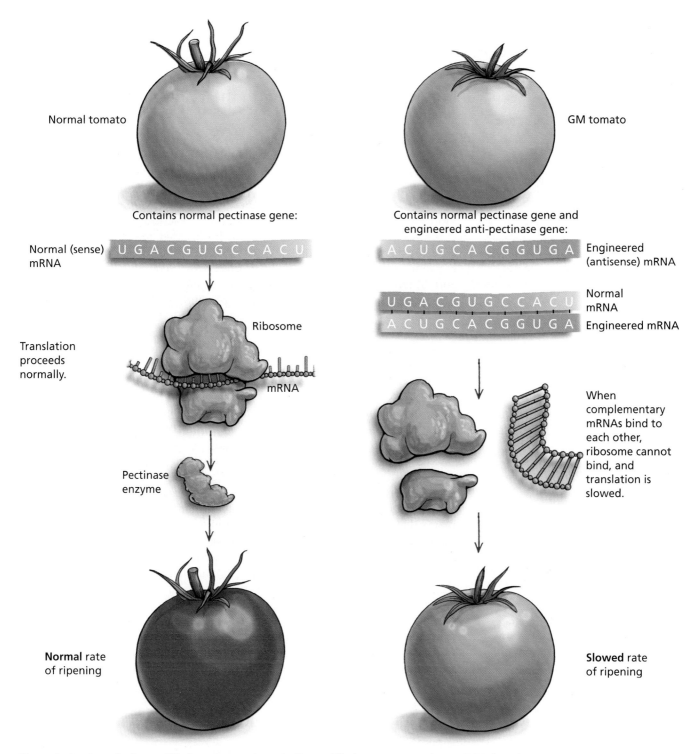

Figure 8.13   Genetically modified tomatoes. Genetically modified tomatoes produce mRNA that decreases the effects of the pectinase enzyme. When the pectinase gene is transcribed and translated, ripening occurs. When the sense pectinase mRNA is bound to the engineered antisense version, translation occurs more slowly, and ripening is slowed.

Genetic engineers may also be able to increase the nutritive value of crops. Some genetic engineers have increased the amount of beta-carotene in rice, a staple food for many of the world's people. Scientists hope that the engineered rice will help decrease the number of people who become blind in underdeveloped nations because cells require beta-carotene in order to synthesize vitamin A, a vitamin required for proper vision. Therefore, eating this genetically modified rice, called golden rice, increases a person's ability to

Figure 8.14 Golden rice. Golden rice has been genetically engineered to produce more β-carotene. The increased concentration of β-carotene makes the rice look more gold in color than unmodified rice does.

synthesize vitamin A (Figure 8.14). However, golden rice is not yet approved for human consumption, and there is debate about how effective the rice will actually be in preventing blindness.

Modifying Crop Plants with the Ti Plasmid and Gene Gun

To modify crop plants, the gene must be able to gain access to the plant cell, which means it must be able to move through the plant's rigid, outer cell wall. One "ferry" for moving genes into flowering plants is a naturally occurring plasmid of the bacterium *Agrobacterium tumefaciens*. In nature, this bacterium infects plants and causes tumors called **galls** (Figure 8.15a). The tumors are induced by a plasmid, called **Ti plasmid** (for tumor-inducing).

Genes from different organisms can be inserted into the Ti plasmid by (1) using the same restriction enzyme to cut the Ti plasmid and the gene, resulting in identical "sticky ends"; (2) connecting the gene and plasmid together; and (3) reinserting the recombinant plasmid into a bacterium. The bacterium, *A. tumefaciens*, with the recombinant Ti plasmid, is then used to infect plant cells. During infection, the recombinant plasmid is transferred into the host-plant cell (Figure 8.15b). For genetic engineering purposes, scientists use only the portion of a plasmid that does not cause tumor formation.

Moving genes into other agricultural crops such as corn, barley, and rice can also be accomplished by using a device called a **gene gun**. A gene gun shoots metal-coated pellets covered with foreign DNA into plant cells (Figure 8.16). A small percentage of these genes may be incorporated into the plant's genome. The gene gun is often used by companies that do not want to pay licensing fees to Monsanto, holder of the *A. tumefaciens* patent.

When a gene from one organism is incorporated into the genome of another organism, a **transgenic organism** is produced. A transgenic organism is more commonly referred to as a **genetically modified organism (GMO)**.

Many people have raised concerns about genetically modified (GM) crop plants. One concern is that large corporations that own many farms, called **agribusiness** corporations, are profiting so much from GM crop production that they will put owners of family farms out of business. Other concerns focus on the impact of GMOs on human health and the environment (Figure 8.17).

Effect of GMOs on Health

Concerns about the potential negative health effects of consuming GM crops have led some citizens to fight for legislation requiring that genetically modified foods be labeled, enabling consumers to make informed decisions about

Figure 8.15 Genetically modifying plants using the Ti plasmid. (a) Plants infected by *Agrobacterium tumefaciens* in nature show evidence of the infection by producing tumorous galls. (b) The Ti plasmid from *A. tumefaciens* serves as a shuttle for incorporating genes into plant cells. The recombinant plasmid is then used to infect developing plant cells, producing a genetically modified plant. When the plant cell reproduces, it may pass on the engineered gene to its offspring.

(a) Gall caused by *A tumefaciens*

(b) Using the Ti plasmid

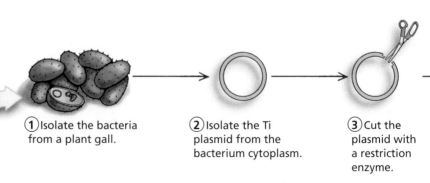

① Isolate the bacteria from a plant gall.

② Isolate the Ti plasmid from the bacterium cytoplasm.

③ Cut the plasmid with a restriction enzyme.

Figure 8.16 Genetically modifying plants using a gene gun. A gene gun shoots a plastic bullet loaded with tiny metal pellets (coated with DNA) into a plant cell. The bullet shells are prevented from leaving the gun, but the DNA-covered pellets penetrate the cell wall, cell membrane, and nuclear membrane of some cells.

the foods they eat. Manufacturers of GM crops argue that labeling foods is expensive and will be viewed by consumers as a warning, even in the absence of any proven risk. Those manufacturers believe that GM food labeling will decrease sales and curtail further innovation.

Genetically Modified Foods in the U.S. Diet.   As the labeling controversy rages, most of us are already eating GM foods. Scientists estimate that over half of all foods in U.S. markets contain at least small amounts of GM foods. Twelve different GM plants have been approved for production in the United States. Over 80% of all soybeans grown in the United States are genetically modified for herbicide resistance. Soybean-based ingredients, including oil and flour, are often produced from genetically modified plants and comprise one or more ingredients in many different processed foods.

Close to 40% of the U.S. corn crop is genetically modified to produce its own pesticide against caterpillar pests. Because GM corn is not separated from non-GM corn by farmers or food processors and because many processed food ingredients are corn-based including corn starch and corn syrup, GM corn is thought to be present in most of our processed foods. The percentage of fresh corn that is GM is thought to be closer to 4%.

Most of the canola oil in the United States is extracted from GM rapeseed plants, which are engineered for herbicide resistance. Canola oil is used in many different products, including vegetable oil, salad dressing, margarine, fried foods, chips, cookies, and pastries.

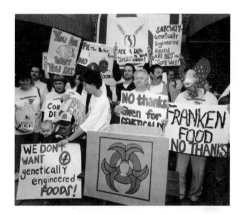

Figure 8.17   Protesters at a World Trade Organization meeting in Seattle. These people are concerned about how GMOs may affect humans and the environment.

④ Use the same enzyme to cut the gene of interest.

⑤ Allow the gene to attach to the plasmid.

⑥ Expose plasmids to young plant cells in culture.

Genetically modified plant contains new gene (and new characteristic).

⑦ Raise plant cells to maturity.

Figure 8.18 "No GMO" labelling. Many food manufacturers and consumers consider the use of unmodified foods to be a selling point for their products.

Genetically modified cotton varieties resistant to caterpillars now account for over 70% of the cotton crop. While cotton is more often used for clothing than for foods, cottonseed oil is used in cooking oils, salad dressing, peanut butter, chips, crackers, and cookies.

Of the 12 different GM plants approved for production, 8 are not commonly grown. Very few farmers are growing GM potatoes, squash, papaya, tomato, sugar beets, rice, flax, and radicchio, likely in response to consumer fears about the health consequences of eating these foods. Products that do not contain GMOs are often labeled to promote that fact (Figure 8.18).

How Are GM Foods Evaluated for Safety? Genetically modified crop plants must be approved by the Environmental Protection Agency (EPA) prior to their release into the environment. The FDA becomes involved in testing the GM crop only when the food from which the gene comes has never been tested, or when there is reason to be concerned that the newly inserted gene may encode a protein that will prove to be a toxin or an allergen.

Allergy is a serious problem for the close to 8% of Americans who experience allergic reactions to foods. Symptoms of food allergy range from a mild upset stomach to sudden death. Genetic engineers must be vigilant about testing foods with known allergens; a person who knows he must avoid eating peanuts may not know to avoid a food that has been genetically modified to contain a peanut gene that may cause a reaction—although no such food currently exists.

If the gene being shuffled from one organism to another is not known to be toxic or cause an allergic reaction, the FDA considers the GM food to be substantially equivalent to the foods from which it was derived; that is, the GM food is GRAS. If a modified crop contains a gene derived from a food that has been shown to cause a toxic or allergic reaction in humans, then it must undergo testing prior to being marketed.

This method of determining potential hazards worked well in the case of a modified soybean that carried a gene from the Brazil nut. This engineering was done in an effort to increase the protein content of soybeans. Since Brazil nuts were known to cause allergic reactions in some people, the modified beans were tested and did indeed cause an allergic reaction in susceptible people. The product was withdrawn, and no one was harmed.

Proponents of genetic engineering cite this as an example of the efficacy of the FDA rules. Opponents of genetically modifying foods wonder whether it will always be possible to predict which foods to test. They point out that it is possible for a protein encoded by a gene with one apparent function to interact with substances in its new environment in unpredictable ways and cause unpredictable effects. For example, those proteins that do not normally cause allergic reactions may be modified in a manner that transforms them into allergens. If a protein originally produced by bacteria (which do not modify proteins in the same manner as plants or other eukaryotic cells do) were inserted into plant cells, the plant cell could modify the protein in such a way that the protein becomes an allergen.

In evaluating toxicity, scientists focus on the protein produced by the modified plant and not the actual gene that is inserted. This is because the gene itself is digested and broken down into its component nucleotides when it is eaten and therefore will not be transcribed and translated inside human cells.

Many plants contain low levels of natural plant toxins. Because plants cannot defend themselves against predation by moving away or physically resisting, they have evolved to rely on these chemical defenses. In fact, the leaves and roots of many plant species are not edible due to the presence of these toxins. When early farmers domesticated plants, selective breeding led to the production of crop plants with reduced levels of toxins. This means that the plants we eat today have a much lower concentration of toxins. This is also part of the reason that modern plants are so susceptible to disease. If an inserted gene were to disrupt the regulation of a toxin gene whose activities had been diminished by selective breeding, it might increase the production of the toxin.

Concern about GM foods is not limited to their consumption. Many people are also concerned about the effects of GM crop plants on the environment.

GM Crops and the Environment

Many genetically modified crops have been engineered in order to increase their yield. For centuries, farmers have tried to increase yields by killing the pests that damage crops and by controlling the growth of weeds that compete for nutrients, rain, and sunlight. In the United States, farmers typically spray chemical pesticides and herbicides directly onto their fields. This practice concerns people worried about the health effects of eating foods that have been treated by these often toxic or cancer-causing chemicals. In addition, both pesticides and herbicides may leach through the soil and contaminate drinking water.

To help decrease farmers' reliance on pesticides, agribusiness companies have engineered plants that are genetically resistant to pests. For example, corn plants have been engineered to kill the European corn borer (Figure 8.19a). To do this, scientists transferred a gene that produces a toxin from the soil bacterium *Bacillus thuringiensis* (Bt) into corn. The *Bt* toxin gene encodes a protein that is lethal to corn borers but not to humans (Figure 8.19b). The idea of using this bacterium for pest control actually came from organic farmers, who have sprayed unengineered *B. thuringiensis* on crop plants for many years. Genetically modified Bt corn has proven to be so successful at resisting the corn borer that close to one-half of all corn currently grown in the United States is engineered with this gene.

Effects on Nontarget Organisms.   Shortly after the arrival of Bt corn, concern arose about its impact on organisms in the surrounding areas that are not pests—that is, nontarget organisms. One laboratory study showed that milkweed, a plant commonly found on the edges of cornfields that had been dusted

(a) Corn plants have been engineered to kill the insects that eat them.

(b) How it works:

(c) Pollen from Bt corn that dusts milkweed might unintentionally kill the butterfly larvae that eat the milkweed.

Corn borer

Bacillus thuringiensis produces a protein that kills corn borers.

Scientist inserts the protein-encoding *Bt* gene into corn

Corn cells produce protein that kills corn borers.

Monarch butterfly caterpillar

Milkweed (common on edges of corn fields)

Figure 8.19   The European corn borer. (a) The European corn borer damages corn and decreases yields. (b) Genetic modification of corn plants provides resistance to the pest. (c) Some researchers have shown that Bt corn may be harmful to other organisms.

with pollen from Bt corn, was lethal to monarch butterfly caterpillars, for which milkweed is the only source of food (Figure 8.19c). This research was performed in a laboratory and has not been shown to occur on farmers' fields, but results of this study indicate there may be cause for concern about how GM crops will affect other organisms.

Modified corn also caused controversy in 2000 when a variety of corn called StarLink™ was found in Taco Bell® taco shells. StarLink, containing a modified gene that was resistant to heat and did not break down during digestion, had not been approved by the EPA for human consumption. As a result, there was a massive recall of the taco shells as well as numerous other cornmeal-based products. This incident raised serious concerns about the ability of regulators, farmers, and food processors to keep unapproved GM products out of the nation's food supply.

Evolution of Resistant Pests.   Critics of Bt corn point out that it is only a matter of time before corn borers evolve resistance to Bt corn. Corn borers with genetic variations that give them a preexisting resistance to the toxin will be more likely to survive and pass on their resistance genes, creating a population of resistant insects. This in turn will require the development of new varieties of genetically engineered corn. The same is true for pesticides applied to crops; pests evolve resistance because application of a pesticide does not always kill all of the targeted organisms. The few pests that have preexisting resistance genes and are not susceptible survive and produce resistant offspring. Eventually, widespread resistance develops, and a new pesticide must be developed and applied.

The problem of accelerated evolution of *Bt* resistance is particularly vexing for the organic farmers, who were the first to use *B. thuringiensis* for controlling the corn borer but who did so in a targeted way. If corn borers develop resistance to *Bt* toxin due to widespread use of Bt corn, organic farmers will have lost a powerful tool for controlling this pest.

The continued need for the development of new pesticides in farming is paralleled by farmers' reliance on herbicides. Herbicide-resistant crop plants, such as Roundup Ready® soybeans, have been engineered to be resistant to Roundup® herbicide, used to control weeds in soybean fields. Farmers can now spray their fields of genetically engineered soybeans with herbicides that will kill everything but the crop plant. Some people worry that this resistance gene will allow farmers to spray more herbicide on their crops since there is no chance of killing the GM plant, thereby exposing consumers and the environment to even more herbicide.

Transfer of Genetic Material.   There is also concern that GM crop plants may transfer engineered genes from modified crop plants to their wild or weedy relatives. Wind, rain, birds, and bees carry genetically modified pollen to related plants near fields containing GM crops (or even to farms where no GM crops are being grown). Many cultivated crops have retained the ability to interbreed with their wild relatives; in these cases, genes from farm crops can mix with genes from the wild crops. This is unlikely to happen with corn or soybeans, which do not have weedy relatives in North America. However, it has already been demonstrated that GM canola has transferred genes to its weedy relatives, and the same is likely to happen with squash and rice. Thus the herbicide is rendered ineffective since both the crop plant and its weedy relative share the same resistance gene. It may become impossible to determine whether weed plants surrounding fields of engineered crops have been pollinated with pollen containing the modified gene, and there could be unintended consequences for the ecology of the surrounding environment. Also, if pollen from GM crop plants drifts to farms that are not growing modified crops, it becomes impossible to determine whether a crop plant has engineered genes. This would be disastrous in the event of a recall of the genetically modified seed.

Additional Problems.   Genetic manipulation could lead to decreasing variation within a species, and this too can have evolutionary consequences. Most GM corn, in addition to carrying the *Bt* resistance gene, has also been selectively bred to mature all at once, produce uniform ears, and have a particular nutrient profile. Because GM varieties do increase production or reduce the cost of inputs, they have often become extremely popular, meaning that most of the nation's corn and soybean crops are nearly genetically identical. If an unforeseen disease or pest were to sweep through an area containing this corn or soybean variety, the disease would probably devastate a large portion of the crop.

Most, but not all, of the genetic engineering that occurs to produce crop plants resistant to pesticides and herbicides is performed by private companies and is designed to maximize profits. For example, Roundup Ready soybeans are purchased by farmers who then apply Roundup herbicide; both the GM soybean and the herbicide are sold by Monsanto. Crops engineered for a more altruistic purpose, such as the golden rice described earlier, were developed in academic research centers and thus far have not proved financially viable for profit-making companies. Someday the techniques pioneered by agribusiness firms may be used to help solve the problem of world hunger and disease, but this has not been the case to date.

While there is hope that genetic engineers will be able to help solve hunger problems by making farming more productive, there are also concerns about any negative health and environmental effects of GM foods. It remains to be seen whether genetic engineering will constitute a lasting improvement to agriculture.

8.5 Genetic Engineers Can Modify Humans

Some genetic engineers are attempting to modify humans. These modifications may one day include replacing defective or nonfunctional alleles of a gene with a functional copy of the gene. If this happens, it might be possible for physicians to diagnose genetic defects in early embryos and fix them, allowing the embryo to develop into an adult without any genetic diseases. Recent developments that have helped scientists to better understand the human genome may make this scenario more likely.

The Human Genome Project

The **Human Genome Project** involves sequencing or determining the nucleotide-base sequence (A, C, G, or T) of the entire human genome and the location of each of the 20,000 to 25,000 human genes. In 1990, the Office of Health and Environmental Research of the U.S. Department of Energy (DOE), along with the National Institute of Health (NIH) and scientists from around the world, undertook this project. At the time, scientists involved in the project proposed to have a complete accounting of all the genes present in humans by the year 2005. However, the race to complete the sequencing of the human genome was drastically accelerated by technological advances and the involvement of a private company named Celera Genomics™. At stake were the rights to patent the gene sequences. Initially, Celera wanted to retain the rights to the DNA sequences, but government scientists were making sequences available to the public. Eventually, the two groups worked together to publish the entire sequence of billions of nucleotide pairs that comprise the human genome in 2003.

The scientists involved in this multinational effort also sequenced the genomes of the mouse, the fruit fly, a roundworm, bakers' yeast, and a common intestinal bacterium named *E. coli*. Scientists thought it was important to sequence the genomes of organisms other than humans because these **model organisms** are easy to manipulate in genetic studies and because important

Sequence from Lab 1

Sequence from Lab 2

ACCGTGTAACCGTATACGCGACCGGTAAG

AGTTTCGTAACCGTAAC⁻

GTAAGCTTACGCGGAATCCGTAACACGATGCTAGTTTC

ACCGTGTAACCGTATACGCGACCGGTAAGCTTACGCGGAATCCGTAACACGATGCTAGTTTCGTAACCGTAAC⁻

Compiled sequence

Figure 8.20   Chromosome walking.
Scientists can determine the location of genes on chromosomes using DNA sequence information uploaded to the Internet from laboratories around the world. Workers search for areas that overlap and fill in the gaps, much like assembling a jigsaw puzzle.

genes are often found in many different organisms. In fact, 90% of human genes are also present in mice; 50% are in fruit flies, and 31% are in bakers' yeast. Therefore, understanding how a certain gene functions in a model organism helps us understand how the same gene functions in humans.

To sequence the human genome, scientists first isolated DNA from white blood cells. They then cleaved the chromosomes into more manageable sizes using restriction enzymes, cloned them into plasmids, and determined the base sequence using automated DNA sequencers. These sequencing machines distinguish between nucleotides based on structural differences in the nitrogenous bases. Sequence information was then uploaded to the Internet. Scientists working on this, or any other project, could search for regions of sequence information that overlapped with known sequences. Using overlapping regions, scientists in laboratories all over the world worked together to patch together DNA sequence information. In this manner, scientists sequenced entire chromosomes by "walking" from one end of a chromosome to the other (Figure 8.20). DNA sequence information obtained by means of the Human Genome Project may someday enable medical doctors to take blood samples from patients and determine which genetic diseases are likely to affect them.

Many people worry about having these types of tests performed because this personal information may get back to their insurance companies or employers, but there is a positive side to having all of this information available. Once the genetic basis of a disease has been worked out—that is, how the gene of a healthy person differs from the gene of a person with a genetic disease—the information can be used to develop treatments or cures.

Gene Therapy

Scientists who try to replace defective human genes (or their protein products) with functional genes are performing **gene therapy**. Gene therapy may someday enable scientists to fix genetic diseases in an embryo. To do so, the scientists would supply the embryo with a normal version of a defective gene; this so-called **germ-line gene therapy** would ensure that the embryo and any cells produced by cell division would replicate the new, functional version of the gene. Thus, most of the cells would have the corrected version of the gene, and when these genetically modified individuals have children, they will pass on the corrected version of the gene. If scientists can fix genetic defects in early embryos, some genetic diseases can be prevented.

Another type of gene therapy, called **somatic cell gene therapy**, can be performed on body cells to fix or replace the defective protein in only the affected cells. Using this method, scientists introduce a functional version of a defective gene into an affected individual cell in the laboratory, allow the cell to reproduce, and then place the copies of the cell bearing the corrected gene into the diseased person.

This treatment may seem like science fiction, but it is likely that this method of treating genetic diseases will be considered a normal procedure in the not-too-distant future. In fact, genetic engineers already have successfully treated a genetic disorder called **severe combined immunodeficiency (SCID)**, a disease caused

Sequence from Lab 3

CTACCGTTACGGATATGCTTACTGTAC

Sequence from Lab 4

TATGCTTACTGTACCTTCAAAACTGACTTTGGCTAACCGTACTCTGTACCTTCAAAACTGACTTTT

CTACCGTTACGGATATGCTTACTGTACCTTCAAAACTGACTTTGGCTAACCGTACTCTGTACCTTCAAAACTGACTTTT

by a genetic mutation that results in the absence of an important enzyme and severely weakens the individual's immune system. Because their immune systems are compromised, people with SCID are incapable of fighting off any infection, and they often suffer severe brain damage from the high temperatures associated with unabated infection. Any exposure to infection can kill or disable someone with SCID, so most patients must stay inside their homes and often live inside protective bubbles that separate them from everyone, even family members.

To devise a successful treatment for SCID, or any disease treated with gene therapy, scientists had to overcome a major obstacle—getting the therapeutic gene to the right place.

Proteins break down easily and are difficult to deliver to the proper cells, so it is more effective to replace a defective gene than to continually replace a defective protein. Delivering a normal copy of a defective gene only to the cell type that requires it is a difficult task. SCID, a disorder that has been treated successfully, was chosen by early gene therapists in part because defective immune system cells could be removed from the body, treated, and returned to the body.

Immune system cells that require the enzyme missing in SCID patients circulate in the bloodstream. Blood removed from a child with SCID is infected with nonpathogenic (non-disease-causing) versions of a virus. This virus is first engineered to carry a normal copy of the defective gene in SCID patients. After the immune system cells are infected with the virus, these recombinant cells, which now bear copies of the functional gene, are returned to the SCID patient (Figure 8.21a).

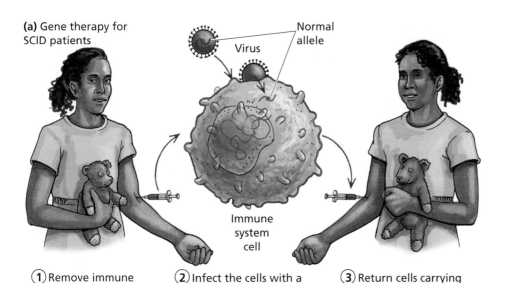

(a) Gene therapy for SCID patients

Normal allele

Virus

Immune system cell

① Remove immune system cells from patient.

② Infect the cells with a virus carrying the normal allele.

③ Return cells carrying the normal allele.

(b) SCID survivor

Figure 8.21   Gene therapy in a SCID patient. (a) A virus carrying the normal gene is allowed to infect immune system cells that have been removed from a person with SCID. The virus inserts the normal copy of the gene into some of the cells, and these cells are then injected into the SCID patient. (b) Ashi DiSilva, the first gene therapy patient.

In 1990, a 4-year-old girl named Ashi DiSilva (Figure 8.21b) was the first patient to receive gene therapy for SCID. Ashi's parents were willing to face the unknown risks to their daughter because they were already far too familiar with the risks of SCID—the couple's two other children also had SCID and were severely disabled. Ashi is now a healthy adult with an immune system that is able to fight off most infections.

However, Ashi must continue to receive treatments because blood cells, whether genetically engineered or not, have limited life spans. When most of Ashi's engineered blood cells have broken down, she must be treated again; thus, she undergoes this gene therapy a few times each year. Since Ashi's gene therapy turned out well, many other SCID patients have been successfully treated and can live normal lives. Unfortunately, Ashi's gene therapy does not prevent her from passing on the defective allele to her biological children because this therapy is not "fixing" the allele in her ovaries.

Although things worked out well for Ashi, successful gene therapy is far from routine. Two of 11 French boys treated with gene therapy for SCID developed leukemia that is thought to be related to their treatment, and an American teenager died from complications of experimental gene therapy meant to cure his relatively mild genetic disorder.

In addition to the risks involved in conducting any experimental therapy, not many genetic diseases can be treated with gene therapy. Gene therapy to date has focused on diseases caused by single genes for which defective cells can be removed from the body, treated, and reintroduced to the body. Most genetic diseases are caused by many genes, affect cells that cannot be removed and replaced, and are influenced by the environment.

Most people support the research of genetic engineers in their attempts to find better methods for delivering gene sequences to the required locations and for regulating the genes once they are in place. A far more controversial type of genetic engineering involves making an exact copy of an entire organism by a process called **cloning**.

Cloning Humans

Human cloning occurs commonly in nature via the spontaneous production of identical twins. These clones arise when an embryo subdivides itself into two separate embryos early in development. This is not the type of cloning that many people find objectionable; people are more likely to be upset by cloning that involves selecting which traits an individual will possess. Natural cloning of an early embryo to make identical twins does not allow any more selection for specific traits than does fertilization. However, in the future it may be possible to select adult humans who possess desired traits and clone them. Since cloning does not actually alter an individual's genes, it is more of a reproductive technology than a genetic engineering technology. However, it may someday be possible to alter the genes of a cloned embryo.

Cloning offspring from adults with desirable traits has been successfully performed on cattle, goats, mice, cats, pigs, rabbits, and sheep. In fact, the animal that brought cloning to the attention of the public was a ewe named Dolly.

Dolly was cloned when Scottish scientists took cells from the mammary gland of an adult female sheep and fused it with an egg cell that had previously had its nucleus removed. Treated egg cells were then placed in the uterus of an adult ewe that had been hormonally treated to support a pregnancy. Scientists had to try many times before this **nuclear transfer** technique worked. In all, 277 embryos were constructed before one was able to develop into a live lamb (Figure 8.22). Dolly was born in 1997.

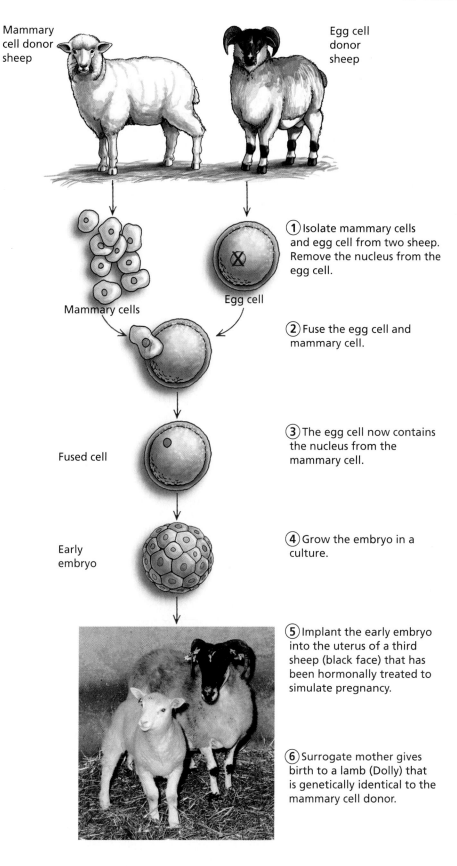

Mammary cell donor sheep

Egg cell donor sheep

Mammary cells

Egg cell

Fused cell

Early embryo

Figure 8.22   Dolly. This sheep is the result of a successful attempt to clone a mammal. She was genetically identical to her cell-donor mother.

(1) Isolate mammary cells and egg cell from two sheep. Remove the nucleus from the egg cell.

(2) Fuse the egg cell and mammary cell.

(3) The egg cell now contains the nucleus from the mammary cell.

(4) Grow the embryo in a culture.

(5) Implant the early embryo into the uterus of a third sheep (black face) that has been hormonally treated to simulate pregnancy.

(6) Surrogate mother gives birth to a lamb (Dolly) that is genetically identical to the mammary cell donor.

The research that led to Dolly's birth was designed to provide a method of ensuring that cloned livestock would have the genetic traits that made them most beneficial to farmers. Sheep that produced the most high-quality wool and cattle that produced the best beef would be cloned.

Essay 8.1 Stem Cells

Genetic engineers in some laboratories are trying to harness the healing powers of human stem cells. These cells, unlike most of the cells in your body, do not perform a specific function; instead, they are able to produce many different kinds of cells and tissues. Because stem cells do not have a particular function, they are said to be **undifferentiated**. Although they are undifferentiated, they can be pressed into service as many different cell types. Imagine that you are remodeling an old home, and you have a type of material that you can mold into anything you might need for the remodeling job—brick, tile, pipe, plaster, and so forth. Having a supply of this material would help you fix many different kinds of damage. Scientists believe that stem cells may serve as this type of all-purpose repair material in the body. If cells are nudged in a particular developmental direction in the laboratory, they can be directed to become a particular tissue or organ. Using stem cells from early embryos to produce healthy tissues as replacements for damaged tissues is called therapeutic cloning. Tissues and organs grown from stem cells in the laboratory may someday be used to replace organs damaged in accidents or organs that are gradually failing due to **degenerative diseases**. Degenerative diseases start with the slow breakdown of an organ and progress to organ failure. Additionally, when one organ is not working properly, other organs are affected. Degenerative diseases include diabetes, liver and lung diseases, heart disease, and Alzheimer's disease.

Stem cells could provide healthy tissue to replace those tissues damaged by spinal cord injury or burns. New heart muscle could be produced to replace muscle damaged during a heart attack. A diabetic could have a new pancreas, and people suffering from some types of arthritis could have replacement cartilage to cushion their joints. Thousands of people waiting for organ transplants might be saved if new organs were grown in the lab.

Stem cells are usually isolated from early embryos that are left over after fertility treatments. **In vitro** (Latin, meaning "in glass") fertilization procedures often result in the production of excess embryos because many egg cells are harvested from a woman who wishes to become pregnant. These egg cells are then mixed with her partner's sperm in a petri dish, resulting in the production of many fertilized eggs that grow into embryos. A few of the embryos are then implanted into the woman's uterus. The remaining embryos are stored so that more attempts can be made if pregnancy does not result or if the couple desires more children. When the couple achieves the desired number of pregnancies, the remaining embryos are discarded or, with the couple's consent, used for stem-cell research.

Early embryonic cells are harvested because stem cells are **totipotent** directly after fertilization; in other words, these stem cells can become any other cell type (Figure E8.1). As the embryo develops, its cells become less and less able to produce other cell types. As a human embryo grows, the early cells start dividing and forming different, specialized cells such as heart cells, bone cells, and muscle cells. Once formed, specialized nonstem cells can divide only to produce replicas of themselves. They cannot backtrack and become a different type of cell.

There is, however, a small supply of stem cells present in adult tissues, probably so that these tissues can repair themselves. Though stem cells exist in adult tissue, they are not present in great numbers, so they can be hard to find and extract for growth. They do not have the limitless replicative potential of embryonic stem cells, making it hard to grow them in large batches. Also, their ability to be transformed into different cell types is limited, making them less useful than embryonic stem cells.

Although embryonic stem cells are easier to work with and have more powerful healing potential than adult stem cells, laws in the United States restrict government funding for embryonic stem cell research. Embryos are destroyed when stem cells are harvested from them—a result that many find objectionable. In 2001, President Bush signed an executive order to ban scientists from using government money for studies involving human embryonic stem cells, unless those cells were created before the 2001 ban.

Figure E8.1 Early human embryo in a petri dish.

This technique is more efficient than allowing two prize animals to breed because each animal gives only half of its genes to the offspring. There is no guarantee that the offspring of two prize animals will have the desired traits. Even when a genetic clone is produced, there is no guarantee that the clone produced will be identical in the appearance and behavior as the original.

No one knows if nuclear transfer will work in humans—or if cloning is safe. If Dolly is a representative example, cloning animals may not be safe. In 2003, at age 6 years, Dolly was put to sleep to relieve her from the discomfort of arthritis and a progressive lung disease, conditions usually found only in older sheep. The fact that Dolly developed these conditions has led scientists to question whether she had aged prematurely. Scientists are watching other cloned animals for similar signs of premature aging.

Another type of cloning technology involves the use of early embryos that can be induced to develop into particular tissues or organs to be used for transplants. Called **therapeutic cloning**, this technique involves **stem cells**, a special type of cells, and has proven to be controversial (Essay 8.1).

The debate about human cloning mimics the larger debate about genetic engineering. As a society, we need to determine whether the potential for good outweighs the potential harm for each application of these technologies (Table 8.2). When it comes to human cloning, the potential for abuse could be substantial. Important questions regarding human cloning will not be resolved by developing human cloning techniques. Ideally, these issues will be discussed and legislation enacted *before* it becomes possible to clone humans.

Table 8.2 Pros and cons of genetic engineering.

Why the work of genetic engineers is important
• GM animals and crops may make farms more productive.
• GM crops may be made to taste better, last longer, or contain more nutrients.
• Genetic engineers hope to cure diseases and save lives.

Why the work of genetic engineers is controversial
• GM crops encourage agribusiness, which may close down some small farms.
• GM animals and crops may cause health problems in consumers.
• GM crops might have unexpected adverse effects on the environment.
• Present research might lead to the unethical genetic modification of humans.
• Lack of genetic diversity of GM crops could lead to destruction of food supply worldwide by pest or environmental change.

CHAPTER REVIEW

Summary

8.1 Genetic Engineers

• Genetic engineers manipulate genes for both nonprofit and for-profit reasons (p. 194).

8.2 Protein Synthesis and Gene Expression

• Genetic engineering techniques allow scientists to use bacterial cells to produce proteins for human use (pp. 194–195).

• Genes carry instructions for synthesizing proteins (p. 195).

• Protein synthesis involves the processes of transcription and translation (pp. 196–197).

• Transcription occurs in the nucleus of eukaryotic cells when an RNA polymerase enzyme binds to the promoter, located at the start site of a gene, and makes a mRNA that is complementary to the DNA gene. RNA differs from DNA in that the sugar in RNA is ribose (not deoxyribose) and the nitrogenous bases are adenine, guanine, cytosine, and uracil (no thymines in RNA) (p. 197).

• Translation occurs in the cytoplasm of eukaryotic cells and involves mRNA, ribosomes, and tRNA. Messenger RNA carries the code from the DNA, and ribosomes are the site where amino acids are assembled to synthesize proteins. Transfer RNA (tRNA) carries amino acids, which bind to triplet nucleotide sequences on the mRNA called codons (pp. 197–198).

• A particular tRNA carries a specific amino acid. Each tRNA has its unique anticodon that binds to the codon and carries instructions for its particular amino acid (p. 198).

• The amino acid coded for by a particular codon can be determined using the genetic code (p. 198).

• The flow of genetic information is from the DNA sequence to the mRNA transcript to the encoded protein (pp. 198–199).

• Mutations are changes to DNA sequences that can affect protein structure and function. Neutral mutations are changes to the DNA that do not result in a different amino acid being incorporated. Insertions or deletions of nucleotides can result in frameshift mutations that change the protein drastically (pp. 200–201).

• A given cell type expresses only a small percentage of the genes that an organism possesses (p. 202).

• Turning the expression of a gene up or down is accomplished in different ways in prokaryotes and eukaryotes. Prokaryotes typically block the promoter with a repressor protein to keep gene expression turned off. Eukaryotes regulate gene expression in any of 5 ways: (1) increasing transcription through the use of proteins that stimulate RNA polymerase binding; (2) varying the time that DNA spends in the uncondensed, active form; (3) altering the mRNA life span; (4) slowing down or speeding up translation; and (5) affecting the protein life span (pp. 202–204).

Web Tutorial 8.1 **Transcription**
Web Tutorial 8.2 **Translation**

8.3 Producing Recombinant Proteins

- Bovine growth hormone is a protein produced by the pituitary glands of cows. To increase the quantity of milk that a cow produces, additional growth hormone is injected into the cow (p. 204).

- Modern genetic engineering techniques enable scientists to produce recombinant BGH in the lab by placing the gene for growth hormone into plasmids, which then clone the gene by making millions of copies of it as they replicate themselves inside their bacterial hosts. Bacteria can then express the gene by transcribing a mRNA copy and translating the mRNA into a protein. The recombinant bovine growth hormone is then isolated and injected into cows (pp. 204–206).

- FDA approval is required for any new food or additive that is not generally recognized as safe (p. 207).

- Basic research is research for which there is not a known commercial application. Applied research typically has a more immediate application from which there can be a profit (p. 207).

Web Tutorial 8.3 **Producing Bovine Growth Hormone**

8.4 Genetic Engineers Can Modify Foods

- Genetic engineering techniques allow for the modification of foods much more quickly than do selective breeding techniques (p. 208).

- Crop plants are genetically modified to increase their shelf life, yield, and nutritive value. Some tomatoes have been modified to slow ripening by the insertion of RNA that is antisense to the ripening-enhancing pectinase gene (pp. 208–209).

- A Ti plasmid or gene gun can be used to insert a particular gene into plant cells (p. 210).

- Although there have been no documented incidents of negative health effects from GM food consumption, there is concern that some GM foods may cause allergic reactions or serve as toxins (p. 212).

- Concerns about GM crops include their impacts on surrounding organisms, the evolution of resistances, transfer of modified genes to wild and weedy relatives, and decreased genetic variation (pp. 213–215).

8.5 Genetic Engineers Can Modify Humans

- Information about genes obtained from the Human Genome Project can be used to help scientists replace genes that are defective or missing in people with genetic diseases (p. 215).

- Gene therapy involves replacing defective genes or their products in an embryo or in the affected adult tissue (p. 216).

- Gene therapy is considered experimental but may hold tremendous promise once scientists determine how to target genes to the right locations and express them in the proper amounts (pp. 216–217).

- Cloning animals with desirable agricultural traits has occurred. It may someday be possible to clone humans, but it is unclear if these humans would be healthy (p. 221).

Learning the Basics

1. List the order of nucleotides on the mRNA that would be transcribed from the following DNA sequence: CGATTACTTA

2. Using the genetic code (Table 8.1 on page 200), list the order of amino acids encoded by the following mRNA nucleotides: CAACGCAUUUUG

3. List the subcellular structures that participate in translation.

4. Transcription _____.
   A. synthesizes new daughter DNA molecules from an existing DNA molecule; **B.** makes a RNA copy of a gene that is to be translated; **C.** pairs thymines (T) with adenines (A); **D.** occurs on ribosomes

5. Transfer RNA (tRNA) _____.
   A. carries monosaccharides to the ribosome for synthesis; **B.** is made of messenger RNA; **C.** has an anticodon region, which is complementary to the mRNA codon; **D.** is the site of protein synthesis

6. During the process of transcription, _____.
   A. DNA serves as a template for the synthesis of more DNA; **B.** DNA serves as a template for the synthesis of RNA; **C.** DNA serves as a template for the synthesis of

proteins; **D.** RNA serves as a template for the synthesis of proteins

7. Translation results in the production of _____.

   A. RNA; **B.** DNA; **C.** protein; **D.** individual amino acids; **E.** transfer RNA molecules

8. The RNA polymerase enzyme binds to _____, initiating transcription.

   A. amino acids; **B.** tRNA; **C.** the promoter sequence; **D.** the ribosome

9. A particular triplet of bases in the coding sequence of DNA is TGA. The anticodon on the tRNA that binds to the mRNA codon is _____.

   A. TGA; **B.** UGA; **C.** UCU; **D.** ACU

10. RNA and DNA are similar because _____.

   A. they are both double-stranded helices; **B.** uracil is found in both of them; **C.** both contain the sugar deoxyribose; **D.** both are made up of nucleotides consisting of a sugar, a phosphate, and a base

Analyzing and Applying the Basics

1. Why are the cells comprising various tissues of your body different, even though they all contain the same genes?

2. Why are Ti plasmids and gene guns used to insert genes into plant cells?

3. Why are some genetic defects more likely to be cured by gene therapy than others are?

Connecting the Science

1. The first "test-tube baby," Louise Brown, was born over 30 years ago. Sperm from her father was combined with an egg cell from her mother. The fertilized egg cell was then placed into the mother's uterus for the period of gestation. At the time of Louise's conception, many people were very concerned about the ethics of scientists performing these in vitro fertilizations. Do you think human cloning will eventually be as commonplace as in vitro fertilizations are now? Why or why not?

2. Do you think it is acceptable to grow genetically modified foods if health risks turn out to be low but environmental effects are high?

CHAPTER 9

Where Did We Come From?

The Evidence for Evolution

Should public school students be taught about alternative hypotheses to evolution?

One idea about the origin of humans: Special creation.

Another idea: Evolution.

Why do biologists insist that only evolution be taught?

L ike many states and school districts, the Kansas State Board of Education recently put significant effort into updating and clarifying the list of concepts that students in grades from kindergarten through 12 are expected to master. In August 1999, the board adopted new statewide educational standards and immediately set off a firestorm of controversy. Conspicuously missing from the list of topics young Kansans were required to know before graduating was a portion of the subject of biological evolution. Specifically, the standards excluded the theory that describes the descent of modern species from extinct ancestors. "We are only being honest with our students," said school board member Steve Abrams. "Evolution is only a theory." In 2000, public backlash against the school board resulted in the defeat of several antievolution members and the reinstatement of evolution as a key component of the science standards. Then, in 2004, many of the ousted members were reelected, and the board took a new approach—mandating the teaching of "alternatives to evolution."

The furor over teaching evolution in Kansas is not an isolated event. In the last decade, legislators in states around the nation, including Michigan, Minnesota, Washington, Hawaii, and Louisiana, have considered requiring science curricula that include discussion of the "challenges" to evolution or even outright rejection of the topic. At the same time, school boards in local districts from Wisconsin to Texas have faced the same issue. In fact, the debate over whether children should learn about evolution is nearly as old as the theory itself.

The subject of evolution probably would not cause so much controversy, in Kansas and elsewhere, if it did not address fundamental questions of human existence: Who are we, and where do we come from? Many religious traditions include as part of their beliefs the understanding that humans were designed by an intelligent supernatural being. In contrast to these beliefs, evolutionary theory argues that humans arose through natural processes and that they are the descendants of ancient apes. On the surface, these ideas appear to be competing hypotheses about the origins of humanity. However, U.S. federal courts have consistently ruled that the First Amendment to the Constitution prohibits instruction of religious beliefs about human origins in American public school science classrooms.

If the belief of creation by a supernatural being is not allowed to be presented in public schools because it is a religious idea, then the theory of evolution should not be presented either, reasoned a majority of Kansas state school board members in 1999. After all, they argued, it seems that the theory of evolution represents a religious belief as well, just one that *rejects* any action of a higher power. In 2005, these same board members voted to allow the teaching of evolution as long as science teachers discussed "gaps" in the theory and offered alternative explanations to explain these gaps.

Do the school board actions in Kansas represent a revolution in science education—the recognition that scientific theories are as much a matter of faith as religious doctrine? Or, as others have said, does it represent a dangerous trend in U.S. education—one that sets religious belief on equal footing with scientific understanding? In this chapter, we examine these questions by exploring the theory of evolution and the origin of humans as a matter of science.

9.1   What Is Evolution?

Evolution really has two different meanings to biologists. The term can refer to either a process or an organizing principle, that is, a theory.

The Process of Evolution

Generally, the word *evolution* means "change," and the process of biological evolution is derived from this definition. **Biological evolution** is a change in the characteristics of a population of organisms that occurs over the course of generations. The changes in populations that are considered evolutionary are those that are inherited via genes. Other changes that may take place in populations due to environmental change are not evolutionary; for instance, the average dress size for women in the United States has increased from 8 to 14 over the past 50 years because of an increase in our average calorie intake and average age, not because of a change in genes.

As an example of the process of biological evolution, consider the species of organism commonly known as head lice. A **species** (Latin, meaning "kind") consists of a group of individuals that can regularly breed together, producing fertile offspring, and that is generally distinct from other species in appearance or behavior. Most species are physically subdivided into smaller groups, or **biological populations**, that are somewhat independent of other populations. As some parents of young children have discovered, some populations of head lice in the United States have become resistant to the pesticide permethrin, found in over-the-counter delousing shampoos. Initially, lice infections were readily controlled through treatment with these products; however, over time, populations of lice changed to become less susceptible to the effects of these chemicals. The evolution of resistance can occur rapidly—a study in Israel demonstrated that populations of head lice were four times less susceptible to permethrin only 30 months (or 40 lice generations) after the pesticide came into widespread use in that country.

Note that in this example, *individual* head lice did not "evolve" or change; instead, the population as a whole changed from one in which most lice were susceptible to one in which most lice were resistant to the pesticide. Individual lice were either susceptible to, or resistant to, permethrin. The variation in resistance among individuals in the population occurred due to variation among them in their genes—in particular, some lice carried gene forms (alleles) that conferred resistance, and others carried nonresistant alleles. Because individuals with the resistant alleles were much more likely to survive permethrin treatment, they passed these alleles on to their offspring. As a result, a population made up primarily of individual lice that carried the susceptible alleles changed into one in which most of the lice carry the resistant alleles. This change in the characteristics of the population took many generations (Figure 9.1). According to the definition of evolution presented earlier, the population of lice has evolved. In this case, the process of **natural selection**, the differential survival and reproduction of individuals in a population, brought about the evolutionary change. Natural selection is the process by which populations adapt to their changing environment. We will discuss natural selection in more detail in Chapter 10. Other forces, including chance, can cause evolutionary changes in the genetic makeup of populations as well. We discuss those forces in Chapter 11.

Most people accept that traits in populations can evolve. Evolutionary change in biological populations, such as the development of pesticide resistance in insects and antibiotic resistance in bacteria, has been observed multiple times. Changes that occur in the characteristics of populations are referred to as **microevolution**. In Chapter 10 we explore the consequences of microevolution resulting from researchers' efforts to combat certain human diseases.

The Theory of Evolution

Few members or supporters of the Kansas state school board disagree that the process of evolution occurs; however, they dispute whether the *theory* of evolution is "scientific truth." In fact, discussion of the theory of evolution is much more controversial in society than discussion of the process. Some of the

Initial lice infestation

After permethrin treatment, most lice are dead, but a few that are slightly more resistant to the pesticide survive.

Reinfestation with the offspring of the resistant lice. The population of lice is now more resistant to permethrin.

Figure 9.1 The process of evolution. The evolution of pesticide resistance in lice occurs as a result of natural selection, one of the mechanisms by which traits in a population can change over time.

controversy is generated by the use of the word *theory*. When people use this word in everyday conversation, they often are referring to a tentative explanation with little systematic support. A sports fan might have a theory about why her team is losing, or a gardener might have a theory about why his roses fail to bloom. Usually, these ideas amount to a "best guess" regarding the cause of some phenomenon. As first described in Chapter 1, a **scientific theory** is much more substantial—it is a body of scientifically acceptable general principles that help explain how the universe works. Scientific theories are supported by numerous lines of evidence and have withstood repeated experimental tests. For instance, the theory of gravity explains the motion of the planets; atomic theory explains the interactions between chemical elements and molecules; and the theory of relativity explains the relationship between mass and energy. The theory of evolution is an organizing principle for understanding how species originate and why they have the characteristics that they exhibit. The **theory of evolution** can thus be stated:

> All species present on Earth today are descendants of a single common ancestor, and all species represent the product of millions of years of microevolution via natural selection and other modes of evolutionary change.

In other words, modern animals, plants, fungi, bacteria, and other living things are related to each other and have been diverging from their common ancestor by various processes since the origin of life on this planet. The origin of modern species from a common ancestor is what most nonscientists think of when they hear the word *evolution*. This part of the theory of evolution, called the theory of common descent, is illustrated in Figure 9.2. However, most biologists understand that the theory of evolution encompasses both the theory of common descent and the processes by which evolutionary change occurs.

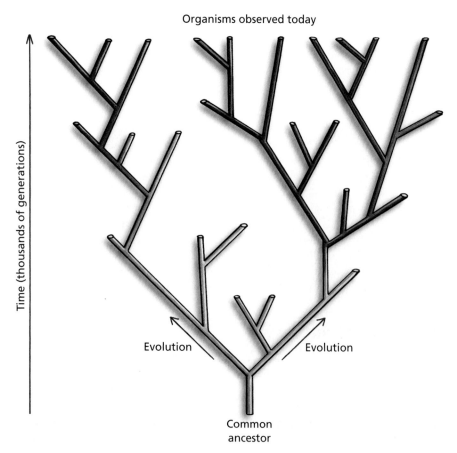

Figure 9.2   The theory of common descent. This theory states that all modern organisms descended from a single common ancestor. Each branching point on the tree represents the origin of new species from an ancestral form.

9.2 Charles Darwin and the Theory of Evolution

Charles Darwin is credited with bringing the theory of evolution into the mainstream of modern science (Figure 9.3). The youngest son of a wealthy physician, Darwin spent much of his early life as a lackluster student. After dropping out of medical school at Edinburgh University and at the urging of his father, Darwin entered Cambridge University to study for the ministry. Darwin barely made it through his classes but did strike up friendships with several members of the scientific community at the college. One of his closest companions at Cambridge was Professor John Henslow, an influential botanist who nurtured Darwin's deep curiosity about the natural world. It was Henslow who secured Darwin his first "job" after graduation. In 1831, at age 22, Darwin set out on what would become his life-defining journey—the voyage of the HMS *Beagle*.

The *Beagle*'s mission was to chart the coasts and harbors of South America. The ship's personnel was to include a naturalist for "collecting, observing, and noting anything worthy to be noted in natural history." Henslow recommended Darwin to be the unpaid assistant naturalist (and socially appropriate dinner companion) to the *Beagle*'s aristocratic, brilliant, but rigid captain, Robert FitzRoy, after two other candidates had turned it down.

Figure 9.3   Charles Darwin. As a young naturalist, Darwin conceived of and developed the theory of evolution. He spent nearly 25 years collecting data and building evidence for his hypothesis before publishing it.

Early Views of Evolution

The hypothesis that organisms, and even the very rock of Earth, had changed over time was not new when Darwin embarked on his voyage. The Greek poet Anaximander (611–546 B.C.) seems to have been the first Western philosopher to postulate that humans evolved from fish that had moved onto land. His ideas were later supplanted by those of Aristotle, who believed that species were fixed and unchanging and formed a progression from "lowest" to "highest" forms of life. Darwin was also familiar with the work of geologists James Hutton and Charles Lyell, who argued that geological features such as gorges and mountains were the result of slow, gradual change over the course of eons.

The first modern evolutionist, Jean Baptiste Lamarck, had published his theory of evolution in 1809, the year of Darwin's birth. Lamarck was the first scientist to clearly state that organisms adapted to their environments. He proposed that all individuals of every species had an innate, inner drive for perfection and that the traits they developed over their lifetimes could be passed on to their offspring. Lamarck used this principle in an attempt to explain why giraffes' necks are long; he proposed that ancient giraffes must have stretched their necks to reach leaves on higher trees and as a result gave birth to calves with longer necks. Lamarck's contemporaries were unconvinced by this proposed mechanism for species change— for instance, it was easily seen that the children of muscular blacksmiths had similar-sized biceps to those of bankers' children and had obviously not inherited their father's highly developed muscles. Lamarck's critics were also unwilling to question the more socially acceptable alternative hypothesis that Earth and its organisms were created in their current forms by God.

The Voyage of the *Beagle*

In the 5 years that he spent on the expedition of the HMS *Beagle*, Darwin spent most of his time on land—luckily for him, as it turns out, he never became accustomed to the ocean's swells and was nearly constantly seasick on board the ship (Figure 9.4 on page 230). The trip was a dramatic awakening for the young man, who was awed at the sight of the Brazilian rain forest, amazed by the scantily clothed natives in the brutal climate of Tierra del Fuego, and intrigued by

Figure 9.4   The voyage of HMS *Beagle*. Darwin's expedition took him to tropical locales from South America to Tahiti.

(a) Tortoise from Santa Cruz Island, an island with abundant vegetation.

(b) Tortoise from Espanola Island, an island with sparse vegetation.

Figure 9.5   Giant tortoises of the Galápagos. The subspecies of giant tortoises on the Galápagos Islands from different environments look distinct from each other. Individuals with dome-shaped shells (a) are common on islands with abundant vegetation, while those with flatter shells (b) are found on islands where vegetation is less abundant. Darwin felt that the existence of relatively similar, but obviously different, tortoises on these islands was best explained by descent from a common ancestor.

the diversity of animals and plants he collected. On the ship, he had ample time to read, including Charles Lyell's book *Principles of Geology*, which put forth the hypothesis that geological processes have not changed through time. In supporting this idea, called **uniformitarianism**, Lyell argued that deep canyons could result from the gradual erosion of rock caused by flowing water, given enough time. The hypothesis of uniformitarianism called into question the belief that Earth was less than 10,000 years old.

The most influential stop Darwin made on his journey was at a small archipelago of volcanic islands off the coast of Ecuador. The Galápagos Islands were not very appealing to Darwin at first; they seemed harsh and nearly lifeless. However, during the month that the *Beagle* spent sailing the islands, Darwin collected an astonishing variety of organisms. Many of the birds and reptiles he observed appeared to be unique to each island. While all islands had populations of tortoises, the type of tortoise found on one island was different from the types found on other islands (Figure 9.5). Darwin wondered why God would place different, unique subtypes of tortoise on islands in the same small archipelago. Upon his return to England, Darwin reflected on these and other observations and concluded that the different subspecies of tortoises on the different islands must have arisen through evolution from a single ancestral tortoise population. He noted a similar pattern of divergence between species on the islands and the closest mainland—prickly pear cacti in Ecuador have the ground-hugging shape familiar to many of us, while on the Galápagos, these plants have diverged into species that are tree-sized (Figure 9.6).

Developing the Hypothesis of Common Descent

Darwin's observations and portions of his fossil collection, sent periodically back to Henslow via other ships, made Darwin a scientific celebrity even before the *Beagle* returned to England. The respect he had gained from his scientific contemporaries provided an important buttress to the intense criticism caused by his interpretation of these observations. As he reflected on his observations and specimens, Darwin began to realize that they fit a pattern. He saw that the evidence supported the view of his own grandfather, Erasmus Darwin, who had hypothesized that all modern organisms are descended from a single common ancestor. This hypothesis, known as **common descent**, had been ridiculed by his grandfather's contemporaries and was still considered radical. Darwin's understanding of the revolutionary nature of this idea prevented him from sharing his hypothesis with all but a few close friends.

Fear of being rejected by his scientific colleagues caused Darwin to carefully document and research support for the hypothesis of common descent. In 1859, when he published his most influential book—*On the Origin of Species by Means of Natural Selection, or the Preservation of Favoured Races in the Struggle for Life*—the evidence he had accumulated was overwhelming. The main point of this text was to put forward a hypothesis proposing how species come about—in other words, the process of evolution via natural selection. Darwin devoted the last several chapters of *The Origin of Species* to describing the evidence for the hypothesis of common descent. He argued that the best explanation for observations of the relationships among modern organisms, and for the existence of fossils of extinct organisms, was that all organisms had descended from a single common ancestor in the distant past.

Darwin had finally been spurred into publishing his work in 1858, after receiving a letter from Alfred Russel Wallace. With Wallace's letter was a manuscript detailing a mechanism for evolutionary change—and it was identical to Darwin's theory of natural selection. Concerned that his years of scholarship would be forgotten if Wallace published his ideas first, Darwin had excerpts of both his and Wallace's work presented in July 1858 at a scientific meeting in London, and the next year he published *The Origin of Species*.

As defined in Chapter 1, a hypothesis is a tentative explanation for an observation. However, the evidence put forth in *The Origin of Species* was so complete, and from so many different areas of biology, that the hypothesis of common descent no longer appeared to be a tentative explanation. In response, scientists began to refer to this idea and its supporting evidence as the theory of common descent.

9.3   Evaluating the Evidence for Common Descent

(a)

(b)

Figure 9.6   Divergence from a common ancestor. (a) Prickly pear cactuses, like this *Opuntia* on continental South America, are shrub-like and low to the ground. (b) On the Galápagos Islands, some prickly pears are tree-sized, such as this variety of *Opuntia echios*. The overall similarity of these two species despite their difference in form supports the hypothesis that they descended from a common ancestral prickly pear.

Most scientists no longer question the theory that modern species evolved from extinct ancestors. Indeed, most biologists would agree that common descent is a scientific fact. However, the acceptance of the theory in the scientific community is in great contrast to the feelings of the nonscientist public. Opinion polls in the United States conducted over the past several decades have consistently indicated that nearly 50% of Americans do not believe that modern species arose from a common ancestor.

Let us explore the statement, "common descent (or evolution) is a fact," more closely. When *The Origin of Species* was published, most Europeans believed that **special creation** explained how organisms came into being. According to this belief, God created organisms during the 6 days of creation described in the first book of the Bible, Genesis. This belief also states that organisms, including humans, have not changed significantly since this beginning. According to some biblical scholars, the Genesis story indicates that creation also occurred fairly recently, within the last 10,000 years. Before Darwin, this story was scientists' best explanation for where living beings came from, and it helped explain the wonderful diversity and complexity of life. The belief in special creation grew from belief in Christianity in Europe. The same creation story is shared by other major religions as well, including Judaism and Islam.

Consider special creation as an alternative to the theory of common descent for explaining how modern organisms came to be. Is it an alternative that should be presented in public high school science classes? Since the idea of special creation requires the action of a supernatural entity, an all-powerful creator, it is not itself a scientific hypothesis. As discussed in Chapter 1, in order for a hypothesis to be testable by science, it must be able to be evaluated through observations of the material universe. Since a supernatural creator is not observable or measurable, there is no way to determine the existence or predict the actions of such an entity via the scientific method. This is the essential problem with any statement that supposes a supernatural cause—including the currently promoted "alternative to evolution" called intelligent design, which argues that some biological processes are too complex to have evolved and must have been supernaturally designed (Essay 9.1). However, the belief of special creation does provide some scientifically testable hypotheses. For instance, the assertion that organisms came into being within the last 10,000 years and that they have not changed substantially since their creation is testable through observations of the natural world. We can call this hypothesis about the origin and relationships among living organisms the static model hypothesis, indicating that organisms are recently derived and unchanging.

Essay 9.1 The Argument from Design

If a person walking through the woods found a watch on the path and examined it closely, he or she would find that it was highly complex, intricate, and accurate. Based on that observation, the person would naturally infer that the watch had been made by a skilled watchmaker. This point, advanced by William Paley, an English priest in 1802, is the basis of the "argument from design." If a complex watch must have been created by a watchmaker, Paley reasoned, then infinitely more complex living organisms must have been created as well. Essentially, the argument from design is an argument based on incredulity. It seems impossible to many of us that structures like the human eye or brain could be the result of an undirected process such as evolution.

The argument from design has two major flaws. First is the assumption that the evolution of structures that increase an organism's chance for survival is based on the random generation of these structures. It is true that the basis for evolutionary change is random—mutations are indeed chance events. However, the selection of mutations (via the survival of individuals who carry them) is not random, a point that we will explore further in Chapter 10. There is a direction in evolutionary change toward better fit to the environment. Thus organisms appear to be "designed" for their environment, but only because natural selection has favored the evolution of traits that increase survival and reproduction in that environment.

The second flaw in the argument from design is the assumption that complex structures must have arisen, fully functional, from formlessness. It seems highly unlikely that something as complex as an eye could have evolved in a series of small adaptations from a simpler structure. Darwin himself was concerned about this very example; so, in *The Origin of Species*, he described examples of less complex eyes that, if put in a series, illustrated how the camera-type human eye might have evolved from simpler structures. Highly complex structures can result from incremental changes to simple beginnings. Darwin's example of progressively complex eyes in nature seemingly put to rest this objection to the theory of evolution.

More recently however, the argument from design has been used when considering molecular structures and biochemical processes of extreme complexity. In his book, *Darwin's Black Box*, Michael Behe argued that many of the interacting parts and processes inside cells are irreducibly complex, meaning that if any one of the parts is not present, the whole system breaks down. This, Behe argues, is evidence that these processes must have been "intelligently designed." However, Behe's thesis has the same flaws as Paley's. He does not acknowledge simpler versions of his irreducibly complex processes and structures, and he assumes that the complexity could only have arisen fully formed, rather than adapted from another use. In fact, Behe's 1996 book posed a challenge to evolutionary biologists, and many have presented rigorous evolutionary analyses that demonstrate how his examples of "irreducibly complex" processes are likely to have evolved.

The essential problem with the argument from design—including the latest version, intelligent design—is that the existence and actions of a supernatural creator cannot be rejected by any observations of the natural world. The "alternative" to evolution that is proposed by intelligent design is not an alternative scientific theory. Instead, it consists of asserting the irreducible complexity of an ever-dwindling set of processes and structures that have not yet been subjected to intense evolutionary analysis.

There are also several intermediate hypotheses between the static model and common descent. One intermediate hypothesis is that organisms were created, perhaps millions of years ago, and that changes have occurred, but brand-new species have not arisen; we will call this the transformation hypothesis. Another intermediate hypothesis is that types of organisms arose separately and since their origin have diversified into numerous species; we will call this the separate types hypothesis. These hypotheses and the theory of common descent are summarized in Figure 9.7. Polls of the American public indicate that many people view all of these alternatives—static model, transformation, separate types, and common descent—as equally reasonable explanations for the origin of biological diversity.

Many nonscientists think that the Kansas State Board of Education made a reasonable choice in 1999 when it decided to remove the theory of evolution (and thus the theory of common descent) from the state's science standards.

Static model
Species arise separately
and do not change over time.

Transformation
Species arise separately but do change over time in order to adapt to the changing environment.

Separate types
Species do change over time, and new species can arise; but each group of species derives from a separate ancestor that arose independently.

Common descent
Species do change over time, and new species can arise. All species derive from a common ancestor.

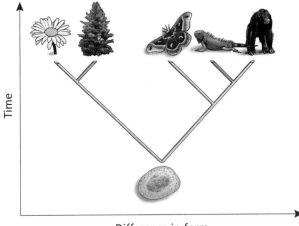

Figure 9.7   Four hypotheses about the origin of modern organisms. A graphical representation of the four hypotheses.

It seems to some that rather than bowing to the arguments of scientists who teach only one of many ideas about the origin of modern organisms, the school board made a justifiable decision to allow school districts to suppress discussion of the topic altogether. After the 2005 decision, many citizens felt that their elected officials were simply trying to be fair when they considered requiring the discussion of other hypotheses in addition to the theory of common descent.

But what about all those scientists who insist that the theory of common descent is fact? Why would so many scientists maintain this position? As you will soon see, the three alternative hypotheses are not equivalent to the theory of common descent. To understand why, we must evaluate the observations that help us test these hypotheses. We will do this as we address one of the most controversial questions underlying this debate: Are humans really related to apes?

Biological Classification Suggests Evolutionary Relationships

Any zookeeper will tell you that the primate house is the most popular exhibit in the park. People love apes and monkeys. It is easy to see why—primates are curious, playful, and agile. In short, they are fun to watch. But something else drives our fascination with these wonderful animals: We see ourselves reflected in them. The forward placement of their eyes and their reduced noses appear humanlike. They have hands with fingernails instead of paws with claws. Some can stand and walk on two legs for short periods. They can finely manipulate objects with their fingers and opposable thumbs. They exhibit extensive parental care, and even their social relations are similar to ours—they tickle, caress, kiss, and pout (Figure 9.8). Humans have long recognized our similarities with the apes. Cultures having close contact with these animals often gave them names that reflect this similarity—such as *orangutan*, a Malay word meaning "person of the forest."

Why are primates, particularly the great apes (gorillas, orangutans, chimpanzees, and bonobos) so similar to humans? Scientists contend that it is because humans and apes are recent descendants of a common biological ancestor. This statement is supported by the biological classification of humans.

Linnaean Classification.   As the modern scientific community was developing in the sixteenth and seventeenth centuries, various methods for organizing biological diversity were proposed. Many of these **classification systems** grouped organisms by similarities in habitat, diet, or behavior; some of these classifications placed humans with the great apes, and others did not.

Into the classification debate stepped Carl von Linné, a Swedish physician and botanist. Von Linné gave all species of organisms a two-part, or binomial, name in Latin, which was the common language of science at the time. In fact, he adopted a Latin name for himself by which he is more universally known— Carolus Linnaeus. The Latin names that Linnaeus assigned to other organisms typically contained information about the species' traits—for instance, *Acer saccharum* is Latin for "maple tree that produces sugar," the tree commonly known as the sugar maple, while *Acer rubrum* is Latin for "red maple."

In addition to the binomial naming system, Linnaeus developed a new way to organize living organisms according to shared physical similarities. His classification system was arranged hierarchically; organisms that shared many traits were placed in the same narrow classification, while those that shared fewer, broader traits were placed in more comprehensive categories. Linnaeus's hierarchy today takes the following form, from broadest to narrowest groupings:

Figure 9.8   Are humans related to apes? Biologists contend that apes and humans are similar in appearance and behavior because we share a common ancestor.

Domain

   Kingdom

      Phylum (or Division)

         Class

            Order

               Family

                  Genus

                     Species

Thus, for example, all organisms that were able to move under their own power, at least for part of their lives, and relied on other organisms for food were placed in the Kingdom Animalia. Within that kingdom, all organisms with backbones (or a related skeletal structure called a notochord) were placed in the same phylum, Chordata, and all chordates that possess fur and produce milk for their offspring were placed in the Class Mammalia, the mammals. Humans are mammals, as are dogs, lions, dolphins, and monkeys. The scientific name of a species contains information about its classification as well. For instance, the black bear, *Ursus americanus*, belongs to the genus *Ursus* along with the grizzly bear, *Ursus arctos*, and a few other bear species.

    Other scientists quickly adopted the logical and orderly Linnaean system of classification, and it became the standard practice for organizing biological diversity (Figure 9.9). To reflect a modern understanding of the tree of life, biologists

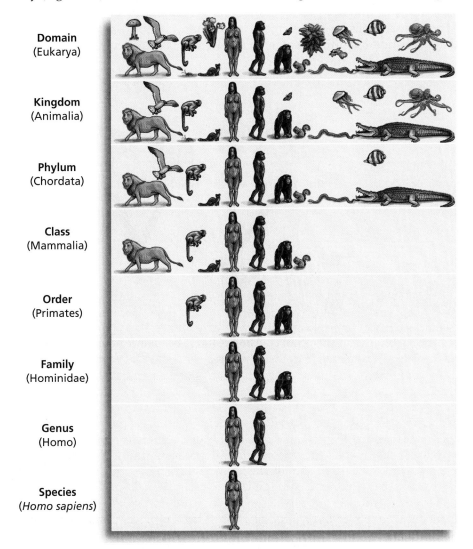

Domain
(Eukarya)

Kingdom
(Animalia)

Phylum
(Chordata)

Class
(Mammalia)

Order
(Primates)

Family
(Hominidae)

Genus
(Homo)

Species
(*Homo sapiens*)

Figure 9.9   The Linnaean classification of humans. This classification system groups organisms into progressively smaller categories. All organisms within a category share basic characteristics, and as the groups become narrower toward the bottom of the figure, the organisms have more similarities.

have added "sub" and "super" levels between these categories—such as super-family between family and order—and have even described new intermediate categories, such as "branch," as a way to best represent the complex evolutionary history within a group of organisms. Modern classification science is discussed more thoroughly in Chapter 12.

Classification of Humans and Apes.   Linnaeus himself did not believe that evolution occurred. His purpose in developing a biological classification was to determine what he called "God's plan" for the living universe. However, Darwin used Linnaeus's classification as a major facet of his argument for the theory of evolution. Darwin argued that Linneaus's system was an effective way to organize biological diversity because it reflected the underlying biological relationships among living organisms. Darwin noted that the levels in Linnaean classification could be interpreted as different degrees of relationship. In other words, all species in the same family share a relatively recent common ancestor, while all families in the same class share a more distant common ancestor. The relationship among species implied by Linnaean classification is typically illustrated with a tree diagram (Figure 9.10).

Using his classification system, Linnaeus placed humans, monkeys, and apes in the same order, which he called Primates, because humans have forward-facing eyes and coordinated hands like other primates. The modern classification of humans reflects only refinements of Linnaeus's ideas. Among living primates, humans are most similar to apes. Humans and apes share a number of characteristics, including relatively large brains, erect posture, lack of a tail, and increased flexibility of the thumb. Scientists now place humans and apes in the same family, Hominidae. Humans and the African great apes (gorillas, chimpanzees, and bonobos) share even more characteristics, including elongated skulls, short canine teeth, and reduced

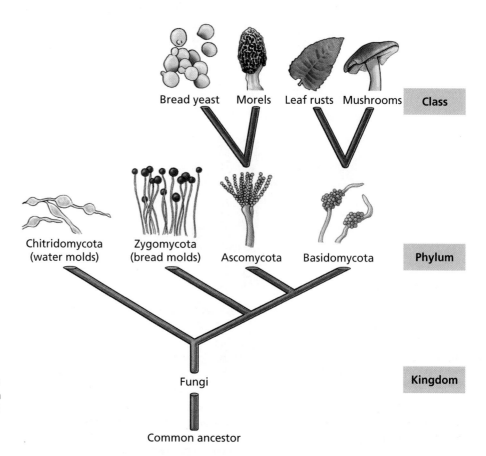

Figure 9.10 Classification implies common ancestry. The shared characteristics among fungal groups that help establish their classification in the Kingdom Fungi can be arranged in the form of a "tree of life."

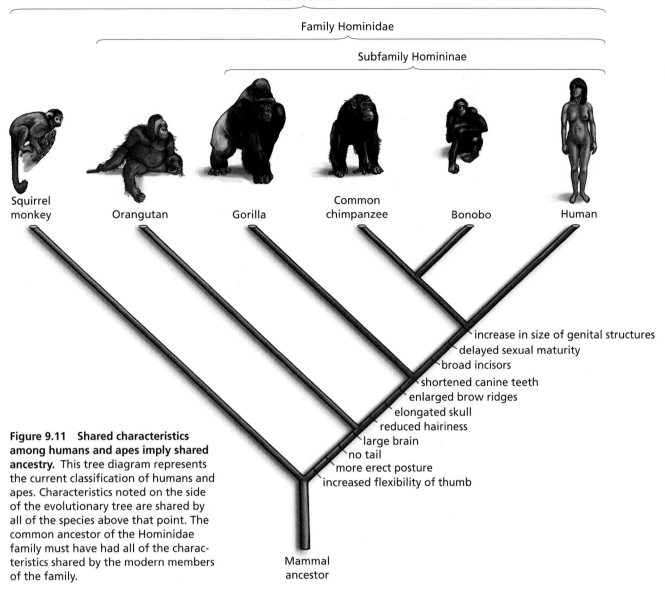

Figure 9.11 Shared characteristics among humans and apes imply shared ancestry. This tree diagram represents the current classification of humans and apes. Characteristics noted on the side of the evolutionary tree are shared by all of the species above that point. The common ancestor of the Hominidae family must have had all of the characteristics shared by the modern members of the family.

hairiness; they are placed together in the same *sub*family, Homininae. When the classification of humans, great apes, and other primates is shown in the form of a tree diagram (Figure 9.11), it is easy to see why Darwin concluded that humans and modern apes may have evolved from the same ancestor.

Evidence of Homology

The fact that different organisms share many traits does not necessarily indicate that they share biological ancestry. Linnaeus's classification alone does not support any of the four competing hypotheses about the origin of modern organisms. However, the tree of relationship implied by Linnaeus's classification, and illustrated in Figure 9.11, forms a hypothesis that can be tested. If modern species represent the descendants of ancestors that also gave rise to other species, we should be able to observe other, less obvious similarities between related modern species in anatomy and genetic material.

Figure 9.12 Homology of mammal forelimbs. The bones in the forelimbs of these mammals are very similar, although they are used for very different functions. Equivalent bones in each organism are shaded the same color. The similarity in basic bone structure is evidence of shared origin.

Comparative Anatomy. Figure 9.12 illustrates the concept of **homology**, which is the similarity in characteristics that has resulted from common ancestry. In this case, the homology is in anatomy, that is, physical structure; however, homology can be observed in other features of organisms from their behavior to their DNA. Each of the mammal forelimbs pictured in Figure 9.12 has a very different function—bat wings are used for flight, sea lion flippers for swimming, lion legs for running, and human arms for grasping and throwing. However, each of these limbs shares a common set of bones that are in the same relationship to each other even if they are quite different in size and relative proportion. The most likely explanation for the similarity in the underlying structure of these limbs is that each species inherited the basic structure from the same common ancestor, and the processes that cause evolution led to their unique modification in each group.

While similarities among organisms in the structure of their functional limbs provides support for the hypothesis that they share a common ancestor, even more compelling evidence comes from similarities between functional traits in one organism and nonfunctional features, or **vestigial traits**, in another. For instance, humans contain a number of traits that appear to have been modified from functional traits found in an ancestor. These traits either do not function in humans, or their function is highly modified from that of other descendants of the same ancestor. In other words, these vestigial traits represent a vestige, or remainder, of our biological heritage. Figure 9.13 provides two examples of vestigial traits in humans. Great apes and humans have a tailbone like other primates; yet neither great apes nor humans have a tail. Additionally, all mammals possess tiny muscles called arrector pili at the base of each hair. When the arrector pili contract under conditions of emotional stress or cold temperatures, the hair is elevated. In furry mammals, the arrector pili help to increase the perceived size of the animal, and they increase the insulating value of the hair coat. In humans, the same conditions produce only goose bumps, which provide neither benefit. Vestigial traits are found in nearly all groups of organisms—for example, flightless birds such as kiwis and ostrich still produce functionless wings, and flowering plants still produce

"Useful" trait in primate relative

Vestigial trait in human

(a) Tailbone

(b) Goose bumps

Figure 9.13   Vestigial traits reflect our evolutionary heritage. (a) Humans and other great apes do not have tails, but they do have a vestigial tailbone, which corresponds to the functional tailbone of a monkey. (b) The ability to elevate their fur helps many mammals seem bigger and provides increased insulation in cold conditions. The vestiges of this trait in humans appear as goose bumps, which arise under similar conditions of cold and intense emotion but serve no known function.

a tiny "second generation" within a developing flower ovule, a vestige of their relationship with ferns and mosses, which produce two independent generations during their life cycle (Figure 9.14).

Darwin maintained that the hypothesis of evolution provided a better explanation for vestigial structures than did the hypothesis of special creation represented by the static model. A useless trait such as goose bumps is better explained as the result of inheritance from our biological ancestors than as a feature that appeared independently in our species.

Homology in Development.   Another striking similarity among organisms that Darwin and his contemporaries observed was the resemblance of different chordates early in their development. All chordates—animals that have a backbone or closely related structure—produce structures called

(a) Fern gametophyte

(b) Flower ovary

Gametophyte generation is found here

Figure 9.14   Vestiges of the evolutionary history of flowering plants. (a) The ancestors of flowering plants were similar to modern ferns, which have two independent stages in their life cycle: the sporophyte stage that we are familiar with and a tiny gametophyte stage. This picture illustrates the fern gametophyte. (b) Flowering plants do not have two independent stages in their life cycle, but they do produce a tiny gametophyte within the ovule of the flower itself.

Snake   Chicken   Possum   Cat   Bat   Human

Early
stage embryo

Intermediate
stage embryo

Late
stage embryo

Pharyngeal slits

Tail

Figure 9.15   Similarity among chordate embryos. Vertebrate embryos are very similar in the first stage of their development, shown here in the top row, evidence that these diverse organisms share a common ancestor that developed along the same pathway. Differences among these species are much greater later in embryonic development.

pharyngeal slits, and most have tails as early embryos (Figure 9.15). These similarities in early development support the hypothesis that humans, bats, chickens, and snakes derived from an ancestor that developed along a similar pathway and that these species thus share an evolutionary relationship with all other chordates. The presence in early human embryos of a tail that is later lost is another piece of evidence that the tailbones of human adults represent a vestige of our evolutionary history. Homology in embryonic development is not limited to animals; for example, every member of the most common class of flowering plants, the eudicotyledons, produces a seedling with two embryonic leaves. Maple trees and sunflowers are essentially identical at "birth" because they share a common ancestor in the first dicotyledon plant (Figure 9.16).

Homology in Biochemistry.   Since Darwin's time, scientists' understanding of the nature of biological inheritance has expanded immensely. Scientists now understand that differences among individuals arise largely from differences in their genes. As we discussed in Chapter 6, these differences are due to randomly occurring mutations that are passed on to descendants. It stands to reason that differences among species must also derive from differences in their genes. If the hypothesis of common descent is correct, then species that appear to be closely related must have more similar genes than do species that are more distantly related.

The most direct way to measure the overall similarity of two species' genes is to evaluate similarities in their DNA. Recall from Chapter 5 that DNA molecules carry genetic information in the sequence of chemical bases making up their linear structures. A single gene on a DNA molecule may be made up of a few hundred to a few thousand bases; this sequence usually contains information about the structure of a protein, the physical manifestation of this stored information.

Many genes are found in nearly all living organisms. For instance, genes that code for histones, the proteins that help store DNA neatly inside cells, are found in algae, fungi, fruit flies, humans, and all other organisms that contain linear chromosomes. Among organisms that share many aspects of structure and function, such as flowering plants, many genes are shared. However, the *sequences* of these genes are not identical. If we compare the sequence of DNA bases for the same gene found in two different flowering plants, we find that the more similar their classification is, the more similar their genes are (Figure 9.17a). In other words, if classification indicates that two plants share a recent common ancestor, then their DNA sequences are more similar than those of two plants that share a more ancient common ancestor.

A comparison of the sequences of dozens of genes that are found in humans and other primates demonstrates the same pattern of relationship (Figure 9.17b). The DNA sequences of these genes in humans and chimpanzees are 99.01% similar, whereas the DNA sequences of humans and gorillas are identical over 98.9% of their length. More distantly related primates are less similar to humans in DNA sequence. This pattern of similarity in DNA sequence exactly matches the biological relationships implied by physical similarity. This result supports the hypothesis of common descent among the primates, as illustrated in Figure 9.11 on page 237. In fact, DNA sequence comparisons in a wide variety of organisms support hypotheses of

Figure 9.16 Embryonic homology in plants. The seedlings of maple trees and sunflowers are nearly identical despite their enormous differences as mature plants.

(a) Maple tree

(b) Sunflower

(a) Comparing DNA sequences

Species A	`ATTGCAACTGGTATCGAGGTTCTAC`
Close relative	`ATTGCCACTGGAATCGAGGTTCTAC`
Distant relative	`ATTGCCACTGGAATCGTGGTTCGAC`

2 differences in 25 nucleotides
2/25 = 8%; or 92% similarity

4 differences in 25 nucleotides
4/25 = 16%; or 84% similarity

(b) Similarity to human DNA sequences

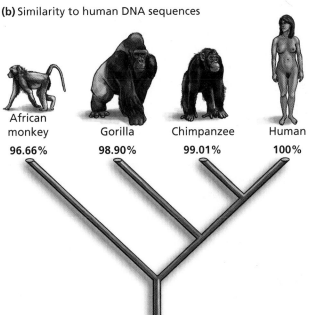

African monkey **96.66%**   Gorilla **98.90%**   Chimpanzee **99.01%**   Human **100%**

Figure 9.17 Similar organisms have similar DNA sequences. (a) The order of DNA bases contains information about the traits of an organism. We can compare the DNA of various organisms by looking at similarities and differences in the DNA sequences for various genes. The number of nucleotide differences between species is exaggerated in this example for illustration purposes. (b) Species that appear to be more similar to humans have more similar DNA sequences for the same genes when compared to species that are less similar.

Figure 9.18   Using DNA evidence to test an evolutionary hypothesis. Scientists comparing several physical features hypothesized a tree of relationships among swallowtail butterflies; a small portion of the tree is illustrated here. When another group of scientists measured similarity in DNA sequences for several genes in these species, the results supported the original hypothesis.

common descent that are implied by physical similarities. Results of one such hypothesis test are described in Figure 9.18.

At first, a finding of similarities in DNA sequence may not seem especially surprising. If genes are like instructions, then you would expect the instructions for building a human and a chimpanzee to be more similar than the instructions for building a human and a monkey. After all, humans and chimpanzees have many more physical similarities than humans and monkeys do, including reduced hairiness and lack of a tail. However, remember that the genes being compared perform the same function in all of these species. For example, one of the genes in this DNA analysis is *BRCA1*, a gene associated in humans with risk for breast cancer, which gives the gene its name. However, *BRCA1* also has the general function of helping repair damage to DNA in all organisms. Given this identical and very basic function in all organisms, there is no reason to expect that differences in the gene sequences for *BRCA1* among different species should conform to any particular pattern—unless the organisms are related by descent. But there is a pattern, and biochemical homology is evident; the *BRCA1* gene of humans *is* more similar to the *BRCA1* gene of chimpanzees than to the *BRCA1* gene of monkeys. The best explanation for this observation is that humans and chimpanzees share a more recent common ancestor than either species shares with monkeys.

Differences in DNA sequence between humans and chimpanzees may allow us to estimate when these two species diverged from their common ancestor. An estimate is based on a **molecular clock** derived from observations of DNA sequence differences in particular genes among several species groups. The principle behind a molecular clock is that the rate of change in certain DNA sequences, due to the accumulation of mutations within a species, seems to be relatively constant. According to one application of the molecular clock hypothesis, the amount of time required to generate a 1% difference in overall DNA sequence (about the difference between modern humans and modern chimpanzees) is approximately 5 to 6 million years.

Evidence from Biogeography

Even without the DNA evidence that modern biologists have, Darwin was able to draw on many examples of anatomical and behavioral homology to support the hypothesis that humans are closely related to chimpanzees and

(a) Charles mockingbird, Champion Island

(b) Hood mockingbird, Hood Island

(c) Longtailed mockingbird, mainland Ecuador

Figure 9.19   Evidence from biogeographic patterns supports common descent. Different mockingbird species with several similarities in appearance are found on different islands in the Galápagos. All species resemble a mockingbird found on mainland Ecuador. This observation supports the hypothesis that the Galápagos mockingbirds all share a common ancestor that came from the mainland.

other great apes. He predicted that additional evidence for this hypothesis would be the discovery of fossils of human ancestors in Africa. Darwin based this prediction on similar patterns he had seen of the distribution of species on Earth, that is, of the **biogeography** of life.

Earlier in this chapter, we described one biogeographical pattern observed by Darwin that supported the hypothesis of common descent—the relationships between different populations of giant tortoises on different islands in the Galápagos. Tortoises on each island are distinct in appearance—so much so that during Darwin's visit, the vice governor of the islands claimed that he could determine with certainty the home island of any tortoise brought to him. However, these varieties of tortoises all belong to the same species and must have the same ancestor. This observation convinced Darwin that all of the populations had changed over time independently on each island. Even more dramatic to Darwin was the appearance of unique species of mockingbirds on each island, all similar in appearance to a different species found on mainland Ecuador (Figure 9.19). Darwin noted that although the unique bird species on the Galápagos appeared to be related to bird species on the nearby mainland, the groups of species on these islands were very different from species found on other, similar islands that he visited, such as New Zealand. These observations support the hypothesis that species in a geographic location are generally descended from ancestors in that geographic location, but they do not support the hypothesis that species arise separately in each location, with each species as a best fit for the environment of that location. If the alternative hypotheses that species appeared independently were true, we would expect to see a set of "tropical-island-adapted" species that are found on all tropical islands.

The hypothesis of common descent helps explain some other puzzling biogeographical patterns, including the diversity of cactus species in the Western Hemisphere and the near absence of this plant family in the Eastern Hemisphere. Deserts are found in both hemispheres, of course, but the absence of these distinctive and dominant plants in African and Asian deserts is difficult to understand—unless the pattern results from the evolution of a large number of cactus species from an ancestor that arose in the Western Hemisphere. Interestingly, African deserts contain several species that superficially resemble cacti but belong to a different family called the euphorbs, which produce a very different flower type (Figure 9.20).

The similarities between cacti and some euphorbs are a result of **convergent evolution**—the evolution of similar adaptations to a common environmental problem in distantly related species. The ability to store water in stems, a column-like shape, and the conversion of leaves to spines

(a) Cactus

(b) Euphorb

Figure 9.20   Unrelated but convergent species. Species in the Cactus family and some in the Euphorb family have similar adaptations to desert conditions. These plants are not closely related, however— a conclusion based on differences in their flower structure. Cacti are found almost exclusively in the Western Hemisphere, while Euphorbs are found in deserts in the Eastern Hemisphere.

are all adaptations to reduce water loss in dry environments and are found in euphorbs and cacti as a result of convergence, not homology. Despite their superficial similarity, the striking differences between cactus flowers and euphorbia flowers is evidence that they are not closely related. However, distinguishing between convergence and homology can often be a challenge for scientists attempting to reconstruct the evolutionary history of a group of organisms, as described in Chapter 12.

We can now see why Darwin predicted, based on biogeographical patterns, that evidence of human ancestors would be found in Africa, the home of chimpanzees and other great apes. If humans and apes share a common ancestor, then highly mobile humans must have first appeared where their less-mobile relatives can still be found.

Evidence from the Fossil Record

Biogeographical patterns and observations of anatomical and genetic similarities among modern organisms provide good evidence to support the theory of evolution. As with nearly all evidence in science, these observations allow us to infer the accuracy of the hypothesis but do not prove the hypothesis correct. This type of evidence is similar to the "circumstantial evidence" presented in a murder trial, such as finding the murder weapon in a car belonging to a suspect or the presence of the suspect's fingerprints on the victim's door. The European scientific community in the nineteenth century was convinced by the circumstantial evidence Darwin had collected and embraced the theory of common descent as the best explanation for the origin of species. Since Darwin's time, scientists have accumulated additional indirect evidence, such as the DNA sequence similarity discussed earlier, to support this theory. But as in a murder trial, direct evidence is always preferred to establish the truth—for instance, the testimony of an eyewitness or a recording of the crime by a security camera. Of course, there are no human eyewitnesses to the evolution of humans, but we have a type of "recording," in the form of the fossil record. The evidence from the fossil record provides even more convincing support for the theory of common descent.

The Fossil Evidence.   **Fossils** are the remains left in soil or rock of living organisms. Most fossils of large animals are rocks that have formed as the organic material in bone decomposed and minerals filled the spaces left behind (Figure 9.21). The process of fossilization requires special conditions because most organisms either quickly decompose after death or are scavenged by other organisms; to form fossils, dead organisms have to be protected from these processes. Therefore, the fossil record is not a complete "recording" of the

Figure 9.21   Fossilization. When bones fossilize, the material that makes up much of their substance slowly decays and is replaced by minerals from water seeping through the sediments in which they are buried. Eventually what remains is a rock "model" of the original bone.

1. An organism is rapidly buried in water, mud, sand, or volcanic ash. The tissues begin to decompose very slowly.

2. Water seeping through the sediment picks up minerals from the soil and deposits them in the spaces left by the decaying tissue.

3. After thousands of years, most or all of the original tissue is replaced by very hard minerals, resulting in a rock model of the original bone.

4. When erosion or human disturbance removes the overlying sediment, the fossil is exposed (as shown here looking from above).

history of life—it is more similar to a security video that captures only a small portion of the action, with many blank segments. Fossils are more likely to form when an organism is rapidly buried by water, mud, or volcanic ash. Fortunately for scientists looking for fossils of **hominins**—humans and human ancestors—these organisms were likely to live near water, so their fossil record is a bit more complete and provides a compelling record.

Fossils were well known by Darwin's time. In the early 1800s, paleontologists—scientists who search for, describe, and study ancient organisms—were beginning to describe the large fossil remnants of dinosaurs. The first complete skeleton of a dinosaur was found in southern New Jersey in 1858, a year before *The Origin of Species* was published. In later books, Darwin noted that convincing evidence for an evolutionary relationship between humans and modern apes would come from the fossils of human ancestors.

One key difference between humans and other apes is our mode of locomotion. While chimpanzees and gorillas use all four limbs to move, humans are bipedal; that is, they walk upright on only two limbs. The reason that bipedalism evolved is unclear. Some scientists, including Darwin, hypothesized that the upright gait was an advantage that freed the hands for tool use. More recently, scientists have hypothesized that an upright posture allowed early hominins to reach fruit in small trees and shrubs. Most likely, bipedalism evolved because it had many benefits to our ancestors.

Whatever its origin, bipedalism evolved through several anatomical changes. The face is now on the same plane as the back instead of at a right angle to it; thus the foramen magnum, the hole in the skull through which the spinal cord passes, is found on the back of the skull in other apes but at the base of the skull in humans. In addition, in bipedal apes, the structures of the pelvis and knee are modified for an upright stance; the foot is changed from being grasping to weight-bearing, and the lower limbs are elongated relative to the front limbs. Thus, a variety of skeletal features can provide clues about the locomotion of a fossil, ape-like creature (Figure 9.22).

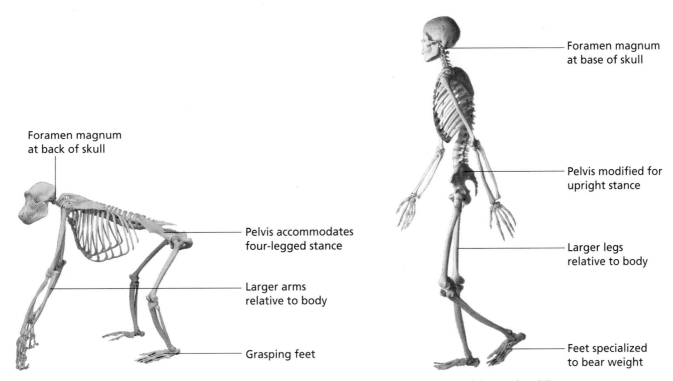

Foramen magnum at back of skull

Pelvis accommodates four-legged stance

Larger arms relative to body

Grasping feet

Foramen magnum at base of skull

Pelvis modified for upright stance

Larger legs relative to body

Feet specialized to bear weight

Figure 9.22 Anatomical differences between humans and chimpanzees. Humans are bipedal animals, while chimpanzees typically travel on all fours. The evolution of bipedalism in hominins resulted from several anatomical changes. If any of these features are present in a fossil primate, the fossil is classified as a hominin.

Figure 9.23   A fossil ancestor. (a) Lucy is still the most complete fossil of *Australopithecus afarensis* ever found. Her pelvis and knee joint provide evidence that she walked on two legs. (b) This artist's conception of what Lucy looked like in life is based on her fossil remains as well as other fossils of the same species.

(a) Lucy's skeleton

(b) Artist's reconstruction

Scientists found the first fossil hominins not in Africa, but in Europe. Johann Fuhlrott discovered a fossil of *Homo neanderthalensis* (Neanderthal man) in 1856, in a small cave within the Neander Valley of Germany. Fuhlrott recognized his find as a primitive human, but the German scientific establishment rejected his interpretation, incorrectly claiming that it was a modern human. Eugene Dubois found fossils of human-like creatures now called *Homo erectus* (standing man) in Java in 1891. It was not until 1924 that Raymond Dart found the first *African* hominin fossil, the Taung child. This fossil was later placed in the species *Australopithecus africanus*. Paleontologists continue to discover new hominin fossils in southern and eastern Africa, including the famous "Lucy," a remarkably complete skeleton of the species *Australopithecus afarensis*, discovered by paleontologist Donald Johanson in 1974 (Figure 9.23). Lucy's fossil skeleton included a large section of her pelvis, which clearly indicated that she walked upright.

When Did Ancient Hominins Live?   Scientists can determine the date when an ancient fossil organism lived by estimating the age of the rock that surrounds the fossil. **Radiometric dating** relies on a natural process that results in change in particular chemical elements, the basic ingredients of matter. This process, called radioactive decay, results as radioactive elements in the rock spontaneously break down into different, unique elements known as 'daughter products.' Each radioactive element decays at its own unique rate. The rate of decay is measured by the element's half-life—the amount of time required for one-half of the amount of the element originally present to decay into the daughter product.

When rock is newly formed from cooled magma, the liquid underlying Earth's crust, it contains a fixed amount of any radioactive element. When the magma hardens, this radioactive element becomes trapped within the resulting

(a)

Radioactive element

Daughter product

Percentage of parent element remaining

100

50

25
19
13
6
0

Decay curve

If 19% of the radioactive element remains in a rock sample, and...

If the half-life of the radioactive element is 1 million years, then...

... the age of the rock sample is 2.5 x 1 million = 2.5 million years

0   1   2   2.5   3   4

Number of half-lives

(b)

Magma 1.8 million years old

Fossils between 1.8–2.5 million years old

Magma 2.5 million years old

Figure 9.24 Radiometric dating. (a) The age of rocks can be estimated by measuring the amount of radioactive material (designated by dark purple circles) with a known half-life and the amount of daughter material (designated by light blue circles) in a sample of rock. In this example, the total amount of still-radioactive material remaining is 19%. By finding that number on the horizontal axis and tracing over to the decay curve, we can determine how many half-lives have passed. Knowing the half-life of the radioactive material, we can calculate the age of the rock. (b) The age of a fossil can be estimated when it is found between two layers of magma-formed rock.

rock. As the element decays over time, the amount of the original element in the rock declines, and correspondingly, the amount of daughter product increases. By determining the ratio of radioactive element to daughter product in a rock sample and knowing the half-life of the radioactive element, scientists can estimate the number of years that have passed since the rock formed (Figure 9.24).

Some critics of the theory of common descent note that different scientists may calculate different dates for the same fossil and that all of these fossil dates represent estimates with a certain degree of potential error. This uncertainty about the exact age of a fossil occurs because the layers of rock containing fossils did not form from magma—meaning that most fossils cannot be dated directly. Instead, fossils are found in rocks formed from the sediments that initially buried the organism. When fossils are found in sedimentary rock between layers of magma-formed rocks, the fossils are assumed to be intermediate in age to the magma-formed rocks. Even though fossils cannot be aged with perfect accuracy, the ages of particular fossil species inferred from the age of surrounding rocks are always within the same general range. Radiometric dating of fossils helps place a timeline on the historical record of living organisms.

Scientists have used radiometric dating to estimate the age of the Earth and the time of origin of various groups of organisms. Using this technique, scientists have also determined that the most ancient hominin fossil, the species *Ardepithecus ramidus*, is 5.2 to 5.8 million years old. (An even older fossil species, *Orrorin tugenensis*, the famous "millennium man," was recently described as a 6-million-year-old human ancestor, but many scientists are reserving judgment about this fossil specimen until more examples are found.) These very early fossils probably represent hominins that are quite similar to the common ancestor of humans and chimpanzees.

Figure 9.25 The evolutionary relationships among hominin species. This tree represents the current consensus among scientists who are attempting to uncover human evolutionary history. Notice that modern humans are the last remaining species of a group that was once highly diverse and consisted of several coexisting species.

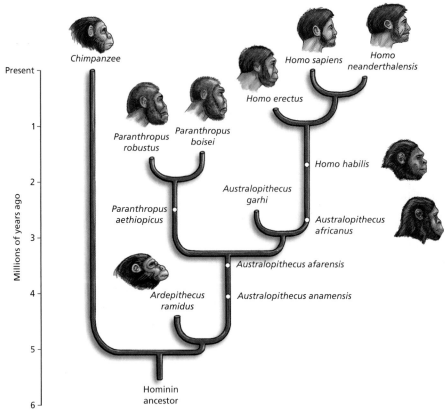

Figure 9.26 The common ancestor of humans and chimpanzees? A "missing link" between humans and chimpanzees would not look half human and half ape, as this cartoon suggests. Both humans and chimpanzees have been evolving separately for at least 5 million years, and their common ancestor would not closely resemble either modern species.

What Is the Missing Link? As the number of described hominin fossils has increased, a tentative genealogy of humans has emerged. The fossil species can be arranged in a pedigree from most ancient to most modern species by determining the age of a fossil and the anatomical similarities among organisms (Figure 9.25). What this pedigree indicates is that modern humans are the last remaining branch of a once diverse group of hominins. The pedigree illustrates a common theme seen in the fossil record of other groups: When a new lifestyle (in this case, bipedal ape) evolves, many different types appear, but only a small number of lineages survive over the long term. Biologists often refer to the lineages that die out as "evolutionary experiments." But does the pedigree provide convincing evidence that modern humans evolved from a common ancestor with other apes?

The common ancestor of humans and chimpanzees is often called the "missing link." Because the common ancestor of humans and chimpanzees has not been identified, some critics of evolution say that the biological relationship between apes and humans remains unproven.

However, finding the fossilized common ancestor between chimpanzees and humans, or between any two species for that matter, is extremely difficult, if not impossible. In order to identify a common ancestor, the evolutionary history of both species since their divergence must be clear. Like humans, modern chimpanzees have been evolving over the 5 million years since they diverged from humans. In other words, a missing link would not look like a modern chimpanzee with some human features, nor like a cross between the two species—as suggested by nineteenth-century satirical cartoons depicting Charles Darwin as an "ape man" (Figure 9.26). While the history of hominins is becoming clearer as new fossils are being identified, much less work has been done on the evolutionary history of chimpanzees.

Furthermore, if we take a closer look at the theory of common descent, we can see that support for this theory does not *require* the identification of a missing link. The theory of evolution is also supported by evidence of intermediate forms between a modern organism and its ancestors. This

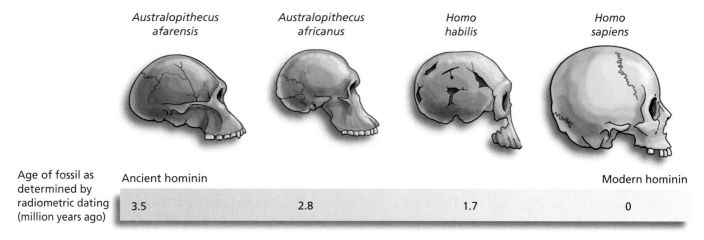

| *Australopithecus afarensis* | *Australopithecus africanus* | *Homo habilis* | *Homo sapiens* |

| | Ancient hominin | | | Modern hominin |
Age of fossil as determined by radiometric dating (million years ago)

| 3.5 | 2.8 | 1.7 | 0 |

Figure 9.27   The ape-to-human transition. Ancient hominins display numerous ape-like characteristics, including a large jaw, small braincase, and receding forehead. More recent hominins have a reduced jaw, larger braincase, and smaller brow ridge, much like modern humans.

evidence should be much easier to locate because these forms have existed since the divergence between chimpanzees and humans about 5 million years ago, and it is the type of evidence provided by the hominin fossil record (review Figures 9.23 and 9.25). Besides being bipedal, humans differ from other apes in having a relatively large brain, a flatter face, and a more extensive culture. The oldest hominins are bipedal but are otherwise similar to other apes in skull shape, brain size, and probable lifestyle. More modern hominins show greater similarity to modern humans, with flattened faces and increased brain size (Figure 9.27). Even younger fossil finds indicate the existence of symbolic culture and extensive tool use, trademarks of modern humans.

The transition from ancient hominin to modern human is not the only fossil record that supports the theory of common descent. Examples of well-described transitions include one between ancient reptiles and mammals and another between ancient and modern horses (Figure 9.28); many others have been found as well.

Figure 9.28   The fossil record of horses. Horse fossils provide a fairly complete sequence of evolutionary change from small, catlike animals with four toes to the modern horse with one massive toe.

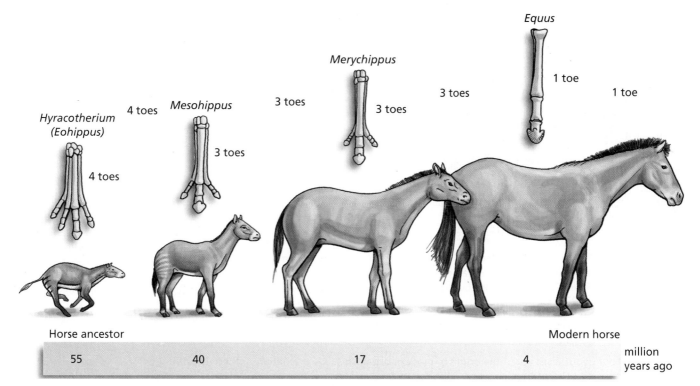

Equus

Merychippus

1 toe

1 toe

Mesohippus   3 toes   3 toes   3 toes

Hyracotherium (Eohippus)   4 toes

3 toes

4 toes

| Horse ancestor | | | | Modern horse |

| 55 | 40 | 17 | 4 | million years ago |

9.4 # Are Alternatives to the Theory of Evolution Equally Valid?

Now we return to the four competing hypotheses: static model, transformation, separate types, and common descent. Do the observations described in the previous section allow us to reject any of these hypotheses? Figure 9.29 summarizes our findings.

The Static Model and Transformation Hypotheses

The physical evidence we have discussed thus far allows us to clearly reject only one of the hypotheses—the static model. Radiometric dating indicates that Earth is far older than 10,000 years, and the fossil record provides unambiguous evidence that the species that have inhabited this planet have changed over time.

Of the remaining three hypotheses, transformation is the poorest explanation of the observations. If organisms arose separately, and each changed on its own path, there is no reason to expect that different species would share structures—especially if these structures are vestigial in some of the organisms. There is also no reason to expect similarities among species in DNA sequence. The hypothesis of transformation predicts that we will find little evidence of biological relationships among living organisms. As our observations have indicated, evidence of relationships abounds.

The Separate Types and Common Descent Hypotheses

Both the hypothesis of common descent and the hypothesis of separate types contain a process by which we can explain observations of relationships. That is, both hypothesize that modern species are descendants of common ancestors. The difference between the two theories is that common descent hypothesizes a single common ancestor for all living things, while separate types hypothesizes that ancestors of different groups arose separately and then gave rise to different types of organisms. Separate types seems more reasonable than common descent to many people. It seems impossible that organisms as different as pine

Static model	Transformation	Separate types	Common descent
Rejected	**Rejected**	**Rejected**	**Supported by a wide variety of evidence, including comparative anatomy, DNA sequences, and the fossil record.**
Earth is far older than 10,000 years, and species have clearly changed over time.	Evidence of relationships among organisms abound.	Universality of DNA, genetic code, and cell components are evidence of a single origin of all life.	

Figure 9.29   Four hypotheses about the origin of modern organisms. An evaluation of these hypotheses leads to the rejection of all of them except for the theory of common descent.

trees, mildew, ladybugs, and humans share a common ancestor. However, several observations indicate that these disparate organisms are all related.

The most compelling evidence for the single origin of all life is the universality of both DNA and of the relationship between DNA and proteins. As noted in Chapter 8, genes from bacteria can be transferred to plants, and the plants will make a functional bacterial protein. This is possible only because both bacteria and plants translate genetic material into functional proteins in a similar manner. If bacteria and plants arose separately, we could not expect them to translate genetic information similarly.

Units One and Two of this text describe biochemical processes and cell structures that are found in nearly all living things. The fact that organisms as different as pine trees, mildew, ladybugs, and humans contain cells with nearly all of the same components and biochemistry is also evidence of shared ancestry. A mitochondrion could have many different possible structural forms and still perform the same function; the fact that the mitochondria in a plant cell and an animal cell are essentially identical implies that both groups of organisms received these mitochondria from a common ancestor.

Pine trees, mildew, ladybugs, and humans *are* very different. Proponents of the hypothesis of separate types argue that the differences among these organisms could not have evolved in the time since they shared a common ancestor. See Essay 9.2 to explore hypotheses about the origin of the common ancestor of all life. But the length of time during which pine trees, mildew, ladybugs, and humans have been diverging is immense—at least 1 billion years. The remaining basic similarities among all living organisms serve as evidence of their ancient relationship (Figure 9.30).

The Best Scientific Explanation for the Diversity of Life

Scientists favor the theory of common descent because it is the best explanation for how modern organisms came about. The theory of evolution—including the theory of common descent—is robust, meaning that it is a good explanation for a variety of observations and is well-supported by a wide variety of evidence from anatomy, geology, molecular biology, and genetics. Evidence for the theory of common descent demonstrates **consilience**, meaning that there is concurrence among observations derived from different sources. Consilience is a feature of all

Figure 9.30   The unity and diversity of life. The theory of evolution, including the theory of common descent, provides the best explanation for how organisms as distinct as pine trees, humans, mildew, and ladybugs can look very different while sharing a genetic code and many aspects of cell structure and cell division.

Essay 9.2 The Origin of Life

The origin of the first living cell is an active research question in biology. There are two scientific hypotheses regarding the source of this first organism:

Hypothesis 1: The common ancestor arose on another planet and was imported to Earth.

Hypothesis 2: The common ancestor arose on Earth abiotically, that is, through natural processes from nonliving materials.

To test hypothesis 1, scientists look for life on more distant planets. Astronomers can use a tool called spectroscopy to analyze light reflecting from the surface of these planets for the chemical "signature" of life. Spectroscopy allows researchers to determine if a planet has oxygen and methane in its atmosphere, two gases that are maintained only by the action of living organisms. None of the planets nearest Earth display this chemical signature.

At least one meteorite has been described as containing evidence of life—the famous "Martian meteorite," ALH84001, found on Antarctic ice in 1984 and described in detail in Chapter 2. Scientists studying this meteorite argued in a 1996 paper that it contained convincing evidence of bacterial life, although not all scientists are as convinced. The recent remote explorations of the Martian surface by robotic rovers confirm that water was once present on the surface of the planet, raising the tantalizing prospect that life once existed there.

Hypothesis 2, the most well-studied hypothesis about the evolution of life, states that the common ancestor arose on Earth through natural processes from nonliving materials. According to this hypothesis, the process can be broken down into three basic steps.

1. Nonbiological processes assembled the simple molecules that were present early in the history of the solar system into more complex molecules.

2. These molecules then assembled themselves into chains that could store information and/or drive chemical reactions.

3. Collections of these complex molecules were assembled into a self-replicating "cell" with a membrane and energy source. This cell fed on other complex molecules.

There is some experimental evidence to support all three steps of hypothesis 2. First, Stanley Miller, a young graduate student working in the lab of his mentor Harold Urey in 1953, attempted to recreate conditions on early Earth within a laboratory apparatus. After allowing the apparatus, which contained very simple molecules and an energy source, to "run" for 1 week, Miller found that complex molecules had formed spontaneously. These molecules included the building blocks of proteins and sugars. His results and others support step 1 of this hypothesis. More recent experiments have demonstrated that these building-block chemicals can be induced to form long chains when put into contact with hot sand, clay, or rock. Long chains of DNA and RNA nucleotides and of amino acids have been created via these methods, providing some experimental support for step 2. Finally, in the early 1980s, two teams of scientists demonstrated that an information-carrying molecule, RNA, could also potentially copy itself, and so at least part of step 3 has experimental support. Although life as we would identify it has not been created in the lab abiotically, these results support the hypothesis that life *could have* formed spontaneously on Earth.

strongly supported scientific theories. Similarities between humans and modern apes are seen in anatomy, behavior, development, and DNA sequence. The age, location, and appearance of hominin fossils also demonstrate a relationship between these species. The theory of common descent is no more tentative than is atomic theory; few scientists disagree with the models that describe the basic structures of atoms, and few disagree that the evidence for the theory of common descent is overwhelming. Most scientists would say that both of these theories are so well-supported that we can call them fact.

Evolutionary theory helps us understand the functions of human genes, comprehend the interactions among species, and predict the consequences of a changing global environment for modern species. Describing evolution as *"just* a theory" vastly understates the importance of evolutionary theory as a foundation of modern biology. Students who do not have a grasp of this fundamental biological principle may lack an appreciation of the basic unity and diversity of life and fail to understand the effects of evolutionary history and change on the natural world and ourselves. For example, we will explore why evolution is important for understanding and treating human disease in Chapter 10. School boards and legislatures do not serve their children well by arguing that "alternative" scientific hypotheses that have been convincingly falsified through systematic observation should be included in their education as possible explanations for the diversity of life.

CHAPTER REVIEW

Summary

9.1 What Is Evolution?

- The process of evolution is the change that occurs in the characteristics of organisms in a population over time (p. 226).

- The theory of evolution, as described by Charles Darwin, is that all modern organisms are related to each other and arose from a single common ancestor (p. 228).

9.2 Charles Darwin and the Theory of Evolution

- Scientists before Charles Darwin had hypothesized that species could change over time. Darwin's voyage on the HMS *Beagle* led him to suspect that this hypothesis was correct. Over the course of 20 years, he was able to gather enough evidence to support this hypothesis and the hypothesis of common descent so that most scientists accepted these theories as the best explanation for the diversity of life on Earth (p. 231).

Web Tutorial 9.1 **Principles of Evolution**

9.3 Evaluating the Evidence for Common Descent

- Linnaeus classified humans in the same order with apes and monkeys based on his observations of physical similarities between these organisms. Darwin argued that the pattern of biological relationships illustrated by Linnaeus's classification provided strong support for the theory of common descent (pp. 236–237).

- Similarities in the underlying structures of a variety of organisms and the existence of vestigial structures are difficult to explain except through the theory of common descent (p. 239).

- Similarities in embryonic development among diverse organisms are best explained as a result of their common ancestry (p. 240).

- Modern data on similarities of DNA sequences among organisms match the hypothesized evolutionary relationships suggested by anatomical similarities and provide an independent line of evidence supporting the hypothesis of common descent (pp. 240–242).

- Biogeographical patterns support the hypothesis of common descent because species that appear related physically are also often close to each other geographically (pp. 243–244).

- Radiometric dating of fossil remains indicates that human-like, bipedal primates appeared about 5 million years ago (pp. 246–247).

- As predicted by the theory of common descent, ancient hominins have more ape-like characteristics than do more modern hominins (pp. 248–249).

9.4 Are Alternatives to the Theory of Evolution Equally Valid?

- Evidence strongly supports the hypothesis that organisms have changed over time and are related to each other (pp. 250–251).

- Shared characteristics of all life, especially the universality of DNA and the relationship between DNA and proteins, provide evidence that all organisms on Earth descended from a single common ancestor rather than from multiple ancestors (pp. 251–252).

Learning the Basics

1. Describe the theory of common descent.

2. What observations did Charles Darwin make on the Galápagos Islands that helped convince him that evolution occurs?

3. What is a vestigial structure, and how does the existence of these structures support the theory of common descent?

4. The process of biological evolution _____.

   A. is not supported by scientific evidence; **B.** results in a change in the features of individuals in a population; **C.** takes place over the course of generations; **D.** b and c are correct; **E.** a, b, and c are correct

5. In science, a theory is a(n) _____.

   A. educated guess; **B.** inference based on a lack of scientific evidence; **C.** idea with little experimental support; **D.** a body of scientifically acceptable general principles; **E.** statement of fact

6. The theory of common descent states that all modern organisms _____.

   A. can change in response to environmental change; **B.** descended from a single common ancestor; **C.** descended from one of many ancestors that originally arose on Earth; **D.** have not evolved; **E.** can be arranged in a hierarchy from "least evolved" to "most evolved"

7. Which of the following lists shows the classification levels in order from broadest grouping to narrowest grouping?

   A. family, phylum, genus, order; **B.** phylum, family, genus, class; **C.** order, genus, species, phylum; **D.** kingdom, order, genus, species; **E.** class, phylum, family, order

8. The DNA sequence for the same gene found in several species of mammals _____.

   A. is identical among all species; **B.** is equally different between all pairs of mammal species; **C.** is more similar between closely related species than between distantly related species; **D.** provides evidence for the hypothesis of common descent; **E.** more than one of the above is correct

9. What characteristics of a fossil can paleontologists use to determine whether the fossil is a part of the human evolutionary lineage?

   A. the position of the foramen magnum; **B.** the structure of the pelvis; **C.** the structure of the foot; **D.** a and c are correct; **E.** a, b, and c are correct

10. The fossil record of hominins _____.

    A. does not indicate a relationship between humans and apes because a missing link has not been found; **B.** dates back at least 5 million years; **C.** indicates that bipedal apes first evolved in Africa; **D.** b and c are correct; **E.** a, b, and c are correct

Analyzing and Applying the Basics

1. The classification system devised by Linnaeus can be "rewritten" in the form of an evolutionary tree. Draw a tree that illustrates the relationship among these flowering species, given their classification (note that "subclass" is a grouping between class and order):

   Pasture rose (*Rosa carolina*, family Rosaceae, order Rosales, subclass Rosidae)

   Live forever (*Sedum purpureum*, family Crassulaceae, order Rosales, subclass Rosidae)

   Spring avens (*Geum vernum*, family Rosaceae, order Rosales, subclass Rosidae)

   Spring vetch (*Vicia lathyroides*, family Fabaceae, order Fabales, subclass Rosidae)

   Multiflora rose (*Rosa multiflora*, family Rosaceae, order Rosales, subclass Rosidae)

2. DNA is not the only molecule that is used to test for evolutionary relationships among organisms. Proteins can also be used, and the sequences of their building blocks (called amino acids) can be compared in much the same way that DNA sequences are compared. *Cytochrome c* is a protein found in nearly all living organisms; it functions in the transformation of energy within cells. The percent difference in amino acid sequence between humans and other organisms can be summarized as follows:

Cytochrome c Sequence of:	Percent Difference from Human Sequence
Chimpanzee	0.0%
Mouse	8.7%
Donkey	10.6%
Carp	21.4%
Yeast	32.7%
Corn	33.3%
Green algae	43.4%

   Draw the evolutionary tree implied by this data that illustrates the relationship between humans and the other organisms listed.

3. Whales and dolphins are sea-dwelling mammals that evolved from land-dwelling ancestors. Describe two pieces of evidence that would help support this hypothesis.

Connecting the Science

1. Humans and chimpanzees are more similar to each other genetically than many very similar-looking species of fruit fly are to each other. What does this similarity imply regarding the usefulness of chimpanzees as stand-ins for humans during scientific research? What do you think it implies regarding our moral obligations to these animals?

2. Creationists have argued that if students learn that humans descended from animals and are, in fact, a type of animal, these impressionable youngsters will take this fact as permission to act on their "animal instincts." What do you think of this claim?

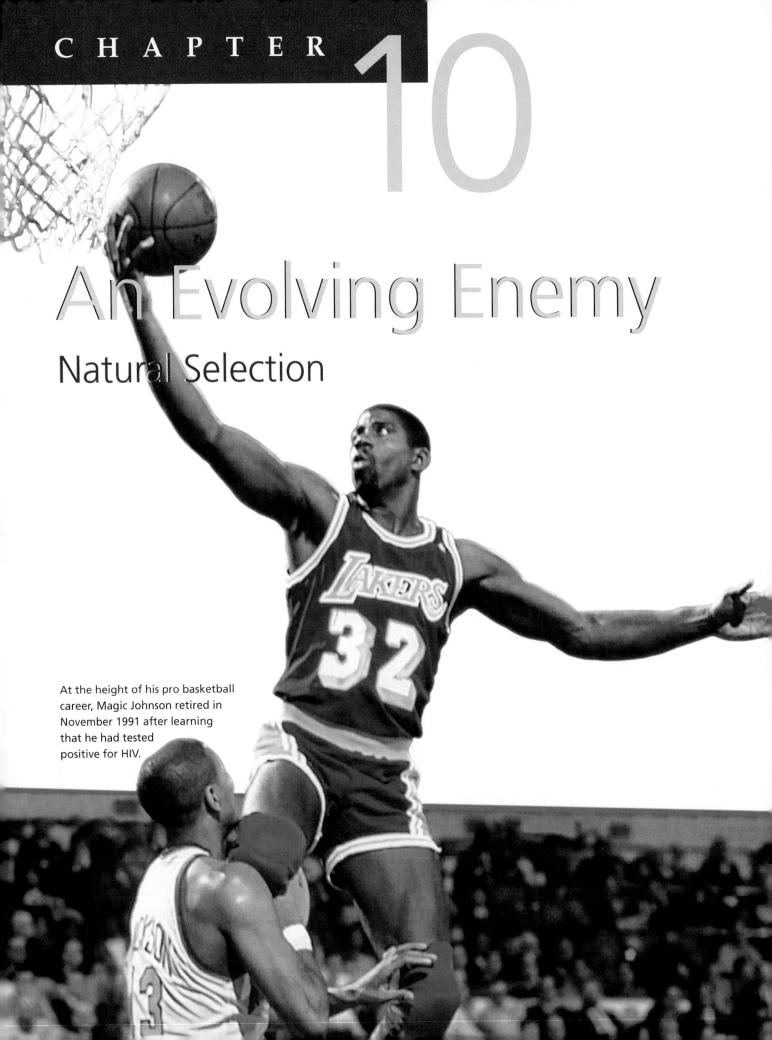

CHAPTER 10

An Evolving Enemy

Natural Selection

At the height of his pro basketball career, Magic Johnson retired in November 1991 after learning that he had tested positive for HIV.

At the time of his retirement, fans expected that Magic would meet the fate of HIV patients such as this man—death from AIDS.

However, Magic Johnson and many other HIV-infected individuals have survived many years without suffering from this debilitating and fatal disease. Why?

In late 1991, basketball fans around the world were hit with devastating news. Earvin "Magic" Johnson, one of the greatest basketball guards ever to play the game, was retiring at age 32, two years after being named the NBA's most valuable player for the third time in 10 seasons. In his relatively short career, Magic broke the record for the most career assists and led the Los Angeles Lakers to five NBA championships. Why was this talented, popular, and successful athlete with a new wife and a baby on the way leaving the game just as he was reaching his physical prime? He had learned only days before that he was infected with HIV, the virus that causes AIDS.

In 1991, a diagnosis of HIV infection was considered to be a death sentence. At that time, the typical length of time between the diagnosis of HIV and death from AIDS was 8 to 10 years. Magic Johnson's fans and other NBA players braced themselves to watch the terrible decline that always occurred with the onset of AIDS.

Now fast forward to November 2005, 14 years after Magic's diagnosis and retirement. The fit and muscular 45-year-old NBA Hall of Fame member is now part owner of the Los Angeles Lakers and head of a company that owns dozens of movie theaters, coffee shops, and a fast-food franchise. In the time since being diagnosed, Magic has won an Olympic gold medal, made two comebacks as an NBA player, coached his former team, celebrated three more

Modern anti-AIDS therapy, consisting of multiple drugs, has disabled a powerful tool of HIV—evolution.

Lakers NBA championships, hosted his own late-night television show, and was inducted into the NBA Hall of Fame. He is about as successful as any former sports star, and just about as healthy. Now many of his friends joke that he will be hit by a bus before he dies of AIDS.

Magic Johnson's survival despite his HIV infection is partly a testament to the huge effort doctors and scientists have invested to control AIDS. The time between the identification of this new disease and the first drug treatments was less than a decade. Although anti-HIV drugs have been available since 1987, five years before Magic's announcement, most people did not remain healthy for long after they started using these drugs. The failure of these early treatment strategies was the result of a single factor—evolution—and the success of current treatments depends on the understanding and management of this powerful process.

In this chapter, we will explore how the process of evolution has shaped HIV and governed our methods for controlling this killer virus. To do so, we must first understand a little of the biology of HIV and AIDS.

10.1   AIDS and HIV

Acquired immune deficiency syndrome, or **AIDS**, was first described in 1981 after dozens of young gay men in New York City and San Francisco were diagnosed with illnesses rarely seen in healthy young people. The types of illnesses seen in affected individuals were more typical of people who had been born with genetic mutations leading to serious defects in the functioning of their immune system. In contrast, this new outbreak of susceptibility to these illnesses appeared to be *acquired* (that is, caused by exposure to some factor) because it was seen suddenly in large numbers of previously normal-functioning people.

AIDS Is a Disease of the Immune System

The increased susceptibility to illness in these young men resulted from a decline in their immune system function. The role of the immune system is to maintain the integrity of the body. The cells of the immune system constantly patrol the tissues and organs of the body for anything that is not clearly produced by the body—that is, anything that is "nonself." Upon encountering a nonself entity, the immune system acts to eliminate it; this is known as an **immune response**.

The virus that causes AIDS primarily kills or disables a particular class of immune system cells called **T4 cells**. T4 cells are also known as helper T cells because they serve as the directors of the immune system's response to specific nonself entities. Thus the loss of T4 cells causes immune deficiency—that is, affected individuals experience diseases that are normally controlled by healthy immune systems. These include infections by organisms commonly found on our bodies in low levels, such as *Pneumocystis carinii*, a fungus that is found in nearly everyone's lungs by age 30. In healthy people, *P. carinii* is held in check by the immune system; but in AIDS patients, this organism often causes pneumonia and extensive lung damage. Diseases like *P. carinii* pneumonia are called opportunistic infections because they occur only when the opportunity arises due to a weakened immune system.

Because individuals with weakened immune systems may have more than one opportunistic infection, each with its own signs and symptoms, there is no single condition that is always associated with AIDS. This is why the disease is called a syndrome—a group of signs and symptoms indicating that an individual has AIDS. Primary among those signs is the depletion of T4 cells.

HIV Causes AIDS

Within months of the initial reports of this new disease, it became clear that AIDS could be transmitted through both sexual intercourse and contact with the blood of affected individuals. By 1983, two independent teams of scientists, led by Luc Montagnier in France and Robert Gallo in the United States, had identified the factor causing the transmission and symptoms of AIDS—the factor was later named the **human immunodeficiency virus**, or **HIV**. The evidence linking HIV to AIDS is outlined in Essay 10.1. Worldwide, most HIV transmission occurs via sexual intercourse without a condom. In the United States, both unprotected sex and the sharing of needles by injection-drug users are primary modes of HIV transmission.

HIV is a virus composed of RNA and proteins, surrounded by a protein coat and membrane envelope. As with all viruses, HIV can reproduce only by using the cellular "machinery" of its **host**, the organism it is infecting, to make copies. The reproductive cycle of HIV is illustrated in Figure 10.1 on page 260.

Essay 10.1 The Evidence Linking HIV to AIDS

Although a small number of scientists have argued that the link between HIV and AIDS is weak, the vast majority of them agree that the statement "HIV causes AIDS" is a fact. Scientists use Koch's postulates, developed by physician Robert Koch in the nineteenth century, as the litmus test for determining the cause of any epidemic disease. The postulates are summarized as follows:

1. *Association:* The suspected infectious agent is found in all individuals suffering from a particular disease.
2. *Isolation:* The supposed infectious agent can be grown outside the host in a pure culture (without any other microorganisms).
3. *Transmission:* Transfer of the suspected pathogen to an uninfected host produces the disease in the new host.
4. *Isolation from new victim:* The same pathogen must be found in the newly infected host.

Does HIV fulfill Koch's postulates as the cause of AIDS? Let us examine the evidence for each assumption.

- *Association:* Studies from around the world show that virtually all AIDS patients carry antibodies that indicate HIV infection.
- *Isolation:* HIV has been isolated from virtually all AIDS patients, as well as in almost all individuals with HIV antibodies. In addition, researchers have documented the presence of HIV genes in both groups of patients.

- *Transmission:* HIV does not readily cause AIDS in other animals, so transmission is difficult to demonstrate. However, this postulate has been fulfilled by a series of tragic incidents. In one case, three laboratory workers with no other risk factors developed AIDS after accidental exposure to concentrated HIV at work. In another case, transmission of HIV from a Florida dentist to six oral surgery patients was documented by genetic analysis of the virus isolated from both the dentist and the patients. The dentist and four of the patients developed AIDS and died. Finally, the Centers for Disease Control (CDC) has documented reports of 57 occupationally-acquired HIV infections among health-care workers, of whom 26 have developed AIDS in the absence of other risk factors.

- *Isolation from new victim:* In the case of the three laboratory workers just described, HIV was isolated from each infected individual, and its RNA sequence was examined. The HIV proved to be the virus that the workers had handled. In the case of the Florida dentist, HIV isolated from his infected patients had very similar RNA sequences, indicating that all of them were infected with the same virus strain.

In short, the link between HIV and AIDS has been firmly established using the standard set of Koch's postulates, and this relationship has been accepted by medical scientists.

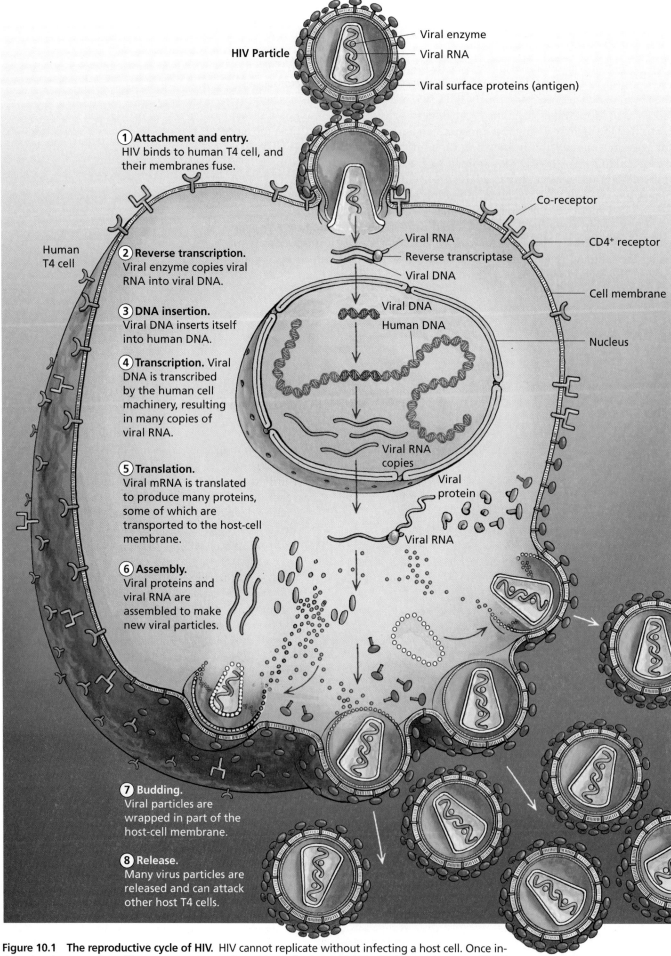

HIV Particle

- Viral enzyme
- Viral RNA
- Viral surface proteins (antigen)

① **Attachment and entry.** HIV binds to human T4 cell, and their membranes fuse.

Co-receptor

Human T4 cell

② **Reverse transcription.** Viral enzyme copies viral RNA into viral DNA.

- Viral RNA
- Reverse transcriptase
- Viral DNA

CD4⁺ receptor

③ **DNA insertion.** Viral DNA inserts itself into human DNA.

- Viral DNA
- Human DNA

Cell membrane

Nucleus

④ **Transcription.** Viral DNA is transcribed by the human cell machinery, resulting in many copies of viral RNA.

⑤ **Translation.** Viral mRNA is translated to produce many proteins, some of which are transported to the host-cell membrane.

- Viral RNA copies
- Viral protein
- Viral RNA

⑥ **Assembly.** Viral proteins and viral RNA are assembled to make new viral particles.

⑦ **Budding.** Viral particles are wrapped in part of the host-cell membrane.

⑧ **Release.** Many virus particles are released and can attack other host T4 cells.

Figure 10.1   The reproductive cycle of HIV. HIV cannot replicate without infecting a host cell. Once inside, the virus uses the cell's components to make copies. Infection with HIV disables and eventually kills the host cell.

HIV initially binds to proteins on the surface of a cell (the CD4$^+$ receptor and a co-receptor) and then releases its own RNA and proteins into the cell (step ①). Once inside the cell, the viral RNA is reverse transcribed into viral DNA by the action of one of the viral proteins, the enzyme called reverse transcriptase (step ②). Transcription is the process in all cells that rewrites the information in DNA into the language of RNA; reverse transcription is simply the converse of that process, in which a strand of RNA is rewritten in the language of DNA.

With the help of another viral enzyme, the viral DNA produced by reverse transcription inserts itself into the cell's genome (step ③), where it commandeers the proteins and cell organelles required for copying genetic material and producing proteins. The cell now makes new copies of the viral RNA (step ④), translates the genes on the viral DNA into the proteins that make up the coat, enzymes, and membrane surface proteins (step ⑤), and assembles new viruses (step ⑥). Then the newly made copies of the virus are released from the cell by budding off the cell membrane and go on to infect other cells that possess the CD4$^+$ receptor and an appropriate co-receptor (steps ⑦ and ⑧). Infection with HIV usually either disables or kills the host cell. Most of the cells infected with HIV are T4 cells, but other cells that carry the CD4$^+$ receptor are susceptible to HIV as well.

The Course of HIV Infection

Early symptoms of HIV infection resemble the flu in about 70% of infected individuals; there are no noticeable symptoms in the remaining 30%. This generalized feeling of fatigue and illness is caused by the initial nonspecific immune response to any viral invader—the same factors that are responsible for the run-down feeling you experience before the onset of a head cold. Because HIV is killing off immune system cells that protect people from a variety of infectious organisms, the run-down feeling may last for a number of weeks. Most people infected with HIV begin to control the virus within 6 to 12 weeks and therefore recover from these flu-like symptoms. This seeming recovery from the infection occurs once the immune system develops a specific response to HIV.

A specific response to a virus or other disease-causing organism develops because the immune system can produce cells and proteins that effectively destroy invaders once the system is exposed to these invaders. Among the billions of cells that make up the immune system, a few individual cells belonging to a particular class (called **B cells**) are capable of recognizing some of the unique proteins that are present in and on HIV particles. A protein or other molecule that is recognized as nonself by immune system cells is called an **antigen**. Proteins called antigen receptors, found on the surface of B cells, bind to an antigen and begin a specific immune system response. In the presence of fragments of HIV, the B cells that have an HIV-antigen receptor are stimulated to divide; these daughter cells will produce HIV **antibodies**, proteins that bind to and help eliminate the virus (Figure 10.2). Within 3 months of initial contact with the virus, 95% of infected individuals have high levels of HIV antibodies circulating in their blood. As a result, the level of HIV in the infected person's bloodstream is greatly reduced.

Once a specific immune response to HIV is fully developed and the number of viruses in the blood has dropped, the levels of T4 cells in an infected individual rebound. At this point, the person is asymptomatic and has a mostly normal immune response. The asymptomatic phase of HIV infection may last for as long as 10 years.

In nearly all HIV-infected individuals who are not receiving drug treatment, the immune system eventually loses control over the virus. At some point, virus levels begin to increase and T4 cell numbers decline, signaling the

HIV-specific antigen receptor

B cell
The binding of HIV fragments to the cell's antigen receptors stimulates the cell to divide.

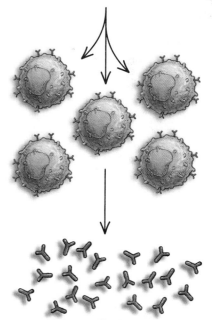

Copies of the B cell release antibodies that destroy HIV particles and disable infected cells.

Figure 10.2   The immune response to HIV. The presence of HIV in the body stimulates the multiplication of immune system cells, some of which produce anti-HIV antibodies. These antibodies help reduce levels of HIV in the bloodstream.

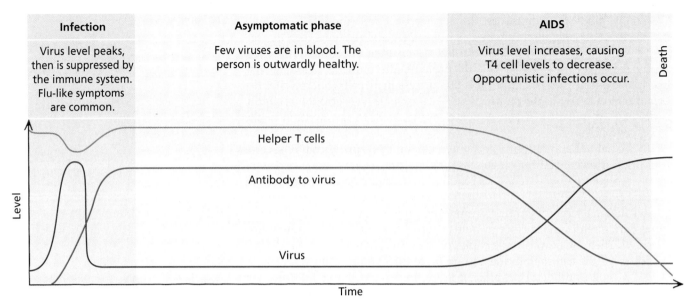

Figure 10.3   The typical course of HIV infection. This graph illustrates the change in HIV levels in the blood, the level of antibodies present, and the level of T4 cells over time. After initial infection, most patients produce enough antibodies to control virus levels for months or years. Eventually, however, nearly all HIV-infected people develop AIDS when the body fails to maintain antibodies to the virus.

onset of AIDS (Figure 10.3). Why does HIV eventually win its battle with the immune system? Primarily due to the evolution of HIV within its host.

Evolution is most simply defined as a change in the genetic characteristics of a population over time. Often these changes occur due to the differential survival of individuals in a population made up of individuals who vary in their traits—that is, by the process of **natural selection**. The natural selection of HIV particles within an infected person's body causes the HIV population within the body to evolve and eventually escape immune system control. Understanding why most people infected with HIV eventually fail to control this virus requires an understanding of this very important idea in biology— any population, whether plant, animal, bacteria, or virus, has the potential to adapt to its environment if it is given enough time and possesses a mechanism to generate variation. Charles Darwin first elaborated the theory of natural selection in the middle of the nineteenth century. That theory has since become one of the most powerful ideas in science.

10.2   The Theory of Natural Selection

In *The Origin of Species*, Charles Darwin put forth two major ideas: the theory of common descent and the theory of natural selection. We discussed the theory of common descent in detail in Chapter 9 and learned that all species living today appear to have descended from a single ancestor that arose in the distant past. Darwin's presentation of this theory was thorough and convincing. Within 20 years of his book's publication, Darwin's principle that all living organisms are related to each other through common descent had been accepted by most scientists. However, it was another 60 years before the scientific community accepted Darwin's ideas about *how* the great variety of living organisms had come about—the process he called natural selection. Today, natural selection is considered one of the most important causes of evolution; although others, such as the processes of genetic drift and sexual selection as described in Chapter 11, also cause populations to change over time.

(a) Variation in coat color

(b) Variation in blooming time

Figure 10.4   Observation 1: Individuals within populations vary. (a) Gray wolves vary in coat color, even within a single litter of animals. (b) Flowers may vary in blooming time, with some individual plants blooming much earlier than others of the same species.

Four Observations and an Inference

The theory of natural selection is elegantly simple. It is an inference based on four general observations:

1. Individuals within Populations Vary. Observations of groups of humans support this statement—people do come in an enormous variety of shapes, sizes, colors, and facial features. It may be less obvious that there is variation in nonhuman populations as well. For example, within a litter of gray wolves born to a single female, some individuals may be black, tawny, or reddish in color (Figure 10.4a); or within a single popula-tion of flowers, some individuals will bloom earlier than the majority, and some will bloom later (Figure 10.4b). We can add all kinds of less obvious differences to this visible variation; for example, the amount of caffeine produced in the seeds of a coffee plant varies among individuals in a wild population. Each different type of individual in a population is termed a **variant**.

2. Some of the Variation Among Individuals Can Be Passed on to Their Offspring. Although Darwin did not understand how it occurred, he observed many examples of the general resemblance between parents and offspring. Farmers regularly take advantage of the inheritance of certain variations. For example, some chickens produce more eggs than others do, and their offspring often produce more eggs than do the offspring of less productive chickens. This variation enables a farmer to select only the off-spring of the best-laying hens as the new flock of egg producers. Darwin noticed that pigeon breeders took advantage of the inheritance of variation when they produced fancy birds—pigeons with fan-shaped tails were more likely to produce offspring with fan-shaped tails than were pigeons with straight tails (Figure 10.5). Darwin hypothesized that offspring tend to have the same characteristics as their parents in natural populations as well.

Figure 10.5   Observation 2: Some of the variation among individuals can be passed on to their offspring. Darwin noted that breeders could create flocks of pigeons with fantastic traits by using as parents of the next generation only those individuals that displayed these traits.

For several decades after *The Origin of Species* was published, the observation that some variations were inherited was the most controversial part of the theory of natural selection. Since scientists could not adequately explain the origin and inheritance of variation, many were unwilling to accept that natural selection could be a mechanism for evolutionary change. When Gregor Mendel's work on inheritance in pea plants (discussed in Essay 6.1) was rediscovered in the 1900s, the mechanism for this observation became clear—natural selection operates on genetic variation that can be passed from one generation to the next.

3. Populations of Organisms Produce More Offspring than Will Survive.

This observation is clear to most of us—the trees in the local park make literally millions of seeds every summer, but only a few of the much smaller number that sprout live for more than a few years. In *The Origin of Species*, Darwin gave a graphic example of the difference between offspring production and survival. In his example, he used elephants, animals that live long lives and are very slow breeders. A female elephant does not begin breeding until age 30, and she produces about one calf every 10 years until around age 90. Darwin calculated that even at this very low rate of reproduction, if all the descendants of a single pair of African elephants survived and lived full, fertile lives, after about 500 years their family would have more than 15 million members—many more than can be supported by all the available food resources on the African continent (Figure 10.6)! Clearly, only a subset of the elephants born in every generation survives long enough to reproduce. The same is true for nearly every species; the capacity for reproduction far outstrips the resources of the environment, and so many individuals do not survive to maturity.

Shelf = Available resources

Generation 0 = 2 elephants
If a female elephant (colored pink) lives a full fertile lifetime, she will bear about six calves in about 90 years.
On average, half of her calves will be female.

Generation 1 = 6 elephants

Generation 2 = 18 elephants

Generation 3 = 54 elephants

Figure 10.6   Observation 3: Populations of organisms produce more offspring than will survive. Even slow-breeding animals like elephants are capable of producing huge populations relatively quickly. Limited resources prevent this breeding from happening.

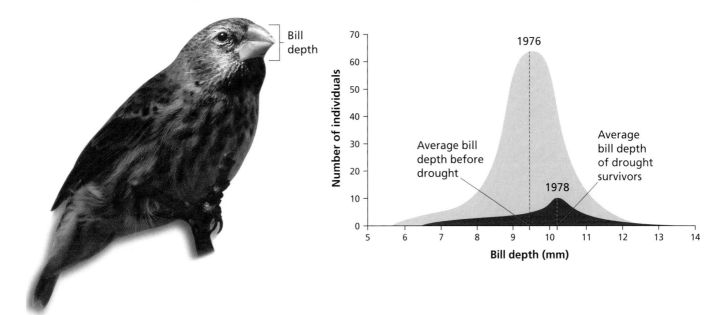

Figure 10.7   Observation 4: Survival and reproduction are not random. Darwin's finches on the Galápagos Islands did not have equal chances of surviving a severe drought. The pale purple curve summarizes bill depth in these birds before the drought. The same population after the drought of 1976 and 1977 (dark purple curve) had an average bill depth of approximately 0.5 mm greater. This illustrates that during the drought years of 1976 and 1977, finches with larger-than-average bills that could crack larger seeds—the only food available during the drought—had higher fitness than did small-billed birds.

4. Survival and Reproduction Are Not Random. In other words, the subset of individuals that survive long enough to reproduce is not an arbitrary group. Some variants in a population have a higher likelihood of survival and reproduction than other variants do. The relative survival and reproduction of one variant compared to others in the same population is referred to as its **fitness**. Traits that increase an individual's fitness in a particular environment are called **adaptations**. Individuals with adaptations to a particular environment are more likely to survive and reproduce than are individuals lacking such adaptations.

Darwin referred to the results of differential survival and reproduction as natural selection. Although Darwin used the word *selection*, which implies some active choice, you should note that natural selection is a passive process. Adaptations are "selected for" in the sense that individuals possessing them survive and contribute offspring to the next generation. For example, among the birds called Darwin's finches, scientists have observed that when rainfall is scarce, a large bill is an adaptation. This is because birds with larger bills are able to crack open large, tough seeds—the only food available during severe droughts. As shown in Figure 10.7, the 300 survivors of a 1977 drought had an average bill depth that was 6% greater than the average bill depth of the original population of 1300 birds.

Adaptations are not only traits that increase survival. Any trait that increases the number of offspring produced relative to others in a population is also an adaptation. For example, flowers in a crowded mountain meadow may have a relatively limited number of potential insect pollinators. More pollinator visits generally results in more seeds being produced by a single flower, and so any trait that increases a flower's attractiveness to pollinators, such as a brighter color or greater nectar production, should be favored by natural selection (Figure 10.8).

You may have heard natural selection described as "survival of the fittest;" however, it is important to recognize that natural selection results in

Figure 10.8   Adaptations are not about survival only. The number of seeds a flowering plant produces depends, in part, on the number of pollinators that visit it. Variations that increase flower's attractiveness to a pollinator can increase its "reproductive success."

the survival and reproduction of those individuals that are best adapted to the current environment given the current variants in the population. The survivors are not "fittest" in an absolute sense, only relatively, meaning that the variants with the highest rates of survival and reproduction are simply better adapted to local conditions than are others in the same population.

Darwin's Inference: Natural Selection Causes Evolution.   The result of natural selection is that favorable inherited variations tend to increase in frequency in a population over time, while unfavorable variations tend to be lost. In other words, adaptations become more common in a population as those individuals who possess them contribute larger numbers of their offspring to the succeeding generation. Natural selection results in a change in the traits of individuals in a population over the course of generations—voilà, evolution.

Testing Natural Selection

Darwin proposed a scientific explanation of how evolution occurs, and like all good hypotheses, it needed to be tested. As Darwin noted in *The Origin of Species*, humans have been testing the hypothesis that selection causes evolution for thousands of years. By imposing selection on domestic animals and plants, humans have changed the characteristics of populations of these organisms.

Artificial selection for dogs with short legs

Artificial selection for dogs with short legs

Artificial selection for dogs with short legs

Dachshunds

Figure 10.9   Artificial selection can cause evolution. When breeders select dogs with certain traits to produce the next generation of animals, they increase the frequency of that trait in the population. Over generations, the trait can become quite exaggerated. Dachshunds are descendants of dogs that were selected for the production of very short legs.

Artificial Selection.   Selection imposed by human choice is called **artificial selection**. It is artificial in the sense that humans control the survival and reproduction of individual plants and animals with favorable characteristics in order to change the characteristics of the population. Dog varieties resulted from artificial selection; they evolved through selection by breeders for various traits (Figure 10.9). Most of the fruits and vegetables we are familiar with also developed as a result of artificial selection. However, due to the direct intervention of humans on the survival and reproduction of these organisms, artificial selection is not exactly equivalent to natural selection.

Natural Selection in the Lab.   Scientists have also observed selection occurring among organisms that are intentionally exposed to different environments. An example of this kind of experiment is one performed on fruit flies placed in environments containing different concentrations of alcohol. High concentrations of alcohol cause cell death. Many organisms, including fruit flies and humans, produce enzymes that metabolize alcohol—that is, they break it down, extract energy from it, and modify it into less-toxic chemicals. There is variation among fruit flies in the rate at which they metabolize alcohol. In a typical laboratory environment, most flies process alcohol relatively slowly, but about 10% of the population possesses an enzyme variant that allows those flies to metabolize alcohol twice as rapidly as the more common variant.

In their experiment, scientists divided a population of fruit flies into two random groups. Initially, these two groups had the same percentage of fast and slow alcohol metabolizers. One group of flies was placed in an environment containing typical food sources; the other group was placed in an environment containing the same food spiked with alcohol. After 57 generations, the percentage of fast-metabolizing flies in the normal environment was the same as at the beginning of the experiment—10%. But after the same number of generations, 100% of the flies in the alcohol-spiked environment were the fast-metabolizing variety (Figure 10.10). Because all of the flies in this environment were now the fast-metabolizing variety, the *average* rate of alcohol

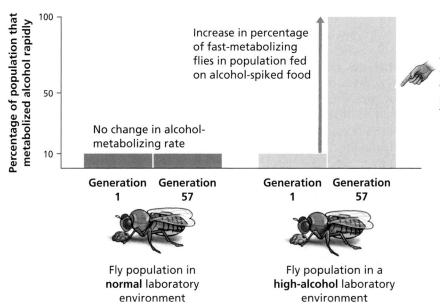

After 57 generations, all flies in the population are fast processors of alcohol. As a result, the **average** rate of alcohol metabolism in this population is twice the rate measured in the unmodified population.

Figure 10.10   An observation of natural selection. When fruit flies are placed in an alcohol-spiked environment, the percentage of flies that can rapidly metabolize alcohol increases over many generations because of natural selection. This causes the average speed of alcohol metabolism in the population to increase as well. In the normal laboratory environment, there is no selection for faster alcohol processing, so the average rate of alcohol metabolism does not change.

metabolism in the population was much higher in generation 57 than in generation 1. The population had evolved.

The evolution of the fruit flies in this experiment was a result of natural selection. In an environment where alcohol concentrations were high, individuals that were able to metabolize alcohol relatively rapidly had higher fitness. Since they lived longer and were less affected by alcohol, they left more offspring than the slow metabolizers did. Thus, each generation had a higher frequency of fast-metabolizing individuals than the previous generation did. After many generations, flies that could metabolize alcohol rapidly predominated in the population.

Natural Selection in Wild Populations. The effects of natural selection have been observed in dozens of wild populations as well. A classic example of natural selection in action is the evolution of bill size in Galápagos finches in response to drought. Figure 10.7 illustrated that a nonrandom subset of the finch population survived a 1977 drought—the survivors tended to be those with the largest bills, which could more easily handle the tough seeds produced by the plants that survived the drought. The survival of this nonrandom subset of birds resulted in a change in the next generation. The population of birds that hatched from eggs in 1978—the descendants of the drought survivors—had an average bill depth of 4% to 5% larger than that of the pre-drought population. Bill size in this population of birds evolved in response to natural selection occurring in a setting uninfluenced by humans.

The Modern Understanding of Natural Selection

One barrier to the acceptance of Darwin's theory of natural selection was a lack of understanding of the origins of variation among individuals and of the mechanism by which variations were passed to the next generation. Without this understanding, it was difficult to see how natural selection could cause a change in the frequency of particular traits in a population. It was not until the early twentieth century, when they began to understand the nature of genes, that most biologists fully accepted the theory of natural selection.

As we discussed in Chapter 6, genes are segments of genetic material (typically DNA, but RNA in some viruses) that contain information about the structure of molecules called proteins. The actions of proteins within an organism help determine its physical traits. Different versions of the same gene are called alleles, and variation in traits among individuals in a population is often due to variation in the alleles they carry.

We can apply these genetic principles to the fruit flies exposed to a high-alcohol environment. In this population there are two variants, or alleles, of the gene that controls alcohol processing. One allele produces an enzyme we will refer to as "fast," and the other produces an enzyme we will call "slow." Flies that make the fast enzyme can metabolize alcohol rapidly. To make this enzyme, they must carry 2 copies of the fast allele. As described in detail in Chapter 6, half of the alleles carried by a parent are passed to their offspring via their eggs or sperm. In the high-alcohol environment, flies with the fast enzyme had more offspring than did flies with the slow enzyme. Since they carry 2 copies of the fast allele, each of the offspring of a fast metabolizer received a copy of this allele. Therefore, in the next generation, a higher percentage of individuals carried the fast allele. We can now describe the evolution of a population as an increase or decrease in the *frequency of an allele* for a particular gene.

Understanding the nature of genes also explains the origin of their variations. Different alleles for the same gene arise through mutation—changes in the DNA sequence. As described in Unit Two, mutations that can be passed on to offspring occur by chance when DNA is copied during the production of eggs and sperm. These mutations can occur anywhere in an organism's genome. If one occurs in noncoding DNA, it may have little effect, but a mutation that occurs in a gene can cause a change in the function of the protein that the gene codes for. If a mutation results in an allele that has a function different from that of the original allele, the resulting variation could become subject to the process of natural selection. The existence of two different alleles for alcohol metabolism in fruit flies suggests that one of these alleles is a mutated version of the other. In the normal laboratory environment, neither of these alleles appears to have a strong effect on fitness. Since the slow metabolizers are more numerous than the fast metabolizers, it appears that there might be a slight disadvantage to carrying the fast enzyme. However, in a different environment, the mutation resulting in the fast allele gives an advantage, and its presence in the population allows for the population's evolution (Figure 10.11).

Scientists now understand that the random process of gene mutation generates the raw material—variations—for evolution, and that natural selection acts as a filter that selects for or against new alleles produced by mutation. We can also see that variation within a population can help ensure the survival of that population when the environment changes—if no individuals in the experiment carried the fast allele, it is possible that no flies would have survived in the alcohol-spiked environment. In Chapter 14, we discuss the importance of genetic variation for the long-term survival of species.

Subtleties of Natural Selection

Natural selection is a fairly simple idea, but it is surprisingly easy to misunderstand. Common misunderstandings of this idea fall into three categories: the relationship between the individual and the population, the limitations on the traits that can be selected, and the ultimate result of selection.

Natural Selection Acts on the Inherited Traits of Individuals.   A common misconception about natural selection can be illustrated by the following erroneous statement: "The Dodo could not adapt to human hunters, so it went extinct." This assertion seems to presume that individuals must change within their lifetimes in order to survive environmental changes. However, evolution can occur only when traits that influence survival are present in a population and have a genetic basis. Again, the example of the

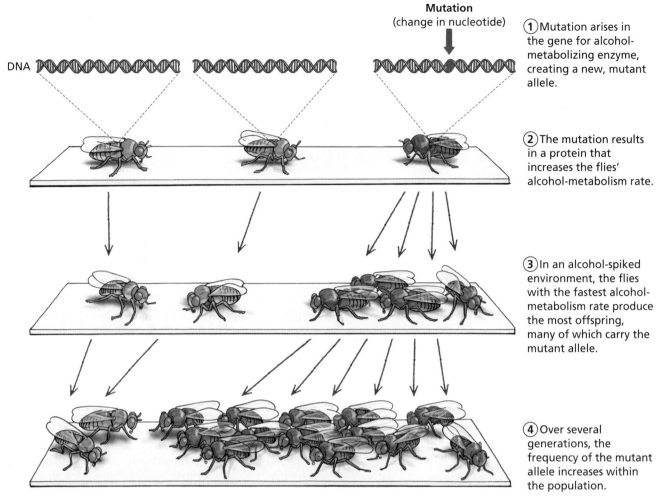

Figure 10.11 Mutation and natural selection. When a gene has mutated, its product may have a slightly different activity. If this new activity leads to increased fitness in individuals carrying the mutated gene, it will become more common in the population through the process of natural selection.

alcohol-metabolizing fruit flies illustrates this point. Selection did not cause change in individual flies; either a fly could rapidly metabolize alcohol, or it could not. It was the differential survival and reproduction of these types of flies in an alcohol-laced environment that caused the population to change.

Natural Selection Does Not Result in Perfection. Natural selection does cause populations to become better fit to their environment, but the result of that process is not necessarily "better" organisms—simply ones that are better adapted to the current situation. Changes in traits that increase survival and reproduction in one environment may be liabilities in another environment. For example, Richard Lenski and his coworkers at Michigan State University found that certain populations of bacteria would adapt to an environment where food levels were low by evolving the production of chemicals that were deadly to other bacteria. Individual bacteria that produced this chemical had an advantage over those that did not; by killing off their nearby competitors, the poisonous bacteria had more of the very limited food available to them and thus became prevalent in the population. However, when the poisonous bacteria were grown in a food-rich environment, they did not grow as well as nonpoisonous bacteria did. This was probably because the poisonous bacteria were using energy to produce their toxin—energy that could have been used for growth and reproduction. Nonpoisonous bacteria in a food-rich environment that

Figure 10.12   The panda's thumb. In addition to the 5 digits on its paw, the giant panda has an opposable "thumb." The thumb is made up of wrist bones that are adapted to help these animals strip leaves from bamboo shoots, their primary food source.

expended the maximum amount of energy on growth had more offspring and thus were better adapted when food was abundant. With nearly all adaptations, there is a trade-off such as this, where increased success in one environment or in one aspect of survival leads to decreased success in another environment or aspect of survival.

The process of natural selection also acts only on the variants that are available within a population. While it might increase our survival and reproduction to have two arms, two legs, and a set of wings, it is not possible to evolve this set of limbs because no such variant exists in the human population. The fact that the adaptations of organisms are constrained by their underlying biology is apparent throughout nature in what evolutionary biologists call *jury-rigged design*, meaning "made using whatever is available." The late Stephen Jay Gould described one of the most famous examples of jury-rigged design—the "thumb" found on a giant panda's front paws (Figure 10.12). These animals apparently have 6 digits: 5 fingers composed of the same bones as our fingers, and a thumb constructed from an enlarged bone equivalent to one found in our wrist. The muscles that operate this opposable thumb are rerouted hand muscles. This structure in pandas appears to be an adaptation that increases their ability to strip leaves from bamboo shoots, their primary food source. A more effective design for an opposable thumb is our own, adapted from one of the basic 5 digits. However, in the panda population, this variation did not exist. Individuals with enlarged wrist bones did exist, so what evolved in giant pandas was a jury-rigged thumb that does its job but is not as flexible as our own thumb.

Natural Selection Results from Current Environmental Conditions.   Selection does not result in the "progress" of a population toward a particular predetermined goal; instead, it is situational. The example of the alcohol-metabolizing flies helps to illustrate this point. Only the population of flies in the high-alcohol environment evolved a faster rate of alcohol metabolism. Without a change in the environment, the alcohol-metabolizing rate of the population of flies in the normal environment did not evolve.

The situational nature of natural selection can lead to evolutionary patterns that defy our sense that species are evolving toward a "more perfect" condition. For example, flowering plants evolved from nonflowering plants; and the flower is, in part, an adaptation to attract bees and other pollinators in order to increase seed production. However, some species of flowering plants, especially grasses, have adapted to environments where wind pollination is particularly effective and the need to attract insect pollinators is much reduced. In these plants, natural selection has favored individuals that have very reduced flower parts and generate primarily pollen-producing and egg-producing structures—much like the reproductive structures of their distant nonflowering ancestors (Figure 10.13). Grasses have not regressed but instead have simply adapted to the environment they experience.

Figure 10.13   Natural selection does not imply progression. Grasses are flowering plants that evolved from ancestors with much showier flowers. The reduction of their flowers as a result of their adaptation to wind pollination does not represent an evolutionary "regression" but instead illustrates the situational nature of natural selection.

While natural selection is often considered a force that causes the traits in a population to change, under certain conditions, selection can cause a population's traits to remain very stable or to split into two species. The type of natural selection experienced by the flies in the alcohol-laden environment is called **directional selection** because it causes the population traits to move in a particular direction, in this case toward higher rates of alcohol metabolism (Figure 10.14a). In certain environments, however, the average variant in the population may have the highest fitness. This results in **stabilizing selection**, in which the extreme variants in a population are selected against and the traits of the population stay the same (Figure 10.14b). For example, in humans, the survival of newborns is correlated to birth weight—both extremely small and extremely large babies

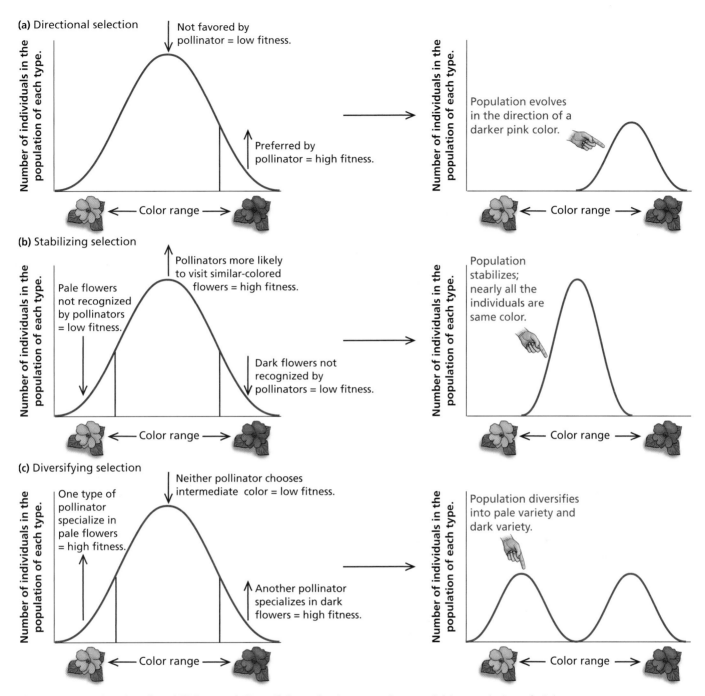

Figure 10.14   Directional, stabilizing, and diversifying selection. Imagine a variable population of pink flowers. If effective pollinators prefer darker-colored flowers as in (a), the population as a whole would be expected to become redder. This is directional selection. (b) If the most "average" color receives the greatest number of pollinator visits, then any extreme variants will likely be lost from the population. This is stabilizing selection. (c) If two populations of pollinators specialize in different ends of the range of colors, the diversity of color in the flower population is expected to increase.

are selected against, causing the average birth weight of babies to be relatively stable over time. Finally, in certain populations, the most common variant may have the lowest fitness, resulting in **diversifying** selection. Diversifying selection causes the evolution of a population consisting of two or more variants (Figure 10.14c). Diversifying selection may contribute to the diversity of HIV particles found in an infected person, a diversity which eventually overwhelms his or her immune system.

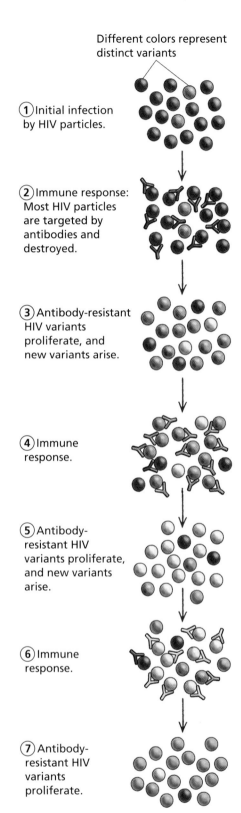

Different colors represent distinct variants

① Initial infection by HIV particles.

② Immune response: Most HIV particles are targeted by antibodies and destroyed.

③ Antibody-resistant HIV variants proliferate, and new variants arise.

④ Immune response.

⑤ Antibody-resistant HIV variants proliferate, and new variants arise.

⑥ Immune response.

⑦ Antibody-resistant HIV variants proliferate.

Figure 10.15 The evolution of HIV. HIV populations evolve in response to changes in the immune system. When the immune system develops a specific response to a strain of HIV, mutants that escape this response proliferate until the immune system develops a response to the mutant strain.

10.3 Natural Selection and HIV

Our understanding of how natural selection causes evolutionary change can help us understand why, before effective drug therapy became available, most people infected with HIV eventually died of AIDS. During the asymptomatic period of HIV infection, the number of HIV virus particles in an infected person's bloodstream is relatively low. However, the immune response to HIV does not completely eliminate the virus. HIV persists inside immune system structures called lymph nodes, where it continues to infect and kill T4 cells. The dying T4 cells release the virus into the bloodstream, where anti-HIV antibodies quickly eliminate them. At the same time, the infected individual maintains a high rate of T4 cell production to replace those lost to HIV. In a sense, the virus and the immune system maintain a balance of power during this period.

HIV Fits Darwin's Observations

The population of HIV is not unchanging during the asymptomatic period; instead, it is evolving in response to the environment created by the infected individual's immune system. HIV will evolve via natural selection because it fulfills all of Darwin's observations.

1. **Viruses in the bloodstream vary.** HIV particles are constantly being reproduced because they continue to infect cells in the lymph nodes; any time there is reproduction, mutation can occur. As a result, during the asymptomatic period, new variants of HIV continually arise. Some of these HIV variants have mutated antigens.

2. **The variation among viruses can be passed on to offspring.** The antigens present on a virus particle are coded for by the particle's RNA. When the virus infects a cell, the RNA is reproduced by the host cell and passed on to a new generation of virus particles.

3. **More viruses are produced than survive.** The HIV antibodies produced by the infected host can eliminate most of the viruses that are produced by infected cells.

4. **Virus survival is not random.** The change in antigens that results from mutation can be great enough that the earlier HIV antibodies do not recognize the new variants. Due to their longer survival in the bloodstream, these new virus variants have higher fitness than do the older variants.

Because of the increased fitness of particular variants, subsequent virus generations consist of a greater percentage of virus particles that possess a new antigen. The population evolves to become resistant to a particular antibody produced by the host.

The Evolutionary Arms Race

At the same time the virus population is evolving, the immune system of the host, and thus the environment to which the virus is exposed, is continually changing. As a new HIV variant becomes more common in an infected person's bloodstream, his or her immune system develops an antibody to the variant's unique antigens. HIV again begins to be cleared from the bloodstream—that is, until the next new HIV antigen variant arises through mutation. In other words, the population of HIV inside the host is continually evolving as natural selection by the host's immune system favors variants with unrecognizable antigens (Figure 10.15).

The evolutionary arms race between HIV and the immune system is somewhat similar to the ongoing battles between computer security professionals and hackers. When a hacker succeeds in accessing protected information, security professionals create a "patch" to cover the flaw that the hacker exploited. A patch keeps hackers at bay until they develop another strategy for breaking into the system—causing the security team to come up with another patch in a seemingly endless cycle. In the case of HIV and the immune system, the patch already exists in the form of one or a few randomly generated B cells that can recognize the new HIV antigens. However, the development of a specific immune response to this new variant by division of these B cells and the production of new antibodies takes several hours to days, and so the overall effect is the same.

Like other RNA viruses that require reverse transcriptase, HIV has a high rate of mutation. Some scientists estimate that every single HIV particle produced has at least one difference from the HIV from which it arose. In addition, HIV has an enormously high rate of reproduction. As the host's immune system focuses its resources on the most common variants, less common types proliferate. The frequent mutation and rapid reproduction of HIV, along with the diversifying selection imposed by the host's immune system, result in a population of the virus within an asymptomatic host that contains about 1 billion distinct variants. With this many variants, it is almost assured that one or more has antigens that are not immediately recognized by the host's immune system. Because these variants avoid the host's antibodies, they can grow to large populations very quickly. Evolution of the HIV population within an infected individual can be extremely fast.

HIV's rapid evolution appears to be the cause of the eventual end of the asymptomatic period in an infected person. The immune system is able to produce antibodies to many different HIV antigen variants, but eventually the sheer number of different HIV variants that the immune system must respond to becomes overwhelming. Finally, one variant arises that escapes immune system control for a long period; large numbers of T4 cells become infected with this variant and are killed or disabled, and the infected individual becomes increasingly immune-deficient. This change initiates the onset of AIDS. In our analogy, imagine millions of hackers, each trying slightly different ways of bypassing the security of a single computer system. As the security team's resources become stretched, the likelihood of one hacker breaking through and crashing the system increases. The relentless evolution of HIV within an infected person's body eventually exhausts his or her ability to control this deadly virus.

10.4 How Understanding Evolution Can Help Prevent AIDS

Immediately after scientists identified and characterized HIV as the virus that causes AIDS, a search began for drugs that would interfere with HIV's ability to replicate. Early drug therapies had rapid failure rates; it was only until scientists incorporated their understanding of natural selection into anti-AIDS treatments that effective, long-term therapies were developed.

Single Drug Therapy Selects for Drug Resistance

One target of anti-HIV drugs is the process of reverse transcription, the rewriting of HIV's genetic information from RNA into DNA. Reverse transcription does not occur in uninfected human cells, and drugs that target this process have the potential for zeroing in on HIV replication without harming normal functions of the human body. One class of drugs used to inhibit reverse transcription is known as the nucleoside analogs. These drugs are similar in structure to one of the four DNA nucleotides described in Chapter 5—A, C, G, and T. Nucleoside analogs inhibit reverse transcription because reverse transcriptase adds one of these analogs to a growing HIV DNA strand in place of the real nucleotide. Once a nucleoside analog is added to a growing DNA strand, replication halts because additional nucleotides cannot be attached to the analog (Figure 10.16a).

One of the first nucleoside analogs approved as treatment for AIDS is known as Azidothymidine, or AZT. While it is not free of side effects, some very severe, AZT first appeared to be a wonder drug—nearly eliminating HIV from the blood of patients who had already progressed to AIDS. However, in all cases, AZT failed to keep virus populations low for an extended period of time. The failure of AZT over time occurred due to the evolution of HIV. Among the virus variants present in an infected person, there are some that do not mistake AZT for a normal nucleotide and never incorporate it into growing HIV DNA strands. These variants are therefore favored by natural selection; they continue to replicate and become the predominant HIV variants in an AZT-treated individual (Figure 10.16b). As a result of natural selection in the presence of AZT, the HIV population evolves to become **drug resistant**—that is, not susceptible to the effects of AZT. The evolution of drug resistance in disease-causing organisms is not new. However, the speed at which AZT resistance arose in AIDS patients was an early clue to HIV's amazing capacity to evolve.

(a) How a nucleoside inhibitor, such as AZT, works.

Reverse transcriptase

Viral RNA

DNA

Nucleotides

AZT

When AZT is incorporated into the DNA strand...

...transcription is terminated and reverse transcriptase falls off.

(b) Drug resistance to AZT

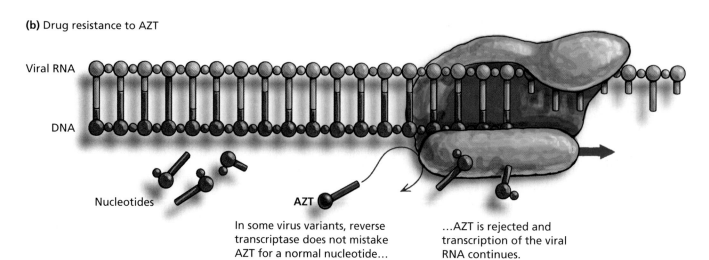

Viral RNA

DNA

Nucleotides

AZT

In some virus variants, reverse transcriptase does not mistake AZT for a normal nucleotide...

...AZT is rejected and transcription of the viral RNA continues.

Figure 10.16   AZT and drug resistance. (a) AZT and other nucleoside inhibitors interfere with HIV replication by tricking reverse transcriptase into adding AZT to the growing DNA strand, stopping replication. (b) Some variants of HIV are resistant to AZT because their reverse transcriptase is mutated and does not mistake AZT for a nucleotide.

Combination Drug Therapy Can Slow HIV Evolution

Since the development of AZT, dozens of new anti-HIV drugs have been made available. In addition to more types of nucleoside analogs, other non-nucleoside analogs that interfere with reverse transcription are now in use. Among these are a powerful class of drugs, called protease inhibitors, that stop HIV replication by interfering with the process that converts inactive viral proteins to active enzymes. In 2003, another class of drugs, called entry inhibitors because they block the HIV receptors on cell membranes, was added to the drug arsenal. However, there are still fewer than 30 anti-HIV drugs available. When patients take only one of any of these drugs, HIV quickly develops resistance to it.

Resistance to Multiple Drugs Is Uncommon.   Understanding that the rapid rate of HIV evolution decreases the effectiveness of these drugs has led doctors to a new standard of care for the infection. This standard is the use of **combination drug therapy**, also commonly called drug cocktail therapy—a combination of at least two reverse transcription inhibitors and a protease inhibitor. This approach has dramatically decreased the number of AIDS cases and deaths due to AIDS in the United States. The effectiveness of combination therapy is based on the following fact: The greater the number of drugs used, the greater the number of changes are required in the virus's genetic material in order for resistance to develop. The likelihood of a virus variant arising that is resistant to a single drug is relatively small but still very possible in a patient with 1 billion different HIV variants. However, the likelihood of a virus variant arising with resistance to all three drugs in a cocktail is extremely small. The chance that an HIV particle exists that is resistant to a single drug is analogous to the likelihood that in 1 billion lottery ticket holders, one person will hold the winning combination—in other words, almost certain. The likelihood of a variant being resistant to several different drugs is analogous to that same ticket holder winning the lottery three times in a row—incredibly unlikely. Just as it is very difficult to win the lottery three times in a row, it is very difficult for HIV to adapt to an environment where it faces three "killer drugs" at once.

Reducing HIV Replication Decreases the Rate of Evolution.   Another key to the effectiveness of combination drug therapy is that when HIV replication is suppressed, new HIV variants arise more slowly. If replication represents the main route by which mutations occur, fewer rounds of replication mean fewer possible mutants. With fewer variants produced, the likelihood also decreases that one contains the combination of mutations that make it resistant to multiple drugs.

Thus, drug cocktails control HIV populations within people by creating an environment that is difficult to adapt to and by slowing the rate at which new adaptations appear. Understanding how HIV evolves to defeat the immune system has allowed scientists to devise ways to interfere with this evolutionary process and prevent HIV infection from progressing into AIDS in most of the treated individuals (Figure 10.17 on page 276). To return to the analogy of computer security professionals and their hacker foes, combination drug therapy not only is equivalent to designing several layers of electronic security around protected information, making it difficult to circumvent all of them at once, but also serves to reduce the number of hackers who are working on ways to defeat the system.

Multiple-Drug-Resistant HIV May Be Less Deadly.   Combination drug therapy greatly reduces the chance of an HIV variant that is resistant to multiple drugs appearing in an infected person, but in about 15% of patients, populations of multiple-drug-resistant variants do evolve. In these cases, HIV infection develops into AIDS and leads to death. However, in 30% to 40% of HIV-infected individuals who host HIV populations resistant to multiple drugs and who thus have high levels of HIV in their blood, there is no decrease in T4 cells. In these patients, the drug-resistant HIV variant in their

Single drug therapy

Combination drug therapy

1. Start with different variants of HIV particles.

2. Single drug reduces fitness of most variants.

3. Resistant variants proliferate.

1. Start with different variants of HIV particles.

2. Combination drug therapy reduces replication and survival of all variants.

3. New mutants are uncommon.

Figure 10.17   Combination drug therapy slows HIV evolution. Using multiple anti-HIV drugs makes the environment much harsher for the virus and decreases the likelihood that a variant with multiple resistances will evolve.

bodies appears to be unable to kill these cells. Patients infected with these meeker HIV variants may live for 3 years or more with high loads of the virus in their bloodstream before progressing to AIDS—much longer than for most people who have high loads of non-drug-resistant HIV. The virus population in this group of patients demonstrates one of the subtleties of natural selection—it rarely results in an organism that is well-adapted for all environments. Recall that for nearly all mutations, there exists a trade-off. In the case of drug-resistant HIV, the mutations that confer this resistance appear to also interfere with the virus's ability to infect large numbers of cells. In some patients with multiple-drug-resistant HIV variants, combination drug therapy has led to the evolution in their bodies of a less deadly virus and has prolonged their lives.

Problems with Combination Drug Therapy

Despite the good news about increased health and prolonged lives of HIV-infected people undergoing combination drug therapy, there are some problems with this approach. Combination drug therapy is expensive;

it often results in severe and unpleasant side effects; and most important, it is difficult to follow. Patients may have to take dozens of pills per day, some of which have very different requirements (for instance, some pills must be taken on an empty stomach, while others need to be taken with food or significant amounts of water). All of this effort is taken to control an infection that initially may not seem any more severe than a mild flu.

The Difficulty of Therapy Creates Drug Resistance.   Due to the difficulty of undergoing combination drug therapy, individuals commonly skip doses or take themselves off the drugs for a period of time. The effect of these breaks in treatment is an increase in HIV replication, which increases the risk of developing drug-resistant varieties. Patients who do not follow the drug treatment schedule carefully can find themselves with large virus populations that are resistant to multiple drugs. Even if this resistant virus is less able to infect cells, few people can live with high levels of HIV for long. To control the virus over the long term, individuals with multiple-drug-resistant variants must change and perhaps increase the number of drugs in their cocktail. For example, some patients take 10 to 15 different drugs several times per day. Many scientists fear that the rapid rate of HIV evolution will eventually outpace their ability to both develop new drugs and prolong the asymptomatic period of HIV infection.

Drug Therapy Makes HIV Seem Less Serious.   Perhaps more troublesome than the evolution of HIV within a patient is the potential evolution of the HIV epidemic in response to combination drug therapy. The rate of transmission of HIV has not significantly changed in the United States for a decade; about 40,000 new infections are reported every year. However, from 10% to 30% of these new infections involve HIV that is already drug resistant. This means that potentially as many as one-third of newly infected people have fewer options for controlling their virus. As the transmission of drug-resistant HIV increases, our ability to control AIDS in the U.S. population will decline. Worse yet, there is some evidence that combination drug therapy has made HIV and AIDS appear to be less of a threat, leading to decreased prevention efforts and an upswing in infection rates. Combination drug therapy does not cure HIV infection; at best, it is an expensive and long-term commitment that increases an individual's ability to live with this disease. Increases in transmission and drug resistance will erode the benefits of this powerful therapy over time.

Drug Therapy Is Expensive.   Combination drug therapy is not available to all of the 42 million HIV-infected individuals around the world. Currently, combination drug therapy costs $1000 to $2000 per month. Worldwide, only about 440,000 of the estimated 6 million people who are in immediate need of anti-HIV therapy receive these drugs. Even in the United States, about 30% of those who take combination drug therapy are subsidized by state and federally funded programs, which are susceptible to budget cuts when tax revenues drop. HIV continues to disproportionately affect the poorest and most vulnerable members of our society and the global community. The gap between the resources of the most-affected populations and the cost of this promising therapy means that for years to come, AIDS will continue to kill people by the millions.

Preventing AIDS

Can Humans Evolve Resistance to HIV?   Scientists have known for many years that a small percentage of people fail to develop an HIV infection despite chronic exposure to the virus. Most of these individuals carry a mutation in the gene for CCR5, one co-receptor for HIV. This variant prevents HIV from binding to immune system cells and thus prohibits its reproduction. About 1% of European-descended whites are homozygous for this mutation and are HIV resistant. Given the existence of this heritable variation and differences in survival among those exposed to this killer virus, can we expect natural selection to cause the human population to evolve resistance to HIV? Eventually, perhaps—but remember that natural selection occurs because of differential survival and reproduction of individuals over time. In short, in order for the human population as a whole to become resistant to this virus, nonresistant variants must die out of the population. Even without anti-HIV drugs, this process would take many generations. Because most nonresistant people will never contract HIV and thus will continue to survive and reproduce, the nonresistant allele will persist for a very long time.

Allowing natural selection to take its course in the human population in order for widespread resistance to evolve would require elimination of all anti-HIV therapy—a prospect unthinkable to most people. The relatively high frequency of the HIV-resistant CCR5 mutation in European populations may itself be a result of unimpeded natural selection. Some scientists have hypothesized that this mutation, which is nearly absent in Asian and African populations, became prevalent in Europeans approximately 700 years ago. The most dramatic event of this time period in Europe was the "Black Death," a plague that killed 33% of the European population in a period of three years from 1347 to 1350. It may be that the CCR5 mutation also confers resistance to infection by the bacteria that causes plague. The death and devastation in Europe caused by the Black Death is nearly unimaginable to us, and yet only 20% of the descendants of that event carry even one copy of the CCR5 mutation. The percentage of people who must die of HIV infection in order to result in a population where 100% carry this mutation is even more dramatic, and it is likely to cause even more serious devastation. Clearly, human evolution is not an acceptable solution to the problem of AIDS.

Preventing Infection.   Despite the success of combination drug therapy, the best "treatment" for AIDS is to avoid becoming infected with HIV at all. HIV is a fragile virus that is transmitted only through direct contact with bodily fluids—primarily blood, semen, vaginal fluid, or occasionally to newborns via breast milk. There is no evidence that the virus is spread by tears, sweat, coughing, or sneezing. It is not spread by contact with an infected person's clothes, phone, or toilet seat. It is not transmitted by an insect bite. And it is unlikely to be transmitted by kissing (although any kissing that allows the commingling of blood could lead to HIV transmission). HIV is frequently spread through needle sharing among injection-drug users, but the primary mode of HIV transmission is via unprotected sex, including oral sex, with an infected partner. So, what is the best way to avoid HIV infection? Do not use addictive injection drugs and avoid sexual activity. If you are sexually active, know your partner's HIV status, drug habits, and sexual activities. The safest relationship is one that is monogamous, where there is no sexual activity outside of the relationship. According to the Centers for Disease Control (CDC), about one-quarter of the approximately 1 million people infected with HIV in the United States do not know that they carry this deadly virus. If your partner might be at risk for HIV infection, practice safer sex—that is, use a condom.

Living with HIV.   Why has Magic Johnson remained free of AIDS for over 14 years since contracting HIV? He has access to the highest-quality medical care to help maintain the effectiveness of his own immune system and the financial and emotional resources that allow him to maintain long-term combination drug therapy. These actions have both reduced the reproduction of HIV in his body and increased the challenges that the virus must overcome in order to escape the control of his immune system. In other words, he has so far successfully limited the chance that natural selection will cause the evolution of HIV in his body. Magic is still infected with HIV; no one knows whether drug therapy will help his body finally eliminate it, or if the therapy will eventually fail and he will lose the battle with this killer virus. Magic's ability to survive, and even thrive, for so long since his diagnosis gives us hope that someday HIV will not be a death sentence for anyone.

CHAPTER REVIEW

Summary

10.1 **AIDS and HIV**

- Infection with HIV eventually leads to collapse of the immune system, resulting in AIDS (p. 258).

- The immune system of an individual infected with HIV can initially control the virus, but the evolution of HIV leads to the eventual loss of immune system cells (pp. 261–262).

Web Tutorial 10.1   HIV: The AIDS Virus

10.2 **The Theory of Natural Selection**

- Individuals in a population vary, and some of this variation can be passed on to offspring (p. 263).

- Not all individuals born in a population survive to adulthood, and not all adults produce the maximum number of offspring possible (p. 264).

- Advantageous traits, called adaptations, increase an individual's fitness, his or her chance of survival and/or reproduction (p. 265).

- The increased fitness of individuals with particular adaptations causes the adaptation to become more prevalent in a population over generations (pp. 265–266).

- Natural selection is a mechanism for evolutionary change in populations (p. 266).

- Artificial selection, when humans control an organism's fitness, causes the evolution of different breeds of animals and varieties of plants (p. 266).

- Populations exposed to environmental changes, both in the lab and in nature, have been shown to evolve traits that make them better fitted to the environment (pp. 266–267).

- The modern definition of evolution is a genetic change in a population of organisms (p. 268).

- The traits of an organism are partially determined by alleles, which arise through the process of mutation (p. 268).

- Alleles that code for adaptations become more common in a population over generations as a result of natural selection (p. 268).

- Natural selection can act only on the variants currently available in the population, results in a population that is better adapted to its environment but usually not perfectly adapted as a result of trade-offs, and does not push a population in the direction of a predetermined "goal" (pp. 268–270).

- Selection can cause the traits in a population to change in a particular direction but in some environments may cause certain traits to resist change and in other environments cause multiple variants to evolve (pp. 270–271).

Web Tutorial 10.2   Natural Selection

10.3 **Natural Selection and HIV**

- HIV eventually overwhelms the immune system because it consists of multiple variants that have differential survival inside the human body; thus, it continually evolves via natural selection (p. 272).

- HIV variants that are immune to the host's anti-HIV antibodies become more common in the host's body because they survive longer and thus reproduce more. As the HIV population consequently changes in form, the immune system develops antibodies to this new form. Eventually, so many different HIV forms exist that the body's attempts to control the virus are overwhelmed (pp. 272–273).

Web Tutorial 10.3   Drug Resistance and Natural Selection

10.4 **How Understanding Evolution Can Help Prevent AIDS**

- HIV variants that are resistant to any single HIV drug are likely in an infected individual with a highly variable HIV population. As a result, HIV evolves resistance to a drug very quickly via natural selection (p. 274).

- Mutants to multiple anti-HIV drugs are relatively unlikely, and drug therapy suppresses HIV replication, thereby reducing the production of mutant varieties (p. 275).

- Varieties of HIV that can survive in an environment containing multiple drugs are sometimes less deadly than nonresistant varieties (p. 275).

- Anti-HIV combination drug therapy selects for drug-resistant viruses, both within patients and in the general population (p. 276).

- Anti-HIV combination drug therapy has major disadvantages, including high cost, difficulty in following the treatment schedule, and severe side effects (p. 277).

- Resistance to HIV infection is unlikely to become commonplace soon in the human population as a result of natural selection (p. 278).

- The best treatment for AIDS is the prevention of HIV infection, which is primarily accomplished through safer sex practices (p. 278).

Learning the Basics

1. Define *fitness* as used in the context of evolution and natural selection.

2. Define *artificial selection*, and compare and contrast it with natural selection.

3. Describe how HIV evolves when it is exposed to a drug that interferes with replication.

4. Which of the following observations is *not* part of the theory of natural selection?

   A. Populations of organisms have more offspring than will survive.; **B.** There is variation among individuals in a population.; **C.** Modern organisms descended from a single common ancestor.; **D.** Traits can be passed on from parent to offspring.; **E.** Some variants in a population have a higher probability of survival and reproduction than other variants do.

5. The best definition of *evolutionary fitness* is _____.

   A. physical health; **B.** the ability to attract members of the opposite sex; **C.** the ability to adapt to the environment; **D.** survival and reproduction relative to other members of the population; **E.** overall strength

6. An adaptation is a trait of an organism that increases _____.

   A. its fitness; **B.** its ability to survive and replicate; **C.** in frequency in a population over many generations; **D.** a and b are correct; **E.** a, b, and c are correct

7. The heritable differences among organisms are a result of _____.

   A. differences in their DNA; **B.** mutation; **C.** differences in alleles; **D.** a and b are correct; **E.** a, b, and c are correct

8. The immune system of an HIV-infected individual _____.

   A. can eliminate HIV entirely; **B.** cannot cause HIV populations to decline; **C.** causes selection for HIV variants that escape immune system control; **D.** evolves quickly in response to HIV infection; **E.** cannot make antibodies to HIV

9. HIV evolves rapidly because it _____.

   A. has a very high rate of reproduction; **B.** has a very high mutation rate; **C.** can detoxify anti-HIV drugs; **D.** a and b are correct; **E.** a, b, and c are correct

10. HIV is transmitted via _____.

    A. sexual intercourse with an infected person; **B.** shaking hands with an infected person; **C.** using the same bathroom as an infected person; **D.** a bite from an insect that has previously bitten an infected person; **E.** all of the above

Analyzing and Applying the Basics

1. The wide variety of dog breeds is a result of artificial selection from wolf ancestors. Use your understanding of artificial selection to describe how a dog breed such as the Chihuahua may have evolved.

2. The striped pattern on zebras' coats is considered to be an adaptation that helps reduce the likelihood of a lion or other predator identifying and preying on an individual animal. The ancestors of zebras were probably not striped. Using your understanding of the processes of mutation and natural selection, describe how a population of striped zebras might have evolved from a population of zebras without stripes.

3. Are all features of living organisms adaptations? How could you determine if a trait in an organism is a product of evolution by natural selection?

Connecting the Science

1. The theory of natural selection has been applied to human culture in many different realms. For instance, there is a general belief in the United States that "survival of the fittest" determines which businesses are successful and which go bankrupt. How is the selection of "winning" and "losing" companies in our economic system similar to the way natural selection works in biological systems? How is it different?

2. Ninety-five percent of worldwide HIV/AIDS cases occur in developing countries, where most of the population cannot afford combination drug therapy. Does the United States have an obligation to provide people in the developing world with low-cost, effective anti-AIDS therapy? In countries where the needs of daily survival often overshadow the requirement to take the drugs in the proper dosage, drug-resistant strains of HIV may be more likely to develop. What do you think will best help to reduce the toll of HIV/AIDS in these regions?

CHAPTER 11

Who Am I?

Species and Races

Individual Census Report

U.S. Department of C
Bureau of the Census

6 **What is your race? Mark** X **one or more races** to indicate what you consider yourself to be.

black or blue pen.

☐ White
☐ Black, African Am., or Negro
☐ American Indian or Alaska
  enrolled or princi

How should a woman who has Asian, black, white, and Native American grandparents respond to this question?

MI

☐ Asian Ind
☐ Chinese
☐ Filipino
☐ Japanese

MOST OF

What race is Indigo?

Do the races on Indigo's census form represent different "basic types" of humans?

Are we more similar to people of the same race than to people of different races?

Indigo pondered the choices in front of her: White, Black, American Indian or Alaskan Native, Vietnamese, Other Asian. As the daughter of a man with African American and Choctaw ancestry, and a woman with a white American father and a Laotian mother, Indigo was not sure what race she should report on the U.S. census form.

The 2000 census was different from all previous censuses in that it allowed people to check multiple boxes under the category, "race." While this change satisfied many multiracial individuals who felt that it was impossible to classify themselves as one particular race, the opportunity to do so was disquieting to others. In Indigo's case, despite having a half-Asian mother, she knew that most people saw her as black, and her connection to her mother's white father was weak at best; she didn't feel a kinship to other whites at all. And while her dad was proud of his Choctaw heritage, Indigo did not know a single Native American. Indigo did not see herself as belonging to any of the races on the census form. Maybe "human" was what she was looking for. She wondered, "Why do I need to specify my racial category? And what does it mean? If I'm part white and part black, am I somehow different from each group?"

Indigo's questions reflect those posed by many people over the years. Why do human groups differ from each other in skin color, eye shape, and stature?

Do these physical differences reveal underlying basic biological differences among these groups?

11.1   What Is a Species?

All humans belong to the same species. Before we can understand the concept of race, we first need to understand both what is meant by this statement and what is known about how species originate.

In the mid-1700s, the Swedish scientist Carolus Linnaeus began the task of cataloging all of nature. As described in Chapter 9, Linnaeus developed a classification scheme that grouped organisms according to shared structural traits. The primary category in his classification system was the **species**, a group whose members have the greatest resemblance. Linnaeus assigned a two-part name to each species—the first part of the name indicates the **genus**, or broader group to which the species belongs; the second part is specific to a particular species within that genus. For example, lions, the species *Panthera leo*, are classified in the same genus with other species of roaring cats such as the leopard, *Panthera pardus*. Linnaeus coined the binomial name *Homo sapiens* (*Homo* meaning "man," and *sapiens* meaning "knowing or wise") to describe the human species. Although Linnaeus recognized the impressive variability among humans, by placing all of us in the same species he acknowledged our basic unity. Linnaeus did classify humans into different varieties within *Homo sapiens*, a point we will return to in Section 11.3.

Modern biologists have kept the basic Linnaean classification, although they have added a **subspecies** name, *Homo sapiens sapiens*, to distinguish modern humans from earlier humans who appeared approximately 250,000 years ago.

The Biological Species Concept

While most people intuitively grasp the differences between most species—lions and leopards are both definitely cats but not the same species—biologists have had difficulty finding a single, objective definition that can be applied to all situations at all times. Several useful concepts have been proposed and used. The concept most commonly used by biologists interested in the process of species formation is called the biological species concept.

Biological Species Are Reproductively Isolated.   According to the **biological species concept**, a species is defined as a group of individuals that, in nature, can interbreed and produce fertile offspring but cannot reproduce with members of other species. In practice, this definition can be difficult to apply. Species that do not overlap in space and so have no opportunity to interbreed, species that do not undergo sexual reproduction (for example, some bacteria), and species known only by their fossils do not easily fit into the biological species concept. However, this definition does provide a basis for us to understand why species generally maintain their distinctness from each other.

Recall that differences in traits among individuals arise partly from differences in their genes. New alleles of a gene occur when the DNA that makes up the gene mutates. By the process of evolution, a particular allele can become more common in a species. If individuals of that species are unable to breed with individuals of another species, then the allele cannot spread from one species to the other. In this way, two species can evolve differences from each other. For example, imagine the common ancestor of lions and leopards. Evidence of the relationship among all large cats indicates that this ancestor had a spotted coat. The allele that eliminated the spots from adult lion coats arose and spread within this species, but it is not found in leopards because lions and leopards cannot interbreed. Scientists refer to the sum total of the alleles found in all the individuals of a species as the species' **gene pool**. Therefore, we can think of a single species as making up an impermeable container for that species' gene pool—a change in the frequency of an allele in a gene pool can take place only within a biological species.

The Nature of Reproductive Isolation.   The spread of an allele throughout a species' gene pool is called **gene flow**. Gene flow cannot occur between different biological species because a pairing between them fails to produce fertile offspring. The inability of pairs of individuals from different species to produce fertile offspring is known as **reproductive isolation**. Reproductive barriers can take two general forms: pre-fertilization barriers or post-fertilization barriers.

Pre-fertilization barriers to reproduction occur when individuals from different species either do not attempt to mate with each other, or if they do, they fail to produce a fertilized egg. The most obvious impediment to mating is that individuals from different biological species simply never contact each other; that is, they are separated by distance, a reproductive barrier known as **spatial isolation**. Among species that are close in space, one barrier to mating is differences in mating behaviors, a mechanism known as **behavioral isolation**. For example, many of the songs and displays produced by birds serve as pre-fertilization barriers. Male blue-footed boobies, sea birds that look almost as goofy as their name implies, perform an elaborate dance for the female before they mate (Figure 11.1a). This dance involves waggling and displaying their electric blue feet and differs from the dances performed by males of other, related booby species. A female blue-footed booby will not respond until she has witnessed several rounds of the dance, at which time she will engage the male in a pointing display (Figure 11.1b). In this display, both birds point their bills skyward, drop their wings, and call out their mating song. The male's dance and the pairs' pointing display presumably provide a way for both birds to recognize that they belong to the same booby species. If a female is courted by a male that cannot perform the "Blue-footed Booby Dance," she will not mate with him. Another barrier to mating results from the physical incompatibility between the sexual organs of two different individuals, a mechanism known as **mechanical isolation**. The genitals of male and female insects of the same species often fit together as specifically as a lock and a key, making matings between two members of different species impossible.

(a) Courting dance

(b) Pointing display

Figure 11.1   A behavioral pre-fertilization barrier to reproduction. (a) Female blue-footed boobies will not mate with males who fail to perform this dance. (b) Male blue-footed boobies will not mate with females who do not engage in the pointing display with them. These behaviors prevent reproduction between unrelated booby species.

(a)

(b)

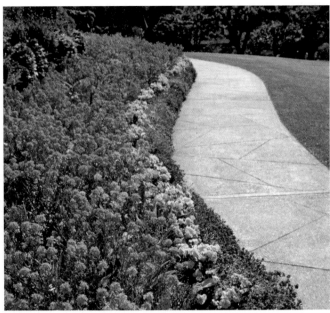

Figure 11.2   A temporal pre-fertilization barrier to reproduction. If different species of flowering plants bloom at different times of the year, they are effectively reproductively isolated. These photos show the completely different suite of species present in (a) spring and (b) summer.

Differences in the timing of reproduction, called **temporal isolation**, can also form a pre-fertilization barrier between species. This is common in flowering plants, different species of which have distinct flowering periods (Figure 11.2). Different species of periodical cicadas, insects that spend most of their lives as larvae in the soil and emerge as adults on a 13- or 17-year cycle, are also isolated temporally. Individuals of the species of cicadas that emerged throughout the eastern United States in the summer of 2004 could not possibly mate with individuals from a species that emerges anytime in the next 12 years.

The most common pre-fertilization barrier between species that *will* mate with each other is an incompatibility between eggs and sperm. For fertilization to occur, a sperm cell must bind to a protein on the surface of an egg cell. If the egg does not recognize the sperm (that is, if the egg does not have a protein that will bind to the sperm), fertilization cannot occur. Among animal species that utilize external fertilization and release their sperm and eggs into the environment—such as fish, amphibians, and sponges—this method of reproductive isolation, called **gamete incompatibility**, is widespread. Plants often have a similar incompatibility—pollen from one species cannot fertilize the ovules of another species.

Post-fertilization barriers occur when fertilization happens as a result of mating between two members of different species, but the resulting offspring does not survive or is sterile. Leopons are the result of matings between a male leopard and a female lion—an event that occurs rarely, if ever, in the wild but has been observed in captive cats. Leopons are apparently sterile, although few have been observed, so it is difficult to know the nature of the reproductive barrier. A better-known example of an offspring of a mating between two species is the mule, resulting from a cross between a horse and a donkey. Mules have a well-earned reputation as tough and sturdy farm animals, but they are also sterile and cannot produce their own offspring. Most instances of post-fertilization barriers are less obvious; most **interspecies hybrids**—that is, the offspring of parents from two different species—do not survive long after fertilization. This inability to develop is primarily a result of an incompatibility between the genes of different species. Since different species have different versions of the genes that direct the development of their bodies, placing these genes in combination often provides a hybrid offspring with incomprehensible information about how to build a body.

In the case of leopons, mules, and other sterile hybrids, the genetic incompatibility is not so large that the offspring cannot develop. Instead, the post-fertilization barrier of hybrid sterility occurs because these hybrids cannot produce proper sperm or egg cells. Recall from Chapters 5 and 6 that during the production of eggs and sperm, compatible genetic sequences called chromosomes pair up and separate during the first cell division of meiosis. Since a hybrid forms from the chromosome sets of two different species, the chromosomes cannot pair up correctly during this process, and the sperm or eggs that are produced in these animals will have too many or too few chromosomes. In the case of mules, the horse parent has 64 chromosomes and therefore produces eggs or sperm with 32; the donkey parent has 62 chromosomes, producing eggs or sperm with 31. The offspring of a cross between a horse and a donkey will therefore have 63 chromosomes and no way to effectively sort these into pairs during the first division of meiosis (Figure 11.3). As a result,

(a) A mule results from the mating of a horse and a donkey.

(b) Why mules are sterile

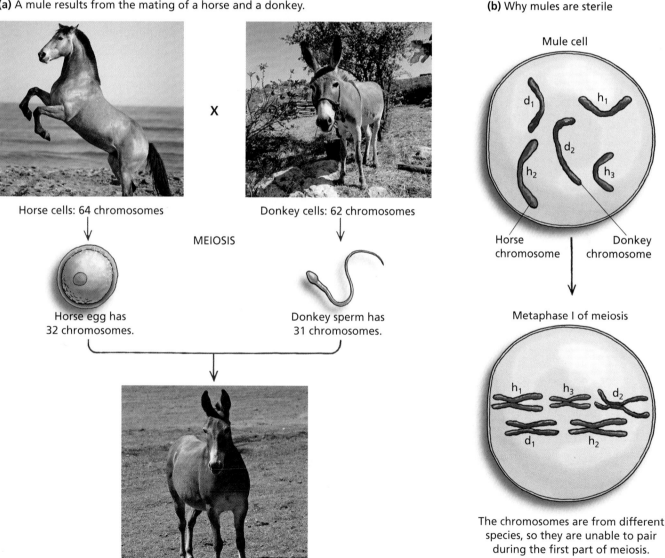

Horse cells: 64 chromosomes

Donkey cells: 62 chromosomes

MEIOSIS

Horse egg has 32 chromosomes.

Donkey sperm has 31 chromosomes.

Mule: 63 chromosomes

Mule cell

Horse chromosome

Donkey chromosome

Metaphase I of meiosis

The chromosomes are from different species, so they are unable to pair during the first part of meiosis.

Figure 11.3 Reproductive isolation between horses and donkeys. (a) A female horse produces an egg carrying 32 chromosomes, and a male donkey produces sperm containing 31 chromosomes. A cross between these two animals produces a mule with 63 chromosomes. (b) Mules produce only very few eggs or sperm because their chromosomes cannot pair properly during meiosis. Fewer chromosomes are illustrated to simplify the drawing.

male mules produce few or no sperm, and females release eggs rarely. Despite this low gamete production, a small number of mules have produced off-spring, but this event is so rare that the gene pools of donkeys and horses have remained separate despite the mule's popularity as a work animal.

Given the definition of species just discussed, we can see that all humans belong to the same biological species. There is no evidence of post-fertilization barriers to reproduction between different human groups, and Indigo's diverse ancestry clearly demonstrates that no pre-fertilization barriers exist that prohib-it mating among the races listed on her census form. To understand the concept of races *within* a species, however, we must first examine how species form.

The Process of Speciation

According to the theory of common descent discussed in Chapter 9, all mod-ern organisms descended from a common ancestral species. This evolution of one or more species from an ancestral form is called **speciation**.

For one species to give rise to a new species, most biologists agree that three steps are necessary:

1. Isolation of the gene pools of subgroups, or **populations**, of the species;
2. Evolutionary changes in the gene pools of one or both of the isolated populations; and
3. The evolution of reproductive isolation between these populations, preventing any future gene flow.

Recall that gene flow occurs when reproduction is occurring within a species. Now imagine what would happen if two populations of a species became *physically* isolated from each other, so that the movement of individu-als between these two populations was impossible. Even without genetic or behavioral barriers to mating between these two populations, gene flow between them would cease.

What is the consequence of eliminating gene flow between two populations? It is identical to what occurs in separate biological species. New alleles that arise in one population may not arise in the other. Thus, a new allele may become common in one population, but it may not exist in the other. Even among existing alleles, one may increase in frequency in one population but not in the other. In this way, each population would be evolving independently. Over time, the traits found in one population begin to differ from the traits found in the other popu-lation. In other words, the populations begin to **diverge** (Figure 11.4). When the

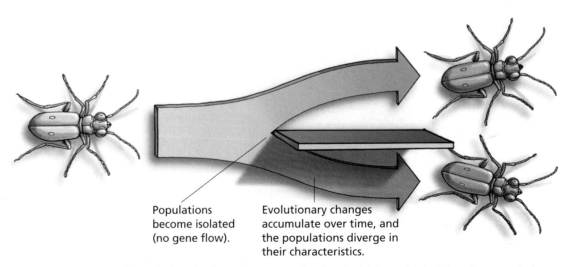

Populations become isolated (no gene flow).

Evolutionary changes accumulate over time, and the populations diverge in their characteristics.

Figure 11.4   Isolation of populations leads to divergence of traits. In this hypothetical situation, populations of beetles diverge as each adapts to its own particular environmental conditions.

Figure 11.5   Migration leads to speciation. The ancestor of Hawaiian silverswords was the much smaller and less dramatic California tarweed. Tarweed seeds were blown or carried by birds to the Hawaiian Islands, creating an isolated population. With no gene flow between the two populations, Hawaiian silverswords evolved into a very different group of species.

divergence is great enough, reproductive isolation can occur. The three steps of speciation are discussed in detail below.

Isolation and Divergence of Gene Pools.   The gene pools of populations may become isolated from each other for several reasons. Often a small population becomes isolated when it migrates to a location far from the main population. This is the case on many oceanic islands, including the Galápagos and Hawaiian islands. Bird, reptile, plant, and insect species on these islands appear to be the descendants of species from the nearest mainland. The original ancestral migrants arrived on the islands by chance. Because it is rare for organisms from the mainland to find their way across hundreds of miles of open ocean to these islands, populations at each site are nearly completely isolated from each other (Figure 11.5). In addition, because migrant populations are often small, their gene pools can change rapidly and dramatically via the process of genetic drift, as described in Section 11.5. Migration of populations from nearby sources appears to have resulted in the evolution of most species in newly emerged or unusual habitats—an idea known as the **founder hypothesis**. According to this hypothesis, the diversity of unique species on oceanic islands, as well as in isolated bogs, caves, and lakes, resulted from colonization of these once "empty" environments by small populations of migrants. Because these habitats had few competitors for resources, variants in a founding population having traits that allowed them to exploit a different resource than that used by other individuals would have had increased fitness. The advantage these different variants had caused the population to diversify and eventually to split into numerous species. For example, the large diversity of fish species once found in Africa's Lake Victoria resulted from rapid divergence from a founding population (Figure 11.6).

Populations may also be isolated from each other by the intrusion of a geologic barrier. This could be an event as slow as the rise of a mountain range or as rapid as a sudden change in the course of a river. The emergence of the Isthmus of Panama between 3 and 6 million years ago represents one such intrusion event. This land bridge connected the formerly separate continents of South and North America but *divided* the ocean gulf between them. Scientists have described at least 6 pairs of biological species of snapping shrimp on both sides of the isthmus that diverged during and after this event. These shrimp species appear to be related to each other because of similarities in appearance, protein structure, and lifestyle. In each case,

Figure 11.6   Support for the founder hypothesis. Lake Victoria contains over 500 species of cichlid fish found nowhere else in the world. A small sampling of this diversity is shown in this tank. This huge diversity appears to have evolved from one or two ancestral species that colonized this "empty" lake less than 20,000 years ago.

Figure 11.7   Physical separation leads to speciation. Species of snapping shrimp on either side of the Isthmus of Panama can be paired according to similarities in appearance and habit. Pairs are numbered for simplicity—the letter *C* before the number indicates the species is found on the Caribbean side of the Isthmus, while a *P* indicates a Pacific species. This pattern indicates that recent speciation in this group occurred during the period that the isthmus was emerging and therefore was dividing formerly continuous populations of ancestral species.

1 member of each of the 6 pairs is found on the Caribbean side of the land bridge, while the other member of each pair is found on the Pacific side (Figure 11.7). This geographic pattern indicates that the two species in each pair descended from a single species. Each original species was most likely found throughout the gulf before the isthmus arose, and each was divided into two isolated populations during the time that the land bridge was arising. Once in isolation, the separated populations of each species diverged into different biological species.

Populations that are isolated from each other by distance or a barrier are known as **allopatric** (meaning "different countries"). However, separation between two populations' gene pools may also occur even if the populations are living in physical proximity to each other, that is, if they are **sympatric** ("same country"). This appears to be the case in populations of the apple maggot fly, a species that may provide one of the clearest examples of speciation "in action."

Apple maggot flies are so named because they are notorious pests of apples grown in northeastern North America. However, apple trees are not native to North America; they were first introduced to this continent less than 300 years ago. Apple maggot flies also infest the fruit of hawthorn shrubs, a group of species that *are* native to North America. Apple-infesting flies appear to have descended from hawthorn-infesting ancestors that began to use the novel food source of apples after the fruit began to be cultivated in their home range. Apples and hawthorns live in close proximity, and apple maggot flies clearly have the ability to fly between apple orchards and hawthorn shrubs. At first glance, it does not appear that the apple maggot flies that eat apples and those that eat hawthorn fruit are isolated from each other.

Figure 11.8 Differences in the timing of reproduction can lead to speciation. This graph illustrates the life cycle of two populations of the apple maggot fly: one that lives on apple trees and another that lives on hawthorn shrubs. The mating period for these two populations differs by a month, resulting in little gene flow between them.

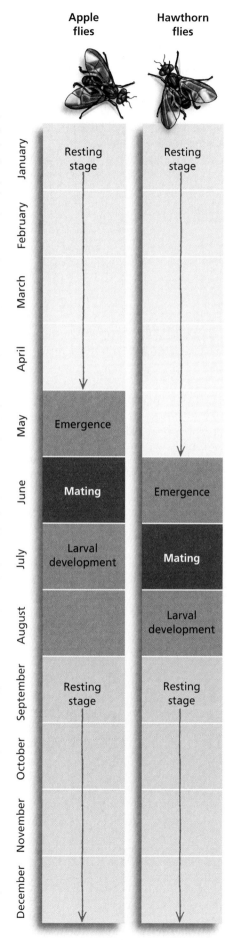

However, upon closer inspection, scientists determined that populations of apple maggot flies on apples and those on hawthorns actually have little opportunity for gene flow between them. Flies mate on the fruit where they will lay their eggs, and hawthorns produce fruit approximately 1 month after apples do. Each population of fly has a strong preference regarding which fruit it will mate on, and flies that lay eggs on hawthorns develop much faster than flies that lay eggs on apples. There appears to be little mixing between the apple-preferring and hawthorn-preferring populations.

Scientists who have examined the gene pools of the two groups of apple maggot flies find that they differ strongly in the frequency of some alleles. Thus it appears that divergence of two populations can occur even if those populations are in contact with each other, as long as some other factor—in this case, the timing of mating and reproduction as a result of variation in fruit preference—is keeping their gene pools relatively isolated (Figure 11.8). While the small amount of mating that still occurs between apple flies and hawthorn flies means that they are still considered the same biological species, the divergence that has occurred between these two populations in the past 300 years indicates that these flies may be headed toward complete reproductive isolation.

Interestingly, in plants, the formation of isolated gene pools can happen instantaneously and without any geographic or temporal barriers between populations. Most plants can undergo **asexual reproduction**; that is, a new plant can form from part of the body of a parent plant. This is often how gardeners propagate their favorite roses, for instance, by taking cutting of a stem and placing the ends of these cuttings in soil, where they will develop roots. Because hybrid plants can perform asexual reproduction, hybrid sterility as a result of an inability to produce eggs or sperm does not doom such plants to only a single generation. A population of many hybrid plants can arise from a single hybrid parent plant via asexual reproduction. However, because the hybrids cannot form eggs and sperm (for the same reason as described earlier for mules), this population is reproductively isolated from its parent populations and can travel its own evolutionary trajectory.

Amazingly, some hybrid plants can become fertile again—if a mistake during mitosis produces a cell containing duplicated chromosomes. Because these cells now contain two of each kind of chromosome, meiosis can proceed, and a plant with these cells can produce eggs and sperm. Since most plants do produce both types of gametes, such an individual can self-fertilize and give rise to a brand new species. One example of such a species is canola, an important agricultural crop grown primarily for the oil that can be extracted from its seeds. Canola developed as a result of chromosome duplication in a hybrid of kale (*Brassica oleracea*) and turnip (*Brassica campestris*); see Figure 11.9 on page 292. In fact, the same process of chromosome duplication, called **polyploidy**, that allows the production of fertile hybrids can also occur in nonhybrid plants. For example, the geneticist Hugo de Vries discovered individual evening primrose plants that had 28 chromosomes—twice as many as other plants in the same population. Upon investigation, he found that these plants were unable to produce viable offspring with evening primrose having only 14 chromosomes. Apparently, a mistake during cell division caused the number of chromosomes to double in an individual plant and led to its immediate reproductive isolation. Recent research suggests that this process of "instantaneous speciation" may have been a key factor in the evolution of diversity in plants. As many as 50% of flowering plant species may have resulted from polyploidy. It appears to occur in some animal groups, such as frogs, as well.

Figure 11.9   Instantaneous speciation.
Canola evolved from a hybrid of kale and
turnip. Although canola was initially a sterile hybrid, a variant that had duplicated
chromosomes must have appeared via
asexual reproduction. This plant could
perform meiosis and is capable of sexual
reproduction. Because canola is a new
species, its pollen cannot fertilize kale
or turnip plant eggs, and vice versa.

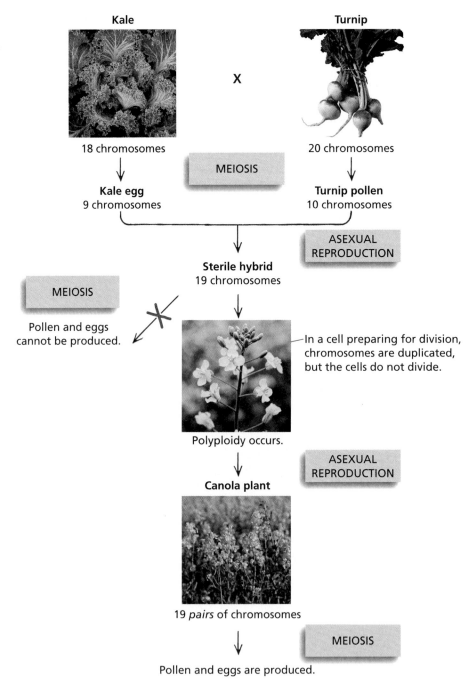

Kale

Turnip

X

18 chromosomes

20 chromosomes

MEIOSIS

Kale egg
9 chromosomes

Turnip pollen
10 chromosomes

ASEXUAL
REPRODUCTION

Sterile hybrid
19 chromosomes

MEIOSIS

Pollen and eggs
cannot be produced.

In a cell preparing for division,
chromosomes are duplicated,
but the cells do not divide.

Polyploidy occurs.

ASEXUAL
REPRODUCTION

Canola plant

19 *pairs* of chromosomes

MEIOSIS

Pollen and eggs are produced.

The Evolution of Reproductive Isolation.   For populations that have diverged
in isolation to become truly distinct biological species, they must become reproductively isolated either by their behavior or by genetic incompatibility. In the
case of canola, genetic incompatibility occurs immediately—a cross between
canola and kale does not result in fertilization. In animals, the process may be
more gradual, occurring when the amount of divergence has caused numerous
genetic differences between two populations. There is no hard-and-fast rule
about how much divergence is required; sometimes a difference in a single gene
can lead to incompatibility, while at other times, populations demonstrating
great physical differences can produce healthy and fertile hybrids (Figure 11.10).
Exactly how reproductive isolation evolves on a genetic level is still unknown
and is an actively researched and intriguing question in biology.

    Of course, once reproductive isolation occurs, each species may take radical-
ly different evolutionary paths because gene flow between the two species is impossible. Once separated, species that derived from a common ancestor can

accumulate many differences, even completely new genes. How rapidly and smoothly different forms evolve is another intriguing question in biology. Darwin assumed that speciation occurred over millions of years as tiny changes gradually accumulated. This hypothesis is known as **gradualism**. Other biologists, most notably the late Stephen Jay Gould, have argued that most speciation events are sudden, result in dramatic changes in form (via natural selection and other mechanisms of evolutionary change) within the course of a few thousand years, and are followed by many thousands or millions of years of little change—a hypothesis known as **punctuated equilibrium**. The hypothesis of punctuated equilibrium is supported by observations of the fossil record, which seems to reflect just this pattern (Figure 11.11).

The period after the separation of the gene pools of two populations but before the evolution of reproductive isolation, could be thought of as a period during which races of a species may form. Determining if the racial groupings on Indigo's census form came about via this process is our focus in the next section.

11.2 The Race Concept in Biology

Biologists do not agree on a standard definition of *biological race*. In fact, not all biologists feel that *race* is a useful term; many prefer to use the term *subspecies* to describe subgroups within a species, and others feel that race is not a useful biological concept at all. When the term is applied, it is often inconsistent. For example, populations of birds with slightly different colorations might be called different races by some bird biologists, while other biologists would argue that the same contrasts in color are meaningless. However, Indigo's question about how the racial group with which she identifies matters leads us to a definition of race that does have a specific meaning. What Indigo wants to know is: If she identifies herself as a member of a particular race, does that mean she is more closely related and thus biologically more similar to other members of the same race than she is to members of other races? The definition of **biological race** that addresses this question is the following: Races are populations of a single species that have diverged from each other. With little gene flow among them, evolutionary changes that occur in one race may not occur in a different race.

(a)

(b)

Figure 11.10 How different are two species? There is no true minimum or maximum amount of divergence that must occur before populations become reproductively isolated. (a) These two species of dragonfly look alike but cannot interbreed. (b) Dog breeds provide a dramatic example of how the evolution of large physical differences does not always result in reproductive incompatibility.

(a) Gradualism   **(b)** Punctualism

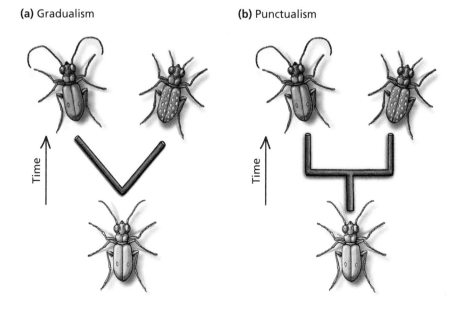

Figure 11.11 Punctuated equilibrium. The pattern of evolutionary change in groups of species may be (a) smooth—representing a constant level of small changes, or (b) more punctuated—with hundreds of thousands of years of seeming stasis followed unpredictably by rapid, large changes occurring within a few thousand years. Paleontologists Stephen Jay Gould and Niles Eldredge argued that evidence from the fossil record indicated that the second pattern—punctuated equilibrium—was predominant. Debate over the "tempo" of evolutionary change continues.

Figure 11.12 Genealogical species. The northern spotted owl belongs to the same biological species as the Mexican and California spotted owl, but its geographic isolation has resulted in the development of a unique gene pool in this population.

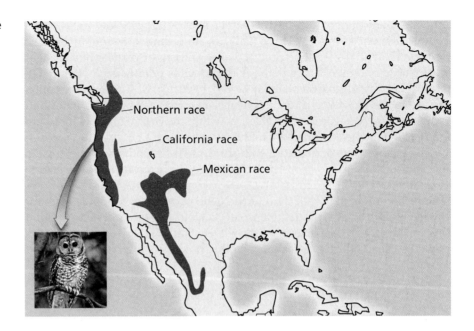

This definition of biological race is actually very similar to another commonly used species definition called the genealogical species concept.

According to the **genealogical species concept**, a species is defined as "the smallest group of reproductively compatible organisms containing all of the known descendants of a single common ancestor." More so than the biological species concept, the genealogical species concept emphasizes unique evolutionary lineages; thus it vastly increases the number of different species that can be identified. For example, the spotted owl, *Strix occidentalis*, is currently described under the biological species concept as a single species with three distinct populations—called northern, California, and Mexican (Figure 11.12). The northern population is the iconic endangered owl of old-growth forests in the northwestern United States. This population is physically isolated from the California and Mexican populations. As a result, its gene pool is separate, and any trait that evolved in the northern spotted owl population is found only within that population. Although spotted owls from Mexico can produce living, fertile offspring with northern spotted owls from Oregon and thus belong to the same biological species, according to the genealogical species concept the northern spotted owl should be identified as a unique genealogical species because it is a unique lineage, representing all the descendants of the first spotted owls to colonize the Pacific Northwest. And while the California and Mexican populations of the spotted owl are not endangered, much effort has been expended to prevent the extinction of the northern population lineage and its unique gene pool.

The advantages of the genealogical species concept are that species are more easily delineated. If the gene pool of a population is consistently different from other related populations, then according to the genealogical species concept, that population represents a different species even if the populations are reproductively compatible. The genealogical concept can apply to all groups of living species whether they reproduce sexually or not. Its disadvantage is that it can be difficult to apply in practice—different populations of the same biological species have to be studied carefully in order to determine if their gene pools differ.

Indigo's question about whether race matters, at least biologically, can now be restated as: Do the racial groups on the census form represent populations of the human species whose gene pools were isolated until relatively recently? In other words, should we consider human races to be different genealogical species? To answer this question, we must first understand how these racial categories came to be identified in American society.

11.3 Humans and the Race Concept

Until the height of the European colonial period in the seventeenth and eighteenth centuries, few cultures distinguished between broad groups of humans based on shared physical characteristics. People primarily identified themselves and others as belonging to particular cultural groups with different customs, diets, and languages. As northern Europeans began to contact people from other parts of the world, being able to set these people "apart" made the process of colonization and subjugation less morally questionable. Thus, when Linnaeus classified all humans as a species, he was careful to distinguish definitive varieties (what we would now call races) of humans. Linnaeus recognized five races of *Homo sapiens*. Not only did Linnaeus describe physical characteristics, he ascribed particular behaviors and aptitudes to each race; reflecting the widespread biases of the scientists of his day, he set the European race as the superior form (Figure 11.13). The classification shown in Figure 11.13 is one of dozens of examples of how scientists' work has been used to legitimize cultural practices—in this case, hundreds of years of injustice based on physical differences among people was supported by a scientific classification that seemed to make "natural" the poverty and oppression experienced by nonwhite groups.

Numerous scientists since Linnaeus have also proposed hypotheses about the number of races that the human species can be divided into. The most common number is 6: white, black, Pacific Islander, Asian, Australian Aborigine, and Native American; although, some scientists have described as many as 26 different races of the human species. The physical characteristics used to identify these races are typically skin color; hair texture; and eye, skull, and nose shape.

To answer Indigo's question about the biological meaning of race, we must determine if the physical characteristics that Linnaeus and other scientists used to delineate their hypothesized human races developed because these groups evolved independently (or mostly independently) of each other. We can test this hypothesis by looking at the fossil record for evidence of isolation during human evolution and by looking at the gene pools of these proposed races for the vestiges of that isolation.

1. HOMO.

Sapiens.   Diurnal; varying by education and situation.

2. Four-footed, mute, hairy.   *Wild Man.*

3. Copper-coloured, choleric, erect.   *American.*
   Hair black, straight, thick; *nostrils* wide, *face* harsh; *beard* scanty; *obstinate*, content free. *Paints* himself with fine red lines. *Regulated* by customs.

4. Fair, sanguine, brawny.   *European.*
   Hair yellow, brown, flowing; eyes blue; gentle, acute, inventive. Covered with close vestments. Governed by laws.

5. Sooty, melancholy, rigid.   *Asiatic.*
   Hair black; *eyes* dark; *severe*, haughty, covetous. *Covered* with loose garments. *Governed* by opinions.

6. Black, phlegmatic, relaxed.   *African.*
   Hair black, frizzled; *skin* silky; *nose* flat; *lips* tumid; *crafty*, indolent, negligent. *Anoints* himself with grease. *Governed* by caprice.

Figure 11.13 Linnaean classification of human variety. Linnaeus published this classification of the varieties of humans in the tenth edition of *Systema Naturae* in 1758. The behavioral characteristics he attributed to each variety reflect a widespread bias among his contemporaries that the European race was superior to other races; this classification was used as justification for the oppression of nonwhite races. Interestingly, one variety of humans recognized by Linnaeus—the *Wild Man* in his classification—is the chimpanzee!

Table 11.1 Comparison of three species concepts.

Species concept	Definition	Pros	Cons
Biological	Species consist of organisms that can interbreed and produce fertile offspring and are reproductively isolated from other species.	Useful in identifying boundaries between populations of similar organisms. Relatively easy to evaluate for sexually reproducing species.	Cannot be applied to organisms that reproduce asexually or to fossil organisms. May not be meaningful when two populations of the same species are separated by large geographical distances.
Genealogical	Species consist of organisms that can interbreed and are all descendants of a common ancestor and represent independent evolutionary lineages.	Most evolutionary meaningful because each species has its own unique evolutionary history. Can be used with asexually reproducing species.	Difficult to apply in practice. Requires detailed knowledge of gene pools of populations within a biological species. Cannot be applied to fossil organisms.
Morphological	Species consist of organisms that share a set of unique physical characteristics that is not found in other groups of organisms.	Easy to use in practice on both living and fossil organisms. Only a few key features are needed for identification.	Does not necessarily reflect evolutionary independence from other groups.

The Morphological Species Concept

The ancestors of humans are known only through the fossil record. We cannot delineate fossil species using the biological species concept. Instead, paleontologists use a more practical definition: A species is defined as a group of individuals that have some reliable physical characteristics distinguishing them from all other species. In other words, individuals in the same species have similar morphology—they look alike in some key feature. The differences among species in these key physical characteristics are assumed to correlate with isolation of gene pools. In the real world, scientists use this **morphological species concept** to distinguish among living organisms since applying the biological and genealogical species concepts can often be nearly impossible in practice. Table 11.1 summarizes the three species concepts.

Natural populations are variable, so a morphological species concept presents a challenge for scientists working with fossil organisms. This challenge is illustrated by the dinosaur, triceratops. In the 1880s, paleontologists working in Wyoming had described at least five different species of triceratops, each different in size or appearance. By the 1990s, scientists were convinced that these five species were actually a single species, *T. horridus*. What had seemed to be differences among species actually reflected variations within *T. horridus* (Figure 11.14). This later analysis was partially based on the close proximity of the original fossils, found in only two Wyoming counties. Paleontologists must use clues about the location, age, and environment of fossils, as well as their morphology, to convincingly group them into different species.

One advantage of the fossil record, however, is that it provides a view of the change in species over time. As described in Chapter 9, the hominin fossil record (the fossil record of humans and their extinct ancestors) consists of a sequence of specimens that are clearly similar to each other but show a pattern of change over time, and are interpreted as making up an evolutionary lineage. The

Figure 11.14 How many species of triceratops? Scientists in the early part of the twentieth century identified several different triceratops species by using the morphological species concept. Scientists who later reexamined the fossil evidence concluded that these several species represent different ages and sexes within a single species.

morphological differences between hominin species are clear, meaning that reconstructing the movement of human ancestors out of Africa is relatively straightforward.

Modern Humans: A History

The immediate predecessor of *Homo sapiens* was *Homo erectus*, a species that first appeared in east Africa about 1.8 million years ago and spread to Asia and Europe over the next 1.65 million years. Fossils identified as early *H. sapiens* appear in Africa in rocks that are approximately 250,000 years old. The fossil record shows that these early humans rapidly replaced *H. erectus* populations in the Eastern Hemisphere.

There is considerable debate among paleontologists about whether *H. sapiens* evolved just once, in Africa (this is called the out-of-Africa hypothesis), or throughout the range of *H. erectus* (known as the multiregional hypothesis). Even if *H. sapiens* evolved in Africa and then migrated to Europe and Asia, it is unclear whether populations of early humans hybridized with *H. erectus* in different areas of the globe (the hybridization and assimilation hypothesis). Because this scientific question is still unresolved, it is difficult to know when the ancestral population of modern humans split into regional populations; it could have been anytime from 150,000 to 1.8 million years ago (Figure 11.15).

Figure 11.15 Three models of modern human origins. (a) The out-of-Africa hypothesis indicates that modern humans arose in Africa and replaced populations of *Homo erectus* that had moved around the globe. (b) The multiregional evolution hypothesis suggests that *H. sapiens* evolved from *H. erectus* multiple times throughout its range. (c) Hybridization and assimilation is an intermediate hypothesis stating that *H. sapiens* evolved in Africa and replaced *H. erectus* populations but that there was some hybridization among local forms of *H. erectus* and incoming *H. sapiens*.

Most paleontologists favor the out-of-Africa hypothesis—that all modern human populations descended from African ancestors within the last few hundred thousand years. This hypothesis is supported by the close genetic similarity among people from very different geographic regions. Humans have much less genetic diversity (measured by the number of different alleles that have been identified for any gene) than any other great ape, which indicates that they are a young species that has had little time to accumulate many different gene variants. The out-of-Africa hypothesis is also supported by evidence that human populations in Africa are more genetically diverse than other human populations around the world. Again, because the amount of genetic diversity within a population or species is a measure of its age, this observation indicates that African populations are the oldest human populations. On balance, the out-of-Africa hypothesis has the strongest support; although with the evidence accumulated to date, none of the other hypotheses can be completely rejected.

If the out-of-Africa hypothesis is correct, the physical differences we see among human populations must have arisen in the last 150,000 to 200,000 years, or in about 10,000 human generations. In evolutionary terms, this is not much time. The recent shared ancestry of human groups does not support the hypothesis that the commonly defined human races are very different from each other.

Genetic Evidence of Divergence

While the fossil evidence discussed thus far indicates that members of the human species have not had much time to diverge and thus are not likely to be very different from each other, Indigo's question about the meaning of race is still relevant. After all, even if two races differ from one another only slightly, if the difference is consistent, then perhaps it is fair to say that people are biologically more similar to members of their own race than to people of a different race.

If a population represents a biological race, there will be a record in its gene pool of its isolation from other groups. Recall that when populations are isolated from each other, little gene flow occurs between them. If an allele appears in one population, it cannot spread to another, and evolutionary changes that occur in one population do not necessarily occur in others. Chapter 9 described the genetic nature of evolutionary change—when a trait becomes more common in a population due to evolution, it is because the allele for that trait has become more common. Evolution results in a change in **allele frequency** in a population, that is, in the percentage of copies of any given gene that are a particular allele. For example, in a population of 50 people, imagine that two individuals carry 1 copy of the allele that codes for blood type B, and the remainder carry 2 copies of the allele that codes for blood type O. Since every person carries 2 copies of each gene, there are actually 100 copies of the blood-type gene in the population of 50 people. Because two of these copies are allele B, the frequency of the B allele is 2 out of 100, or 2%, in this population. An evolutionary change in this population would be seen as an increase or decrease in the frequency of the B allele in the next generation.

Because evolution leaves a genetic record, we can make two predictions to test a hypothesis of whether biological races exist within a species. If a race has been isolated from other populations of the species for many generations, it should have these two traits:

1. Some unique alleles
2. Differences in allele frequency for some genes relative to other races

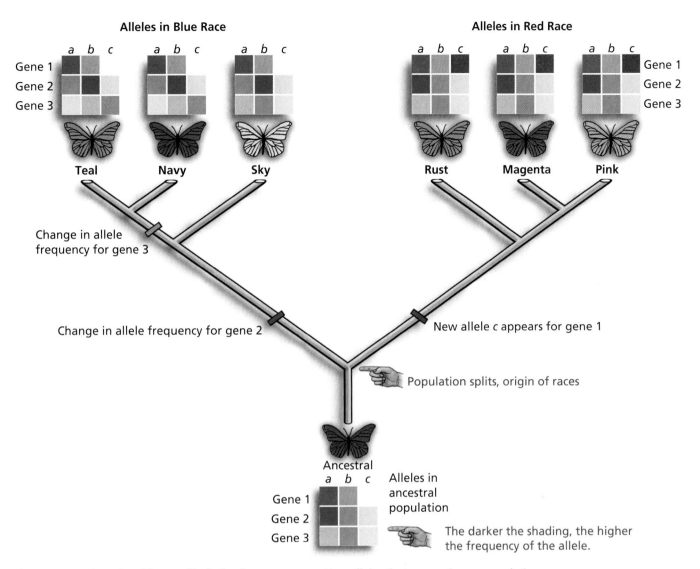

Figure 11.16   Genetic evidence of isolation between races. New alleles that appear in one population will not spread to the other populations because gene flow is restricted; thus, each race should have unique alleles. The frequency of alleles for various genes will also be more similar among populations within a race than between races.

The tree diagram in Figure 11.16 illustrates these predictions of the hypothesis of biological races. In the figure, butterfly populations colored teal, navy, and sky are all part of the same race ("blues"), and populations colored rust, magenta, and pink are part of a separate race ("reds"). The grid at the bottom of the tree illustrates the frequency of alleles for three genes in the ancestral butterfly population. For instance, there are two alleles for gene 1—one that is very common (allele *a*) and thus high in frequency, and one that is rare (allele *b*) and low in frequency. The two races described at the top of the tree originated when the ancestral population split, and the two resulting populations became isolated from each other. Notice the following patterns:

- **A race-specific allele.** Not long after the divergence between the blue and red races, mutation causes a new allele for gene 1 (that is, allele *c*) to arise but only in the red race. Because the two races are isolated, this allele does not spread to the gene pool of the blue race—there is no allele *c* for gene 1 in any of the blue populations. Additionally, because the populations colored rust, magenta, and pink diverge *after* this allele appears in reds, all of these populations contain individuals that carry allele *c* for gene 1.

- **Similar allele frequencies in populations within races.** Also not long after the divergence between reds and blues, natural selection results in a change in the allele frequency of gene 2. Perhaps the environment inhabited by blues favors individuals that carry allele *b*—this results in these individuals having more offspring than do the individuals that carry only allele *a*. Thus allele *b* becomes more common in the blue race. This evolutionary change occurred before the divergence of the populations colored teal, navy, and sky. Therefore, all of these blue race populations have a similar pattern of allele frequency for this gene, but the pattern differs from that in the populations that make up the red race.

As a result of the evolutionary independence of the blue race and the red race, if you compare the allele frequency grids of all the populations, you will notice that populations colored teal, navy, and sky are similar (although not identical) to each other, and populations colored rust, magenta, and pink are similar to each other. However, the allele frequencies for the genes in the teal population are distinctly different from those in the rust, magenta, and pink populations.

Observing a pattern of unique allele frequencies in different populations of the same species is one piece of supporting evidence showing that the populations have been isolated from each other. For example, scientists have observed that certain alleles are more common in apple-eating populations of apple maggot flies than in the hawthorn-eating populations. This observation has led researchers to conclude that these populations of flies are genetically isolated from each other and should be considered different races.

Human Races Are Not Biological Groups

Recall the six major human races described by many authors: white, black, Pacific Islander, Asian, Australian Aborigine, and Native American. Do these groups show the predicted pattern of race-specific alleles and unique patterns of allele frequency? In a word, no.

No Race-Specific Alleles Have Been Identified.   Let us first examine whether any alleles that are unique to a race have been identified. Sickle-cell anemia is a condition we discussed in Chapter 6, and one that has long been thought to be a "black" disease. This illness occurs in individuals who carry 2 copies of the sickle-cell allele, resulting in red blood cells that deform into a sickle shape under certain conditions. The consequences of these sickling attacks include heart, kidney, lung, and brain damage. Many individuals with sickle-cell anemia do not live past childhood.

Nearly 10% of African Americans and 20% of Africans carry 1 copy of the sickle-cell allele. However, if we examine the distribution of the sickle-cell allele more closely, we see that the pattern is not quite so simple. Just as we can divide the human species into populations that share similarity in skin color and eye shape (the typical races), we can divide these races into smaller populations that live in a defined geographic area and share cultural and language similarities. When we do this, we find that not all populations classified as black have a high frequency of the sickle-cell allele. In fact, in populations from southern and north-central Africa, which are traditionally classified by race as black, this allele is very rare or absent. Among populations that are classified as white or Asian, there are some in which the sickle-cell allele is relatively common, such as among white populations in the Middle East and

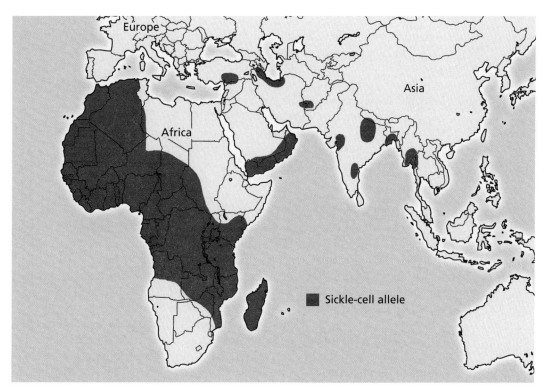

Figure 11.17   The sickle-cell allele: Not a "black gene." The map illustrates where the sickle-cell allele is found in human populations. Note that it is not found in all African populations but is found in some European and Asian populations.

Asian populations in northeast India (Figure 11.17). Thus, the sickle-cell allele is not a characteristic of all black populations nor is it unique to a supposed "black race."

Similarly, cystic fibrosis, a disease that results in respiratory and digestive problems and early death, was often thought of as a disease of the "white race." Cystic fibrosis occurs in individuals who carry 2 copies of the cystic fibrosis allele. As with sickle-cell anemia, it has become clear that the allele that causes cystic fibrosis is not found in all white populations and is found, in low frequency, in some black and Asian populations. Thus, the cystic fibrosis allele is not a characteristic of all white populations, nor is it unique to a supposed "white race."

These examples of the sickle-cell allele and cystic fibrosis allele demonstrate the typical pattern of gene distribution. Scientists have not identified a single allele that is found in all (or even most) populations of a commonly described race and that is not found in other races. The hypothesis that human races represent mostly independent evolutionary groups is not supported by these observations.

Populations Classified in the Same Race Do Not Have Similar Allele Frequencies.   What about the second prediction of the hypothesis that human racial groups are biologically independent—that we should observe unique patterns of allele frequency within these different races? Until the advent of modern techniques allowing scientists to isolate genes and the proteins they produce, there was no way of directly measuring the frequency of alleles for most of the genes in a population. However, scientists could evaluate the racial categories already in place and assume that their average physical differences reflected genetic differences among them. Thus populations with dark skin were assumed to have a high frequency of "dark skin" alleles, while populations with light skin were assumed to have a low frequency of these alleles. Similar assumptions were made about a range of physical differences—eye shape, skull shape, and hair type all clearly have a genetic basis, and all clearly differ among racial categories. These observations appear to support the hypothesis that different races have unique allele frequencies. However, physical characteristics such as skin color, eye shape, and hair type are each influenced by several different genes, each with a number of different alleles. Because skin color is affected by numerous genes, each of them affecting the amount and distribution of skin pigment, two human populations with fair skin could have completely different gene pools with respect to skin color.

If the physical characteristics that describe races illustrate biological relationships, then the allele frequency for *many different* genes should also be more similar among populations within a race than between populations of different races. In the last half-century, scientists have been able to directly measure the allele frequency of different genes in a variety of human populations. Essay 11.1 on page 304 describes how we can calculate allele frequency from the frequency of genotypes.

Let us examine a few examples of data collected on allele frequencies for various genes in different human populations. Figure 11.18a shows the frequency of the allele that interferes with an individual's ability to taste the chemical phenylthiocarbamide (PTC) in several populations. People who carry 2 copies of this recessive allele cannot detect PTC, which tastes bitter to people who carry 1 or no copies of the allele.

Figure 11.18b lists the frequency of one allele for the gene *haptoglobin 1* in a number of different human populations. Haptoglobin 1 is a protein that helps scavenge the blood protein hemoglobin from old, dying red blood cells.

Figure 11.18c illustrates variation among human populations in the frequency of a repeating DNA sequence on chromosome 8. Repeating

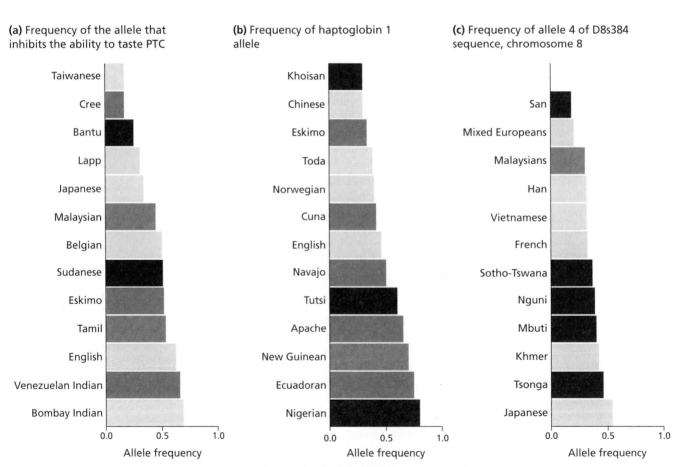

(a) Frequency of the allele that inhibits the ability to taste PTC

(b) Frequency of haptoglobin 1 allele

(c) Frequency of allele 4 of D8s384 sequence, chromosome 8

Figure 11.18   Do human races show genetic evidence of isolation? The bars on each of these histograms illustrate the frequency of the described allele in many different human populations. Bars with the same color represent populations within the same commonly defined race. These histograms illustrate that populations within these "races" are not necessarily more similar to each other than they are to populations in different races.

sequences are common in the human genome, and differences among individuals in the number of repeats create the unique signatures called DNA fingerprints (described in Chapter 8). The frequency of one pattern of repeating sequence in a segment of chromosome 8, called allele 4 of the D8s384 sequence, is illustrated for a number of populations.

Notice that the human populations in each part of Figure 11.18 are listed by increasing frequency of the allele in the population. If the hypothesis that human racial groups have a biological basis is correct, then populations from the same racial group should be clustered together on each bar graph. The color coding of each population group in each of the graphs corresponds to the racial category in which they are typically placed. To help us evaluate the hypothesis, we are using stereotypical colors for the races—pale brown for white, dark brown for black, yellow for Asian, red for Native American, and medium brown for Pacific Islanders.

What we see in the three graphs is that allele frequencies for these genes are *not* more similar within racial groups than between racial groups. In fact, in two of the three graphs, the populations with the highest and lowest allele frequencies belong to the same race—for these genes, there is more variability *within* a race than there are average differences *among* races. Scientists have observed this same pattern for every gene they have studied in the human population. These observations do not match the prediction made by the hypothesis that morphological similarity among populations reflects an underlying close genetic relationship. Both the fossil evidence and genetic evidence indicate that the six commonly listed human racial groups do *not* represent biological races.

Human Races Have Never Been Truly Isolated

The genetic analysis that caused scientists to reject the hypothesis that human populations within the same races are very similar to each other, and consistently different from other races, has shown that the exact opposite is true. The evidence that human populations have been "mixing" since modern humans first evolved is contained within the gene pool of human populations.

For instance, the frequency of the B blood group decreases from east to west across Europe (Figure 11.19). The I^B allele that codes for this blood type apparently evolved in Asia, and the pattern of blood group distribution seen in Figure 11.19 corresponds to the movement of Asians into Europe beginning

Frequency of the B blood group

0%   5%   10%   15%   20%   25%   30%

Figure 11.19 The map of blood types indicates mixing between populations. The frequency of the type B blood group in Europe declines from east to west across the continent. This pattern reflects the movement of alleles from Asian populations into European populations over the past 2000 years.

Essay 11.1 The Hardy-Weinberg Theorem

Reginald Punnett, who developed the Punnett square, was a scientist at Cambridge University during the early 1900s. Punnett is considered to be one of the fathers of modern genetics. His verification of Mendel's work helped establish Cambridge University as a center of genetic research. Among Punnett's accomplishments was his dissemination of Mendel's work to a wide and somewhat skeptical scientific audience.

One of these skeptics was George Udny Yule, a statistician at University College, London. Yule argued that inheritance could not work by Mendelian principles—as an example, he used the dominant trait of brachydactyly (having extra fingers). Since this allele is dominant over the five-fingered condition, Yule asserted that we would expect it to eventually become more common. He based this assertion on the observation that a cross between heterozygotes results in three-fourths of the offspring expressing the dominant trait and one-fourth expressing the recessive trait; thus, you should get three brachydactylys for every five-fingered person.

Punnett intuitively knew that Yule's assertion was false, but he did not have the mathematical background to prove his intuition. For help, he turned to another Cambridge scientist named Godfrey Hardy. Hardy was a renowned "pure mathematician" whose teaching revolutionized mathematics education. Legend has it that Hardy wrote the mathematical proof on his shirt cuff during a dinner party, and he felt that it was so simple it fell below his standards of publication. He did eventually publish it as a letter to the editor in the journal *Science* at nearly the same time as Wilhelm Weinberg's identical proof was published in a German journal. Hardy's letter was his only contribution to the field of biology; in his autobiography, *A Mathematician's Apology*, it receives no mention.

The Hardy-Weinberg theorem states that allele frequencies will remain stable in populations that are large in size, randomly mating, and experiencing no migration or natural selection. Subsequent geneticists used this theorem as a baseline for predicting how allele frequencies would change if any of the theorem's assumptions were violated. In other words, the Hardy-Weinberg theorem enables scientists to quantify the effect of evolutionary change on allele frequencies. Today, the Hardy-Weinberg theorem forms the basis of the modern science of population genetics.

In the simplest case, the Hardy-Weinberg theorem (which we abbreviate to Hardy-Weinberg) describes the relationship between allele frequency and genotype frequency for a gene with 2 alleles in a stable population. Hardy-Weinberg labels the frequency of these two alleles, p and q.

Imagine that we know the frequency of alleles for a particular gene in a population; let us say that 70% of the alleles in the population are dominant (A), and 30% are recessive (a). Thus, $p = 0.7$ and $q = 0.3$. Each gamete produced by members of the population carries 1 copy of the gene. Therefore, 70% of the gametes produced by this entire population will carry the dominant allele, and 30% will carry the recessive allele. The frequency of gametes produced of each type is equal to the frequency of alleles of each type (Figure E11.1a).

For the purposes of Hardy-Weinberg, we assume that every member of the population has an equal chance of mating with any member of the opposite sex. In other words, there is no relationship between the alleles that an individual carries for the gene and the alleles of her or his partner. The fertilizations that occur in this situation are analogous to the result of a lottery drawing. In this analogy, we can imagine individuals in a population each contributing an equal number of gametes to a "bucket." Fertilizations result when one gamete drawn from the sperm bucket fuses with another drawn from the egg bucket. Since the frequency of gametes carrying the dominant allele in the bucket is equal to the frequency of the dominant allele in the population, the chance of drawing an egg that carries the dominant allele is 70%.

In Figure E11.1b, a modified Punnett square illustrates the relationship between allele frequency in a population and genotype frequency in a stable population. On the horizontal axis of the square, we place the two types of gametes that can be produced by females in the population (A and a), while on the vertical axis we place the two types of gametes that can be produced by males. In addition, on each axis is an indication of the frequency of these types of egg and sperm in the population: 0.7 for A eggs and A sperm, 0.3 for a eggs and a sperm. Used like the typical Punnett square in Chapter 6, the grid of the square also shows the frequency of each genotype in this population. The frequency of the AA genotype in the next generation will be equal to the frequency of A sperm being drawn (0.7) times the frequency that A eggs will be drawn (0.7), or 0.49. This calculation can be repeated for each genotype. The frequency of the AA genotype is $p \times p \, (= p^2)$, the aa genotype $q \times q \, (= q^2)$, and the Aa genotype $p \times q \times 2 \, (= 2pq)$, because an Aa offspring can be produced by an A sperm and an a egg, or an a sperm and an A egg. Yule was proven wrong: The dominant-recessive relationship among alleles does not determine the frequency of genotypes in a population. Hardy and Weinberg mathematically proved that the frequency of genotypes in one generation of a population depends on the frequency of genotypes in the previous generation of the same population. The dominant trait of brachydactyly is rare in human populations because the allele for the trait is very low in frequency—in the absence of any factor that will cause finger number in the population to evolve, it should remain rare.

Scientists rarely have information about allele frequency; however, they often have information about genotype frequency. When scientists know the frequency

(a) Parental generation: 70% **A** ($p = 0.7$) and 30% **a** ($q = 0.3$)

Gametes: 70% **A** ($p = 0.7$) and 30% **a** ($q = 0.3$)

(b)

Eggs

A ($p = 0.7$)   **a** ($q = 0.3$)

Sperm

A ($p = 0.7$)

AA	**Aa**
Frequency: $= p \times p$ $= 0.7 \times 0.7$ $= 0.49$	Frequency: $= p \times q$ $= 0.7 \times 0.3$ $= 0.21$

a ($q = 0.3$)

Aa	**aa**
Frequency: $= p \times q$ $= 0.7 \times 0.3$ $= 0.21$	Frequency: $= q \times q$ $= 0.3 \times 0.3$ $= 0.09$

AA Aa aa
$$p^2 + 2pq + q^2 = 1$$

If $p = 0.7$
and $q = 0.3$ then...
$0.49 + 0.42 + 0.09 = 1$

Figure E11.1 The relationship between allele frequency and gamete frequency. (a) The frequency of any allele in a population of adults is equal to the frequency of that allele in gametes produced by that population. (b) Knowing the allele frequency of gametes allows us to predict the frequency of various genotypes in the next generation. In the absence of any process causing evolutionary change, allele frequency should remain constant from one generation to the next.

of a phenotype produced by a recessive allele, they know the frequency of that genotype. They can then use Hardy-Weinberg to calculate the allele frequency in a population. For instance, if the frequency of individuals with sickle-cell anemia is 1 in 100 births (0.01), we know that q^2 —the frequency of homozygous recessive individuals in the population—is equal to 0.01. Therefore, q is simply the square root of this number, or 0.1.

about 2000 years ago. As the Asian immigrants mixed with the European residents, their alleles became a part of the European gene pool. Populations that encountered a large number of Asian immigrants (that is, those closest to Asia) experienced a large change in their gene pools, while populations that were more distant from Asia encountered a more "diluted" immigrant gene pool made up of the offspring between the Asian immigrants and their European neighbors. Other genetic analyses have led to similar maps—for example, one indicates that populations that practiced agriculture arose in the Middle East, migrated throughout Europe and Asia, and interbred with resident populations about 10,000 years ago.

These data indicate that there are no clear boundaries within the human gene pool. Interbreeding of human populations over hundreds of generations has prevented the isolation required for the formation of distinct biological races.

11.4   Why Human Groups Differ

As you learned in the previous section, human races such as those indicated on Indigo's census form do not represent mostly evolutionarily independent groups; that is, they are not true biological races. However, as is clear to Indigo and to all of us, human populations do differ from each other in many traits. In this section, we explore what is known about why populations share certain superficial traits and differ in others.

Natural Selection

Recall the distribution of the sickle-cell allele in human populations as shown in Figure 11.17 (page 301). It is found in some populations of at least three of the typically described races. The frequency of the sickle-cell allele in these populations is much higher than scientists would predict if its only effect was to cause a life-threatening disease when it is homozygous; that is, when an individual carries 2 copies of the allele. If causing disease was this allele's only effect, then most individuals who carry 1 copy of it would have lower fitness (or, fewer surviving offspring) than would individuals who have no copies, because at least some of the carriers' offspring would have sickle-cell disease. Natural selection, the process described in Chapter 10 that results in a higher frequency of alleles that increase fitness and a lower frequency of alleles that decrease fitness, therefore should cause the sickle-cell allele to become rare in a population. The reason that the sickle-cell allele is common in certain populations has to do with the advantage it provides to heterozygotes—individuals who carry 1 copy of the sickle-cell allele—in particular environments.

The sickle-cell allele has the highest frequencies in populations that are at high risk for malaria. Malaria is caused by a parasitic, single-celled organism that spends part of its life cycle feeding on red blood cells, eventually killing the cells. Because their red blood cells are depleted, people with severe malaria suffer from anemia, which may result in death. When individuals carry a single copy of the sickle-cell allele, their blood cells deform when infected by a malaria parasite. These deformed cells quickly die, reducing the parasite's ability to reproduce and infect more red blood cells and therefore reducing a carrier's risk of anemia.

The sickle-cell allele reduces the likelihood of severe malaria, so natural selection has caused it to increase in frequency in susceptible populations. The protection that the sickle-cell allele provides to heterozygote carriers is demonstrated by the overlap between the distribution of malaria and the distribution of sickle-cell anemia (Figure 11.20). The sickle-cell anemia allele is an adaptation, a feature that increases fitness, within populations in malaria-prone areas. The allele for sickle-cell disease is not associated with a particular racial category; instead, it is associated with populations that live in particular environments.

(a) Malaria parasite, *Plasmodium falciparum*   **(b)** Malaria sickle-cell overlap

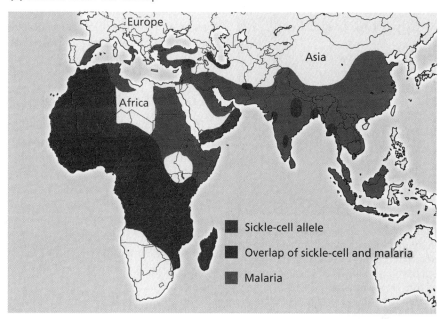

Another physical trait that has been affected by natural selection is nose form. In some populations, most individuals have broad, flattened noses; in others, most people have long, narrow noses. The pattern of nose shape in populations generally correlates to climate factors—populations in dry climates tend to have narrower noses than do populations in moist climates. Long, narrow noses appear to increase the fitness of individuals in dry environments, serving to increase the water content of inhaled air before it reaches the lungs. A narrower nose has a greater internal surface area, exposing inhaled air to more moisture. For instance, among tropical Africans, people living at drier high altitudes have much narrower noses than do those living in humid rain-forest areas (Figure 11.21). Interestingly, our preconception puts these two populations of

Figure 11.20 The sickle-cell allele is common in malarial environments. (a) Malaria is caused by a microscopic infectious organism, *Plasmodium falciparum* (colored yellow in photo), that is transmitted from person to person by mosquitoes. (b) This map shows the distributions of the sickle-cell allele and malaria in human populations. The overlap is one piece of evidence that the sickle-cell allele is in high frequency in certain populations because it provides protection from malaria.

(a) Ethiopian with a narrow nose   **(b)** Bantu with a broad nose

Figure 11.21 Nose shape is affected by natural selection. (a) Long, narrow noses are more common among populations in cold, dry environments. (b) Broad, flattened noses are more common in warm, wet environments.

Figure 11.22   Convergence. The similarity in shape between sharks (above) and dolphins (below) results from similar adaptations to life as an oceanic predator of fish. However, sharks are more closely related to stingrays, and dolphins are more closely related to land mammals.

Africans in the same race and explains differences in their nose shape as a result of natural selection, but we place white and black populations into different races and explain their skin color differences as evidence of long isolation from each other. However, like nose shape, skin color is a trait that is strongly influenced by natural selection.

Convergent Evolution

Traits that are shared by populations because they share similar environmental conditions rather than sharing ancestry are termed *convergent*. For example, the similarity in shape between white-sided dolphins and reef sharks is a result of convergence; we know by their anatomy and reproductive characteristics that sharks are most closely related to other fish, and dolphins to other mammals (Figure 11.22). The pattern of skin color in human populations around the globe also appears to be the result of **convergent evolution**. When scientists compare the average skin color in a native human population to the level of ultraviolet (UV) light to which that population is exposed, they see a nearly perfect correlation—the lower the UV light level, the lighter the skin (Figure 11.23).

UV light is light energy in a range that is not visible to the human eye. Among its many effects, this high-energy light interferes with the body's ability to maintain adequate levels of the vitamin folate. Folate is required for proper development in babies and for adequate sperm production in males. Men with low folate levels have low fertility, and women with low folate levels are more likely to have children with severe birth defects. Therefore, individuals who maintain adequate folate levels have higher fitness than individuals who do not. Darker-skinned individuals absorb less UV light and thus have higher folate levels in high-UV environments than light-skinned individuals do. In other words, in environments where UV light levels are high, dark skin is favored by natural selection—it is an adaptation in these environments.

Human populations in low-UV environments face a different challenge. Absorption of UV light is essential for the synthesis of vitamin D. Vitamin D is crucial for the proper development of bones. Women are especially harmed by low vitamin D levels—inadequate development of the pelvic bones can make safely giving birth impossible. There is no risk of not making enough vitamin D when UV light levels are high, regardless of skin color. However, in areas where levels of UV light are low, individuals with lighter skin are able to maximize their absorption of what light is available and thus have higher levels of vitamin D. In these environments, light skin has been favored by natural selection

Figure 11.23   There is a strong correlation between skin color and exposure to UV light. The average reflectance of skin in various human populations is correlated to the average UV radiation that these populations experience. Reflectance is an indication of color—higher reflectance indicates lighter skin. The color of the dots on the graph specifies the racial category of each population.

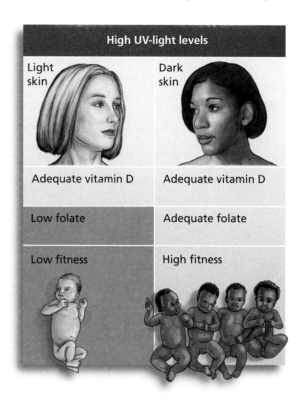

Figure 11.24   The relationship between UV light levels, folate, vitamin D, and skin color. Populations in regions where UV light levels are high experience selection for darker, UV-resistant skin. Populations in regions where UV light levels are low experience selection for lighter, UV-transparent skin.

(Figure 11.24). An exception to this pattern is for populations living in low-UV environments but have high levels of vitamin D in their diet; in this case, being able to make vitamin D is less important, and skin color may be darker.

Because UV light has important effects on human physiology, it has served as a mechanism for natural selection for skin color in human populations. Where UV light levels are high, dark skin is an adaptation, and populations become dark-skinned. Where UV light levels are low, light skin is usually an adaptation, and populations evolve to become light-skinned. The pattern of skin color in human populations is a result of the convergence of different populations in similar environments, not evidence of separate races of humans. Natural selection has caused differences among human populations, but it has also resulted in some populations superficially appearing more similar to some other human populations. Convergence in skin color and other physical characteristics has contributed to the commonly held hypothesis that people with similar skin color are more alike than people with different skin colors. As we saw in the previous section, there is no evidence to support this hypothesis. In fact, populations that appear to be similar could be quite different from each other, simply by chance.

Genetic Drift

As we have seen, differences among populations may arise through the effect of natural selection in various environments. However, differences may also arise through chance processes. A change in allele frequency that occurs due to chance is called **genetic drift**. Human populations tend to travel and colonize new areas, and so we seem to be especially prone to evolution via genetic drift.

Founder Effect.   A common cause of genetic drift occurs when a small sample of a larger population establishes a new population. The gene pool of this sample is rarely an exact model of the source population's gene pool. The difference between a subset of a population and the population as a whole is called sampling error. As discussed in Chapter 1, sampling error is more severe for smaller subsets of a population, such as those that typically found new settlements. This type of sampling error is often referred to as the **founder effect** (Figure 11.25a).

Genetic diseases that are at unusually high levels in certain populations often result from the founder effect. For example, the Amish of Pennsylvania are descended from a population of 200 German founders established approximately 200 years ago. Ellis-van Creveld syndrome, a recessive disease that causes dwarfism (among other effects), is 5000 times more common in the Pennsylvania Amish population than in other German American populations. This difference is a result of a single founder in that original population who carried this very rare recessive allele. Since the Pennsylvania Amish usually marry others within their small religious community, this allele has stayed at a high level—1 in 8

(a) Founder effect: A small sample of a large population establishes a new population.

Frequency of red allele is low in original population.

Several of the travelers happen to carry the red allele.

Frequency of red allele much higher in new population.

(b) Population bottleneck: A dramatic but short-lived reduction occurs in population size.

Frequency of red allele is low in original population.

Many survivors of tidal wave happen to carry red allele.

Frequency of red allele much higher in new population.

(c) Chance events in small populations: The carrier of a rare allele does not reproduce.

Frequency of red allele is low in original population.

The only lizard with red allele happens to fall victim to an eagle and dies.

Red allele is lost.

Figure 11.25   The effects of genetic drift. A population may contain a different set of alleles because (a) its founders were not representative of the original population; (b) a short-lived drop in population size caused a change in allele frequency in one of the populations; (c) one population is so small that low-frequency alleles are lost by chance.

Pennsylvania Amish are carriers of the Ellis-van Creveld allele, compared to less than 1 in 100 non-Amish German Americans.

Plants with animal-dispersed seeds appear to be especially prone to the founder effect. For example, cocklebur, a widespread weed that produces hitchhiker fruit (Figure 11.26), consists of populations that are quite variable in form. The variation among populations appears to have been caused by the subset of burrs that were carried from an ancestral location to new colonies.

Population Bottleneck.   Genetic drift may also occur as the result of a **population bottleneck**, a dramatic but short-lived reduction in population size followed by a rapid increase in population (Figure 11.25b). Bottlenecks often occur as a result of natural disasters. As with the founder effect, the new population differs from the original because the gene pool of the survivors is not an exact model of the source population's gene pool.

A sixteenth-century bottleneck on the island of Puka Puka in the South Pacific resulted in a human population that is clearly different from other Pacific island populations: The 17 survivors of a tsunami on Puka Puka were all relatively petite, and their modern descendants are significantly shorter in stature compared to populations found on other islands. Bottlenecks are experienced by non-human populations as well; the genetic similarity among individuals in a large population of Galápagos tortoises on the island of Isabela seems to suggest that most of these animals were wiped out during a volcanic eruption about 88,000 years ago and that the current population descended from a tiny group of survivors. Many less common breeds of dogs, cats, and other pet animals are constantly at risk of experiencing a severe genetic bottleneck, especially if their popularity declines for a period of time.

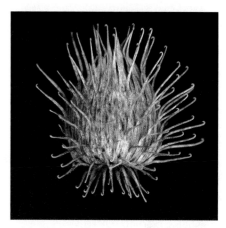

Genetic Drift in Small Populations.   Even without a population bottleneck, allele frequencies may change in a population due to chance events. When an allele is low in frequency within a small population, only a few individuals carry a copy of it. If one of these individuals fails to reproduce, or passes on only the more common allele to surviving offspring, the frequency of the rare allele may drop in the next generation (Figure 11.25c). When the population is very small, there is a relatively high probability that a rare allele will fail to be passed on to the next generation because of chance events. If the population is small enough, even relatively high-frequency alleles may be lost after a few generations by this process.

A human population that illustrates the effects of genetic drift in small populations is the Hutterites, a religious sect with communities in South Dakota and Canada. Modern Hutterite populations trace their ancestry back to 442 people who migrated from Russia to North America between 1874 and 1877. Hutterites tend to marry other members of their sect, and so the gene pool of this population is small and isolated from other populations. Genetic drift in this population over the last century has resulted in a near absence of type B blood among the Hutterites, as compared to a frequency of 15% to 30% in other European migrants in North America. Genetic drift in populations that remain small for many generations can lead to a rapid loss of alleles, which may be very harmful to a population. While this problem is uncommon in humans, the effects of genetic drift on small populations of endangered species can lead to extinction, which will be explored in Chapter 14.

Humans are a highly mobile species, and we have been founding new populations for millennia. Most early human populations were also probably quite small. These factors make human populations especially susceptible to the founder effect, population bottlenecks, and genetic drift, and have contributed to the differences among modern human groups. However, in addition to natural selection and random genetic change, humans' highly social nature and extensive culture have contributed to superficial differences in the physical appearance of various human populations.

Figure 11.26   An example of the founder effect in plants. Cocklebur disperses by hooking the spikes on its fruits to the fur (or sock) of a passing animal. When these burrs are removed at a distant location, the plant can found a new population that is a genetic subset of the parent population. As a result, the founder effect is very common in cockleburs and may contribute to the large morphological differences in fruit size and shape among cocklebur populations.

(a) Peacock

(b) Lion

(c) Blue morpho butterfly

Figure 11.27   The effects of sexual selection. Sexual selection is responsible for many unique and fantastic characteristics of organisms from (a) the peacock's tail to (b) the male lion's mane to (c) the bright colors of butterflies.

Sexual Selection

Men and women within a population may have preferences for particular physical features in their mates. These preferences can cause populations to differ in appearance. When a trait influences the likelihood of mating, that trait is under the influence of a form of natural selection called **sexual selection**. Darwin hypothesized sexual selection in 1871 as an explanation for many of the differences between males and females within a species. For instance, the enormous tail on a male peacock results from female peahens that choose mates with showier tails. Because large tails require so much energy to display, and males with these tails are more conspicuous to their predators, peacocks with the largest tails must be both physically strong and smart in order to survive. Peahens can use the size of the tail, therefore, as a measure of the "quality" of the male. Tail length does appear to be a good measure of overall fitness in peacocks. Research has demonstrated that the offspring of well-endowed males are more likely to survive to adulthood than are the offspring of males with scanty tails. When a peahen chooses a male with a large tail, she is ensuring that her offspring will receive high-quality genes. The same phenomenon seems to account for the differences between males and females in many species (Figure 11.27). In humans, there is some evidence that the difference in overall body size between men and women is a result of sexual selection—namely, a widespread female preference for larger males—perhaps again because size may be an indication of overall fitness. In these cases, female choice for particular mates has led to the evolution of the population.

Some apparently sexually selected traits seem to have little or no relationship to fitness and reflect simply a "social preference." In our highly social species, this type of sexual selection may be common. For example, some scientists have hypothesized that a trait common in the Khoikhoi people of South Africa evolved because of a male preference. Women in this population store large amounts of fatty tissue in their buttocks and upper thighs, giving them a body shape that is considerably different from other African populations. Men in these populations prefer women with this body shape and appear to have caused selection for this trait in the population, although this pattern of fat storage appears to have little effect on a woman's fitness. Other scientists have suggested that lack of facial and thick body hair in many Native American and Asian populations resulted from selection by both men and women for less hairy mates. Some even hypothesize that many physical features that are unique to particular human populations evolved as a result of these socially derived preferences. While the hypothesis that sexual selection was a key process for creating differences among human populations is intriguing, there is as yet little evidence to support these ideas and no simple way to test them.

Assortative Mating

Some differences between human populations may be reinforced by the ways in which people choose their mates. Individuals usually prefer to marry someone who is like themselves, by a process called positive **assortative mating**. For example, there is a tendency for people to mate assortatively by height—that is, tall women tend to marry tall men—and by skin color. When two populations differ in obvious physical characteristics, the number of matings between them may be small if the traits of one population are considered unattractive to members of the other population. Assortative mating has been observed in other organisms as well; for instance, sea horses choose mates that are similar in size to themselves, and in some species of fruit flies, females will mate only with males who have the same body color. Positive assortative mating tends to maintain and even exaggerate physical differences between populations. In highly social humans, assortative mating may be an important cause of differences between groups.

While human populations may show superficial differences due to natural selection in certain environments—genetic drift, sexual selection, and assortative mating—many of these differences are literally no more than skin deep. Beneath a veneer of physical differences, humans are basically the same.

11.5   Race in Human Society

The discussion in this chapter may still leave Indigo unsatisfied. Scientific data indicate that the racial categories on her census form are biologically meaningless. Races that were once thought of as unitary groups have been revealed to be hugely diverse collections of populations. Two unrelated individuals of the "black race" are no more likely to be biologically similar than a black person and a white person. Yet everywhere she looks, Indigo sees evidence that the racial categories on her census form matter to people—from the existence of her college's Black Student Association to the heated discussions in her American Experience class about immigration policies in the United States.

Part of the disparity between what recent science has revealed and what our common experience tells us about the reality of race comes from the fact that racial categories are *socially* meaningful. In the United States, we all learn that skin color, eye shape, and hair type are the primary physical characteristics that denote meaningful differences among groups. These physical characteristics have this significance due to the history of European colonization, slavery, immigration, and Native American oppression. In other words, race is a social construct—a product of history and learned attitudes. The construction of racial groups allowed some "races" to justify unethical and inhumane treatment of other "races." Thus, human races were described in the seventeenth century primarily to support **racism**, the idea that some groups of people are naturally superior to others. The United States government collects information about race on the census form as part of its effort to measure and ameliorate the lingering effects of historical, state-supported racism, but the Census Bureau acknowledges that the races with which people identify "should not be interpreted as being primarily biological or genetic."

It may be easier to see that racial categories are socially constructed if you imagine what might have happened if Western history had followed a different path. If the origin of American slaves had been from around the Mediterranean Sea, we might now identify racial groups on the basis of some other physical difference besides skin color—perhaps height, weight, or the presence of thick facial hair. Alternatively, compare the racial groupings in modern North America to those in modern Rwanda, where individuals are identified with different racial groups (Hutu and Tutsi) based on physical stature only. This classification reflects the differential social status attained by the typically taller Tutsi tribe and the typically shorter Hutu tribe under European colonization in the nineteenth century. In the United States, we would classify Hutu and Tutsi together in the same "black race"—an assignment that many members of these two groups would vigorously reject. In every society, children learn from birth which physical differences among people are significant in distinguishing "us" from "them." Even if a child is never explicitly taught racial categories, the fact that many communities are highly segregated into racial enclaves provides a lesson about which physical characteristics mark someone as "different from me."

When socially constructed racial categories are considered biologically meaningful, they become traps that are extremely difficult for individuals to escape. The most important insight that has come from studies of human diversity is that grouping human populations on the basis of skin color and eye shape is as arbitrary as grouping them on the basis of height and weight. However, arbitrary groupings are not necessarily bad. We all group ourselves into social categories: Christian or Muslim, baseball or football fan, cat or dog person. Even if the racial categories on the census form were once part of a racist system, when people identify themselves as members of a particular race, they are acknowledging a shared history with others who also identify themselves as members of that race. This self-identification can be important for realizing individual and group goals of equality and self-determination as well as continuing the fight against the real and serious vestiges of state-supported racism. The biological evidence tells Indigo that she is able to choose her racial category based on her own history and relationships—and that she should feel free to choose "none of the above" if she desires.

CHAPTER REVIEW

Summary

11.1 What Is a Species?

- All humans belong to the same biological species, *Homo sapiens sapiens*. A biological species is defined as a group of individuals that can interbreed and produce fertile offspring. Biological species are reproductively isolated from each other, thus separating the gene pools of species (p. 284).

- Reproductive isolation is maintained by pre-fertilization or post-fertilization factors (pp. 285–286).

- Speciation occurs when populations of a species become isolated from each other. These populations diverge from each other, and reproductive isolation between the populations evolves (pp. 288–292).

Web Tutorial 11.1 The Process of Speciation

11.2 The Race Concept in Biology

- Biological races are populations of a single species that have diverged from each other but have not become reproductively isolated. This definition of race corresponds to the definition of the genealogical species concept (pp. 293–294).

11.3 Humans and the Race Concept

- The fossil record provides evidence that the modern human species is approximately 200,000 years old, which is not much time for major differences between human groups to have evolved (pp. 297–298).

- The genetic evidence for biological races includes alleles that are unique to a particular race; as well as similar allele frequencies for a number of genes among populations within

races but differences in allele frequencies among populations in different races (pp. 298–300).

- Modern human groups do not show evidence that they have been isolated from each other and formed distinct races (p. 305).

- Genetic evidence indicates that human groups have been mixing for thousands of years (p. 306).

11.4 Why Human Groups Differ

- Similarities among human populations may evolve as a result of natural selection. The sickle-cell allele is more common in populations where malaria incidence is high, and light skin is more common in areas where the UV light level is low; both adaptations are a result of natural selection in these environments (p. 306).

- Human populations may show differences due to genetic drift, which is defined as changes in allele frequency due to chance events such as founder effects or population bottlenecks (pp. 309–311).

- Sexual selection, whereby individuals—typically females—choose mates that display some "attractive" quality, may also be responsible for creating differences among human populations (p. 312).

- Positive assortative mating, in which individuals choose mates who are like themselves, can reinforce differences between human populations (p. 312).

11.5 Race in Human Society

- While race in the human species has no biological meaning, it is an important social construct based on shared history and self-identity (p. 313)

Learning the Basics

1. Describe the three steps of speciation.

2. Can speciation occur when populations are not physically isolated from each other? How?

3. Describe three ways that evolution can occur via genetic drift.

4. Which of the following is an example of a pre-fertilization barrier to reproduction?

   A. A female mammal is unable to carry a hybrid offspring to term.; **B.** Hybrid plants produce only sterile pollen.; **C.** A hybrid between two bird species sings a song that is not recognized by either species.; **D.** A male fly of one species performs a "wing-waving" display that does not convince a female of another species to mate with him.; **E.** A hybrid embryo is not able to complete development.

5. According to the most accepted scientific hypothesis about the origin of two new species from a single common ancestor, most new species arise when _____.

    A. many mutations occur; **B.** populations of the ancestral species are isolated from each other; **C.** there is no natural selection; **D.** a Creator decides that two new species would be preferable to the old one; **E.** the ancestral species decides to evolve

6. For two populations of organisms to be considered separate biological species, they must be _____.

    A. reproductively isolated from each other; **B.** unable to produce living offspring; **C.** physically very different from each other; **D.** a and c are correct; **E.** a, b, and c are correct

7. The statement that "human populations classified in the same race appear to be more genetically similar than

human populations placed in different races" is _____.

    A. true; **B.** false

8. Similarity in skin color among different human populations appears to be primarily the result of _____.

    A. natural selection; **B.** convergence; **C.** shared ancestry among these populations; **D.** a and b are correct; **E.** a, b, and c are correct

9. The tendency for individuals to choose mates who are like themselves is called _____.

    A. natural selection; **B.** sexual selection; **C.** positive assortative mating; **D.** the founder effect; **E.** random mating

10. When you identify yourself as a member of a particular human group based on shared customs, language, and recent history, you are using the _____.

    A. biological definition of race; **B.** social construct definition of race; **C.** psychological definition of race; **D.** incorrect definition of race; **E.** morphological definition of race

Analyzing and Applying the Basics

1. Wolf populations in Alaska are separated by thousands of miles from wolf populations in the northern Great Lakes of the lower 48 states. Wolves in both populations look similar and have similar behaviors. However, the U.S. government has treated these two populations quite differently, listing the Great Lakes populations as endangered until recently but allowing hunting of wolves in Alaska. Some opponents of wolf protection have argued that the "wolf" should not be considered endangered at all in the United States because of the large population in Alaska, while supporters of wolf protection state that the Great Lakes population represents a unique population that deserves special status. Should these two populations be considered different races or species? What information would you need to test your answer?

2. The frequency of phenylketonuria (PKU), which results from the inability to metabolize the amino acid phenylalanine, in Irish populations is 1 in every 7000 births, while the frequency in urban British populations is 1 in 18,000 and only 1 in 36,000 in Scandinavian populations. Give two reasons that this allele, which results in severe mental retardation in homozygous individuals, may be found in different frequencies in these populations. (If you have read Essay 11.1, you should also be able to use Hardy-Weinberg to calculate the frequency of this allele in each population.)

3. Medical researchers have often excluded particular racial groups from their studies to minimize variability among the subjects of their studies. Given the biological understanding of race, does this policy make sense?

Connecting the Science

1. Black people in the United States have higher rates of hypertension (high blood pressure), heart disease, and stroke than do white people. Is this difference likely to be biological? How could you test your hypothesis?

2. The only information that was collected from *every resident* of the United States in the 2000 census was name,

place of residence, sex, age, and race. (Note that some residents received a "long form" with many additional questions, but everyone had to answer these five basic questions.) Do you think it is important to collect race data from all citizens? Why or why not? Is there some other piece of information that you think is more useful to the government?

CHAPTER 12

Prospecting for Biological Gold

Biodiversity and Classification

This strange Yellowstone hotspring ...

… is a source of this extremely valuable chemical, produced by microorganisms.

Scientists are interested in finding other useful products in Yellowstone.

Are they likely to succeed? Is there much left to discover about life?

At the end of a narrow foot trail in Yellowstone National Park lies a natural curiosity—Octopus Spring. The boiling hot water of the spring is colored an otherworldly blue. A gooey white crust encircles its main pool, and along the banks of the drainage streams radiating in all directions from it are brightly colored mats and streamers of pink, yellow, green, and orange. Although Octopus Spring is certainly not the most beautiful or dramatic feature of Yellowstone, this relatively small spring looms large in the history of biological discovery and represents a source of continued controversy.

The brilliant colors of Octopus Spring result from large numbers of microscopic organisms living in the water and on nearby surfaces. In the 1960s, Dr. Thomas Brock of the University of Wisconsin was the first to describe this biological community. One of the species he discovered, which he named *Thermus aquaticus* for its affinity for hot water, contained a protein new to science. This protein, an enzyme called *Taq* polymerase, can help produce long chains of DNA at much higher temperatures than other organisms can. *Taq* polymerase is now an integral part of the polymerase chain reaction (PCR), a high-temperature process used by many laboratories to prepare DNA samples for research or for use in the process of DNA fingerprinting. PCR has helped to revolutionize DNA research and has made many of the recent advances in genetic technology possible.

317

Until it was ruled invalid in 1999, the patent for *Taq* polymerase was held by the Swiss pharmaceutical firm Hoffman-LaRoche, and licensing agreements with other companies that used and produced *Taq* polymerase netted the company over $100 million every year. Of this substantial sum, Yellowstone National Park, the National Park Service, and the U.S. Treasury received . . . nothing in royalty payments. Even a small share of the royalties for *Taq* polymerase would have provided funds to improve and manage this heavily used national park.

The managers of Yellowstone Park do not want to miss out on the financial rewards that may come with protecting other valuable species within their borders. To capitalize on future discoveries in the park, Yellowstone entered into an agreement in 1997 with Diversa Corporation to identify and describe some of the microscopic species in the park. In return, Diversa agreed to make a one-time payment of $100,000 to Yellowstone Park and to provide several thousand dollars for research services. Diversa also agreed to share an undisclosed percentage of royalties from any profitable products that result from their research. The announcement of this deal set off a flurry of criticisms— from environmentalists who fear the disruption of biological communities in the park, to government watchdogs concerned about a few private stockholders profiting from resources taken from a park maintained for the entire public. Diversa has yet to begin exploration in Yellowstone's hot springs, pending the resolution of several legal challenges to their agreement.

Even without the legal challenges, the agreement between Yellowstone's managers and Diversa is a calculated risk by both parties. Diversa is investing nearly a million dollars in this venture, and Yellowstone faces potential damage to the wild and scenic resources that the park was designed to protect for the public good. Why are the parties to this agreement willing to take these risks? What is the likelihood of success in Diversa's search for valuable species within Yellowstone Park? Can there be many organisms that humankind has yet to discover? What do we know about the organisms we have identified? And how can we learn quickly about the traits of newly discovered species? We can answer these questions by applying evolutionary theory to investigations of the amazing variety of life on Earth.

12.1 Biological Classification

Diversa Corporation's proposed hunt for new organisms and new uses of known organisms in Yellowstone is called **bioprospecting**. Bioprospectors seek to strike biological "gold" by finding the next penicillin (originally discovered in a fungus), aspirin (produced by willow trees), or *Taq* polymerase in the living world. Yellowstone isn't the only potentially rich source of biological gold— drug companies are also investing in bioprospecting in the vast Amazonian rain forests, the strange hydrothermal vents of the ocean depths (Figure 12.1), and the bleak expanses of Antarctic ice. Other scientists are also surveying more commonly encountered organisms such as airborne molds and the bacteria that cause tooth decay. To understand the challenges associated with this survey, we must know something about the diversity of life on Earth.

How Many Species Exist?

The company name "Diversa" reflects the promise and challenge of looking for new drugs and other useful chemicals in the natural world. A characteristic of life on Earth is that it is full of variety—that is, the living world is diverse. Scientists refer to the variety within and among living species as **biodiversity**. Understanding the evolutionary origins of biodiversity and discovering the role of other species in the health of the planet is an essential aspect of biological science, but Diversa and other bioprospectors are interested in biodiversity

Figure 12.1   A source of biological riches? The organisms surrounding this deep-sea volcanic site have been known to science for less than 30 years. They represent an intriguing source of unique biological chemicals.

for a more utilitarian reason—they are banking on the variety of life to give them biochemical "solutions" to human problems.

To bioprospectors, the promise of biodiversity is its great variety, but great variety is also a source of challenge. The number of species described by science is between 1.4 and 1.8 million; even if prospectors spent only a single day screening each species for valuable products, examining them would take more than 5,000 years. The variety of different species also greatly underestimates the biodiversity within a species. Just as the species of tomato can come in many shapes, sizes, and flavors, there are species in which individuals differ greatly in the amount and potency of a particular biological molecule (Figure 12.2). Bioprospectors could miss a valuable molecule if they test only a small number of individuals from one population of a species.

Biologists disagree about the total number of species that have already been identified and described. Much of this uncertainty stems from the method of storing and cataloguing known species. When biologists identify what appears to be a new species, they collect individual specimens of the organism for storage in specialized museums. Most animal collections are found in natural history museums, while plant repositories are called herbaria (Figure 12.3); many types

Figure 12.2   Biological diversity. These varieties of the garden tomato illustrate diversity within a species.

(a) Natural history museum

(b) Herbarium

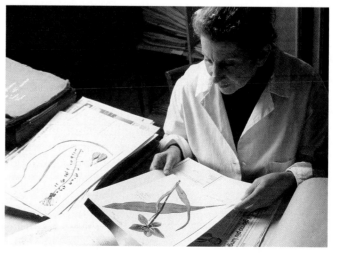

Figure 12.3   Biological collections. (a) A collection of animals in a natural history museum. Most collections contain several examples of each species to show the range of variation within the species. (b) Plant specimens are stored in herbaria.

of microbes and fungi are kept in specialized facilities called type-collection centers. **Systematists** are biologists who specialize in describing and categorizing a particular group of organisms. For an organism to be considered a new species by the scientific community, a systematist must create a description of the species that clearly distinguishes it from similar species, and she must publish this description in a professional journal. Because there are numerous large natural history museums and herbaria all over the world, along with many different journals, it is often unclear whether a species has already been described. Systematists evaluate collections to see if there is any overlap, but this process is slow and further complicated by the continual discovery and description of new species. The lack of a central resource for species collections and descriptions means that the total number of described species is only an estimate.

The number of known species represents a fraction of the total number of species on Earth. Some estimates of the actual number are as large as 100 million unique species. Most biologists agree that our planet is home to at least 10 million distinct species. Biologists are a long way from knowing all there is to know about the diversity of life. However, it is extremely rare for scientists to describe an organism that appears to have numerous features found in no other species. In fact, living species can be grouped into a few broad categories based on shared characteristics. The most general categories are kingdoms and domains.

Kingdoms and Domains

Systematists work in the field of **biological classification**, in which they attempt to organize biodiversity into discrete and logical categories. The task of classifying life is much like categorizing books in a library—books can be divided into "fiction" or "nonfiction," and within each of these divisions, more precise categories can be made (for example, nonfiction can be divided into biography, history, science, etc.). The book-cataloguing system used in most public libraries, the Dewey decimal system, is only one way of shelving books. For instance, academic and research libraries use a different system, developed by the U.S. Library of Congress. Librarians use the cataloguing system that is appropriate to the collection of books owned by the library and the needs and interests of the library's users; just as there are alternative methods of organizing books, there is more than one way to organize biodiversity to meet differing needs.

Biologists have traditionally subdivided living organisms into great groups that share some basic characteristic. Fifty years ago, most biologists divided life into two categories: plants, for organisms that were immobile and apparently made their own food; and animals, for organisms that could move about and relied on other organisms for food. When it became clear that too many organisms did not fit easily into this neat division of life, some scientists began to argue for a system of five **kingdoms**, in which organisms were categorized according to the type of cell they possessed and their method of obtaining energy. Table 12.1 provides an overview of this system.

The five-kingdom system is not perfect either; for instance, the Protista kingdom contains a wide diversity of life-forms, from amoebas to seaweeds, that have only superficial similarities. More recently, many biologists have argued that the most appropriate way to classify life is according to evolutionary relationships among organisms. Recall that the theory of evolution states that all modern organisms represent the descendants of a single common ancestor that existed nearly 4 billion years ago. Evidence for this theory includes the universality of the genetic code, many cell structures, and certain biochemical pathways. Separate populations of this ancestor diverged as natural selection and genetic drift occurred in each group, resulting in evolutionary lineages as described in Chapter 11.

Table 12.1   The classification of life. Until recently, most biologists used the five-kingdom system to organize life's diversity. Now many use a six-category system, which better reflects evolutionary relationships by acknowledging the existence of three major domains as well as four of the kingdoms.

Kingdom Name	Kingdom Characteristics	Examples	Approximate Number of Known Species	Domain Name and Characteristics
Plantae	Eukaryotic, multicellular, make own food, largely stationary	Pines, wheat, moss, ferns	300,000	**Eukarya** All organisms contain eukaryotic cells.
Animalia	Eukaryotic, multicellular, rely on other organisms for food, mobile for at least part of life cycle	Mammals, birds, fish, insects, spiders, sponges	1,000,000	
Fungi	Eukaryotic, multicellular, rely on other organisms for food, reproduce by spores, body made up of thin filaments called hyphae	Mildew, mushrooms, yeast, *Penicillium*, rusts	100,000	
Protista	Eukaryotic, mostly single–celled forms, wide diversity of lifestyles, including plant–like, fungus–like, and animal–like types	Green algae, *Amoeba*, *Paramecium*, diatoms, chytrids	15,000	The orange boxes indicate the six categories currently used to classify the diversity of life.
Monera	Prokaryotic, mostly single–celled forms, although some form permanent aggregates of cells	*Escherichia coli*, *Salmonella*, *Bacillus anthracis*, *Anabena*, sulfur bacteria	4,000	**Bacteria** Prokaryotes with cell wall containing peptidoglycan. Wide diversity of lifestyles, including many that can make their own food.
		Thermus aquaticus, *Halobacteria halobium*, methanogens	1,000	**Archaea** Prokaryotes without peptidoglycan and with similarities to Eukarya in genome organization and control. Many known species live in extreme environments.

Five-Kingdom System

Three-Domain System

Figure 12.4   The tree of life. This tree is a simplification of the current state of knowledge regarding evolutionary relationships among living organisms. Note that the branch tips are all on the same plane, representing the present time. The tree of life has been heavily pruned through nearly 4 billion years of evolution. Living organisms represent a small remnant of all the species that have appeared over Earth's history.

The process of divergence from early ancestors into the diversity of modern species has resulted in the modern "tree of life" (Figure 12.4).

When life is classified according to the relationships among organisms, major groupings correspond to divergences that occurred very early in life's history, and minor groupings correspond to more recent divergences. Classifying life according to evolutionary relationships may be especially useful to bioprospectors if a close relationship indicates a similarity in the compounds produced by living organisms.

Determining the evolutionary relationship among *all* living organisms requires comparisons of their DNA. Since each species is unique, the DNA sequence—that is, the sequence of nucleotides within the genetic instructions—of each species is unique. However, because all species share a common ancestor, all organisms also have basic similarities in their DNA sequences. As evolutionary lineages diverged from each other, mutations in DNA sequences occurred independently in each lineage and appear now as a record of evolutionary relationship among living organisms. In other words, as described in Chapter 9, the DNA sequences of closely related organisms should be more similar than the DNA sequences of more distantly related organisms (Figure 12.5).

To determine the evolutionary relationship among all modern species, scientists must compare sequences for a gene that performs a similar function among organisms as diverse as humans, willow trees, and *Thermus aquaticus*. The DNA sequence that best fits these criteria is one containing the instructions for making ribosomal RNA (rRNA), which functions as a structural part of ribosomes. As discussed in Chapter 8, the function of ribosomes—the fundamental "factories" found in all cells—is to translate genes into proteins. Each ribosome contains several rRNA molecules. The ones primarily used by scientists who are interested in the relationships among living organisms are found in the small subunit of the ribosome. A comparison of the DNA coding for small-subunit rRNAs from myriad organisms yielded the tree diagram in Figure 12.4.

You should note from Figure 12.4 that three of the kingdoms (Fungi, Animalia, and Plantae) represent relatively recently diverged groups of organisms. What was formerly called the kingdom Monera is actually made up of two groups of organisms that are quite distinct, the Archaea and Bacteria; and the kingdom Protista is a hodgepodge of many, very different organisms. To better reflect such biological relationships, biologists categorize life into three **domains** (represented by the three main branches on the tree—Bacteria, Archaea, and Eukarya), each containing several kingdoms. These three domains represent the descendants of the most ancient divergence of living organisms.

Figure 12.5 **Determining the relationships among organisms.** Comparisons of the DNA sequences that code for ribosomal RNA in various organisms have helped scientists construct the tree of life. The actual comparison consisted of thousands of base pairs; only a few are shown here for simplicity.

Systematists have begun delineating kingdoms in the Bacteria and Archaea domains using DNA sequence comparisons, while other biologists are revising the Eukarya kingdoms so that these categories more accurately reflect evolutionary history. However, just as the Library of Congress cataloguing system is not the most appropriate organization for a public library, an evolutionary classification is not the most effective organization for the purpose of surveying diversity. In this chapter we use a common hybrid of the five-kingdom system and three-domain system that specifies six categories: the domains Bacteria and Archaea and the four Eukarya kingdoms—Protista, Fungi, Plantae, and Animalia.

12.2 The Diversity of Life

In reality, the number of kingdoms and domains into which life is appropriately classified is of minor importance to a bioprospector, but understanding more recent evolutionary relationships may be *very* important, as we will discuss in section 12.3. However, dividing life into six categories—those described in Table 12.1—simplifies *our* discussion of biodiversity and of where bioprospectors may find valuable resources within the variety of life. In this section we describe the six categories: bacteria, archaea, protists, animals, fungi, and plants.

Bacteria and Archaea

Life on Earth arose at least 3.6 billion years ago, according to the fossil record. (The origin of this life remains an intriguing scientific mystery, described in detail in Essay 9.2.) The most ancient fossilized cells (Figure 12.6) are remarkably similar in external appearance to modern **bacteria** and **archaea**. Both bacteria and archaea are **prokaryotes**; this means they do not contain a nucleus, which provides a membrane-bound, separate compartment for the DNA in

Figure 12.6 **The oldest form of life.** This photograph of a fossil is accompanied by an interpretive drawing showing the fossil's living form. It was found in rocks dated at 3.465 billion years old.

(a) *Escherichia coli*

(b) *Streptomyces venezuelae*

(c) *Halobacterium*

Figure 12.7 A diversity of prokaryotes. (a) *Escherichia coli*, the "lab rat" of basic genetic studies, lives on the partially digested food in our intestines. (b) A colony of billions of cells of *Streptomyces venezuelae*, source of the antibiotic chloramphenicol and a member of the group of bacteria that live in soil and decompose dead plant matter. (c) *Halobacterium*, a salt-loving archaean, is found in high populations in this salty pond. The red pigment in this bacterium's cells is very similar in structure to the light-sensing pigments in our eyes and is used to convert energy from sunlight into cellular energy.

other cells. Prokaryotes also lack other internal structures bounded by membranes, such as mitochondria and chloroplasts, which are found in more complex **eukaryotes**. Although some species may be found in chains or small colonies as in Figure 12.6, most prokaryotes are **unicellular**, meaning that each cell is an individual organism. The discarded kingdom name "Monera" (from the Greek *moneres,* meaning "single" or "alone"), which once applied to all prokaryotes, refers to this trait. Individual prokaryotic cells are hundreds of times smaller than the cells that make up our bodies; for this reason, they are often called microorganisms or **microbes**, and biologists who study these organisms (as well as single-celled eukaryotes) are known as **microbiologists**. Their small size, easily accessible DNA, and simple structure make prokaryotes very attractive to bioprospectors because the process of studying, growing, and manipulating these organisms is typically less difficult than with eukaryotes.

The relatively simple structure of prokaryotes belies their incredible chemical complexity and diversity. Some prokaryotes can live on petroleum, others on hydrogen sulfide emitted by volcanoes deep below the surface of the ocean, and some simply on water, sunlight, and air (Figure 12.7). Prokaryotes are ubiquitous—they are found in and on nearly every square centimeter of Earth's surface, including very hot and very salty places, and even thousands of feet below ground. Prokaryotes are also incredibly numerous; for instance, there are more prokaryotes living in your mouth right now than the total number of humans who have ever lived!

Domain Bacteria. Although most are harmless to humans, the majority of the well-known bacteria have been identified because they cause disease in humans or crops. While enormous efforts are expended to control these organisms, the fact that they can live in and on other living creatures is remarkable. To survive within a host organism, bacteria must escape eradication by their host's infection-fighting system. Therefore, the molecules that allow bacteria to effectively colonize in or on living humans could be useful in treating diseases of the human immune system. Disease-causing organisms thus, surprisingly, represent one source of bacterial "biological gold."

Many of the known bacteria obtain nutrients by decomposing dead organisms. Bacterial species that function as decomposers often have competitors for their food sources—other species that also consume the same food source. When many individuals compete for the same resource, natural selection will cause the evolution of traits that provide an edge over the competition. In the case of competing bacterial decomposers, this edge is often in the form of chemicals such as **antibiotics** that kill or disable other bacteria with which the decomposers may be competing. Today, more than half of commercial antibiotics are derived from bacterial prokaryotes. Another class of valuable molecules that has evolved as a result of competition among bacteria is restriction enzymes, proteins that can chop up DNA at specific sequence sites and thus interfere, or restrict, the growth of other organisms. Restriction enzymes are used in the production of DNA fingerprints and were crucial to the success of the Human Genome Project, an effort to describe the entire DNA sequence of humans. Bioprospectors are very interested in finding more of both of these extremely valuable compounds—antibiotics and restriction enzymes—within the domain Bacteria.

Domain Archaea. Although superficially similar to domain Bacteria because of its prokaryotic cells, the domain Archaea differs from Bacteria in many fundamental ways, including the structure of cells and membranes. The known Archaea encompass numerous organisms found in extreme environments, including high-salt, high-sulfur, and high-temperature habitats. *Thermus aquaticus*, the source of *Taq* polymerase and a hot-spring dweller, belongs to Archaea. *Taq* polymerase is valuable because it operates at a high temperature—making it, along with compounds from other hot-spring archaeans, potentially useful in industrial settings. Natural selection of archaea in extreme environments has likely caused the evolution of other unique and useful biological molcules that can operate at high temperatures, high pressures, or in extremely salty environments in this group of organisms.

Scientists and bioprospectors still have much to learn about bacteria and archaea, beginning with a basic understanding of exactly how diverse they are. For instance, although most archaeans are known from extreme environments, members of this domain are found everywhere they are sought. Some scientists estimate that the number of undescribed prokaryotic species could range up to 100 million. Diversa Corporation's focus on the microscopic organisms of Yellowstone Park reflects the effort that drug companies are putting into microbial bioprospecting. However, prokaryotes are not the only microscopic organisms; most species in the diverse kingdom Protista also cannot be seen with the naked eye.

Protista

The kingdom **Protista** is made up of the simplest known eukaryotes—organisms composed of cells containing a nucleus and other membrane-bound internal structures. Most protists are single-celled creatures, although several have enormous multicellular (many-celled) forms.

The most ancient fossils of eukaryotic cells are approximately 2 billion years old, nearly 1.5 billion years *younger* than the oldest prokaryotic fossils. According to the **endosymbiotic theory** for the origin of protists, eukaryotes were most likely the descendants of a "confederation" of cells. At least some eukaryotic cell structures, including mitochondria, the cell's power plants, appear to have descended from bacteria that took up residence inside (*endo-*) ancestral eukaryotic cells. When organisms live together, the relationship is known as a *symbiosis*. In this case the relationship was mutually beneficial, and over time, the cells became inextricably tied together (Figure 12.8). When biologist Lynn Margulis first popularized the endosymbiotic hypothesis in the United States in 1981, many of her colleagues were skeptical, but an examination of the membranes, reproduction, and ribosomes of mitochondria shows clear similarities to the same features in certain bacteria. Even more convincingly, the sequence of DNA found in mitochondria (mtDNA) is most similar to the DNA sequence found in a particular group of bacteria. Mutually beneficial symbioses between eukaryotic cells and photosynthetic bacteria appeared after the evolution of mitochondria. These endosymbioses led to the evolution of chloroplasts—one relationship led to green algae and land plants, while several other instances of endosymbiosis led to other modern groups, including red algae and brown algae.

Not long after the first eukaryotes appeared, a wide diversity of eukaryotic forms established themselves on Earth. The modern kingdom Protista contains organisms resembling animals, fungi, and plants. There is no agreement among scientists regarding how many **phyla**, that is, groups below the level of kingdom, are contained within Protista. Some argue as few as eight, and others

Figure 12.8 The evolution of eukaryotes. The leading hypothesis regarding the evolution of eukaryotes is endosymbiosis. Mitochondria and chloroplasts appear to be descendants of once free-living bacteria that took up residence within an ancient nucleated cell.

Ancestral free-living, **oxygen-consuming** prokaryote   Ancestral nucleated eukaryote

① Ancestral prokaryotes and eukaryotes coexisted.

② Ancestral eukaryote engulfed but did not kill prokaryote.

Modern eukaryotic cell   **Mitochondrion**

③ The prokaryote survived inside the eukaryote, and each evolved dependence on the other.

Ancestral free-living photosynthetic prokaryote

④ In the ancestors of algae and land plants, photosynthetic prokaryotes were engulfed but not killed.

Chloroplast

⑤ The cells evolved dependence on each other. Multiple different symbiosis led to different algal groups.

Table 12.2   The diversity of Protista. Protista contains animal-like, fungus-like, and plant-like organisms. A sampling of protistan phyla is described here.

Kingdom Protista: Common Names and Characteristics of Select Phyla		Example
Animal-like Protists	**Ciliates** Free-living, single-celled organisms that use hair-like structures to move.	Paramecium
	Flagellates Use one or more long whip-like tail for locomotion. Most are free-living but some cause disease by infecting human organs.	Giardia
	Amoebas Flexible cells that can take any shape and move by extending pseudopodia ("false feet").	Amoeba
Fungus-like Protists	**Slime molds** Feed on dead and decaying material by growing net-like bodies over a surface or by moving about as single amoeba-like cells.	Physarum
Plant-like Protists	**Diatoms** Single cells encased in a silica (glass) shell.	Diatom
	Brown algae Large multicellular seaweeds.	Kelp
	Green algae Closest relatives to land plants. Single-celled to multicellular forms.	Volvox

propose as many as 80. Table 12.2 lists a few of the more common groups of organisms within the kingdom. As a result of its diversity, kingdom Protista may contain a plethora of useful organisms and compounds (Figure 12.9). As with Bacteria and Archaea, most members of kingdom Protista remain unknown.

The group of protists that bioprospectors have investigated in the most detail are **algae**, the only members of this kingdom with the ability to manufacture food. As with plants, algae make food with the aid of sunlight via the process of photosynthesis. The photosynthetic production of carbohydrates by algae represents a rich and tempting food source to non-photosynthetic organisms that will eat, or prey on, photosynthesizers. As a result, natural selection in

(a) Animal-like protist

Mattesia oryzaephili, a parasite that infests and kills beetles in stored grain. *M. oryzaephili* may be useful in controlling these pests.

(b) Fungus-like protist

Lagenidium giganteum, used to control mosquito populations.

(c) Plant-like protist

Gonyaulax polyedra, a luminescent alga used to measure levels of toxic materials in ocean sediments.

(d) Plant-like protist

Chondrus crispus, a red alga that is the source of carageenan.

Figure 12.9   Protista. Protista is the most diverse of life's kingdoms. Valuable organisms and compounds have been found in many major groups, including animal-like forms (a), fungus-like forms (b), and plant-like forms (c) and (d).

most photosynthetic organisms including algae has favored the evolution of defensive chemicals. These molecules make the algae distasteful or even poisonous to a potential predator; therefore, we humans can use these chemicals to control *our* predators. For example, extracts of red algae stop the reproduction of several different viruses in human tissues grown in laboratory dishes; this effect is presumably related to the algae's defensive chemicals. As yet, no drugs derived from algae are available to consumers, although there is significant interest in developing these drugs as treatments for HIV, influenza, and severe acute respiratory syndrome (SARS), as well as other dangerous viruses.

The group of organisms commonly referred to as algae is actually made up of several distinct, quite divergent, categories of organisms. Each of these algal phyla have methods of producing and storing food that is quite different from the others, as each represents the descendants of unique endosymbiosis. Some of the unique compounds produced by different algal phyla are potentially useful to humans. For example, carageenan is a slimy carbohydrate produced by red algae (see Figure 12.9d). Red algae is commercially harvested from ocean algal beds for its carageenan, which is then used as a stabilizer and thickener in foods, medicines, and cosmetics.

Figure 12.10 Ediacaran fauna. This reconstruction of multicellular organisms that lived before the Cambrian explosion is based on 580-million-year-old fossil remains. Note some of the bizarre forms that preceded the ancestors of modern multicellular organisms.

The animal-like and fungus-like protists are currently less interesting to most bioprospectors, presumably because animals and fungi have not been as rich a source of useful biological products as photosynthesizers have. However, as we shall see, both the kingdom Animalia and the kingdom Fungi have some intriguing characteristics and can be a source of useful biochemicals. It may be that within these categories of protists, bioprospectors also will find natural products that humans can use.

Animalia

From the origin of the first prokaryote until approximately 600 million years ago, life on Earth consisted only of single-celled creatures. Then, multicellular organisms first began to appear in the fossil record. (Note that there is some disagreement about when multicellular organisms first appeared. In 2002, Australian scientists announced that they had found evidence of multicellular life from at least 1.2 billion years ago. This date is nearly twice as old as most other estimates.) The ancient, many-celled creatures of 600 million years ago, called the Ediacaran fauna, were organisms unlike any modern species, including giant fronds and ornamented disks (Figure 12.10). Biologists are unsure which of these species is the common ancestor of modern animals—defined as multicellular organisms that make their living by ingesting other organisms and are motile (have the ability to move) during at least one stage of their life cycle. Within about 40 million years of the first appearance of the early multicellular organisms pictured in Figure 12.10, *all* modern animal groups had emerged.

The apparent relatively sudden emergence of the modern forms of animals—a period comprising little more than 1% of the history of life on Earth—is referred to as the **Cambrian explosion**, named for the geologic period during which it occurred. One of the most compelling questions in biology is the source of this explosion of biodiversity. The relatively rapid evolution of the immense diversity of life from simpler ancestors is remarkable. Some scientists hypothesize that the evolution of the animal lifestyle itself—that is, as predators of other organisms—led to the Cambrian explosion.

It can be difficult to conceive of an animal as complex as a human ever evolving from a simple eukaryotic ancestor. However, humans are actually not very different from other eukaryotic organisms. When the first cell containing a nucleus appeared, all of the complicated processes that take place in modern cells, such as cell division and cellular respiration, must have evolved. When the first multicellular animals appeared, many of the complex processes required to maintain these larger organisms, such as communication systems among cells and the formation of organs and organ systems, arose. Although a human and a starfish appear to be very different, the way they develop and the structures and functions of their cells and common organs are nearly identical. In fact, there appears to be surprisingly little genetic difference between humans and starfish; most of that difference occurs in a group of genes that control **development**, the process of transforming from a fertilized egg into an adult creature. Additionally, the amount of time since the divergence of the major evolutionary lineages of animals is still quite long—so long that it can be difficult to grasp.

While members of the kingdom **Animalia** (Latin *anima*, meaning "breath," or "soul") share basic characteristics such as multicellularity and motility, there is still significant diversity within the kingdom as described by **zoologists**, the biologists who study these organisms (Table 12.3). Most people typically picture mammals, birds, and reptiles when they think of animals, but species with backbones (including mammals, birds, reptiles, fish, and amphibians) represent only 4% of the total species in the kingdom. A small number of these **vertebrates** have traits interesting to a bioprospector. For instance, poison dart frogs (Figure 12.11) secrete high levels of toxins onto their skin.

Figure 12.11 Poison dart frog. This brightly colored frog contains glands in its skin that release poison when the frog is handled. The frog's bright colors warn potential predators of its toxicity.

Table 12.3 Phyla in the kingdom Animalia. A sampling of the diversity of animals. The rows are arranged generally in order of appearance in evolutionary time—from the more ancient sponges to the more recent chordates.

Kingdom Animalia: Major Phyla	Description	Example	
Porifera	The most ancient animal group. Fixed to underwater surface and filter bacteria from water that is drawn into their loosely organized body cavity.	Sponge	
Cnidaria	Radially symmetric (like a wheel) with tentacles. Some are fixed to a surface as adults (e.g. corals), while others are free floating in marine environments.	Jellyfish	
Platyhelminthes	Flatworms with a ribbon-like from. Live in a variety of environments on land and sea, or as parasites of other animals.	Tapeworm	
Nematoda	Roundworms with a cylindrical body shape. Very diverse and widespread in many environments. Earliest animal to evolve a complete digestive tract including a mouth and anus.	Roundworm	
Mollusca	Soft-bodied animals often protected by a hard shell. Body plan consists of a single muscular foot and body cavity enclosed in a fleshy mantle. Phylum includes snails, clams, and squid.	Octopus	
Annelids	Segmented worms. Body divided into a set of repeated segments.	Earthworm	
Arthropods	Segmented animals where the segments have become specialized into different roles (such as legs, mouthparts, and antennae). Body completely enclosed in an external skeleton that molts as the animal grows. Phylum includes insects and spiders as well as crabs and lobsters.	Shrimp	
Echinoderms	Slow moving or immobile animals without segmentation and with radial symmetry. Internal skeleton with projections gives the animal a spiny or armored surface.	Sea urchin	
Chordates	Animals with a spinal cord (or spinal cord-like structure). Includes all large land animals, as well as fish, whales, and salamanders.	Duck-billed platypus	

(a) Ant (*Pseudomyrmex triplarinus*)

Source of arthritis treatment.

(b) Jellyfish (*Aequorea victoria*)

Source of fluorescent protein, a useful labeling tool in microbiology.

(c) Horseshoe crab (*Limulus polyphemus*)

Source of blood proteins used to test for pathogens in humans.

Figure 12.12   Examples of invertebrates. Most animals are invertebrates, including insects, jellyfish, and horseshoe crabs.

These toxins are nerve poisons that cause convulsions, paralysis, and even death to their potential predators. In fact, the name of these frogs derives from their traditional use by humans as a source of toxins to coat the tips of hunting darts. The nerve toxins produced by these frogs are potentially valuable to bioprospectors as sources of potent, nonaddictive painkillers—in low doses, of course.

Most of the bioprospecting work in kingdom Animalia focuses on the remaining 96% of known organisms—the **invertebrates** (animals without backbones, as shown in Figure 12.12). The vast majority of multicellular organisms on Earth are invertebrate animals, and most of these animals are insects. Many invertebrates contain chemical compounds not found elsewhere in nature. Many species of beetles, ants, bees, wasps, and spiders produce venom to repel predators and competitors; these venoms are sources of potential drugs. For example, the tropical ant *Pseudomyrmex triplarinus* produces venom that appears to be useful for treating the joint swelling and pain associated with arthritis. Ants, bees, wasps, and termites that manage to flourish in crowded colonies have evolved protective molecules that reduce the spread of disease in these environments. These organisms may prove to be a source of compounds for reducing the spread of disease in human populations as well.

The animals that inhabit the oceans' incredibly diverse coral reefs are especially interesting to bioprospectors (Figure 12.13). These biological communities are very crowded with life, and the individuals within them continually interact with predators and competitors. As a result of this challenge, many successful coral-reef organisms contain defensive chemicals that might be useful as drugs. One mantra of reef bioprospectors is, "If it is bright red, slow-moving, and alive, we want it," reflecting the assumption that in order for a marine animal to survive despite being so conspicuous and easy to catch, it must have evolved powerful deterrents to predators. The number of unknown invertebrate species, especially in the oceans, is estimated to be anywhere from 6 to 30 million.

Figure 12.13   A coral reef. This extremely diverse biological community is a rich source of interesting biological chemicals.

While our ignorance about the diversity of animals is great, another kingdom of multicellular eukaryotes is even less well known, although no less important to the functioning of ecosystems and as a source of molecules useful to humans—the fungi.

Fungi

Early classifications that separated the biological world into two kingdoms, plants and animals, placed **fungi** in the plant kingdom. Like plants, fungi are immobile, and many produce organs that function like fruit by dispersing **spores**, cells that are analogous to plant seeds in that they can germinate into new individuals. However, the mushroom you think of when you imagine fungi is a misleading image of the kingdom. Most of the functional part of fungi is made up of very thin, stringy material called **hyphae**, which grows over and within a food source (Figure 12.14).

Fungi feed by secreting chemicals that break down the food into small molecules, which they then absorb into the cells of the hyphae. The string-like form of fungal hyphae maximizes the surface over which feeding takes place, so the vast majority of the "body" of most fungi is microscopic and diffuse. Fungal food sources typically include dead organisms, and the actions of fungi are key to recycling nutrients from these organisms. Fungi are more like animals than plants in that they rely on other organisms for food. In fact, DNA sequence analysis by **mycologists**, biologists who study fungi, indicates that Fungi and Animalia are more closely related to each other than either kingdom is to the plants.

The phyla of fungi are distinguished by their method of spore formation (Table 12.4). However, convergent evolution has led to the evolution of similar body shapes and lifestyles—which we call "fungal forms" among these different phyla. One of the most commerically important fungal forms is **yeast**, a single-celled type of fungi, found in at least two different fungal phyla, that inhabits liquids such as plant sap, soaked grains, or fruit juices. The activity of yeasts in oxygen-poor but sugar-rich environments results in

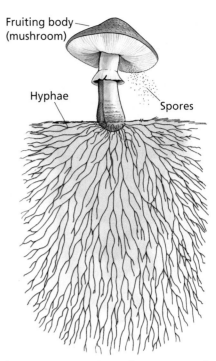

Figure 12.14   Fungi. Hyphae can extend over a large area. The familiar mushroom, as well as the fruiting structures of less-familiar fungi, primarily function only as methods of dispersing spores.

Table 12.4   Fungal diversity. Fungi are classified into phyla based on their mode of spore production. The most common phyla are listed here.

Kingdom Fungi: Major Phyla	Description	Example
Zygomycota	Sexual reproduction occurs in a small resistant structure called a zygospore. Most reproduction is asexual— directly via mitosis.	*Rhizopus stolonifera*— bread mold
Ascomycota	Spores are produced in sacs on the tips of hyphae in fruiting structures.	Morel
Basidiomycota	Spores are produced in specialized club-shaped appendages on the tips of hyphae in fruiting structures.	*Amanita muscaria*— the poisonous Fly Agaric

Figure 12.15 Antibiotics from fungi. The pink-and-gray mold in the center of this petri dish is a colony of *Penicillium*. The red dots on the edges of the dish are colonies of bacteria. The pink dots surrounding the *Penicillium* colony are bacterial colonies that are dying from contact with antibiotic secretions from the fungi.

the formation of alcohol, which explains the important role of these fungi in beer brewing, wine making, and the production of other forms of alcoholic beverages. The metabolism of yeasts within flour batter leads to the production of carbon dioxide bubbles, which are constrained by wheat protein fibers and thus allow the dough to "rise" during bread making. The fungal form known as **mold** is found in all phyla (another example of convergent evolution, described in Chapter 11) and is also commercially important, both because this quickly reproducing, fast-growing organism can spoil fruits and other foods and because it provides the essential activity for converting milk into certain types of cheese, including blue and Camembert.

About one-third of the bacteria-killing antibiotics in widespread use today are derived from fungi. Fungi produce antibiotics because natural selection favored the development of these chemicals as a tool to reduce populations of their main competitors for food—bacteria. Penicillin, the first commercial antibiotic, is produced by a fungus (Figure 12.15). Its discovery is one of the great examples of good fortune in science.

Before he went on vacation during the summer of 1928, British bacteriologist Alexander Fleming left a dish containing the bacteria *Staphylococcus aureus* on his lab bench. While he was away, this culture was contaminated by a spore from a *Penicillium* fungus that may have come from a different laboratory in the same building. When Fleming returned to his laboratory, he noticed that the growth of *S. aureus* had been inhibited on the fungus-contaminated culture dish. Fleming had been searching for a method to control bacterial growth, and this chance discovery provided a clue. He inferred that some chemical substance had diffused from the fungus, and he named this antibiotic penicillin, after the fungus itself. The first batches of this bacteria-slaying drug became available during World War II. Many historians believe that penicillin helped the Allies win the war by greatly reducing the number of deaths from infection in wounded soldiers. Since the discovery of penicillin, hundreds of other antibiotics have been isolated from different fungus species.

Some fungi infect living animals and thus also have potential as sources of drugs. Cyclosporin is a molecule produced by a number of different fungi as an adaptation that allows these species to infect live hosts. Cyclosporin has the effect of suppressing a host's immune response, and humans have used this effect to prevent the immune systems of organ-transplant recipients from attacking lifesaving, but foreign, transplanted organs. Fungi are also the source of a powerful class of anticholesterol drugs, called statins, that help treat and prevent heart disease. The fungus *Claviceps purpurea*, also known as ergot, has long been known to have powerful effects on the human body. Midwives throughout the nineteenth century used this pest of rye and wheat to stimulate uterine contractions and speed labor, and farmers throughout Europe knew that consuming grain infected with ergot could lead to neurological effects, including burning pain in the limbs ("St. Anthony's Fire"), hallucinations, and convulsions. Biologist Linda Caporael has suggested that the symptoms of "demonic possession" that led to the Salem Witch Trials in 1691–92 were caused by widespread consumption of ergot in contaminated grain. More recently, the illegal drug LSD has been derived from this fungus in order to produce the same hallucinogenic effects in recreational users.

While fungi have been the third most important source of molecules useful to humans—right after bacteria in terms of numbers and impacts of derived compounds—the source of most naturally derived drugs has been the plant kingdom.

Plantae

The kingdom **Plantae** consists of multicellular eukaryotic organisms that make their own food via photosynthesis. Plants have been present on land for over 400 million years, and their evolution is marked by increasingly effective

Table 12.5 Plant diversity. The four major phyla of plants are listed here in order of their appearance in evolutionary history.

Kingdom Plantae: Major Phyla	Description	Example	
Bryophyta	Mosses. Lacking vascular tissue, these plants are very short and typically confined to moist areas. Reproduce via spores.	Moss	
Pteridophyta	Ferns and similar plants. Contain vascular tissue and can reach tree size. Reproduce via spores.	Staghorn fern	
Coniferophyta	Cone-bearing plants resembling the first seed-producers.	Cycad	
Anthophyta	Flowering plants. Seeds produced within fruits, which develop from flowers. Advances in vascular tissues and chemical defenses contribute to their current dominance on Earth.	Orchid	

adaptations to the terrestrial environment (Table 12.5). The first plants to colonize land were necessarily small and close to the ground, for they had no way to transport water from where it is available in the soil to where it is needed, in the leaves. The evolution of **vascular tissue**, made up of specialized cells that can transport water and other substances, allowed plants to reach tree-sized proportions and to colonize much drier areas. The evolution of **seeds**, structures that protect and provide a food source for young plants, represented another adaptation to dry conditions on land. However, most modern plants belong to a group that appeared only about 140 million years ago, the **flowering plants**. Like their ancestors, flowering plants possess vascular tissue and produce seeds, but in addition, these plants evolved a specialized reproductive organ, the flower. Over 90% of the known plant species are flowering plants (Figure 12.16).

(a) Foxglove (*Digitalis purpurea*)   **(b)** Aloe (*Aloe barbadensis*)   **(c)** Willow (*Salix alba*)

Source of heart drug digitalis.   Source of aloe vera, used to treat burns and dry skin.   Source of aspirin.

Figure 12.16 Diversity of flowering plants. A few of the enormous variety of plants that provide important medicines.

From about 100 million to 80 million years ago, the number of distinct groups, or families, of flowering plants increased from around 20 to over 150. During this time, flowering plants became the most abundant plant type in nearly every habitat. The rapid expansion of flowering plants is called **adaptive radiation**—the diversification of one or a few species into a large and varied group of descendant species. Adaptive radiation typically occurs either after the appearance of an evolutionary breakthrough in a group of organisms or after the extinction of a competing group. For example, the radiation of animals during the Cambrian explosion is hypothesized by some scientists to be a result of the evolution of the predatory lifestyle, and the radiation of mammals beginning about 65 million years ago occurred after the extinction of the dinosaurs. Essay 12.1 describes the chronicle of life as a history of successive adaptive radiations of organisms. The radiation of flowering plants must be due to an evolutionary breakthrough—some advantage they had over other plants allowed them to assume roles that were already occupied by other species.

Plant biologists, or **botanists**, are still debating which traits of flowering plants give them an advantage over nonflowering types. Some botanists believe that the unique reproductive characteristics of flowering plants led to their radiation. These unique characteristics include preventing the allocation of nutrients to a developing embryo until successful fertilization of the egg (a process called "double fertilization"), as well the assistance of animals in transferring male gametes to the female reproductive structure

Essay 12.1 Diversity's Rocky Road

Paleontologists study fossils and other evidence of early life, and they have been able to piece together the history of life on Earth from these data. Early in this reconstruction, they recognized distinct "dynasties" of groups of organisms that appeared during different periods. The rise and fall of these dynasties allowed scientists to subdivide life's history into geologic periods. Each period is defined by a particular set of fossils. Table E12.1 gives the names of major geologic periods, their age and length, and major biological events that occurred during each period.

The dominance of a biological dynasty often ends due to mass extinctions—species losses that are rapid, global in scale, and affect a wide variety of organisms. For instance, the mass extinction of the dinosaurs (and of 60% to 80% of *all* organisms) distinguishes the division between the Cretaceous and Tertiary periods. Mass extinctions are most probably the result of a global catastrophe. Paleontologists believe that the mass extinction of the dinosaurs was most likely due to an enormous asteroid strike that occurred off the coast of what is now the Yucatán Peninsula in Mexico. This strike appears to have caused not only the incineration of large areas of forest in both North and South America but a massive, global tidal wave. It also probably threw up an enormous cloud of debris that blocked the sun's light for up to 3 months and led to a decade of severe acid rain. Organisms that were fortunate enough to survive this cataclysm, including our mammal ancestors, formed the basis of modern species. The adaptive radiation of these survivors has led to the current dynasty—the Age of Mammals.

Currently, Earth appears to be experiencing another mass extinction, this one caused by human activity. The current mass extinction is the topic of Chapter 14. The state of biological diversity—and the fate of humans—after this modern mass extinction is in doubt. But if the history of life is any indication, the next great era will be as different as the ones preceding it.

Table E12.1 Geological periods. The history of life is divided into four major eras, with all but the first era divided into several periods. Periods are marked by major changes in the dominant organisms present on Earth.

Era	Period	Millions of Years Ago	Features of Life on Earth
Cenozoic	Quaternary	0	Most modern organisms present.
	Tertiary	1.8	After the extinction of the dinosaurs. Mammals, birds, and flowering plants diversify.
Mesozoic	Cretaceous	65	Massive carnivorous and flying dinosaurs are abundant. Large cone-bearing plants dominate forests. Flowering plants appear.
	Jurassic	144	Huge plant-eating dinosaurs evolve. Forests are dominated by cycads and tree ferns.
	Triassic	206	Early dinosaurs, mammals, and cycads appear on land. Life "restarts" in the oceans.
Paleozoic	Permian	251	Early reptiles appear on land. Seedless plants abundant. Coral and trilobites abundant in oceans. Permian ends with extinction of 95% of living organisms.
	Carboniferous	290	Land is dominated by dense forests of seedless plants. Insects become abundant. Large amphibians appear.
	Devonian	354	Known as the age of fishes. Sharks and bony fish appear. Large trilobites are abundant in the oceans.
	Silurian	408	Life begins to invade land. The first colonists are small seedless plants, primitive insects, and soft-bodied animals.
	Ordovician	439	Life is diverse in the oceans. Cephalopods appear, and trilobites are common.
	Cambrian	495	All modern animal groups appear in the oceans. Algae are abundant.
Pre-Cambrian		543	Life is dominated by single-celled organisms in the ocean. Ediacaran fauna appear at the end of the era.
		4500	

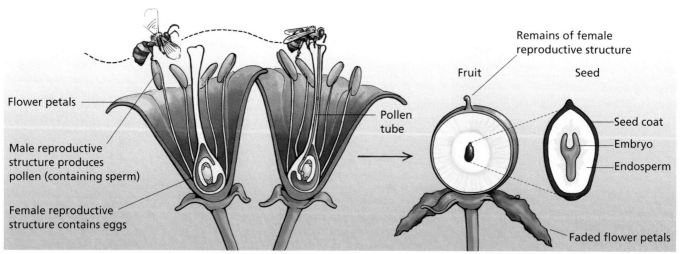

(1) Flower petals attract insects that move pollen from one flower to another, helping fertilization to occur.

(2) Double fertilization occurs. The pollen tube carries two sperm. One fertilizes the embryo, and the other fuses with two nuclei in another cell to produce the endosperm, a tissue that nourishes the embryo.

(3) Fruit consists of seeds packaged in a structure that aids their dispersal, such as tasty flesh or a parachute.

(4) Seeds contain an embryo and endosperm, and are highly resistant to drying.

Flower petals

Male reproductive structure produces pollen (containing sperm)

Female reproductive structure contains eggs

Pollen tube

Remains of female reproductive structure

Fruit

Seed

Seed coat

Embryo

Endosperm

Faded flower petals

Figure 12.17   Sexual reproduction in flowering plants. Flowering plants reproduce quite differently from nonflowering plants. These differences, including the production of fruit for dispersal, may have provided their advantage over other plant types and possibly led to their adaptive radiation.

(Figure 12.17). Other botanists think that chemical defenses in these plants reduced their susceptibility to predators and provided their edge over other plant groups.

The diversity of chemical defenses in flowering plants makes them particularly interesting to bioprospectors. Because plants cannot physically escape from their predators, natural selection has favored those plants that produce predator-deterring toxins via side reactions to primary biochemical pathways. For instance, curare vines produce toxins that block the ability of nerves to control muscle tissue. Organisms that get this toxin, called curarine, in their bloodstreams become paralyzed and can do little damage to the vine. Curarine is produced via a secondary pathway of the process of amino acid synthesis; in this case, natural selection must have favored ancestors to the curare vine that had genetic variations leading to production of not only normal amino acids, but this toxic by-product as well. Today, doctors use curarine as a muscle relaxant during surgery.

The kingdom Plantae is the source of many other well-known, naturally derived drugs. Aspirin from willow, the heart drug digitalis from foxglove, the anticancer chemical vincristine from the rosy periwinkle, morphine from opium poppies, caffeine from coffee, and dozens of other pharmaceutical products are derived directly from plants. Hundreds of other drugs based on plant chemicals are now reproduced via manufacturing processes. Many botanists believe that the number of unknown plant species is relatively small—probably a few thousand—but the potential of even the known species as sources of drugs is still mostly unknown.

12.3   Learning About Species

The living world is amazingly diverse, and our knowledge of it is only fragmentary. Many biologists are attracted to the study of biological diversity for just this reason—because the variety of life is remarkable, fascinating, and largely an unexplored frontier in human knowledge. As we observed in the previous section, however, nonhuman species can also represent an enormous resource for humans. Our survey of life's diversity illustrates the essential problem for a bioprospector seeking to tap this diverse resource—there are many more potentially valuable species than there are resources to find and evaluate them. This challenge has led many drug companies to abandon most of their bioprospecting programs in favor of a strategy called "rational drug design," which allows their scientists to use an understanding of the causes of illness to create synthetic drugs used to treat a particular disease. However, many scientists have noted that some of the most effective plant-based drugs are so strange in structure that they would never have been designed through this process. Companies like Diversa are banking on finding some of these strange compounds in nature. What tools can biologists offer to bioprospectors who seek to mine biological gold?

Fishing for Useful Products

The National Cancer Institute (NCI) has taken a brute-force approach to screening species for evidence of cancer-suppressing chemicals. NCI scientists receive frozen samples of organisms from around the world, chop them up, mix them with various chemical solvents, and separate them into a number of extracts, each probably containing hundreds of components. These extracts are tested against up to 60 different types of cancer cells to evaluate their efficacy in stopping or slowing growth of the cancer. Promising extracts are then further analyzed to determine their chemical nature, and chemicals in the extract are tested singly to find the effective compound. This approach is often referred to as the "grind 'em and find 'em" strategy.

To date, this strategy has been effective in identifying one major anticancer chemical—paclitaxel, also known by the trade name Taxol, from the Pacific yew. Paclitaxel continues to be produced via extraction from the needles of other species of yew trees and is effective against ovarian cancer, advanced breast cancers, malignant skin cancer, and some lung cancers. Dozens of other less well-known anticancer drugs have been identified by this route as well.

Understanding Ecology

The grind 'em and find 'em approach works best when researchers are seeking treatments for a specific disease or set of diseases, such as cancer. But most bioprospectors are much more speculative—they are interested in determining whether an organism contains a chemical that is useful against *any* disease. Doing this effectively requires a more thoughtful approach, taking into account the biology of the species.

One aspect of an organism's biology that can be illuminating to the bioprospector is its **ecology**—that is, its relationship to the environment and other living organisms. Our survey of diversity illustrated some ecological characteristics that increase the likelihood of a species containing valuable chemicals. Some of these characteristics include high levels of competition with bacteria and fungi, susceptibility to predation, ability to live in and on other living organisms, and high population density. In each of these cases, natural selection of species in a particular ecological situation can lead to the evolution of antibacterial, antifungal, or antiviral compounds; molecules that suppress or modify the effects of the immune system; and chemical defenses that may have physiological effects. An understanding of ecology is useful even within a species. For instance, populations of plants experiencing high levels of insect attacks may produce more defensive compounds than do populations of the same species that are not under attack. Screening organisms whose ecology indicates the probability of defensive or antibiotic compounds is one method bioprospectors use to increase their success.

Reconstructing Evolutionary History

One clue to an organism's chemical traits can come from understanding its relationship to other species and knowing the traits found in its closest relatives. This is one reason some scientists argue that a classification system reflecting evolutionary relationships is more useful than one based on more superficial similarities. The classification of certain birds helps illustrate this point. Vultures (Figures 12.18a and 12.18b) are birds that specialize in feeding on dead animals. These birds spend a large amount of time soaring on broad, flat wings, have sharp beaks for tearing meat, and regurgitate food to feed their offspring. A nonevolutionary classification places all vultures together. However, research published in the 1970s demonstrates that New World vultures in the Western Hemisphere (Figure 12.18b) appear to be more closely related to storks (Figure 12.18c)—long-legged birds with long beaks that specialize in catching fish in shallow waters—than they are to Old World vultures from the Eastern Hemisphere (Figure 12.18a). Even though species of New World vulture *look* like Old World vultures, they share a more recent common ancestor with storks and are thus much more similar to storks anatomically, physiologically, and genetically.

An **evolutionary classification** can be quite useful in the study of living organisms; for instance, if scientists wish to know more about the basic biology of New World vultures, then they might start by learning what is known about the biology of storks, their closest relatives. And if bioprospectors want to look for new valuable biological compounds, they could start by investigating the chemicals found in relatives of organisms with already known valuable chemicals.

Developing Evolutionary Classifications. Evolutionary classifications are based on the principle that the descendant species of a common ancestor should share any biological trait that first appeared in that ancestor. For example,

(a) Old World vulture

(b) New World vulture

(c) Stork

Hooded Vulture (*Necrosyrtes monachus*)   Turkey Vulture (*Cathartes aura*)   Wood Stork (*Mycteria americana*)

Figure 12.18   The challenge of biological classification. (a) Old World vulture; (b) New World vulture; and (c) stork. The evolutionary relationship between New World vultures and storks is not evident from their appearance.

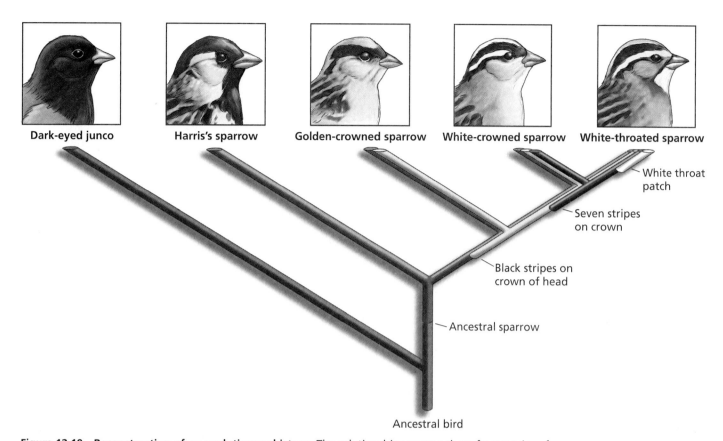

Dark-eyed junco　　**Harris's sparrow**　　**Golden-crowned sparrow**　　**White-crowned sparrow**　　**White-throated sparrow**

White throat patch

Seven stripes on crown

Black stripes on crown of head

Ancestral sparrow

Ancestral bird

Figure 12.19　Reconstruction of an evolutionary history. The relationships among these four species of sparrow can be illustrated by their shared physical traits compared to a more distant relative, the dark-eyed junco. White-throated sparrows and white-crowned sparrows share a very distinct black-and-white head pattern, indicating that they are more closely related to each other than to golden-crowned sparrows, which have fewer stripes on their crowns. These three birds apparently share a more recent common ancestor with each other than with Harris's sparrow, however, which lacks crown stripes.

this principle has been used in an attempt to uncover the evolutionary relationship among different species of sparrow belonging to the genus *Zonotrichia*.

Figure 12.19 illustrates a hypothesized **phylogeny**—the evolutionary relationship—of species in the genus *Zonotrichia*. Scientists have used a technique called **cladistic analysis**, an examination of the variation in these sparrows' traits relative to a closely related species, to determine this phylogeny. For example, if we examine just the heads of the four sparrow species compared to their relative, the dark-eyed junco, we see that three—all but the Harris's sparrow—have dark and light alternating stripes on the crown of their heads. This observation seems to indicate that crown striping evolved early in the radiation of these sparrows. Among the three species with crown stripes, two have 7 stripes while the golden-crowned sparrow has only 3 stripes. An increase in the number of stripes appears to have evolved after the original striped crown pattern and contributes to the radiation among birds in this group. Finally, of the two species with 7 crown stripes, only one has evolved a white throat—the aptly named white-throated sparrow. In the case of these four sparrows, it appears that every step in their radiation involved a visible change in their appearance. A phylogeny of humans produced in a very similar way is presented in Chapter 9.

Unfortunately, reconstructing evolutionary relationships is not as simple as the sparrow example suggests. Descendant species may lose a trait that evolved in their ancestor, or unrelated species may acquire identical traits via convergent evolution, as described in Chapter 11. You can even see convergence in the phylogeny in Figure 12.19; golden-crowned sparrows, like the white-throated species, have a patch of golden feathers on their heads that appears to have evolved independently. The existence of convergent traits complicates attempts to determine the accurate evolutionary classification of organisms.

Testing Evolutionary Classifications.   Any classification developed by a biologist can be considered a hypothesis of the evolutionary relationship among organisms. It is difficult to test this hypothesis directly—scientists have no way of observing the actual evolutionary events that gave rise to distinct organisms. However, scientists *can* test their hypotheses by using information from both fossils and living organisms. By examining the fossils of extinct organisms, scientists can gather clues about the genealogy of various groups. For example, fossils of vulture-like birds clearly indicate that this lifestyle evolved independently in both the Old World and the New World.

Information from living organisms can provide an even finer level of detail about evolutionary relationships. As illustrated earlier in Figure 12.5, closely related species should have similar DNA. If the pattern of DNA similarity matches a hypothesized evolutionary relationship among species, the phylogeny is strongly supported. This is the case with the hypothesized relationship between New World vultures and storks; DNA sequence comparisons indicate that the DNA of New World vultures is more similar to the DNA of storks than it is to that of Old World vultures. These data allow scientists to strongly infer that New World vultures are closely related to storks and that their similarities to Old World vultures are a result of convergent evolution. In contrast, DNA sequence comparisons do not support the sparrow phylogeny presented in Figure 12.19. Here, the data suggest that white-crowned birds and golden-crowned birds are closely related, and white-throated sparrows are a more distant relative. In this case, more observations are needed to discern the true evolutionary relationships among the *Zonotrichia*. In Chapter 9 we described multiple supportive tests of another phylogeny—in that case, the evolutionary relationship among humans and apes.

The examples of phylogeny reconstruction and testing as described here and in Chapter 9 provide nice illustrations of the process, but the species they employ are not likely to contain biological molecules that are valuable to humans. The evolutionary reconstructions that bioprospectors are most interested in are those performed on groups of species that already contain valuable members. For instance, relatives of the curarine-producing curare vine are likely to contain similar secondary biochemical pathways that produce slightly different muscle relaxants. Once a hypothesis of evolutionary relationship *is* reasonably well supported by additional data, bioprospectors can use the information gathered about one species in a classification group to predict the characteristics of other species in that group. This helps them to identify related species that could be additional sources of biological gold.

Although an evolutionary classification of groups that are likely to contain valuable chemicals would probably increase the speed of discovery, the current slow pace of reconstructing and testing phylogenies means that bioprospectors do not always have this information. Some bioprospectors have turned to the humans who have the most intimate knowledge of the usefulness of organisms in their environments—healers in cultures that extensively use their natural landscapes as sources of medicines.

Learning from the Shamans

To this point, we have been describing the search for biologically active compounds in living organisms as a process of scientific exploration whereby bioprospectors "discover" new compounds from nature. This is true for organisms in Yellowstone's hot springs; chemicals derived from these organisms were probably truly unknown to humans. For many other species, however, people have known of their usefulness for thousands of years. This knowledge is maintained in the traditions of indigenous people in biologically diverse areas, people who use native organisms as medicines, poisons, and foods. In many cultures, the repository of this traditional knowledge is the medicine man or woman. A shaman, as aboriginal healers are often called, can help direct bioprospectors to useful compounds by teaching about their culture's traditional methods of healing. Shamans employ many remedies that are highly effective against disease. Several bioprospectors have consulted with shamans to increase their chances of finding useful drugs (Figure 12.20).

Using the knowledge of native people in developing countries to discover compounds for use in wealthy, developed countries is highly controversial. This process is often referred to as **biopiracy** because organisms and active compounds discovered by traditional healers can be patented in the United States and Europe, potentially providing enormous financial rewards to the bioprospector with no return to a shaman or his people. The United Nations Convention on Biodiversity has sought to alleviate biopiracy by asserting that each country owns the biodiversity within its borders. However, because the U.S. government has not signed this legally binding document, companies in the United States are not required to abide by its terms. Additionally, even when a country makes a bioprospecting agreement with a pharmaceutical firm, it is unlikely that the indigenous community within the country will benefit in any way from a new drug developed from its store of knowledge. Indigenous peoples recently have begun questioning the ethics of bioprospecting via shamans, and several proposed agreements between developing countries and pharmaceutical firms have come under criticism.

The bioprospecting agreement between Diversa and Yellowstone National Park has not escaped charges of biopiracy. While Diversa is not relying on information from indigenous people to help locate valuable organisms, critics have charged that the managers of Yellowstone are essentially "selling off" organisms and chemical compounds that belong to the American public. In addition, they argue that the action of bioprospecting itself will damage the very resource that provides these remarkable discoveries. The federal courts have dismissed lawsuits against Yellowstone National Park and Diversa that address these points according to current law, but the issue remains an ethical dilemma: What is the responsibility of individuals and corporations profiting from biological diversity to the source and survival of that diversity?

Biological diversity represents an enormous resource for humans, but it also comes with an awesome responsibility. Actions of the U.S. Congress protected Yellowstone National Park and perhaps ultimately enabled the discovery of *Thermus aquaticus* and *Taq* polymerase. But thousands of useful organisms are lost every year through the destruction of native habitat, and our ability to use these organisms is diminished by the loss of indigenous cultures and their shamans. The dramatic rate of biodiversity loss not only denies humans still-undiscovered biological molecules but also diminishes Earth's ability to sustain our population and robs future generations of the diverse wonder and beauty of nature that our generation may be the last to truly enjoy. Chapter 14 discusses the causes, consequences, and possible solutions for the current biodiversity crisis. Humans can help to reduce the rate at which biodiversity is being lost but only if we begin to appreciate the value of the diversity that surrounds and sustains us.

Figure 12.20 Indigenous knowledge. This shaman of the Matses people of the Amazon rain forest is collecting plants for use in medicines. His intimate knowledge of the natural world is the product of the long history of his people in this diverse environment.

CHAPTER REVIEW

Summary

12.1 Biological Classification

- Bioprospectors seek to discover new drugs and other useful chemicals from the diversity of living organisms on Earth (p. 318).

- The number of known living species is estimated to be between 1.4 and 1.8 million, but the total number of species may be as high as 100 million (p. 319).

- Organisms are classified in domains according to evolutionary relationships and in kingdoms based on similarities in structure and lifestyle (pp. 320–322).

Web Tutorial 12.1 The Tree of Life

12.2 The Diversity of Life

- Life on Earth began about 3.5 billion years ago with simple prokaryotes, but it would be 1.5 billion years before eukaryotes evolved (p. 323).

- Bacteria and Archaea are prokaryotes, simple single-celled organisms without a nucleus or other membrane-bound organelles. They are abundant, found in a variety of habitats, and rely on a variety of food sources. Prokaryotes may produce antibiotics or have chemicals that function in extreme conditions (pp. 323–324).

- Eukaryotes, cells with nuclei and other membrane-bound organelles, probably evolved from symbioses among ancestral eukaryotes and prokaryotes (p. 325).

- The kingdom Protista is a hodgepodge of organisms that are typically unicellular eukaryotes. Algae are protists that are especially interesting to bioprospectors because they make defensive chemicals against predators and produce unique food-storage compounds (pp. 325–327).

- Multicellular organisms did not appear until approximately 600 million years ago, and this advance in form led to the diversity of species on Earth today (p. 328).

- Animals are motile, multicellular eukaryotes that rely on other organisms for food. Animal groups evolved in a short period of time known as the Cambrian explosion. Bioprospectors are interested in animals that produce venom or defensive chemicals (pp. 328–330).

- Fungi are immobile, multicellular eukaryotes that rely on other organisms for food and are made up of thin, threadlike hyphae. Fungi often produce antibiotics that kill their competitors, and some can escape detection by their living host's immune system (pp. 331–332).

- Plants are multicellular, photosynthetic eukaryotes. They have become increasingly adapted to land habitats over time. The diversity of flowering plants may be due partly to their production of defensive chemicals (pp. 332–334).

Web Tutorial 12.2 Endosymbiotic Theory

12.3 Learning About Species

- Some bioprospectors look for useful products by screening as many compounds as possible against a particular disease (pp. 337–338).

- An understanding of the ecological relationships of organisms provides clues to the likelihood and nature of possible chemical compounds in organisms (p. 338).

- Determining the evolutionary relationships among living organisms can help provide clues about an organism's traits. Phylogenies are created and tested by evaluating the shared traits of different species that indicate they shared a recent ancestor (pp. 339–340).

- Studying how indigenous healers called shamans use organisms can help bioprospectors identify species that may have useful chemicals (p. 341).

- Biopiracy occurs when a small group of people benefit from the knowledge of an indigenous culture; this practice may also undermine society's efforts to protect biodiversity (p. 341).

Learning the Basics

1. What characteristics of flowering plants may have driven the diversification of this group of organisms?

2. How is knowledge of the ecology of an organism useful for predicting what types of valuable chemicals it may possess?

3. How are hypotheses about the evolutionary relationships among living organisms tested?

4. Which of the following kingdoms or domains is a hodgepodge of different evolutionary lineages?

   A. Bacteria; **B.** Protista; **C.** Archaea; **D.** Plantae; **E.** Animalia

5. Comparisons of ribosomal RNA among many different modern species indicate that _____.

   A. there are two very divergent groups of prokaryotes;
   B. the kingdom Protista represents a conglomeration

of very unrelated forms; **C.** fungi are more closely related to animals than to plants; **D.** a and b are correct; **E.** a, b, and c are correct

6. Which of the following characteristics distinguishes prokaryotes from eukaryotes?

   A. Eukaryotes have a nucleus, while prokaryotes do not.; **B.** Prokaryotes lack ribosomes, which are found in eukaryotes.; **C.** Prokaryotes do not contain DNA, but eukaryotes do.; **D.** Eukaryotic organisms are much more widespread than prokaryotes.; **E.** Prokaryotes produce antibiotics, and eukaryotes do not.

7. The mitochondria in a eukaryotic cell _____.

   A. serve as the cell's power plants; **B.** probably evolved from a prokaryotic ancestor; **C.** can live independently of the eukaryotic cell; **D.** a and b are correct; **E.** a, b, and c are correct

8. Most animals _____.

   A. are insects; **B.** lack a backbone; **C.** are still unidentified; **D.** a, b, and c are correct; **E.** b and c are correct

9. Fungi feed by _____.

   A. producing their own food with the help of sunlight; **B.** chasing and capturing other living organisms; **C.** growing on their food source and secreting chemicals to break it down; **D.** filtering bacteria out of their surroundings; **E.** producing spores

10. Phylogenies are created based on the principle that all species descending from a recent common ancestor _____.

    A. should be identical; **B.** should share characteristics that evolved in that ancestor; **C.** should be found as fossils; **D.** should have identical DNA sequences; **E.** should be no more similar than species that are less closely related

Analyzing and Applying the Basics

1. Unless handled properly by living systems, oxygen can be quite damaging to cells. Imagine an ancient nucleated cell that ingests an oxygen-using bacterium. In an environment where oxygen levels are increasing, why might natural selection favor a eukaryotic cell that did not digest the bacterium but instead provided a "safe haven" for it?

2. Imagine you have found an organism that has never been described by science. The organism, made up of several hundred cells, feeds by anchoring itself to a submerged rock and straining single-celled algae out of pond water. What kingdom would this organism probably belong to, and why do you think so?

3. Imagine two fungi. Both weigh the same; however, one consists of a few short, very thick hyphae, and the other consists of many long, thin hyphae. Can they both absorb the same amount of food? If not, which fungus is more effective?

Connecting the Science

1. Scientists initially ridiculed the hypothesis that eukaryotic cells evolved from a set of cooperating independent cells. Most biologists still believe that competition for resources among organisms is the primary force for evolution. Do you think biologists' dismissal of the role of cooperation in evolution is a reflection of how life really "works," or do you think that it is a function of scientists' immersion in a culture that values competition over cooperation? Explain your choice.

2. Do we have an obligation to future generations to preserve as much biodiversity as possible, considering that many organisms may contain currently unknown "biological gold"? Would simply preserving the information contained in an organism's genes (in a zoo or other collection) be good enough, or do we need to preserve organisms in their natural environments?

Is the Human Population Too Large?

Population Ecology

From space, Earth doesn't look too crowded.

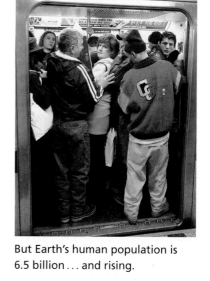

But Earth's human population is 6.5 billion . . . and rising.

Is this Ethiopian child hungry because the planet is overpopulated?

In its most recent estimate in 2005, the United Nations (UN) reported that the human population on Earth is approximately 6.5 billion—double the number of people alive in 1960. The UN also predicted that the population would continue to grow for several more decades before stabilizing at as high as 10.6 billion by about 2050. As is usually the case, many observers greeted the report as another piece of bad news. While the UN's population projection is lower than past predictions (previous reports forecast a population of over 12 billion by 2050), many scientists and environmentalists wonder if our planet can support the current population for very long, let alone an additional 4.1 billion people.

Other commentators, such as the late economist Julian Simon, a former senior fellow at the influential Cato Institute, are skeptical of environmentalists' statements about population growth. They point to predictions made in the best-selling book *The Population Bomb* (1968), in which author Paul Ehrlich forecast worldwide food and water shortages by the year 2000. In fact, most measures of human health have become more upbeat since 1970, including global declines in infant mortality rates, increases in life expectancy, and a 20% increase in per capita income—despite a near doubling in population since the publication of Ehrlich's book. By most measures, the average person is better off today than in 1970. Paul Ehrlich was clearly

Or can Earth support everyone at the same level as that of the average North American family?

345

wrong in 1968; why should we believe his doom-and-gloom predictions about the future now?

Ehrlich and his colleagues counter that while they were wrong about how soon it would happen, there are some indications that the large human population is rapidly reaching a real limit to growth. For example, the UN previously released another report—*The State of Food and Agriculture, 2003–2004*—describing numerous food crises around the world. According to the UN, as of August 2003, thirty-eight countries and over 62 million people were facing food emergencies, meaning that starvation could be imminent. Worldwide, 842 million people—including 150 million children under the age of 5—do not get enough food regularly for a healthy existence. A staggering 55% of the nearly 12 million deaths each year among children under 5 in the developing world are associated with inadequate nutrition. Despite years of international attention and billions of dollars spent to address this problem, the situation has not improved dramatically—there are only 10% fewer children suffering from malnutrition today than there were in 1980.

So what is the truth? Is the human population larger than Earth can support for much longer? Are we headed into a global food crisis and massive famine? Or are we gradually moving toward an era where all people on Earth will be as well-fed, long-lived, and affluent as the average North American?

13.1 A Growing Human Population

Ecology is the field of biology that focuses on the interactions among organisms as well as between them and their environment. The relationship between organisms and their environments can be studied at many levels—from the individual, to populations of the same species, to communities of interacting species, and finally to the effects of biological activities on the nonbiological environment, such as the atmosphere. The three chapters in this unit present basic ecological principles obtained from the study of ecology at all of these levels.

From an ecological perspective, a **population** is defined as all of the individuals of a species within a given area. Populations exhibit a structure, which includes the spacing of individuals (that is, their distribution) and their density (abundance). Much of the science of ecology is concerned with the factors influencing the distribution and abundance of the individuals within populations. The interactions among species described in Chapter 14 make up one set of influences, but another set is the dynamics of the population, including the relative numbers of individuals of different sexes and ages and the numbers that are born or die in a given time period.

Population Structure

The first task of a population ecologist is to understand how many individuals make up the population of interest. Certain populations can be counted directly, as in a census tabulating the number of humans in an area or a survey identifying all individuals of a particular tree species in a forest tract. The size of more mobile and inconspicuous species can be estimated by the **mark-recapture method**. In this technique, researchers capture many individuals, mark them in some way (for instance, with an ear tag) and release them back into the environment. At some later time, the researchers capture another group of individuals and calculate the proportion of previously marked individuals in this group. This proportion can be used to estimate the size of the total population. For example, imagine that a researcher captured, marked, and released 100 beetles. If he returns a week later and finds that 10% of the beetles he caught on the second round are marked,

(a) Clumped

(b) Uniform

(c) Random

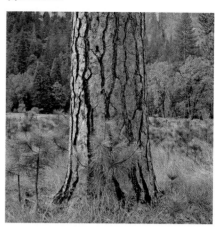

Figure 13.1   Patterns of population dispersion. Individuals in a population may be (a) clumped, like these cattails growing in soil with the correct water content; (b) uniformly distributed, like these birds at a feeder; or (c) randomly dispersed, like these seedlings in a forest.

he can assume that the 100 beetles he marked originally represented 10% of the entire beetle population. According to this mark-recapture survey, the total population is approximately 1000 beetles.

Another basic aspect of population structure is dispersion—that is, how organisms are distributed in space. Many species show a **clumped distribution**, with high densities of individuals in certain resource-rich areas and low densities elsewhere. Plants that require certain soil conditions and the animals that depend on these plants tend to be clumped (Figure 13.1a). On a global scale, humans show a clumped distribution, with high densities found around transportation resources such as rivers and coastlines. This clumped distribution masks a more **uniform distribution** on a local scale; for instance, the spacing between houses in a subdivision or strangers in a classroom tends to equalize the distances among individual property owners or people. Species that show a uniform distribution are often territorial—they defend their own personal space from intruders. Human territoriality has a social component; for instance, you may have noticed the variation among cultures, even within the same country, regarding how much space between two conversing people is appropriate. However, spacing between humans has a biological component, just as in other species—we all react strongly, and physically, to invasions of our socially delineated personal space. We can observe these same strong reactions among certain species of birds at bird feeders (Figure 13.1b). Nonsocial species with the ability to tolerate a wide range of conditions typically show a **random distribution**, wherein no compelling factor is actively bringing individuals together or pushing them apart. The distribution of seedlings of trees with windblown seeds is often random (Figure 13.1c).

A population's distribution and abundance provides a partial snapshot of its current situation. The dispersion of the human population—and recent changes in that pattern—profoundly affects the natural environment, as discussed in Chapter 15. However, to better understand how a population is responding to its environment, we need to determine how it is changing through time.

Population Growth

Historians have been able to use archaeological evidence and written records to determine the size of the human population on Earth at various times during the past 10,000 years. This record, presented in Figure 13.2, dramatically illustrates the pattern of population growth. For most of our history, the human population has remained at very low levels. At the beginning of the agricultural era, about 10,000 years ago, there were approximately 5 million humans. There were 100 million people during the Egyptian Empire (7000 years later) and about 250 million at the dawn of the Christian religion in 1 C.E. (C.E. refers to Common Era, the year designation used by most Western countries). The population was growing, but at a very slow rate—approximately 0.1% per year. Beginning around 1750, the rate at which the human population was growing jumped to about 2% per year. The human population reached 1 billion in 1800, had doubled to 2 billion by 1930, and then doubled again to 4 billion by 1970. Although the current growth rate is slower, about 1.2% per year, the rapid increase in population looks quite dramatic on a graph of human population over time.

The graph of human population growth is a striking illustration of **exponential growth**—growth that occurs in proportion to the current total. In other words, populations growing exponentially do not add a fixed number of offspring every year; instead, the quantity of new offspring is an ever-growing number. Exponential growth results in the J-shaped growth curve seen in Figure 13.2. The larger a population is, the more rapidly it grows because an increase in numbers depends on individuals reproducing in the population. So, while a growth rate of 1.2% per year may seem rather small, the number of individuals added to the 6.5-billion-strong human population every year at this rate of growth is a mind-boggling 77 million (approximately the entire population of Germany). Put another way, three people are added to the world population per second, and about a quarter of a million people are added every day.

What has fueled this enormous increase in human population? The annual **growth rate** of a population is the percent change in population size over a single year. Growth rate is a function of the birth rate of the population (the number of births averaged over the population as a whole) minus the death rate (the number of deaths averaged over the population as a whole). For example, 22 babies are born per year, on average, in a group of 1000 people—that is, the birth rate for the population is 2.2%:

$$\frac{22}{1000} = 0.022 = 2.2\%$$

In addition, each year 10 individuals die out of every 1000 people, resulting in a death rate of 1.0%:

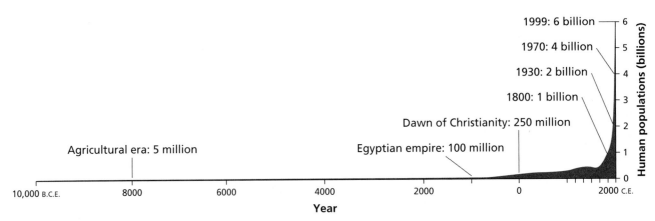

Figure 13.2   Human population growth. Estimates of human populations indicate that the number of people on Earth grew relatively slowly from the origin of agriculture through the eighteenth century. Beginning around the time of the Industrial Revolution, growth rates and population numbers began to soar.

$$\frac{10}{1000} = 0.01 = 1\%$$

This results in the current growth rate of 1.2%:

$$\text{growth rate} = \text{birth rate} - \text{death rate}$$

$$1.2\% = 2.2\% - 1.0\%$$

Today's relatively high growth rate, compared to the historical average of 0.1%, is the result of a large difference between birth rates and death rates.

In human populations, the tendency has been for decreases in death rate to be followed by decreases in birth rate. The speed of this adjustment helps to determine population growth in the future.

The Demographic Transition

Prior to the Industrial Revolution, both birth rates and death rates were high in most human populations. Although women gave birth to many children, relatively few children lived to reach adulthood. The rapid increase in population growth rate that occurred in the eighteenth century resulted from a dramatic decrease in infant mortality (the death rate of infants and children) in industrializing countries. In particular, new knowledge of how deadly infectious diseases could be prevented greatly reduced the number of children who suffered from these illnesses. With birth rates high and death rates declining, the population growth rate increased. Not long after death rates declined in these countries, birth rates followed suit, lowering growth rates again. Scientists who study human population growth refer to the period when birth rates are dropping toward lowered death rates as the **demographic transition** (Figure 13.3). The length of time that a human population remains in the transition has an enormous effect on the size of that population. Countries that pass through the transition swiftly remain small, while those that take longer can become extremely large. Countries that began the process of industrialization in the eighteenth century and that now have a high per capita income are called more developed countries. These include countries in Western Europe, North America, and Japan. Nearly all more developed countries have already passed through the demographic transition and have low population growth rates.

However, global human population growth rates have remained high because the least developed countries (countries that are early in the process of industrial development and have low per capita incomes) remain in the

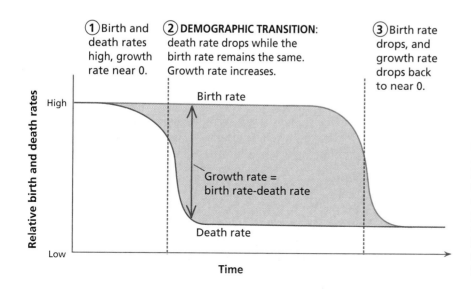

Figure 13.3 The demographic transition. As improvements in sanitation and medical care in human populations cause a decrease in infant mortality, death rates drop and growth rates soar. Eventually, people in these populations respond by decreasing the number of children they have. The longer a country remains in the transitional period, the larger its population becomes.

demographic transition. In addition, several recent changes have decreased infant mortality even more dramatically. These changes include the use of pesticides to reduce rates of mosquito-borne malaria, immunization programs against cholera, diphtheria, and other fatal diseases, and the widespread availability of antibiotics. While birth rates are gradually declining in less developed countries, they still remain high, contributing to high growth rates. The vast majority of future population growth will occur within populations in the less developed world, especially those in Africa and Asia, but these countries are where the vast majority of food crises are occurring. Are the populations in these countries already too large to support themselves? Answering that question requires an understanding of the factors that limit population growth.

13.2 Limits to Population Growth

In their study of nonhuman species, ecologists see clear limits to the size of populations. They can also observe the sometimes awful fates of individuals in populations that outgrow these limits. For this reason, many professional ecologists express grave concern about the consequences of a rapidly growing human population.

You may know of several instances of nonhuman populations outgrowing their food supplies. The elk population in Yellowstone National Park suffered enormous mortality throughout the 1970s after it grew so large that it degraded its own rangeland. The massive migrations of Norway lemmings that occur every 5 to 7 years and lead to many deaths result from population crowding; while these animals do not commit "mass suicide," as often assumed, the loss of high-quality food in an area as populations increase incite the lemmings to disperse and often meet their death in the process. Even yeast in brewing beer grow large populations that eventually use up their food source and die off during the fermenting process. Let us explore what ecology can tell us about the likelihood of human populations suffering the same fate as elk, lemmings, or yeast.

Carrying Capacity and Logistic Growth

The examples of the elk in Yellowstone and the Norway lemmings illustrate a basic biological principle. While populations have the capacity to grow exponentially, their growth is limited by the resources—food, water, shelter, and space—that individuals need to survive and reproduce. The maximum population that can be supported indefinitely in a given environment is known as the environment's **carrying capacity**.

The growth of a population in an environment where resources are limited is exponential at first, but the effects of declining resources gradually take their toll on growth rate. A simplified graph of population size over time in resource-limited populations is S-shaped (Figure 13.4). This model shows the growth rate of a population declining to zero as it approaches the carrying capacity. In other words, birth rate and death rate become equal, and the population stabilizes at its maximum size. Not long after ecologists first predicted this pattern of growth, called logistic growth, populations of organisms as diverse as flour beetles, water fleas, and single-celled protists were shown in laboratory studies to conform with this projected growth curve.

The declining growth rate near a population's carrying capacity is caused by **density-dependent factors**, which are population-limiting factors that increase in intensity as the population increases in size. Density-dependent factors include limited food supplies, increased risk of infectious disease in more crowded conditions, and an increase in toxin concentration caused by in-

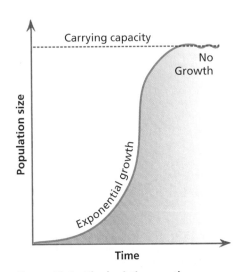

Figure 13.4 The logistic growth curve. This graph illustrates the change in size of an idealized population over time. The S-shaped curve is due to a gradual slowing of the population growth rate as it approaches the carrying capacity of the environment.

(a) Fruit flies   **(b)** Water fleas   **(c)** White-tailed deer   **(d)** Humans

Figure 13.5   Limits to growth. Populations of (a) fruit flies in a laboratory culture, (b) water fleas in an aquarium, and (c) white-tailed deer in the northeastern United States all experience high death rates and/or low birth rates as their populations approach the carrying capacity of the environment. (d) Do human populations face these same limits?

creased waste levels. Density-dependent factors cause declines in birth rate or increases in death rate. In organisms such as fruit flies growing in laboratory culture bottles, high populations lead to increased mortality of the flies as food supplies dwindle and wastes accumulate. Water fleas living in crowded aquariums do not have enough food to support egg production, and so birth rates drop. Females of white-tailed deer populations living in crowded natural habitats are less likely to be able to carry a pregnancy to term than deer in less-crowded environments. Density-dependent factors can be contrasted with **density-independent factors** that influence population growth rates—for instance, severe droughts that increase the death rate in plant populations regardless of their density, or increased temperatures that increase the birth rate in cold-limited insects. However, density-independent factors do not occur in a vacuum; they can have more or less severe effects depending on the size of a population. For example, a density-independent factor such as an unusually cold winter can be deadly to individuals in a white-footed mouse population, but the likelihood of survival is also a function of how much food each individual has stored for the winter. How much food is stored depends on the density of mice competing for food sources during the autumn.

Are density-dependent factors beginning to reduce growth rates in a human population? That is, are humans nearing the carrying capacity of Earth for our population? If we are, will death rates increase as food resources dwindle and more people starve? Or will birth rates decline because fewer women will have enough food to support themselves *and* a developing baby (Figure 13.5)?

Earth's Carrying Capacity for Humans

One way to determine if the human population is reaching Earth's carrying capacity is to examine whether, and how rapidly, the growth rate is declining. As we saw in Figure 13.4 on page 350 the S-shaped curve of population size over time results from a gradually declining growth rate as the population approaches carrying capacity.

Human population growth rates were at their highest in the early 1960s, about 2.1% per year, but they have since declined to the current rate of 1.2%. This steady decline is one indication that the population, though still currently growing, is nearing a stable number. Uncertainty about the future rate of growth has led the UN to produce differing estimates of this number and how soon population stability will be reached (Figure 13.6). However, the unique characteristics of humanity make it difficult to determine exactly which population size represents Earth's carrying capacity for humans.

Signs That the Population Is Not Near Carrying Capacity. The rates of population increase of fruit flies and water fleas in the laboratory have slowed as these populations neared carrying capacity because their growth rates were forced down by density-dependent factors; lack of resources caused increased death

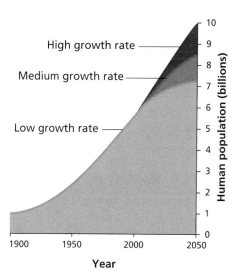

Figure 13.6   Projected human population growth. The United Nations' report predicting the eventual size of the human population is based on a number of uncertainties and leads to three projections: a low-growth scenario of 7.2 billion people; medium growth resulting in 8.6 billion; or even a high-growth estimate of 10.3 billion.

rates or decreased birth rates. However, this is not the case in human populations. Even as the human population has rapidly increased, death rates continue to decline—an indication that people are not limited by food resources. Growth rates are declining because birth rates are falling faster than death rates. Unlike the water fleas and white-tailed deer, whose females are unable to have offspring when populations are near carrying capacity, birth rates in human populations are falling because women and families, even those with adequate resources, are *choosing* to have fewer children.

Although the human population's growth rate is slowing as might be expected near carrying capacity, rising living standards indicate that the population is not currently experiencing a density-dependent factor. Another way to determine if we are near Earth's carrying capacity is to estimate the amount of resources that are currently being used by humans and use that estimate to approximate the theoretical limit to population size. The amount of food energy available on the planet is referred to as the **net primary production (NPP)**. NPP is the amount of solar energy captured via plant photosynthesis minus the amount of energy that plants need to support themselves. In other words, NPP is a measure of plant growth, typically over the course of a single year. Several different analyses of the global extent of agriculture, forestry, and animal grazing estimate that humans use roughly one-third of the total land NPP. If we accept these rough estimates, we can approximate that the carrying capacity of Earth is three times the present population, or approximately 19 billion people. This theoretical maximum is the total number of humans that could be supported by all of the photosynthetic production of the planet—leaving no resources for millions of other species. Given the dependence of humans on natural systems (explored in Chapter 14), it is unlikely that our species could survive on a planet where no natural systems remained. However, even the largest population projection by the UN, 10.6 billion, falls well short of this theoretical maximum.

Signs That the Population Is Near Carrying Capacity.   Ecologists caution that the resources required to sustain a population include more than simply food, and so the carrying capacity deduced from NPP estimates may be much too high. Humans also need a supply of clean water, clean air, and energy for essential tasks such as heating, food production, and food preservation. The relationship between population size and the supply of these resources is not as straightforward as the relationship between population and food. For instance, every new person added to the population requires an equivalent amount of clean water, but every new person also introduces a certain amount of pollution to the water supply. We cannot simply divide the current supply of clean water by 10.6 billion to determine if enough will be available in the future, since increased population leads to increased pollution and therefore less total clean water.

Furthermore, many essential supplies that sustain the current human population are **nonrenewable resources**, meaning that they are a one-time stock and cannot be easily replaced. The most prominent nonrenewable resource is fossil fuel, the buried remains of ancient plants transformed by heat and pressure into coal, oil, and natural gas. The use of fossil fuel and other nonrenewable resources is a function not only of the number of people but also of average lifestyles, which vary widely around the globe. For example, Americans make up only 5% of Earth's population but are responsible for 24% of global energy consumption. The average American uses as many resources as 2 Japanese or Spaniards, 3 Italians, 6 Mexicans, 13 Chinese, 31 Indians, 128 Bangladeshis, 307 Tanzanians, or 370 Ethiopians. Americans also consume a total of 815 billion food calories per day—about 200 billion calories more than is required, or enough to feed an additional 80 million people. Much of modern food production relies on the energy provided by fossil fuel. When these resources begin to run out, we might find that we need far more of Earth's NPP than we do now to sustain abundant food production. In other words, the actual carrying capacity of our planet may be much lower than our approximations.

The question posed at the beginning of this section remains unanswered; there is no agreement among scientists concerning the carrying capacity of Earth for the human population. Given that uncertainty, what can ecologists tell us about the risks facing the human population that may result from massive, rapid population growth?

13.3 The Future of the Human Population

Unlike nearly all other species, human populations are not simply at the mercy of environmental conditions. With its ability to transform the natural world, human ingenuity has helped populations circumvent seemingly fixed natural limits. However, ingenuity has a dark side, in that it can lull people into believing that nature has an almost infinite capacity to support their ever-growing needs. Managing the growth of human populations before even the most secure of them face environmental and economic disaster requires an understanding of the risks of continued rapid growth and the strategies that help reduce it.

A Possible Population Crash?

The use of nonrenewable resources creates a risk of the human population overshooting a still unknown carrying capacity. Ecologists have long known that when populations have high growth rates, they may continue to add new members even as resources dwindle. This causes the population to grow larger than the carrying capacity of the environment. The members of this large population are then competing for far too few resources, and the death rate soars while the birth rate plummets. This results in a **population crash**, a steep decline in number (Figure 13.7). For instance, in some species of water flea, healthy offspring continue to be born for several days after the food supply becomes inadequate because females can use their fat stores to produce additional young. The size of the population continues to rise even when there is no food left to graze on; however, when these young water fleas run out of stored fat, most individuals die. For many species with high birth rates, rapid growth followed by dramatic crashes produce a **population cycle** of repeated "booms" and "busts" in number.

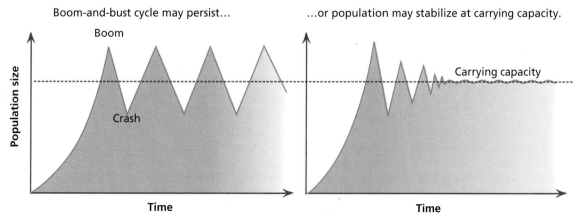

Figure 13.7 Overshooting and crashing. These graphs illustrate rapid population growth followed by a population crash. Over time, the population may stay in a "boom-and-bust" cycle, or it may stabilize at its carrying capacity.

Figure 13.8   The crash of a human population. On Rapa Nui, also known as Easter Island, the human inhabitants created these large statues. Soon after completely deforesting this small island, the large population suffered a severe crash.

A population overshoot and subsequent crash affected the human population on the Pacific island of Rapa Nui (also known as Easter Island) during the eighteenth century (Figure 13.8). This 150-square-mile island is separated from other landmasses by thousands of miles of ocean; therefore, its people were limited to using only the resources on or near their island. Archaeological evidence suggests that at one time, the human population on Rapa Nui was at least 7000—apparently a number far greater than the carrying capacity of the island. By 1775, the subsequent overuse and loss of Rapa Nui's formerly lush palm forest had resulted in a rapid decline to fewer than 700 people, a population likely much lower than the initial carrying capacity of the island. It is possible that humanity's use of the stored energy in fossil fuels may be allowing us to overshoot Earth's true carrying capacity.

Biological populations may also overshoot carrying capacity when there is a time lag between when the population approaches carrying capacity and when it actually responds to that environmental limit. Scientists who study human populations note a lag between the time that humans reduce birth rates and when population numbers respond. They call this lag **demographic momentum**. The momentum occurs because while parents may be reducing their family size, their children will begin having children before the parents die, causing the population to continue growing. Even when families have an average of two children, just enough to replace the parents, demographic momentum causes the human population to grow for another 60 to 70 years before reaching a stable level. The potential demographic momentum of a population can be estimated by looking at its **population pyramid**, a summary of the numbers and proportions of individuals of each sex and each age group. As Figure 13.9 illustrates, the potential for high levels of demographic momentum occurs when the age structure most closely resembles a true pyramid, with a large proportion of young people. In more stable populations, the proportion that is young is

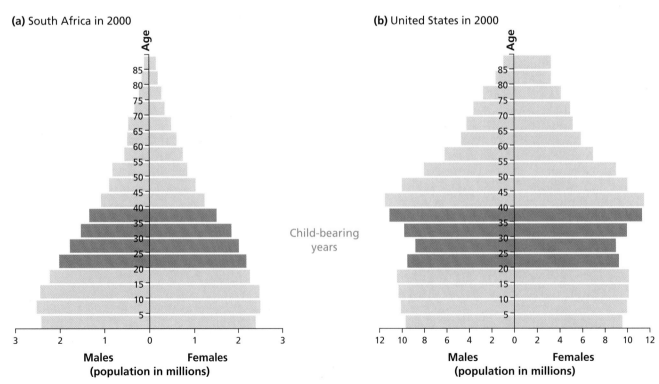

(a) South Africa in 2000

(b) United States in 2000

Figure 13.9   Demographic momentum. In a rapidly growing human population like that of (a) South Africa in 2000, most of the population is young, and the population will continue to grow as these children reach child-bearing age. In a slower-growing or stable human population, the ages are more evenly distributed, as in (b) the United States in 2000.

not significantly larger than the proportion that is middle aged, and the pyramid looks more like a column.

Whether or not our reliance on stored resources and the potential demographic momentum in human populations will result in an overshoot of Earth's carrying capacity—followed by a severe crash, as on the island of Rapa Nui—remains to be seen. But human ecologists already know what factors help to slow population growth so that a crash may become less likely.

Avoiding Disaster

As discussed earlier in the chapter, when death rates drop in human population, birth rates eventually follow. Unlike any other species known to science, humans will voluntarily limit the number of babies they produce. When more opportunities become available outside of child rearing, most women delay motherhood and have fewer children. In fact, birth rates are lowest in countries where income is high and women are provided with education (Figure 13.10). This information provides a clear direction for public policies attempting to decrease population growth rates: improve conditions for women, including increasing access to education, health care, and the job market, and provide them with the information and tools that allow them to regulate their fertility.

Slowing growth rates before the human population reaches some environmentally imposed limit has additional benefits. Determining Earth's carrying capacity for humans as simply a function of whether food and water will be available also ignores quality-of-life issues, or what some scientists call cultural carrying capacity. An Earth that was wholly given over to the production of food for the human population would lack wild, undisturbed places and the presence of species that nurture our sense of wonder and discovery. With human populations at the limits of growth, much of our creative energy would be used for survival, taking away our ability to make and enjoy music, art, and literature. Limiting human population growth also leaves room for nonhuman species. As we discuss in Chapter 14, human activity is posing a direct threat to the survival of a significant percentage of Earth's biodiversity—a threat that

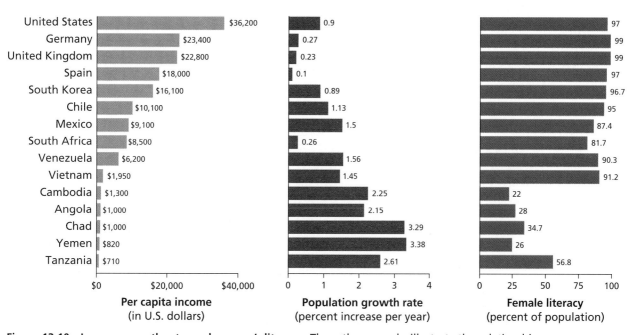

Figure 13.10 Income, growth rate, and women's literacy. These three graphs illustrate the relationships among income, population growth, and female literacy. Note that higher income and literacy are correlated with decreased birth rates and thus decreased population growth in most countries.

increases in direct proportion to the size and affluence of the planet's human population.

What we have learned is that scientists cannot tell us exactly how many people Earth can support, partly because humans make unpredictable choices and partly because humans have the capacity to innovate and adjust seemingly fixed biological limits. Ultimately, the question of how many people Earth should support—and at what quality of life, or including support for nonhuman species—is a question not solely of science but also of values and ethics.

CHAPTER REVIEW

Summary

13.1 A Growing Human Population

- A population is defined as a group of individuals of the same species living in a fixed area. The structure of a population can be described by the number of individuals and their dispersion (p. 346).

- The human population has grown very rapidly over the last 150 years and exhibits a pattern of exponential growth, which is an increase in numbers as a function of the current population size (p. 348).

- Human population growth is spurred by decreases in death rate caused by decreased infant mortality. In most populations, this decrease is followed by a decrease in birth rates. The gap between when death rates drop and birth rates follow is called the demographic transition (pp. 348–349).

  Web Tutorial 13.1 Population Growth

13.2 Limits to Population Growth

- Nearly all populations eventually reach the carrying capacity of their environment. Near carrying capacity, density-dependent factors cause an increase in death rate (p. 350).

- The growth rate of the human population is declining, but not as a result of density-dependent factors. Instead, birth rates are dropping because women are choosing to have fewer children (p. 351).

- Rough calculations indicate that the energy received from the sun each year could support a population of 19 billion people, well over the largest population projections (p. 352).

- Humans' reliance on nonrenewable resources may be temporarily inflating the actual carrying capacity of Earth (pp. 352–353).

13.3 The Future of the Human Population

- Fast-growing populations that overshoot their environment's carrying capacity may experience a crash or go through periodic booms and busts (p. 353).

- It is possible that the human population will overshoot Earth's carrying capacity because of our reliance on nonrenewable resources and due to demographic momentum (p. 354).

- Human population growth rates decline when women are empowered to seek an education and may choose to work outside the home (p. 355).

Learning the Basics

1. What factors have led to the explosive increase in the human population over the past 150 years?

2. Explain why a decrease in population growth rate is expected as a nonhuman population approaches carrying capacity.

3. Describe why demographic momentum may cause the human population to overshoot Earth's carrying capacity.

4. When individuals in a population are evenly spaced throughout their habitat, their dispersion is termed as _____.

   A. clumped; **B.** uniform; **C.** random; **D.** excessive; **E.** exponential

5. The growth of human populations over the past 150 years has increased primarily due to _____.

   A. increases in death rate; **B.** increases in birth rate; **C.** decreases in death rate; **D.** decreases in birth rate; **E.** increases in net primary production

6. According to Figure 13.11, the carrying capacity for fruit flies in the environment of the culture bottle is _____.

   A. 0 flies; **B.** 100 flies; **C.** 150 flies; **D.** between 100 and 150 flies; **E.** impossible to determine

7. In contrast to nonhuman populations, human population growth rates have begun to decline due to _____.

   A. voluntarily increasing death rates; **B.** voluntarily decreasing birth rates; **C.** involuntary increases in death rates;

Figure 13.11

   D. involuntary decreases in birth rates; **E.** voluntarily increasing birth rates

8. Populations that rely on stored resources are likely to overshoot the carrying capacity of the environment and consequently experience a _____.

   A. demographic momentum; **B.** cultural carrying capacity; **C.** decrease in death rates; **D.** population crash; **E.** exponential growth

9. Demographic momentum refers to the tendency for _____.

   A. low population growth rates to continue to decline; **B.** high population growth rates to continue to increase; **C.** populations to continue to grow in number even when growth rates reach zero; **D.** populations to continue to grow in number even when women are reducing the number of children they bear; **E.** women to continue to have children even though they no longer wish to

10. Which of the following factors is associated with declines in a country's population growth rate?

   A. an increase in per capita income; **B.** an increase in female educational attainment; **C.** an increase in women's social status; **D.** a and c are correct; **E.** a, b, and c are correct

Analyzing and Applying the Basics

1. A researcher captures 50 penguins, marks them with a spot of paint on their bills, and releases them. One month later she returns, captures another 50 penguins, and notes that only 1 has a previous mark. What is the likely size of the total penguin population in the researcher's study area?

2. Review Figure 13.11 above. How would you expect the carrying capacity of the population to change if the flies are supplied with a greater amount of food? What other factors might influence the carrying capacity in this environment?

3. Imagine two human populations, each one made up of 5 million individuals. In one population, over 50% of the members are in the age group of 0 to 20 years and about 2% are over 65. In the other population, about 20% are from 0 to 20 years old and about 20% are over 65. Which of these populations will probably stabilize at a larger number, and why?

Connecting the Science

1. Review your answer to Question 2 in "Analyzing and Applying the Basics." How are the factors that limit fruit-fly populations in a culture bottle similar to the factors that limit human populations on Earth? How are they different?

2. Africa is the only continent where increases in food production have not outpaced human population growth. Many of the most severe food crises are in African countries. Should those of us in the more developed world assist African populations? How? What factors influence your thoughts on this question?

CHAPTER 14

Is Earth Experiencing a Biodiversity Crisis?

Community Ecology, Ecosystem Ecology, and Conservation Biology

The Lost River sucker
faces extinction...

... but saving the fish has angered these farmers.

Who has the right to use this lake for their survival—the farmers or the fish?

I n the summer of 2001, the typically quiet, conservative community of Klamath Falls, Oregon, suddenly began seething with revolutionary passion. Anger at federal authorities was widespread and palpable; signs along Route 39 outside the city read, "Please thank the U.S. Fish and Wildlife Service for destroying the Klamath Basin's economy," and "Crime Scene . . . by the U.S. Federal Government." The residents' fury reached a boiling point in late June and July, when distraught farmers repeatedly confronted and threatened federal officials and eventually destroyed a Bureau of Reclamation facility using chainsaws, pry bars, and blowtorches.

The wrath of the people of Klamath Falls and surrounding communities was generated by the federal government's legal requirement to protect species that are recognized as in danger of extinction. In the case of the Klamath crisis, the species at risk are two fish—the Lost River sucker and the shortnose sucker. In the midst of a multiyear drought and dangerously low water levels in Upper Klamath Lake, home to these endangered fish, the U.S. Fish and Wildlife Service stopped the outflow of water from the lake in April 2001. The irrigation canals that had fed barley, potato, and alfalfa fields in the high desert of the Klamath Basin since the early 1900s suddenly went dry. Without irrigation, thousands of farmers were unable to produce crops and faced the prospect of bankruptcy, foreclosure, and loss of their livelihood.

Why should we care about the fate of such a controversial endangered species—or any endangered species?

Lost River and shortnose suckers are dull-colored fish that feed on the mucky bottoms of lakes and streams in the region. These fish have not represented a viable economic resource for humans for several decades. In contrast, the crops produced annually by irrigated fields in the region produce millions of dollars in income. Ty Kliewer, a student at Oregon State University whose family farms in the basin, summarized the feelings of many when he told his senator in 2001 that he had learned the importance of balancing mathematical and chemical equations in school. "It appears to me that the people who run the Bureau of Reclamation and the U.S. Fish and Wildlife Service slept through those classes," Kliewer said. "The solution lacks balance, and we've been left out of the equation." The Klamath crisis of 2001 was not unique; thousands of people all over the United States have had their jobs threatened or eliminated by the government's attempts to protect endangered species.

Why should the survival of one or a few species come before the needs of humans? Many biologists and environmentalists say that the Klamath Falls bumper sticker "Fish or Farmers?" misstates the dilemma. Instead, they argue, humans depend on the web of life that creates and supports natural ecosystems, and they worry that disruptions to this web may become so severe that our own survival as a species will be threatened. In this view, protecting endangered species is not about pitting fish against farmers; it is about protecting fish to ensure the survival of farmers. In this chapter, we explore the causes and consequences of the loss of biological diversity.

14.1 The Sixth Extinction

The government agencies that stopped water delivery to the Klamath Basin farmers were acting under the authority of the **Endangered Species Act (ESA)**, a law passed in 1973 with the purpose of protecting and encouraging the population growth of threatened and endangered species. Lost River and shortnose suckers were once among the most abundant fish in Upper Klamath Lake—at one time, they were harvested and canned for human consumption. Now, with populations of fewer than 500 and minimal reproduction, these fish are in danger of **extinction**, defined as the complete loss of a species. Critically imperiled species such as the Lost River and shortnose suckers are exactly the type of organisms that legislators had in mind when they enacted the ESA.

The ESA was passed because of the public's concern about the continuing erosion of **biodiversity**, the entire variety of living organisms. The fate of the whooping crane, one of only two cranes native to North America, prompted the passage of the ESA. Biologists estimate that more than 1000 whooping cranes were alive in the mid-1860s. By 1938, as a result of hunting and the loss of nesting areas, the whooping crane had disappeared from much of the continent; only two small flocks were left. By 1942, only 16 birds remained in the wild. Unfortunately, the near extinction of the whooping crane is not a unique event. Bald eagles, peregrine falcons, gray wolves, and elephant seals—once abundant species—are or recently have been pushed close to extinction as a result of human activity. Even one of the most numerous bird species on the planet, the passenger pigeon (Figure 14.1), was not safe. This species was driven to extinction in North America nearly 100 years ago by familiar causes—habitat loss and overhunting. The ESA was drafted because humans appear to be triggering an unprecedented and rapid rate of species loss.

Critics of the ESA argue that the goal of saving all species from extinction is unrealistic. After all, extinction is a natural process—the approximately 10 million species living today constitute less than 1% of the species that have ever existed—and trying to save rare species, as we have seen in the Klamath Basin, can be detrimental to humans. In the next section, we

Figure 14.1   An extinct species. Passenger pigeons were once the most common bird in North America, but they were driven to extinction by human activity.

explore the scientific questions posed by ESA critics: How does the rate of extinction today compare to the rates in the past? Is the ESA just attempting to postpone the inevitable, natural process of extinction?

Measuring Extinction Rates

If ESA critics are correct in stating that the current rate of species extinctions is "natural," then the extinction rate today should be roughly equal to the rate in previous eras. The rate of extinction in the past can be estimated by examining the fossil record.

Figure 14.2 illustrates what examinations of the fossil record tell scientists about the history of biodiversity on Earth. Since the rapid evolution of a wide variety of animal groups approximately 580 million years ago, the number of families of organisms has generally increased. However, this increase in biodiversity has not been smooth or steady. The history of life on Earth has been punctuated by five **mass extinctions**—species losses that are global in

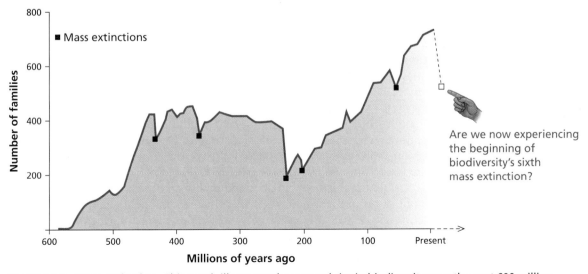

Are we now experiencing the beginning of biodiversity's sixth mass extinction?

Figure 14.2   Mass extinction. This graph illustrates the general rise in biodiversity over the past 600 million years, as indicated by an increase in the number of marine families present in the fossil record. However, this rise has been punctuated by five mass extinctions (marked here with black squares), each resulting in a global decline in biodiversity. The proportion of species lost during these mass extinctions appears to be even greater than the proportion of families lost because families that were especially species-rich died out.

Youngest fossil of this species

Oldest fossil of this species

Total life span of species:
13.1 − 12.3 = 0.8 million years

Figure 14.3   Estimating the life span of a species. Using a process called radiometric dating, the ages of these fossil shells are estimated from the age of the rocks in which they are embedded. Fossils of the same species are arranged on a timeline from oldest to youngest. The difference in age between the oldest and youngest fossil of a species is an estimate of the species' life span.

scale, affect large numbers of species, and are dramatic in impact. During mass extinctions, 20% to 50% of living families, containing 50% to 90% of all living species, were lost over the course of a few thousand to a few hundred thousand years. Past mass extinctions were probably caused by massive global changes—for instance, changes in sea levels brought about by climate fluctuations, shifts in ocean and land forms caused by continental drift, or widespread destruction and climate disruption caused by an asteroid impact. Many scientists argue that we are now seeing biodiversity's sixth mass extinction, this one caused by the massive global changes resulting from human activity.

Determining whether the current rate of extinction is unusually high requires some knowledge about the **background extinction rate**, the rate at which species are lost through the normal evolutionary process. Normal extinctions occur when a species lacks the ability to adapt to environmental change—for instance, if a species does not have the right combination of alleles to survive in a new climate condition, compete with new species for the same resources, or escape new predators. In many cases, a new species arises due to the evolution of populations of the old species. When individuals in a population possess unique traits that allow them to survive environmental changes, they will give rise to populations of descendants that display these traits. If the differences between the original population and its descendants are great enough, scientists will identify fossils of the ancestral and descendant populations as separate species. In other cases, the extinction of one species increases the resources for a population of a different species, which may adapt and change in form as it fills the now "open" role in the system. The fossil record can provide clues about the background extinction rate that results from this continual process of species turnover.

The span of geological time in which fossils of an individual species are found represents the life span of that species (Figure 14.3). Biologists have thus estimated that the "average" life span of a species is around 1 million years (although there is tremendous variation) and that the overall rate of extinction is about one species per million (0.0001%) per year. Some scientists have argued that these estimates are too low because they are based on observations of fossils, a record that may be biased toward long-lived species. However, the estimates are currently scientists' best approximation of background extinction rates.

Current rates of extinction are calculated from actual recorded extinctions. This is a challenge because extinctions are surprisingly difficult to document. The only way to conclude that a species no longer exists is to exhaustively search all areas where it is likely to have survived. In the absence of a complete search, most conservation organizations have adopted this standard: To be considered extinct, no individuals of a species must have been seen in the wild for 50 years.

A few searches for specific species give hints to the recent extinction rate. In Malaysia, a 4-year search for 266 known species of freshwater fish turned up only 122. In Africa's Lake Victoria, 200 of 300 native fish species have not been seen for years. On the Hawaiian island of Oahu, half of 41 native tree snail species have not been found, and in the Tennessee River, 44 of the 68 shallow-water mussel species are missing. Despite these results, few of the missing species in any of these searches is officially considered extinct.

The most complete records of documented extinction occur in groups of highly visible organisms, primarily mammals and birds. Since 1600, eighty-three out of an approximate 4500 identified mammal species have become extinct, while 113 of approximately 9000 known bird species have disappeared. The known extinctions of mammals and birds, spread out over the 400 years of these records, correspond to a rate of 0.005% per year. Compared to the background rate of extinctions calculated from the fossil record, the current rate of extinction is 50 times higher. If we examine the past 400 years more closely, we see that the extinction rate has actually increased since the start of this historical record (Figure 14.4) to about 0.01% per year, making the current rate 100 times higher than the calculated background rate.

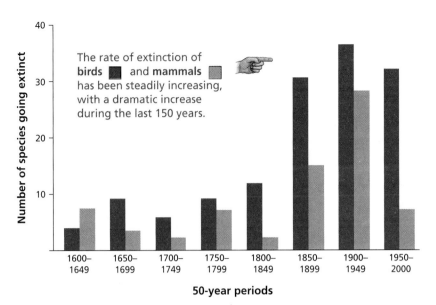

The rate of extinction of **birds** ■ and **mammals** ■ has been steadily increasing, with a dramatic increase during the last 150 years.

50-year periods

Figure 14.4   Rate of extinction. This graph illustrates the number of species of mammals and birds known to have become extinct since 1600.

In addition, there are reasons to expect that the current elevated rate of extinction will continue into the future. The World Conservation Union (known by its French acronym, IUCN), a highly respected global organization composed of and funded by states, government agencies, and nongovernmental organizations from over 140 countries, collects and coordinates data on threats to biodiversity. According to the IUCN's most recent assessment, 11% of all plants, 12% of all bird species, and 24% of all mammal species (the three best-studied groups of organisms) are in danger of extinction, and human activities on the planet pose the greatest threat to most of these species.

Habitat Loss and Food Chains

A variety of human activities can put species at risk of extinction. The most severe threats belong to one of four general categories: loss or degradation of habitat, introduction of nonnative species, overharvesting, and effects of pollution. However, these four categories are not equal; the IUCN estimates that 83% of endangered mammals, 89% of endangered birds, and 91% of endangered plants are directly threatened by damage to or destruction of the places where they live.

Habitat Destruction.   The dramatic reduction in numbers of shortnose and Lost River suckers in Upper Klamath Lake is almost entirely due to human modification of these species' **habitat**, the place where they live and obtain their food, water, shelter, and space. At one time, 350,000 acres of wetlands regulated the overall quality and amount of water entering into the lake. Most of these wetlands have been drained and converted to irrigated agricultural fields now. The disruption of natural water flows into and out of the lake has interfered with sucker reproduction and has reduced the number of offspring they produce by as much as 95%.

The outright loss of habitat experienced by the Lost River and shortnose suckers is commonly called **habitat destruction**, and it is not limited to species in the more developed world (Figure 14.5). Rates of habitat destruction caused by agricultural, industrial, and residential development accelerated throughout the twentieth century as Earth's total human population more than tripled from less than 2 billion in 1900 to over 6 billion today. This trend will likely continue as the human population continues to increase and become more affluent. As the amount of natural landscape declines, the number of species supported by the habitats in these landscapes naturally also decreases.

The relationship between the size of a natural area and the number of species that it can support follows a general pattern called a **species-area curve**. A species-area curve for reptiles and amphibians on a West Indian archipelago is

Figure 14.5   Lost habitat = lost species. Lemurs are the most highly endangered primates in the world. These acrobatic animals are found only on the island of Madagascar, first settled by humans 1500 years ago. Today only 10% of natural forest remains there. Of the 48 species of lemur present on the island 2000 years ago, 16 have become extinct, and 15 are at risk of extinction.

(a) Species diversity increases with area.

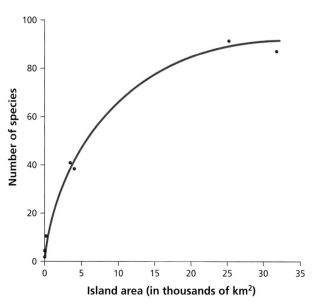

(b) Habitat reduction is predicted to result in loss of species.

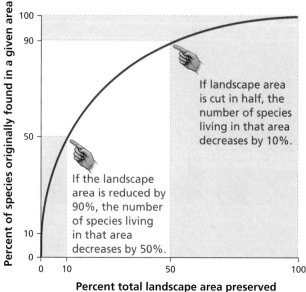

Figure 14.6   Predicting extinction caused by habitat destruction. (a) This curve demonstrates the relationship between the size of an island in the West Indies and the number of reptile and amphibian species living there. (b) We use a generalized species-area curve to roughly predict the number of extinctions in an area experiencing habitat loss.

illustrated in Figure 14.6a. Similar graphs have been generated in studies of different groups of organisms in a variety of habitats. Although the precise relationship between habitat area and number of species found in that habitat varies, the general pattern is that the number of species in an area increases rapidly as the size of the area increases, but the rate of increase slows as the area becomes very large. This rule of thumb, the approximation derived from the studies, is shown in Figure 14.6b. From the graph, we can estimate that a 90% decrease in landscape area will cut the number of species living in the remaining area by half.

Applying species-area curves to estimate extinction rates requires that we calculate the amount of natural landscape that has been lost in recent decades—a difficult task. Most studies have focused on tropical rain forests, which cover a broad swath of land roughly 20 degrees north and south of the equator. Tropical rain forests contain, by far, the greatest number of species of any habitat type on Earth. In 1988, biologist Edward O. Wilson estimated the rate of habitat destruction in the rain forest; he calculated that about 1% of the tropical rain forest is converted to agricultural use every year. Conservatively estimating the number of species in the rain forest at 5 million, Wilson applied the generalized species-area curve and projected that nearly 20,000 to 30,000 species (about 0.5%) are lost each year due to rain forest destruction.

More modern studies using images from satellites (Figure 14.7a) indicate that approximately 20,000 square kilometers (about 7722 square miles, an area the size of Massachusetts) of rain forest are cut each year in South America's Amazon River basin. This is a rate of 2% per year, or double Wilson's estimate. At this rate of habitat destruction, tropical rain forests will be reduced to 10% of their original size within about 35 years. If we apply the species-area curve, the habitat loss translates into the extinction of about 50% of species living in the Amazonian rain forest. Most of these species are small, and most are not even known to science; but if this prediction proves accurate, the extinct species in the rain forest would include about 50,000 of all known 250,000 species of plants, 1800 of the known 9000 species of birds, and 900 of the 4500 species of mammals in the world.

Of course, habitat destruction is not limited to tropical rain forests. When all of Earth's biomes are evaluated, freshwater lakes and streams, grasslands, and temperate forests are also experiencing high levels of modification. According to the IUCN, if habitat destruction around the world continues at

Figure 14.7 The primary causes of extinction.

(a) Habitat destruction

Humans are rapidly destroying tropical rain forests. This 1999 satellite photo illustrates the extent of destruction in an area of Brazilian rain forest that, until 30 years ago, contained no agricultural lands. The lighter parts of the photo are agricultural fields; the darker regions are intact forest.

(b) Habitat fragmentation

This "island" of tropical forest was created when the surrounding forest was logged. Scientists have documented hundreds of localized extinctions within fragments such as this.

(c) Introduced species

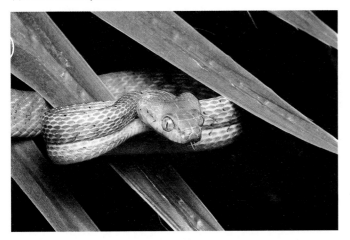

The introduced brown tree snake is responsible for the extinction of dozens of native bird species on the Pacific island of Guam.

(d) Overexploitation of species

These tiger skins represent a small fraction of the illegal harvest of tigers in Asia, primarily for the Chinese market.

(e) Pollution

Pollution from herbicides appears to be responsible for the increase of deformities in frogs in the midwestern United States and may partially explain the worldwide decline in frog species.

(f) Global warming

Polar bears hunt for seals, their primary prey, from sea ice. The extent of sea ice in the Arctic Ocean has been steadily declining over the past 20 years, threatening the bears' survival.

its present rate, nearly one-fourth of *all* living species will be lost within the next 50 years.

Some critics have argued that these estimates of future extinction are too high because not all groups of species are as sensitive to habitat area as the curve in Figure 14.6b suggests. Many species may still survive and even thrive in human-modified landscapes. Other biologists contend that there are other threats to species, including habitat fragmentation, and therefore the rate of species loss is likely to be even higher than these estimates.

Habitat Fragmentation.   Habitat destruction rarely results in the complete loss of a habitat type. Often what results from human activity is **habitat fragmentation**, in which large areas of intact natural habitat are subdivided (Figure 14.7b). Habitat fragmentation is especially threatening to large predators, such as grizzly bears and tigers, because of their need for large hunting areas.

Large predators require large, intact hunting areas due to a basic rule of biological systems: Energy flows in one direction within an ecological system along a **food chain**, which typically runs from the sun to **producers** (photosynthetic organisms) to the **primary consumers** that feed on them, to **secondary consumers** (predators that feed on the primary consumers), and so on. Along the way, most of the calories taken in at one **trophic level** (that is, a level of the food chain) are used to support the activities of the individuals at that level and therefore are not available for use by organisms at the next level. In other words, a substantial amount of the solar energy initially fixed by producers is dissipated—that is, given off as heat—at each level within a food web. You can see this in your own life; an average adult needs to consume between 1600 and 2400 Calories per day simply to maintain his or her current weight. The flow of energy along a food chain leads to the principle of the **trophic pyramid**, the bottom-heavy relationship between the **biomass** (total weight) of populations at each level of the chain (Figure 14.8). Habitat destruction and fragmentation can cause the lower levels of the pyramid to shrink, depriving the top predators of adequate calories for survival.

Habitat fragmentation also exposes wide-ranging predators to additional dangers. For example, grizzly bears need 200 to 2000 square kilometers of habitat to survive a Canadian winter, but the Canadian wilderness is increasingly bisected by roads built for tree harvesting. Each road represents an increased chance of grizzly-human interaction. Every interaction between grizzly bears and humans represents a greater danger to the bears than to humans. For example, of the 136 grizzlies that died in Canada's national parks between 1970 and 1995, seventeen died of natural causes and 119 were killed by humans.

The species that do remain in small fragments of habitat are more susceptible to extinction because populations in each fragment become isolated. Habitat fragmentation often makes it impossible for individuals to move from an area that has become unsuitable because of natural environmental changes to an area that is suitable. Isolated populations are also subject to genetic problems that threaten their long-term survival, as we discuss in detail later in the chapter.

Figure 14.8   A trophic pyramid. The relationship between producers, primary consumers, and secondary consumers in a biological community is illustrated here. Because most of the energy consumed by a trophic level is used within that level for maintenance, biomass decreases as position in the food chain increases. As a result, for secondary consumers to survive, a habitat must be large enough to support large producer and primary consumer populations.

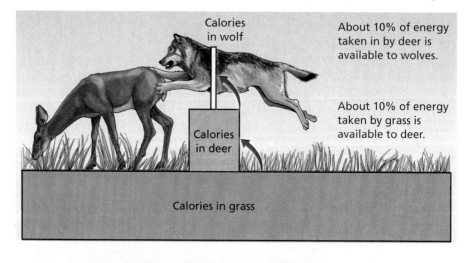

Calories in wolf

About 10% of energy taken in by deer is available to wolves.

Calories in deer

About 10% of energy taken by grass is available to deer.

Calories in grass

These genetic threats accrue to species whose populations are made small by other causes as well.

Other Human Causes of Extinction

According to the IUCN, the remaining threats to biodiversity posed by human activity play a role in about 40% of all cases of endangerment (Figure 14.7c–f on page 365).

Introduced Species. **Introduced species** include organisms brought by human activity to a region where they previously had never been found. Introduced species are often dangerous to native species because they have not evolved together; for instance, many birds on oceanic islands such as Hawaii and New Zealand are unable to defend themselves from introduced ground hunters. On Hawaii, one of these introduced predators, the Pacific black rat, became very adept at raiding eggs from nests and contributed to the extinction of dozens of species of honeycreepers, birds found nowhere else on Earth. Even domestic cats can take an enormous toll on wildlife when introduced into new habitats. A recent study in Wisconsin estimated that free-ranging cats in that state kill approximately 39 million birds every year.

Introduced species may also compete with native species for resources, causing populations of the native species to decline. For example, zebra mussels in the Great Lakes crowd out native mussel species as well as other organisms that filter algae from water, and the introduced vine known as kudzu shades over native trees and vines in the Southeast.

When introduced species crowd out native species, it is not only their direct competitors that suffer. Other species that rely on the struggling natives may decline as well. For example, opposum shrimp were introduced into Montana's Flathead Lake as a potential food source for game fish, primarily kokanee salmon. Unfortunately, the shrimp competed with the kokanee for the same food source, and introduction of the shrimp led to sharp declines in kokanee abundance. As a result, bald eagles that relied on kokanee populations in Flathead Lake also declined.

Humans continue to move species around the planet. As the global trade in agriculture and other goods continues to expand, the number of species introductions are likely to increase over the next century without a concerted effort to prevent them.

Overexploitation. When the rate of human destruction or use of a species outpaces its ability to reproduce, the species is subject to **overexploitation**. This often occurs when particular organisms are highly prized by humans, which is the case with animals and plants whose parts are used as medicinal therapies in some societies. For instance, three of the planet's eight tiger species have become extinct, and the remaining species are gravely endangered, in part due to the demand for their bones for their purported ability to treat arthritis and their genitals, which are erroneously believed to reverse male impotence. Popular wild-grown herbal medicines—including echinacea, the common cold remedy discussed in Chapter 1—are also at risk of overexploitation. The Pacific yew, a tree that produces the anticancer chemical taxol described in Chapter 5, faced the potential risk of extinction via overharvesting until production processes were developed to extract taxol from more common and faster-growing species of yew.

Overexploitation is also likely when an animal competes directly with humans—as in the case of the gray wolf, which was nearly exterminated in the United States by human hunters determined to protect their livestock. Species that cross international boundaries during migration or live in the world's oceans are also highly susceptible to overexploitation since it can be difficult to maintain a healthy population when no single government can regulate the total harvest. The near extinction of numerous whale species occurred in the nineteenth and early twentieth centuries due to unregulated harvest by many nations. Stocks of cod, swordfish, and tuna are now similarly threatened.

Pollution.   The release of poisons, excess nutrients, and other wastes into the environment—a practice otherwise known as **pollution**—poses an additional threat to biodiversity. For example, the herbicide atrazine poses a risk to frogs and salamanders in agricultural areas of the United States; and nitrogen pollution caused by fertilizer, automobile exhaust, and industrial emissions has led to drastic declines in populations of sensitive plant species within native grasslands in Europe. Fertilizer pollution from farms in the Klamath Valley poses a risk to the shortnose and Lost River suckers. Increased levels of nitrogen and phosphorus from fertilizer that runs off into waterways have increased the growth of algae in Upper Klamath Lake; when these algae explode in numbers, the bacteria that feed on them flourish and rapidly use up the available oxygen in the water. This process of oxygen-depleting **eutrophication** results in large fish kills, not just in Upper Klamath Lake but also in rivers, ponds, and oceans that receive fertilizer runoff from farms. Eutrophication threatens dozens of fish and invertebrate species in the Gulf of Mexico, where fertilizer draining from farm fields in the midwestern United States creates a low-oxygen "dead zone" the size of New Jersey at the mouth of the Mississippi River.

Perhaps the most serious pollutant released by humans is carbon dioxide, also a principal cause of global climate change (see Chapter 4). Computer models that link predicted changes in climate to known ranges and requirements of over 1000 species indicate that 15% to 37% of these plants will face extinction in the next century as the climate changes.

While determining the exact causes for why certain species are disappearing can sometimes be difficult, the evidence just discussed suggests that ESA critics who describe modern extinction rates as "natural" are incorrect. Over the past 400 years, humans have caused the extinction of species at a rate that appears to far exceed past rates, and it is clear that human activities continue to threaten thousands of additional species around the world. Earth appears to be on the brink of a sixth mass extinction of biodiversity—and the pervasive global change causing this extinction is human activity.

Many people who feel a moral responsibility to minimize the human impact on other species continue to support actions that conserve species despite the cost. However, in addition to supporting rights for nonhuman species, there is a practical reason to prevent the sixth extinction from occurring—the loss of nonhuman species can cause human suffering as well.

14.2   The Consequences of Extinction

Concern over the loss of biodiversity is not simply a matter of an ethical concern for nonhuman life. Humans have evolved with and among the variety of species that exist on our planet, and the loss of these species often results in negative consequences for us. In addition, the fossil record illustrated in Figure 14.2 (page 361) reveals that it takes 5 to 10 million years to recover the biological diversity lost during a mass extinction. The species that replaced those lost in previous mass extinctions were also very different; for instance, after the mass extinction of the reptilian dinosaurs, mammals replaced them as the largest animals on Earth. We cannot predict what biodiversity will look like after another mass extinction. In other words, the mass extinction we may be witnessing today will have consequences felt by people in thousands of generations to come.

Loss of Resources

The Lost River and shortnose suckers were once numerous enough to support fishing and canning industries on the shores of Upper Klamath Lake. Even before the arrival of European settlers, the native people of the area relied heavily upon these fish as a mainstay of their diet. The loss of these species represents a tremendous impoverishment of wild food sources. The biological resources that are harvested directly from natural areas are numerous and diverse—for example, wood for fuel and lumber, shellfish for protein, algae for gelatins, and herbs for medicines. The loss of any of these species affects human populations economically; one estimate places the value of wild species in the United States at $87 billion a year, or about 4% of the gross domestic product.

Wild species can also provide resources for humans in the form of unique biological chemicals. One example of a natural origin for valuable chemicals is the rosy periwinkle (*Catharanthus roseus*), which evolved on the island of Madagascar and has been exported around the world as a garden plant (Figure 14.9a). In the 1950s, scientists interested in the medicinal properties of plants noted that several African populations used extracts of rosy periwinkle as a treatment for diabetes. While they did not find an effective diabetes treatment when they screened this plant, two different laboratories found that chemicals derived from it appeared to be quite effective at stopping the growth of cancer cells. These two drugs, vincristine and vinblastine, have contributed to major gains in the likelihood of survival from leukemia and Hodgkin's disease. If we are unable to screen living species due to their extinction, we will never know which ones might have provided compounds that would improve human lives. In Chapter 12, we describe a small fraction of the thousands of species that have provided valuable biochemicals to humans.

Wild relatives of plants and animals that have been domesticated, such as agricultural crops and cattle, are also important resources for humans. Genes and gene variants that have been "bred out" of domesticated species are often still found in their wild relatives. These genetic resources represent a reservoir of traits that could be reintroduced into agricultural species through breeding or genetic engineering. Agricultural scientists who are attempting to produce better strains of wheat, rice, and corn look to the wild relatives of these crops as sources of genes for pest resistance and for traits that improve yields in specific environmental conditions. For example, the Mexican teosinte species *Zea diploperennis* (Figure 14.9b), discovered in 1977, appears to be an ancestor of modern corn. This species of teosinte is resistant to several viruses that plague cultivated corn; some genes that confer this resistance have been transferred via hybridization to produce disease-resistant varieties. Additionally, there is value in preserving wild relatives of domesticated crops in their natural habitats; often the wild organisms in these communities provide the key to reducing pest damage and

(a) Rosy periwinkle

(b) Teosinte, ancestor of modern corn

(c) Boll weevil wasp

Figure 14.9 Resources from nature. (a) Anticancer drugs vincristine and vinblastine were first isolated from rosy periwinkle (*Catharanthus roseus*), a species of flower native to Madagascar. (b) Teosinte is the ancestor of modern corn, first cultivated in Central America. This species, *Zea diploperennis*, was discovered in a remote Mexican site in 1978. (c) *Catolaccus grandis* is a predator of boll weevils, which are one of the most damaging cotton pests. *C. grandis* was discovered preying on pests of wild cotton in southern Mexico.

disease on the domestic crop. For example, the wasp *Catolaccus grandis* consumes boll weevils and is used to control infestations of these pests in cotton fields (Figure 14.9c). *C. grandis* was discovered in the tropical forest of southern Mexico, where it parasitizes a similar pest in wild cotton populations. Of course, introducing an insect such as *C. grandis* into a new environment carries risk, even if the introduction is meant to reduce environmental damage. There are many examples of environmental disasters caused by introduced predators—for example, the cane toad, which was brought to Australia to control beetles in sugarcane crops but now threatens the survival of several native frog species as well as their predators, which are poisoned by the toad's skin toxins. Often, a less risky approach to reducing pest damage to crops is to preserve nearby habitats and the ecological interactions that persist there.

Disruption of Ecological Communities

Although humans receive direct benefits from thousands of species, most threatened and endangered species are probably of little or no use to people. Even the Lost River and shortnose suckers, as valuable as they once were to the

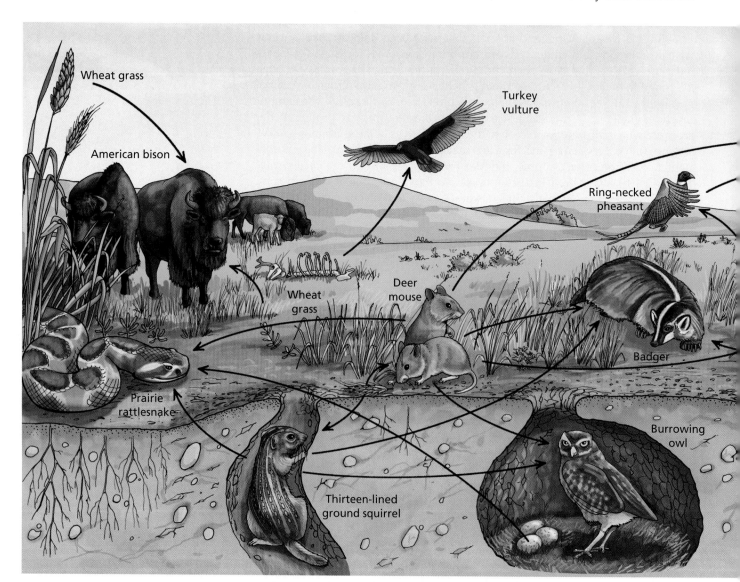

Figure 14.10 The web of life. Species are connected to each other and to their environments in various, complex ways. This drawing shows some of the important relationships among organisms and their environments in a North American prairie. Black arrows represent feeding relationships; for example, thirteen-lined ground squirrels eat wheat grass and in turn are eaten by badgers.

native people of the Klamath Basin, are not especially missed as a food source; no one has starved simply because these fish have become less common.

In reality, most species are beneficial to humans because they are connected to other species and natural processes in a biological **community**, consisting of all the organisms living together in a particular habitat area. The complex linkage among organisms in a community is often referred to as a **food web** (Figure 14.10). Because relationships between species may be based on requirements other than food, ecologists often use the phrase "web of life" to illustrate all of these interactions. As with a spider's web, any disruption in one strand of the web of life is felt by other portions of the web. Some tugs on the web cause only minor changes to the community, while others can cause the entire web to collapse. Most commonly, losses of strands in the web are felt by a small number of associated species. However, some disruptions caused by the loss of seemingly insignificant species have the potential to be felt by humans, especially when we rely on a functional community for particular products or services. Some examples of this phenomenon are described next.

Mutualism: How Bees Feed the World.   An interaction between two species that benefit each other is called **mutualism**. Cleaner fish that remove and

Figure 14.11 A close partnership. The relationship between plants and fungi in mycorrhizae provides benefits to both partners and is a clear example of mutualism. Most of the root-like threads in this soil are fungal hyphae.

consume parasites from the bodies of larger fish, fungi called mycorrhizae that increase the mineral absorption of plant roots while consuming the plant's sugars (Figure 14.11), and ants that find homes in the thorns of acacia trees and defend the trees from other insects are examples of mutualism. The mutualistic interaction between plants and bees is perhaps the most often described example.

Bees are the primary pollinators of many species of flowering plants; that is, they transfer sperm, in the form of pollen grains, from one flower to the female reproductive structures of another flower. The flowering plant benefits from this relationship because insect pollination increases the number and vigor of seeds that the plant produces. The bee benefits by collecting excess pollen and nectar to feed itself and its relatives in the hive (Figure 14.12).

Wild bees pollinate at least 80% of all the agricultural crops in the United States, providing a net benefit of about $8 billion. In addition, populations of wild honeybees have a major and direct impact on many more billions of dollars of agricultural production around the globe.

Unfortunately, bees have suffered dramatic declines in recent years. According to the U.S. Department of Agriculture, we are facing an "impending pollination crisis" because both wild and domesticated bees are disappearing at alarming rates. These dramatic declines are believed to result from an increased level of bee **parasites** (infectious organisms that cause disease or drain energy from their hosts), competition with the invading Africanized honeybees ("killer bees"), and habitat destruction. The extinction of these inconspicuous mutualists of crop plants would be extremely costly to humans.

Predation: How Songbirds May Save Forests.   A species that survives by eating another species is typically referred to as a **predator**. The word conjures up images of some of the most dramatic animals on Earth: cheetahs, eagles, and killer whales. You might not picture wood warblers, a family of North American bird species characterized by their small size and colorful summer plumage, as predators; however, these beautiful songsters are voracious consumers of insects (Figure 14.13a). The hundreds of millions of individual warblers in the forests of North America collectively remove literally tons of insects from forest trees and shrubs every summer. Most of these insects prey on plants. By reducing the number of insects in forests, warblers reduce the damage that insects inflict on forest plants. The results of a study that excluded birds from white oak seedlings showed that the trees were about 15% smaller because of insect damage over 2 years, as compared to trees from which birds were not excluded. However, other studies have shown less dramatic benefits.

Although scientists still disagree about how important warblers and other insect-eating birds are to the survival of trees, most agree that reducing the number of forest pests increases the growth rate of the trees. Harvesting trees for paper and lumber production fuels an industry worth over $200 billion in the United

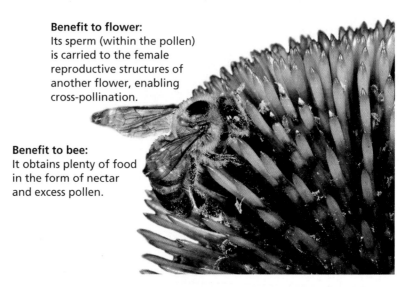

Benefit to flower:
Its sperm (within the pollen) is carried to the female reproductive structures of another flower, enabling cross-pollination.

Benefit to bee:
It obtains plenty of food in the form of nectar and excess pollen.

Figure 14.12 Mutualism. Honeybees transfer pollen, allowing one plant to "mate" with another plant of some distance away.

(a) Black-throated blue warbler, predator of insects.

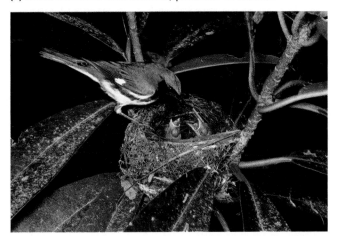

(b) Forests suffer when insects are unchecked by predators.

Figure 14.13   Predation. (a) The black-throated blue warbler is one of many warbler species native to North American forests. These colorful birds are among the most active predators of plant-eating insects in these forests. (b) Insects can kill many trees, as seen in this photo of a spruce budworm infestation. Warblers and other insect-eating birds probably reduce the number and severity of such insect outbreaks.

States alone. At least some of the wood harvested by the timber industry was produced only because warblers were controlling insects in forests (Figure 14.13b).

Many species of forest-inhabiting warblers appear to be experiencing declines in abundance. The loss of warbler species has several causes, including not only habitat destruction in their summer habitats in North America and their winter habitats in Central and South America but also increased predation by human-associated animals such as raccoons and house cats. Although other, less-vulnerable birds may increase in number when warblers decline, the warblers' effects on insect pest populations may not be completely replaced by less insect-dependent birds. If smaller warbler populations correspond to lower forest growth rates and higher levels of forest disease, then these tiny, beautiful birds definitely have an important effect on the human economy.

Competition: How a Deliberately Infected Chicken Could Save a Life.   When two species of organisms both require the same resources for life, they will be in **competition** for the resources within a habitat. We may imagine lions and hyenas fighting over a freshly killed antelope or weeds growing in our vegetable gardens as typical examples of competition, but most competitive interactions are invisible.

In general, competition limits the size of competing populations. To determine whether two species that use similar resources are competing, we remove one from an environment. If the population of the other species increases, then the two species are competitors. One of the least visible forms of competition occurs among microorganisms. Competitive interactions among microbes may be among the most essential factors for maintaining the health of people and communities.

Salmonella enteritidis is a leading cause of food-borne illness in the United States. Between 2 million and 4 million people in this country will be infected by this bacterium in the coming year, and they will experience fever, intestinal cramps, and diarrhea as a result. In about 10% of cases, the infection results in severe illness requiring hospitalization. If it is not treated promptly, the infected individuals may die. Nearly 400 to 600 Americans die as a result of *S. enteritidis* infection every year.

Most *S. enteritidis* infections result from consuming raw or improperly cooked poultry products, especially eggs. The U.S. Centers for Disease Control estimate that as many as 1 in 50 consumers are exposed to eggs contaminated with *S. enteritidis* every year. Most of these eggs have had their shells disinfected and do not look damaged in any way; the bacteria were deposited in the egg by the hen when the egg was forming inside her. Thus, the only way to prevent *S. enteritidis* from contaminating eggs is to keep it out of hens.

A common way to control *S. enteritidis* is to feed hens antibiotics—chemicals that kill bacteria. However, like most microbes, *S. enteritidis* strains can evolve drug resistance if some members of the population contain genes that allow them to survive the effects of the antibiotic. Another way to reduce the chance of *S. enteritidis* infection in poultry is to reduce the amount of resources available for the bacteria's growth. Most *S. enteritidis* infections originate in an animal's digestive system. If another bacterial species is already monopolizing the food and available space in a hen's digestive system, then *S. enteritidis* has trouble colonizing there. Some poultry producers now intentionally infect hens' digestive systems with harmless bacteria, via a practice called **competitive exclusion**, to reduce *S. enteritidis* levels in their flocks. This technique involves feeding cultures of benign bacteria to one-day-old birds. When the harmless bacteria become established in their intestines, the chicks will be less likely to host large *S. enteritidis* populations (Figure 14.14). There is evidence that this practice is

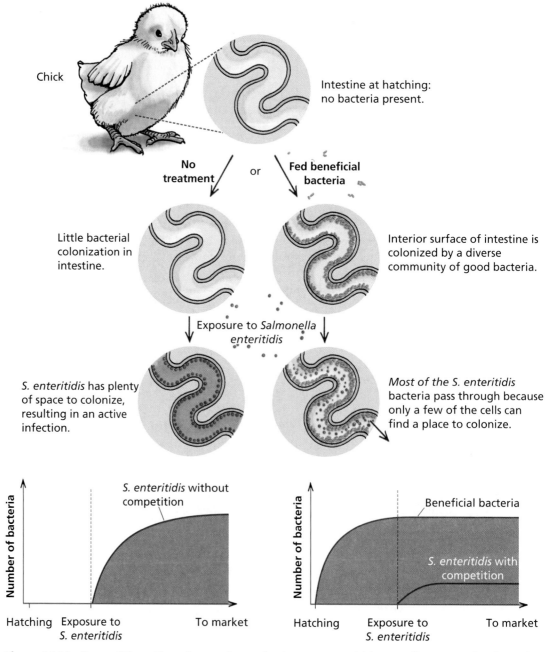

Figure 14.14   Competition. If poultry producers feed very young chicks non-disease-causing bacteria, the bacteria take up the space and nutrients in the intestine that would be used by *S. enteritidis*; thus, they will have no site to colonize and increase their population.

working; *S. enteritidis* infections in chickens have dropped by nearly 50% in the United Kingdom, where competitive exclusion in poultry is common practice.

The competitive exclusion of *S. enteritidis* in hens mirrors the role of some human-associated bacteria, such as those that normally live within our intestines and genital tracts. For instance, many women who take antibiotics for a bacterial infection will then develop vaginal yeast infections because the antibiotic kills noninfectious bacteria as well, including species in the genital tract that normally compete with yeast. Maintaining competitive interactions between larger species can be important for humans as well. For instance, in temporary ponds, the main competitors for the algae food source are mosquitoes, tadpoles, and snails. In the absence of tadpoles and snails, mosquito populations can become quite large—potentially with severe consequences since these insects may carry deadly diseases such as malaria, West Nile virus, and yellow fever. With frogs and their tadpoles increasingly endangered, this risk is a real one.

Keystone Species: How Wolves Feed Beavers.   Table 14.1 summarizes the major types of ecological interactions among organisms. However, this table emphasizes the effects of each interaction on the species directly involved; it does not illustrate that many of these interactions may have multiple indirect effects. Look again at the food web pictured in Figure 14.10 on pages 370–371. None of the species in the prairie's biological community is connected to only one other species—they all eat something, and most of them are eaten by something else. You can imagine that badgers, by preying on deer mice, have a negative effect on rattlesnakes, which they compete with for these mice, and a more indirect positive effect on ground squirrels, which compete with mice for grass seeds. The existence of indirect effects of varying importance has led ecologists to hypothesize

Table 14.1   Types of species interactions and their direct effects.

Interaction	Example	Effect on Species 1	Effect on Species 2
Mutualism: Association increases the growth or population size of both species.	1. Ants 2. Acacia tree	**+** The swollen thorns of the acacia provide shelter for the ants. The acacia leaves provide "protein bodies" that the ants harvest for food.	**+** Ants kill herbivorous insects and destroy competing vegetation, benefiting the acacia.
Predation and Parasitism: Consumption of one organism by another.	1. Brown bear 2. Salmon	**+** The brown bear catches the salmon and eats it, obtaining nourishment.	**−** The salmon does not survive.
Competition: Association causes a decrease or limitation in population size of both species.	1. Dandelion 2. Tomato plant	**−** The dandelion weed does not grow as well in the presence of the tomato plant. Dandelion produces fewer seeds and fewer offspring.	**−** The tomato plant does not grow optimally in the presence of the weed. Tomato plant produces fewer flowers and fruit.

(a)

(b)

Figure 14.15   Keystone species. (a) The keystone in an archway helps to stabilize and maintain the arch. (b) A keystone species, such as wolves in Yellowstone National Park, help to stabilize and maintain other species in an ecosystem.

that in at least some communities, the activities of a single species can play a dramatic role in determining the composition of the system's food web. These organisms are called **keystone species** because their role in a community is analogous to the role of a keystone in an archway (Figure 14.15a). Remove the keystone, and an archway collapses; remove the keystone species, and the web of life collapses. Biologists can point to several examples of keystone species, including the population of gray wolves in Yellowstone National Park.

Gray wolves were exterminated within Yellowstone National Park by the mid-1920s because of a systematic, highly effective campaign to rid the American West of this occasional predator of livestock. However, by the 1980s, increased understanding and appreciation of these animals by the American public led to renewed interest in returning wolves to their historical homeland. In the mid-1990s, thirty-one wolves originally trapped in Canada were released into Yellowstone National Park. By the end of 2003, this number had grown to at least 301 wolves in over 30 packs, living both in the park and in surrounding public lands. During the time that wolves were extinct in the park, biologists noticed dramatic declines in populations of aspen, cottonwood, and willow trees. They attributed this decline to an increase in the consumption of these trees by elk, especially during winter when grasses become unavailable. However, just a few years after wolf reintroduction, aspen, cottonwood, and willow tree growth has rebounded in some areas of the park (Figure 14.15b). Besides the regions near active wolf dens, these areas include places on the landscape where elk have limited ability to see approaching wolves or to escape. Thus, they will stay away from these areas to avoid wolf predation. Wolves, by both reducing elk numbers and changing elk behavior, appear to be an important factor in maintaining large populations of hardwood trees in Yellowstone Park.

The rebound of aspen, cottonwood, and willow populations in Yellowstone has effects on other species as well. Beaver rely on these trees for food, and their populations appear to be growing in the park after decades of decline. Warblers, insects, and even fish that depend on shelter, food, and shade from these trees are increasing in abundance as well. Wolves in Yellowstone appear to fit the profile of a classic keystone species, one whose removal had numerous and surprising effects on biodiversity in the park. Since biologists cannot usually predict which species, if any, will act as keystones in a community, it is often impossible to know whether the rippling effects of one extinction will change that community forever.

Changed Ecosystems

As the examples in the previous section illustrate, the extinction of a single species can have sometimes surprising effects on other species in a habitat. What may be even less apparent is how the loss of seemingly insignificant species can change the environmental conditions on which the entire community depends.

Ecologists define an **ecosystem** as all of the organisms in a given area, along with their nonbiological environment. The function of an ecosystem is described in terms of the rate at which energy flows through it and the rate at which nutrients are recycled within it. The loss of some species can dramatically affect both of these ecosystem properties.

Energy Flow.   In nearly all ecosystems, the primary energy source is the sun. As discussed in Chapter 4, producers convert sun energy into chemical energy during the process of photosynthesis, and the energy is passed through trophic levels making up a food chain; energy is then partitioned among trophic levels in a bottom-heavy trophic pyramid (review Figure 14.8 on page 366). The amount of sunlight reaching the surface of the Earth and the availability of water at any given location are the major determiners of both trophic pyramid structure and energy flow through it. (Chapter 15 provides a summary of how variance in sunlight and water availability leads to differences in Earth's ecosystem types.)

However, the biodiversity found in an ecosystem can also have strong effects on energy flow within it.

Studies in grasslands throughout the world have provided convincing evidence that loss of species can affect energy flow. By comparing experimental prairie "gardens" planted with the same total number of individual plants but with different numbers of species, scientists at the University of Minnesota and elsewhere have discovered that the overall plant biomass tends to be greater in more diverse gardens. This research indicates that a decline in diversity, even without a decline in habitat, may lead to less energy being made available to organisms higher on the food chain, including people who depend on wild-caught food.

Nutrient Cycling.   When essential mineral nutrients for plant growth pass through a food web, they are generally not lost from the environment—hence the term **nutrient cycling**. Figure 14.16 illustrates the nitrogen nutrient cycle in a natural prairie. Nitrogen is a major component of protein, and abundant protein is essential for the proper growth and functioning of all living organisms. Nitrogen is therefore often the nutrient that places an upper limit on production in most ecosystems—more nitrogen generally leads to greater production, while areas with less available nitrogen can support fewer plants (and therefore animals).

You should note as you review Figure 14.16 that plants absorb simple molecules from the soil and incorporate them into more complex molecules. These complex molecules move through the food web with relatively minor changes until they return to the soil. Here, complex molecules are broken down into simpler ones by the action of **decomposers**, typically bacteria and fungi. Changes in the soil community can greatly affect nutrient cycling and thus the survival of certain species in ecosystems. Scientists investigating the effects of introduced earthworms that have replaced native soil communities in forests throughout the northeastern United States have observed dramatic

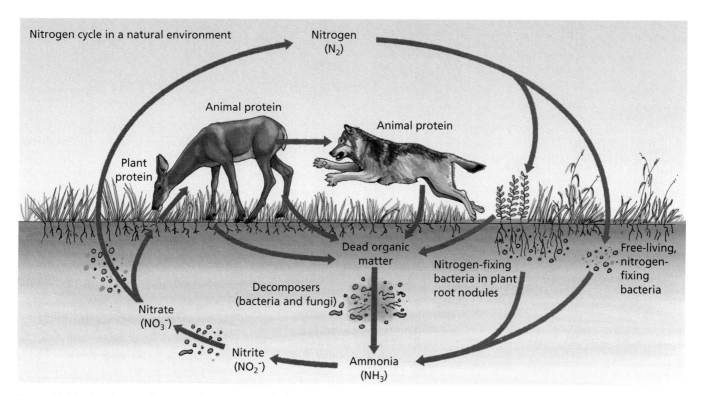

Figure 14.16   Nutrient cycling. Nutrients are recycled in an ecosystem, flowing from soil to producers to consumers and then back into the soil, where complex nutrients are decomposed into simpler forms.

Figure 14.17 Changes in ecosystem function. In forests in northern North America, the introduction of earthworms and loss of native species have changed the plant community. Notice how barren the worm-infested forest floor, at bottom, appears, compared to the worm-free forest at top. One reason for this dramatic change may be a disruption in the native nutrient cycle.

reductions in the diversity and abundance of plants on the forest floor (Figure 14.17). These changes in the plant community may be related to the effects on nutrient cycling resulting from this change in the soil community.

The loss of biodiversity can clearly have profound effects on the health of communities and ecosystems on which humans depend. However, controversy exists over whether the current extinction may negatively affect our own psychological well-being.

Psychological Effects

Some scientists argue that the diversity of living organisms sustains humans by satisfying a deep psychological need. One of the most prominent scientists to promote this idea is Edward O. Wilson, who calls this instinctive desire to commune with nature **biophilia**.

Wilson contends that people seek natural landscapes because our distant ancestors evolved in similar landscapes (Figure 14.18). According to this hypothesis, ancient humans who had a genetic predisposition driving them to find diverse natural landscapes were more successful than those without this predisposition, since diverse areas provide a wider variety of food, shelter, and tool resources. Wilson claims that we have inherited this genetic imprint of our preagricultural past.

While there is no evidence of a gene for biophilia, there is evidence that our experience with nature has powerful psychological effects. Studies in dental clinics indicate that patients viewing landscape paintings experience a 10- to 15-point decrease in blood pressure. Patients in a Philadelphia hospital who could see trees from their windows recovered from surgery more quickly and required fewer pain medications than did patients whose views were of brick walls. Individual experiences with pets and houseplants indicate that many people derive great pleasure from the presence of nonhuman organisms. Although not conclusive, these studies and experiences are intriguing since they suggest that a continued loss of biodiversity could make life in human society less pleasant overall.

Figure 14.18 Is our appreciation of nature innate? Humans evolved in a landscape much like this one in East Africa. Some scientists argue that we have an instinctive need to immerse ourselves in the natural world.

14.3 Saving Species

So far in this chapter, we have established the possibility of a modern mass extinction occurring, and we have described the potentially serious costs of this loss of biodiversity to human populations. Since current elevated extinction rates are largely a result of human activity, reversing the trend of species loss requires mostly political and economic, rather than scientific, decisions. But what *can* science tell us about how to stop the rapid erosion of biodiversity?

Protecting Habitat

Without knowing exactly which species are closest to extinction and where they are located, the most effective way to prevent loss of species is to preserve as many habitats as possible. The same species-area curve that Wilson used to estimate the future rate of extinction also gives us hope for reducing this number. According to the curve in Figure 14.6b (page 364), species diversity declines rather slowly as habitat area declines. Thus, in theory, we can lose 50% of a habitat but still retain 90% of its species. This estimate is optimistic because habitat destruction is not the only threat to biodiversity, but the species-area curve tells us that if the rate of habitat destruction is slowed or stopped, extinction rates will slow as well.

Protecting the Greatest Number of Species. Given the growing human population, it is difficult to imagine a complete halt to habitat destruction. However, biologist Norman Myers and his collaborators have concluded that 25 biodiversity "hotspots," making up less than 2 percent of Earth's surface, contain up to 50 percent of all mammal, bird, reptile, amphibian, and plant species (Figure 14.19). Hotspots occur in areas of the globe where favorable climate conditions lead to high levels of plant production, such as rain forests, and where geological factors have resulted in the isolation of species groups, allowing them to diversify. Stopping habitat destruction in these hotspots could greatly reduce the global extinction rate. By focusing conservation efforts on hotspot areas at the greatest risk, humans can very quickly prevent the loss of a large number of species. Of course, preserving these biodiversity hotspots is not easy. It requires the concerted actions of a diverse community of nations and

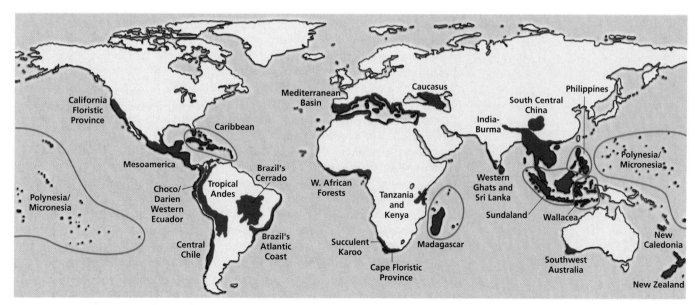

Figure 14.19 Diversity "hotspots." This map shows the locations of 25 identified biodiversity hotspots around the world. Notice how unevenly these regions of high biodiversity are distributed.

people, some of whom must also address pressing concerns of poverty, hunger, and disease. Even with habitat protection, many species in these hotspots will likely become extinct anyway for other human-mediated reasons.

In the long term, we must find ways to preserve biodiversity while including human activity in the landscape. One option is **ecotourism**, which encourages travel to natural areas in ways that conserve the environment and improve the well-being of local people. Some hotspot countries, such as Costa Rica and Kenya, have used ecotourism to preserve natural areas and provide much-needed jobs; other countries have been less successful.

While preserving hotspots may greatly reduce the total number of extinctions, this approach has its critics, who say that by promoting a strategy that focuses intensely on small areas, we risk losing large amounts of biodiversity elsewhere. These critics promote an alternative approach—identifying and protecting a wide range of ecosystem *types*—designed to preserve the greatest diversity of biodiversity rather than just the largest number of species.

Protecting Habitat for Critically Endangered Species.   Although preserving a variety of habitats ensures fewer extinctions, already endangered species require a more individualized approach. The ESA requires the U.S. Department of Interior to designate critical habitats for endangered species; that is, areas in need of protection for the survival of the species. The amount of critical habitat that becomes designated depends on political as well as biological factors. The biological part of a critical habitat designation includes conducting a study of habitat requirements for the endangered species and setting a population goal for it. The U.S. Department of Interior's critical habitat designation has to include enough area to support the recovery population. However, federal designation of a critical habitat results in the restriction of human activities that can take place there. The U.S. Department of Interior has the ability to exclude some habitats from protection if there are "sufficient economic benefits" for doing so—a decision that is in some part political.

Decreasing the Rate of Habitat Destruction.   Preserving habitat is not simply the job of national governments that set aside lands in protected areas or of private conservation organizations that purchase at-risk habitats. All of us can take actions to reduce habitat destruction and stem the rate of species extinction. Conversion of land to agricultural production is a major cause of habitat destruction, and so eating lower on the food chain and reducing your consumption of meat and dairy products from animals that are fed field crops is one of the most effective actions you can take. Reducing your use of wood and paper products, and limiting your consumption of these products to those harvested sustainably (that is, in a manner that preserves the long-term health of the forest), can help slow the loss of forested land.

Other measures to decrease the rate of habitat destruction require group effort. For instance, increased financial support for developing countries and the reduction of their international debt may help slow the rate of habitat destruction. These actions allow countries to invest money in technologies that decrease their use of natural resources. Strategies that slow the rate of human population growth offer more ways to avoid mass extinction. You can participate in group conservation efforts by joining nonprofit organizations focused on these issues, writing to politicians, and educating others.

Although protecting habitat from destruction can reduce extinction rates for species on the brink of extinction—like the shortnose and Lost River suckers—preserving habitat is not enough. Populations can become so small that they can disappear, even with adequate living space. Recovery plans for both the Lost River and shortnose suckers set a short-term goal of one stable population made up of at least 500 individuals for each unique stock of suckers. To understand why at least this many individuals are required to protect the species from extinction, we need to review some of the special problems faced by small populations.

(a) Lost River sucker

(b) California condor

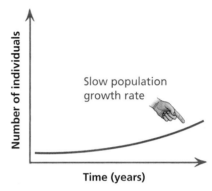

Figure 14.20 The effect of growth rate on species recovery. (a) This graph illustrates the rapid growth of a hypothetical population of quickly reproducing Lost River suckers. (b) The slow growth rate of the California condor has made the recovery of this species a long process. Today, nearly 30 years after recovery efforts began, the population of wild condors is still only in the dozens. Two wild populations of 150 condors each must be established for the bird to be removed from endangered status.

Protection from Environmental Disasters

A species' growth rate is influenced by how long the species takes to reproduce, how often it reproduces, the number of offspring produced each time, and the death rate of individuals under ideal conditions. For instance, species that reproduce slowly take longer to grow in number than do species that reproduce quickly. Thus the growth rate of an endangered species influences how rapidly that species can attain a target population size. Shortnose and Lost River suckers have relatively high growth rates and will meet their population goals quickly if the environment is ideal (Figure 14.20a). For slower-growing species, such as the California condor (Figure 14.20b), populations may take decades to recover. The rate of recovery is important because the longer a population remains small, the more it is at risk of experiencing a catastrophic environmental event that could eliminate it entirely. The story of the heath hen is a classic example of the dangers facing small populations.

The heath hen was a small wild chicken that once ranged on the East Coast of the United States, from Maine to Virginia (Figure 14.21). In the eighteenth century, the heath hen population numbered in the hundreds of thousands of individuals, and the birds were a favorite wild meal of European settlers. Continued settlement of the eastern seaboard of the United States resulted in the loss of heath hen habitat and increased hunting, causing the rapid and dramatic decline of the birds' population. By the end of the nineteenth century, the only remaining heath hens lived on Martha's Vineyard, a 100-square-mile island off the coast of Cape Cod. Farming and settlement on the island further reduced the habitat for heath hen breeding. By 1907, only 50 heath hens were present on the island.

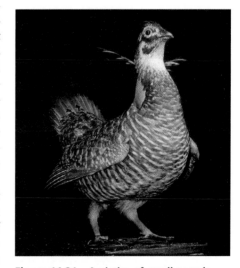

Figure 14.21 A victim of small population size. The heath hen was once abundant throughout the eastern United States. Although it was protected when its population fell to nearly 50 individuals, a series of unexpected disasters caused its extinction.

Recognizing the precariousness of the population, Massachusetts established a 2.5-square-mile reserve for the remaining birds on Martha's Vineyard in 1908. The establishment of this sanctuary seemed to be the solution to the heath hen's precipitous decline. By 1915, the population was recovering and had grown to nearly 2000 individuals. However, beginning in 1916, a series of disasters struck the remaining birds. First, fire destroyed much of the species' breeding habitat on the island. The following winter was long and cold, and an invasion of starving predatory goshawks from the north further reduced the heath hen population. Finally, a poultry disease brought to Martha's Vineyard in imported domestic turkeys wiped out much of the remaining population. By 1927, only 14 heath hens remained—almost all males. The last surviving member of the species was spotted for the last time on March 11, 1932.

The final causes of heath hen extinction were natural events—fire, harsh weather, predation, and disease. But it was human-caused habitat loss and human hunting that initially caused the population to become more vulnerable to these relatively common challenges. A population of 100,000 individuals can weather a disaster that kills 90% of its members but leaves 10,000 survivors; but a population of 1000 individuals will be nearly eliminated by the same circumstances. Even when human-caused losses to the heath hen population were halted, the species' survival was still extremely precarious.

Small populations of endangered species can still be protected from the fate that befell the heath hen. Having additional populations of the species at sites other than Martha's Vineyard would have nearly eliminated the risk that *all* members of the population would be exposed to the same series of environmental disasters. This is the rationale behind placing captive populations of endangered species at several different sites. For instance, captive whooping cranes are located at the U.S. National Biological Service's Patuxent Wildlife Research Center in Maryland, the International Crane Foundation in Wisconsin, the Calgary Zoo in Canada, and the Audubon Center for Endangered Species Research in New Orleans. However, if populations of endangered species remain small in number, they are subject to a more subtle but potentially equally devastating situation—the loss of genetic variability.

Protection from Loss of Genetic Diversity

A species' **genetic variability** is the sum of all of the alleles and their distribution within the species. Differences among alleles produce the variety of traits within a population. For example, the gene that determines your ABO blood type comes in three different forms, and the combination of alleles that you possess determines whether your blood type is O, A, B, or AB. Thus a population containing all three blood-type alleles contains more genetic variability for this gene than does a population with only two alleles.

The loss of genetic variability in a population is a problem for two reasons: (1) On an individual level, low genetic variability leads to low fitness; and (2) on a population level, rapid loss of genetic variability may lead to extinction.

The Importance of Individual Genetic Variability.   As discussed in Chapter 10, fitness refers to an individual's ability to survive and reproduce in a given set of environmental conditions. Low individual genetic variability decreases fitness for two reasons. We can use an analogy to illustrate them. First, imagine that you could own only two jackets (Figure 14.22a). If both are blazers, then you would be well prepared to meet a potential employer. However, if you had to walk across campus to your job interview in a snowstorm, you would be pretty uncomfortable. If you own two parkas, then you will always be protected from the cold, but you would look pretty silly at a dinner party. However, if you own one warm jacket and one blazer, you are ready for slush and snow as well as a nice date. In a way, individuals experience the same advantages when they carry two different functional alleles for a gene—that is, when they are heterozygous for codominant alleles. In this case, since each allele codes for a functional protein, a heterozygous individual produces two, slightly different proteins for the same function. If the protein produced by each allele works best in different environments (for instance, if one works best at high temperatures and the other at moderate temperatures), then heterozygous individuals are able to function efficiently over a wider range of conditions than are homozygotes, who only make one version of the protein.

The second reason high individual genetic variability increases fitness is that, in many cases, one allele for a gene is **deleterious**—that is, it produces a protein that is not very functional. In our jacket analogy, a nonfunctional, deleterious allele is equivalent to a badly torn jacket. If you have this jacket and an intact one, at least you have one warm covering (Figure 14.22b). In this case, heterozygosity is valuable because a heterozygote still carries one functional copy of the gene. Often these deleterious alleles are recessive, meaning that the activity of the functional allele in a heterozygote masks the fact that a deleterious allele is present (see Chapter 6). An individual who is homozygous (carries two identical copies of a gene) for the deleterious allele will have low fitness—in our analogy, two torn jackets and nothing else. For both of these reasons, when individuals are heterozygous for many genes, (or, in our analogy, have two choices for all clothing items), the cumulative effect is often greater fitness relative to individuals who are homozygous for many genes.

In a small population, where mates are more likely to be related to each other than in a very large population simply because there are fewer mates to choose from, heterozygosity declines. When related individuals mate—known as **inbreeding**—the chance that their offspring will be homozygous for any allele is relatively high. The negative effect of homozygosity on fitness is known as **inbreeding depression**. This is seen in humans as well as in other species; numerous studies consistently show that the children of first cousins have higher mortality rates (thus lower fitness) than children of unrelated

(a) Heterozygote has higher fitness than either homozygote.

Homozygote: Relatively low fitness
(only one type of jacket in wardrobe)

Homozygote: Relatively low fitness
(only one type of jacket in wardrobe)

Heterozygote: Relatively high fitness
(two types of jackets in wardrobe)

(b) Heterozygote masks the deleterious allele.

Homozygote: Relatively high fitness
(two functional jackets in wardrobe)

Homozygote: Relatively low fitness
(two nonfunctional jackets in wardrobe)

Heterozygote: Relatively high fitness
(one functional jacket in wardrobe)

Figure 14.22   The benefits of heterozygosity. In this analogy, each jacket represents an allele. (a) Heterozygotes may be better prepared for a diversity of life experiences than homozygotes are. (b) Heterozygotes may have higher fitness than some homozygotes because certain alleles are deleterious and recessive. In this case, homozygotes for the normal allele also have higher fitness than homozygotes for the recessive allele.

parents. Because these children are the offspring of close relatives, they are more likely to have high homozygosity as compared to the children of nonrelatives. In a small population of an endangered species, low rates of survival and reproduction that are associated with high rates of inbreeding can seriously hamper a species' ability to recover from endangerment.

How Variability Is Lost via Genetic Drift.   Small populations lose genetic variability because of **genetic drift**, a change in the frequency of an allele within a population that occurs simply by chance. Although in Chapter 11 we discussed genetic drift as a process for causing evolutionary change, in a small population, genetic drift can have detrimental consequences also.

Imagine a human population in which the frequency of blood-type allele A is 1%; that is, only 1 out of every 100 blood-type genes in the population is the A form (we use the symbol I^A). In a population of 20,000 individuals, we calculate the total number of I^A alleles as follows:

total number of blood-type alleles in population =

total population $\times$ 2 alleles/person

$20,000 \times 2 = 40,000$

total number of I^A alleles =

total number of alleles in population $\times$ frequency of allele I^A

$40,000 \times 1\% = 40,000 \times 0.01$

$= 400$

If a few of the individuals who carry the I^A allele die accidentally before they reproduce, the number of copies of the allele drops slightly in the next generation of 20,000 people—down to about 385 out of 40,000 alleles. The chance occurrences that lead to this drop result in a new allele frequency:

frequency of I^A alleles in population =

total number of I^A alleles/total number of blood-type alleles in population

$\dfrac{385}{40,000} = 0.0096$

$= 0.96\%$

The change in frequency from 1% to 0.96% is the result of genetic drift.

A change in allele frequency of 0.04% is relatively minor. Hundreds of individuals will still carry the I^A allele. It is likely that in a subsequent generation, a few individuals carrying allele I^A will have an unusually large number of offspring, thus increasing the allele's frequency within the next generation.

Now imagine the effects of genetic drift on a small population. In a population of only 200 individuals and with an I^A frequency at 1%, only 4 of the individuals in the population carry the allele. If 2 of these individuals fail to pass it on, the frequency will drop to 0.5%. Another chance occurrence in the following generation could completely eliminate the two remaining I^A alleles from the population. Thus genetic drift occurs more rapidly in small populations and is much more likely to result in the complete loss of alleles (Figure 14.23). In most populations, alleles that are lost through genetic drift have relatively small effects on fitness at the time. However, many alleles that appear to be nearly neutral with respect to fitness in one environment may have positive fitness in another environment. When this is the case, their loss may spell disaster for the entire species.

The Consequences of Low Genetic Variability in a Population.   Populations with low levels of genetic variability have an insecure future for two reasons.

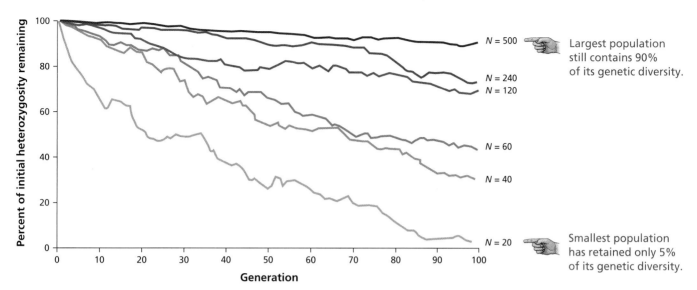

Figure 14.23 Genetic drift affects small populations more than large populations. In this graph, each line represents the average of 25 computer simulations of genetic drift for a given population size. After 100 generations, a population of 500 individuals still contains 90% of its genetic variability. In contrast, a population of 20 individuals has less than 5% of its original genetic variability.

First, when alleles are lost, the level of inbreeding depression in a population increases, which means lower reproduction and higher death rates, leading to declining populations that are susceptible to all the other problems of small populations. Second, populations with low genetic variability may be at risk of extinction because they cannot evolve in response to changes in the environment. When few alleles are available for any given gene, it is possible that no individuals in a population will possess an adaptation that allows them to survive an environmental challenge. For example, there is some evidence that individuals with type A blood are more resistant to cholera and bubonic plague than are people with type O or B blood. Therefore, possessing the I^A allele may be neutral relative to other blood-type alleles in places where these diseases are rare, but it could be an advantage where the diseases are common. If blood-type A really does protect against some infectious diseases, those individuals who carry an I^A allele will survive if a population is exposed to bubonic plague, and the population will persist. If a population has lost diversity and contains no individuals carrying the I^A allele, then it may become extinct upon exposure to bubonic plague.

As is often the case, there are some exceptions to the "rules" just described. For example, widespread hunting of northern elephant seals in the 1890s reduced the population to 20 individuals, thus probably wiping out much of the genetic variation in the species. However, elephant seal populations have rebounded to include about 150,000 individuals today. The dramatic recovery of elephant seals is apparently unusual. There are many more examples of populations that suffer because of low genetic variability. The Irish potato is perhaps the most dramatic example of this condition.

Potatoes were a staple crop of rural Irish populations until the 1850s. Although the population of Irish potatoes was high, it had remarkably low genetic variability. First, potatoes are not native to Ireland (in fact, they originated in South America), so the crop was limited to just a few varieties that were originally imported. In fact, most of the potatoes grown on the island were of one variety, called Lumper for its bumpy shape. Second, new potato plants are grown from potatoes that are produced by the previous year's plants and thus are genetically identical to their parents. This agricultural practice ensured that all of the potatoes in a given plot had identical alleles for every gene. All available evidence indicates that the genetic variability of potatoes grown in Ireland during the nineteenth century was extremely low.

(a) *Phytophthora infestans*, the protozoan that causes blight

(b) A potato plant with blight

Figure 14.24   Potato blight. (a) The organism that causes late blight in potatoes, *Phytophthora infestans*. (b) Infected potato tubers.

When the organism that causes potato blight arrived in Ireland in September 1845, nearly all of the planted potatoes became infected and rotted in the fields (Figure 14.24). The few potatoes that by chance escaped the initial infection were used to plant the following year's crops. Some varieties of potatoes in South America carry alleles that allow them to resist potato blight and escape an infestation unaffected. However, apparently very few or no Irish potatoes carried these alleles; and in 1846, the entire Irish potato crop failed. Because of this failure and another in 1848, along with the ruling British government's harsh policies that inhibited distribution of food relief, nearly 1 million Irish peasants died of starvation and disease, and another 1.5 million peasants emigrated to North America.

Irish potatoes descended from a small group of plants that were missing the allele for blight resistance, so even an enormous population of these plants could not escape the catastrophe caused by this disease. Similarly, since small populations lose genetic variability rapidly through genetic drift, preventing endangered species from declining to very small population levels may be critical for avoiding a similar genetic disaster. These historical situations support the current need to preserve adequate numbers of Lost River and shortnose suckers, even at the expense of crop production in the Klamath Basin, in order to save these species from extinction.

14.4   Protecting Biodiversity Versus Meeting Human Needs

Saving the Lost River and shortnose suckers from extinction requires protecting all of the remaining fish and restoring the habitat they need for reproduction. These actions cause economic and emotional suffering for humans who make their living in the Klamath Basin. In fact, many actions necessary to save endangered species result in immediate problems for people. As pointed out earlier by the Oregon State University student, we need to balance the costs and benefits of preserving endangered species.

A provision of the Endangered Species Act allows members of a committee to weigh the relative costs and benefits of actions taken to protect endangered species. The Endangered Species Committee, which includes the U.S. Secretaries of Agriculture and the Interior as well as the Chairman of the Council of Economic Advisors, has convened a number of times for this purpose. This so-called God Squad decides if they should overrule a federal action meant to save an endangered species in order to protect the livelihoods of people. Farmers in the Klamath Basin have advocated for a God-Squad ruling on the diversion of water from Upper Klamath Lake, but history suggests that a decision is not likely to be in their favor. The Endangered Species Committee convened only four times from 1973 through 2001, and it has granted two exemptions, one of which was essentially overturned by the subsequent presidential administration.

If the debate in the Klamath Basin follows the pattern set by other ESA controversies, a political solution that causes some economic hardship while ensuring the immediate survival of the fish will prevail. Biologists working on the problem agree that the recovery goal for shortnose and Lost River suckers is high enough to ensure short-term survival (50 years), but it is not high enough to ensure both species' long-term survival (500 years). The biologists' assessment is based on computer models predicting how the population will respond to predicted environmental changes. The recovery population size of 500 is large enough to withstand environmental catastrophes in the short term, but in the long term, continued loss of genetic variability results in the extinction of populations of only 500 fish in their models.

The risk to the long-term survival of the fish helps balance the cost to the farmers of the Klamath Basin. The short-term cost to farmers was somewhat alleviated when they received federal disaster assistance to help them adjust to the loss of lake-derived irrigation water. While recent increases in rainfall have helped provide enough water for both farmers and fish and reduced the level of conflict, the federal government continues to purchase farmland from willing sellers in the Basin in order to protect and restore the fishes' habitat. The U.S. Fish and Wildlife Service hopes that this long-term solution will help provide adequate habitat for the survival of both the shortnose and Lost River suckers.

The ESA has been a successful tool for bringing species such as the peregrine falcon, American alligator, and gray wolf back from the brink of extinction, but all of these successes have come with some cost to citizens. If the solution to these and other endangered species controversies is any guide, many Americans are willing to devote tax dollars to efforts that balance the needs of people and wildlife in order to protect our natural heritage.

As with any challenge that humans face, the best strategy for preserving biodiversity is to prevent species from becoming endangered to begin with. Table 14.2 provides a list of actions that can help preserve biodiversity. Meeting this challenge requires some creativity, but it is often possible to provide for the needs of people while preserving our natural heritage. However, it will take all of us to help keep the equation in balance.

Table 14.2 Taking action to preserve biodiversity.

Objective	Why do it?	Actions
Reduce fossil fuel use.	• Mining, drilling, and transporting fossil fuels modifies habitat and leads to pollution. • Burning fossil fuels contributes to global climate change, further degrading natural habitats.	• Buy energy-efficient vehicles and appliances. • Walk, bike, carpool, or ride the bus whenever possible. • Choose a home near school, work, or easily accessible public transportation. • Buy "clean energy" from your electric provider, if offered.
Reduce the impact of meat consumption.	• The primary cause of habitat destruction and modification is agriculture. • Modern beef, pork, and chicken production relies on grains produced on farms. One pound of beef requires 4.8 pounds of grains, or about 25 square meters of agricultural land.	• Eat one more meat-free meal per week. • Make meat a "side-dish" instead of the main course. • Purchase grass-fed or free-range meat.
Reduce pollution.	• Pollution kills organisms directly or can reduce their ability to survive and reproduce in an environment.	• Do not use pesticides. • Buy products produced without the use of pesticides. • Replace toxic cleaners with biodegradable, less harmful chemicals. • Consider the materials that make up the goods you purchase, and choose the least-polluting option. • Reuse or recycle materials instead of throwing them out.
Educate yourself and others.	• Change happens most rapidly when many individuals are working for it.	• Ask manufacturers or store owners about the environmental costs of their goods. • Talk to family and friends about the choices you make. • Write to decision makers to urge action on effective measures to reduce human population growth and curb habitat destruction and species extinction.

CHAPTER REVIEW

Summary

14.1 The Sixth Extinction

- The loss of biodiversity through species extinction is exceeding historical rates by 50 to 100 times (p. 362).

- Species-area curves help us predict how many species will become extinct due to human destruction of natural habitat (pp. 363–364).

- Species at the top of the food chain are more susceptible to extinction because less energy is available for survival at higher trophic levels (p. 366).

- Additional threats of habitat fragmentation, introduced species, overexploitation, and pollution also contribute to species extinction (pp. 367–368).

    Web Tutorial 14.1 **Tropical Deforestation and the Species Area Curve**
    Web Tutorial 14.2 **Habitat Destruction and Fragmentation**

14.2 The Consequences of Extinction

- Species are important to us as resources, either directly as consumed products or indirectly as organisms used to provide potential medicines or genetic resources (p. 369).

- Species are members of communities; their loss as mutualists, predators, competitors, and keystone species may change a community, making it less valuable or even harmful to humans (pp. 371–376).

- Species also play a role in ecosystem function, including effects on energy flow and nutrient cycling. Changes to the biological components of an ecosystem may change its non-biological properties as well (pp. 376–377).

- Biodiversity may fulfill a human need to experience natural landscapes (p. 378).

14.3 Saving Species

- If habitat protection is focused on a few well-defined biodiversity hotspots, then the number of organisms becoming extinct can be markedly reduced (p. 379).

- When species are already endangered, restoring larger populations is critical for preventing extinction (p. 380).

- Small populations are at higher risk for extinction due to environmental catastrophes (p. 381).

- Small populations are at risk when individuals have low fitness due to inbreeding and thus are less able to increase population size (p. 383).

- Genetic variability is lost in small populations because of genetic drift—the loss of alleles from a population due to chance events. Therefore, small populations may be less able to evolve in response to environmental change (p. 384).

14.4 Protecting Biodiversity Versus Meeting Human Needs

- The political process enables people to develop plans for helping endangered species recover from the brink of extinction while minimizing the negative effects of these actions on people (pp. 386–387).

Learning the Basics

1. How is the estimate of historical rates of extinction generated? What are the criticisms of these estimates?

2. Describe how habitat fragmentation endangers certain species. Which types of species do you think are most threatened by habitat fragmentation?

3. Compare and contrast the species interactions of mutualism, predation, and competition.

4. Current rates of species extinction appear to be approximately _____ historical rates of extinction.

    A. equal to; **B.** 10 times lower than; **C.** 10 times higher than; **D.** 50 to 100 times higher than; **E.** 1000 to 10,000 times higher than

5. The relationship between the size of a natural habitat and the number of species that the habitat supports is described by a(n) _____.

    A. habitat fragmentation measure; **B.** inbreeding depression matrix; **C.** species-area curve; **D.** overexploitation scale; **E.** ecosystem services cost

6. A mass extinction _____.

    A. is global in scale; **B.** affects many different groups of organisms; **C.** is caused only by human activity; **D.** a and b are correct; **E.** a, b, and c are correct

7. The web of life refers to the _____.

    A. evolutionary relationships among living organisms; **B.** connections between species in an ecosystem;

C. complicated nature of genetic variability; **D.** flow of information from parent to child; **E.** predatory effect of humans on the rest of the natural world

8. According to many scientists, the most effective way to reduce the rate of extinction is to _____.

   A. preserve habitat, especially in highly diverse areas; **B.** focus on a single species at a time; **C.** eliminate the risk of genetic drift; **D.** produce less trash by recycling more; **E.** encourage people to rely more on agricultural products and less on wild products

9. The risks faced by small populations include _____.

   A. erosion of genetic variability through genetic drift; **B.** decreased fitness of individuals as a result of inbreeding; **C.** increased risk of experiencing natural disasters; **D.** a and b are correct; **E.** a, b, and c are correct

10. One advantage of preserving more than one population and more than one location of an endangered species is _____.

    A. a lower risk of extinction of the entire species if a catastrophe strikes one location; **B.** higher levels of inbreeding in each population; **C.** higher rates of genetic drift in each population; **D.** lower numbers of heterozygotes in each population; **E.** higher rates of habitat fragmentation in the different locations

Analyzing and Applying the Basics

1. Review Figure 14.6a on page 364. The graph depicts the relationship between island size and the number of amphibian and reptile species found on an island chain in the West Indies. How many species of reptiles and amphibians would you expect to find on an island that is 15,000 square kilometers in area? Imagine that humans colonize this island and dramatically modify 10,000 square kilometers of the natural habitat. What percentage of the species that were originally found on the island would you expect to become extinct?

2. Examine the web of relationships among organisms depicted in Figure 14.10 on pages 370–371. Which of the following species pairs are likely competitors? In each case, describe what they compete for.

   A. badger, jackrabbit; **B.** bison, coyote; **C.** rattlesnake, badger; **D.** ground squirrel, deer mouse; **E.** jackrabbit, prairie dog

   How could you test your hypothesis that these animals are in competition with each other?

3. The piping plover is a small shorebird that nests on beaches in North America. The plover population in the Great Lakes is endangered and consists of only about 30 breeding pairs. Imagine that you are developing a recovery plan for the piping plover in the Great Lakes. What sort of information about the bird and the risks to its survival would help you to determine the population goal for this species as well as how to reach this goal?

Connecting the Science

1. From your perspective, which of the following reasons for preserving biodiversity is most convincing? (1) Nonhuman species have roles in ecosystems and should be preserved in order to protect the ecosystems that support humans; or (2) nonhuman species have a fundamental right to existence. Explain your choice.

2. If a child asks you the following question 20 or 30 years from now, what will be your answer, and why?

   "When it became clear that humans were causing a mass extinction, what did you do about it?"

Where Do You Live?

Climate and Biomes

What is your "biological address?"

Who are your neighbors?

It can be difficult to identify a biological address in a human-designed landscape.

How do you answer the question, "Where do you live?" Most of us would respond with a neighborhood or street address to someone from our community, a city or town name to someone from elsewhere in our state, or a state or country name to someone who lives far away. Not too many of us would give a reply such as "the Sonoran Desert" or "the boreal forest." These descriptions of the natural environment in which we live can be thought of as biological addresses. Our biological addresses include the native vegetation and the resident animals, fungi, and microbes that share, or once shared, our living space.

In this chapter, we will explore factors that help to determine the qualities of a particular biological address by guiding you to learn more about your own natural neighborhood. See if you can answer the following questions:

- Is the native vegetation of the place where you live forest, grassland, or desert?

- What are the seasonal weather changes in your area?

- Can you describe the physical characteristics of three plant species that are native to your area?

And it is sometimes difficult to determine where the resources on which we rely come from.

- What is the largest mammalian predator that lives, or once lived, in the habitat that is now your neighborhood?

- Can you describe three native bird species that breed in your region?

Is it important to know the answers to questions like these? Consider that many Americans would have a difficult time answering at least some of these questions. Human settlements around the United States are remarkably similar, and many of us have little experience with local habitats in their pre-settlement condition. Knowing the answers to these questions implies bioregional awareness, an understanding of local environmental conditions. One consequence of a general lack of bioregional awareness that you may have experienced is local summertime water shortages caused by the intense thirst of suburban lawns, especially in areas where water-loving vegetation is not native. Other costs of lack of bioregional awareness may be more severe, including the construction of homes in areas where periodic fires are common, or building on sandy coastlines, which are very unstable in their natural condition. Having a solid awareness of one's own bioregion may allow people to build human settlements that are better both for humans and the natural environment. Taking into account one's bioregion when developing human habitats can occur in many ways; for example, in the southwest desert regions of the United States, environmentally sensitive housing developers use xeriscaping, a kind of landscaping that involves native, drought-tolerant plants. Xeriscaping not only prevents the overconsumption of water but also provides a habitat for resident wildlife.

Human populations, like all natural populations, require resources and produce waste. Despite our familiarity with human settlements in the United States, surprisingly few Americans have an understanding of where our resources come from and where our waste goes. In other words, we do not have a clear picture of the ecology of the human environment. See if you can answer these questions:

- What body of water serves as the source of your tap water?

- What primary agricultural crops are produced in your area?

- How is your electricity generated?

- Where does your garbage go?

- How is the waste handled when you flush the toilet or pour liquids down the drain?

Why is knowing the answers to these questions important? Humans remain dependent on the natural world for our resources and waste disposal. An understanding of the capacity of the natural environment in our bioregion can help us design ways of living that take advantage of a region's natural gifts and respect its limits. The consequences of a lack of understanding about how our human communities fit into the surrounding biological community can include air and water pollution and the negative effects of that pollution on ourselves and our biological neighbors. In this chapter, we explore how the ecology of a bioregion intersects with the biology of human habitats.

15.1 Global and Regional Climate

Why is Buffalo, New York so much snowier than frigid Winnipeg, Manitoba? Why is Miami, Florida hot and humid while Tuscon, Arizona is hot and dry? Why does much of India experience monsoons? Why are there greater temperature differences between winter and summer in Moscow, Russia than in Dublin, Ireland? Why is the daily high temperature on tropical Pacific islands always near 80°F? The answers to all of these questions require an understanding of **climate**, the average weather of a place as measured over many years. The climate of a place comprises various measures such as average temperature, average rainfall, and the average number of severe weather events in a given time period such as a month or a year. Climate should be distinguished from **weather**, which can be thought of as the current conditions in terms of temperature, cloud cover, and **precipitation** (rain or snowfall). Put another way, the weather in a place will tell you if you have to shovel snow tomorrow morning, while the climate in a place will tell you if you even need to own a snow shovel.

The major components of climate in a geographic area are temperature, precipitation, and the variability of these two factors. As you read this section, think about the climate of your hometown and see if you can identify the global and local factors that influence it.

Temperature

You already know that temperatures are warmer at Earth's equator than at its poles. You have also probably noticed that some areas of your region are consistently cooler or warmer than other areas. The temperature in any site depends on both its location on the globe and the local environment.

Global Temperature Patterns. On a broad scale, the average temperature of any spot on the globe is determined by the amount of **solar irradiance** it receives—that is, the amount of light energy per unit area hitting that spot. Locations that receive large amounts of solar irradiance in a year have a higher average temperature than places receiving small amounts. The amount of sun energy striking Earth's surface varies because of the planet's spherical shape. Figure 15.1 illustrates two streams of solar

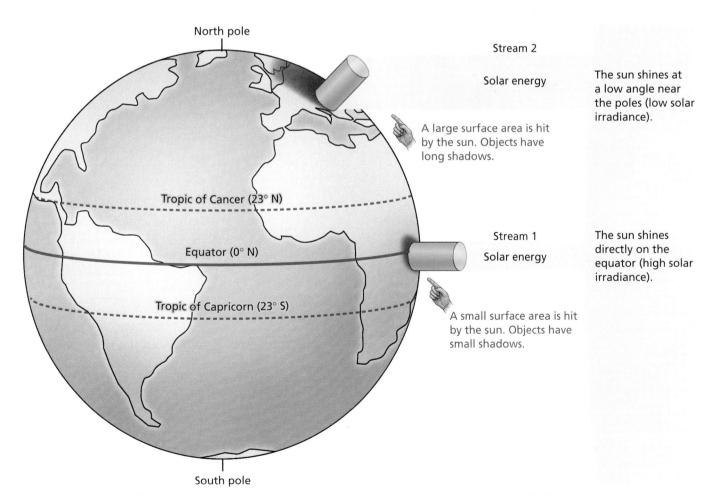

Figure 15.1 Solar irradiance on Earth's surface. The annual average temperature in a location on Earth's surface is most directly determined by its solar irradiance. Areas near the equator receive the greatest amount of solar energy (stream 1), while areas near the poles receive the least (stream 2).

energy flowing from the sun; both are the same diameter and equivalent in light energy content. Energy stream 1 strikes Earth's surface where it is facing directly toward the stream, while energy stream 2 strikes the surface where it is "curving away" from the sun. As you can see in the figure, the surface area hit by stream 1 is much smaller than the surface area hit by stream 2. In other words, the solar irradiance is greatest in areas that directly face the sun and lower in areas that curve away from the sun.

Earth's axis is roughly perpendicular to the flow of energy from the sun. The extremes of this axis are called **poles**, while the circle around the planet that is equidistant to both poles is called the **equator**. The position of the axis in relation to the sun's energy means that solar irradiance is greatest near the equator and declines gradually until it reaches its lowest levels at the poles. This is why southern Florida has a warmer climate than northern Vermont.

Solar irradiance also varies in a particular location on an annual basis. This occurs because Earth's axis is not exactly perpendicular to the sun's rays. In fact, the axis tilts approximately 23.5° from perpendicular (Figure 15.2). Due to this tilt, as Earth orbits the sun, the Northern Hemisphere (the region of the globe between the equator and the North Pole) is tilted toward the sun during the northern summer and away from the sun

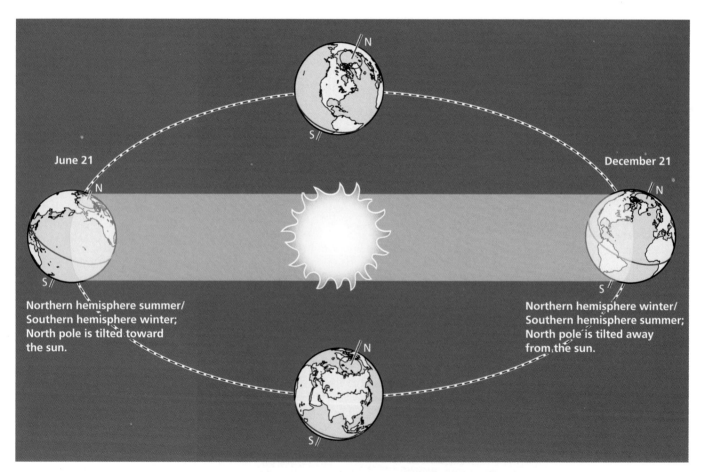

June 21

December 21

Northern hemisphere summer/
Southern hemisphere winter;
North pole is tilted toward
the sun.

Northern hemisphere winter/
Southern hemisphere summer;
North pole is tilted away
from the sun.

Figure 15.2   Earth's tilt leads to seasonal temperature patterns. Since Earth's axis is 23° from perpendicular to the sun's rays, as Earth orbits the sun during a year, the poles appear to move toward and away from the sun. Solar irradiance is increased at and near a pole when it tilts toward the sun and is decreased when that pole tilts away. The difference in solar energy causes temperatures to vary throughout the year.

Figure 15.3 The sun travels across the sky during the year.
This image is produced by taking a picture of the sun at the
same location and time once each week throughout a year.
The sun is highest in the sky at the summer solstice and lowest
at the winter solstice.

during the northern winter. The tilt also helps explain why the position of
the sunrise changes over the course of a year, moving from south to north
as winter turns to summer (Figure 15.3). Note that wintertime in the Northern Hemisphere is summertime in the Southern Hemisphere, and vice
versa. Solar irradiance is at its annual maximum in the Northern Hemisphere during the summer **solstice**, the point at which the sun reaches its
northern maximum and the North Pole is tilted closest to the sun. In
addition, Earth's tilt affects day length, such that the hours of sunlight
in the north are greatest at the summer solstice and least at the winter
solstice. For example, in Chicago, day length (from sunrise to sunset) is
approximately 15 hours at the summer solstice and 9 hours at the winter
solstice. The closer a region is to a pole, the greater the variance in day
length over the course of a year because the distance traveled by the poles
toward and away from the sun is greater than the distances traveled by
spots closer to the equator. So while day length in Chicago varies by 6 hours
from winter to summer, in Fairbanks, Alaska, it varies by 18 hours from a
low of less than 4 hours at the winter solstice to a high of nearly 22 hours at
the summer maximum.

Even though the summer solstice is the day of highest solar irradiance in
the Northern Hemisphere, it is not the warmest day of the year. Instead,
the warmest days of the year are about one month after the solstice. This
occurs because Earth stores the light energy it gains from the sun as heat and
releases it gradually into the atmosphere. You can think of the solar heat
stored in the earth as a bank account. Heat is always dissipating, much like
money being withdrawn regularly from a bank account. If money is added
to a bank account at the same rate as it is withdrawn, the balance remains
the same. Similarly, if heat is added to a location on Earth's surface at the
same rate that it is dissipating, the temperature there remains constant. As
day length increases, the amount of heat deposited is greater than the
amount withdrawn, and the bank account gets larger—that is, the temperature starts to rise. On the longest day of the year, the amount deposited is
greatest; but even after that day, the heat deposits are greater than the withdrawals, and the balance (temperature) continues to increase. Similarly, the
coldest days of the year are typically a month or so after the winter solstice.
In this case, withdrawals remain larger than deposits for several weeks after

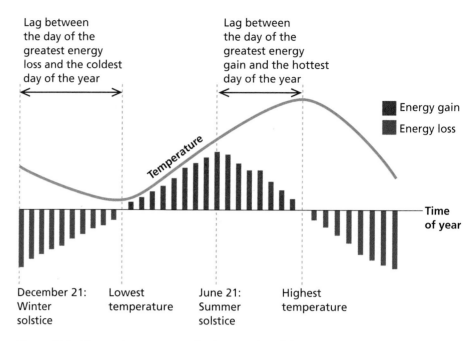

Figure 15.4 Thermal momentum. The longest day of the year—the summer solstice—is not the warmest day of the year at a geographic location because the amount of heat deposited in Earth's "bank" is larger than the amount lost for several weeks following the summer solstice.

the winter solstice (Figure 15.4). The same analogy can help explain why the warmest part of the day is in the middle to late afternoon, even though the solar irradiance is greatest at noon when the sun is directly overhead. In the first few hours of the afternoon, heat gain remains greater than heat loss, increasing temperature.

Because day length changes more dramatically near the poles than near the equator, the bank account balance also changes more dramatically closer to the poles. The amount of balance change explains why seasonal temperature swings are more pronounced closer to the poles than near the equator. The average low temperature for January in Tampa, Florida is 10°C (50°F), and the average high in July is 32°C (90°F), a difference of 22 degrees. On the contrary, the January low in Missoula, Montana is −10°C (14°F), and the July high is 29°C (84°F), a difference of 39 degrees.

Local Factors That Influence Temperature. Solar irradiance is not the only factor that determines average temperature and seasonal variation in a geographic location. Three other characteristics of a location's setting have an effect on its temperatures: (1) altitude, (2) the proximity of a large body of water, and (3) characteristics of the land's surface and vegetation.

Temperature drops as altitude—the height above sea level—increases. Temperature differences due to elevation are dramatic; the summit of Mt. Everest, 8.8 km (5.5 mi) above sea level, averages −27°C (−16°F), while nearby Kathmandu, Nepal, at 1.3 km (4385 feet), averages 18°C (65°F). However, smaller differences in elevation within a region have a converse effect on air temperature. Because cold air is denser than warm air, cold air tends to "drain" to the relatively lowest point on a landscape. Thus valleys and other low spots will often remain colder than nearby hilltops.

Temperatures in areas near oceans, seas, and large lakes are influenced by the thermal properties of water. The temperature of water rises and falls slowly in response to solar irradiation, when compared to the rate of temperature change

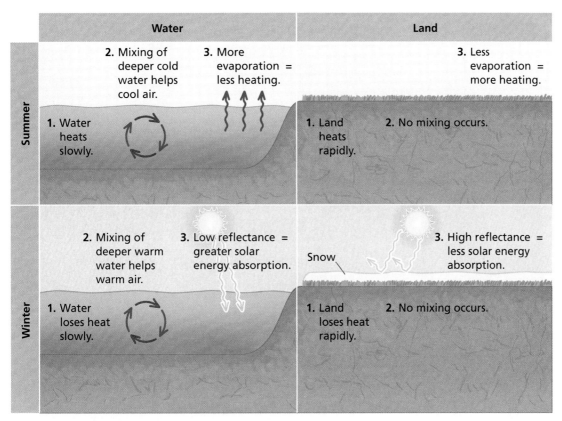

Figure 15.5   The moderating influence of water. Because water heats slowly, large bodies of water increase in temperature more slowly than land regions. Winds blowing across water in spring will cool nearby landmasses. Conversely, water loses heat more slowly than land, so it will send warmer breezes over landmasses in winter.

of land surfaces. Thus, air masses over water are cooler in summer and warmer in winter than are air masses over land (Figure 15.5). Land areas close to large bodies of water experience more moderate temperatures than do regions further inland. Because of this phenomenon, the growing season on the shores of the Niagara peninsula in Canada, bordered by Lake Erie and Lake Ontario, is as much as 30 days longer than the season is further from shore. Another reason that Tampa experiences a much less dramatic temperature range than Missoula is that it is located on the Gulf of Mexico, while Missoula is far from oceans or large lakes.

Oceans also have an effect on local climates because heat is transferred around these large masses of water via currents. Currents that run north and south can carry heat, in the form of warm water, thousands of kilometers around the globe. The warm water heats the air mass above it and thus the nearby landmasses. The Gulf Stream is a current that carries water from the tropical Atlantic Ocean to the shores of northern Europe, producing a much milder climate there than in other areas at the same distance from the equator. The warmth of the Gulf Stream makes Dublin, Ireland as warm as San Francisco, even though Dublin is 1600 km (1000 mi) closer to the North Pole.

The amount of light absorbed or reflected by the land's surface will also influence surrounding air temperature. A surface that reflects most light energy will have a lower nearby air temperature compared to a surface that absorbs most of that energy, heats up, and radiates that heat into the air. Snow reflects more light than a forest does, so air over a snowpack remains cold. The low reflectance of asphalt pavement and most building materials contributes to the urban heat island effect, which is the tendency for

Less vegetation = less energy used to evaporate water, and more energy converted to heat

Air conditioning = heat vented into outside areas

Dark surfaces = high albedo, solar energy absorbed rather than reflected.

Fossil fuel use = waste heat produced

Figure 15.6   Urban heat island. Urban areas are significantly warmer than nearby rural areas for a variety of reasons.

cities to be from 1° to 6°F warmer than the surrounding areas (Figure 15.6). In addition to the heat absorbed by streets and buildings, the increased amount of human activity influences air temperatures in urban areas. Waste heat given off as a result of gas and oil burning or vented from air conditioners can contribute up to one-third of the excess heat in city centers. Cities are also warmer because they contain little vegetation, which tends to reduce air temperature. Much of the solar energy absorbed by plants converts liquid water inside the plant to vapor, which also prevents the energy from being converted to heat.

Precipitation

As with temperature patterns, the amount of rain or snowfall experienced in any location on Earth is a product of both global and local factors.

Distribution of Precipitation on Earth's Surface.   Energy from the sun is one of the primary drivers of rain and snowfall patterns on Earth. To understand how sunlight causes rainfall, we must first understand some of the properties of water vapor. Air is composed primarily of nitrogen, oxygen, and water vapor. For water to remain as vapor, evaporation must exceed condensation; that is, the number of water molecules that randomly clump together to form liquid droplets must be less than the number of water molecules that escape from droplets. The rate of evaporation depends on temperature; at high temperatures, evaporation rate is high, and vice versa. Thus when air cools and the rate of evaporation decreases, water molecules clump into larger and larger droplets. When the droplets are large enough,

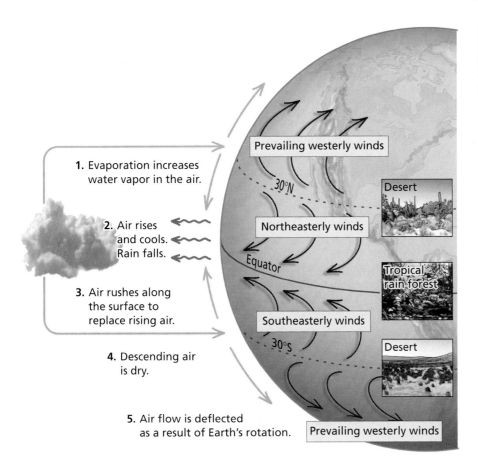

Figure 15.7 Global precipitation and wind patterns. High levels of solar irradiance at the solar equator lead to high levels of evaporation and rainfall. This phenomena drives massive movements of air near the tropics into the temperate zones.

concentrations of them can be seen as clouds. As clouds grow even larger, droplets can become heavy enough to fall as rain. If the temperature inside the cloud is cold enough, droplets will freeze into ice crystals, which may fall as snow.

Rainfall patterns in tropical regions result from the air cycling near Earth's surface to high in the atmosphere and back down, as illustrated in Figure 15.7. Where solar irradiation is highest, at or near the equator, air temperatures rise quickly during the day. Hot air is less dense than cold air. Therefore, air at the equator rises and leaves near Earth's surface an area of low air pressure that is filled by breezes blowing from north and south of this point, which is called the intertropical convergence zone. As the air rises, it cools; water vapor condenses to form clouds and falls to Earth as rain. The now-dry air flows in the upper atmosphere toward the poles, where pressure is lower because of the air flow in the lower atmosphere. The once warm, wet air mass continues to cool and release water and finally drops back to Earth's surface at about 30° north and south latitude. The falling air displaces the ground-level air at these latitudes, and the ground-level air flows toward the poles. As the air mass drops, it is affected by Earth's rotation and thus deflects to the east or west. The movements of these vast air masses create the prevailing winds in various regions of the globe. The pattern of air flow helps explain the band of rain forests near the equator, the great deserts at 30° north and south of the equator, and the tendency for weather patterns in North America to come from the west.

Global rainfall patterns exhibit seasonality as well. The area of maximum solar irradiance travels from 23°N to 23°S over the course of a year due to the 23° tilt in Earth's axis. Therefore, the intertropical convergence zone also moves, creating distinct rainy seasons wherever it is located. Rainy seasons occur in desert areas when the movement of the convergence zone results in prevailing winds that accumulate water vapor from nearby oceans and fall as rain. The monsoon seasons in India and southern Arizona are both associated with these wind shifts.

Local Precipitation Patterns.   The amount of precipitation that falls in a given land region is highly dependent on the context of that area—in particular, the land's proximity to a large body of water. Wind blowing across warm water accumulates water vapor that condenses and falls when it reaches a cooler landmass. Communities surrounding the Great Lakes provide a dramatic example of this effect. Because the prevailing winds are from the west, areas immediately to the north and west of these lakes accumulate much less snow than do regions on the southern and eastern sides, which receive winds that have crossed over the warmer lakes. For example, Toronto, on the northwest side of Lake Ontario, averages about 140 cm (55 in) of snow per year, while Syracuse, New York, on the southeast side averages almost twice that—274 cm (108 in).

Precipitation amounts are also affected by the presence of mountains or mountain ranges. When an air mass traveling horizontally approaches a mountain, it follows the landscape's contours and is forced upward. Cooling as it moves upward, the water vapor within it condenses to form clouds, rain, or snow, which falls on the windward side of the mountain. Warming again as it drops down the other side, the dry air mass causes water to evaporate from land on the sheltered or leeward side of the mountain. The dry area that results is often referred to as the mountain's "rain shadow." The Great Basin of North America, encompassing nearly all of Nevada and parts of Utah, Oregon, and California, is a desert because of the rain shadow cast by the Sierra Nevada mountains.

The temperature and rainfall patterns in a geographical region play a major part in determining not only the primary vegetation in that area but also the human activities that are most successful there.

Know Your Bioregion: Describe the climate of the place where you live, including the annual average temperature, average precipitation, average number of sunny days, and names of seasons. What global and local factors influence the climate where you live?

15.2   Terrestrial Biomes

Climate plays the greatest role in determining the physical appearance of the vegetation in a particular geographic area. Plants (and animals) native to a region have adapted, via the process of natural selection, to the water availability and temperatures experienced there. In general, the size of the vegetation is limited by water availability—large trees require large amounts of water. Water availability is obviously a function of total precipitation, but it is also influenced by temperature; frozen water cannot be taken up by plants.

Four basic land **biome** categories, or primary vegetation types, are typically recognized: forest, grassland, desert, and tundra. Each of these biome categories may contain several biome types; for instance, a grassland may be either prairie, steppe, or savanna. The relationship between climate and biome type is illustrated in Figure 15.8. Which of these biomes is your home? And how has the human population changed the environment of your biological neighborhood?

Forests and Shrublands

Forests are vegetation communities dominated by trees and other woody plants. They occupy approximately one-third of Earth's land surface, and when all forest-associated organisms are included, contain about 70% of the

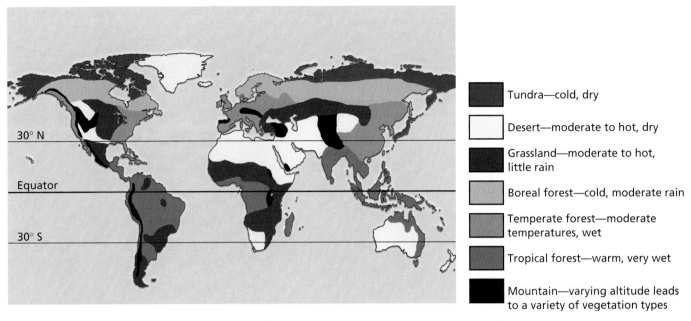

Figure 15.8   The distribution of Earth's biomes. The primary vegetation type in a given area is determined by the region's climate.

biomass, or total weight of living organisms, found on Earth's land surface. Forests are generally categorized into three groups based on their distance from the equator: (1) tropical forests at or near the equator, (2) temperate forests from 23° to 50° north and south of the equator, and (3) boreal forests close to the poles (Figure 15.9).

Tropical Forests. Extensive areas of **tropical forest** were once found throughout Earth's equatorial region—in Central and South America, central Africa, India, southeast Asia, and Indonesia. Tropical forests contain a large amount of biological diversity; one square kilometer may contain as many as 100 different tree species. One hypothesis regarding why tropical forests are so diverse is that these areas receive the largest amount of solar irradiance over the course of a year. Because the energy level is high, adequate populations of many varying species can be supported in a relatively small area. Think of the available energy as analogous to a pizza—the larger the pizza, the greater the number of individuals who can be fed to satisfaction. High energy and water levels also support the growth of very large trees. Because the trees are large and produce many leaves, most of the sunlight is absorbed

(a) Tropical rain forest

(b) Temperate forest

(c) Boreal forest

Figure 15.9   Forest biomes. (a) Tropical rain forests are highly diverse and efficient at capturing light before it reaches the ground. (b) Temperate forests contain mostly deciduous trees that drop all of their leaves annually. (c) Boreal forests are often highly uniform, made up of one or two species of coniferous trees.

Figure 15.10   Life in the canopy. Biologists estimate that 70% to 90% of species in the rain forest are found primarily in the trees. This bromeliad, often found more than 60 meters from the forest floor, is vase-shaped as an adaptation to acquire water and nutrients from rainfall rather than soil.

before it hits the ground, meaning that most living organisms in these forests are adapted to survive high in the treetops. Animals are able to fly, glide, or move freely from branch to branch, while small plants are able to obtain nutrients and water even while living on the upper branches of larger plants (Figure 15.10).

With warm temperatures, abundant water, and high energy levels, the process of **decomposition**, the breakdown of waste and dead organisms, is rapid in tropical forests. Dense vegetation quickly reabsorbs the resulting simple nutrients. As a result, few nutrients are stored in the soils of tropical forests. When vegetation is cleared and burned in preparation for agricultural crops in these areas, the ash-fertilized soils can support crop growth for only 4 to 5 years. Among human populations in tropical forest areas, this slash-and-burn (or swidden) agricultural system is common. Once soils are depleted of nutrients, the plots are abandoned, and a new field is cleared using the same method. The history of human settlement and agriculture in tropical rain forests indicates that tropical forests can indefinitely support swidden agriculture if the population is small enough that abandoned plots have several decades to recover. However, increasing human population levels in tropical countries may be too large to allow for adequate recovery before the land is needed again, and road building into areas with previously small human populations has forever changed the nature of these interior forests—from ancient and mostly undisturbed to swidden fields (Figure 15.11).

Temperate Forests.   Some areas of tropical forest, especially those further from the equator, may demonstrate seasonal changes between annual wet and dry periods. But most people associate major seasonal change with forests in temperate areas, where winter temperatures can drop well below the freezing point of water. Large areas of **temperate forest** appear in eastern North America, and only remnants of these forests remain in western and central Europe and eastern China. In temperate forests, water is abundant enough throughout the year to support the growth of large trees, but cold temperatures during the winter limit photosynthesis and freeze water in the soil. Trees with broad leaves can grow faster than narrow-leaved trees, so they have an advantage in the summer. However, broad leaves allow tremendous amounts of water to evaporate from a plant, leading to potentially fatal

Figure 15.11   Clearing for agriculture. The red dots trailing white smudges of smoke in this satellite image of the Amazonian rain forest in Brazil are active fires, set by newly arrived farmers in order to prepare the land for farming. The amount of tropical forest land converted to agriculture is greater now than at any other time in human history.

dehydration when water supplies are limited. To balance these two seasonal challenges, most trees in temperate forests have evolved a **deciduous** habit, meaning that they drop their leaves every autumn. In preparation for shedding its leaves, a deciduous tree reabsorbs their chlorophyll, the green pigment essential for photosynthesis. The colorful fall leaves that grace eastern North America in September and October result from secondary photosynthetic pigments and sugars that are left behind.

The short time lag between the onset of warm temperatures and the re-leafing of deciduous trees provides an opportunity for plants on the temperate forest floor to receive full sunlight. Spring in these forests triggers the blooming of wildflowers that flower, fruit, and produce seeds quickly before losing the competition for light and water to their towering companions (Figure 15.12). The lighter leaves of deciduous trees allow more sunlight through than do their tropical forest equivalents. Therefore, temperate forests typically have a shrub layer that may be missing in the tropics. Also, animals in temperate forests are more evenly distributed throughout the forest, not primarily concentrated in the treetops.

Since their soils are relatively easy to turn over and are rich in nutrients, nearly all of the forested lands in the eastern United States were converted to farmland within 100 years of the American Revolution. However, in the late nineteenth and early twentieth centuries, farms in the eastern United States were abandoned as inexpensive transportation networks made produce grown elsewhere cheaper for consumers and more reliable for suppliers. Now, many of these abandoned farms have grown back into forest. However, these second-growth forest sites are once again threatened—this time by expanding urban development. The World Wildlife Fund estimates that worldwide, only 5% of temperate deciduous forests remain relatively untouched by humans.

Boreal Forests.   The largest biome on Earth is the **boreal forest**, covering vast expanses of northern North America, northern Europe and Asia, and high-altitude areas in the western United States. Unlike tropical and temperate forests, which are dominated by trees from the phylum of flowering plants called angiosperms, **coniferous** trees from the phylum of plants that produce seed cones (gymnosperms) dominate boreal forests (Figure 15.13). In fact, boreal forests are among the only land areas where flowering plants are not the dominant vegetation type. Climate conditions for boreal forests include

Figure 15.12   Springtime in a temperate forest. The time between soil thawing and the production of leaves by the canopy trees is a window of opportunity for plants growing on the temperate forest floor. They are adapted to take advantage of this limited sunlight by emerging from the soil, flowering, and producing fruit rapidly.

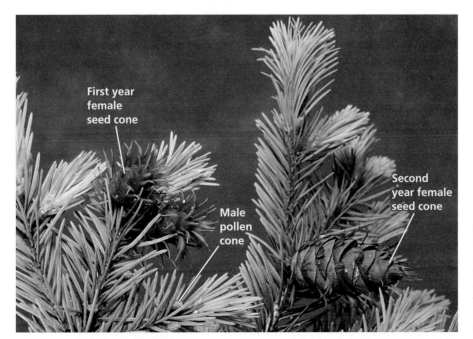

First year female seed cone

Male pollen cone

Second year female seed cone

Figure 15.13   Coniferous trees. Cone-bearing plants produce sperm and eggs on cones instead of in flowers. Sperm is packaged in pollen grains, which are typically carried by wind to a female cone that contains the egg. The fertilized egg and its surrounding tissue develops into a seed inside the cone.

Figure 15.14   Diversity in the boreal forest. Plant diversity in the boreal forest is relatively low due to difficult winter conditions, but bird diversity during summer is surprisingly high. This Blackburnian warbler is one of many bird species breeding in the boreal forest during summer, when a combination of abundant sunshine and few permanent resident birds provides enormous numbers of available insects to feed their growing offspring.

very cold, long, and often snowy winters, and short, moist summers. The dominant conifers in these forests are evergreens, meaning that like many tropical forest trees, they maintain their leaves throughout the year. However, coniferous tree leaves are needle shaped and coated with a thick, waxy coat—both adaptations to reduce water loss when water freezes in the soil. The evergreen habit likely explains conifers' dominance over flowering trees in boreal forests—the growing season is so short that an evergreen's ability to begin photosynthesizing as soon as water is available in spring gives it a great advantage over a faster-growing but slower-to-start deciduous tree.

Since they occupy regions with such climate challenges, boreal forests tend to be lightly populated by humans. These areas represent some of the "wildest" landscapes on Earth—home to moose, wolves, bobcat, beaver, and a surprising diversity of summer-resident birds (Figure 15.14). Trees in these landscapes are valuable for both building materials and paper products, and logging in the boreal forest is extensive. There are increasing concerns that the boreal forest in North America is being cut at rates faster than it can be replaced. Certainly, as logging operations push deeper into previously untouched areas, this biome is changing rapidly, putting the iconic wild species of the boreal forest at increased risk of becoming endangered or extinct.

Chaparral.   One major biome is dominated by woody plants but is not a forest. This landscape is known as **chaparral**, and its vegetation consists mostly of spiny evergreen shrubs (Figure 15.15). Chaparral is found extensively in areas surrounding the Mediterranean Sea and in smaller patches in southern California, South Africa, and southwest Australia. Long, dry summers and frequent fires maintain the shrubby nature of chaparral. In fact, chaparral vegetation is uniquely adapted to fire. Natural selection in this climate favored species that respond well to fire (Essay 15.1). Several species have seeds that will germinate only after experiencing high temperatures. Many **perennial**, or long-lived, chaparral plants have extensive root systems that quickly resprout after aboveground parts are damaged. Chaparral will grow into temperate forest when fire is suppressed. As a result, natural selection has favored chaparral shrubs that actually encourage fire—such as rosemary, oregano, and thyme, which contain fragrant oils that are also highly flammable.

Fire can be an important contributor to the structure and function of many biomes. In southern California, the flammability of chaparral vegetation has come directly into conflict with rapid population growth and urbanization. In fall 2003, this area experienced its most extensive and expensive fire season ever as 750,000 acres of shrubland and surrounding forest burned—consuming over 3200 buildings and killing 17 people in San Diego County alone. In response to this disaster, recommendations by a state task force have called for a policy of putting wildfire protection ahead of protection of wildlife and wildlife habitat.

Figure 15.15   Chaparral. Much of the scrubby landscape of southern California is made up of this biome type.

Essay 15.1 Wildfire!

Since 1944, Smokey Bear's message has been the same: "Only YOU can prevent forest fires." At that time in history, forest fires seemed only to be destructive, sinister forces that could cause widespread death and destruction of people, buildings, and wildlife. In fact, in 1905, the U.S. Forest Service's official policy became total fire suppression—all wild land fires would attempt to be snuffed out, and prevention, as symbolized by Smokey, became everyone's business.

Unfortunately, Forest Service biologists of the time had little understanding of the role of periodic fire in maintaining healthy biological communities in various regions of the United States. Six vegetative communities in North America are fire-adapted, meaning that the species they contain have evolved strategies for surviving periodic fires. In fact, in the absence of fire, these communities may change dramatically. Fire-dependent systems include chaparral, prairies, and pine-dominated habitats in the Great Lakes, Southeast, Northwest, and Rocky Mountains.

Adaptations to fire include thick bark (Ponderosa Pine bark is 4 inches thick at the soil's surface), the ability to quickly resprout from extensive underground roots and stems (as in the case of prairie grasses and fireweed), cones that require intense heat in order to release seeds (such as Jack Pine), and seeds that require exposure to smoke in order to germinate (several flowering herbs in the western United States, for example). In fire-adapted systems, periodic fires help to increase overall plant growth by converting downed woody material into ash that fertilizes the soil and also serves to increase diversity by creating patches of different ages within larger regions.

After a fire, communities go through **succession**, the progressive replacement of different suites of species over time. Generally, the first colonizers of vacant habitat are fast-growing species that produce abundant, easily dispersed seeds. These pioneers are replaced by plants that grow more slowly but are better competitors for light and nutrients. Eventually, a habitat patch is dominated by a set of species—called the **climax community**—that cannot be displaced by other species without another environmental disturbance. For instance, in boreal forests, the fireweed and lupine that dominate immediately after the fire are replaced by deciduous aspen and birch forest, which is then replaced by coniferous spruce and fir forest (Figure E15.1). Without periodic fires, the boreal forest system would consist solely of the climax community and thus be significantly less diverse.

The century-long policy of fire suppression has changed fire-adapted ecosystems dramatically. Now many forests contain large amounts of "fuel" in the form of dead vegetation, and fires that start in these forests quickly become intense conflagrations that can cover thousands of acres, killing trees and other organisms that are adapted to smaller, cooler fires. The Forest Service now struggles with the consequences of their policy of fire suppression, not only fighting more dangerous, destructive fires, but also fighting the perception that they fostered for many years—that forest fires are only bad.

(a) 1 year

(b) 5 years

(c) 25 years

Figure E15.1   Succession after fire in a boreal forest.

Because of this policy, and in combination with the long-term human modification of chaparral in the Mediterranean basin, this unique biome is one of the most threatened on the planet.

Grasslands

Grasslands are regions dominated by nonwoody grasses; they contain few or no shrubs or trees. These biomes occupy geographic regions where precipitation is too limited to support woody plants. Grasslands can be further categorized into tropical and temperate categories.

Figure 15.16 Savanna. Grasslands in tropical areas can support huge herds of grass-eating mammals—and their associated predators.

Tropical grasslands are known as **savannas** and are characterized by the presence of scattered individual trees (Figure 15.16). Savanna covers about half of the African continent as well as large areas of India, South America, and Australia. Savannas are maintained by periodic fires or clearing. In regions where wet and dry seasons are distinct, yearly grass fires during the dry season kill off woody plants; in regions where elephants and other large grazing animals are present, the damage these animals do to trees helps to maintain the grass expanse (Figure 15.17). Because grazing animals eat the tops of plants, natural selection in these environments has favored plants, like grasses, that keep their growing tip at or below ground level. Because grass grows from its base, grazing (or mowing) is equivalent to a haircut—trimming back but not destroying. Woody plants grow from the tips, so intense grazing can destroy these plants.

Grass leaves are very fibrous and thus difficult to digest. Grazers that could obtain more nutrients from these plants were favored by natural selection in grassland environments. The most common adaptation is a relationship with bacteria that live in a specialized chamber of the grazer's digestive system and partially digest the grass so that more nutrients are available to the animal. Although grazing mammals are characteristic of savannas, in some savanna areas where domesticated grazers such as cattle have been introduced, the landscape has been transformed from grassland to bare, sandy soil. In the Sahel region of central Africa, hundreds of acres of savanna are lost each year due to overgrazing—that is, allowing more livestock on the grassland than can be sustained.

Temperate grassland biomes include tallgrass **prairies** and shortgrass **steppes**. Generally, the height of the vegetation corresponds to the precipitation— greater precipitation can support taller grasses. Prairies and steppes are found in central North America, central Asia, parts of Australia, and southern South America. These landscapes are generally flat to slightly rolling and contain no trees. In the cooler temperate regions, decomposition is relatively slow, and the soil of prairies and steppes is rich with the partially decayed roots of grasses and small nonwoody plants (Figure 15.18). These soils provide an excellent base for agriculture, and most native prairies and steppes have been plowed and replanted to crops. In North America, less than 1% of native prairie remains. Where precipitation is too low to support plant crops, cattle may be grazed or groundwater pumped to irrigate "dry land" crops such as wheat.

Figure 15.17 Savanna maintenance. Some of the grasslands in Africa are maintained by the enormous appetites and destructive power of elephants. Elephants severely prune acacia trees when feeding on them, keeping the trees' size relatively small; they actively destroy thorny myrrh bushes that they do not eat, which provides additional habitat for grass to grow.

Desert

Where rainfall is less than 50 centimeters (20 inches) per year, the biome is called **desert**. This biome can be found throughout the world, but the world's great deserts include the Sahara in northern Africa, the Gobi of central Asia, and the deserts of the Middle East, central Australia, and the southwestern United States. Most of these deserts are close to 30° north or 30° south of the equator, where the air masses that were "wrung" of water in the rain-forest regions around the equator fall back to Earth's surface as hot and dry winds. Although the image of a desert is often hot and sandy, some deserts can be quite cold; most deserts have vegetation, although it can be sparse.

Plants and animals in desert regions have evolutionary adaptations to retain and conserve water. For example, plants are often thickly coated with waxes to reduce evaporation, contain photosynthetic adaptations that reduce water loss through leaf pores, and may be protected from predators by spines and poisonous compounds (Figure 15.19). Deserts are also home to many fast-growing **annual** herbs—flowering plants that complete their entire life cycle from seed to seed in a single season. In deserts, the wet season, the time during which a shallow-rooted annual plant can survive, is quite short; many desert-adapted, annual plants can germinate, flower, and produce seeds in a matter of 2 or 3 weeks. The seeds they produce are hardy and adapted to survive in the hot, dry soil for many years until the correct rain conditions return. Some animals in these dry environments have physiological adaptations that allow them to survive with little water intake. The most amazing of these animals are the various species of kangaroo rat, which apparently never consume water directly. Instead, they conserve water produced during the chemical reactions of metabolism and have kidneys that produce urine 4 times more concentrated than our own.

The sunny, warm, and dry climate of the deserts in Arizona and New Mexico is appealing to people; this region has the greatest rate of human population growth in the United States. The increasing population is putting stress on water supplies in these dry states, threatening the ability to support the water needs of cities and agricultural production as well as depleting water sources for native animals. The Colorado River is so extensively used that often it cannot sustain a flow through its historic outlet at the Gulf of California. Recent surveys indicate that of the species once found in the delta region of the Colorado River, which a century ago was one of the most biologically productive areas in North America, up to 95% are now gone.

Figure 15.18   A prairie in bloom. Although the dominant plants on prairies are grasses, some of the less numerous plants produce large, colorful flowers. This is a midwestern prairie in late summer.

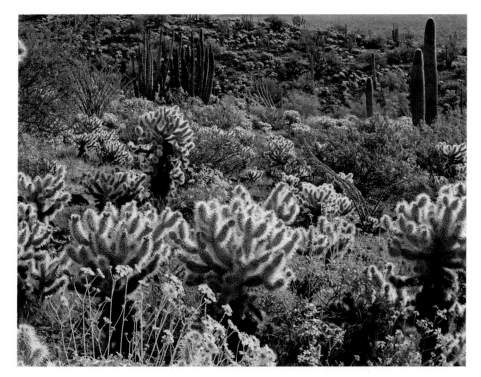

Figure 15.19   Adaptations to constant drought. Desert plants have several strategies for conserving water: thick, whitish, and waxy coatings to reflect light and reduce evaporation; a column-like form to reduce exposure to the high-intensity sun during the middle of the day; the ability to store water in their stems; and the modification of leaves to spines that discourage predators from accessing their water or damaging their protective surface.

Figure 15.20 Tundra. A marmot feeds on lush green tundra vegetation in Alaska. The short growing season and the year-round chance of below-freezing temperatures limit the vegetation of the tundra to ground-hugging plants.

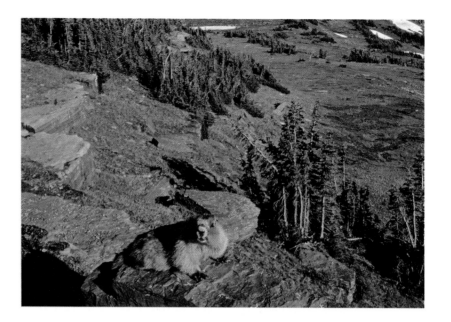

Tundra

The biome type where temperatures are coldest—close to Earth's poles and at high altitudes—is known as **tundra** (Figure 15.20). Here plant growth can be sustained for only 50 to 60 days during the year, when temperatures are high enough to melt ice in the soil. High temperatures are not sustained long enough to melt all of the soil's ice. Therefore, places like the arctic tundra near the North Pole are underlaid by **permafrost**, icy blocks of gravel and finer soil material. The permafrost layer impedes water drainage, and soils above permafrost are often boggy and saturated. Plants in tundra regions are low to the ground and adapted to windswept expanses and freezing temperatures. For example, tundra plants often grow in multispecies "cushions" where all individuals are the same height, obtaining shelter from and providing shelter for each other. Surprisingly, this low vegetation supports a large and diverse community of grazing mammals, such as caribou and musk oxen, and their predators, such as wolves. Animals in tundra regions have evolved to survive long winters with adaptations, such as storing fat and producing extra fur or feathers. Other animals, such as ground squirrels, have adapted by evolving a behavior called hibernation; they enter a sleeplike stage during which their metabolism is extremely low in order to maximize energy conservation. Grizzly bears and female polar bears also spend much of the coldest months in a deep sleep; although this is not true hibernation, these bears are so lethargic that females give birth in this state without fully awakening. Other animals survive by migrating south to avoid the hardships of a long, frigid winter. Hundreds of millions of birds, migrating from the arctic tundra or other polar regions toward the equator and back again, take part in one of the great annual dramas of life on Earth (Figure 15.21).

Tundra is very lightly settled by humans, but it is threatened by our dependence on fossil fuels—oil, natural gas, and coal that formed from the remains of ancient plants. Some of the largest remaining untapped oil deposits are found in tundra regions, including northern Alaska, Canada, and Siberia. However, use of fossil fuels appears to be causing Earth to warm (see Chapter 4), and global warming has been greatest at the poles, where tundra is predominant. Winters in Alaska have warmed by 2° to 3°C (4°–6°F), whereas elsewhere they have warmed by about 1°C. As climate conditions change, areas that were once tundra have begun to support shrub and tree growth and are changing into boreal forest.

Know Your Bioregion: Describe the native vegetation of the place where you live. How much native vegetation remains? What ecological factors (climate, fire, etc.) have influenced the native vegetation type?

Figure 15.21   Migration. Birds that breed in the Arctic, such as these red knots massed on the beach at Delaware Bay undertake a fatiguing and perilous journey in order to take advantage of the abundant daylight and insect life available during summer on the tundra.

15.3   Aquatic Biomes

Nearly all human beings live on Earth's land surface; but most people also live near a major body of water and are both influenced by, and have an influence on, these **aquatic** systems. Aquatic biomes are typically classified as either freshwater or saltwater. As you read this section of the chapter, consider the water bodies that affect you most and how you affect them.

Freshwater

Freshwater is characterized as having a low concentration of salts—typically less than 1% of total volume. Scientists usually describe three types of freshwater biomes: lakes and ponds, rivers and streams, and wetlands.

Lakes and Ponds.   Bodies of water that are inland, meaning surrounded by land surface, are known as **lakes** or **ponds**. Typically, ponds are smaller than lakes, although there is no set guideline for the amount of surface area required for a body of water to reach "lake" status. Some ponds, however, are small enough that they dry up seasonally. In temperate regions, these are called vernal ponds, from the Latin word *vernalis*, meaning "of the springtime." Vernal ponds are often crucial to the reproductive success of frogs and salamanders as well as a variety of insects that spend part of their lives in water.

The aquatic environment of lakes and ponds can be divided into different zones: the surface and shore areas, which are typically warmer, brighter, and thus full of living organisms; and the deepwater areas, which are dark, low in oxygen, cold, and home to mostly decomposers. The biological productivity of lakes in temperate areas is increased by seasonal turnovers—times of the year when changes in air temperature and steady winds lead to water mixing,

(a) Lake in summer

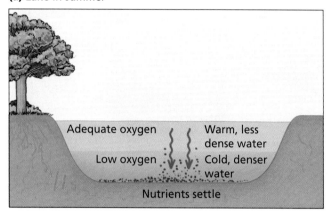

(b) Lake in fall and spring

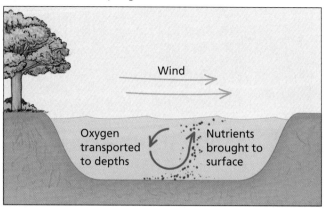

(c) Algal bloom in lake

Figure 15.22   Nutrients in lakes. (a) A lake is stable during the summer, and oxygen levels in deeper regions diminish as nutrient levels near the surface decline. (b) Nutrients and oxygen are mixed throughout a lake during fall and spring "turnover," providing the raw materials that enable photosynthetic production. (c) Large inputs of nutrients during summer may occur when nearby agricultural and landscaping activity fertilizes a lake. High nutrient levels lead to "blooms" of algae growth, severe oxygen depletion, and a risk of fish kills.

redistributing nutrients that had settled on the bottom of the lake and bringing fresh oxygen to deeper water (Figure 15.22).

Nutrients applied to agricultural lands and residential lawns near lakes and ponds can also increase their algae populations as the nutrients leach into the water; ironically, too many nutrients added in this way can lead to the "death" of these water bodies. Their degradation occurs because large populations of algae (and the microorganisms that feed on them) can lead to low oxygen levels and thus the death of fish.

Rivers and Streams.   Rivers and **streams** are flowing water moving in one direction. These waterways can be divided into zones along their lengths. At the headwaters of a river—often a lake, an underground spring, or near a melting snowpack—the water is clear, cold, and fast-flowing; it is thus high in oxygen, providing an ideal habitat for cold-water fish such as trout (Figure 15.23a). Near the middle reaches of a river, the width typically increases. Also increasing is the diversity of fish, reptile, amphibian, and insect species that the river supports because the warming water provides a better habitat for photosynthetic plants and algae. At the mouth of a river, where it flows into another body of water, the speed of water flow is often slower, and the amount of sediment—soil and other particulates carried in the water—is high. High levels of sediment reduce the amount of light in the water and therefore the diversity of photosynthesizers that survive there. Oxygen levels are also typically lower near the mouth since the activities of decomposers increase relative to photosynthesis. Many of the fish

(a) Headwaters

(b) River mouth

Figure 15.23   River habitats. (a) The Colorado River is fed by numerous headwater streams that originate in the mountain snowpacks of Colorado and Wyoming. The river here is cold, fast flowing, and rich in oxygen. (b) With sluggish water flow, warmer temperatures, and high nutrient levels, the mouth of a river is a very different habitat.

found at the mouth of a river are bottom-feeders such as carp and catfish, which eat the dead organic matter that flowing water has picked up along its way (Figure 15.23b).

Rivers and streams are threatened by the same pollutants that damage lakes. But their habitats also face wholesale destruction with the development of dams and channels—dams to provide hydropower or reservoirs for cooling fossil-fuel-powered, electricity-generating plants; and channels to simplify and expedite boat traffic.

Wetlands. Areas of standing water that support emergent, or above-water, aquatic plants are called **wetlands** (Figure 15.24). In numbers of species supported, wetlands are comparable to tropical rain forests. The high biological productivity of wetlands results from the high nutrient levels found at these interfaces between the aquatic and land environments. Besides their importance as biological factories, wetlands provide health and safety benefits by slowing down the flow of water. Slower water flows reduce the likelihood of flooding and allow sediments and pollutants to settle before the water flows into lakes or rivers.

Since the European settlement of the continental United States, over 50% of wetlands have been filled, drained, or otherwise degraded. Although extensive efforts by environmental organizations over the past 25 years have led to legislation that has greatly slowed the rate of wetland loss in the United States, about 58,000 acres are still destroyed every year.

Saltwater

About 75% of Earth's surface is covered with saltwater, or **marine**, biomes. Marine biomes can be categorized into three types: oceans, coral reefs, and estuaries.

Oceans. Like lakes, **oceans** can be divided into distinct zones. Areas near the surface with abundant light support the largest number of phytoplankton—microscopic algae suspended in the water—and thus provide food for most organisms that inhabit the ocean. Deeper, darker areas are colder, have less oxygen, and contain mostly decomposers, which filter dead organic matter that drops from the surface.

Unlike lakes, oceans experience tides—regular fluctuations in water level caused by the gravitational pull of the moon on these large bodies of water. As a result of tides, oceans contain unique habitats known as **intertidal zones**, which are underwater during high tide and exposed to air during low tide. Organisms

Figure 15.24   Wetlands. This dragonfly rests on a cattail emerging from a lakeside marsh. Wetlands provide an important habitat for a wide variety of creatures.

Figure 15.25   Intertidal zones. Tatoosh Island off the coast of Washington state is the site of the longest-running ecological research program in the United States. Robert Paine, the scientist who began studying this area in 1960, has used this site as a natural laboratory to study ecological interactions among the organisms living in the island's tidal pools.

in intertidal zones must be able to survive the daily fluctuations and rough wave action they experience. Adaptations such as strong anchoring structures, burrow building, and water-retaining, gelatinous outer coatings allow animals and seaweeds to take advantage of the high-nutrient environment found along the shore. Because they experience such dramatic environmental changes and are so rich in nutrients, intertidal zones have proven to be excellent habitats for studies of important ecological interactions (Figure 15.25).

Oceans also contain a habitat known as the abyssal plain; in very deep water where sunlight never penetrates, temperatures can be quite cold, and the weight of the water above creates enormous pressure. Once thought to be lifeless because of these conditions, the abyssal plain is surprisingly rich in life, supported primarily by the nutrients that rain down from the upper layers of the ocean. Animals living in these dark recesses are often blind, with highly developed senses of touch that can detect the tiniest movements of potential food items. In this case, natural selection favored animals that had devoted more of their nervous system to senses that are useful in the dark, relative to animals that still committed their resources to producing eyes. In the 1970s, researchers studying the deep ocean discovered an entire ecosystem supported by bacteria that use hydrogen sulfide escaping from underwater volcanic vents as an energy source. Animals in this ecosystem either use the bacteria as a food source directly or have evolved a mutualistic relationship with them, providing living space for the bacteria while benefiting from the bacteria's metabolism. The abyssal plain represents the last major unexplored frontier on Earth.

Photosynthetic organisms in the open oceans function as Earth's lungs, taking in carbon dioxide and releasing enormous amounts of oxygen (the opposite of our lungs, which take in this oxygen and release carbon dioxide). In fact, about 50% of the oxygen in Earth's atmosphere is generated by single-celled photosynthetic plankton in the oceans. Photosynthetic plankton serve as the base of a food chain, providing energy to microscopic animals called zooplankton, which in turn feed fish, sea turtles, and even large marine mammals such as blue whales. These predators provide a source of food for yet another group of predators, including sharks and other predatory fish, as well as ocean-dwelling birds such as albatross. Oceans also generate most of Earth's freshwater. The water molecules evaporating from their surface condense and fall on adjacent landmasses as rain and snow.

Coral Reefs.   Coral reefs are unique biomes in that the structure of the habitat is composed of the skeletons of the dominant organism in the habitat: coral animals. Coral animals are very simple in structure but have a unique lifestyle—they filter dead organic material from the water and harbor photosynthetic algae inside their bodies. Up to 90% of a coral's nutrition is provided by the algae, which receives as its benefit a protected site for photosynthesis and easy access to nutrients and carbon dioxide from the coral animal. Reef-building coral live in large colonies, and each individual coral secretes a limestone skeleton that protects it from other animals and from wave action (Figure 15.26). Coral reefs are made up of the accumulations of billions of these skeletons, as well as the remains of other organisms that use the reef as a home.

Coral reefs are found throughout the tropics, in warm and well-lit water, providing ample resources for abundant plankton and algae growth. Their complex structure and high biological productivity make them the most diverse aquatic habitats, rivaling terrestrial tropical rain forests in number of species per unit area. Coral reefs are sensitive to environmental conditions and prone to "bleaching," which occurs when host coral animals expel their algae companions. Bleaching can occur for various reasons, but recent episodes seem to be associated with high ocean temperatures. Although coral can recover from bleaching, increased global temperatures due to global warming (see Chapter 4) may lead to more frequent bleaching and the death of especially sensitive reef systems.

Estuaries.   The zone where freshwater rivers drain into salty oceans is known as an **estuary** (Figure 15.27). The mixing of fresh and salt water that occurs in

Figure 15.26   Coral reefs. These coral animals have their filtering appendages extended and are straining small pieces of organic matter from the water. Most of their calories are provided from the algae that live in their bodies and give the animals in this picture their orange color.

estuaries, combined with water-level fluctuations produced by tides, creates a unique habitat that is extremely productive, just as in freshwater wetlands. Estuaries provide a habitat for up to 75% of commercial fish populations and 80% to 90% of recreational fish populations; they are sometimes called the "nurseries of the sea." Estuaries are also rich sources of shellfish—crabs, lobsters, and clams. Vegetation surrounding estuaries, including extensive salt marshes consisting of wetland plants that can withstand the elevated saltiness compared to freshwater, provides a buffer zone that stabilizes a shoreline and prevents erosion. Some familiar and economically important estuaries are Tampa Bay, Puget Sound, and Chesapeake Bay.

Estuaries benefit people economically by providing shellfish and ocean fish production, protection from erosion, and recreational resources. Unfortunately, estuaries are threatened by human activity as well, including eutrophication from increasing fertilizer pollution and outright loss as a result of housing and resort development.

As is clear from our discussion of both aquatic and terrestrial biomes, no habitat on Earth has escaped the effects of humans. Consequently, in order to truly understand our biological addresses, we must learn how human populations

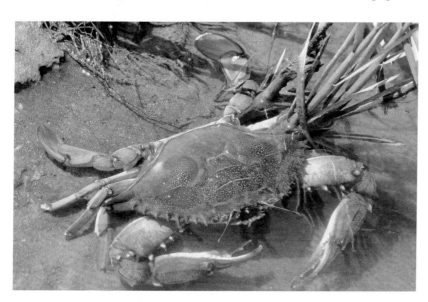

Figure 15.27   Estuary. The estuary in Chesapeake Bay is the largest in North America; it supports an enormous number of organisms, some with significant commercial value, such as the blue crab.

use the environment. Preserving our biological neighborhoods requires being able to meet human needs and at the same time respecting the needs of other species with which we coexist.

Know Your Bioregion: What are the aquatic habitats nearest to you? Can you name some of the dominant species in these habitats? What threats do these habitats face in your area?

15.4 Human Habitats

According to the United Nations Food and Agriculture Organization, humans have modified 50% of Earth's land surface for our own use. Most of this modification has resulted from agriculture and forestry, but a surprisingly large amount—2% to 3% of Earth's land surface, or 3 to 4.5 million square kilometers (larger than the area of India)—has been modified for human habitation (Figure 15.28). And this is only direct conversion; a human settlement has environmental effects far beyond its geographic boundaries, including the conversion of natural landscapes to agriculture and the changes to Earth's atmosphere, which receives some of our waste. In fact, because our activities have changed the very atmosphere of Earth, it is fair to say that no natural habitat has escaped human-caused alteration.

At the beginning of the nineteenth century, 98% of the human population lived in rural areas, where they were dependent on agriculture or harvesting natural resources for survival. As the Industrial Revolution accelerated and expanded, that pattern began to change. By 1950, one-third of the 2.5-billion human population lived in cities. Today, half of the human population of over 6 billion lives in cities. If current trends continue—which appears likely—by the year 2050, two-thirds of the population, nearly 6 billion people, will live in urban areas. Clearly, cities are going to become increasingly important in determining the human impact on native biomes.

Energy and Natural Resources

Consider the requirements of a forest. The only energy required for its growth is the solar irradiance striking the area, and the only nutrients available to support this growth are what is already present in the soil and slowly being increased by the activities of bacteria. In addition, nearly all of the "waste" produced by the organisms in a forest biome is processed on site; that is, it becomes part of the soil or air and is recycled endlessly. Thus forests are limited by the solar energy they receive and are enmeshed in material cycles that tend to preserve the nutrients needed for growth in a given area. In contrast, cities rely extensively on energy

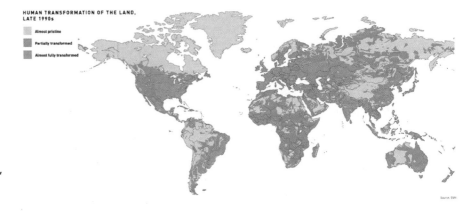

Figure 15.28 Human modification of Earth's land surface. About half of Earth's land surface remains relatively untouched by humans; most of the pristine land lies in the vast boreal forest covering much of Canada and Siberia, the great deserts of Africa and Australia, and the dense Amazonian rain forest in north and central South America.

imported from elsewhere and tend to be the central link in a linear flow of materials—from natural or agricultural landscapes, through the cities, and into waste disposed of elsewhere. Cities are very different from most typical biomes.

Energy Use.   In more developed countries, much of the energy required to power the activities within cities—from running buses and heating and cooling buildings to lighting traffic signals and energizing wireless phones—is derived from fossil fuels. These fuels, extracted from wells and mines around the globe, include the region surrounding the Persian Gulf, source of much of the world's oil, and the Appalachian Mountains, a rich source of coal. For most people in developed countries, the energy we use is not associated with the bioregion in which we live. The environmental impacts of acquiring and transporting fossil fuel can be substantial, ranging from the degradation of oceans and estuaries resulting from oil spills to the wholesale dismantling of forested mountains during coal mining (Figure 15.29).

In less developed countries, fossil fuel use is still relatively small, and energy sources are more directly tied to the surrounding natural environment. A primary source of energy for heating and cooking in these countries is wood and other plant-based materials (including the dung of plant-eating animals). As urban areas in less developed countries grow, surrounding forests are stripped of trees, sometimes at a faster rate than they can be replaced. In both less developed and more developed countries, an awareness of cleaner sources of energy within a given bioregion—such as abundant solar or wind resources—could help to reduce the environmental costs of energy consumption in the region's urban areas.

Know Your Bioregion: What is the primary source for the electricity you use in your home? Is your bioregion rich in any less environmentally damaging energy resources?

Natural Resources.   In addition to energy, human settlements require materials for survival—food from agricultural production and harvesting of natural resources, metals and salts from mines, fresh water for human and industrial consumption, petroleum for asphalt and manufactured goods, and trees for processing into paper and packaging. Developing and extracting all these raw materials changes biomes far from a city's center. The amount of material needed to support a city can be tremendous. A year 2000 estimate calculated that the **ecological footprint** of the city of London—that is, the amount of land needed to support the human activity there—was 293 times the actual size of the city, equal to twice the entire land surface of the United Kingdom. Clearly, the resources to support London must come from other countries around the world—from the boreal forests of Russia and Scandinavia that supply wood for building and paper production to the tropical forests of India and China that supply Londoners' daily spot of tea. There is no reason to believe that London is exceptional; most cities in more developed countries likely have similarly-sized footprints. Cities' footprints can be reduced by their citizens, especially if they reduce the amount of energy used for transportation and make sensitive consumer choices, including buying locally produced foods and less meat.

Know Your Bioregion: What agricultural products are produced in your bioregion? How easy is it to buy local produce? What other natural resources does your bioregion supply?

Waste Production

The ecological footprint calculated for London includes not only the land needed to provide resources but also that required to handle the city's waste. The waste produced by a city presents a special problem. The density of the human population is so great in cities that specialized systems are required to handle the enormous amounts of human waste generated in a small area.

Figure 15.29   The cost of fossil fuel extraction. Much of the remaining coal in West Virginia is found hundreds of feet below the surface of steep mountains. In the 1970s, mining companies developed a technique known as "mountaintop removal"—using explosives to blast off the rock above the coal deposit and dumping the waste rock into nearby valleys—to access this coal. Since 1981, more than 500 square miles of West Virginia have been destroyed by this method of mining.

Figure 15.30   Treatment of human waste. In more developed countries, human waste is treated in expensive and technologically advanced sewage treatment plants. Water entering the plant is first treated with oxygen to increase decomposition activity, sent to settling ponds to remove solids, and treated with chlorine or another disinfectant before being released into a nearby waterway. Solids are compacted and either sent to a nearby landfill or composted at high temperatures to produce a soil fertilizer.

Human wastes can be liquid, in the form of wastewater; solid, in the form of garbage; or gaseous, in the form of air pollution.

Wastewater.   In developed countries, cities have sewage treatment systems to handle the **wastewater** that drains from sinks, tubs, toilets, and industrial plants (Figure 15.30). These systems typically treat water by removing semisolid wastes through settling, using chemical treatment to kill any disease-causing microorganisms, and eventually discharging the treated water to lakes, streams, or oceans. Unfortunately, older treatment systems can be overwhelmed by storm water, causing the discharge of large volumes of wastewater directly into waterways. Untreated wastewater can cause nutrient levels to spike and algae and bacteria growth to increase, thus leading to beaches being closed to swimming and other water-based recreation. Huntington Beach, one of the most famous surf spots in California, was closed for nearly the entire summer of 1999 because of untreated wastewater flowing into the ocean there.

Semisolid wastes, called sludge, are often composted (that is, allowed to decompose via the action of bacteria and fungi) and applied to land as fertilizer or trucked to landfills for burial. Land application of sewage sludge has its own problems; this material contains not only human waste—a valuable fertilizer if properly composted—but also industrial wastewater, which can contain a wide variety of toxic chemicals. A major challenge of the wastewater treatment industry in developed countries is the safe and effective disposal of sewage sludge.

In less developed countries, safe disposal of treated sewage from rapidly growing cities may be a distant dream. In many rapidly industrialized regions, the emigration of people to urban areas has overwhelmed antiquated and inadequate sewer systems. Large numbers of people in these cities live in slums where running water is scarce, and untreated human waste flows in open gutters. The consequences of inadequate disposal of human waste can be severe—intestinal diseases due to contact with waste-contaminated drinking water cause the preventable deaths of over 2 million children under 5 years old every year.

Know Your Bioregion: Where does your wastewater go? Does your community have any problems handling wastewater and sewage?

Figure 15.31   Disposal of garbage. Solid waste contained in a sanitary landfill is locked within a thick liner and surrounded by a drainage system that collects all water releasing from or leaching through the waste. Garbage is compacted and "capped" with soil or other material as soon as it is dumped. Much of the waste sent to sanitary landfills in the United States could be recycled or composted.

Garbage and Recycling.   In addition to dealing with wastewater, people in urban areas also must find ways to control their **solid waste**—that is, their garbage. In more developed countries, most of the solid waste finds its way into sanitary landfills, which are pits lined with resistant material such as plastic. Landfills have systems for collecting liquid that drains through the waste, and exhaust pipes to vent dangerous gases that may build up as a result of decomposing waste (Figure 15.31). As nearby communities fight the development of new landfills, and older ones reach their maximum capacity, landfill space is becoming more and more limited and farther removed from cities. To stave off a looming "garbage crisis," many states and communities mandate high levels of recycling of paper, glass, and metal; and many have or are considering community-wide composting programs to reduce the amount of food and yard waste trucked to landfills. Unfortunately, even in cities and states where recycling rates are relatively high, household garbage production continues to increase. In less developed countries, the problem of solid waste disposal is more severe. Many large cities in these areas have large, open dumps in which unstable, fire-prone piles of garbage provide living space for desperately poor immigrants in the city.

Know Your Bioregion: Where does your garbage go? What is the rate of recycling in your community, and can it be improved?

Air Pollution.   Because of human dependence on fossil fuels, urban areas produce large amounts of gaseous waste. These air emissions include carbon dioxide, a chief contributor to global warming (see Chapter 4), as well as combustion by-products including nitrogen and sulfur oxides, small airborne particulates, and fuel contaminants such as mercury. Exposure to sunlight and high temperatures can cause some of these by-products to react with oxygen in the air to form ground-level ozone or **smog** (Figure 15.32). For individuals with asthma, heart disease, or reduced lung function, increased ozone levels can lead to severe illness, even death. When the gaseous pollution produced by human settlements enters the upper reaches of Earth's atmosphere, it can be carried on air currents throughout the globe. Air emissions from coal-fired power plants throughout the Midwest cause severe acid rain, even in lightly settled regions of the northeast United States, and airborne toxins such as benzene and PCB have been found in high levels in animals in the seemingly pristine Arctic.

Air pollution is a problem in both developed and less developed countries. In less developed countries, pollution control is weak or lacking altogether; in more developed countries, the sheer volume of fossil fuel use contributes to poor air quality. In the United States, the number of miles driven by car per household has nearly doubled in the last 25 years, partially due to an increase in the distance that individuals live from their workplaces. Development of suburban settlements outside the geographical limits of cities has been termed **urban sprawl**. Urban sprawl not only contributes to an increase in fossil fuel consumption by causing increased transportation needs but also affects wildlife through habitat destruction and fragmentation (see Chapter 14). Urban sprawl also impairs water quality via the destruction of wetlands and the increasing amount of paved and built surfaces, which then funnel pollutants and warmed water into urban lakes, streams, and rivers.

Figure 15.32   The effect of waste gas emissions. The brownish haze over downtown Los Angeles, California in this picture is smog, resulting from automobile emissions and other air pollutants causing the conversion of oxygen to ozone in sunny, still conditions. Days like this one can be sickening or deadly for people with impaired lung function.

Know Your Bioregion: How is the air quality in your community? What are the major air pollutants and their sources?

The effect of human settlements on surrounding natural biomes can be significant and severe, as we have discovered in this chapter. However, many of these impacts can be mitigated with thoughtful planning and the use of improved technology. In the United States, laws such as the Clean Air Act and the Clean Water Act—passed in 1970 and 1972, respectively—have greatly reduced air and water pollution and have contributed to the recovery of once severely impaired habitats (Figure 15.33). Cities throughout the more developed world are supporting projects aimed at creating sustainable communities that are both economically vital and environmentally "intelligent." Perhaps by getting to know our biological neighbors and understanding how our choices affect these organisms, we can be inspired to help create communities that are safe and healthy for humans and other species.

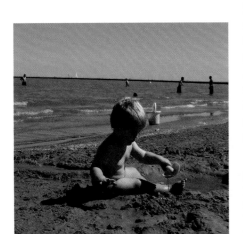

Figure 15.33   Fitting human needs into the bioregion. These bathers are enjoying Lake Erie; once so polluted that it was considered "dead" in the 1960s, the lake has now recovered and provides clean swimming beaches and a healthy fish population. The lake's recovery provides hope that other threatened biomes and habitats can support sustainable human settlements with our thoughtful efforts.

CHAPTER REVIEW

Summary

15.1 Global and Regional Climate

- Climate is the weather of a place as measured over many years. Climate in an area is determined by global temperature patterns, which are driven by solar irradiance and local factors such as proximity to large bodies of water (pp. 392–393).

- Temperatures are warmer at the equator than at the poles because solar irradiance is greater at the equator (pp. 393–394).

- Seasonal changes are caused by the tilt of Earth's axis. For example, during summer in the Northern Hemisphere, total solar irradiance during the day is much greater than it is during winter (pp. 394–395).

- Local temperatures are influenced by altitude, proximity to a large body of water, nearby ocean currents, the light reflectance of the land surface, and the amount of surrounding vegetation (p. 396).

- Global precipitation patterns are driven by solar energy, with areas of heavy rainfall near the solar equator and dry regions at latitudes of 30° north and south of the equator (p. 399).

- Local precipitation patterns are influenced by the presence of a large body of water, which tends to increase rainfall, and the presence of a mountain range, which tends to reduce rainfall on its leeward or sheltered side (p. 400).

Web Tutorial 15.1 **Tropical Atmospheric Circulation and Global Climate**

15.2 Terrestrial Biomes

- Categories of primary vegetation types found on Earth's land surface are called biomes (p. 400).

- Forests are dominated by trees and categorized by distance from the equator into tropical, temperate, and boreal types (p. 401).

- The chaparral biome is dominated by woody shrubs (p. 404).

- The major vegetation type on grasslands is nonwoody grasses. These biomes are categorized by distance from the equator into tropical and temperate types (p. 405).

- Deserts are found where precipitation is 50 cm (20 in) per year or less and contain mostly drought-resistant plants (pp. 406–407).

- Tundra occurs in areas where the growing season is less than 60 days, both near the poles and at high altitudes (p. 408).

15.3 Aquatic Biomes

- Freshwater biomes include lakes, rivers, and wetlands (p. 409).

- Marine biomes include oceans, coral reefs, and estuaries (p. 411)

15.4 Human Habitats

- Humans have modified over 50% of Earth's land surface; much of this modification now supports the activities of cities (p. 414).

- Cities must rely on imported energy and other resources to survive; extraction of these resources carries an environmental cost felt by other bioregions (p. 415).

- Waste disposal is a significant challenge for large urban areas. Sewage must be treated to avoid contaminating drinking water; garbage must be disposed of effectively, and air emissions must be controlled. More developed countries have more resources to consume and handle the vast amounts of waste that they produce, while less developed countries struggle with inadequate systems (pp. 415–416).

• Increasing the bioregional awareness of citizens and communities can help them devise methods for supporting human activities in a more sustainable manner (pp. 417–418).

Learning the Basics

1. Explain why the northern United States experiences a cold season in winter and a warm season in summer.

2. Why does proximity to a large body of water moderate climate on a nearby landmass?

3. Compare and contrast tropical, temperate, and boreal forests.

4. The solar equator, the region of Earth where the sun is directly overhead, moves from 23° N to 23° S latitudes and back over the course of a year. Why?

   A. Earth wobbles on its axis during the year.; **B.** The position of the poles changes by this amount annually.; **C.** Earth's axis is 23° from perpendicular to the sun's rays.; **D.** Earth moves 23° toward the sun in summer and 23° away from the sun in winter.; **E.** Ocean currents carry heat from the tropical ocean north in summer and south in winter.

5. A land area's climate is determined by _____.

   A. annual average solar irradiance; **B.** whether or not it is near a large body of water; **C.** the amount of variation in solar irradiance over the course of a year; **D.** the area's altitude; **E.** all of the above

6. The biome type characterized by evergreen coniferous trees and low average temperature is _____.

   A. chaparral; **B.** desert; **C.** temperate forest; **D.** tundra; **E.** boreal forest

7. Tundra is found _____.

   A. where average temperatures are low, and growing seasons are short; **B.** near the poles; **C.** at high altitudes; **D.** a and b are correct; **E.** a, b, and c are correct

8. Which statement best describes the desert biome?

   A. It is found wherever temperatures are high.; **B.** It contains the largest amount of biomass per unit area than any other biome.; **C.** Its dominant vegetation is adapted to conserve water.; **D.** Most are located at the equator.; **E.** It is not suitable for human habitation.

9. Which of the following biomes has a structure made up primarily of the remains of its dominant organisms?

   A. coral reefs; **B.** freshwater lakes; **C.** rivers; **D.** estuaries; **E.** oceans

10. In more developed countries, wastewater containing human waste _____.

   A. is dumped directly into waterways; **B.** is applied directly to farm fields; **C.** is piped into sanitary landfills; **D.** is treated to remove sludge and disease-causing organisms and released into nearby waterways; **E.** is a major source of infectious disease in human populations

Analyzing and Applying the Basics

1. Consider the following geographic factors, and predict both the climate and biome type found in the location described. Explain the reasoning that you used to determine your answer. This small city is:

   • On the coast of the Pacific Ocean
   • 20° north of the equator

   • Altitude of 200 meters above sea level
   • At the base of a mountain range

2. One prediction of global climate change models is that significant amounts of melting ice will change the salt content of the ocean, causing the Gulf Stream in the Atlantic Ocean to stop altogether. How will this change likely affect Europe?

Connecting the Science

1. How many biomes do you rely on to supply your food? Many grocery stores label the origin of their produce. The next time you go to the grocery store, try to determine the number of different countries from which your produce comes. Could you easily change your diet and shopping habits to rely on locally produced food? Why or why not?

2. Consider the setting of your home. What kind of changes to the environment of your home would help it fit better into the bioregion? Are these changes feasible? Are they desirable?

Organ Donation

Tissues, Organs, and Organ Systems

A motorcycle accident results in severe head injuries to a young man.

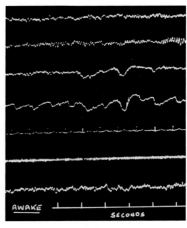

He is rushed to the hospital, where it is determined that his brain has stopped functioning.

Medical personnel search his wallet, looking for information on relatives to contact and seeing if he carried a signed organ donor card.

The drama plays out in emergency rooms around the country thousands of times every year. A formerly healthy young man is rushed to the hospital in an ambulance after having been involved in a motorcycle accident. After many failed attempts to find brain activity, emergency room physicians decide that the young man is brain dead. In the meantime, his family has assembled in the waiting room, overcome by the shock of the loss they are facing. In this time of terrible emotional upheaval, the family is asked to answer a question: will they donate their loved one's tissues and organs?

The family is understandably confused and has many questions. Is there no chance that their loved one will ever be okay—even many years from now? If they agree, will the medical staff work less hard to keep their loved one alive? What will happen to the tissues and organs if they do decide to donate them? The family knows that their loved one would have liked the idea of helping others, but they have never really discussed organ donation. Even though his driver's license lists him as a donor, the family is not sure what to do.

This chapter will help you understand the types of tissues and organs that can be donated and why properly functioning tissues and organs are necessary to maintain life.

His family is asked if they will donate his organs.

16.1 Tissues

Groups of similar cell types that perform a common function comprise a **tissue**. Damage to tissues can sometimes prevent them from recovering their function, which may be restored by a tissue transplant. We will first learn about the structure and function of different tissue types and then see some situations where transplants can help replace damaged tissues. The human body has four main types of tissue: epithelial, connective, muscle, and nervous.

Epithelial Tissue

Epithelial tissue, or **epithelia**, are tightly packed sheets of cells that cover organs and outer surfaces and line hollow organs, vessels, and body cavities (Figure 16.1a).

The sheets of epithelial tissue are usually not attached on all surfaces, as in other tissue types. Instead, epithelial tissues are anchored on one face of the tissue but free on the other tissue face. The unattached tissue can be exposed to body fluids or the external environment. For example, the epithelial tissue that lines blood vessels is attached to the wall of the blood vessel, but its outer surface is exposed to the bloodstream. The epithelium that comprises the outer layer of skin, called the epidermis, is anchored to the underlying tissue but has a surface that is exposed to air. This is also true of the epithelia that line other parts of the body, including respiratory surfaces and the digestive and urinary and genital, or urogenital, tracts.

Epithelial cells function in protection, secretion, and absorption. Epithelial cells in the skin (Figure 16.1b) help protect the body from injury and ultraviolet light. Some glands in the skin are formed by outgrowths of epithelial tissue.

Figure 16.1 Epithelial tissues. (a) Epithelial tissues, or epithelia, line organs including the heart and blood vessels, respiratory tract, and digestive and urogenital tracts. (b) Epithelial tissues are tightly packed cells that usually have one unattached surface as seen in the epidermis of the skin.

These glands secrete substances such as mucus, oils, and sweat. Epithelial tissues can also function to protect the body from water loss and pathogens. Epithelial cells that line blood vessels and intestines function in absorbing nutrients.

Epithelial tissues constantly slough off cells, which are replaced by cell division. Skin cells are rubbed off by clothing, and cells that line the mouth and intestines are scraped off by food. These cells are replaced, resulting in a replacement of the epidermis every 4–6 weeks.

Some epithelial tissues can be transplanted, as we will see at the end of this section.

Connective Tissue

Connective tissues function, as their name implies, to form connections. They usually bind organs and tissues to each other. Connective tissue is loosely organized, not tightly packed like epithelium. Our bodies contain six different types of connective tissue, all of them composed of cells embedded in a **matrix** made of protein fibers and an amorphous, gel-like ground substance. The matrix can be likened to Jell-O® with carrot shavings in it: The Jell-O is analogous to the ground substance, and the carrots are analogous to the protein fibers. The proportions of these components vary from one part of the body to another, depending on structural requirements. In some areas, the connective tissue is highly cellular; in others, the protein fibers or ground substance of the matrix is the predominant feature. The consistency of the matrix is a function of the types of fibers and ground substances present. Blood, for example, has a liquid matrix. Adipose or fat tissue has a thick fluid, or viscous, matrix; and bone has a solid matrix. Specific compositions and functions of the six types of connective tissue follow.

Loose Connective Tissue.   Loose connective tissue (Figure 16.2a on page 424) is the most widespread connective tissue in animals. It connects epithelia to underlying tissues, holds organs in place, and acts as padding under the skin and elsewhere. The tissue is called loose connective tissue due to the loose weave of its constituent fibers. The cells of loose connective tissue, called **fibroblasts**, secrete the proteins comprising the matrix. The matrix of loose connective tissue is its chief structural feature. Two proteins are important constituents of this matrix—collagen and elastin. Collagen fibers give connective tissue tremendous strength, and the elastin fibers allow connective tissue to stretch without breaking. The degradation of collagen and elastin in skin that occurs with age, exposure to sunlight, and cigarette smoke causes the weakening of the connection between the skin and the underlying muscle, leading to wrinkles.

Adipose Tissue.   Adipose (fat) tissue (Figure 16.2b) connects the skin to underlying structures and insulates and protects organs. Adipose cells are specialized for the synthesis and storage of energy-rich reserves of fat (lipids). A fat droplet fills the cytoplasm of these cells and shrinks when the fat is used for energy. Cells are the predominant constituent of this connective tissue; only a small amount of matrix is associated with adipose tissue.

Blood.   Blood (Figure 16.2c) is a type of connective tissue that circulates throughout the body through arteries, veins, and capillaries and transports oxygen and nutrients to cells. The bloodstream allows the connection of one part of the body to another. The cellular component of blood tissue includes red cells, which carry oxygen; white cells, which help fight infection; and cell fragments called platelets, which function in clotting. Blood cells are suspended in a liquid- and protein-rich matrix called plasma.

Fibrous Connective Tissue.   Fibrous connective tissue forms the tendons that connect muscles to bones and the ligaments that connect bones to each other at joints. As in loose connective tissue, the cells of this tissue are called fibroblasts.

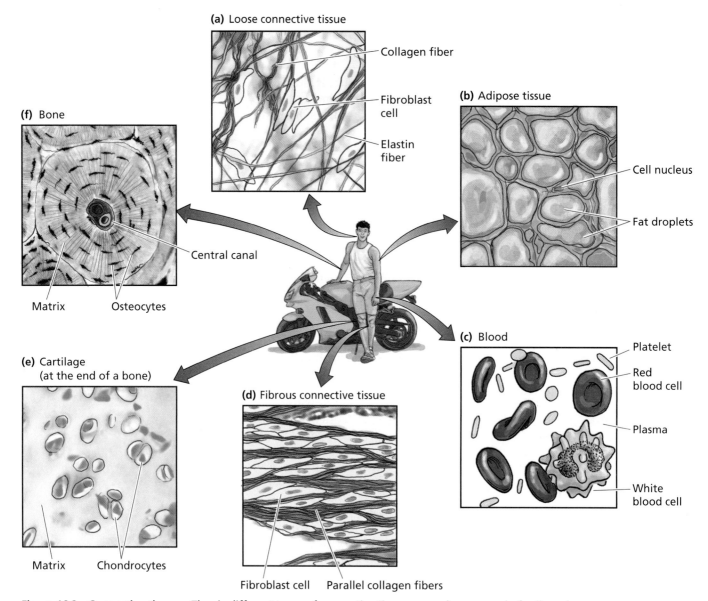

Figure 16.2   Connective tissues. The six different types of connective tissue are each composed of cells and a specialized matrix.

The densely packed collagen fibers of the matrix, arranged in parallel, are the most conspicuous feature of fibrous connective tissue (Figure 16.2d).

Cartilage.   Cartilage is a type of connective tissue composed of cells, called **chondrocytes**, that secrete substances to form the dense matrix surrounding them (Figure 16.2e). The matrix is rich in collagen and other structural proteins. Cartilage connects muscles with bones, provides flexible support for the ears and nose, and allows for shock absorption. At joints, the ends of bones are covered in cartilage, which helps permit smooth gliding where joint surfaces contact each other.

Bone.   Bones that make up the skeleton are connected to each other at the joints. In addition to being a framework for the body, bones support and protect other tissues and organs. Marrow, inside of the bone, produces blood cells.

Bone is a rigid type of connective tissue whose branched cells, called **osteocytes**, secrete substances that harden into a solid matrix of collagen and calcium and other minerals (Figure 16.2f). Bone serves as a reservoir of calcium and minerals that the body can use if dietary levels of these substances are low. Central canals inside bones house nerves and blood vessels.

Many connective tissues can be transplanted from tissue donors. Some of these tissues, including blood and bone (marrow), can be transplanted from living donors.

While connective tissue serves to support and connect, another important tissue is muscle tissue. In fact, muscle is the most abundant type of tissue in animals. Anything we think of as "meat" is actually muscle tissue.

Muscle Tissue

Muscle is a highly specialized tissue capable of contracting. Most muscle tissue is composed of bundles of long, thin cylindrical cells called muscle fibers. Muscle fibers contain specialized proteins, **actin** and **myosin**, that cause the cell to contract when signaled by nerve cells.

Muscle tissues are differentiated according to whether their function requires conscious thought—such as walking—and is **voluntary**, or whether their function requires no conscious thought, such as the beating of the heart, and is **involuntary**. Muscle tissues are also differentiated by the presence of bands that look like stripes under a microscope and are called **striated muscle**, or the lack of obvious bands, called **smooth muscle** tissues. The striated appearance is due to the banding pattern formed by actin and myosin deposits in the cell. Muscle tissue that lacks these striations also contains actin and myosin deposits, just not in a banded pattern. Humans and other vertebrates have three types of muscle tissues. These contracting tissues are cardiac, skeletal, and smooth muscle.

Cardiac Muscle.   **Cardiac muscle** is found only in heart tissue. This involuntary striated tissue undergoes rhythmic contractions to produce the heartbeat. Cardiac muscle cells are highly branched and interwoven (Figure 16.3a); this

(a) Cardiac muscle (e.g., heart)

Muscle fiber

Nucleus

(b) Skeletal muscle (e.g., biceps)

Muscle fiber   Nucleus

(c) Smooth muscle (e.g., intestine)

Muscle fiber   Nucleus

Figure 16.3   Muscle tissue. Cardiac (a) and skeletal (b) muscles are striated, while smooth muscle (c) is not.

allows a contraction signal to be propagated among many muscle cells at once, producing a coordinated heartbeat to help pump blood throughout the body.

Skeletal Muscle.   **Skeletal muscle** is usually attached to bone and produces all the movements of body parts in relation to each other. Skeletal muscle is responsible for voluntary movements such as walking. This striated muscle tissue is shown in Figure 16.3b. Exercise causes an increase in the size of skeletal muscle cells, not an increase in the number of cells.

Smooth Muscle.   **Smooth muscle**, as its name implies, is not striated (Figure 16.3c). This involuntary muscle is composed of spindle-shaped cells that comprise the musculature of internal organs, blood vessels, and the digestive system. Smooth muscle contracts more slowly than skeletal muscle, but it can remain contracted for a long time. Consider the difference between the contractions of the smooth muscle lining the digestive tract and helping to move food along as opposed to a biceps contraction. Contractions of the smooth muscles of the digestive tract are involuntary and are held for long periods of time. Contractions of the biceps are voluntary and occur more quickly.

Cardiac muscle tissue can be transplanted, as long as the heart has not been damaged. The last of the four types of tissues, and the least likely to be involved in a transplant, is nervous tissue.

Nervous Tissue

Nervous tissue is composed mainly of cells called neurons (Figure 16.4) that can conduct and transmit electrical impulses. The main function of nervous

Figure 16.4   Nervous tissue. The brain and spinal cord are composed of nervous tissues. Nerves are composed of neurons that transmit signals to and from the brain.

tissue is to help the body sense stimuli, process the stimuli, and transmit signals to and from the brain out to the rest of the body. For example, if a car door suddenly opens in front of you while you are driving, you react quickly by trying to swerve and maneuver your vehicle out of the way. This action requires your body to recognize the threat and coordinate a response to it very quickly. It is the nervous tissue that makes this possible. Nervous tissue is found in the brain, spinal cord, and nerves. In Chapter 21, we cover nervous tissue in greater detail.

Hopefully, you can now see the importance of the various tissue types. For an organism to be fully functional, damage to these tissues must be repaired, or the damaged tissue must be replaced.

Tissue Donation

If a family consents to tissue donation, tissues can be removed from a person who has been declared brain dead or from a person who has suffered cardiac death. Brain death occurs when the cerebrum and brain stem of the brain cease to function. The cerebrum, the largest and most sophisticated part of the brain, is composed of two hemispheres. It controls movement and higher mental functions, such as thought, reasoning, emotion, and memory. The brain stem, located at the base of the brain, governs such automatic functions as heartbeat, respiration, and swallowing. When the brain stem ceases to function, a person has to rely on machines to perform these functions in order to remain alive.

Head injuries caused by motor vehicle and cycling accidents, ruptured blood vessels, and drowning are all common causes of brain death. When a person suffers this type of injury, their brain cells die because swelling and damage to the tissues prevents them from being supplied with oxygen and nutrients. Once dead, brain cells cannot recover. A person who is brain dead has a different prognosis than someone who is in a coma or vegetative state. A person in a coma or vegetative state might recover some functions, while a brain dead person will not.

Organ donations can come from individuals who have suffered a cardiac death as well. Cardiac death occurs when the heart and lungs stop functioning.

In the case of the young motorcyclist discussed at the beginning of the chapter, the family was asked whether they would donate his tissues and organs because testing showed that all activity had ceased in his cerebrum and brain stem. If the family grants permission for tissue donation, a specially trained team of surgeons will carefully remove any tissues that the family agrees to donate. These tissues can include bones, tendons, ligaments, cartilage, veins, and skin.

The motorcyclist's bones may be donated to someone who has had bone cancer and would require an amputation without the transplant. His tendons, ligaments, and cartilage may be used to reconstruct and stabilize the damaged joints of many different people, helping them resume normal activities after accidents and injuries. Veins from the young man's legs can be used to help recipients with circulatory problems, and they can be used as a bypass vein for people undergoing heart bypass surgery. Donated skin can be used for burn patients as a type of biological dressing, not only to help prevent infections but also to decrease blood loss and pain due to exposed nerve endings until the patients' own skin can grow back.

It is also common to transplant the transparent cornea of the eye from a brain-dead individual to a living person who has experienced vision loss due to corneal damage. The cornea consists of a thin layer of surface epithelium overlaying a layer of fibrous connective tissue. In a healthy eye, the cornea serves as a barrier that shields the eye from dust and germs. Damage to the cornea caused by injury, infection, or chemical burn can result in loss of vision.

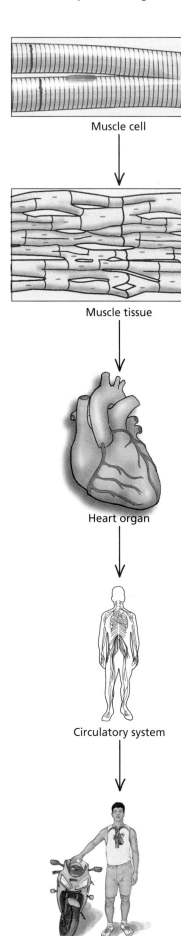

Muscle cell

Muscle tissue

Heart organ

Circulatory system

Organism

Once removed, the young man's tissues will be checked for the presence of bacteria and viruses that could cause disease in the recipients. Many of his tissues will then be treated with chemicals to remove any cells or proteins that a recipient's immune system might recognize as foreign, and thus attack. Because of these treatments, tissue recipients do not normally reject tissue transplants and usually do not need to take drugs to suppress their immune system. For this reason, tissue recipients need not be genetically similar to the donor. After being tested and treated, the tissues will be frozen and sent to tissue and eye banks, where they will be catalogued and stored until they are needed. Most tissues can be stored for up to 5 years.

Tissue removal and treatment takes a few hours and can be performed during the time that the family begins to make plans for a funeral. If the young man's family decides to donate his tissues, they can be used to dramatically improve the lives of as many as 50 different people. In addition to improving people's lives by donating the young man's tissues, the family may save several lives by donating his organs.

16.2 Organs and Organ Systems

Many structures in the body are actually assemblages of different tissue types. **Organs** are structures composed of two or more tissues packaged together, working in concert to produce the organ's specific function. When many organs interact to perform a common function, these organs are said to be part of an **organ system**. All of the organ systems in a body work together to form the functional organism.

This activity can be likened to the manner in which the engine (or organ) of a motor vehicle (organism) functions due to the combined actions of its different component parts, such as valves and pistons (tissues). Many different systems in the vehicle, such as the electrical system (organ system), or the fuel injection system (organ system), work together to keep the vehicle (organism) functioning.

Living things display increasingly complex levels of organization, from cells, to tissues, to organs, to organ systems, to an entire organism (Figure 16.5). Body parts at each level of the hierarchy act in concert to regulate the organism's bodily functions.

The regulation of the motorcyclist's body temperature while riding through different climates requires several levels of organization. The brain (an organ) sends signals through the cells of the nervous system (an organ system) to the epidermis (tissue) that result in changes in individual units (cells) that comprise sweat glands and blood vessels, resulting in an overall change in the amount of sweat secreted and in the dilation of blood vessels.

There are many different organs in humans and other animals. Some of these organs, such as the heart, lungs, kidney, and brain, are covered in more detail in subsequent chapters of this unit. In the following section, we will use the liver as a model organ to illustrate how an organ's ability to perform its required function is made possible by the actions of different tissue types and to see how an organ can function as part of an organ system.

Figure 16.5 Levels of organization. Cells give rise to tissues. Tissues with a common function comprise organs. Organs working together make up an organ system, and an entire organism often has many organ systems that communicate with each other.

(a) Human liver   **(b)** Liver tissue   **(c)** Liver lobules

Figure 16.6   The liver. (a) The human liver and greenish gallbladder. (b) The liver is subdivided by connective tissue into (c) lobules. A central vein allowing blood flow to the tissue is lined with epithelia.

Organs: The Liver as a Model Organ

The **liver** is a large, reddish-brown organ found on the right side of the abdominal cavity below the diaphragm. The human liver is divided into four lobes. A greenish sac called the gallbladder is held in close contact with the liver (Figure 16.6a).

Structurally, the epithelial tissue comprising the liver is covered with connective tissue that branches and extends throughout the body of the liver (Figure 16.6b). This connective tissue acts as a sort of scaffolding to support the tissue and subdivides the tissue into hexagonal structures called **lobules**. In the middle of each lobule is a central vein that allows blood to reach all parts of the liver (Figure 16.6c). Blood flows through the liver and is filtered by liver cells called **hepatocytes**. Filtering results in the removal of toxic materials, dead cells, pathogens, drugs, and alcohol from the bloodstream.

As described above, one organ (in this case, the liver) is composed of different tissues (epithelial tissue and connective tissue). These tissues work together to help the organ perform a function (such as filtering).

The liver is an important part of both the circulatory and the digestive systems. As a part of the circulatory system, the liver synthesizes blood-clotting factors, detoxifies harmful substances in the blood, regulates blood volume, and destroys old blood cells. As an accessory part of the digestive system, the liver helps metabolize fats and other nutrients.

Most organs of the digestive system are not suitable for transplantation. After you have gained an understanding of the digestive system as a whole, we will discuss the few digestive system organs that can be transplanted.

Organ Systems: The Digestive System

The digestive system is one of 12 organ systems in the human body. Table 16.1 on page 430 briefly outlines the functions and component organs involved in each system.

Table 16.1 Organ systems. Organs and functions of the human organ systems.

Organ system	Function	Organ system	Function
Digestive — Esophagus, Stomach, Liver, pancreas, gallbladder, Small and large intestine	Ingests and breaks down food so that it can be absorbed by the body	Cardiovascular — Blood vessels, Heart	Enables the transport of nutrients, gases, hormones, and wastes to and from cells of the body
Excretory — Kidney, Ureter, Bladder, Urethra	Eliminates liquid wastes; regulates water balance	Endocrine — Pituitary gland, Thyroid/Parathyroid, Thymus, Gonads, others	Secretes hormones into bloodstream for regulation of body activities
Respiratory — Trachea, Bronchi, Lung	Enables gas exchange, supplying blood with oxygen and removing carbon dioxide	Nervous — Brain, Spinal cord, Nerves	Senses environment, communicates with and activates other parts of the body
Skeletal — Cartilage, Bone	Provides mechanical support for the body; mineral storage; red blood cell production	Lymphatic and Immune — Thymus, Lymph nodes, Lymphatic vessels, Spleen	Protects against infections
Muscular — Skeletal muscles	Enables movement, posture, and balance via contraction and extension of muscles	Reproductive—Female — Ovary, Uterus, Cervix, Vagina	Produces eggs, and supports the development of offspring
Integumentary — Hair, Nails, Skin	Protects body from environment, injury, and infection; stores fat	Reproductive/Male — Prostate, Testicle, Penis	Produces and delivers sperm and associated fluids

The job of the digestive system is to break down or metabolize food so that the body can utilize the energy stored in its chemical bonds. Food is broken down into successively smaller units as it is passed through a long tube with openings at the mouth and anus. Called the **alimentary canal** or **digestive tract**, this tube allows food to be exposed to the actions and secretions of various organs and glands (Figure 16.7). Essay 16.1 outlines the evolutionary trends resulting in development of the digestive tract in both humans and other animals.

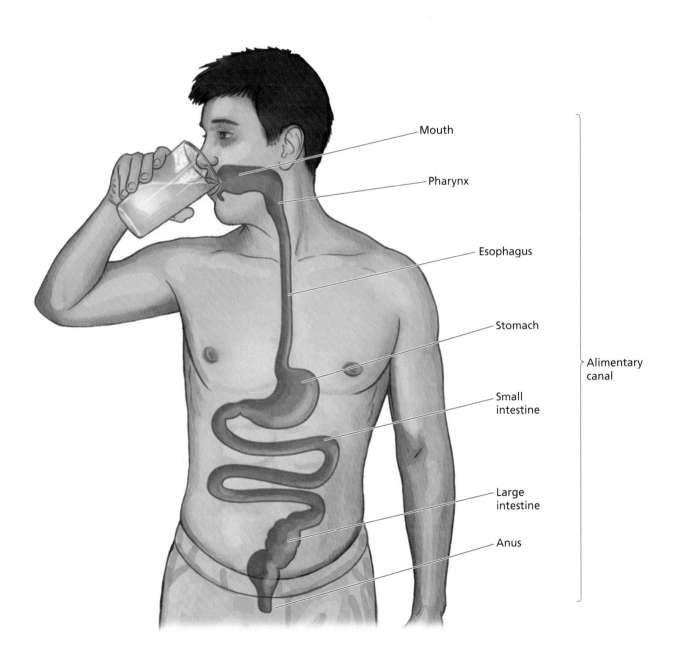

Figure 16.7   The alimentary canal. Food enters the body when it is absorbed by cells lining the alimentary canal.

Essay 16.1 Focus on Evolution: The Digestive System

The evolution of the vertebrate digestive system can be illustrated by comparing the digestive structures in more simple to evolutionarily more complex organisms. The trend in the development of digestive systems is from digestion inside structures called food vacuoles, to the development of a saclike digestive cavity, to the development of a one-way tube with a mouth and anus to help maximize nutrient gathering, to the development of specialized organs and glands for digestion.

The simplest eukaryotes, the protozoans, do not have a true digestive system made up of specialized cell types. However, these single-celled creatures do contain specialized organelles called food vacuoles, which contain enzymes capable of breaking down biological molecules. Paramecia use projections of their membranes, called cilia, to sweep food into their "mouth." Once inside the cytoplasm, the food is packaged into vacuoles. The engulfed food is digested by enzymes in the vacuole. Wastes are released to the outside via an anal pore (Figure E16.1a).

Hydra are simple animals that have a digestive system consisting of a saclike structure with a single opening that serves as both the mouth and anus (Figure E16.1b). Because it does not have two openings, this type of digestive system is called an **incomplete digestive system**. Hydra feed on entire organisms, which they first paralyze by stinging them with their tentacles. The paralyzed organism is then brought into the hydra's cavity, where cells secrete enzymes to aid in digestion. This digestive sac has its limitations; all cells of the body must be in close proximity to this cavity in order to receive nutrients.

Animals with this type of digestive system, by necessity, have very simple body plans.

While an incomplete digestive system is adequate for organisms with very simple body structures, animals with more complex internal anatomy require a more efficient mechanism for delivering food energy to the body's tissues. The evolution of a **complete digestive system**, consisting of two openings (a mouth and an anus), paired with a circulatory system or some other mechanism for transporting nutrients, permitted the other internal organs of animals to become more elaborate. The one-way flow of food through a complete digestive tract permits more efficient and thorough digestion, and close proximity of the digestive tract and the circulatory system allows food molecules to be carried throughout the body. The earthworm has a one-way digestive tract with specialized organs able to carry out digestion (Figure E16.1c). Food enters the mouth and moves through the digestive system. Nutrients are absorbed through the intestine, and undigested foods are released through the anus.

An even greater degree of organ specialization can be seen in organisms having accessory organs that help support the function of the main digestive tract. Digestion and absorption of nutrients from food is extremely efficient in organisms with such specialized digestive systems but at a cost—up to 30% of the energy obtained from the diets of animals with these complex systems is required simply to support the activities of digestion. One reason animals with simpler digestive systems survive is that these simpler strategies are the most appropriate adaptations in certain environments.

(a) Food vacuoles in paramecium

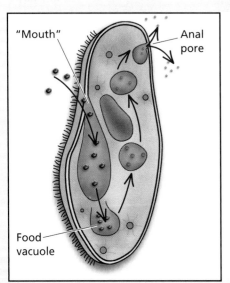

(b) Incomplete digestive system of hydra

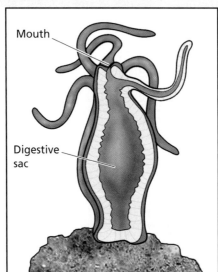

(c) Complete digestive system in an earthworm

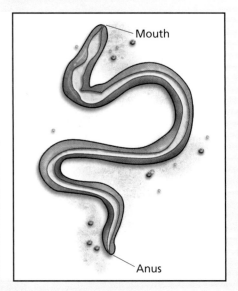

Figure E16.1

The breakdown of food begins in the mouth, or oral cavity, where chewing fractures food into smaller pieces (Figure 16.8). Mechanical digestion of food occurs as teeth grind down food, which increases the surface area exposed to enzymes within the mouth. Chemical digestion also begins in the mouth with the secretion of saliva from salivary glands located underneath the tongue. Saliva contains the enzyme **salivary amylase**, which breaks down sugars.

The carbohydrate breakdown that begins in the mouth is the first step in the conversion of polymers that we consume (carbohydrates, proteins, and fats) into their component subunits (monosaccharides, amino acids, glycerol and fatty acids, respectively). This conversion is necessary because the body cannot absorb the polymers you ingest or utilize them to build cellular structures directly. Instead, the polymers are broken down and rebuilt according to your body's specifications.

The **tongue** has taste buds that help you taste food. This muscular structure shapes food into a ball and pushes the ball of food, called the **bolus**, to the back of the mouth. From there it is swallowed and enters the digestive system (Figure 16.9).

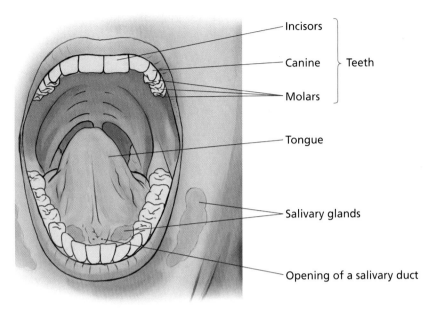

Figure 16.8   The human oral cavity. Food is ground into smaller pieces by the 32 adult teeth. Incisors are specialized to enable biting off pieces of food. Canine teeth are sharp enough to help rip food apart. Molars have broad surfaces that enhance grinding.

Figure 16.9   The digestive system. The digestive system consists of the alimentary canal and accessory organs.

When you swallow chewed food, it moves from your mouth to your **pharynx**. The pharynx, forming the back of your throat, branches to feed into both the trachea (which leads to the lungs) and the esophagus (which leads to the stomach). A flap of tissue called the **epiglottis** keeps swallowed food from entering the trachea. The **esophagus** brings food to a large digestive organ called the **stomach**. The movement of the bolus of food is hastened by rhythmic waves of smooth muscle contraction down the esophagus, through a process called **peristalsis**.

Further breakdown of food occurs in the fluid of the stomach, which has a very low pH. This acidic pH hastens the metabolism of proteins. In addition, many digestive enzymes are secreted by stomach epithelia. Stomach epithelial cells also secrete mucus in order to prevent the digestive enzymes from damaging the stomach.

Digested food mixes into a slurry with digestive enzymes, becoming something called **chyme**. Chyme is slowly allowed to enter the **small intestine**, where it is neutralized by other secretions so that the small intestine is not damaged by low pH. The small intestine of an adult is about 6 meters, or close to 20 feet, long and is the major organ for chemical digestion and the absorption of nutrients into the bloodstream via its epithelium. Many of the digestive enzymes used in the small intestine are produced by an organ called the **pancreas**. Secretions from the pancreas neutralize stomach acids that enter the small intestine and contain enzymes that break down carbohydrates, fats, proteins, and nucleic acids. While in the small intestine, chyme is exposed to a substance called bile, which is synthesized by the liver.

The liver is considered an **accessory organ** along with the pancreas and gallbladder. In a sense, they are accessories to the alimentary canal because they are outside of the tube but produce or secrete substances required for digestion. **Bile** is produced by the the liver and is secreted into the small intestine. Bile contains no digestive enzymes but does contain water, pigments, and salts that help dissolve fats. Bile salts cause emulsification, which is a process that prevents fats from clumping. Fats that have been emulsified are smaller and have more surface area exposed than do large clumps of fat. Thus, secretion of bile into the small intestine facilitates digestion because smaller, emulsified fats are easier for enzymes to gain access to and digest.

In addition to helping the body digest fats by secreting bile into the small intestine, the liver also helps to metabolize and store nutrients. Portal veins carry nutrients from the intestines to the liver, where they can be metabolized. The liver is able to convert excess carbohydrates and proteins into fat, which is then exported and stored in adipose tissue. The liver also serves as a storage site for some vitamins and excess glucose. Glucose monomers are joined to each other to produce a polymer of glucose called glycogen. The liver removes glucose from the blood when glucose levels are high, such as after a meal, and stores them as glycogen. When a meal has not been eaten recently, blood glucose levels fall, and the liver breaks the bonds between glucose molecules in glycogen and releases glucose monomers into the bloodstream.

The **gallbladder** is an organ that stores and concentrates bile, which will be released into the small intestine to help dissolve fats. Concentrating bile involves removing water from it. When the gallbladder removes too much water, the bile can crystallize, causing gallstones.

The products of digestion can be absorbed across the epithelia of the small intestine into the bloodstream (Figure 16.10). Absorption of the products of digestion in the small intestine is facilitated by projections of

Figure 16.10 Absorption of nutrients. The foldings of the small intestine to produce finger-like projections called villi and microvilli increase the surface area across which nutrients can enter the bloodstream. Once in the bloodstream, nutrients can be transported to cells throughout the body.

the small intestine called **villi** that increase the small intestine's surface area. The surface area of the small intestine is larger than that of a typical dorm room floor! Surface area is increased even further because each villus is also covered with smaller projections of the plasma membrane, called **microvilli**, that transport nutrients into the blood vessels inside each villus.

After traversing the small intestine, most polymers have been converted into monomers and can move across the small intestine into the bloodstream, where they are transported to individual cells. Materials that are not absorbed are passed through the **large intestine** (also known as the **colon**) and will exit the body as feces. Fecal matter consists largely of indigestible plant fibers.

We have now seen how a body system, in this case the digestive system, consists of many organs working together to perform a common function. The failure of some of these organs can compromise the system's ability to function properly.

Several hundred intestinal transplants are performed each year to replace intestines that have been damaged by infection or cancer. Pancreatic transplants are becoming more common for people with diabetes. When a pancreatic transplant takes place, the diabetic's original pancreas is left in place. That way, if the transplant fails, the diabetic's original pancreas can still perform its digestive functions. The gallbladder is not usually transplanted because it is possible to live a normal life without this organ. This is because the gallbladder does not synthesize bile; it only secretes it. Likewise, stomach transplantation is not commonly performed because a person can survive without a stomach. In that case, the person obtains nutrition from digestion and absorption in the small intestine only.

Failure of an individual organ can affect an entire organ system, or more than one organ system. Liver failure, for instance, would have a large impact not only on the digestive system but also on the circulatory system. In addition to affecting organ systems, organ failure compromises the body's overall ability to maintain a steady state. The ability to maintain a consistent internal environment is imperative for proper health.

16.3   Regulating the Internal Environment

An organism's ability to maintain relatively constant internal conditions under a wide array of environmental conditions is called **homeostasis**. Maintaining this steady state requires the pooled effort of the cells, tissues, organs, and organ systems. Each cell engages in basic metabolic activities that will ensure its own survival. Cells of a tissue perform activities that contribute to the proper functioning of organs; organs perform required functions alone, or in conjunction with organ systems, to contribute to the maintenance of a stable internal environment.

Animals must keep their heart rate, blood pressure, water and mineral balance, and temperature and blood glucose levels within a narrow range in order to maintain a stable internal environment. Regulation of temperature, called **thermoregulation**, is an example of a bodily function that, in many organisms, is tightly regulated. Humans and all other mammals and birds are **endotherms**, that is, organisms with the ability to maintain a body temperature that is warmer than the surrounding environment. Endotherms derive most of their body heat from metabolism. Most invertebrates, fish, amphibians, and reptiles are **ectotherms**, or organisms that obtain their body heat primarily by absorbing it from their surroundings.

Keeping physical and chemical properties within tolerable ranges utilizes feedback from one system to another. Feedback is information that is sent to a control center, which in turn directs a cell or tissue or organ to respond by turning up or turning down a given process. Feedback can be negative or positive. In the following examples, we will illustrate that *negative* feedback *negates* change and *positive* feedback *promotes* change.

Negative Feedback

Negative feedback occurs when the product of the process inhibits the process. Thermoregulation in endotherms occurs by negative feedback, in a manner very similar to the way that temperature is controlled in a house. Temperature in a house is regulated by a thermometer that functions as a sort of control center to turn the heater on when the room becomes too cold. When the room becomes too hot, the heater is turned off. Therefore, the negative feedback of the temperature change regulates heating systems.

Body temperature is controlled in a similar manner. When the body temperature increases, a control center in the brain sends signals to the body to dilate blood vessels near the skin and activate sweat glands, allowing heat to escape. When the body temperature cools, blood vessels near the surface of

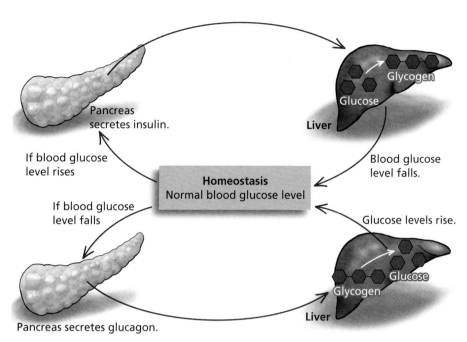

Figure 16.11   Regulation of glucose levels in the blood. When blood glucose is high, such as after a meal, the amount of glucose in the blood increases. In response to the increase in blood glucose, the pancreas secretes the hormone insulin, which stimulates the liver to remove glucose from the blood and store it as glycogen. When blood glucose is low, the pancreas secretes a hormone called glucagon, which stimulates the liver to break down glycogen and secrete glucose units into the bloodstream.

the skin constrict; muscles shiver, and body temperature increases. These negative feedback mechanisms keep the body temperatures of endotherms within a very narrow range.

Regulation of blood glucose levels by the liver provides another example of negative feedback (Figure 16.11). When food is digested, the amount of glucose in the blood increases. This stimulates the pancreas to produce insulin, a hormone that stimulates the liver (along with fat and muscle tissues) to remove glucose from the blood and store it as glycogen.

When most of the glucose has been removed from the blood, the pancreas secretes a hormone called glucagon, which stimulates the liver to break down glycogen and secrete glucose units into the bloodstream. Consequently, negative regulation of blood glucose levels reverses the direction of diet-induced changes in glucose levels, keeping glucose levels from swinging wildly after a meal and confining glucose levels to within a very narrow range— thus promoting homeostasis.

Positive Feedback

Positive feedback occurs when the product of the process intensifies the process. For example, during labor and childbirth, hormones cause the muscles of the uterus to contract, pushing the baby out of its mother's body. The uterine contractions stimulate the release of more hormones, which cause the contractions to intensify in rate and duration. This in turn causes even more hormones to be released and results in even more and longer contractions. In animals, most homeostatic regulation is via negative feedback because positive feedback amplifies a response and can actually drive the process away from homeostasis.

Damaged and diseased organs do not respond properly to signals that help promote homeostasis. Transplanting healthy organs to replace defective ones can restore, at least temporarily, homeostasis to a diseased individual.

Organ Donation

The young motorcyclist was a good candidate for organ donation because he died from a brain injury. People who die from cardiac death can be tissue donors, but the lack of oxygen experienced during cardiac death causes organs to deteriorate, making such people less-suitable organ donors. Since many more people will die of cardiac death than brain injury, tissues can be banked for future use, but there is a shortage of organs for use in transplant operations.

If the family agrees, the young man will be kept on a ventilator so that his organs continue to be nourished with oxygen and blood until his organs can be surgically removed. While the organs are being removed, medical personnel will attempt to find the best recipient. To select the recipient, a search of a computerized database is performed. Starting at the top of the list, medical staff will narrow down possible recipients, based in part on how long the recipient has been waiting. People at the top of the waiting list for a given organ tend to be very ill and will die if a transplant does not become available in a matter of days or weeks. These people are often hospitalized or waiting at home or in hotels near the hospital, hoping to receive a call telling them that an organ has become available. Recipients must be located close to the hospital since donor organs cannot be preserved indefinitely; they must be transplanted as soon as possible. Lungs, for instance, can be preserved for around 6 hours before the transplant operation must begin.

Organ donors and recipients must have the same blood type so that the recipient's immune system does not react against the organ. Likewise, donors and recipients are matched for the presence of certain markers on the surface of their tissues. When it has been determined that blood and marker types of donor and recipient are a close enough match, a transplant may occur. The most commonly transplanted organs are the liver, kidney, heart, and lungs.

Liver transplants are performed when the liver is failing due to infection by the hepatitis virus; when the patient has a condition called cirrhosis, usually caused by chronic alcohol abuse; and in cases where the liver has become cancerous. Liver donors can be either living or brain-dead individuals. Living donors can be used for liver transplants because the liver has a remarkable characteristic that other adult organs do not share—it can regenerate itself. If a piece of the liver is removed from a living donor and placed in a recipient who has the same or similar surface markers, then both the donor and the recipient's liver will grow to full size. In a given year, approximately 17,000 people are waiting for a liver transplant; approximately 10% of them will die due to lack of donor organs.

Kidney transplants are performed when the kidneys fail due to kidney disease. The kidneys are a pair of bean-shaped organs located under the rib cage by the small of the back. Each kidney is a little smaller than a fist and functions as a filter to remove toxins and wastes from the blood and excrete them, along with excess water, as urine. When kidneys fail, waste products build up in the blood, which can be toxic. The two most common causes of kidney disease are diabetes and high blood pressure.

About half of all kidney donations come from living donors. Living donors can donate one of their kidneys and survive with the remaining kidney. In a typical year, over 50,000 people are on waiting lists for kidneys; close to 10% of those people will die due to a lack of donated kidneys.

Heart transplants are performed when a person's heart has been irreversibly damaged by genetic disease, heart disease, or infection of the heart, all of which can irreparably damage the cardiac muscle of the heart. In the United States, over 3000 people are on waiting lists for a heart transplant, and over 10% of them will die before an organ becomes available.

Lung disease is one of the leading causes of death in the United States. Lungs can be damaged by genetic disease, infection, and emphysema, which is often caused by cigarette smoking (Chapter 17). In the United States, around 4000 people are on waiting lists for lung transplants; close to 25% of them will die while waiting for a donor.

If you would like to be an organ donor, discuss your plans with your family. Even if you carry a signed organ donor card, your family will need to agree to donate your tissues and organs. If you should suffer an accidental death, like the young motorcyclist, your family will be forced to make many difficult decisions quickly. Having a discussion about organ donation now will relieve them of the burden of trying to make this decision for you.

CHAPTER REVIEW

Summary

16.1 Tissues

- Tissues are composed of similar cell types that perform a common function (p. 422).

- Epithelia line and cover organs, vessels, and body cavities. They are tightly packed tissues with one free surface. Outgrowths of epithelia form some glands. These tissues function in protection, secretion, and absorption (p. 422).

- Loosely packed connective tissues bind tissues and organs to each other. The six types of connective tissues are each composed of a characteristic cell type, embedded in a characteristic matrix (p. 423).

- Loose connective tissue connects epithelia to underlying tissues and holds organs in place. The cellular component is fibroblasts; and the matrix is rich in the proteins collagen and elastin, which provide this tissue with tensile strength and elasticity (p. 423).

- Adipose tissue connects the skin to underlying structures and insulates and protects organs. Cells of this tissue synthesize and store fat and have very little extracellular matrix (p. 423).

- Blood is a type of connective tissue that transports oxygen and nutrients to body cells. Blood cells have a liquid matrix called plasma (p. 423).

- Fibrous connective tissue forms the tendons and ligaments. The cells of this tissue are fibroblasts, and the matrix is rich in collagen (pp. 423–424).

- Cartilage is a flexible, shock-absorbing tissue composed of cells called chondrocytes that secrete a dense, collagenous matrix (p. 424).

- Bone tissue provides support for the body. The cells of this tissue are called osteocytes, and the matrix is rich in collagen and minerals (p. 424).

- Muscle tissues are composed of fibers that contract and conduct electrical impulses. The three types of muscle tissue can be either voluntary or involuntary and smooth or striated (p. 425).

- Cardiac muscle comprises the heart; it is involuntary and striated (p. 425).

- Skeletal muscle is attached to bones, responsible for voluntary movements, and striated (p. 426).

- Smooth muscle comprises the musculature of internal organs and blood vessels. Smooth muscle is involuntary and not striated (p. 426).

- Nervous tissue, found in the brain and spinal cord, is composed of neurons; it senses stimuli and transmits signals throughout the body (pp. 426–427).

16.2 Organs and Organ Systems

- Organs are groups of tissues working in concert (p. 428).

- Organ systems are suites of organs working together to perform a function or functions (p. 428).

- The liver, like all organs, is composed of different tissues whose combined functions give rise to the organ's overall function. The epithelial tissue of the liver is divided into lobes. Lobes are divided by connective tissue into lobules. Lobules contain a central vein that allows blood to reach the hepatocytes, which function as filters to remove toxins and pathogens. Other functions of the liver include the production of proteins and cholesterol, secretion of bile, and storage of vitamins and glycogen (p. 429).

- The digestive system is a group of organs and glands working together to break foods into their component parts for reassembly into forms that the body can use, or for use in generating energy (p. 431).

- Food moves from the mouth through the esophagus to the stomach to the small intestine. Digested nutrients are absorbed into the bloodstream across the small intestine and brought to cells (pp. 433–434).

- The pancreas, liver, and gallbladder are accessory organs that secrete substances which aid in digestion. Digestive enzymes produced by the pancreas help break down most food molecules. Bile produced by the liver facilitates the breakdown of fats, and the gallbladder stores and concentrates bile before its release into the small intestine (p. 434).

- Undigested materials move through the large intestine and out the anus (p. 435).

Web Tutorial 16.1 **The Digestive System**

16.3 Regulating the Internal Environment

- Homeostasis is the ability to maintain a constant internal environment even when the external environment changes (p. 436).

- Negative feedback is one mechanism that helps facilitate homeostasis. When negative feedback occurs, the product of the process acts to turn down the process. Negative feedback negates change (p. 436).

- Positive feedback is a less commonly used homeostatic mechanism. When positive feedback occurs, the product of the process intensifies the process. Positive feedback promotes change (p. 437).

Learning the Basics

1. Describe the four tissue types found in animals.

2. What are the differences between the six separate types of connective tissue?

3. What are the differences between the three separate types of muscle tissue?

4. Epithelia _____.

   A. are loosely packed tissues that hold organs in place; **B.** can be free-floating tissues such as blood; **C.** line organs and cavities and have one surface exposed; **D.** include the tissues that hold joints together; **E.** are tissues that conduct nerve impulses

5. Connective tissues include _____.

   A. cartilage; **B.** blood; **C.** fat; **D.** bone; **E.** all of the above

6. Muscle tissues _____.

   A. contain actin and myosin, whether or not they are striated; **B.** that facilitate movements requiring conscious thought are involuntary; **C.** increase in cell number with exercise; **D.** in the heart are smooth; **E.** all of the above

7. Which of the following is not a function of the liver?

   A. storing bile; **B.** filtering toxins; **C.** producing cholesterol; **D.** storing glycogen

8. The flow of food in the human digestive system moves as follows:

   A. mouth → stomach → small intestine → pharynx → large intestine; **B.** mouth → stomach → large intestine → pharynx → small intestine; **C.** mouth → pharynx → small intestine → stomach → large intestine; **D.** mouth → pharynx → stomach → small intestine → large intestine; **E.** mouth → large intestine → small intestine → pharynx → stomach

9. When a woman is breastfeeding, the more her infant drinks, the more milk she produces. This is an example of _____.

   A. negative feedback; **B.** positive feedback; **C.** thermoregulation; **D.** independent regulation

10. Bile _____.

    A. is stored in the pancreas; **B.** helps break down glycogen; **C.** emulsifies fats; **D.** removes water from indigestible materials in the large intestine; **E.** all of the above

Analyzing and Applying the Basics

1. Describe a nonbiological example of both negative and positive feedback.

2. Describe what happens to food as it moves through the alimentary canal.

3. What are the roles of accessory organs in the digestive system?

Connecting the Science

1. Why might families refuse to donate the tissues and organs of a loved one?

2. How do you think your family would react to being asked to donate your tissues and organs?

CHAPTER 17

Clearing the Air

Respiratory, Cardiovascular, and Excretory Systems

Many communities have banned smoking in all public places, including restaurants and bars.

Supporters of these bans say that secondhand tobacco smoke is a public health threat.

Opponents include smokers, who feel discriminated against while engaging in a lawful activity . . .

. . . as well as bar owners, who worry that smoking bans will cost them business.

L ike many trends, it started in California. In 1977, Berkeley became the first city in the United States to limit smoking in public places, including restaurants. In 1983, the city of San Francisco passed the first ban on smoking in private workplaces. In 1995, the state banned smoking in all indoor spaces but gave bars and restaurants a few years to come up with strategies for effectively protecting employees and nonsmoking customers from the smoke generated by their cigarette-using patrons. By 1998, a total ban went into effect. States and cities across the country—from Delaware to Florida; from Boulder, Colorado, to St. Paul, Minnesota—have followed California's lead and enacted their own bans on indoor smoking. The trend has even carried overseas, where Ireland became the first smoke-free country and Bhutan became the first country to ban the sale of all tobacco-containing products in 2004.

While many nonsmokers celebrate the enforcement of smoking bans as a mechanism to remove what they see as annoying odor and haze from their favorite restaurants, bars, and bowling alleys, ban supporters have also justified these restrictions as essential to protecting public health. Over the last dozen years, an emerging medical and scientific consensus about the dangers of "secondhand smoke" has linked the airborne particles and gases that are emitted by the lit end of cigarettes, cigars, and pipes to illnesses from lung cancer to heart disease.

443

The passage of smoking bans has never been without controversy, however. While nonsmokers demand a right to smoke-free air, smokers claim a right to engage in a lawful activity where and when they wish. While restaurant employees advocate for work environments that are healthful, bar owners worry about the effects of a ban on their bottom line. Citizens in every community that is considering a ban try to find the right balance between protecting public health and the rights of private individuals and property owners.

Is there a public place for smoking, or do the risks demand that we protect all nonsmokers from the consequences of secondhand smoke? In this chapter, we examine the pathway of secondhand smoke into a nonsmoker's lungs and body and investigate the evidence that links exposure to this smoke to preventable diseases that affect the lungs, heart, and other organs.

17.1 Effects of Smoke on the Respiratory System

The "secondhand" smoke emitted by the lit end of a cigarette, combined with the smoke exhaled by active smokers, is more accurately referred to as **environmental tobacco smoke**, or **ETS**. This smoke affects not only the **active smoker**, who is holding and inhaling from the cigarette, but also nonsmoking individuals in the environment. A nonsmoker in an environment high in ETS is sometimes called a **passive smoker**. ETS is similar to the smoke inhaled by an active smoker, but it also appears to have some key differences.

According to the National Cancer Institute, more than 4500 chemicals have been identified in the smoke inhaled by an active smoker, and there may be as many as 100,000 chemicals in this mix of gases and particles. Passive smokers are exposed to the same mix of chemicals that active smokers are, although typically in much lower concentrations. Surprisingly, certain chemical compounds are actually higher in concentration within ETS than within actively inhaled smoke. These compounds are produced when the tobacco smolders and does not burn completely on the end of a cigarette or in the bowl of a pipe. When a smoker takes a drag on a cigarette, oxygen in the air is pulled through the smoldering tobacco, causing it to burn more vigorously and thus lowering the concentration of these products of incomplete burning. If you have ever built a campfire, you are probably familiar with this phenomenon. A smoldering fire is a smoky fire, and blowing on a smoldering fire helps bring in oxygen, relights some of the smoldering material, and reduces the total amount of visible smoke produced.

The most abundant gas in ETS is carbon monoxide, a by-product of the incomplete combustion of any carbon-containing compound. You may have heard of carbon monoxide as a potential source of sometimes fatal air pollution inside homes resulting from faulty furnaces or other gas-burning appliances. Carbon monoxide is approximately five times more abundant in ETS than it is in the smoke inhaled by active smokers. While the levels of carbon monoxide in ETS alone are not fatal in the short term or even the long term, effects of exposure to elevated levels of carbon monoxide can lead to serious consequences.

ETS also contains high concentrations of airborne particulates, particles with a diameter less than half the width of a human hair. These particles are commonly known to cigarette smokers as tar. Because most active smokers inhale cigarette smoke through a filter tip, the amount of particles acquired by primary smoking is reduced. However, because neither active nor passive smokers breathe through a filter when breathing normally, they are both exposed to the

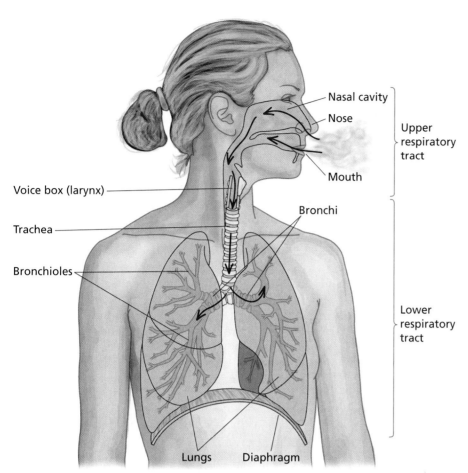

Figure 17.1 The human respiratory system. Smoke enters the upper respiratory system through the nasal cavity or mouth, then passes through the larynx into the lower respiratory system. Here, it flows into a series of smaller and smaller tubes within the lungs, until it comes into close contact with circulating blood in the lungs' alveoli.

full concentration of particles present in ETS. Chronic exposure to airborne particulates can lead to a number of negative health effects, including emphysema and cancer.

Airborne particulates enter the human body via the mouth and nose and settle in the **lungs** (Figure 17.1), where they cause some of their most severe consequences. However, some of the chemicals in these particles, as well as carbon monoxide gas, have the ability to cross from the lungs into the bloodstream and thus can be carried throughout the body. To understand how chemicals pass from the air into our bodies, first we must understand the structure and function of lungs.

What Happens When You Take a Breath?

You take a breath approximately 12 times per minute, or over 6 million times per year, without thinking much about it. In each normal breath, you take in about 500 milliliters—approximately 2 cupfuls—of air and release the same amount. Over the course of a single minute, about 4 liters of air come into direct contact with the surfaces of your lungs where the exchange of gases occurs. Normal breathing requires the action of the diaphragm to bring air into and expel it from the lungs.

Diaphragm.   In contrast to many other animals, respiration in humans and other mammals is an active process (see Essay 17.1 on pages 446–448). The human respiratory system is separated from the organs that make up the bulk of the digestive and reproductive systems by a strong, dome-shaped muscle called the **diaphragm**. When we inhale at rest, the diaphragm contracts and

Essay 17.1 Focus on Evolution: The Respiratory and Circulatory Systems

The majority of known organisms on Earth today generate energy by breaking down carbon-containing molecules via aerobic respiration. This form of metabolism requires oxygen as a reactant and generates carbon dioxide as a waste product. Because aerobic respiration occurs in every cell, multicellular organisms must be able to deliver oxygen and organic carbon to, and vent carbon dioxide molecules from, all parts in their bodies. Animals have active lifestyles and extremely high energy needs; as a result, organ systems that allow for gas exchange and for the distribution of materials throughout their bodies have evolved in these organisms. The following paragraphs detail the evolution of respiratory and circulatory systems in the animal kingdom.

Simple Diffusion

Small animals have a large surface-to-volume ratio—that is, their surface area is great compared to their internal volume (Figure E17.1a). As a result, most of their body cells are in close proximity to the external environment. Flatworms such as planaria are an example of animals small enough to allow diffusion of gases across body walls. Food molecules pass directly from the digestive system to body cells; therefore, these organisms have neither a specialized respiratory system nor a circulatory system.

Increased Surface Area and Open Circulation

Larger animals have a lower surface-to-volume ratio, and body cells may be far removed from the external environment (Figure E17.1b). Insects and spiders have solved this dilemma by having pores called spiracles on their surface. Spiracles allow gases to flow in and out through a complex branching system of tracheae. This strategy, which essentially increases the surface of the animal, brings outside air into close contact with every cell of the body. Food molecules are distributed via an **open circulatory system**, in which vessels empty fluid directly into the body's tissues. The fluid carried by these vessels, called hemolymph, is pumped into the body by a simple, tube-shaped heart and returns to the heart via pores that close when the heart contracts, allowing a one-way flow of material.

Gills for Gas Exchange

Larger water-dwelling animals like clams, sea stars, and fish have gills—thin sheets of tissue that rely on the constant movement of water past them to bring in oxygen and diffuse carbon dioxide (Figure E17.1c). Like the alveoli in lungs, gills have an enormous surface area and network of capillaries. Clams pair gills with an open circulatory system and tubular heart in order to deliver oxygenated blood to their tissues.

Figure E17.1a Flatworm

Figure E17.1b Insect

Countercurrent Gas Exchange in Water and Closed Circulatory Systems

Fish are especially efficient at employing gills to remove oxygen from relatively oxygen-poor water via a process called countercurrent exchange (Figure E17.1d). In this process, water flows in a single direction over the surface of the gills, while blood flows in the opposite direction. As a result, oxygen is able to diffuse out of water over the entire surface of the gills because the blood always contains a lower concentration of oxygen. Fish, as well as reptiles, mammals, birds, and even squid, have **closed circulatory systems** in which the circulating fluid, blood, is maintained in vessels. Among closed systems, fish have the simplest circulation with a single loop from heart to gills, to the rest of the body, and then back to the heart. Because blood loses pressure as it passes through capillaries, circulation of oxygenated blood is relatively slow in fish, and oxygen delivery to tissues is limited. Squid and their relatives have evolved a multiple-pump system in which two hearts force blood into each set of gills, and a third systemic heart pumps this oxygenated blood to the rest of the body.

Lungs and Double Circulation

Large animals on land evolved internal air sacs, called lungs, that allow the respiratory surface to remain moist for gas exchange (Figure E17.1e on page 448). The advent of lungs permitted the **double circulation** system as seen in humans, with a compact heart that pumps blood to the lungs for gas exchange and then back to the heart, where it can be pumped at high pressure to the rest of the body. In amphibians, the heart allows some mixing of oxygenated blood from lungs with deoxygenated blood heading there. This system is less efficient than the one for mammal and bird hearts but sufficient for ectotherms—animals who do not generate much internal body heat. Reptiles have a single ventricle, like amphibians, but

Figure E17.1c   Mussel

Figure E17.1d   Fish

(*Continued*)

a muscular septum between the two halves of the ventricle provides more separation between oxygenated and deoxygenated blood.

Four Chambered Hearts and Countercurrent Gas Exchange in Air

Birds and mammals are endothermic animals that maintain a regular body temperature and must use significant energy to do so. They are able to maintain this lifestyle because of adaptations that increased the efficiency of respiratory and circulatory systems. Birds have the most efficient respiratory system of all, using countercurrent exchange in a one-way flow of air, allowing for maximum oxygen "capture" by the blood (Figure E17.1f). The asymmetrical, four-chambered hearts of birds and mammals ensure that blood with high oxygen content can be pumped vigorously to the body's tissues, providing ample oxygen for maintaining regular body temperature as well as for engaging in physical activity.

Figure E17.1e Amphibian

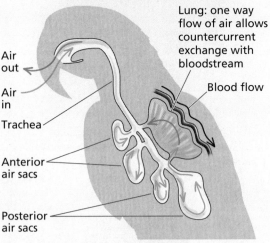

Figure E17.1f   Bird

flattens out, increasing the volume of our chest cavity (Figure 17.2a). At the same time that the diaphragm is contracting, the rib cage is lifted up and outward, further increasing the cavity's volume. Increasing the volume decreases the air pressure inside the chest, causing air to flow from the nasal cavity through the **larynx** (voice box) and **trachea** (windpipe) and into the lungs. The change in pressure in our chest cavity during inhalation is similar to the change in the air pressure inside a syringe when the plunger is pulled back. If the syringe is in a liquid solution, we can see the liquid flow toward the

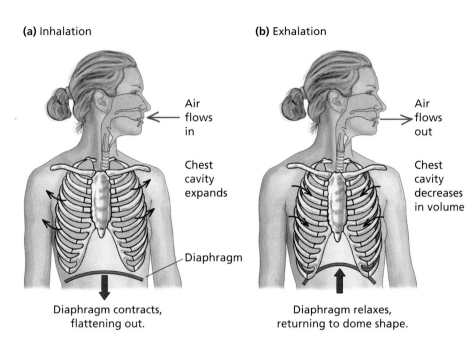

(a) Inhalation

Air flows in

Chest cavity expands

Diaphragm

Diaphragm contracts, flattening out.

(b) Exhalation

Air flows out

Chest cavity decreases in volume

Diaphragm relaxes, returning to dome shape.

Figure 17.2 How the diaphragm works. (a) Contraction of the dome-shaped diaphragm flattens it out, increasing the volume of the chest cavity. This lowers air pressure inside the cavity, causing air to rush into the lungs. (b) Relaxation of the diaphragm causes the chest cavity to lose volume, increasing pressure and thus forcing air out of the nose and mouth.

lower pressure inside the syringe, just as air flows to the lower-pressure environment inside the lungs.

When we are resting, exhalation requires no muscle contraction. The diaphragm relaxes, and the volume of the chest cavity decreases (Figure 17.2b). The air in the lungs is now squeezed into a smaller volume, increasing the pressure and causing air to flow back out of the trachea and the nasal cavity. Passive exhalation is similar to the release of air from an expanded balloon. Because the elastic walls of the balloon are exerting pressure on the contents, air rushes out even without actively squeezing the balloon. Exhalation becomes an active process during intense aerobic exercise, when the body must actively rid itself of carbon dioxide. In this case, muscles in the abdomen contract, forcing the diaphragm upward, and muscles in the chest wall tighten, squeezing the chest cavity further. Forcing liquid out of a syringe by depressing the plunger and reducing the volume is much like the process of active exhalation.

Our rate of breathing is regulated by control centers in the most ancient part of the brain, called the brain stem. The brain stem signals the diaphragm to contract in response to carbon dioxide levels in the blood. Breathing rate increases when carbon dioxide levels in the body increase, such as during exercise. As discussed in Chapter 4, the activities of cells require energy in the form of ATP, which is produced when carbon-containing food molecules are broken down, resulting in the production of carbon dioxide. Rapid breathing moves this carbon dioxide out of the body more quickly while also providing oxygen for optimal ATP production.

Contraction of the diaphragm can also be controlled voluntarily. Inhaling from a cigarette requires stronger contraction of the diaphragm than normal breathing does, in order to pull air through the burning tobacco and into the mouth and lungs. Holding your breath requires consciously overriding the signals from the brain stem—at least for a time. Speaking or making any vocal noises requires active exhalation, as air forced out past the vocal cords in the larynx causes these structures to vibrate and produce sound. The amount of air forced past the vocal cords determines the volume of our speech, while muscles that control the length of the vocal cords help to determine the pitch of our speech. The shape of our mouths, lips, tongue, and the position of our teeth determines the actual sound that is produced. Sustained exposure to tobacco smoke can cause parts of the larynx to become covered with scar tissue, often making long-time smokers sound quite hoarse. A study by researchers at the University of Cincinnati found that nonsmoking women regularly exposed to

(a) Human lungs

(b) Tennis court

Figure 17.3   The respiratory surface of the human body. (a) Lungs contain the surfaces over which gases are exchanged between the body and the environment. This photo also shows a heart, the pear-shaped structure between and below the two lungs. (b) The total surface area for gas exchange contained within a pair of healthy human lungs is approximately the size of a tennis court.

environmental tobacco smoke also had some changes to their vocal cords relative to other nonsmokers, although these changes had relatively little effect on their voices.

The diaphragm also may contract involuntarily for other reasons, for instance when we inhale deeply before a cough. Coughing is a reflex that helps remove irritants from the trachea. Keeping air passageways into the lungs clear is crucial for maintaining adequate oxygen uptake.

Lungs.   Healthy lungs are pink, rounded, and very spongy in texture (Figure 17.3a). Even at rest, these organs almost completely fill the chest cavity. The external surface of the lungs is linked to the lining of the chest cavity by two thin membranes, one covering the lungs and one lining the chest wall. These membranes stick together because they are moist, much like two wet pieces of plastic wrap that stick together because of the cohesive attraction among water molecules. Thus, when the chest cavity enlarges, the lungs expand. If the two membranes become separated—for instance, as a result of injury to the chest wall—then air will enter the space between them, sometimes causing the underlying lung to collapse.

Once air enters the lungs, it flows through tubes called **bronchi** that produce finer and finer branches (**bronchioles**), each of which dead-ends at structures called **alveoli**. These 300 million tiny sacs within our lungs contain the **respiratory surface** of our bodies, the tissue across which gases from inside the body are exchanged with gases in the air. The area of this respiratory surface in humans is approximately 160 square meters (1725 square feet), about the size of a tennis court (Figure 17.3b). A large surface area is required to supply oxygen and remove carbon dioxide from large-bodied, active mammals. Each grape-like cluster of alveoli is surrounded by a net of tiny blood vessels, called **capillaries**, connecting the gases exchanged in these structures with the entire body (Figure 17.4).

Exposure to smoke can damage and eventually destroy the respiratory surfaces within alveoli. A reduction in the total amount of respiratory surface results in shortness of breath, wheezing, and an inability to participate in even moderately vigorous activity. These symptoms are all caused by the lung's reduced capacity for gas exchange.

Gas Exchange

The primary function of the lungs is **gas exchange**, the process that allows us to acquire oxygen for cellular respiration and to expel carbon dioxide, one of the waste products of respiration (Figure 17.5). Exchange of gases occurs between the air spaces of the alveoli and the blood contained in neighboring capillaries. The

(a)

Blood in:
Low oxygen,
high carbon dioxide

Air

Blood out:
High oxygen,
Low carbon dioxide

(b)

Figure 17.4   Alveoli. (a) Alveoli occur in clusters at the end of bronchi. The clusters are enmeshed in a dense network of thin-walled capillaries, which bring blood into close proximity with the gas in the alveoli. (b) An electron micrograph of alveoli in lungs.

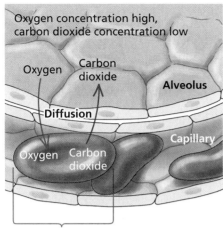

Oxygen concentration high,
carbon dioxide concentration low

Oxygen

Carbon
dioxide

Alveolus

Diffusion

Oxygen Carbon
dioxide

Capillary

Oxygen concentration low,
carbon dioxide concentration high

Figure 17.5   Gas exchange at a respiratory surface. The movement of oxygen and carbon dioxide gas between air spaces in the lungs and the blood occurs via simple diffusion. Oxygen in high concentration in the alveoli diffuses to deoxygenated blood, while carbon dioxide in high concentration in the blood diffuses into the air spaces.

exchange works by simple diffusion across the thin membranes surrounding the blood vessels and the alveoli walls. Carbon dioxide in high concentration within the capillaries passes from these structures into the alveoli, where it is maintained in low concentration by exhalation. Oxygen in high concentrations, maintained by inhalation, passes from the alveoli to the deoxygenated blood in the capillaries.

Carbon dioxide and oxygen gas cannot directly cross membranes but must be dissolved in fluid that coats the respiratory surface. Because the size of the alveoli increases and decreases over the course of a single breath, this layer of fluid must be somewhat slippery to maintain constant coverage of the surface and to prevent the alveoli walls from sticking together. A soap-like substance called surfactant provides this slippery consistency. The composition of surfactant can be negatively affected by tobacco smoke, and its degradation can reduce levels of gas exchange and make breathing more laborious.

The Role of Hemoglobin in Gas Exchange.   Carbon dioxide is in relatively low concentrations in the atmosphere, and so it readily moves from the blood into the alveoli and out into the atmosphere. However, oxygen does not dissolve as readily into the blood from the air; thus all animals (and even some plants) produce specialized proteins called respiratory pigments that can bind oxygen. Respiratory pigments contain a metal atom that produces a color when oxygen binds to it. For instance, the pigment in insects contains copper, which produces the bluish "blood" you can see when a large insect is crushed. When these proteins are located in a circulatory system, they quickly whisk oxygen away from the respiratory surface. The oxygen-removing action of these proteins helps to maintain the maximum concentration difference between the outside air and circulatory fluid, increasing the rate of oxygen diffusion.

Hemoglobin is the respiratory pigment in humans that acquires and transports oxygen. Hemoglobin is actually made up of four separate protein chains, each containing a single iron atom that can bind to a single oxygen molecule (Figure 17.6). When oxygen attaches to the iron atom, the hemoglobin becomes red, giving our blood its distinctive color. The relationship among the four protein chains is such that if one picks up an oxygen molecule, the remaining three are more likely to bind to oxygen. As with all respiratory pigments, hemoglobin must be able to release oxygen where it is needed, and the physical relationship among the protein chains means that losing a single oxygen atom increases the likelihood that the remain-

Heme group

Oxygen (O_2)

Iron (Fe^{+++})

Figure 17.6   Hemoglobin, the blood's oxygen shuttle. This drawing illustrates the relationship between the four protein chains in a single hemoglobin molecule. Where oxygen levels in the surrounding solution are high, the iron-containing binding site (called a heme group) on all four protein chains will pick up an oxygen molecule. Where surrounding oxygen levels are low, the oxygen molecules are rapidly unloaded.

ing three will be lost. Each of our red blood cells contains approximately 250 million hemoglobin molecules; therefore, each cell can carry about 1 billion oxygen atoms. In areas of the body where oxygen levels are low and carbon dioxide production is high, hemoglobin releases oxygen molecules. In this way, red blood cells are loaded with oxygen in the lungs and will drop their load in areas of the body where high levels of cellular respiration are occurring.

Hemoglobin picks up oxygen effectively, but surprisingly, it binds to the smoke by-product carbon monoxide about 200 times more strongly. Even when in a mixture of gases containing oxygen, carbon monoxide inhaled in tobacco smoke is preferentially loaded into red blood cells at the respiratory surface. Because the binding of carbon monoxide is so strong, hemoglobin is slow to release it. Even small amounts of carbon monoxide can tie up large amounts of hemoglobin in the body, starving body tissues of oxygen. While the amounts of carbon monoxide in ETS are not high enough to cause death from lack of oxygen, chronic low levels of oxygen deprivation can damage tissues and organs. Carbon monoxide is especially damaging to developing embryos and fetuses, since they must acquire the oxygen they need through exchange with their mother's blood supply. The lower-than-average birth weights of babies born to smoking mothers may be due to their relative oxygen deprivation. There is some evidence from animal studies that long-term exposure to the carbon monoxide levels found in ETS may contribute to diminished brain function in infants and children for the same reason—oxygen deprivation.

The biggest threat ETS poses to respiratory function does not stem from the interaction between hemoglobin and carbon monoxide. Instead, the threat comes from damage caused to lungs by microscopic particles of smoke.

Smoke Particles and Lung Function

Much of the visible smoke rising from the lit end of a cigarette consists of tiny particles of partially burned tobacco and its by-products. These **particulates** are a significant hazard to infants and children as well as to anyone with impaired lung function, including individuals who have been chronically exposed to particulate-laden air.

Bronchitis and Asthma.   Our lungs have various mechanisms for expelling and removing foreign objects. Coughing is often our first response to larger particles or droplets that enter the trachea. Particles too small to trigger a cough can settle in the upper respiratory tract, trachea, or larger bronchi. These particles become trapped in mucus produced by the tissue lining these structures. The mucus and particles are swept upward toward the mouth and nose by the actions of tiny hair-like projections called cilia on the tissue surface (Figure 17.7). When a glob of mucus reaches the tip of the larynx, it is

Figure 17.7   The surface of the bronchi. This highly magnified image from an electron microscope shows the hairs on the surface of epithelial cells within a nonsmoker's bronchi. These hairs continually sweep small particles and mucus out of the respiratory passages. The same surface in a smoker would have fewer and shorter hairs than in a nonsmoker.

coughed up, expelled out of the nose, or swallowed. Airborne particles in to-bacco smoke increase the production of mucus, but they also damage the cilia lining the bronchi, making it more difficult to expel these particles. This damage in turn can lead to inflammation of the bronchi, known as **bronchitis**.

The tiniest particulates in ETS remain airborne and can be drawn deeply into the lungs, even into the alveoli. Because alveoli lack cilia, the movement of foreign materials out of these structures is much more limited. Small particles can remain in the alveoli for long periods, promoting inflammation and fluid accumulation as well as generally reducing gas exchange. Particulates are also known to exacerbate **asthma**, an allergic response that results in the constriction of bronchial walls and an overproduction of mucus. According to the U.S. Environmental Protection Agency (EPA), the number of additional cases of asthma caused by ETS among children in the United States is likely to be 26,000 per year.

Emphysema.   The physical consequences of chronic bronchitis and asthma can lead to the formation of scar tissue in the lungs, permanently blocking bronchi. The inflammation also causes damage to alveoli walls, causing small alveoli to merge into fewer, larger sacs (Figure 17.8). This merging of alveoli reduces the overall area of the respiratory surface, interfering with gas exchange. The sacs themselves become surrounded by thick scar tissue, which interferes with the passage of oxygen into the capillaries. The buildup of scar tissue also makes the lungs less elastic, meaning that the passive process of exhalation is ineffective. As more "dead air" remains in the lungs, they become overinflated, gradually increasing the size of the chest. This is a characteristic physical sign of the underlying disease called **emphysema**.

Because alveoli cannot be regenerated, the lung damage that results in emphysema is permanent and irreversible. Individuals with emphysema are chronically short of breath and unable to participate in vigorous activity. The most severely affected individuals require supplemental oxygen simply to engage in the activities of daily living. The EPA estimates that adult non-smokers exposed to ETS have a 30% to 60% higher occurrence of emphysema, asthma, and bronchitis when compared to unexposed nonsmokers.

Lung Cancer.   Many components of tobacco smoke are carcinogens, chemicals that are known to cause cancer. Carcinogens cause changes to the structure or sequence of DNA molecules; that is, they cause mutations. As discussed in Chapter 5, cancer results from an accumulation of mutations that affect genes controlling the cell cycle.

Small particulates in tobacco smoke contain carcinogens. When these particulates are drawn deeply into the lungs, they may stay on the respiratory surface for months or even years without being cleared. We can see the accu-

(a) Normal alveoli   **(b)** Effects of smoke on alveoli

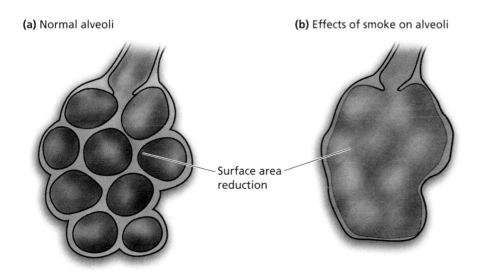

Surface area reduction

Figure 17.8   The effect of chronic smoke exposure on alveoli. (a) Normal alveoli appear as clusters of grapes in the lung. Gas exchange occurs across the "skin" of each of these grapes. (b) The walls of the alveoli are destroyed by exposure to particulates. The total surface area for gas exchange is therefore much reduced.

Figure 17.9   Tobacco tar trapped in lungs. The lung on the left is the typical healthy pinkish color, while the one on the right shows the dark staining caused by the accumulation of tiny airborne particles in tobacco smoke. The lighter-colored mass on the smoker's lung is a cancerous tumor.

Healthy lung   Smoker's lung

Cancerous tumor

mulation of particles in the lungs of a long-time smoker, which are often black with tar trapped inside the alveoli (Figure 17.9). Because these tobacco smoke carcinogens are retained in the lungs, the DNA of lung cells remains under the threat of cancer-causing mutations long after smoke inhalation has stopped.

About 170,000 new cases of lung cancer are diagnosed in the United States each year, 90% in current or former active smokers. However, for the 17,000 cases of lung cancer not due to active smoking, researchers at the National Institutes of Health (NIH) have estimated that 3000 cases are directly attributable to exposure to ETS. After decades of scientific study, the link between exposure to environmental tobacco smoke and lung cancer has been demonstrated conclusively.

How Do We Know That Tobacco Smoke Kills?

The description of a cause-and-effect relationship between two factors—such as "Exposure to secondhand smoke causes cancer in humans"—is called a hypothesis. As described in Chapter 1, a hypothesis is equivalent to "an educated guess," which must be tested to determine whether it is likely to be true. Testing hypotheses about the distribution and causes of diseases in human populations is called **epidemiology**. To test epidemiological hypotheses, researchers may use several different scientific strategies, ranging from highly controlled experiments to simple observations of populations (Table 17.1). Because the effects of exposure to secondhand smoke are potentially very serious and because they occur after many years of exposure, experimental studies on humans testing this hypothesis are impossible. As a result, the relationship between cancer and ETS exposure is studied using correlational studies. After combining and analyzing the results of dozens of different studies of this type, epidemiologists have determined that nonsmokers exposed to ETS have a 20% to 30% higher risk of lung cancer than do unexposed nonsmokers.

As discussed in Chapter 1, all correlational studies have the same basic problem—it is impossible to eliminate the chance that some *other* difference between individuals in the study is, in reality, causing a difference in lung cancer rates. In the case of ETS and lung cancer, the hypothesized link is also supported by the presence of known carcinogens in secondhand smoke, by experiments using animals that show higher rates of cancer in ETS-exposed individuals, and by a link established by hundreds of studies between active smoking and lung cancer. The relationship between exposure to secondhand smoke and diseases in other organs and organ systems is less clear from the data collected. However, given the connection between the lungs and the bloodstream, one would expect that passive smokers have a higher risk of all the diseases that are more common in active smokers.

To understand how tobacco smoke breathed into the lungs can affect the entire body, we must understand first how blood and its components are moved around via the bloodstream.

Table 17.1   Types of epidemiological studies.

Name	Description	Pros	Cons
Ecological studies	Examine specific human populations for unusually high levels of various diseases (e.g., documenting a "cancer cluster" around an industrial plant).	Inexpensive and relatively easy to do	Unsure whether exposure to environmental factor is actually related to onset of the disease
Cross-sectional surveys	Question individuals in a population to determine amount of exposure to an environmental factor and whether disease is present.	More specific than ecological study	• Expensive. • Subjects may not know exposure levels • Cannot control for other factors that may be different among individuals in survey • Cannot be used for rare diseases
Case-control studies	Compare exposures to specific environmental factors between individuals who have a disease and individuals matched in age and other factors but who do not have a disease.	• Relatively fast and inexpensive • Best method for rare diseases	• Does not measure absolute risk of disease as a result of exposure • Difficult to select appropriate controls to eliminate alternative hypotheses • Examines just one disease possibly associated with an environmental factor
Cohort studies	Follow a group of individuals, measuring exposure to environmental factors and disease prevalence.	Can determine risk of various diseases associated with exposure to particular environmental factor	• Expensive and time-consuming • Difficult to control for alternative hypotheses • Not feasible for rare diseases

17.2   Spreading the Effects of Smoke: The Cardiovascular System

The risks of active smoking, in addition to lung and airway damage, include increased rates of throat, bladder, and pancreatic cancer; higher rates of heart attack, stroke, and high blood pressure; and even premature aging of skin. All of these effects occur because many of the components of tobacco smoke can cross the thin walls of alveoli into the bloodstream and move throughout the body.

Structure of the Cardiovascular System

Gases and other materials are distributed around the body of most animals by a **cardiovascular system**, which consists of three major components: a circulating fluid (often called blood), a pump (in most cases, this is called a heart), and a vascular system (blood vessels and capillaries).

Blood.   The 5 liters (or 11 pints) of **blood** in the vascular system of an adult human is made up of both liquid and solid (that is, cellular) portions. The liquid portion, called **plasma**, consists of water and dissolved proteins, salts, and gases. The cellular or solid portion is primarily made up of red blood cells (giving blood

Figure 17.10   Components of blood.
Blood consists of the fluid plasma (a) as
well as large amounts of red blood cells
(b) and lesser amounts of white blood
cells (c) and platelets (d).

(a) Plasma

(b) Red blood cells

(c) White blood cells

(d) Platelets

its reddish color) with much smaller numbers of white blood cells and platelets
(Figure 17.10). In adults, all of the cellular components of blood are produced by
cells in the **bone marrow**, tissue found in the cavities of certain bones. **Red blood
cells** are uniquely adapted to their primary task—shuttling oxygen from the
lungs to the rest of the body. These cells are packed with hemoglobin molecules
and have hardly any internal structure; in other words, they lack a nucleus and
other organelles. Red blood cells are also small and pinched in the middle, which
makes their surface area large relative to their volume, ensuring rapid diffusion of
oxygen into and out of the cells. **White blood cells** come in many different
varieties and are the essential component of immune system function, attacking
invading organisms as well as removing toxins, wastes, and damaged cells
throughout the body. The principal function of **platelets** is to prevent blood loss.
These cell-like structures are actually membrane-bound fragments of larger cells
that have been produced in the bone marrow.

Platelets contain enzymes and other proteins that are essential to **blood
clotting**, a process that stems the flow of blood out of damaged blood vessels.
Immediately after a break occurs, muscles in the walls of blood vessels contract
to restrict the flow of blood. Within seconds, sticky platelets become attached to
the damaged tissue ends. Chemical signals released by platelets at the site of vessel damage initiate the complex clotting process, while masses of these cells form
a temporary plug to help restrict blood flow. A clot consists of a net made up of
the protein fibrin, which forms a patch over the damaged area (Figure 17.11).
Cell division in the walls of the damaged vessel and surrounding tissues eventually seals the cut, and the patch dissolves. A substance in tobacco smoke, perhaps
nicotine, increases the stickiness of platelets and promotes production of fibrino-
gen, the precursor to fibrin, making blood clots form more readily. When a clot
forms *within a* blood vessel, it can block blood flow where it forms (a condition
called a thrombosis) or break free and travel throughout the bloodstream until it
becomes lodged in another artery or capillary bed (in which case it is called an

Figure 17.11   A blood clot. A blood clot
consists of a net of protein fibers that
trap red blood cells, forming a temporary
patch over a damaged blood vessel.

embolism). When a thrombus or embolus becomes lodged within the blood vessels of the heart or any other organ, it can cause cells to die and severe damage.

Heart.   We often think of the heart in poetic terms as the seat of our emotions, intuition, and spirituality; but in physiological terms, the heart is a fairly simple and very utilitarian organ. The fist-sized **heart** consists of two muscular pumps that are coordinated but also somewhat independent. One pump on the right side of the heart receives oxygen-poor blood from the body and sends it to the lungs, while the left pump receives the oxygen-rich blood from the lungs and sends it into general circulation within the body. The two pumps are each divided into two chambers, a relatively thin-walled **atrium** and a thick-walled **ventricle** (Figure 17.12).

The heart muscle contracts with an intrinsic rhythm determined by a small patch of muscle tissue in the wall of the right atrium. This patch, called the **sinoatrial (SA) node**, acts as the heart's **pacemaker** by sending out electrical signals that first cause both left and right atria to contract and then, one-tenth of a second later, cause both ventricles to contract. The complete sequence within the heart of filling with blood and then pumping is called the **cardiac cycle**, alternating between a relaxed period (called **diastole**) and a contraction phase (called **systole**). During systole, the **AV valves** between the atrium and the ventricle close as the ventricle contracts ("lubb"), and the **semilunar valves** that connect the ventricles to the rest of the circulatory system close as the ventricle relaxes ("dupp"). This cycle leads to that distinctive heartbeat sound ("lubb-dupp, lubb-dupp"). The speed of the cardiac cycle, also known as the heart rate, is regulated at the SA node by the internal pacemaker (whose function can be replaced by an implanted electric pacemaker) and by signals from the brain stem and spinal cord while we are at rest. Input from other parts of the brain can also affect the heart's rhythm, speeding up the heart rate in response to experiencing danger or intense emotion, or to taking certain drugs such as nicotine.

Compared to muscles elsewhere in the body, the heart is much more dependent on oxygen to generate energy for its activities. A thrombus or embolus that restricts blood flow to the heart muscle can cause portions of the muscle to die

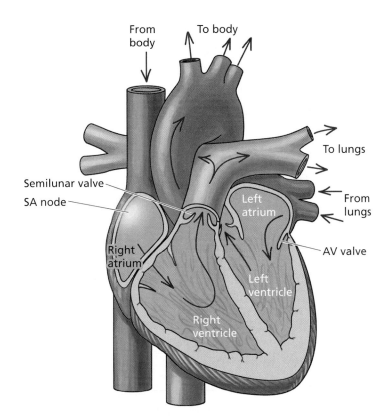

Figure 17.12   The human heart. The heart consists of four chambers making up two mostly independent pumps. One pump sends deoxygenated blood to the lungs, and the other sends blood returning from the lungs to the rest of the body. Note that this drawing illustrates the orientation of a person's heart when facing you. The left and right sides are relative to that person, not to you.

because of oxygen starvation. Formation of these blockages depends not only on clotting factors in the blood but also on the condition of the blood vessels.

Blood Vessels.   The system of tubes that carries blood to and from the heart is generally referred to as the **vascular system**. Components of the vascular system, broadly termed the **blood vessels**, include arteries, veins, and capillaries (Figure 17.13). **Arteries** are the branching blood vessels that carry blood from the heart, and **veins** are the converging vessels that bring it back. As in the lungs, capillaries throughout the body are tiny, thin-walled blood vessels that create a net of channels between the smallest arteries and the smallest veins.

Arteries have thick, elastic walls that balloon out as contraction of the ventricles causes a mass of blood to flow into the system and that snap back to resting size once the blood passes by. The wave of blood is called a **pulse**; you can measure your heart rate by feeling the pulse as it passes through an artery close to the surface of the skin.

Exchange of gases and other materials between the circulating fluid and the body's tissues and organs occurs across the porous walls of the capillaries. Capillaries are extremely diffuse and abundant; with a combined length in an adult human of an estimated 160,000 kilometers (100,000 miles) of capillaries, no living cell is more than 0.1 mm (about the thickness of a sheet of paper) from one of the 10 to 40 billion capillaries we possess. Liquid and materials in these vessels are forced out of capillaries due to higher blood pressure near the arterial end of a **capillary bed**. Materials and liquid from the body tissues then flow back into the capillaries as a result of concentration differences at the venous (vein) end of the capillary bed (Figure 17.14). Muscles surrounding the arterial ends of capillary beds can contract to cut off blood flow to less-needy organs in order to

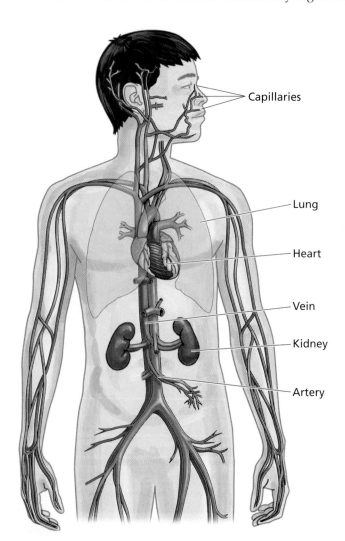

Figure 17.13   The circulatory system. Arteries carry blood away from the heart, while veins carry blood to the heart. The two types of blood vessels are connected by nets of capillaries.

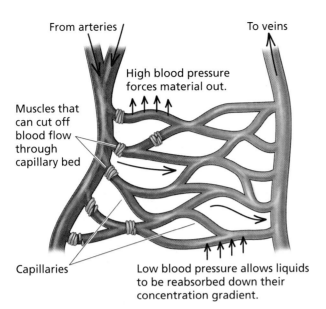

From arteries

To veins

High blood pressure forces material out.

Muscles that can cut off blood flow through capillary bed

Capillaries

Low blood pressure allows liquids to be reabsorbed down their concentration gradient.

Figure 17.14 Capillary function. Liquid forced out of the upstream end of a capillary bed by blood pressure carries oxygen and nutrients to the tissues. Pressure is much lower at the downstream end, allowing liquid carrying wastes to be reabsorbed into the blood vessels. Muscles on the upstream side can contract to restrict flow through the bed.

effectively deliver blood and nutrients to more essential regions. Capillaries that nourish the skin often become constricted in response to a crisis in another organ, which is why a common symptom of illness is paler skin (most apparent in the tissues lining the eyes and inside the mouth.) Skin capillaries also play an essential role in the regulation of body temperature, opening when temperature rises to allow body heat to dissipate and closing to conserve heat in cold temperatures.

Veins have much thinner, less-elastic walls than arteries, and the pressure of the blood is much lower once it reaches these vessels. Blood tends to pool within veins, a fact you can easily see by lowering your hand to your side. Blood pooling in the veins on the back of the hand will make these veins become distended and stand out. Movement of blood from the veins back to the heart is facilitated by the contraction of skeletal muscles, which compress the veins and squeeze the blood through them. The blood flows in only one direction, toward the heart, due to the presence of one-way valves within the veins (Figure 17.15).

To heart

Backflow prevented

Muscle contraction squeezes blood toward heart.

Blood pools in veins when muscles relax; one way valves prevent backflow.

Figure 17.15   Flow of blood in veins. Blood in the veins is under low pressure and returns to the heart due to the contraction of skeletal muscle. The blood flows in one direction thanks to the presence of one-way valves.

The efficient flow of oxygen and materials to the body's tissues requires that the blood is maintained under enough pressure to move it throughout the vascular system. **Blood pressure**, the force of the blood against blood-vessel walls, is created in part by the pulse of blood from the contracting heart and in part by the diameter of the arteries. Blood pressure rises when arteries become constricted by the action of muscles in the vessel walls. For instance, this happens when release of the stress hormones adrenaline or epinephrine indicates that muscles need oxygen delivered quickly. Unless a person is under chronic stress and overproducing the adrenaline signal, blood pressure will rise in response to vigorous activity or excitement but return to normal under restful conditions. Chronic high blood pressure, called **hypertension**, may be caused by arteries that are narrowed by constant psychological stress (resulting in stress hormone release) or by the accumulation of fatty material within the walls of the arteries, which occurs as a result of aging, a high-fat diet, or exposure to tobacco smoke. Hypertension can lead to damage of the walls of blood vessels, and it necessitates that the heart work much harder to push blood through the arteries.

The components of tobacco smoke can exert effects on all organ systems because materials crossing the respiratory surface enter the bloodstream, which then circulates these materials throughout the body.

Movement of Materials Through the Cardiovascular System

The lungs and the cardiovascular system are connected by a double circulation system in which blood flows in two distinct but related circuits. The **pulmonary circuit** circulates the blood into the lungs, where oxygen and other components inhaled in the air are picked up, and returns this blood to the heart, where it enters the second circuit. This **systemic circuit** pumps blood to the rest of the body, where the blood drops off its oxygen and picks up carbon dioxide. Blood flowing through the systemic circuit also picks up and distributes materials from the digestive system and filters non-gaseous wastes via the excretory system.

The path of circulation between the heart, lungs, and body is detailed in Figure 17.16. From this, you can see that oxygen, carbon monoxide, and other materials picked up by the bloodstream at the lungs (step ① in the figure) return to the heart via a pair of large veins. These veins empty into the left atrium (step ②), where the beginning of a heartbeat forces the blood into the left ventricle (step ③). There is a slight lag between when the atrium contracts and when the contraction signal reaches the ventricles. This lag allows the left venticle to fill with oxygenated blood. Once this part of the heart contracts, the blood is forced at high pressure into the arteries, which repeatedly branch out, become smaller in diameter, and lead into capillary beds (step ④). In the capillaries, oxygen and other components in high concentration within the blood diffuse out, and carbon dioxide and wastes diffuse in (step ⑤). The deoxygenated blood then travels to the systemic veins, which empty into the right atrium (step ⑥). Contraction of the right atrium forces blood into the right ventricle (step ⑦). Contraction of the right ventricle sends blood into the arteries and capillary beds of the pulmonary circuit (step ⑧). A single red blood cell carrying a load of oxygen—or carbon monoxide—can travel this entire circuit in approximately 1 minute. Thus the chemicals present in ETS have immediate effects on the body. These immediate effects can cause more severe long-term damage as well.

Smoke and Cardiovascular Disease

Most people believe that lung cancer and other lung diseases are the primary risk associated with exposure to tobacco smoke. In reality, most of the deaths due to smoking result from heart and blood-vessel damage, or **cardiovascular disease**.

Most of the cardiovascular damage that results from tobacco smoke is caused by **nicotine**, the primary active ingredient in tobacco. Nicotine is a

Carbon dioxide out

Oxygen in

Gas exchange in lungs

① Oxygenated blood from the lungs travels through veins to the left atrium of the heart.

② Contraction of the heart muscle forces blood from the left atrium to the left ventricle.

③ The slightly delayed contraction of the left ventricle forces blood into arteries at high pressure.

④ Blood flows through the arteries and arterioles into capillaries.

⑤ In the capillaries, oxygen and other components in high concentration in the blood diffuse out and carbon dioxide and wastes diffuse in.

⑥ Capillaries empty into the venules, which carry blood to the veins and eventually to the right atrium.

⑦ Blood is forced into the right ventricle by contraction of the right atrium.

⑧ Contraction of the right ventricle sends blood into the arteries and capillary beds in the lungs.

Carbon dioxide in

Oxygen out

Gas exchange in capillary beds throughout body tissues

Figure 17.16   Connecting the respiratory and circulatory system. Blood picks up oxygen and other inhaled materials in the lungs and carries it to the heart, where it is pumped to the rest of the body. The blood then returns to the heart, where it is pumped back to the lungs to dump a load of waste gases and to pick up more oxygen.

chemical produced by tobacco plants as a natural pesticide—in fact, nicotine sulfate spray is sold commercially for use by organic gardeners to combat aphids and other insects. The nicotine in tobacco smoke is readily absorbed across the alveoli walls and enters the bloodstream almost immediately after smoke is drawn into the lungs.

In high doses, nicotine is toxic to humans and other mammals. In low doses, nicotine interacts with cells in the brain to stimulate the release of adrenaline, which in turn increases heart rate and blood pressure. (See Chapter 19 for a more detailed description of adrenaline and other homones.) Some of the brain cells activated by nicotine include those in the brain's "reward pathways," regions whose activity is associated with feelings of well-being and happiness. The reward pathways of the brain are an adaptation that reinforces behaviors important for our survival and reproduction; for instance, these pathways are activated in response to tasting delicious food or engaging in sexual activity. By hijacking this pathway, nicotine and other drugs such as cocaine and methamphetamine become addictive, driving the user to seek out these drugs again and again to get the same response. Nicotine addiction is part of what keeps smokers smoking, despite the extensive financial and physical costs associated with tobacco use.

(a) Normal artery

(b) Atherosclerotic artery

Figure 17.17   Atherosclerosis. Fat and cholesterol accumulate in the walls of arteries, reducing their diameter and their ability to carry blood. The formation of these plaques is facilitated by changes in cholesterol levels caused by exposure to tobacco smoke.

Nicotine has effects on other parts of the body besides the brain. It appears to increase production of LDL, the "bad cholesterol" discussed in Chapter 3, and reduce production of HDL, the "good cholesterol." As a result, individuals exposed to tobacco smoke have higher levels of circulating fats (lipids) and thus higher risk for blockages in blood vessels. These blockages are caused by **atherosclerosis**, the accumulation of fats and other debris on the interior walls of arteries, reducing their interior diameter (Figure 17.17). Recall that nicotine stimulates blood-clot formation. When a blood clot becomes trapped in arteries narrowed by atherosclerosis and by nicotine-induced constriction, it can block blood flow. When a blockage occurs in an artery that brings blood to the brain, it results in a **stroke**, which causes the death of brain tissue. When a blockage occurs in an artery that supplies blood to the heart muscle, it can result in the death of part of that muscle in a **heart attack**. Stroke and heart attack are the primary causes of death from cardiovascular disease.

Combining the artery-damaging and clot-inducing results of nicotine consumption with the oxygen-robbing increase in inhaled carbon monoxide, exposure to tobacco smoke represents a serious risk to heart function. According to the U.S. Centers for Disease Control (CDC), heart disease accounts for about 147,000 deaths per year among active smokers. A smoker has a 200% to 300% greater chance of dying from heart disease than does a nonsmoker. However, the CDC also estimates that 35,000 deaths per year from heart disease among non-smokers result from exposure to environmental tobacco smoke. There is good reason to suspect that the artery-damaging effects of smoke exposure also lead to higher rates of stroke and other vascular system damage among smokers and individuals breathing in ETS, but epidemiological studies have not convincingly demonstrated this link.

Unlike the permanent lung changes caused by tobacco smoke, the effects of nicotine and carbon monoxide on cardiovascular health are much more reversible. According to the World Health Organization (WHO), the risk of heart disease decreases by 50% in smokers within 1 year of quitting, and the risk to smokers by 15 years after quitting is the same as the risk to people who have never smoked. The effects of nicotine are reversed over time because this drug is metabolized, that is, changed into simpler forms and excreted. To understand how the body rids itself of nicotine and other toxins, we must first understand the structure and function of the excretory system.

17.3   Removing Toxins from the Body: The Excretory System

The metabolism of nicotine results in waste products that circulate through the bloodstream until they are expelled from the body. The **excretory system** has the task of efficiently removing these and other wastes while retaining valuable materials that can be reused and recycled. In humans, the major organs of the excretory system include the **kidneys**, which filter and cleanse circulating blood before sending the waste through **ureters** to the **bladder**, where it is stored until it is expelled via the **urethra** (Figure 17.18).

Kidney Structure and Function

The kidneys are paired, approximately fist-sized organs that sit behind the liver and stomach in the upper abdominal cavity. These compact organs are packed with a dense network of looped tubules, called **nephrons**. Each kidney contains approximately 1,250,000 nephrons, with a combined length in an adult human of about 145 km (85 miles). Networks of capillaries surround the nephrons, allowing wastes to diffuse out of the blood and into these tubules for excretion.

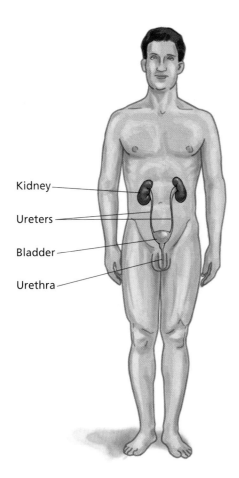

Figure 17.18   The human excretory system. Kidneys filter the blood and transfer waste urine to the bladder, where it is held until voluntarily released through the urethra.

Kidney

Ureters

Bladder

Urethra

The total volume of blood passes through the kidneys hundreds of times per day, such that each kidney filters about 1000 liters of blood every 24 hours.

The processing of waste in kidneys has four distinct phases, which we can follow on a diagram of a single, loop-shaped nephron imbedded in the body, or **cortex** of a kidney. (Figure 17.19 on page 464). Step ①, **filtration**, occurs when some of the plasma portion of the blood is forced through tiny holes in the capillary walls by blood pressure and into the upstream end of the nephron. In step ②, **reabsorption,** valuable substances such as sugars, proteins, and water initially forced out into the filtrate are reabsorbed by the kidney cortex and returned to the bloodstream. Water passively flows out of the nephron and into the kidney cortex as it descends into this salty environment. On the ascending limb, the nephron walls are impermeable to water but actively secrete salt into the cortex. The structure of the nephron allows for a gradient of salt concentration to be maintained, from relatively low at the top of the loop to relatively high at the base. The last segment of the nephron permits active **secretion**, step ③, of certain wastes that are in low concentration in the plasma, including toxins such as the products of nicotine metabolism. The fluid then moves from the nephron into a collecting duct. As the duct crosses the salty kidney cortex, more water diffuses from the filtrate into the kidney tissue. Water and other materials removed from the filtrate are eventually drawn into the capillaries and returned to the bloodstream. During **excretion**, step ④, the remaining concentrated fluid, called **urine**, leaves the nephron and ultimately flows to the bladder. Release of urine from the bladder, called **micturation**, is primarily under our conscious control.

Smoking and the Excretory System

Smoking appears to damage the filtration capacity of kidneys, probably due to increased blood pressure and blood-vessel wall damage. As a result, wastes are not removed as rapidly as they are from a healthy body; more valuable

Figure 17.19   Nephron function. The functional unit of a kidney is the nephron, which first filters plasma from the blood-stream and then allows the reabsorption of essential substances, such as water and salt, from the filtrate. The kidney also actively secretes toxins into the urine, which is then excreted into the bladder.

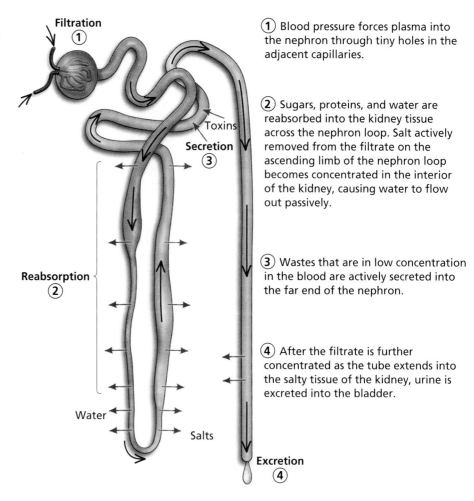

Filtration ①

Toxins

Secretion ③

Reabsorption ②

Water

Salts

Excretion ④

① Blood pressure forces plasma into the nephron through tiny holes in the adjacent capillaries.

② Sugars, proteins, and water are reabsorbed into the kidney tissue across the nephron loop. Salt actively removed from the filtrate on the ascending limb of the nephron loop becomes concentrated in the interior of the kidney, causing water to flow out passively.

③ Wastes that are in low concentration in the blood are actively secreted into the far end of the nephron.

④ After the filtrate is further concentrated as the tube extends into the salty tissue of the kidney, urine is excreted into the bladder.

materials, such as proteins, that would have been reabsorbed are not. Constant exposure to nicotine and its breakdown products may also explain why cohort studies show that the risk of kidney cancer is at least 38% greater among smokers than among those who have never smoked. The toxins in tobacco smoke are even more dangerous to the bladder, where they are retained for long periods of time. Bladder cancer is nearly three times more common in smokers than in nonsmokers. As with many of the health consequences of ETS, it is difficult to determine from current cohort and case-control studies whether kidney and bladder cancers are more common in passive smokers than in nonsmokers. However, given our understanding of kidney function, there is certainly reason to believe that exposure to the toxins in environmental tobacco smoke causes damage to these important organs.

The links between our respiratory, cardiovascular, and excretory systems provide a mechanism whereby what we inhale affects all the organs of our bodies (Figure 17.20). While many of the links between environmental tobacco smoke and disease have not been "proven beyond a reasonable doubt," there is accumulating evidence from both epidemiology and the known risks of active smoking that ETS is dangerous. As citizens, we are often called upon to make decisions in the absence of perfect knowledge. In most of the communities that have considered imposing smoking bans, voters have decided that the risk of ETS is large enough to justify restricting smokers. Given that nonsmokers outnumber smokers by almost 4 to 1, clearing the air in formerly smoky eating establishments and entertainment venues has proven to be mostly good for business. In this case at least, resolution of the controversy has been a win for both business owners and the public's health.

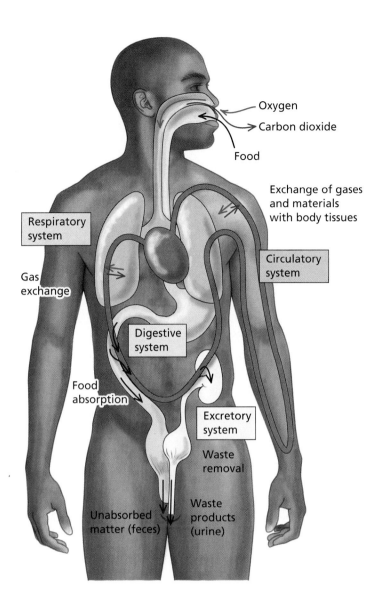

Oxygen

Carbon dioxide

Food

Exchange of gases
and materials
with body tissues

Respiratory
system

Gas
exchange

Circulatory
system

Digestive
system

Food
absorption

Excretory
system

Waste
removal

Unabsorbed
matter (feces)

Waste
products
(urine)

Figure 17.20   Digestion, respiration, circulation, and excretion. This figure illustrates the relationship between these four functions by simplifying the actual structure of their associated organs. These organ systems provide the primary routes of materials into, out of, and throughout the body.

CHAPTER REVIEW

Summary

17.1 Effects of Smoke on the Respiratory System

- Environmental tobacco smoke (ETS) consists of thousands of chemicals, including carbon monoxide, and particles that arise from the smoldering end of a lit cigarette, cigar, or pipe (p. 444).

- The movement of air into and out of the lungs depends largely on the action of the diaphragm (pp. 445–446).

- Air in the lungs flows through bronchi and into alveoli, where gas exchange occurs (p. 450).

- Oxygen for aerobic respiration diffuses from the alveoli into blood inside the surrounding capillaries, while carbon dioxide, a waste product of aerobic respiration, diffuses out of blood into the alveoli (pp. 450–451).

- Hemoglobin in the blood reversibly binds to oxygen molecules as blood flows through the lungs and gives up its load of oxygen in working tissues (p. 451).

- Small particles in tobacco smoke land in the larger bronchi, causing cell damage and eventually leading to chronic bronchitis. Chronic bronchitis in turn can lead to emphysema, in which inelastic lungs trap air and impede gas exchange (pp. 452–453).

- The tiniest smoke particles are drawn into the alveoli, where they may remain for long periods, exposing the alveolar cells to carcinogens within the smoke particles (pp. 453–454).

- Epidemiologists have linked exposure to environmental tobacco smoke to a risk of lung cancer by looking at lung cancers in nonsmokers as well as by following large populations of individuals who have different exposures to ETS and measuring their health outcomes (pp. 454–455).

Web Tutorial 17.1 **Gas Exchange in the Lung**

17.2 Spreading the Effects of Smoke: The Cardiovascular System

- Blood is made up of a liquid portion, called plasma, and a cellular portion, consisting of red blood cells, white blood cells, and platelets (pp. 455–456).

- Platelets are involved in blood clotting. Exposure to nicotine increases the production and "stickiness" of platelets (p. 456).

- The heart is a double pump consisting of four chambers—two atria and two ventricles. The right-side pump sends oxygen-poor blood to the lungs; the left-side pump sends oxygen-rich blood to the body (p. 457).

- Blood leaves the heart via arteries, exchanges materials with other tissues of the body in capillaries, and returns to the heart via veins (p. 458).

- Blood from the lungs flows to the heart, where it is pumped into the systemic circulation. After dropping off a load of oxygen and picking up carbon dioxide in the body tissues, the blood returns to the heart and is pumped back to the lungs (p. 460).

- The nicotine in smoke increases heart rate and blood pressure, putting strain on the heart muscle (pp. 460–461).

- Nicotine also increases the production of LDL ("bad cholesterol"), causing atherosclerosis (p. 462).

Web Tutorial 17.2 **The Cardiovascular System**

17.3 Removing Toxins from the Body: The Excretory System

- The excretory system consists of the kidneys, bladder, ureters, and urethra (p. 462).

- Plasma is filtered from the blood within nephrons. These structures help return water and other valuable components to the blood while allowing excretion of the waste products of metabolism and other toxins (pp. 462–463).

- Toxins from tobacco smoke are stored in urine within the bladder, leading to a greatly increased risk of bladder cancer among smokers (p. 464).

Web Tutorial 17.3 **The Mammalian Kidney**

Learning the Basics

1. Describe how contraction of the diaphragm allows the lungs to fill, and how oxygen from inhaled air enters the lungs.

2. Describe the flow of blood through the circulatory system, heart, and lungs in humans.

3. List the three processes that occur within the nephrons of kidneys.

4. Compared to smoke inhaled from the filter end of a cigarette, ETS _____.

   A. is less concentrated; **B.** contains more products of incomplete combustion; **C.** contains more particles; **D.** a and b are correct; **E.** a, b, and c are correct

5. Alveoli are _____.

   A. small air sacs at the ends of bronchi in the lungs; **B.** the respiratory surface in mammals; **C.** surrounded by a net of capillaries; **D.** subject to damage because of exposure to tobacco smoke; **E.** more than one of the above is correct

6. Emphysema leads to overinflation of the lungs because _____.

   A. affected individuals inhale too much; **B.** exposure to tobacco smoke leads to uncontrolled production of additional alveoli; **C.** the bronchial tubes are filled with thick mucus, which traps air; **D.** buildup of scar tissue makes lungs less elastic, trapping air; **E.** lungs are subject to chronic, allergic inflammation

7. The function of platelets is _____.

   A. carrying oxygen; **B.** defending the body from infection; **C.** aiding in the production of blood clots; **D.** carrying water and dissolved salts; **E.** recycling hemoglobin

8. Blood flowing from body tissues to the heart _____.

   A. is pumped via veins to the rest of the body; **B.** is pumped via arteries to the lungs; **C.** is under relatively high pressure compared to blood leaving the heart; **D.** returns via millions of capillaries tied directly to the heart's two atria; **E.** has used up all of its hemoglobin;

9. After blood is filtered in the nephrons of the kidneys, _____.

   A. water and nutrients are reabsorbed; **B.** toxins are actively secreted into the urine; **C.** carbon dioxide is actively loaded into unfiltered blood cells; **D.** a and b are correct; **E.** a, b, and c are correct

10. Which of the following diseases is associated with exposure to tobacco smoke?

    A. heart disease; **B.** lung cancer; **C.** bladder cancer; **D.** asthma; **E.** all of the above

Analyzing and Applying the Basics

1. Some endurance athletes, such as cyclists, engage in the practice of "blood doping," in which packed red blood cells are transfused into their bloodstream immediately before an event. How might increasing the volume of red blood cells provide an advantage to an athlete? What do you think could be the risks of this practice?

2. About 17,000 babies born in the United States each year have a heart defect that allows blood from the right side of the heart to mix with blood in the left side of the heart. How do you think a "hole in the heart," as this condition is commonly termed, affects the function of the cardiovascular system?

3. Why are valves found in veins but not in arteries?

Connecting the Science

1. Exposure to tobacco smoke causes early death, but it also causes years or even decades of disabling, chronic health problems such as emphysema and lung cancer. Some people have argued that the best way to reduce the number of smokers is to make smoking, and the health care costs associated with it, more expensive. This strategy entails placing high taxes on cigarettes and other tobacco products and restricting current and former smokers' access to subsidized medical care. What do you think of these strategies? Do you think they would be successful? Do you think they might be unfair?

2. Although tobacco causes more cases of long-term disability, alcohol is more deadly to young men and women than cigarettes because of automobile and other accidents. Do you think the success of tobacco bans means that alcohol will become more strongly restricted? Do you think this is a good policy? Why or why not?

CHAPTER 18

Will Mad Cow Disease Become an Epidemic?

Immune System, Bacteria, Viruses, and Other Pathogens

Scientists are trying to
prevent the spread of
mad cow disease.

The brains of affected individuals look spongy.

In deer, this disease is called chronic wasting disease.

In the late 1980s, dairy farmers in Great Britain began noticing unusual behaviors in a few of their cows. Cows that had previously shown no signs of illness became progressively less coordinated. They would shake and tremble and rub parts of their bodies against walls or fences. Farmers watched with alarm as their cows staggered around fields. Even their dispositions seemed to change. Many of the cows appeared to be irritated or even crazy, leading to the name "mad cow disease."

Several months after the onset of symptoms, the cows inevitably died. Autopsies revealed that the brains of the "mad" cows had holes in them characteristic of a class of diseases known as spongiform encephalopathies. An encephalopathy is a pathology, or disease, of the brain. The diseased cows' brains resembled porous, natural sponges filled with holes. Mad cow disease is a type of spongiform encephalopathy that affects only cows, so it is called bovine spongiform encephalopathy, or BSE.

Spongiform encephalopathies have been diagnosed in sheep, whose skin becomes so itchy that they scrape their bodies against fences. In fact, the disease name, scrapie, describes this behavior. When the disease is present in elk and deer, it is called chronic wasting disease, due to the emaciated appearance of affected animals.

This disease can also be spread to humans.

Humans can also become ill with spongiform encephalopathies. An obscure disease called kuru has long been known to occur in natives of the eastern highlands of New Guinea. The disease results from the tribal custom of honoring the dead by eating their brains. Individuals with this disease lose coordination and often become demented. They, too, inevitably die.

A long-recognized but previously very rare condition called Creutzfeldt-Jakob Disease (CJD) is another spongiform encephalopathy of humans. Like the diseased cows, affected humans become very agitated, dizzy, and short-tempered. They experience short-term memory loss, lack of coordination, and slurred speech. As in the case with cows, this disease is lethal in all the affected humans. Typically, only elderly people display the disease symptoms. However, a recent and alarming trend has been an increase in the rate of diagnosis of CJD in young British patients.

On average, these patients live just over a year after being diagnosed, and when examined at autopsy, their brains look more similar to those of BSE-infected cattle than CJD-infected humans. The increased number of people diagnosed with the disease, the structure of their brains after death, and the younger age of those infected led scientists to believe that BSE was, in some way, being transmitted from infected cows to these individuals. This new, transmissible form of the disease was named new-variant CJD (nvCJD).

Scientists, doctors, veterinarians, hunters, and patients and their families all over the world are concerned about preventing the spread of these diseases. Several recent cases of mad cow disease in Canada—while the United States currently considers lifting the ban on importing beef from that country—bring concerns about the spread of this disease to U.S. citizens. As a first step toward understanding this disease and preventing its spread in cows and humans, scientists had to determine what type of agent was causing it.

18.1   Infectious Agents

Infectious diseases result from the activity of infectious agents. An infectious agent gains access to the body and uses the body's resources for its own purposes. Infectious diseases differ from genetic diseases in that they are caused by organisms rather than caused by malfunctioning genes—although malfunctioning genes can make an organism more susceptible to infection.

Disease-causing organisms are called **pathogens**. When a pathogen can be spread from one organism to another, it is said to be **contagious**. When it finds a tissue inside the body that will support its growth, it becomes **infectious**. Some organisms can cause an infection in an individual but are not contagious if the infected individual cannot pass the infection to other individuals. Organisms that can be seen only when viewed under a microscope are called microscopic organisms or **microbes**. Bacteria and viruses are the most common infectious microbes.

Bacteria

Bacteria (Figure 18.1) are a diverse group of single-celled organisms. They are tiny and numerous. In fact, there are more bacteria in your mouth than there are humans on Earth. These rod-shaped (bacilli), spherical (cocci), or spiral (spirochetes) cells are prokaryotic and do not contain organelles that are surrounded by membranes, such as the endoplasmic reticulum or the Golgi apparatus (see Chapter 2). Prokaryotes do not have a nucleus. Unlike their eukaryotic counterparts, which contain their DNA in a nucleus, prokaryotes have DNA that is coiled up inside the **nucleoid region**. The bacterial genome consists of a double-stranded, circular DNA molecule. In addition to the large DNA chromosome, bacteria may also contain small,

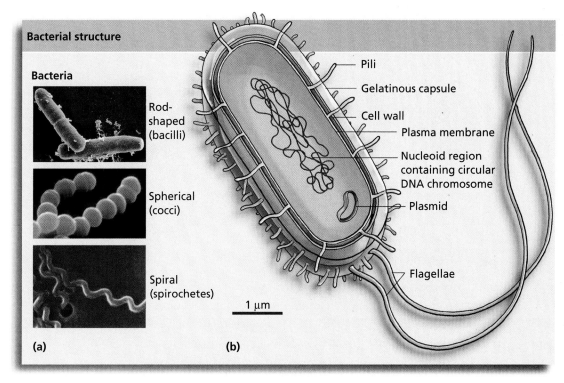

Figure 18.1   Bacterial structure. The structure of a typical bacterium (a) as viewed under a microscope and (b) illustrated.

circular extrachromosomal DNA (DNA that is separate from the chromosome) called **plasmids**. In Chapter 8, you learned that scientists use plasmids during genetic engineering. Some of these plasmids also carry antibiotic-resistant genes that allow bacteria to resist the drugs that would normally kill them. Essay 18.1 addresses the growing problem of antibiotic resistance in pathogenic bacteria.

Most bacterial cells are surrounded by a **cell wall** that provides rigidity and protection; it is composed of carbohydrate and protein molecules. The cell walls of many bacteria are surrounded by a gelatinous **capsule**, which helps the bacteria not only to attach to cells within tissues they will infect but also to escape destruction by cells of the immune system. Bacteria also may have external **flagella** to aid in motility, and **pili**, which help some bacterial cells pass genes to each other.

Bacteria reproduce by a process called **binary fission** (see Figure 18.2 on page 472). When a bacterial cell divides by binary fission, the single, circular chromosome that is attached to the plasma membrane inside the cell wall is copied. The copy is attached to another site on the plasma membrane, and the membrane between the attachment sites grows and separates the two copies of the original chromosome until it eventually produces two separate daughter cells. This is an example of asexual cell division because one parent cell gives rise to identical daughter cells.

Bacteria can reproduce rapidly under favorable conditions, doubling their population every 20 minutes or so. For example, after 8 hours at room temperature, a single *Salmonella* bacterium in a chicken salad sandwich will give rise to over 33 million bacterial cells. This exponential growth occurs because one cell gives rise to two, and those two yield four, the four divide to become eight, then 16, and so on (see Figure 18.3 on page 472). *Salmonella* reproduces quite quickly in food because the bacteria have access to all the nutrients they need. They reproduce even more quickly at room temperature than when kept in a refrigerator.

Bacterial replication

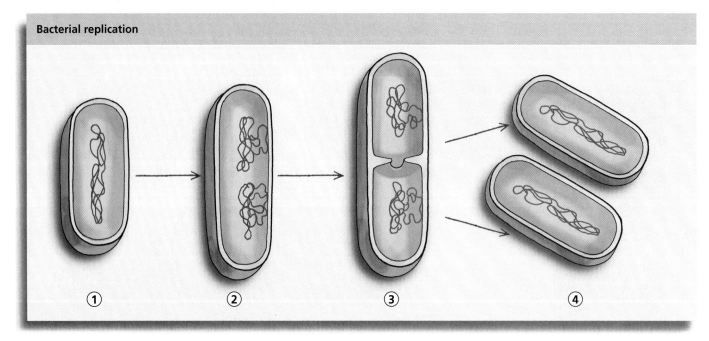

Figure 18.2   Bacterial replication. (1) The bacterial cell begins with one copy of the circular DNA chromosome in the nucleoid region. (2) The circular chromosome is copied, and each copy is attached to the plasma membrane. (3) Continued growth separates the two chromosomes. The plasma membrane grows inward in the middle, and a new cell wall is constructed. (4) Daughter cells separate, and chromosomes return to the nucleoid region.

When an infection occurs in your body, the rapidly increasing bacteria support their growth by using your cells' nutrients, effectively preventing your cells from functioning properly. However, most cases of disease-causing bacterial infection result from more than just the large numbers of bacteria in your system. The symptoms of a disease arise due to the effects of biological molecules secreted by the bacterial cells, called **toxins**, that block cellular processes.

Anthrax is an infectious disease caused by the bacterium *Bacillus anthracis*. This microorganism lives mainly in soil and produces structures called **spores** that can be lethal when inhaled. This disease is rarely contracted by humans; weapons-grade anthrax requires difficult and expensive processing to make the spores airborne. Anthrax is an example of a disease

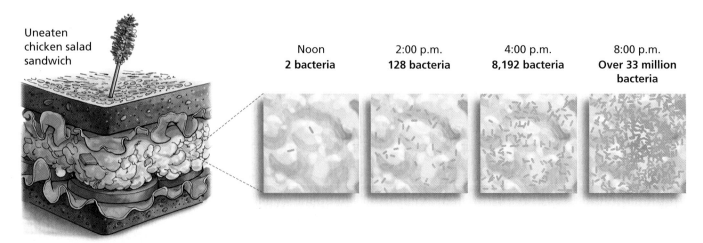

Figure 18.3   Exponential growth of bacteria. *Salmonella*, the bacterium that causes food poisoning, reproduces every 20 minutes. If a sandwich with two bacteria is left out for 20 minutes, each bacterial cell will make a copy of itself, yielding 4 bacteria. After 40 minutes, there will be 8 bacteria; within 8 hours, there will be 33 million bacteria.

Essay 18.1 Focus on Evolution: Antibiotic Resistance

Antibiotics are medications that doctors prescribe for people with bacterial infections. When you were a child and had an ear infection, your parents probably gave you spoonfuls of a pink medicine (Figure E18.1). Now when you have a sinus infection, your doctor prescribes other antibiotics. These infections are easily treated and cause you little inconvenience besides a trip to the doctor and pharmacy. Unfortunately, the days of easily treatable bacterial infections may soon be over because the bacteria that cause many diseases—including tuberculosis, ear infections, and gonorrhea—have become resistant to the antibiotics developed to cure them.

Antibiotic resistance develops as a result of natural selection on a bacterial population. Variation exists within any population of individuals, including within a population of bacterial cells. Some bacteria, even before exposure to the antibiotic, carry genes that enable them to resist it. You can develop a drug-resistant infection by selecting for the resistant bacteria within your own body or by contracting the resistant bacteria from someone else.

Tetracycline is an antibiotic used to kill the bacteria that cause acne. It works to prevent translation (the synthesis of proteins) in prokaryotes. Since the workbenches of translation, the ribosomes, differ structurally in prokaryotes and eukaryotes, this antibiotic—like all effective antibiotics—selectively kills bacterial cells and not eukaryotic host cells. One bacterial cell in a population of millions may have a different DNA sequence in the region of DNA that encodes a protein involved in translation. If this protein enables the bacterial cell to withstand tetracycline treatment, then it will survive and pass on the resistant sequence to its offspring. In this manner, those bacteria that are not killed are selected for, and resistant infections develop.

The problem of resistance in bacteria originated from medical and agricultural overuse of antibiotics. Some patients ask their doctors to give them antibiotics for a cold or the flu, both of which are caused by viruses. Viruses are not affected by antibiotics, since viruses do not have the organelles and other structures that antibiotics target. For example, tetracycline would not be able to kill a viral infection because the virus is using the patient's eukaryotic ribosomes for translation and is thus unaffected by the treatment. Adding to the problem of medical overuse is that people often discontinue their prescribed antibiotics when they feel better. A person who does not finish all of the prescribed medication allows resistant bacteria to proliferate; the infection will come back in a few weeks and be resistant to that antibiotic, and another course of medication will be required.

The use of antibiotics in agriculture is also a problem. Animals such as cows and chickens are given antibiotic drugs in their feed to prevent them from becoming ill. Animals who are fighting infections do not gain as much weight and must then be sold for less money. Antibiotic-treated animals may harbor resistant bacteria, and eating undercooked meat from an animal infected by a resistant population of bacteria can give you a drug-resistant infection.

Figure E18.1 Amoxicillin is a familiar antibiotic.

that is infectious without being contagious because infected individuals do not pass the disease on to other people. When anthrax bacteria are inhaled or lodged in the skin, the disease results from the secretion of toxins that produce an accumulation of fluid inside white blood cells called **macrophages**, which then swell and release the toxic chemicals that they normally use to kill bacteria. The toxic chemicals released by macrophages kill the host organism unless antibiotics have been administered soon after exposure; this is why anyone suspected of having come in contact with anthrax spores is given antibiotics even before an infection is confirmed.

While many different types of bacteria cause infections that are troublesome or even fatal to humans, bacteria also play many important roles in our bodies and in nature. Viruses are another category of microorganisms that serve as infectious agents.

Viruses

Viruses (Figure 18.4) are not considered to be living organisms in part because they cannot replicate (copy) themselves without the aid of a host cell. Hosts for viral infections include many organisms other than humans, such as bacteria and plants. Viruses lack the enzymes for metabolism and contain no ribosomes; therefore, they cannot make their own proteins. They also lack cytoplasm and membrane-bound organelles. Viruses are really nothing more than packets of nucleic acid (DNA or RNA) surrounded by a protein coat.

The genetic material, or **genome**, of a virus can be DNA or RNA; it can be double stranded or single stranded, and it can be linear or circular. For example, the herpes virus has a double-stranded DNA genome, and the polio virus has a single-stranded RNA genome. The genes of a virus can code for the production of all the proteins required to produce more viruses.

The protein coat surrounding a virus is called its **capsid**. Many of the viruses that infect animals develop an additional structure outside the capsid called the **viral envelope**. The envelope is derived from the cell membrane of the host cell and may contain additional proteins encoded by the viral genome. Viruses that are surrounded by an envelope are called enveloped viruses.

Infection by an enveloped virus occurs when the virus gains access to the cell by fusing its envelope with the host's cell membrane (Figure 18.5). An unenveloped virus uses its capsid proteins to bind to receptor proteins in the plasma membrane of a host cell. Some capsid proteins function as enzymes that digest holes in the plasma membrane, thereby allowing the viral genome to enter into cells. Once inside the host cell, the capsid is removed.

Whether the virus is enveloped or not, after the genome enters the host cell, the infection continues when the virus makes copies of itself. First the genome is copied; then the virus uses the host-cell ribosomes and amino acids

Figure 18.4   Viral structure. The structure of a typical virus (a) as viewed under a microscope and (b) illustrated.

Viral replication

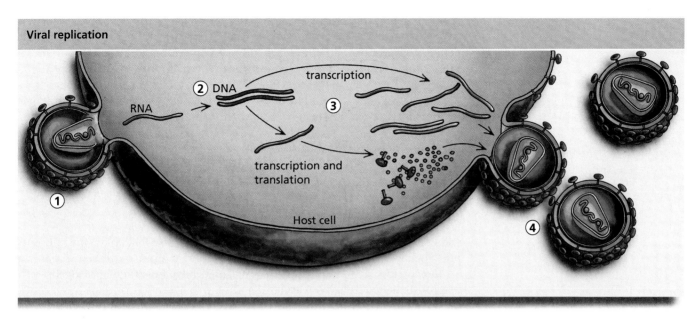

Figure 18.5 Viral replication. (1) Viral membrane fuses with the host cell; the capsid is removed, and the genome enters. (2) The genome is copied and used to synthesize many copies of double-stranded DNA. (3) The DNA is transcribed and translated by the host cell. More viral proteins and copies of the viral genome are produced. (4) Once assembled, new viruses leave the host cell and infect other cells.

to make viral proteins for building new capsids and synthesizing some of the envelope proteins. Once assembled, the new virus exits the cell; leaving behind some viral proteins in the host's cell membrane, and moves to another cell to spread the infection.

Some viruses, called **latent viruses**, enter a state of dormancy. During that time, the virus is not being replicated. The herpes virus, for example, can undergo long periods of dormancy when it is still present in the body but not causing symptoms such as painful sores and blisters.

Viruses whose incidence has increased within the past two decades or those that seem likely to increase substantially in the near future are called **emerging viruses**. One example of a disease caused by an emerging virus is severe acute respiratory syndrome (SARS), which caught the world's attention in 2003 (Figure 18.6). SARS causes a high fever, chills, and body aches. SARS patients can die if the infection causes pneumonia. The World Health Organization (WHO) reported in 2003 that over 8,000 people were infected with the virus that causes SARS; about 10% of those infected died.

Viral infections may emerge for many reasons. Emergence can be caused by the proliferation of a new agent or by detection of an agent that has been present but was unidentified. Emergence can also be due to the reappearance of a previously known infection. Many organisms in the population may have developed immunity to one strain of a virus. If that virus undergoes genetic mutation that allows it to evade detection by the immune system, it would be better at replicating itself—and thus selected for—and reemerge.

Emergent viruses can also arise due to the spread of an existing virus from one host species to another. Some new human diseases originate in other animals. The virus that causes Ebola is one such example. Ebola virus, which has caused outbreaks in Africa, first appeared in 1976. Symptoms of Ebola infection include fever, headache, and muscle pain. The virus

Figure 18.6 SARS. A highly magnified photograph of the virus that causes SARS.

contains genes that code for the production of proteins that are very similar to proteins of the immune system that help stop blood flow—a process called **coagulation**. In the early stage of infection, the virus overstimulates coagulation so much that later in the infection, when these proteins are really needed, the supply has been exhausted, and blood begins to leak from every orifice. Humans' encroachment on monkey habitat led to increased contact between humans and monkeys, allowing Ebola virus to spread.

While many infectious diseases are caused by bacteria or viruses, some are also caused by eukaryotic organisms.

Eukaryotic Pathogens

Many pathogenic organisms are **parasites**, or organisms that obtain the nutrients and shelter required for growth and development from a different organism while contributing nothing to the survival of its host. Eukaryotic pathogens such as protists (single-celled eukaryotes) and worms cause tremendous human suffering and death worldwide.

Malaria is caused by several different protists of the genus *Plasmodium* (Figure 18.7a) This disease is transmitted when a mosquito bites an infected person, carries the parasite, and then passes the parasite on to an uninfected person by a subsequent bite. In humans, the parasites grow and reproduce in the liver and red blood cells. The growth of parasites in the red blood cells destroys the cells and releases more parasites into the bloodstream. If a female *Anopheles* mosquito bites a human who is infected with the parasite and later bites an uninfected person, the parasite is passed on. Symptoms include shaking and chills, high fever, and severe headache. Damage to the kidney and red blood cells can cause kidney failure and anemia, respectively. Malaria affects 300 to 500 million people annually (that's more than the population of the United States) and results in nearly 3 million deaths. This disease is most common in tropical and subtropical climates.

Schistosomiasis is caused by worms of the genus *Schistosoma* (Figure 18.7b). Infection results from contact with contaminated water in which certain types of freshwater snails that carry schistosomes are living. Freshwater becomes contaminated by *Schistosoma* eggs when infected people urinate or defecate in the water. Eggs present in urine and feces hatch, and the young worms grow and develop inside snails. At adulthood, the parasite leaves the

(a) *Plasmodium falciparum* **(b)** *Schistosoma* worm

Figure 18.7 Eukaryotic pathogens. The *Plasmodium* pathogen that causes malaria is shown in part (a), and the pathogen that causes schistosomiasis is shown in part (b).

snail and enters the water. The worms can then penetrate the skin of persons who are swimming, bathing, or washing. Once a person is infected, the worms grow inside the person's blood vessels and then produce their eggs. Some of these eggs reach the bladder or intestines, where they are expelled along with urine and feces and can make their way into the water to set up the cycle anew. Symptoms of this infection include a rash, itching, chills, fever, cough, and muscle aches. This infection is highly prevalent in developing countries where plumbing is lacking. Roughly 200 million people worldwide are infected.

Bacteria, viruses, and eukaryotic parasites are the most common infectious agents. In Table 18.1, you can read about more diseases caused by these organisms. However, the spongiform encephalopathies are not caused by any of these organisms. Scientists were able to make this determination because the infectious agent causing these diseases does not respond to the treatments—such as heat, chemicals, and radiation—that usually kill bacteria, viruses, and eukaryotic parasites. Instead, the spongiform encephalopathies seem to be caused by a novel class of infectious agents called prions.

Table 18.1   Infectious agents and the diseases they cause.

Agent	Disease	Agent	Disease
Bacteria 1 µm	• Salmonellosis • Anthrax • Gonorrhea • Botulism • Impetigo • Leprosy • Lyme disease • Tuberculosis • Strep throat • Tetanus • Diphtheria • Bacterial meningitis	**Viruses** 0.01 µm	• Chicken pox • Measles • Mumps • Influenza • Rabies • Shingles • Smallpox • AIDS • Hepatitis • Viral pneumonia • Viral meningitis • Polio
Prions 0.001 µm	Spongiform encephalopathies: • Mad cow disease • Scrapie • Chronic wasting disease • Kuru • Creutzfeldt–Jakob disease (CJD)	**Eukaryotic Pathogens** Worms, Giardia, Yeast cells	• Worms cause schistosomiasis, trichinosis, tapeworm • Protists cause amoebic dysentery, giardia, sleeping sickness, and malaria • Fungi cause athlete's foot, oral thrush

Prions

After a cell synthesizes a protein, the protein folds into its characteristic shape. If a protein is folded incorrectly, it can no longer perform its job properly. A **prion** (Figure 18.8) is a normally occurring protein produced by brain cells that, when misfolded, causes spongiform encephalopathy.

The word *prion* is a shortened form of the term *proteinaceous infectious particle*. Normal prions are present in the brains of all mammals that have been studied. When highly magnified, the normal shape of a prion resembles a coil. The misfolded version of this protein resembles a sheet of paper that has been accordion-folded lengthwise several times.

The normal role of the prion in the brain is not yet clear, but experiments in mice lacking the prion gene, and therefore unable to make the normal version of the protein, indicate that it may protect mammals against dementia and other degenerative disorders associated with aging. The very rare CJD seen in elderly people is believed either to arise spontaneously when a prion is mistakenly misfolded, or to be caused by a mutation to the gene that encodes the prion protein in humans, leading to the production of a misfolded prion protein.

In contrast, the newly emerging form of the disease that affects younger people, nvCJD, results when an individual is *infected* by misfolded prion proteins. Remarkably, the misfolded protein refolds the properly folded proteins into the mutant, disease-causing version (Figure 18.9). This is very unusual behavior for a protein; no other known protein has this capability.

Over time, the nerve cells in the brain become clogged with the misfolded prions, causing them to improperly transmit normal impulses and eventually stop functioning altogether. Ultimately the cells burst, freeing their misfolded prions to find and refold normal prions in other nerve cells. Finally, the brain becomes riddled with empty spaces formerly occupied by normal cells, producing the spongelike character of the diseased brain.

Prion structure

Prions

(a) Normal prion protein has more helical regions.

(b) Misfolded prion protein has more pleated regions.

0.001 μm

Amino acid chain

Figure 18.8   Prion structure. (a) Normal and (b) misfolded prions are illustrated.

Prion replication

Figure 18.9   Prion replication. (1) A misfolded prion protein refolds a normal protein into the mutant form. (2) The number of misfolded prions is now doubled. Each refolds a normal prion protein into the mutant form. (3) The number of misfolded prions continues to double, eventually bursting the host cell and spreading infection to other cells.

Unlike viruses, bacteria, and eukaryotic pathogens, prions have no RNA or DNA, which were once thought to be necessary for any infectious agent to multiply. Furthermore, although most proteins within cells are easily broken down, prions resist degradation, allowing these rogue disease agents to propagate relentlessly.

Prior to the discovery of prions in the early 1980s, all known infectious agents had their own genomes, and many researchers did not believe that proteins themselves could be infectious. In fact, very little money was available for the study of prions because this type of infectious agent is so rare and poorly understood. Dr. Stanley Prusiner, a neurologist at the University of California, San Francisco, was the first to systematically study prions and to link them with disease (Figure 18.10). Although many scientists remained skeptical about his findings, Prusiner was awarded the 1997 Nobel Prize in Medicine or Physiology for his research. However, understanding the nature of prions as infectious agents is only the first step toward preventing spongiform encephalopathies from becoming widespread in humans.

18.2   Epidemics

An **epidemic** is a contagious disease that spreads rapidly and extensively among members of a population. An **epidemiologist** is a scientist who attempts to determine who is prone to a particular disease, where risk of the disease is highest, and when the disease is most likely to occur. Epidemiologists try to answer these questions by determining what the victims of the disease have in common. By identifying what factors increase the risk of a disease, epidemiologists help formulate public health policy. When it is difficult to pinpoint what is causing a disease or how it is spread, deadly epidemics can result. For an epidemic to occur, the infectious agent must cause disease and must be transmissible from one organism to another.

Figure 18.10   Dr. Stanley Prusiner. Dr. Prusiner, at left, won the 1997 Nobel Prize for his work with prions.

Table 18.2 Transmission of infectious agents. Infectious agents are transmitted in four ways, allowing pathogenic organisms to gain access to a host.

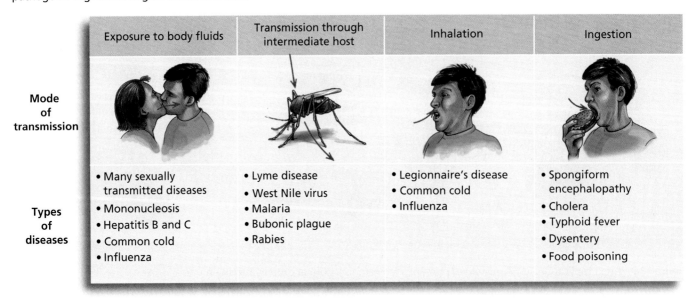

	Exposure to body fluids	Transmission through intermediate host	Inhalation	Ingestion
Mode of transmission				
Types of diseases	• Many sexually transmitted diseases • Mononucleosis • Hepatitis B and C • Common cold • Influenza	• Lyme disease • West Nile virus • Malaria • Bubonic plague • Rabies	• Legionnaire's disease • Common cold • Influenza	• Spongiform encephalopathy • Cholera • Typhoid fever • Dysentery • Food poisoning

Infectious diseases are transmitted via contact with the disease-causing organism. This can include contact with infected fluids such as blood, saliva, and semen as well as exposure facilitated by an intermediate host. Transmission also occurs via inhalation or ingestion of the disease organism or its toxins (Table 18.2).

Exposure to Infected Body Fluids

As discussed in Chapter 10, AIDS is caused by exposure to infected blood. Hepatitis is another disease caused by contact with infected blood. There are at least six different hepatitis viruses, but the most dangerous is the virus known as hepatitis C. Exposure to hepatitis C leads to chronic liver diseases such as cirrhosis (irreversible, potentially fatal scarring of the liver), liver cancer, and liver failure. Hepatitis C ranks second to alcoholism as a major cause of liver disease and is the leading reason for liver transplants in the United States.

It is estimated that about 3% of the world's population carries this virus, but they are unaware of it because there are often no symptoms until the liver damage is severe. However, people who donate blood may find out that they have the disease before it causes serious damage because blood banks routinely test for the hepatitis C virus and inform donors who test positive. Prior to 1990, there were no tests for hepatitis C, and people who had transfusions prior to that are at risk for the disease. Intravenous drug use and possibly some skin-piercing practices, notably tattooing and body piercing, may also be contributing to the spread of hepatitis C.

Saliva is another body fluid that transmits infectious disease. Mononucleosis is spread when an infected individual kisses or shares a drinking glass with an uninfected person. This disease is caused by a virus that infects cells of the immune system; symptoms include sore throat, fever, and excessive fatigue.

The exchange of semen and vaginal fluids facilitates the spread of sexually transmitted diseases (STDs), which are discussed in more detail in Chapter 20.

Transmission Through an Intermediate Host

Transmission of microorganisms can occur through an intermediate organism such as an insect. The organism that carries disease-causing microorganisms from one host to another is called a **vector**.

(a) Deer tick

(b) Western black-legged tick

Figure 18.11 Ticks that carry the bacteria that causes Lyme disease. (a) Deer tick.
(b) Western black-legged tick.

Lyme disease, caused by the bacterium *Borrelia burgdorferi*, is transmitted via
the bite of an infected tick. Not all kinds of ticks carry the bacteria. In the eastern
United States, the deer tick is usually responsible for transmitting the disease
(Figure 18.11a). In the western states, the western black-legged tick is the culprit
(Figure 18.11b). A rash with a clear center that resembles a bull's-eye may but does
not always appear anytime from one day to one month after the infected tick's
bite. Symptoms of the disease include numbness and pain in the arms or legs,
paralysis of the facial muscles, fever, fatigue, stiff neck, and headaches. Antibiotics
are now available that can cure this disease when it is diagnosed early.

The mosquito serves as a vector for the West Nile virus. Mosquitoes pick up
the virus when they feed on birds that have died due to West Nile infection.
Infected mosquitoes can then transmit the virus to humans with a bite. Once the
West Nile virus is in a human's bloodstream, it multiplies and can move
through the bloodstream into the brain, causing a swelling of the brain called
encephalitis.

Inhalation

Other pathogens cause disease when they enter the body along with the air we
breathe. The flu is transmitted when a person who has it coughs, sneezes, or
speaks and sends the virus into the air, and other people then inhale the virus.
The flu also may be spread when people touch a surface that has flu viruses on it
and then touch their nose or mouth. A person can spread the flu even before he
or she feels sick, starting from one day prior to the onset of symptoms until about
seven days after symptoms start. These symptoms usually begin suddenly and
may include fever, headache, tiredness, dry cough, sore throat, and body aches.
The influenza virus is the pathogen that causes the flu. The flu is with us every
winter because it mutates rapidly and is able to evade the immune system.
Those viruses that undergo mutations that allow them to escape the immune
system are more likely to replicate themselves and therefore are selected for.

Ingestion

The transmissible types of spongiform encephalopathies are spread by ingesting food containing misfolded prions. Scientists believe that the cows with
mad cow disease ingested misfolded proteins when they ate the remnants of
diseased animals that were present in their feed.

After a cow has been slaughtered, what is left of the carcass is cooked and then ground up, producing a type of cattle feed called meat-and-bone meal. Feed mills buy the meal and make it into a toasted, crunchy, cereal-like substance that is fed to cows. When feed mills unwittingly use cows with mad cow disease in making meat-and-bone meal, the infectious prions from those cows are fed to other cows, which become infected and thus spread BSE to more cows.

Humans become infected with misfolded prions when they eat meat from diseased cows (or from other humans, as is the case with kuru). Although the pathogen that causes spongiform encephalopathies in cows and humans infects the brain and spinal cord—parts of the cow that are normally not consumed—meat can be contaminated during the rendering process. For example, hamburger containing parts of the spinal cord that were stripped off the carcass may be infectious. If people eat contaminated meat, it is difficult for their immune systems to fight the infection.

18.3   The Body's Response to Infection: The Immune System

Organisms have evolved defenses against pathogens gaining access to the body. In humans and most vertebrates, these consist of three different lines of defense (Figure 18.12).

First Line of Defense: Skin and Mucous Membranes

The skin, secretions, and mucous membranes comprise the first line of defense against pathogens. These external physical and chemical barriers are

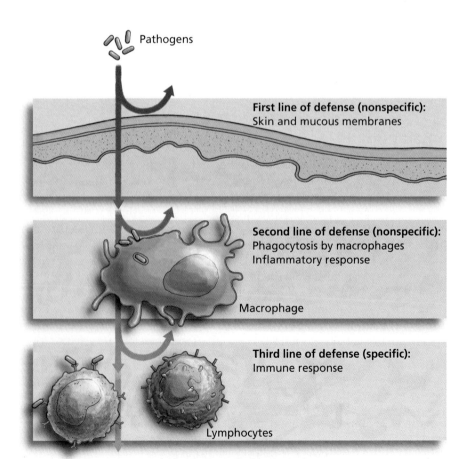

Pathogens

First line of defense (nonspecific): Skin and mucous membranes

Second line of defense (nonspecific): Phagocytosis by macrophages Inflammatory response

Macrophage

Third line of defense (specific): Immune response

Lymphocytes

Figure 18.12   The body's defense against infection. The body has three lines of defense: The skin and mucous membranes try to keep infection out; macrophages attack pathogens that slip through and cause an inflammatory response; and lymphocytes target specific pathogens and attempt to contain them.

nonspecific defenses; that is, they do not distinguish one pathogen from another. Organisms living on the body's surface are kept out by the skin. Glands in the skin even secrete chemicals that slow down the growth of bacteria. Tears and saliva secrete enzymes that break down bacterial cells. Mucous membranes lining the respiratory, digestive, urinary, and reproductive tracts also secrete mucus that traps pathogens, which are washed away or prevented from entering the body. Pathogens that are able to evade the first line of defense next encounter the internal, or second, line of defense.

Second Line of Defense: White Blood Cells, Proteins, and Inflammation

If a pathogen is able to get past the first line of defense and start an infection, the second line of defense, which is also nonspecific, can often stop the infection. This internal defense system consists partly of white blood cells that indiscriminately attack and ingest invaders. The ingestion of pathogens by cells is called **phagocytosis**. Phagocytosis is not limited to ingesting invading organisms. Phagocytosis is a type of endocytocis that some cells can use when they need to take in materials that are too large to pass through the cell membrane. Phagocytic white blood cells called **macrophages** move throughout the lymphatic fluid, cleaning up dead and damaged cells. To destroy the cells, macrophages extend their long **pseudopodia** (cellular extensions used for eating and moving), grab the invading organism, and engulf it; enzymes inside the macrophage help break the invader apart. Much of the destruction of the offending cells occurs in the lymph nodes, which is why those nodes swell when we are ill with an infection.

Additional white blood cells that circulate through the blood and lymph destroying invaders are **natural killer cells**. These cells attack virus-invaded body cells by penetrating the plasma membranes of infected cells and causing them to burst.

In addition to white blood cells, some proteins act as nonspecific defenders. **Interferons** are proteins produced by virus-infected body cells to help uninfected cells resist infection. When infected cells die, they release interferons that bind to receptors on uninfected cells and stimulate the healthy cells to produce proteins that inhibit viral reproduction.

Another type of defensive protein, called **complement proteins**, circulate in the blood. These proteins can coat the surfaces of microbes, making them easier for macrophages to engulf. Complement proteins can also poke holes in the membranes surrounding microbes, causing them to break apart. These proteins also amplify the inflammatory response.

The **inflammatory response** is a reaction producing redness, pain, swelling, and fever that is induced by chemicals released from macrophages. These substances increase blood flow to the infected area, causing redness. They cause leakiness in the surrounding blood vessels, allowing cells and fluids to enter tissues and cause pain and swelling. Fever can also be caused by chemicals released from macrophages. It is thought that the high temperatures associated with fever may help slow bacterial growth.

While the second line of defense carries out its activities, the third line of defense often begins to function as well.

Third Line of Defense: Lymphocytes

If a pathogen makes it past the first two nonspecific defense systems, the immune system presents a third line of defense. Cells of the immune system identify and attack specific microorganisms that are recognized as foreign. This **specific defense** system consists of millions of white blood cells, called **lymphocytes**. Lymphocytes travel throughout the body by moving through spaces between cells and tissues or being transported via the blood and lymphatic systems (Figure 18.13).

The specific response generated by the immune system is triggered by proteins and carbohydrates on the surface of pathogens, or on cells that have been infected by pathogens. Molecules that are foreign to the host and stimulate the immune system to react are called **antigens**. Examples of antigens include molecules found on invading viruses, bacteria, fungi, protozoans, and worms, as well as molecules on the surface of foreign substances such as dust, pollen, or transplanted tissues. Antigens are present on tissues transplanted from one person to another because every individual's cells have a characteristic set of proteins and sugars on their surfaces. This is why people receiving organ transplants are given drugs to suppress their immune systems.

Regardless of the source, when an antigen is present in the body, the production of two types of lymphocytes is enhanced: the **B lymphocytes (B cells)** and **T lymphocytes (T cells)**. Like macrophages, these lymphocytes circulate throughout the blood and lymph system and are concentrated in the spleen and lymph nodes. Because lymphocytes recognize specific antigens, they are said to display **specificity**.

B and T cells both recognize and help eliminate antigens, but they do so in different ways. B cells secrete proteins called **antibodies** that bind to and inactivate antigens. T cells, in contrast, do not produce antibodies but instead directly attack invaders.

Lymphocytes differ from most other cells due to their ability to recognize foreign molecules. This recognition is based on the presence of proteins whose shape is complementary to a portion of the foreign molecule, called **antigen receptors**. The receptor is able to bind to the antigen in a manner similar to the way that a key fits into a lock. These receptors are either attached to the

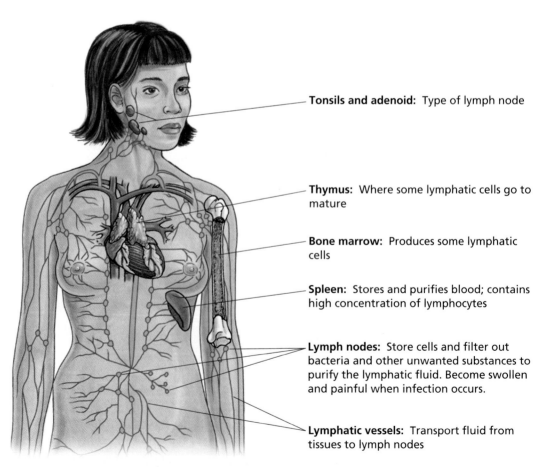

Tonsils and adenoid: Type of lymph node

Thymus: Where some lymphatic cells go to mature

Bone marrow: Produces some lymphatic cells

Spleen: Stores and purifies blood; contains high concentration of lymphocytes

Lymph nodes: Store cells and filter out bacteria and other unwanted substances to purify the lymphatic fluid. Become swollen and painful when infection occurs.

Lymphatic vessels: Transport fluid from tissues to lymph nodes

Figure 18.13   The lymphatic system. Pathogens trigger a response by the lymphatic system. The various organs of this system work to eliminate infections.

surface of the lymphocyte or secreted by the lymphocyte. Both B and T cells have receptor molecules on their surfaces, but B cells also secrete antibodies into the surrounding body fluids to locate and destroy antigens.

The ability of the body's B and T cells to respond to specific antigens begins before humans are born. Thus, we are able to respond to infectious agents the very first time we are exposed. This ability continues into adulthood because these cells are manufactured, at the rate of about 100 million per day, throughout our lives.

The ability to respond to an infection, the **immune response**, actually results from the increased production of B and T cells. B and T cells recognize different types of antigens: B cells recognize and react to small, free-living microorganisms such as bacteria and the toxins they produce (Figure 18.14a). T cells recognize and respond to body cells that have gone awry, such as cancer cells or cells that have been invaded by viruses. T cells also respond to transplanted tissues and larger organisms such as fungi and parasitic worms (Figure 18.14b).

Making B and T Cells

Lymphocytes are produced from special cells, called **stem cells**, that have the ability to become any other cell type (see Essay 8.1). Many parts of the body, including the bone marrow, retain a supply of stem cells that can develop into more specialized cells. Bone marrow stem cells enable the bone marrow to produce blood cells throughout a person's lifetime. Lymphocytes are produced from the stem cells of bone marrow and released into the bloodstream.

Some lymphocytes continue their development in the bone marrow and become B cells. Others take up residence in the thymus gland. The thymus gland, located behind the top of the sternum, stimulates T cells to develop. When immune cells finish their development in the **b**one marrow, they

(a) B lymphocytes
B cells recognize and react to small free-living microorganisms such as bacteria and the toxins they produce.

(b) T lymphocytes
T cells recognize and react to body cells that have gone awry, such as cancer cells or cells that have been invaded by viruses. T cells also respond to larger organisms, such as fungi and parasitic worms.

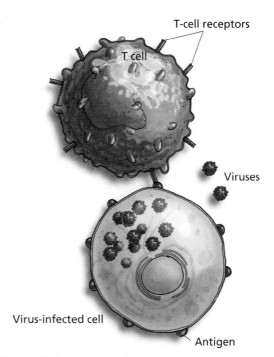

Figure 18.14   B cells and T cells. (a) The receptors produced by B cells function as cell-surface antigen receptors or as secreted antibodies that are present in bodily fluids. (b) T cells also have antigen receptors. These receptors help B and T cells recognize and destroy invaders.

Figure 18.15   Lymphocyte development. Lymphocytes develop from cells in the bone marrow. Lymphocytes that continue their development in the bone become B cells. Those that move to the thymus to continue their development become T cells.

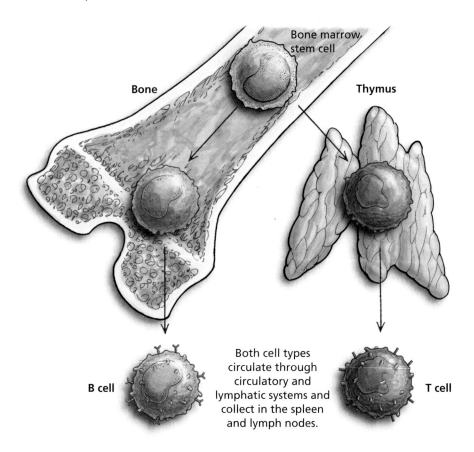

are called B cells; when they mature in the thymus, they are called T cells (Figure 18.15).

Lymphocytes can recognize trillions of different antigens. This extraordinary ability results from the trillions of different antigen receptors that are produced by B and T cells. How is it that such an incredible diversity of receptors can be produced?

Antigen Diversity.   If every antigen receptor were encoded by its own gene, there would need to be trillions of genes to code just for these immune system proteins. Since the entire human genome contains approximately 20,000 genes, this cannot be the case. Instead, B and T cells have evolved a mechanism to generate an enormous variety of receptors from a limited number of genes. As B and T cells develop, each cell's DNA segments that code for the production of antibodies do something very unusual: They rearrange themselves. Some portions are cut out, and the remaining DNA segments are joined together (Figure 18.16). Each unique arrangement of DNA produced by a given cell encodes a different receptor protein. Once synthesized, the proteins move to the surface of the B or T cell and act as antigen receptors. Thus, from a pool of only hundreds of genes, an unlimited variety of diverse antigen receptors can be created.

Another interesting facet of the immune system is its ability to determine whether a molecule is part of the host cell or foreign.

Self Versus Nonself.   While B and T cells are maturing, their antigen receptors are tested for potential self-reactivity. The cells of a given individual have characteristic proteins on their surfaces, and developing lymphocytes are tested in the thymus to determine whether they will bind to self-proteins. Any developing lymphocyte whose antigen receptors bind to self-proteins is eliminated, making an immune response against one's self less likely. Lymphocytes whose receptors do not bind are then allowed to develop to maturity (Figure 18.17). Thus the body normally has no mature lymphocytes that react against self-proteins, and the immune system exhibits self-tolerance. When this testing fails, the cells of the

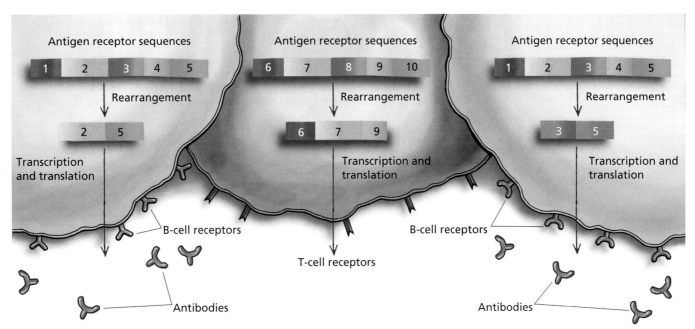

Figure 18.16 Genetic rearrangements allow for the production of millions of different antigen receptors. The genetic information encoding antibody structures is stored in bits and pieces. These bits and pieces are arranged in various orders during development of the antibody-producing cells. Removal of various DNA segments produces a wide variety of antigen receptors and increases the number of antigens to which each of us can respond.

immune system attack normal body cells. Diseases that result when a person's immune system is attacking itself are called **autoimmune diseases**.

Multiple sclerosis is an autoimmune disease that occurs when T cells specific for a protein on nerve cells attack these cells in the brain. In insulin-dependent diabetes, T and B cells attack cells that produce the hormone insulin in the pancreas. Lupus is a disease that occurs when self-antibodies to the nuclei of *all*

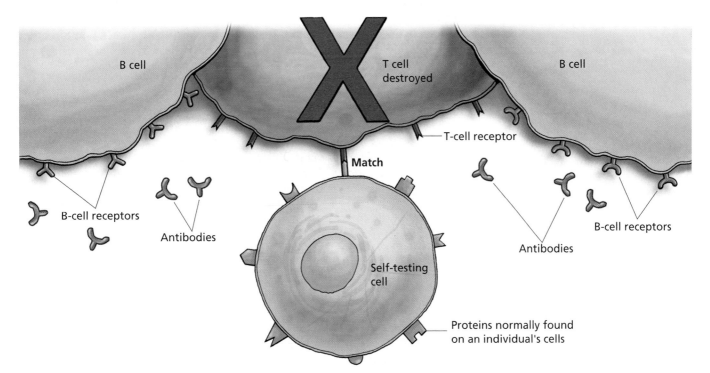

Figure 18.17 Testing developing lymphocytes for self-proteins. Cells have characteristic proteins on their surfaces. Developing lymphocytes are tested to determine whether they are self or nonself. Lymphocytes that bind to antigen receptors on testing cells are destroyed; those that do not bind are allowed to develop.

cells are formed. These antinuclear antibodies build up in cells, causing inflammation of many tissues in the body.

Even though we have a single immune system, it is diversified into two subsystems so that we can combat the multitude of infectious agents encountered in our lifetimes. This diversification is a result of the B and T cells' differing approaches to ridding the body of infectious agents once they are found.

Humoral and Cell-Mediated Immunity

Since antibodies circulate through the body fluids (which in medieval times were called humours), the protection afforded by B cells is called **humoral immunity**. T cells provide immunity that depends on the involvement of cells rather than antibodies and is called **cell-mediated immunity**.

Humoral Immunity.   When a B cell encounters an antigen, it immediately makes copies of itself, resulting in a population of identical cells able to help fight the infection (Figure 18.18). This population of cells is called a **clonal population**. The sheer number of cells in the clonal population strengthens the immune system's ability to rid the body of the infectious agent.

The entire clonal population has the same DNA arrangement. Therefore, all the cells in a clonal population carry copies of the same antigen receptor on their membrane and secrete the same antibody. Some cells of the clonal population, called **memory cells**, will help the body respond more quickly if the infectious agent is encountered again. Should subsequent infection occur, the large number of memory cells facilitates a quicker immune response. Elimination of the infectious agent occurs via the actions of complement proteins that cause lysis (bursting) of the foreign and infected cells. The complement proteins circulate in the blood and lymph in an inactive state until they come in contact with antibodies bound to the surface of microorganisms, at which time they are activated.

Cell-Mediated Immunity.   T cells also respond to infection by undergoing rapid cell division to produce memory cells and then becoming specialized cells (Figure 18.19). However, unlike B cells, T cells do not secrete antibodies; instead, they directly attack other cells. Two of these attacking cell types are the cytotoxic T cells and helper T cells. **Cytotoxic T cells** attack and kill body cells that have become infected with a virus. When a virus infects a body cell,

Humoral Immunity

B-cell receptors   Antigen   Clonal population

B cell

Antibodies

Figure 18.18   Humoral immunity. B lymphocytes do not kill cells bearing antigens directly. Instead they make and secrete antibodies specific to antigens. The antibodies circulate in the blood and lymph. The antibody-antigen complex binds with complement proteins in the blood, and together the antibody and complement proteins can burst bacterial plasma membranes.

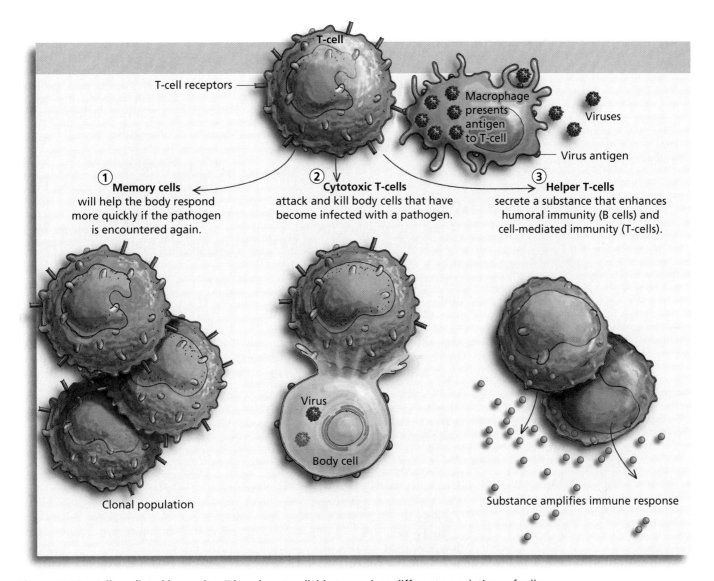

Figure 18.19   Cell-mediated immunity. T lymphocytes divide to produce different populations of cells: (1) Memory cells carry the specific antigen receptor; (2) Cytotoxic T cells attack and kill cells; and (3) Helper T cells boost the immune response.

viral proteins are placed on the surface of the host cell. Cytotoxic T cells recognize these proteins as foreign, bind to them, and destroy the entire cell. They do this before the virus has had time to replicate by releasing a chemical that causes the plasma membranes of the target cell to leak.

Helper T cells, also called T4 cells, can be thought of as boosters of the immune response. These cells detect invaders and alert both the B and T cells that infection is occurring. Without helper T cells, there can be almost no immune response. Helper T cells also secrete a substance that greatly increases the level of cytotoxic T-cell response. The AIDS virus, HIV, infects helper T cells, thus crippling the body's ability to respond to any infection (see Chapter 10).

Macrophages also help fight infection by transferring an antigen to their own plasma membranes, thus alerting the immune system (via the T cell) to a foreign antigen present in the body. When the macrophage presents an antigen to a T cell that has the correct receptor, the T cell replicates itself to produce more memory cells, more cytotoxic T cells, and more helper T cells.

As you can see, the various cells of the immune system work in concert to strengthen the overall response of the immune system. However, even with all the complexity of the immune system, some substances, such as prions, are able to invade the body without eliciting an immune response.

There Is No Immune Response to Prions

An inability to mount an immune response can be the result of an infectious agent carrying antigens for which no amount of DNA rearrangement can yield the proper receptor. In other cases, one group of people may carry DNA segments that can be rearranged to produce the right receptors, while other groups do not. This partly explains why some people do not become ill when exposed to an infectious disease, and others die from the same exposure.

Sometimes an infectious agent is effectively combated for a period of time by the immune system, only to reemerge in a form that is newly pathogenic. Pathogens are under evolutionary pressure to change in order to evade the immune system. We see evidence of changing pathogens every winter with the flu virus. In preparing a vaccine against the flu virus, scientists must try to predict which of the many strains of virus will make their way to the United States and then include parts of several different viruses in the vaccine. **Vaccines** are made of components of the disease-causing organisms, such as proteins from the plasma membrane of a bacterial cell, parts of a virus, or a whole virus that has been inactivated. The immune system responds to the introduced vaccine's challenge by producing the clonal population of memory cells that will be prepared for a real infection by the virus, should it occur. Some vaccines require multiple doses before a sufficient response is generated; others require booster shots to maintain protection. Since only parts of the virus are present in the vaccine, being vaccinated does not result in the flu but should give the immune system a head start. Flu vaccines must be given every year because the flu virus rearranges its genetic sequences rapidly. This results in different proteins being encoded and placed on the surface of the virus, thereby preventing existing memory cells from recognizing a virus they have already encountered. Because the proteins on the surface of a flu virus change so quickly, a vaccine that was prepared to protect you from last year's flu virus will not necessarily work against this year's flu virus.

In the case of spongiform encephalopathies, the immune system does not mount a response to misfolded prions at all because they are just refolded versions of the normal prion protein. It seems that the refolded version of the protein is similar enough to the normal version that the immune system recognizes them both as self. Therefore, the infection, which bypasses the first two lines of defense when it is consumed, skates through the third line of defense as well. This ability to evade the immune system is quite unusual. A healthy immune system normally can stop millions of different pathogens. However, because this infectious agent is a protein resembling the cell's normal proteins, it does not evoke an immune response.

After being ingested by humans, the misfolded bovine prions move to the spleen and lymph nodes and refold all the prions there. From there, the refolded proteins move to the brain and spinal cord, where they refold even more normal proteins. The lack of normal proteins damages the brain and eventually causes the patient to die.

Treatment of prion diseases does not conform to conventional methods. Since prions are proteins, not bacteria, viruses, or eukaryotic pathogens, conventional treatments for infectious diseases will not prevent or cure the spongiform encephalopathies. An anti-prion agent would have to be able to determine which prions were properly folded and which were not. This makes treating spongiform encephalopathies quite difficult. Because heat and radiation—generally effective in destroying pathogens—do not work against prions, scientists hope to discover how to stabilize the normal prion protein so that it will maintain its normal shape in brain cells. Until this happens, epidemiologists and other scientists are focusing on preventing the spread of these diseases.

18.4 Preventing the Spread of Prion Diseases

It is now illegal to import cows, sheep, goats, and deer to the United States and the practice of feeding these animals meat-and-bone meal produced during the rendering of other mammals is banned. Because certain parts of cattle are at higher risk for harboring prions, materials obtained from cow skulls, brains, eyes, vertebral columns, and spinal cords are prohibited from use in human foods, as is the use of this high-risk material in cosmetics.

Despite all these preventative measures, scientists do not know whether mad cow disease—and its human counterpart, nvCJD—will become an epidemic in the United States. The outcome will be decided, in large part, by the level of stringency practiced by the makers of animal feed, by the farmers responsible for feeding cows and observing their behavior, and by the governmental regulatory agencies charged with protecting human health.

CHAPTER REVIEW

Summary

18.1 Infectious Agents

- Infectious diseases are usually caused by pathogens such as bacteria, viruses, and eukaryotic parasites (p. 470).

- Bacteria are single-celled organisms with DNA genomes contained in a nucleoid region. A cell wall surrounds and protects most bacteria, which in turn can be surrounded by a capsule. Flagella help bacteria move (pp. 470–471).

- Bacteria reproduce asexually by binary fission (p. 471).

- Bacteria cause disease by using host resources to rapidly reproduce and by releasing toxins into the host (p. 472).

- Viruses can reproduce only inside host cells. They are composed of nucleic acids encased in a capsid and sometimes an envelope (p. 474).

- Viruses cause disease by using host-cell resources and by destroying host cells as part of their infectious cycle (p. 474).

- Viruses replicate their genome inside host cells. The viral genome then directs the synthesis of viral components (pp. 474–475).

- Eukaryotic parasites, including insects and worms, cause many diseases worldwide (p. 476).

- Prions are novel infectious agents comprised of protein only. They cause disease by refolding the host cell's normally occurring prion proteins (p. 478).

- Misfolded prions can arise spontaneously, as the result of a mutation to the gene that encodes the prion protein, or as a result of ingesting misfolded proteins from a diseased organism. Once inside the body, the misfolded prions refold normal proteins so that they assume the disease-causing shape (pp. 478–479).

Web Tutorial 18.1 **Structure and Reproduction of Bacteria**
Web Tutorial 18.2 **Structure and Reproduction of Viruses**

18.2 Epidemics

- An epidemic is a contagious disease that spreads rapidly and extensively among a population (p. 479).

- The spread of infectious disease results from physical contact with the disease-causing organism. Contact can occur via exposure to body fluids, an intermediate host, inhalation, or ingestion (pp. 480–481).

18.3 The Body's Response to Infection: The Immune System

- The skin, its secretions, and mucous membranes are nonspecific defenses that comprise the first line of defense against infection (pp. 482–483).

- The second line of defense consists of nonspecific internal defenses (p. 483).

- White blood cells, such as phagocytic macrophages and natural killer cells, remove infectious agents (p. 483).

- Proteins are also involved in the second line of defense. Interferons, released by infected cells, help protect uninfected cells from becoming infected. Complement proteins coat microbes and make them easier for macrophages to ingest and amplify the inflammatory response (p. 483).

- Chemicals released by macrophages cause inflammation (p. 483).

- Lymphocytes help comprise the third, specific, line of defense in response to antigens on the surface of pathogens (pp. 483–484).

- Exposure to antigens causes increased production of B cells and T cells. B cells secrete antibodies against pathogens, and T cells attack invaders (p. 484).

- Lymphocytes are produced during development and allow the immune system to respond to trillions of different antigens (p. 485).

- Antigen receptors on the surface of immune cells are produced by rearrangements of gene segments (p. 486).

- The antigen receptors of B and T cells are tested for self-reactivity, and those that react against self are eliminated (pp. 486–487).

- The humoral response of the immune system involves B cells, which divide to produce cells that carry and secrete antibodies when exposed to an antigen. Complement proteins burst foreign and infected cells (p. 488).

- The cell-mediated response of the immune system involves T cells, which become specialized into different cell types upon exposure to an antigen. These cells speed up the immune response and destroy virus-infected cells (pp. 488–489).

- Prions do not evoke an immune response because the immune system recognizes them as self (p. 490).

Web Tutorial 18.3 **The Body's Response to Infection**

18.4 Preventing the Spread of Prion Diseases

- Regulating the import and use of animals may decrease the likelihood of a U.S. epidemic (p. 491).

Learning the Basics

1. Describe the structure of a typical bacterium and a typical virus.

2. Describe how bacteria and viruses reproduce.

3. What roles do B cells and T cells have in the immune response?

4. Binary fission _____.
   A. allows bacterial cells to produce more plasmids; **B.** is a type of sexual cell division that viruses undergo; **C.** allows prions to replicate themselves; **D.** is the method by which bacteria replicate

5. Transmission of infectious diseases *does not* occur _____.
   A. via contact with infected body fluids; **B.** by transmittance of a defective gene from parent to offspring; **C.** through exposure to an intermediate host; **D.** by inhalation of infectious agents; **E.** by ingestion of infectious agents

6. The immune system _____.
   A. has a gene for each antigen; **B.** undergoes genetic rearrangement in response to different antigens; **C.** is able to make many antigen receptors by rearranging the DNA of immune cells; **D.** can always devise an antibody that will bind to an antigen

7. Autoimmune diseases result when _____.
   A. a person's endocrine system malfunctions; **B.** liver enzymes malfunction; **C.** B cells attack T cells; **D.** the immune system fails to differentiate between self and nonself cells

8. Which of the following cell types divides to produce cells that make antibodies?
   A. helper T cells; **B.** B cells; **C.** cytotoxic T cells; **D.** all of the above

9. Helper T cells secrete substances that _____.
   A. help prevent leukemia; **B.** prevent bacteria from entering cells; **C.** boost B-cell and cytotoxic T-cell response; **D.** inhibit reverse transcriptase; **E.** stimulate the thymus to make more B cells

10. The immune system can recognize a virus you have been exposed to once because _____.
    A. you harbor the virus for many years; **B.** we have genes to combat every type of virus; **C.** a cell that makes receptors to a virus multiplies upon exposure to it and produces memory cells; **D.** a copy of the viral genome is inserted into a memory cell

Analyzing and Applying the Basics

1. Why do you need a flu shot every year but only one inoculation against some other diseases?

2. Why is your immune system usually more effective at fighting off infection the second time you are exposed to a pathogen?

3. Why might one person die of an infectious disease while another person recovers?

4. How do prions differ from other infectious agents? Why don't animals generate antibodies to them?

Connecting the Science

1. Some parents refuse to have their children vaccinated. Should these parents be forced to allow their children to be vaccinated to protect society against the reemergence of once-deadly killers such as polio? Defend your answer.

2. In 1796, Edward Jenner intentionally exposed a child to smallpox in order to test his hypothesis that the child had already acquired immunity via exposure to cowpox virus. Why might Jenner have felt this was a necessary risk? Would such treatment be ethical? Why or why not?

Sex Differences and Athleticism

Endocrine, Skeletal, and Muscular Systems

Many young athletes dream of
playing professional sports.

Ila Borders is the first woman to win a professional baseball game.

Annika Sorenstam has had success with competing in golf against men.

The scene on the baseball field is one that young athletes dream about all over the world. For one young baseball pitcher, the dream has come true. On a warm Friday evening late in July, the sun is beginning to set over left field in a beautiful, open-air, brick baseball stadium. On the mound, a 23-year-old lefthander stands motionless, squinting to better see the catcher's signals.

The scoreboard above the centerfield wall shows that it is the top of the sixth inning, and the Duluth-Superior Dukes are leading the Sioux Falls Canaries by the score of 2 to 0. The Dukes pitcher has given up only three hits this game; but the Canaries may be putting together a rally, with one runner on base and their number three hitter at bat.

The pitcher glances at the runner on second base, turns to face the batter, and in one strong, fluid motion, rears back and hurls the ball across the inside corner of home plate. "Strike three. The batter is out," yells the home-plate umpire.

The stadium erupts as the fans jump to their feet cheering wildly, and the 5'-10" pitcher walks off the mound toward the dugout to a standing ovation. For the next half-inning, the hometown fans chant the pitcher's name in unison, until the southpaw finally comes out of the dugout to salute the cheering fans. When the pitcher raises the black-and-purple Dukes baseball

Danica Patrick has had success in auto racing against men.

495

cap toward the fans behind home plate, her long hair spills across the back of her uniform, obscuring her name and number.

The pitcher is Ila Borders, the first woman in the starting lineup of a men's professional baseball game and on that warm 1997 July evening, the first woman to win a men's professional baseball game. This accomplishment was so impressive that *Sports Illustrated* honored Borders as one of the top 100 female athletes in history; the uniform and glove that served her in this historic victory adorn the prestigious Baseball Hall of Fame in Cooperstown, New York, and the national television stations ESPN and CNN broadcasted footage of her pitching that night.

Will Ila Borders be the first of many women to play major league baseball? Will women infiltrate professional football and hockey leagues next? Will the successes of women golfers and auto racers be replicated in other sports as opportunities increase for women to compete? Or, are the bodies of women and men simply constructed so differently that they will be able to compete against each other only rarely?

To answer these questions, we need to delve into the study of sex differences—average differences between males and females. They can be biological, such as differences in reproductive anatomy, or cultural, such as differences in opportunities to practice and to access top-quality coaching.

One of the main ways that sex differences originate and are maintained has to do with the sexes secreting different amounts of sex-specific hormones. Hormones are secreted by specialized organs and glands that comprise the endocrine system.

19.1   The Endocrine System

The **endocrine system** is an internal system of regulation and communication involving hormones, the glands that secrete them, and the particular cells that respond to the hormones. A given hormone will elicit a response only in some cell types. The cells that respond to a particular hormone are called its **target cells**.

Hormones

Hormones are chemicals that travel through the circulatory system and act as signals to elicit a response from target cells. There are two general mechanisms by which hormones elicit a specific response: (1) Hormones can bind to receptors on the target cell's surface and trigger a change inside the cell, or (2) hormones can diffuse across the cell membrane and bind to receptors inside the cell to trigger the response. Protein hormones (Figure 19.1) typically bind to a receptor on the surface of the target cell, and that receptor stimulates a series of other proteins to become involved in relaying the message to the inside of the cell, eliciting a specific cellular response. The chain reaction that relays or transduces a signal from the outside of the cell to the inside is called **signal transduction**.

Hormones that can cross cell membranes can bind to receptors inside the target cell and cause cells to turn specific genes on or off. The hormones most involved in producing anatomical sex differences are called **sex hormones**. These include the male hormone **testosterone** and the female hormone **estrogen**. Structurally, sex hormones are a type of steroid hormone whose synthesis requires the steroid cholesterol as a precursor or starting material. Steroid hormones are fat soluble and move readily across cell membranes, where they exert their sex-specific effects on target cells that contain receptors for them. Once a sex hormone passes into the cell and binds to a receptor, the hormone and receptor bind to the DNA in the

(a)
Protein hormone
Hormone receptor
Relay molecules
Cellular response

(b)
Hormone receptor with no hormone bound
No signal transduced

Figure 19.1   Protein hormones. (a) When a protein hormone is secreted into the bloodstream, it can bind to a receptor on the cell membrane of its target cell. This binding can induce a shape change in the receptor molecule, serving as a way of transmitting a signal to the inside of the cell that some response is required. This signal can be transduced or relayed through many relay molecules. (b) When no protein hormone is available, no signal is transduced.

cell's nucleus and either enhance or inhibit the expression of different genes (Figure 19.2).

In summary, the dual roles of hormones include acting as chemical messengers when they are carried in the blood to all parts of the body and helping to regulate many processes, including metabolism, reproduction, and development. The role of a particular hormone, whether it be acting as a messenger or regulator, begins when it is synthesized and secreted by a particular endocrine gland.

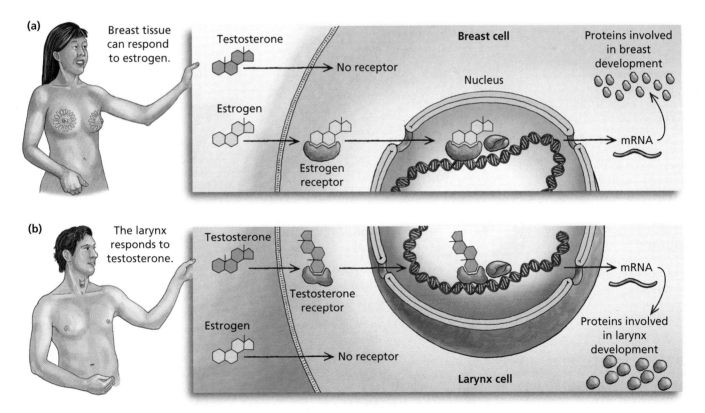

(a)
Breast tissue can respond to estrogen.
Testosterone → No receptor
Estrogen → Estrogen receptor
Breast cell
Nucleus
Proteins involved in breast development
mRNA

(b)
The larynx responds to testosterone.
Testosterone → Testosterone receptor
Estrogen → No receptor
mRNA
Proteins involved in larynx development
Larynx cell

Figure 19.2   Steroid hormones. (a) Estrogen produced by females enters the cells that compose the breast tissue, where it binds to its receptor. The estrogen and its receptor together bind to DNA to increase the expression (transcription) of genes required for the growth and development of the breast at puberty. Males have lower levels of estrogen and fewer estrogen receptors in their breast tissue, so the breasts do not develop to the extent that they do in females. (b) In males, testosterone enters the cells of the larynx and finds receptors, which increase the production of proteins that change the structure of the larynx, altering the sounds it produces and resulting in voice deepening.

Endocrine Glands

Organs that secrete hormones are called **endocrine glands**. Many glands are involved in the endocrine system (Figure 19.3). For example, the **thyroid gland** secretes hormones that stimulate metabolism. Along with the **parathyroid glands**, the thyroid gland also helps to regulate blood-calcium levels. The **thymus** produces some immune cells and is part of the lymphatic system, and the **pancreas** helps regulate blood glucose levels. Five additional endocrine organs are involved in the production of biological sex differences and are investigated more thoroughly below: the hypothalamus, the pituitary gland, the adrenal glands, the ovaries, and the testes.

Hypothalamus and Pituitary Gland. The **hypothalamus**, located deep inside the brain, regulates body temperature and affects behaviors such as hunger, thirst, and reproduction. In the reproductive system, the hypothalamus secretes a hormone that stimulates the activities of the **gonads** (testes or ovaries); it is called **gonadotropin-releasing hormone (GnRH)**. GnRH moves through a complex series of veins from the hypothalamus to the **pituitary gland**, located centrally in the brain. The pituitary gland secretes many different hormones; two in particular are involved with producing sex differences: **follicle-stimulating hormones (FSH)** and **luteinizing hormones (LH)**. In males, FSH stimulates sperm production, and LH

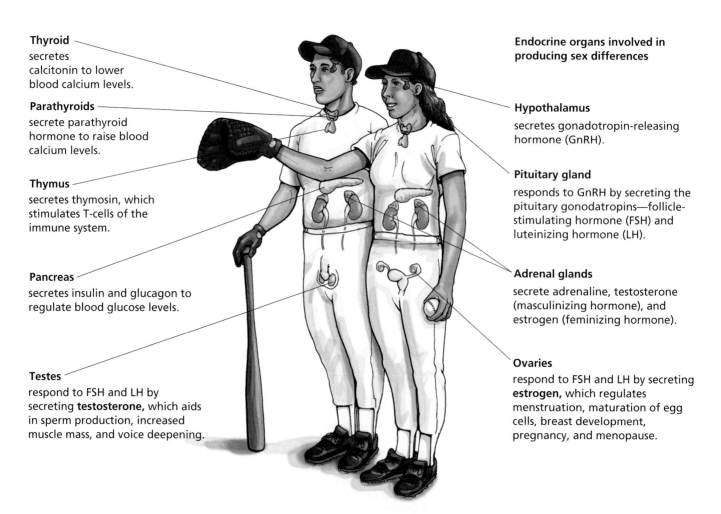

Thyroid
secretes
calcitonin to lower
blood calcium levels.

Parathyroids
secrete parathyroid
hormone to raise blood
calcium levels.

Thymus
secretes thymosin, which
stimulates T-cells of the
immune system.

Pancreas
secretes insulin and glucagon to
regulate blood glucose levels.

Testes
respond to FSH and LH by
secreting **testosterone,** which aids
in sperm production, increased
muscle mass, and voice deepening.

**Endocrine organs involved in
producing sex differences**

Hypothalamus
secretes gonadotropin-releasing
hormone (GnRH).

Pituitary gland
responds to GnRH by secreting the
pituitary gonodatropins—follicle-
stimulating hormone (FSH) and
luteinizing hormone (LH).

Adrenal glands
secrete adrenaline, testosterone
(masculinizing hormone), and
estrogen (feminizing hormone).

Ovaries
respond to FSH and LH by secreting
estrogen, which regulates
menstruation, maturation of egg
cells, breast development,
pregnancy, and menopause.

Figure 19.3   Major endocrine glands. The locations and functions of the major endocrine glands are shown.

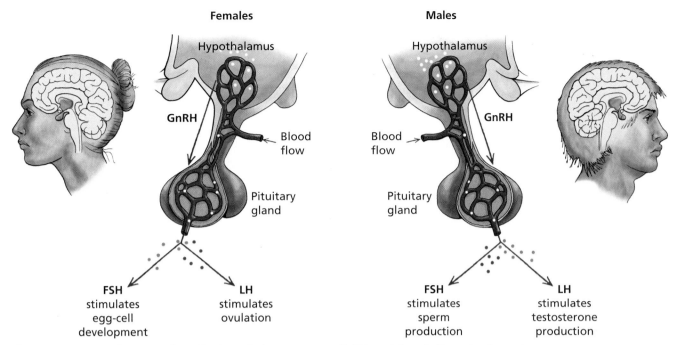

Figure 19.4   FSH and LH secretion. The hypothalamus secretes GnRH, and the pituitary gland secretes FSH and LH. FSH and LH act on the testes of a male, resulting in sperm and testosterone production; but in females, they stimulate the maturation and release of egg cells from the ovary.

stimulates testosterone production. In females, FSH stimulates egg-cell development, and LH stimulates the release of an egg cell during ovulation (Figure 19.4).

Adrenal Glands.   An adrenal gland sits atop each kidney. These glands secrete the hormone **adrenaline** (epinephrine) in response to stress or excitement. They also secrete **androgens**, which include masculinizing hormones such as testosterone and feminine hormones such as estrogen.

The cells of the outer covering of each adrenal gland synthesize sex hormones from cholesterol. While cholesterol can be synthesized in many body tissues, it is converted into sex hormones only within the adrenal glands, testes, and ovaries. Thus the adrenal glands of both males and females secrete small amounts of both testosterone and estrogen. Most sex-hormone production and secretion takes place in the testes and ovaries.

Testes.   The paired **testes** are oval organs suspended in the scrotum of human and other mammalian males. Testes secrete **testosterone**, the hormone that aids in sperm production, hair thickness and distribution, increased muscle mass, and voice deepening. Beginning at puberty, sperm are produced by the cells of the testes. Sperm production is most efficient at temperatures that are lower than body temperature. Therefore, the testes are situated outside the body cavity, in the scrotum.

Ovaries.   The paired **ovaries** are about the size and shape of almonds in the shell. They produce and secrete estrogen. Estrogen regulates many functions in the female body, including menstruation, the maturation of egg cells, breast development, pregnancy, and the cessation of menstruation after reproductive age, called **menopause**. Inside the ovaries are all of the cells that can mature into the egg cells for ovulation. Production of egg cells begins while a female is developing inside her mother's uterus, pauses at birth, resumes at puberty, and continues until menopause, at around 50 years of age.

The hormones of the female endocrine system regulate pregnancy and menstruation, as we will detail in Chapter 20. Many people have questioned whether a woman's athletic performance or capabilities are altered during any portion of the menstrual cycle. The answer is an unequivocal *no*. Researchers who measured speed and strength among sprint swimmers and weight lifters found that the women's performances remained consistent throughout their menstrual cycles. Other research has shown no differences in maximum bicycling and jumping power at different stages of the menstrual cycle. In fact, study after study has shown no loss of physical function during or prior to menstruation.

Puberty.   The endocrine systems of both males and females increase the secretion and synthesis of hormones around the time of puberty. **Puberty** marks the beginning of sperm production in males and the beginning of egg-cell maturation and menstruation in females. Boys typically begin puberty at the age of 9 to 14 years, with the average being 13 years. Puberty in males includes enlargement of the penis and testes; an overall growth spurt; growth of muscles and the skeleton, resulting in wide shoulders and narrow hips; and changes in hair growth, including the production of pubic, underarm, chest, and facial hair. In addition, the larynx enlarges and vocal cords lengthen to produce a deeper voice.

For girls, the first signs of puberty occur between ages 8 and 13, most commonly around age 11. These signs include breast development, increased fat deposition, a growth spurt, pubic and underarm hair growth, and the commencement of menstruation.

It is at puberty that biologically induced sex differences in athleticism begin to emerge, largely because the endocrine system begins to affect skeletal and muscle development.

19.2   The Skeletal System

The skeletal system provides support for the body, protects internal organs, aids in movement, and stores minerals (Essay 19.1). In humans, this internal framework is composed of the skeleton, which in turn is composed of bones and cartilage. Bones are rigid body tissues that function to provide support for the actions of muscles and protection for soft parts of the body—for example, what the skull does for the brain. Bones also help produce blood cells within the bone marrow and serve as a reservoir for minerals like calcium that circulate in the body. Cartilage protects the ends of bones from degradation.

The Human Skeleton

The human skeleton is composed of 206 bones and is organized into two basic units, the axial and the appendicular skeleton (Figure 19.5 on page 502). The **axial skeleton** supports the trunk of the body and consists largely of the bones making up the vertebral column or spine and much of the skull. The **appendicular skeleton** is composed of the bones of the hip, shoulder, and limbs.

Essay 19.1 Focus on Evolution: The Skeletal System

Natural selection has provided a variety of means by which organisms move. In animals that actively travel through the environment, the skeleton provides a surface on which muscles function as well as a protective cage for internal organs that could be damaged during movement. Three types of skeletal systems have evolved in animals: hydrostatic skeletons, endoskeletons, and exoskeletons. Each type of skeletal system has advantages for organisms in certain ecological roles but may be ill-suited for others.

The simplest skeleton consists of fluid held under pressure in a closed body compartment. These hydrostatic skeletons are found in jellyfish, flatworms, and earthworms, among other creatures. For these organisms, movement is controlled by contraction of muscles against the fluid-filled compartment. Because water, the typical fluid found in a hydrostatic skeleton, cannot compress much, muscles can "push off" against the compressed cylinder, forcing the body forward. For instance, some worms hold fluid in their body cavity at high pressure and contract longitudinal muscles against it to produce a whiplike movement. In other worms, different segments of the body can change shape individually by using circular and longitudinal muscles to produce rhythmic contractions of the hydrostatic skeleton. This produces waves of head-to-tail movement (Figure E19.1a). Hydrostatic skeletons are adapted for life in aquatic environments or in soil. However, these fluid-filled cylinders cannot maintain an animal's shape above ground, as anyone who has ever picked up an earthworm knows. Animals that move through the air by walking, running, or flying require a stiffer type of skeleton.

One solution that has evolved for animals that require a stiff form is an external skeleton. An exoskeleton is a hard encasement, on the surface of the animal, to which muscles attach. Clams and many other mollusks produce a hard shell that serves primarily as protection but also as a point of attachment for muscles that open and close their shells (Figure E19.1b). These shells grow throughout the life of the animal. Arthropods such as lobsters and many insects have evolved a more flexible exoskeleton that helps them move more efficiently and provides more limited protection than a shell. Because an arthropod's exoskeleton is like a form-fitting coat, it must be periodically shed as the animal grows.

Endoskeletons, which consist of hard supporting elements such as bone, are located inside the soft body tissues and can grow with the animal. However, they provide very little protection from predators. Endoskeletons may be made primarily of a mixture of protein and inorganic material, as in sponges and sea stars; cartilage, as in sharks and rays; or bones, as in humans. Harder endoskeletons must be made flexible by joints, which are gaps between bones held together by ligaments.

Figure E19.1a Earthworms move by contracting their hydrostatic skeletons.

Figure E19.1b The clam's shell is an exoskeleton.

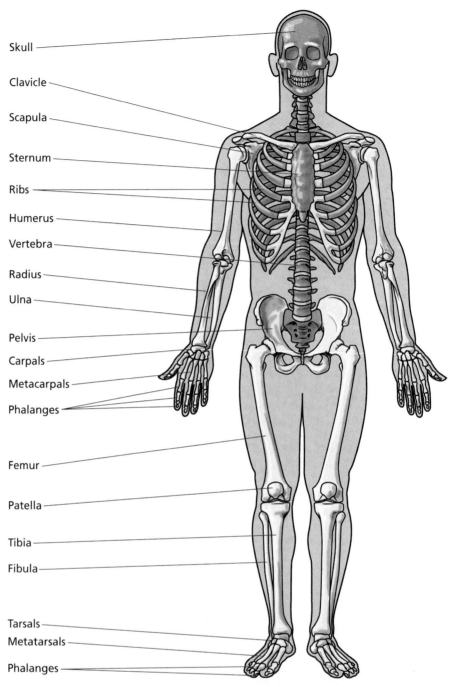

Skull

Clavicle

Scapula

Sternum

Ribs

Humerus

Vertebra

Radius

Ulna

Pelvis

Carpals

Metacarpals

Phalanges

Femur

Patella

Tibia

Fibula

Tarsals

Metatarsals

Phalanges

Figure 19.5 The human skeleton. The human skeleton can be divided into those bones that support the trunk (shaded red in the figure), called the axial skeleton, and those that compose the limbs, called the appendicular skeleton. Cartilage between joints is shaded blue.

To allow for a variety of movements, humans have different types of movable joints (Figure 19.6). **Ball-and-socket joints** in the hips and shoulders enable arms and legs to move in three dimensions. **Hinge joints** such as those in the knee and elbow allow back-and-forth movement similar to opening and closing a door. The **pivot joint** in the neck allows the head to turn from side to side. To facilitate movement, joints are lined with cartilage and lubricated by fluids. The bones are kept in place by ligaments and moved by muscles.

The bones of the skeleton are living tissues composed of cells that actively reshape the tissue. Calcium is absorbed as bone is built and released when bone breaks down.

(a) Ball and socket joint

Hip

Femur

(b) Hinge joint (knee)

Femur

Tibia

Fibula

(c) Pivot joint (neck)

Atlas

Axis

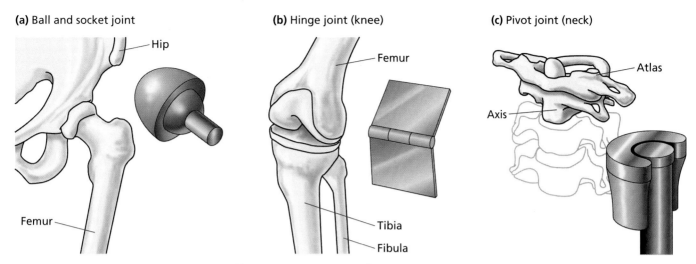

Figure 19.6   Joints in the human skeleton. (a) The hip is an example of a ball-and-socket joint. The ball-shaped end of the femur fits into a cup-shaped socket, allowing a wide range of motion. (b) The hinge joint of the knee allows movement in a single plane. Hinge joints are formed when a projection on one bone fits into a depression within another. (c) The pivot joint of the neck allows for freedom of movement of the head.

Bone Structure and Remodeling

Bone tissue can be tightly or loosely packed. **Compact bone** forms the hard outer shell of bones, and **spongy bone** is the porous, honeycomb-like inner bone (Figure 19.7). Because bone is a living tissue, it has blood vessels throughout. The interior of bone contains **bone marrow**, which helps to produce blood cells.

Osteoblasts, Osteoclasts, and Calcium.   The cells that make up bone are called **osteoblasts** and **osteoclasts**. Osteoblasts help bone tissue regenerate itself by a process called bone deposition. Osteoclasts are involved in breaking down or reabsorbing bone tissue. The delicate balance between deposition and reabsorption is integral to the maintenance of bones. This process is regulated by a negative feedback loop, which is in turn controlled by the

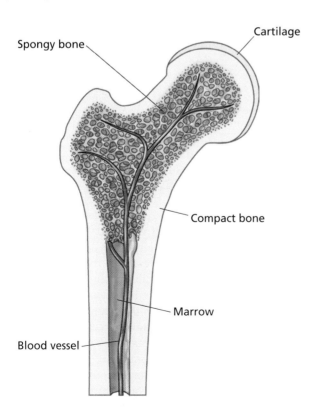

Cartilage

Spongy bone

Compact bone

Marrow

Blood vessel

Figure 19.7   Bone structure. Bone is composed of compact and spongy bone. There are blood vessels running through bone, and the marrow inside bones helps produce blood cells.

Figure 19.8 Bone remodeling. Bone is in a constant state of remodeling, whereby old bone is removed by osteoclasts and new bone is laid down by osteoblasts. (a) Bone removal is stimulated by the parathyroid hormone, resulting in the release of calcium. (b) Bone formation is stimulated by the hormone calcitonin, and results in the uptake of calcium.

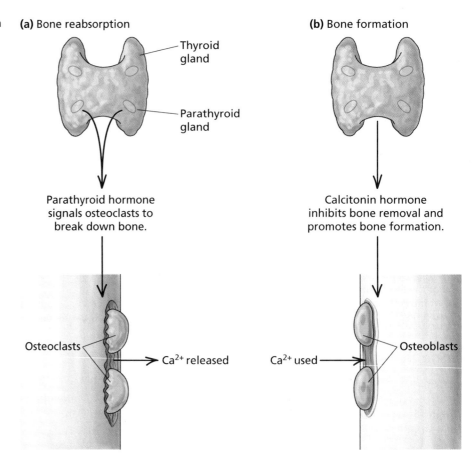

(a) Bone reabsorption

Thyroid gland

Parathyroid gland

Parathyroid hormone signals osteoclasts to break down bone.

Osteoclasts

Ca²⁺ released

(b) Bone formation

Calcitonin hormone inhibits bone removal and promotes bone formation.

Ca²⁺ used

Osteoblasts

endocrine system (Figure 19.8). Recall from Chapter 16 that negative feedback negates change.

Calcium is delivered to the bloodstream in response to the effects of the parathyroid hormone, which is produced by the parathyroid glands. Release of this hormone signals the kidneys to decrease the amount of calcium secreted in the urine and stimulates the breakdown of bone, so that stored calcium is released into the blood. The thyroid gland secretes another hormone, **calcitonin**, which decreases the concentration of calcium within the blood when the level rises too high.

When calcium levels in the blood are high, osteoblasts remove calcium from the blood and use it to make new bone. When calcium levels in the bloodstream are low, osteoclasts break down bone and release calcium into the bloodstream. The maintenance of bloodstream calcium levels is important in many physiological processes, including blood clotting, muscle contraction, production of nerve impulses, and the activity of many enzymes.

Sex Differences in Bone Structure. Average differences in male and female skeletons arise due to the timing of puberty. From infancy to adulthood, humans grow in overall size; but a larger proportion of growth occurs in the extremities (legs and arms) as compared to the torso—for example, consider how short toddlers' legs look in relationship to their bodies. A growth spurt occurs during adolescence, with a larger proportion of growth occurring in the arms and legs than in the torso. In part because puberty both occurs later and lasts longer in boys than in girls, the average adult man has longer arms and legs than the average adult woman. Overall, the average man is 15 cm (5.9 inches) taller than the average woman.

The differences between women's and men's skeletons contribute to average differences in success at performing certain physical activities. Because women generally have long torsos relative to their leg length, their center of gravity (the point on the body where the weight above equals the weight below) is lower than in men. Individuals with a lower center of gravity are better able to maintain balance. Competitive gymnastics recognizes this difference in center of gravity— women, but not men, compete on the balance beam apparatus. Ice skating, ballet, and other forms of dance also require a lot of balance, and the lower center of gravity may enhance a woman's ability to stay on her feet.

Men, with their generally longer legs and arms, have more difficulty balancing, but they bring more power to activities that rely on the lever action of their extremities. A longer arm or leg transmits more force; therefore, men have faster and stronger slap shots in hockey, stronger kicks in soccer, and faster swings and throws in baseball. Longer legs also have longer stride lengths, which means that men will usually be faster runners than women are.

Aside from differences in leg and arm length, there are few differences in the human skeleton that affect athleticism. (Figure 19.9a). In fact, there are only six bones that differ between same-sized males and females. Four of the bones that differ are found on the head. These are the **mandible**, or jawbone, which is larger in males; the paired **temporal bones** found near the temple, which have a larger opening in males to allow for the connection of thicker muscles to support the large jaw; and the **frontal bone** of the cranium, or forehead, which is generally more rounded in females and has a less-pronounced ridge above the eyes. The other two bones that show general differences between men and women are the two **ossa coxae**, which form the bony pelvis.

The differences between the pelvises of men and women evolved in response to different selective pressures. The evolution of upright-walking early humans

(a) Skeletal differences

Women have:
- Smaller frontal bone
- Smaller temporal bones
- Smaller mandible

- Flatter, broader pelvis (two ossa coxae)
  - pelvis is tipped forward, causing greater curvature of lower spine
  - larger, rounder pelvic inlet

Centers of gravity

- Shorter long bones

(b) Q angle

- Increased Q angle between femur and knee

Figure 19.9 Sex differences in the skeleton. Human male and female skeletons are very hard to distinguish. (a) Consistent differences are found in the frontal bone, temporal bones, mandible, and the two ossa coxae. The female pelvic inlet is larger and rounder than the male pelvic inlet. (b) Q angles in males are typically smaller than Q angles in females.

required that these changes take place. However, the changes to the ossa coxae, combined with our large head size, results in one of the most difficult births of any mammal. Early human females with rounder pelvic inlets may have had a selective advantage because they were less likely to die during childbirth, since rounder inlets generally permit the easier passage of a child's head through the birth canal.

The round pelvic inlet that is typical of most women is produced by a bony pelvis that is flatter and broader than a man's pelvis, which has an inlet that resembles an elongated oval. A flatter pelvis also requires that the bony pelvis be tipped forward to bring the hipbones to the front of the body, which is maintained by the curvature of the lower spine. The spinal curvature and greater pelvic tilt in individuals with broad pelvises elevates the buttocks and gives a curvy appearance to the profile of women. Conversely, the male pelvis lowers the buttocks and gives a flat appearance to a man's profile.

When you bring your feet together, your femurs, or thighbones, extend diagonally to your knees from where they attach at the hips. A broader pelvis means that the femurs are farther away from each other at the point of attachment than are femurs attached to a narrower pelvis. Therefore, the **Q angle** (Figure 19.9b) formed between the kneecap and femur increases as the broadness of the bony pelvis increases.

An increased Q angle has been thought to be a sex-influenced risk factor for knee injury. For instance, knee injuries among basketball players are much more common in women, who generally have greater Q angles than men do. However, many scientists who study sports injuries are quick to point out that social differences in early physical activity and conditioning may have prevented development of the leg-muscle strength that supports the knee-stressing movements typical of basketball. To accurately discern the effects of greater Q angles, studies must compare men and women who have followed similar conditioning and training regimens.

While the role of the Q angle may or may not influence the rate of knee injury, the male body has one clear advantage in sports requiring greater strength. A taller, heavier skeletal frame supports more muscle, and muscle mass does affect athleticism.

19.3 The Muscular System

The major function of the muscular system is movement. Skeletal muscle is one of three types of muscle in humans; the other two types are cardiac and smooth muscle. Skeletal muscle attaches to bones at tendons and produces voluntary body movements by interacting with the skeleton. Skeletal muscle is nourished by blood vessels that supply oxygen and nutrients and controlled by nerves that help signal the process of muscle contraction.

Muscle Interaction with Bones

Figure 19.10 illustrates how muscles and tendons interact with bone. The ability to move the arm to pitch requires that two muscle groups perform opposite tasks. Contraction of the biceps muscle shortens the muscle while the triceps muscle is relaxed. Likewise, when the biceps muscle is relaxed, the triceps muscle is contracted. Such **antagonistic pairs** of muscles, in which each muscle is paired with a muscle of the opposite effect, are a common feature of the muscular system.

Muscle Structure and Contraction

The structure of a muscle involves smaller and smaller parallel arrays of filaments (Figure 19.11a). A muscle contains bundles of parallel **muscle fibers**.

Figure 19.10 Antagonistic muscle pairs. Movements are often produced through the actions of two opposing muscles. (a) Contraction of the biceps raises the forearm, and (b) contraction of the triceps lowers the forearm. Tendons attach muscles to bones.

Each muscle fiber is a single cell. Within each muscle fiber is a parallel array of threadlike filaments called **myofibrils**. Bands of all myofibrils in a cell are aligned together, or in register, like pipes of the same length stacked neatly with their ends aligned. This gives the skeletal muscle its characteristic striated or striped appearance (Figure 19.11b). Each myofibril is a linear arrangement of **sarcomeres**, which are the unit of contraction of a muscle fiber. A sarcomere is composed of thin filaments of the protein **actin**

(a) Skeletal muscle structure

(b)

Figure 19.11   Skeletal muscle structure. (a) Muscle cells (fibers) are arranged in parallel bundles inside the muscle's outer sheath. (b) This photomicrograph shows a skeletal muscle cell with its myofibrils in register, producing the characteristic banded or striated pattern. (c) The myofibrils inside a muscle cell are linear arrangements of sarcomeres. Sarcomeres are bounded by Z discs.

and thick filaments of the protein **myosin**. The sarcomere is the region between two dark lines, called **Z discs**, in the myofibril (Figure 19.11c).

When a shortstop moves to field a ball, his bones move because the skeletal muscles attached to them shorten. To contract the muscles, the sarcomeres shorten. This shortening takes place through a series of steps outlined in the next section.

Sliding-Filament Model of Muscle Contraction. Sarcomere shortening involves coordinated sliding and pulling motions within the sarcomere (Figure 19.12). Actin filaments are anchored at the Z discs at the ends of a

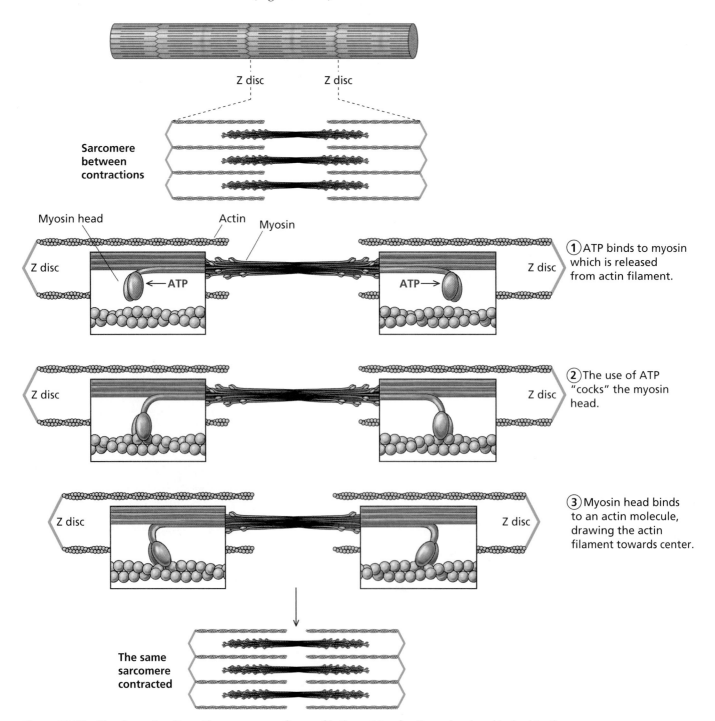

Figure 19.12 Muscle contraction. The sarcomere of a myofibril consists of actin molecules attached to the Z disc and myosin molecules. Using energy from ATP, the myosin head binds to actin and pulls it toward the center. Since this is happening at both ends of the sarcomere, the sarcomere shortens, allowing contraction of the muscle cell and movement of the muscle.

sarcomere. They overlap a parallel, stationary set of myosin molecules that are not attached to Z discs. During contraction, actin filaments slide over the fixed myosin filaments, pulling the ends of the sarcomere with them. The sarcomere ends are brought toward the center when the head of the myosin molecule attaches to binding sites on the actin molecules and pulls toward the center. Contraction of the sarcomere can be likened to the hand-over-hand action of pulling a rope. The stationary hands (myosin) contract and release the sliding rope (actin).

The sliding-filament mechanism allows the sarcomere to shorten or contract in a stepwise manner. First, the high-energy compound ATP binds to the myosin head, causing the myosin head to detach from the binding site on the actin filament. Second, the breakdown of ATP provides energy to the myosin head, and it changes shape or "cocks." Third, the myosin head binds to another actin-binding site; pulls in short, powerful strokes; releases the actin; and reattaches further along. This process serves to pull the Z discs inward, shortening the sarcomere.

Sex Differences in Muscle.   Testosterone released by the testes at puberty increases the size of muscle fibers. Muscle fibers contain testosterone receptors in their cytoplasm, and the presence of testosterone stimulates the cells to increase in mass—in part by retaining dietary nitrogen, which is used to produce more muscle proteins.

Although the ovaries and adrenal glands do release small quantities of testosterone, males have more of this hormone than do females, and the subsequent difference in muscle mass accounts for an overall difference in strength between males and females. In sports that require brute strength, such as football, men have an advantage; but in sports like baseball, where success depends more on agility and hand-eye coordination, the effect of sex differences is reduced.

Even if a man and a woman had similar amounts of muscle mass, the man would look more muscular because women have a different pattern of body-fat deposition. Women have more body fat under their skin, which tends to smooth out the appearance of muscles beneath the skin.

We have seen that sex differences in the endocrine system can lead to differences in skeletal structure and muscle mass, all of which can affect athleticism. In addition to these differences, a few other sex differences can influence a person's athletic abilities.

19.4 Other Sex Differences That Influence Athleticism

As we have seen, some of the endocrine, skeletal, and musculature differences between the sexes may affect athleticism. In this section, we will investigate a few other sex differences and then turn our attention to what these differences really mean.

Body-Fat Differences

Differences in body-fat percentages and locations may affect athleticism. These differences arise at puberty, when a young woman begins to store fat in her torso, abdomen, hips, thighs, and buttocks. This body fat, in addition to skeletal differences, gives women's bodies their curvy shape. Men usually carry most of their fat in the abdomen. Overall, women have about 10% more body fat than men do; this body fat is necessary to maintain fertility. When a female does not have enough body fat, the hormones that regulate menstruation (estrogen, progesterone) are altered or not secreted, and menstruation ceases. Lack of menstruation can be permanent, resulting in sterility, and it can result

Figure 19.13 Swimmer in streamline position. Fat makes a woman's body smoother and more buoyant.

in bone damage because estrogen helps bones stay strong. Excessive exercise or starvation that leads to the cessation of menstruation, called **amenorrhea**, causes permanent bone damage when the estrogen that normally increases prior to ovulation is not produced.

In addition, women may metabolize fat differently than men do. Females seem to utilize more fat to produce energy than do males, who tend to use the body's stored sugars more readily. In women, this has the effect of slowing down glucose metabolism, meaning that more sugars are available for prolonged exercise. This difference may result in a greater tolerance for endurance events in women. In fact, women win endurance events such as ultra-marathons as often as men do.

In standard marathon events, women have also improved their performance when compared to men in recent years. For example, in 1972—the first year that women were officially allowed to compete in marathons—the first-place woman won the Boston Marathon in a time of 3:10:26 (3 hours, 10 minutes, and 26 seconds). That same year, the first man to cross the finish line did so with a time of 2:15:39. In 2004, the first-place woman finished with a time of 2:24:27, over 45 minutes faster than the top 1972 female finisher. The first-place male in 2004 finished with a time of 2:10:37—only a 4-minute difference. In 30 years, women have made substantial athletic gains, bringing the finishing times of men and women closer together. This could simply be due to access and exposure to training; historically, men began training before women did and had access to better coaching and practice facilities. Whether this trend continues, ultimately making women as fast as—or faster than—men at endurance events, remains to be seen.

The extra body fat that gives women increased endurance may also be a physical advantage in long-distance swimming events—fat increases buoyancy and enables women to maintain the most energy-conserving streamlined position with less effort (Figure 19.13). In addition, fat not only provides increased insulation, which slows the rate of body-heat loss, but also stores energy that can be converted to ATP during endurance events.

While women's physiology may be favored in endurance events, men have the advantage when it comes to events requiring a high level of intensity over a shorter period of time; this advantage is partially due to sex differences within the cardiovascular system.

Cardiovascular Differences

Women have smaller hearts and less blood volume relative to body size, and they have fewer blood cells per unit volume than similarly sized men do. This means that men are better able to supply oxygen and nutrients to their greater muscle mass. Women also have smaller lungs that work harder to transport oxygen to the cells and tissues that need it. If a man and woman of similar size are running side by side at the same speed, the woman's respiratory and cardiovascular system are working harder to bring oxygen to her tissues.

Group Differences and Individual Differences

While we can determine average differences between males and females in terms of skeletal size, muscle mass, body fat, and cardiovascular activity, these differences are often of little value in the real world. Average differences hide the fact that there are a wide *range* of values for each of these characteristics influencing athleticism, and the ranges for males and females typically overlap a great deal. Consider the example of body fat: The average healthy woman has 22% body fat, and the average healthy man has 14% body fat. This seems like a large difference, but most males and females fall in the same range (Figure 19.14). Healthy women have from 12% to 32% body fat, and the range

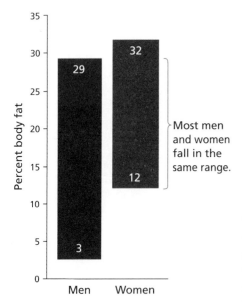

Figure 19.14 Male and female body-fat percentages. This graph shows that there is a lot of overlap in body-fat percentages between the sexes.

is between 3% and 29% for men. Most males and females fall within the 12% to 29% range of body fat.

In fact, it is impossible to determine a person's gender based on their body-fat percentage. If all you knew about a person was that his or her body-fat percentage was 18%, then you would have no way to predict whether they were male or female. In the same way, it would be very difficult to predict a person's body-fat percentage solely on the basis of his or her gender.

The range of body-fat percentages for normal women shows a 20% difference from lowest to highest; for men, the range varies by 26%. However, the average difference between men and women is only 7%. As is often the case when we try to study differences between various human groups, the differences *within* a group (individual-based) are greater than the differences *between* two groups (group-based).

The overlap of ranges also limits assumptions we can make about athletic performance. If we randomly picked 10 men and 10 women from a college course and timed them running 100 meters, we might see a slight difference in average speed between the males and the females. However, the large overlap of ranges would make it impossible either to predict a person's gender based on his or her time or to predict a person's time based on his or her gender. If we also timed 10 women from the track team, they would surely be at the fastest end of the distribution—probably much faster than most or all of the randomly selected men. So, in this example, average differences can tell us something about athleticism, but they are certainly not the whole story.

To understand the whole story of athleticism and gender, we must evaluate the impact that our culture has on intensifying average differences in athletic performance. Strong cultural forces often determine the amount of time athletes spend practicing as well as the amount of support they receive for doing so.

As a young girl, Ila Borders was prevented from registering for Little League baseball when she first tried to play at age 10. In college, where she was the first woman to be given a baseball scholarship, and the only woman on the team, she was taunted by her own teammates as well as by members of the opposing team. In a television interview with Mike Wallace of CBS's *60 Minutes*, Borders spoke of how her own teammates would throw baseballs at her when she had her back turned, and how players from opposing teams yelled obscenities at her.

Today, more than 100,000 women participate in intercollegiate athletics—a fourfold increase since 1971. This is due in part to Title IX federal legislation, which mandates equal athletic opportunities for schools receiving federal monies. Financial support for female athletes, and their social acceptance, has grown tremendously.

Physical activity is important to overall health, regardless of gender. Studies have shown that children of both sexes who participate in sports are less likely to drop out of high school, smoke, or drink. Participation in sports teaches values such as teamwork, discipline, pride in accomplishment, drive, and dedication.

Children who exercise are also more likely to continue exercising in adulthood than children who do not exercise. In a country where most adults do not get the recommended 30 to 60 minutes of exercise most days of the week, it makes sense to encourage everyone to become more athletic. When good exercise habits are carried into adulthood, there is a decreased risk of heart disease, obesity, diabetes, and many cancers. Additional benefits include lowered cholesterol, and studies suggest that exercise may decrease anxiety and depression. Sex differences that lead to differences in strength, speed, balance, and endurance are meaningless for recreational athletes. What really matters to most athletes is finding an activity they enjoy, so that they can keep healthy for as long as possible.

CHAPTER REVIEW

Summary

19.1 The Endocrine System

- Glands in the endocrine system release hormones that are involved in regulating many bodily functions (p. 496).

- Hormones are secreted by endocrine glands and travel through the circulatory system to their target organs. To trigger a response, hormones can bind to receptors on the target cell's surface or diffuse into the cell and bind to receptors inside the cell (p. 496).

- Testosterone and estrogen are sex hormones that move across cell membranes and bind to receptors inside target cells. The hormone and receptor move together to the cell nucleus and control the expression of target genes (pp. 496–497).

- There are several different kinds of endocrine glands. The thyroid gland secretes hormones that stimulate metabolism, and along with the parathyroid gland, regulates calcium levels. The thymus helps the immune system form, and the pancreas helps to regulate blood glucose levels (p. 498).

- The hypothalamus regulates body temperature, hunger, thirst, and reproduction. The hypothalamus secretes GnRH, which stimulates the pituitary gland to release FSH and LH (pp. 498–499).

- Adrenal glands secrete adrenaline, androgens, and estrogens (p. 499).

- Testes secrete testosterone, which affects the development of male characteristics (p.499).

- Ovaries secrete estrogen, which fosters the development of female characteristics (pp. 499–500).

Web Tutorial 19.1 **The Endocrine System**

19.2 The Skeletal System

- The skeletal system is composed of bones and cartilage. It provides support, protection, movement, and mineral storage.

- The skeleton supports body and limbs. Joints are regions where bones come together. Different types of joints allow for different kinds of movements. Ligaments hold bones together at joints (pp. 500–502).

- Bones are composed of cells that are encased in a strong matrix. Bone remodeling is under the control of the endocrine system. Hormones stimulate the osteoblasts to use calcium to build bone and the osteoclasts to break down bone and release calcium (p.503).

Web Tutorial 19.2 **Contraction of Skeletal Muscle**

19.3 The Muscular System

- The muscular system functions in movement, which is coordinated by the combined actions of antagonistic pairs of muscles. Each muscle in the antagonistic pair has an opposite effect (p. 506).

- Muscles are bundles of parallel fibers. Muscle fibers are bundles of parallel myofibrils. Myofibrils are linear arrays of sarcomeres, which are the unit of muscle contraction (pp. 506–507).

- Sarcomeres are composed of actin and myosin. Actin filaments attached to the ends of a sarcomere are pulled toward the middle of the sarcomere by the movement of myosin heads. These myosin heads then attach to the filaments, pull them toward the sarcomere, release the filaments, and reattach further down the filament. This process is called the sliding-filament model of muscle contraction (pp. 508–509).

19.4 Other Sex Differences That Influence Athleticism

- Women, on average, have more body fat than do men. Women store fat on their hips and thighs; men store fat in their abdomens (p. 509).

- Men have larger hearts, more blood volume, and more blood cells than do women (p. 510).

Learning the Basics

1. List the functions of the endocrine organs involved in producing sex differences.

2. Describe bone structure.

3. Describe muscle structure.

4. Steroid hormones act by _____.

   A. causing neurons to fire; **B.** increasing muscle contraction; **C.** diffusing across membranes and regulating the expression of genes; **D.** binding to receptors on the cell surface and transducing a response without entering the cell

5. The endocrine organ that sits atop a kidney is _____.

   A. the pituitary gland; **B.** the hypothalamus; **C.** the ovary; **D.** the adrenal gland; **E.** the testicle

6. Bones of the limbs form part of the _____.

A. appendicular skeleton; **B.** axial skeleton; **C.** spongy bone; **D.** compact bone

7. Osteoblasts _____.

A. are found in the marrow of bone; **B.** regulate bone deposition; **C.** regulate bone reabsorption; **D.** are more active when calcium is low

8. The antagonistic pairs of arm and leg muscles _____.

A. allow healthy muscles to compensate for injured ones; **B.** allow muscles to produce opposing movements; **C.** allow different types of rotations of joints; **D.** allow myofibrils to fire in response to different stressors

9. During muscle contraction, _____.

A. the actin heads pull the sarcomere closed; **B.** myosin attaches to actin, pulls it toward the center of the sarcomere, releases it, and reattaches further along; **C.** myofibrils shrink due to the actions of testosterone; **D.** muscle fibers contract under the actions of testosterone

10. Which of the following is a true statement regarding muscle contraction?

A. Actin filaments are stationary, and myosin heads can move.; **B.** Z discs are pulled toward the center of the sarcomere.; **C.** Actin is used up during this process.; **D.** No ATP is required for movement to occur.

Analyzing and Applying the Basics

1. How would a chemical that blocks testosterone production affect muscle development?

2. Why are most men taller than most women?

3. In what ways do skeletal differences between males and females benefit each sex?

Connecting the Science

1. Provide several examples of cultural influences that sway our views of biological differences between males and females.

2. Describe other situations in which males and females behave differently but in which that difference does not necessarily make one sex better than the other.

CHAPTER 20

Is There Something in the Water?

Reproductive and Developmental Biology

Alligators with altered reproductive structures.

Fish that are all female.

Is there something in the water …

Is something wrong with the water? People in many communities are concerned about the health of organisms that live in their local lakes, rivers, and streams. In 1980, a pesticide spill in Lake Apopka, in central Florida, killed most of the resident alligators. Today, the male alligators in the lake have abnormally small penises. Fish near a waste treatment plant in Colorado's Boulder Creek have both male and female reproductive organs. Populations of chinook salmon in the Pacific Northwest are almost all female, and most of these female fish have chromosomes typical of males of this species. The salamander population in Tecumseh, Missouri, has decreased by 75%, and their eggs and larvae are no longer found in local rivers.

Some scientists and environmental activists connect reproductive changes in wildlife to similar effects in people. Consider the following: According to some scientific studies and the popular press, human males have ever-decreasing sperm counts and reproductive organ abnormalities. Human females are reported to be experiencing more fertility problems and menstrual irregularities than in the past. If several women working in the same office or living in close proximity become pregnant at the same time, people will often joke that "it must be something in the water." Current fertility problems have turned that notion on its head, and now scientists and laypersons alike are wondering if there *is* something in the water that is causing a decrease in fertility among both humans and animals.

…that is altering animal and human reproduction?

515

Studies of aquatic animals with certain reproductive problems seem to implicate a class of chemicals called **endocrine disruptors**. Endocrine disruptors disrupt the actions of the hormone-producing endocrine system. Evidence of endocrine disruption in aquatic animals has led to concern about the water humans are drinking. Might there be endocrine disruptors in the water supply? Are these endocrine disruptors tricking the human endocrine system into releasing hormones in situations when no hormone should be released? Are endocrine disruptors blocking the actions of hormones that have been released?

To answer these questions, a more thorough understanding of reproduction and development is required. We will begin by looking at basic reproductive processes.

20.1   Principles of Animal Reproduction

Reproduction is the process whereby organisms produce offspring, allowing for the continuation of their species. Organisms compete with other individuals for the resources required to survive and reproduce. During reproduction, cells such as sperm and eggs that are nonessential to the animal's own survival, but indispensable to the perpetuation of the species, are produced and can help give rise to offspring. Because reproduction begins with the production of single cells, any environmental factor that affects these cells can dramatically influence the offspring produced. This is why an early clue to the presence of a biologically active pollutant in an area is often a decline in the survival and health of offspring produced in that environment.

Animals employ a wide variety of strategies for producing offspring. However, the two basic schemes for reproduction are asexual and sexual reproduction.

Asexual Reproduction

If you have ever propagated plants from cuttings, you know that these organisms can reproduce without the aid of a partner. This process also occurs in some animals. Called **asexual reproduction**, this method of reproduction occurs when one parent produces offspring that are genetically identical to the parent and to each other. In effect, the parent clones itself.

Prokaryotic organisms such as bacteria produce two genetically identical cells, called daughter cells, by first copying their single chromosome and then splitting the single cell into two cells. This process is called **binary fission**. Some single-celled eukaryotic organisms, such as amoebas, divide by the asexual process of mitosis to produce two genetically identical offspring. Sometimes multiple fissions can occur, fragmenting one individual into many individual daughter cells that can develop into adults. Sponges are multicellular animals that reproduce by fission, during which masses of cells break away from the parent to form offspring (Figure 20.1a).

Budding is a form of asexual reproduction that is similar to fission. A yeast cell budding in bread dough undergoes cell division to produce a daughter cell that remains attached to the parent until it achieves sufficient growth to break away. Hydra are simple animals that reproduce by budding (Figure 20.1b).

A variant of budding is a process called **vegetative reproduction**. Certain plants and aquatic animals, such as sea squirts, send out runners on which buds form. These buds then develop into separate individuals (Figure 20.1c).

Asexual reproduction has many advantages as well as disadvantages. There is no need to find a partner in order to reproduce, which is particularly advantageous for organisms that do not move or are isolated from other

(a) Sponge   **(b)** Hydra   **(c)** Sea squirt

Figure 20.1   Asexual reproduction in animals. Asexual reproduction produces genetically identical off-spring. There are many ways in which animals can reproduce asexually. (a) Sponges can reproduce by fission when cells that were once part of the parent's body break off and form new offspring. (b) Hydra reproduce by budding, which involves undergoing cell division to produce daughter individuals that eventually break away from the parent. (c) Sea squirts reproduce by vegetative reproduction, which involves the formation of many identical offspring on runners.

organisms. When conditions are favorable, asexual reproduction also allows for the rapid production of large numbers of offspring. However, because all offspring are identical to the parent, there is not much ability to adapt to environmental change. Populations of individuals that are not identical are more likely to have a few members whose genetic makeup will allow them to survive and perpetuate the species when conditions change. When a population consists of identical individuals, an environmental change could decimate the entire population.

Environmental variability favors another form of reproduction, called **sexual reproduction**. Organisms using this strategy produce offspring that differ genetically from each other and from the parents.

Sexual Reproduction

Creation of offspring by the fusion of sex cells from two different individuals—for example, a mating between one male and one female parent—is required for sexual reproduction. Individuals that produce large sex cells or **gametes** are classified as females, and individuals that produce small gametes are classified as males. Essay 20.1 (page 518) discusses why natural selection has resulted in two classes of different-sized gametes.

Offspring are produced after the gametes from two different individuals combine genetic information at fertilization. The gamete-producing structures are called **gonads**. In males, sperm are produced by reproductive organs called testes. Eggs are produced by the ovaries of the female.

Because two different individuals contribute genetic information to their gametes, the offspring will be a genetic mixture of both parents. Each gamete is said to be haploid (n), and contains half the genetic information of the individual that produced it. Fusion of two haploid gametes produces a diploid ($2n$) **zygote** or fertilized egg cell. Because a mixing of parental genetic information occurs in the offspring, sexual reproduction produces an almost infinite variety of genetically unique individuals. Most commonly, an organism will produce either male or female gametes but not both. However, there are exceptions.

Essay 20.1 Focus on Evolution: The Development of Two Different-Sized Gametes

Males produce small gametes, and females produce large gametes. In birds, a female produces 3 to 5 eggs that each weigh 15% to 20% of her body weight. In the same breeding season, a male bird may produce trillions of sperm that in aggregate weigh less than 5% of his body weight. In humans, women produce eggs that are 1 million times larger than a single sperm.

Biologists agree on one, fairly simple, explanation as to why there are sex differences in male and female gamete size. Once sexuality evolved as a method of reproduction, the formation of two different sexual types (i.e., female and male) was probably inevitable. This diversification occurred through the process of natural selection and is diagrammed in Figure E20.1.

When a range of gamete sizes are produced in a population, individuals producing gametes near either the small or large ends of the size distribution will be more successful than individuals who produce medium-sized gametes. Imagine three different ways of distributing the same amount of energy into gametes—one could produce many small gametes, few large gametes, or a moderate number of medium-sized gametes. Individuals that produce few, but large, gametes provide a large supply of nutrients in each cell. These individuals will therefore have larger, healthier offspring at birth. Their offspring, because of their size and hardiness, will be better able to compete for food than will offspring produced from smaller gametes. Accessing more food allows these larger offspring to have a large number of their own offspring.

Thus individuals that produce just a few large gametes will have more grandchildren than will individuals who produce a greater number of medium-sized gametes. By the process of natural selection, individuals with large gametes should therefore become prevalent in the population.

On the other hand, individuals who produce many small gametes can potentially parent more offspring than can those who produce fewer gametes because they can distribute these abundant gametes more widely. In addition, small gametes may be more likely than larger gametes to win the race to fertilization sites because they can move faster. Natural selection would favor individuals who produce many small gametes over individuals who produce fewer, larger ones.

The two successful gamete production strategies described previously coexist in populations because each strategy alone is unstable. In a population where all gametes were large, individuals who produced smaller gametes would be successful at parenting offspring because they could fertilize many other gametes. Where all gametes were small, individuals that appeared with larger gametes would be more successful at producing grandchildren. The most stable situation is a population where 50% of the individuals produce a few large gametes and 50% produce many small gametes—in other words, where the population is 50% female and 50% male.

(a) Initial frequency

Many and faster gametes results in increased numbers of fertilizations

Slower than smaller gamete and less resources than larger gamete so selected against.

More resources so greater offspring survival.

Frequency in population

Gamete size

(b) Frequency after selection

Frequency in population

Gamete size

Figure E20.1 The evolution of gamete size. Natural selection resulting in two successful gamete production strategies.

Hermaphrodites are animals that have both male and female reproductive systems. Earthworms, for example, can both donate and receive sperm during mating. Because the production of male and female gametes by one individual is very common in the plant kingdom, the term *hermaphrodite* applies only to animals.

The mechanism of fertilization can differ among sexually reproducing animals. Fertilization can occur inside or outside the body. Depositing sperm in or near the female reproductive tract through **copulation**, or sexual intercourse, allows **internal fertilization** to occur. Sharks, reptiles, birds, and mammals reproduce via internal fertilization. Many aquatic invertebrates and most fish and amphibians use **external fertilization**, during which the parents need not have physical contact with each other. The female lays eggs in the water, and the male releases sperm over the eggs. Because these eggs are fertilized and develop outside the body, in water, they are very susceptible to aquatic contaminants.

Figure 20.2 Deformed frogs. This frog may have been exposed to chemicals that altered its development.

Environmental Contaminants and Sexual Reproduction

Frogs are sexually reproducing organisms that use external fertilization. The fertilized eggs of frogs give rise to tadpoles that develop into adults in the water. Because so much of a frog's development occurs in the water, and because chemicals in the water can be absorbed directly by the developing embryo, the embryos of frogs and other amphibians are highly sensitive to environmental contaminants. In fact, you have probably seen pictures of frogs with extra limbs, missing eyes, and other obvious deformities (Figure 20.2). Many scientists are studying whether these deformities are caused by man-made chemicals that have found their way into the frog's habitat. In a sense, frogs can be used as environmental early warning signals in the same way that canaries were once used to warn of high concentrations of carbon monoxide in coal mines. When canaries that were brought into mines showed distress or died, the miners knew to get out of the mine quickly before the carbon monoxide levels were high enough to kill them too.

Because problems in frog reproduction and development can foreshadow problems for other organisms, the results of work done by Dr. Tyrone Hayes (Figure 20.3) concern many scientists. Hayes's research has shown that a widely used pesticide called atrazine, which makes its way from farmers' fields into the groundwater, marshes, and ponds, is causing male frogs to grow eggs inside their testes.

Why would male frogs produce egg cells inside their testes? Because these "feminized" frogs were collected from atrazine-contaminated ponds and marshes, Hayes hypothesized that this pesticide might be an endocrine disruptor that affects the development of the frog gonad.

To test the hypothesis that atrazine was responsible for the conversion of sperm-producing tissue into egg-producing tissue, Hayes moved his research from ponds and marshes into his laboratory. He injected normal male tadpoles with atrazine. He found that this did indeed disrupt the development of testicular tissue, resulting in gonads that could also produce egg cells. Further research by Hayes also suggests that atrazine does this by activating the enzyme **aromatase**, which converts testosterone, a hormone normally produced in higher quantities within males, into estrogen, a hormone normally produced in higher quantities within females (see Figure 20.4 on page 520). This resulted in the "feminization" of the male testicle. Alarmingly, Hayes found that this endocrine disruption will occur with doses of atrazine that are lower than those found in our drinking water.

Frogs are not the only organisms that have been shown to undergo changes to reproductive organs after exposure to endocrine disruptors. For

Figure 20.3 Dr. Tyrone Hayes. Dr. Hayes at work collecting frogs from an atrazine-contaminated site.

Figure 20.4 Converting testosterone to estrogen. The enzyme aromatase converts testosterone into estrogen. Scientists believe that the herbicide atrazine causes this enzyme to be overactive in male frog testes.

3 years, scientists in Canada have placed birth-control pills into the remote Pearl Dace Lake in Ontario. A remote lake was selected for this study so that no other known sources of contamination would be present. Researchers selected the birth-control pill for this experiment because many scientists are concerned that the estrogen from these pills is making its way into the water supply. Women using the birth-control pill secrete estrogen in their urine, which is eventually deposited into waterways. The amount of estrogen placed in the lake was equivalent to what is found in a typical urban waterway.

When scientists compared fish exposed to estrogen with those from the same lake before estrogen treatment, they found significant increases in the numbers of "feminized" males from all species of fish. From minnows to trout, the scientists found male fish that were producing eggs. They found no increase in such fish in two nearby untreated lakes that were used as controls.

The work of these and other researchers shows that organisms living in waters contaminated by endocrine-disrupting chemicals can develop altered reproductive structures. While humans do not grow and develop in contaminated waters, they might be drinking contaminated water. Are human reproductive structures affected by exposures to endocrine disruptors?

20.2 Human Reproduction

To understand how human reproduction might be affected by environmental contaminants, we will start by considering the anatomical structures that comprise the human male and female reproductive systems.

Human Reproductive Systems

The reproductive systems of males and females consist of external and internal structures. These structures are designed to allow the production and maturation of gametes, signal the synthesis and secretion of substances required for reproductive function, and provide a route to deliver the gametes through ducts.

Male Reproductive Anatomy. The external and internal components of the male reproductive system are shown in Figure 20.5.

The **penis** delivers sperm to the female reproductive tract during sexual intercourse. The penis is composed of spongy erectile tissue. During sexual arousal, this tissue fills with blood. Pressure from the increased volume of blood in the penis seals off the veins that drain blood from it. This causes the penis to become engorged with blood, keeping it erect. The erection is essential for insertion of the penis into the vagina, which facilitates delivery of sperm to the egg. A tube inside the penis, called the

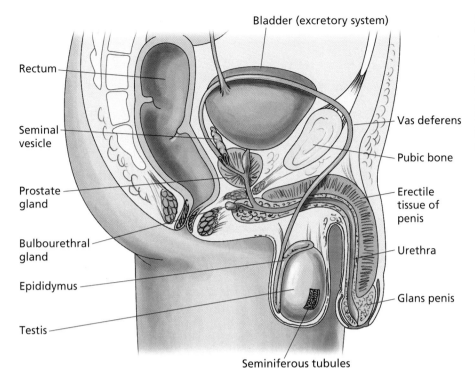

Bladder (excretory system)

Rectum

Seminal vesicle

Prostate gland

Bulbourethral gland

Epididymus

Testis

Seminiferous tubules

Vas deferens

Pubic bone

Erectile tissue of penis

Urethra

Glans penis

Figure 20.5   Male reproductive anatomy. The male external genitalia consist of the penis and scrotum. The scrotum houses the testicles. The testicles of a male produce sperm, which is stored in the epididymis and carried by vas deferens through the urethra to the outside of the body. Secretions from the seminal vesicle, prostate, and bulbourethral glands help sperm mature and gain motility.

urethra, provides passage out of the body for both sperm and urine. The head, or **glans penis**, consists of highly sensitive skin that is covered by a fold of skin called the foreskin.

The **scrotum** is a pouch below the penis that contains the **testes** or **testicles**. The skin of the scrotum is thin and devoid of fatty tissue. It tends to be folded or wrinkled with little hair. Below the skin is a layer of involuntary smooth muscle that regulates the position of the testes relative to the body in order to keep them at a temperature that maximizes sperm production. The scrotum contracts in the cold in order to keep the sperm-producing testes closer to the body.

In addition to producing sperm, testes also produce male hormones called **androgens**. The testes consist of many highly coiled tubes, called **seminiferous tubules**, where sperm form. The seminiferous tubules are held in place by connective tissues that are rich in hormone-producing cells called **Leydig cells**. The process of sperm formation takes about 60 days and begins in the seminiferous tubules. From the seminiferous tubules, sperm pass through the coiled **epididymis**, which is an approximately 6-meter (20-foot) tube that rests atop each testis. During the 20 or so days that sperm pass through the epididymis, the sperm become able to move and capable of fertilizing an egg cell.

During ejaculation, sperm are propelled from the epididymis through sperm-carrying ducts called the **vas deferens**. Each duct is covered in smooth muscle and is capable of peristaltic contractions (contractions in waves) that move sperm through the duct.

Several glands add secretions to the developing sperm as they make their way through the ducts of the reproductive system. **Seminal vesicles** are glands that secrete mucus and fructose (a sugar) for the sperm to use as energy. The **prostate** gland secretes a thin, milky white fluid into the urethra. These secretions contain nutrients for sperm. The **bulbourethral glands** are a pair of small glands that lie below the urethra between the prostate and the penis. Before ejaculation takes place, these glands secrete clear mucus that helps neutralize any acidic urine present in the urethra. Sperm and these secretions make up **semen**.

Figure 20.6   Female reproductive anatomy.
The female external genitalia consist of the vulva and the clitoris. Openings to the outside allow for the expulsion of urine (via the urethra) and menstrual tissue (vagina). The ovaries of the female produce egg cells and hormones required for reproduction. Egg cells move through the oviducts to the uterus. The uterus can house a developing fetus. The cervix is the bottom portion of the uterus. The cervix widens during birthing. The vagina is the passageway into and out of the uterus.

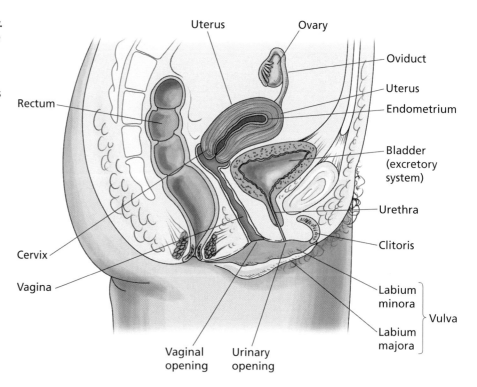

Female Reproductive Anatomy.

Figure 20.6 shows the female reproductive anatomy. The most obvious feature of the female external genitalia is a structure called the **vulva**. The vulva consists of two sets of lips, or labia: the outer **labia majora**, which are fatty and have a hairy external surface, and the inner **labia minora**, which contain neither fat nor hair. At the front of the vulva, the labia minora divide around the **clitoris**, an important organ for female sexual arousal.

Between the folds of the labia, there is an opening for the urethra, which in females serves as a passageway for urine from the bladder. The urethra in women is short (4 cm on average) compared to the urethra in men (18 cm on average). Partially due to this difference in length, women are more susceptible to bladder infections because bacteria have a shorter distance to travel from the outside of the body. A few centimeters below the urinary opening is the vaginal opening.

The organs that make up the internal genitalia are the ovaries, vagina, uterus, and oviducts. The **ovaries** are the female gonads. They produce the gametes and sex hormones that circulate in a woman's body during her reproductive years. Ovaries are about the size of a walnut; but at birth, they contain about 2 million ova (eggs) each. During nearly every menstrual cycle, one or more eggs mature and are released from a structure called a **follicle**. The follicle, a fluid-filled sac containing the developing egg, secretes the ovarian hormone estrogen. During ovulation, the follicle ruptures, much like a blister popping, to release the egg. The remnant of the follicle, called the **corpus luteum**, secretes both estrogen and progesterone.

The **vagina** is a muscular organ that acts primarily as a passageway into and out of the uterus. The **uterus** is about the size of a fist. The wall of the uterus is thick (about 1 cm) and is composed of some of the most powerful muscles in the human body. These muscular walls contract rhythmically during labor, childbirth, and orgasm. The internal surface of the uterine wall is called the **endometrium**, which changes in thickness during the course of the menstrual cycle. The lower third of the uterus becomes narrower than the upper portion and is called the **cervix**.

The **oviducts** are actually an extension of the top surface of the uterus. These tubes extend from the body of the uterus toward the ovaries, which are suspended within the abdominal cavity. The oviducts are not attached to the ovaries directly. Instead, they end in brushy structures that move over the surface of the ovary. These movements, along with suction inside the oviducts, direct the eggs released by the ovary into the oviducts.

The pathway from the vagina to the uterus to the oviduct represents the only natural pathway in humans directly from the outside of the body to the inside of the abdominal cavity. This means that women can experience bacterial infections inside their abdominal cavity without experiencing an injury that punctures the cavity wall. These infections are often sexually transmitted. For a description of common sexually transmitted diseases in both men and women, see Table 20.1 on pages 524–526.

Sexually transmitted diseases can cause permanent **infertility**, defined as the failure to achieve pregnancy after 1 year of unprotected intercourse. Infertility can result from physical blockages of ductal structures caused by scarring in response to infection. For pregnancy to occur, the oviducts must be open, allowing the sperm to reach the egg and the fertilized egg to travel back down the oviduct to the uterus.

Likewise, the sperm must be able to move from the testes through the vas deferens and urethra to be delivered to the egg. Physical blockages or obstructions to the male ducts can occur also, but this is less often a cause of infertility in males than it is in females.

STDs are not the only cause of infertility. Endocrine-disrupting chemicals can also cause changes to reproductive structures that lead to infertility.

Endocrine Disruptors, Reproductive Anatomy, and Infertility. A well-studied example of an endocrine disruptor that affects human fertility is provided by a chemical called diethylstylbestrol or DES. Between 1938 and 1971, many pregnant women were given this synthetic drug to help prevent miscarriage. However, the physicians who prescribed this drug did not know that it behaved as a mimic of the female hormone estrogen and altered the development of the uterus, resulting in a misshapen uterus less able to sustain a pregnancy. Daughters of women who took this drug have a higher than average rate of infertility; approximately 24% of these women are infertile.

The most common cause of male infertility is the inability to produce enough healthy sperm. Many scientific studies have shown that sperm counts are declining in developed countries. Although no one knows for sure why this is happening, some scientists are studying whether endocrine disruptors are interfering with human males' ability to produce sperm, an effect these chemicals have been shown to have on other animals. The production of gametes is a multistep process allowing many different opportunities for endocrine disruptors to act.

Gametogenesis

The production of gametes is called **gametogenesis**. In human males and females, gametogenesis involves the process of meiosis (discussed in Chapter 5), which is a type of cell division that reduces the chromosome number by one half. Since human body cells contain 46 chromosomes, meiosis in humans helps to produce gametes that contain 23 chromosomes. Human body cells with 46 chromosomes are diploid, and the gametes produced after meiosis are haploid.

Meiosis alone is not enough to produce a functional gamete. Other changes to sex cells occur as they go through gametogenesis to enable them to mature into gametes capable of being involved in fertilization. For instance, sperm cells gain motility, and egg cells increase in size and nutrient content to allow for the development of the early embryo.

Table 20.1 Sexually transmitted diseases. The causes, symptoms, prevention, and prevalence of common STDs are discussed.

Bacterial pathogens	Mode of infection	Symptoms	Treatment and prevention	Incidence
Chlamydia Cause—*Chlamydia trachomatis* (bacterium)	• This sexually transmitted disease infects the urethra, cervix, and oviducts of women and the urethra of men.	• Pelvic pain, fluid discharge • If untreated, can lead to pelvic inflammatory disease and infertility • Many cases are undiagnosed; there may be no symptoms until years after infection.	• Treated with antibiotics • Condoms prevent transmission.	• Most common bacterial STD • Approximately 3 million cases each year in the United States.
Gonorrhea Cause—*Neisseria gonorrhoeae* (bacterium)	• This disease, also known as "the clap," can be sexually transmitted or spread from infected mothers to babies during birthing.	• Thick discharge from penis or vagina. • 80% of gonorrhea cases are asymptomatic. • Untreated gonorrhea can cause infertility in women if bacteria spreads to the oviducts and causes pelvic inflammatory disease.	• Treated with antibiotics • Condoms prevent transmission.	• Over 500,000 cases of gonorrhea are diagnosed in the United States each year.
Pelvic Inflammatory Disease (PID) Cause—Gonorrhea or chlamydia	• This infection of the female reproductive tract occurs when sexually transmitted bacteria ascend from the vagina into the uterus and oviducts.	• Pelvic pain • Difficulty becoming pregnant due to scarring and blockage of reproductive organs	• Antibiotics kill PID–causing organisms but do not heal damaged reproductive organs. • Condoms prevent transmission.	• About 1 million United States women diagnosed each year • Approximately 100,000 become infertile from PID. • Thousands suffer complications with pregnancy.
Syphilis Cause—*Treponema pallidum*	• Sexually transmitted	• Immediate symptoms include a small sore, a chancre, at the site of infection. • Weeks later, a rash, fever, fatigue, and joint soreness will follow. • If untreated, neurological problems, paralysis, and death can occur.	• Antibiotics can cure if caught early.	• 70,000 cases diagnosed annually in the United States.

Viral pathogens	Mode of infection	Symptoms	Treatment and prevention	Incidence
AIDS Cause—human immunodeficiency virus (HIV)	• Spreads through oral, anal, and vaginal sex, as well as via blood transfusion.	• Over time, HIV infection weakens the immune system so much that infections that are typically controlled cause severe damage.	• Combination-drug therapies halt progression of disease. • Cannot be cured •Condoms reduce the likelihood of transmission.	• Worldwide, about 42 million people are living with AIDS/HIV.
Genital Warts Cause—human papilloma virus (HPV)	• Sexual contact (intercourse or oral-genital contact) with an infected partner.	• Growths or bumps on the pubic area, penis, vulva, or vagina • Some types of HPV cause genital warts; other types can cause abnormal cell changes on a woman's cervix, lead to cervical cancer.	• No current cure for HPV • Warts can be removed surgically, burned off with lasers, or frozen off with liquid nitrogen. • Condoms reduce the likelihood of infection.	• Some studies estimate that the majority of the sexually active population has been exposed to at least one or more of the over 70 different types
Hepatitis B Cause—hepatitis B virus (HBV). Hepatitis can also be caused by noninfectious agents such as chemical poisons, drugs, and alcohol.	• Transmitted through blood and other bodily fluids • Newborns can get infected from their mothers during delivery.	• Inflammation and scarring (cirrhosis) of the liver, which may be fatal.	• Preventable through vaccination • A few antiviral drugs are effective for treating chronic HBV infection. • Condoms prevent transmission.	• Approximately 200,000 U.S. cases are diagnosed annually. • About 4,000–5,000 people die from the infection.
Herpes Simplex Cause—herpes simplex virus type 1 (HSV-1) or herpes simplex type 2 (HSV-2)	• Spread by direct skin-to-skin contact, usually by kissing or by oral, vaginal, or anal intercourse.	• Cold sores or fever blisters on mouth or face • Can also cause similar symptoms in the genital area, known as genital herpes.	• Antiviral medications lessen the duration and discomfort of herpes outbreaks but do not provide a cure. • Condoms reduce the likelihood of transmission.	• Nearly 20% of the population is infected with genital herpes.

Insect, protozoan, and fungal pathogens	Mode of infection	Symptoms	Treatment and prevention	Incidence
Pubic Lice Cause— *Pediculus pubis* (insect)	• This sexually transmitted disease, also known as "crabs," is transmitted via skin-to-skin contact or contact with an infected towel or clothing.	• Itching of the pubic area caused by an allergic reaction to the bites usually starts about 5 days after initial infection.	• Cured by washing the affected area with a delousing agent	• Approximately 3 million U.S. cases of crabs every year

(*table continued on next page*)

(Table 20.1   continued)

Insect, protozoan, and fungal pathogens	Mode of infection	Symptoms	Treatment and prevention	Incidence
Trichomoniasis Cause— *Trichomonas vaginalis* (protozoan)	• This disease, also known as "trich," is transmitted by sexual intercourse.	• In women, vaginal itching with a frothy yellow-green vaginal discharge. • Most men do not have symptoms. • Some men experience irritation in urethra after urination or ejaculation.	• Treated with antibiotics • Condoms decrease the likelihood of transmission.	• An extremely common STD, infecting up to 15% of sexually active women in the United States (over 2 million per year)
Yeast Cause—*Candida* (fungi)	• Normal inhabitants of the female reproductive tract that increase in number during illness or stress • Antibiotics, taken for bacterial infections, can kill vaginal bacteria and allow the yeast to grow further. • Also passed from person to person, such as via sexual intercourse	• Thick whitish discharge from vagina and vaginal itching	• Antifungal medicines can cure yeast infections.	• Nearly 75% of all adult U.S. women have had at least one yeast infection.

Spermatogenesis. The production of sperm, called **spermatogenesis** (Figure 20.7a), begins to occur in the testes at puberty. Cells that will undergo gametogenesis line the walls of the seminiferous tubules. During spermatogenesis, each parent cell first duplicates by mitosis, and then one of the two daughter cells undergoes meiosis. The other daughter cell maintains the function of the parent cell. Since each round of mitosis produces two cells, only one of which goes on to complete meiosis, a man will never exhaust the supply of cells that can be used to make sperm cells.

The diploid cell that begins meiosis is called a **primary spermatocyte**. After the first meiotic division, which separates the 23 homologous pairs of chromosomes, the cell is called a **secondary spermatocyte**. Secondary spermatocytes, with their 23 duplicated but unpaired chromosomes, are haploid. These cells then undergo meiosis II to produce haploid cells that no longer have duplicated chromosomes; these haploid cells are called **spermatids**.

Cells that aid the developing sperm, called **Sertoli cells**, are located also in the seminiferous tubules. These cells secrete the substances required for further sperm development and convert the spermatids into **spermatozoa** by removing some of their excess cytoplasm. The mature sperm is composed of a small head containing the DNA, a midpiece that has mitochondria to provide energy for the journey to the oviduct, and a tail (or flagellum) to propel the sperm (Figure 20.7b). At the tip of the sperm's head lies the **acrosome**. This structure contains digestive enzymes that help a sperm cell gain access to the egg cell.

(a) Spermatogenesis

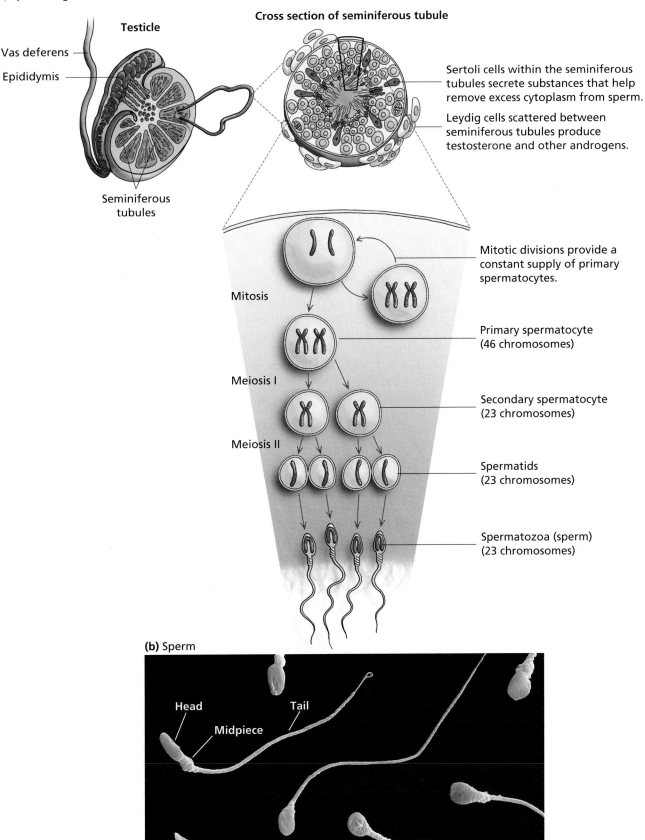

Testicle

Vas deferens

Epididymis

Seminiferous tubules

Cross section of seminiferous tubule

Sertoli cells within the seminiferous tubules secrete substances that help remove excess cytoplasm from sperm.

Leydig cells scattered between seminiferous tubules produce testosterone and other androgens.

Mitosis

Meiosis I

Meiosis II

Mitotic divisions provide a constant supply of primary spermatocytes.

Primary spermatocyte (46 chromosomes)

Secondary spermatocyte (23 chromosomes)

Spermatids (23 chromosomes)

Spermatozoa (sperm) (23 chromosomes)

(b) Sperm

Head

Midpiece

Tail

Figure 20.7 Spermatogenesis. (a) Sperm are produced in the seminiferous tubules of each testicle. Cells in the testes first undergo a round of mitosis, producing two identical daughter cells with 46 chromosomes. The primary spermatocyte begins meiosis with 46 chromosomes. After meiosis I, the two resultant daughter cells are the secondary spermatocytes. Meiosis II results in the production of four haploid daughter cells containing 23 chromosomes, called spermatids. Spermatids are then modified to become mature spermatozoa. (b) The head of the sperm contains the DNA. The midpiece contains mitochondria to provide energy for the flagellum or tail.

Oogenesis.   The formation and development of female gametes, called **oogenesis**, occurs in the ovaries and results in the production of egg cells (Figure 20.8a). A small percentage of these egg cells will be released over a woman's reproductive life span, and an even smaller percentage may be fertilized. While spermatogenesis begins at puberty, oogenesis actually begins while the female is still in her mother's uterus and then pauses until puberty. At puberty, development of preexisting eggs continues each month until the cessation of menstruation or **menopause**. Amazingly, the egg cell that helped produce you started developing inside your mother while she was developing inside your grandmother's uterus!

Developing egg cells, called **oogonia**, begin meiosis but pause at prophase I, at which stage they are called **primary oocytes**. Each primary oocyte is surrounded by a single layer of flattened cells called **follicle cells**. The primary oocyte surrounded by follicle cells is called the **primary follicle**. A female produces around 2 million of these potential egg cells prior to her birth, but some of these cells degenerate from about 700,000 at birth to around 350,000 at puberty.

When maturation of egg cells commences at puberty, the follicle cells begin dividing, forming layers and secreting estrogen. During the menstrual cycle, one primary follicle continues its development from prophase I to metaphase II of meiosis within the ovary. At this point, there are two cells—one small cell, called the first **polar body**, that will not be released from the ovary; and a much larger cell, called the **secondary oocyte**, that will give rise to the egg cell. About 12 hours after this secondary oocyte forms, it is expelled from the ovary. The release of an egg cell from the ovary is called **ovulation**. The secondary oocyte is released on around the 14th day of the menstrual cycle. It takes about 3 days for the secondary oocyte—an object about the size of a pinhead—to move from the ovary through the oviduct, uterus, and cervix and then exit the body through the vagina. If sperm are present in the oviduct and the secondary oocyte is fertilized, meiosis resumes from metaphase II, producing the ovum and a second polar body. Due to its small size and lack of nutrients, the polar body does not function as an egg cell and is not thought to be capable of producing a child. When fertilization does occur, chromosomes are inherited in equal proportions from both parents, but most of the cytoplasm and its organelles are inherited along with the egg cell (Figure 20.8b).

Women are fertile, or able to produce gametes, only a few days per month from puberty until menopause. Men are fertile throughout their lifetimes, with a slow but progressive decrease in sperm count as they age. Many different chemicals, potentially including endocrine disruptors, can artificially hasten this decline in sperm counts.

Endocrine Disruptors, Gametogenesis, and Infertility.   The vast majority (around 90%) of male infertility cases are due to problems with sperm production and formation including low numbers of sperm, improperly shaped sperm, or sperm with impaired motility. The likelihood of male infertility increases with alcohol and drug use, including cigarette smoking. In addition to these factors, many scientists are investigating whether exposures to endocrine disruptors can cause altered sperm production in humans.

One study that compared reproductive health of men from around the United States showed that sperm counts of men from rural Missouri were lower than they were for men from many large cities across the country. Scientists who performed this 2003 study found higher levels of three commonly used agricultural pesticides (including atrazine) in the bodies of men with lower sperm counts than were present in the bodies of men with higher sperm counts. Most of the men with decreased sperm counts did not work directly with these chemicals, leading researchers to question whether these men may have been exposed to these chemicals in their drinking water.

Another important reproductive function in humans may be affected by the water we drink. That function is the menstrual cycle, which we discuss next.

(a) Oogenesis

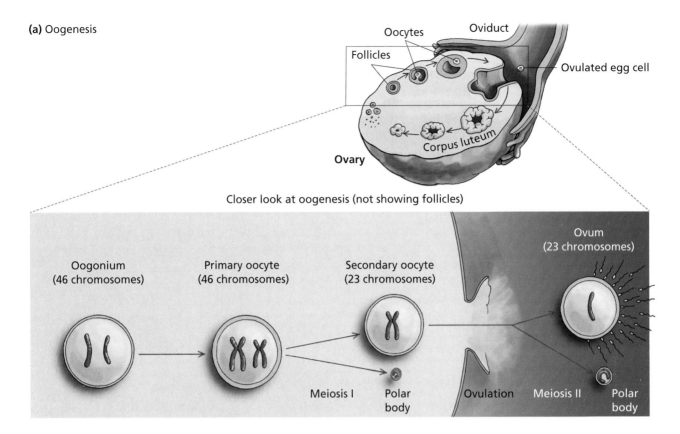

Oocytes

Oviduct

Follicles

Ovulated egg cell

Corpus luteum

Ovary

Closer look at oogenesis (not showing follicles)

Ovum
(23 chromosomes)

Oogonium
(46 chromosomes)

Primary oocyte
(46 chromosomes)

Secondary oocyte
(23 chromosomes)

Meiosis I   Polar
body

Ovulation   Meiosis II   Polar
body

(b) Egg (shown with sperm)

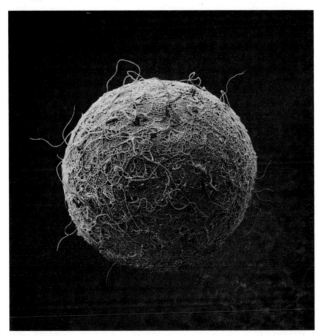

Figure 20.8   Oogenesis. (a) Through meiosis, human females produce egg cells that will be ovulated. From birth to puberty, the process is paused at prophase I. At puberty, primary oocytes move from prophase I to metaphase II. The secondary oocytes produced will move through telophase II only if fertilized. Cell division results in the production of polar bodies and one ovum, thus producing one functional gamete, not four as in males. (b) While only one sperm cell can actually fertilize an egg cell, many try. The fertilized cell contains all the materials required to produce a human.

The Menstrual Cycle

The term **menstrual cycle** refers specifically to cyclic changes that occur in the uterus. This cycle depends on intricate interrelationships among the brain, ovaries, and lining of the uterus or endometrium. During the course of a single menstrual cycle, a woman's body prepares an egg for potential fertilization and her uterus for a potential pregnancy. If no pregnancy has occurred, the uterine lining is excreted during **menstruation**, and a new cycle begins.

Figure 20.9 illustrates the changes in hormone levels, ovaries, and condition of the endometrium that occur throughout the 28-day cycle. Most women's bodies do not adhere precisely to a 28-day cycle—some women have longer cycles, and others have shorter cycles. The first day of a menstrual cycle is considered to be the first day of actual bleeding.

Like many other biological processes, the menstrual cycle is self-regulating, operating by responding to feedback. Negative feedback slows or stops the process; positive feedback speeds up the process.

Figure 20.9 The menstrual cycle. Changes in hormone levels are linked to uterine condition and the state of ovarian follicles over the course of the menstrual cycle.

Recall from Chapter 19 that the hypothalamus, located in the brains of both males and females, secretes gonadotropin-releasing hormone (GnRH), which stimulates another endocrine structure in the brain—the pituitary gland—to release follicle-stimulating hormone (FSH) and luteinizing hormone (LH). In males, FSH and LH are involved in the production of sperm; in females, these hormones help regulate the development of egg cells and ovulation.

High levels of estrogen provide positive feedback to the hypothalamus. The hypothalamus secretes GnRH, which acts on the pituitary gland to increase the secretion of FSH and LH. Conversely, high levels of the hormone **progesterone** have a negative feedback effect on the hypothalamus; GnRH secretion is decreased, and FSH and LH levels decline (Figure 20.10).

When a woman menstruates, the endometrial tissue is shed and released through the cervix and vagina. Inside the ovary, the primary follicle begins to grow under the influence of FSH. As the follicle grows, it produces estrogen. After menstruation ends, the growing estrogen levels once again stimulate the cells of the endometrium to grow and divide.

When the primary follicle is large enough, it produces enough estrogen to stimulate GnRH release. This leads to a spike in both FSH and LH levels, which lasts for about 24 hours. Ovulation occurs 10 to 12 hours after the LH peak, around 14 days before menstruation in a typical 28-day cycle. The follicle then literally bursts open, releasing the egg, which is drawn into the oviducts.

After ovulation, the ruptured follicle is called the **corpus luteum**. The corpus luteum is a hormone-producing tissue that produces progesterone and estrogen.

Progesterone helps maintain the blood flow to the uterine lining in order to support early fetal development. In addition, progesterone inhibits continued

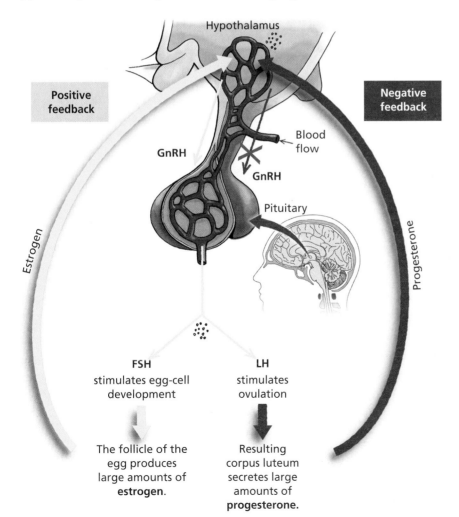

Figure 20.10 Feedback during the menstrual cycle. High levels of estrogen provide positive feedback to the hypothalamus; conversely, high levels of the hormone progesterone have a negative feedback effect on the hypothalamus. These two feedback loops do not occur simultaneously because follicle development takes place before ovulation and production of the corpus luteum.

LH production, so the LH surge is inhibited by the ovarian hormones that it stimulates—an example of negative feedback.

If fertilization does not occur, the corpus luteum degenerates in about 12 to 14 days after ovulation. Because the corpus luteum is no longer secreting them, progesterone and estrogen levels fall, triggering the arteries that supply the uterus to spasm. This allows the lining of the uterus to be shed and causes menstrual cramps. Decreasing levels of progesterone also serve to release the hypothalamus from inhibitory control. Therefore, LH and FSH levels rise, and the cycle starts over.

If fertilization has occurred, the process of endometrial breakdown stops. The endometrium and an endocrine organ that forms during pregnancy, called the **placenta**, release a hormone called human chorionic gonadotropin (HCG) that extends the life of the corpus luteum. Women test for the presence of this hormone in most over-the-counter pregnancy tests—a positive test indicates that a placenta has formed. With the corpus luteum intact, progesterone and estrogen levels remain high, and the endometrium is maintained. In a pregnant woman, the corpus luteum finally disintegrates after about 6 or 7 weeks of pregnancy, when the placenta begins to produce enough progesterone to maintain the uterine lining on its own.

The birth-control pill supplies the body with low continuous doses of synthetic versions of estrogen and progesterone. The estrogen present in the pill is at a level low enough to prevent follicle development. The progesterone contained in the pill serves as a backup (in case ovulation does occur) by making the mucus that the cervix secretes more resistant to the ascent of sperm. This is in contrast to the cervical mucus secreted during ovulation, which is thin and stringy and has the consistency of raw egg white. This type of cervical mucus actually facilitates fertilization by providing channels through the cervix for the sperm to swim through and gain access to the egg. The low hormone levels present in the birth-control pill also prevent the uterine lining from developing enough to support a pregnancy. See Table 20.2 on pages 533–535 for a list of other methods of birth control.

Endocrine Disruptors, Menstruation, and Infertility.   Some cases of female infertility can be related to a condition called **endometriosis**, which occurs when menstrual tissues that normally line the uterus begin to grow in other areas such as the oviducts and ovaries. These tissues continue to respond to the hormonal cycle, growing and shedding in sync with the lining of the uterus. In addition to being very painful, endometriosis can lead to scarring and inflammation of the oviducts and disruption of ovulation. An endocrine disruptor, the chemical Di(2-ethylhexyl)phthalate (DEHP), is under investigation to determine whether it plays a role in causing endometriosis. DEHP, a type of chemical called a plasticizer, is added to plastics to keep them soft and pliable. Some scientists believe that drinking water from plastic bottles or heating food in plastic containers might increase an individual's exposure to this and other plasticizers that leach out of the plastic and into the water or food. Researchers at the University of Siena in Italy found that women with endometriosis have higher levels of DEHP in their blood than do women who do not have endometriosis. These studies provide no direct evidence that chemicals present in plastics are causing reproductive problems. These women might also have high levels of other, untested for, chemicals that are actually causing the problem. The increased level of environmental chemicals also may be due to some other factor. Research on the hormone-disrupting effects of plasticizers is still in its infancy; before drawing any definitive conclusions, scientists must wait until many studies that provide direct evidence of human reproductive problems caused by plasticizers have been completed.

Reproductive problems can arise because of exposures to toxins that a person experiences during his or her lifetime. Such problems also can result from exposures experienced in the womb during development, when reproductive and other structures are forming.

Table 20.2 Birth-control methods. The merits and efficacy of various birth control methods are outlined.

	Method	Mode of action	Risks
Abstinence	**Abstinence**	• Sperm and egg never have contact.	• No associated risks. • 100% effective*
Hormonal methods	**Combination birth control pill**	• Synthetic estrogen and progesterone given at continuous doses (versus the cyclic fluctuations of a menstrual cycle) prevent ovulation.	• Increased risk of heart disease and fatal blood clots for women over age 35 who smoke. May have slightly increased risk of breast cancer. Decreased risk of uterine and ovarian cancers. • Does not protect against STDs • 99% effective
	Minipill	• The minipill contains progesterone only. Since there is no estrogen to combat the effects of progesterone, cervical mucus is thickened, sperm ascent is impeded, and the uterine lining is not prepared to support a pregnancy.	• Increased risk of ovarian cysts. • Does not protect against STDs • 95% effective
	Depo-Provera	• This method requires the user to have progesterone injections every 3 months. Same mode of action as the minipill.	• Irregular vaginal bleeding • Does not protect against STDs • 99% effective
	Implantables	• Norplant, recently taken off the market, delivered progesterone to thicken cervical mucus. Another implant, Implanon, may be available soon.	• Under investigation • Does not protect against STDs • Effectiveness unknown
Sterilization	**Female sterilization**	• Permanently blocking oviducts with a tightly coiled tube.	• Does not protect against STDs • 99.8% effective
	Male sterilization (vasectomy)	• Cutting each vas deferens prevents sperm from being ejaculated in the semen.	• Does not protect against STDs • 99.9% effective

*Percent effectiveness assumes correct usage. Pregnancy rate without birth control is 85%.

(*table continued on next page*)

Method	Mode of action	Risks
Barrier methods		
Cervical cap	• When inserted against cervix before intercourse, prevents sperm and egg contact.	• No known risks • Does not protect against STDs.
Diaphragm	• When inserted into vagina before intercourse, prevents sperm and egg contact.	• No known risks • Does not protect against STDs.
Female condom	• Held against cervix by a flexible ring, the female condom prevents sperm and egg contact.	• No known risks
Male condom	• Prevents sperm and egg contact.	• No known risks
Other methods		
Spermicides	• When inserted into vagina 1 hour before intercourse, kills sperm.	• No known risks • Does not protect against STDs.
Fertility awareness	• Abstinence for the 4 days before and 4 days after predicted time of ovulation.	• None • Does not protect against STDs.
Intrauterine device	• When this small plastic device is inserted into the uterus by a physician, it prevents fertilization and prevents the uterus from supporting a pregnancy.	• May increase risk of pelvic inflammatory disease.

*Percent effectiveness assumes correct usage. Pregnancy rate without birth control is 85%.

Method	Mode of action	Risks
Patch	• When applied to skin, delivers progesterone and estrogen to prevent ovulation and thicken cervical mucus.	• Same as combination pill • Does not protect against STDs • 99% effective
Vaginal ring	• When inserted in vagina, delivers progesterone and estrogen to prevent ovulation and thicken cervical mucus.	• Same as combination pill • Does not protect against STDs • 99% effective
Emergency birth control pills	• High doses of progesterone and or estrogen prevent sperm from reaching egg or prevent fertilized egg from attaching to wall of uterus. Can also prevent ovulation for one cycle if it has not occurred. These so-called "morning after pills" are not the same as the mifepristone abortion pill. Emergency birth control pills prevent pregnancy, and mifepristone terminates an established pregnancy.	• Nausea, vomiting, abdominal pain, fatigue, headache • Does not protect against STDs • 80% effective

(Left side vertical label: Hormonal methods)

*Percent effectiveness assumes correct usage. Pregnancy rate without birth control is 85%.

20.3 Human Development

Development is the series of events that take place after fertilization; it leads to the formation of a new multicellular organism. Fertilization sets the stage for development to occur.

Fertilization

Of the roughly 300 million sperm ejaculated, only about 200 reach the site of fertilization in the oviduct. Initially, sperm must survive the acidic (low pH) conditions of the upper vagina. The natural acidity of the upper vagina works to help prevent bacterial infections in women but also can kill sperm. Some sperm get stuck in folds of the cervix and in the uterine wall, and only half of the sperm that make it to the uterus enter the oviduct that contains the ovulated egg.

The oviducts themselves undergo muscular contractions during the time of ovulation, and these contractions allow the sperm to move toward the upper portion of the oviduct to wait for an egg cell to be ovulated. It takes about half an hour to move a sperm from the upper vagina to the oviduct.

How Fertilization Takes Place.   When both sperm and egg are present in the oviduct, fertilization can occur (Figure 20.11). First, the sperm cells must move through the follicle cells that surround and nourish the egg. Then they must pass through a translucent covering on the egg cell called the **zona pellucida**. This protective layer shields the egg from mechanical damage and acts as a species-specific barrier. Only sperm produced by a male of the same species as the female that produced the egg can traverse the zona pellucida. This is because getting through the zona pellucida requires binding of a specific receptor on the sperm head. This binding of the sperm head to the zona pellucida triggers the release of enzymes present in a structure, located at the head of the sperm, called the acrosome. The acrosomal enzymes interact with the egg cell's zona pellucida to allow the formation of a tunnel through the zona pellucida toward the egg cell's plasma membrane. Once this sperm cell makes it through the zona pellucida, its plasma membrane and the egg's plasma membrane fuse, and the egg cell draws the sperm cell's nucleus inward. The fertilized egg is now ready to begin the divisions of early development.

Chemicals in the environment can affect the ability of sperm to bind to the egg during fertilization. Some scientists are concerned that this could be a cause of infertility.

Endocrine Disruptors and Fertilization.   Trichloroethylene (TCE) is an industrial solvent that is used to remove grease from metal and is found in many household products. Large amounts of this chemical can get into the water supply via air emissions from metal degreasing plants and via wastewater from factories that produce metals, paints, cleaning products, electronics, and rubber products. A 1996 study at the National University of Singapore showed that males exposed to TCE while working as mechanics or dry cleaners had a larger number of abnormally shaped sperm than did men who had not been exposed

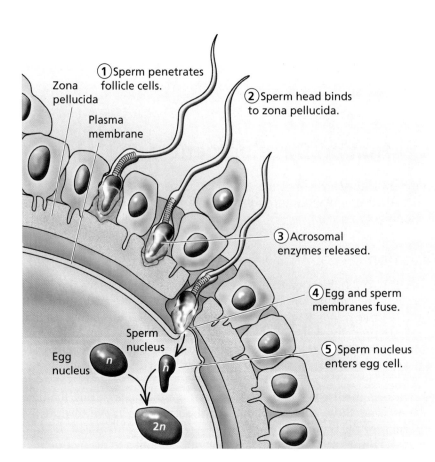

Figure 20.11 Fertilization. The steps involved in fertilization are illustrated in this figure.

to TCE, and these abnormal sperm were less likely to be able to fertilize normal egg cells. Likewise, a 2004 study at the University of California Davis showed that male mice exposed to TCE had a dose-dependent decrease in their ability to fertilize eggs. In other words, high doses of TCE resulted in high levels of infertility, while low doses of TCE resulted in lower levels of infertility in this population of mice.

Studies showing an effect of endocrine-disrupting chemicals on reproductive health, as we described earlier in this and the previous sections, seem to support the hypothesis that exposures during adulthood can decrease fertility. But these studies provide no definitive evidence. In the next section we describe something scientists are more sure of—that minute chemical exposures during critical windows of development can have tremendous impacts on a developing fetus.

Human Embryonic Development

After fertilization, the fertilized egg cell, or zygote, undergoes a series of changes to produce a multicellular structure capable of producing adult tissues (Figure 20.12). In humans, the term **embryo** is used to describe the stage of development from the first divisions of the zygote until body structures begin to appear at about the ninth week. After that time, the developing human is referred to as a **fetus**. Early development of the embryo involves a series of rapid cell divisions, called **cleavage**, that begin while the zygote is still in the oviduct. These divisions continue until the now multicellular embryo is propelled down the oviduct toward the uterus.

When the embryo reaches the uterus, a few days after fertilization, it is in the form of a hollow ball of cells known as the **blastocyst**. Inside the blastocyst, a cluster of cells called the **inner cell mass** forms. Some cells of the inner cell mass give rise to the embryo. The outer cells of the blastocyst form part of the placenta. The inner cell mass undergoes a dramatic reshuffling of cells to produce a structure with three layers of cells, called the **gastrula**. The cells comprising the three layers of the gastrula will begin to specialize, or **differentiate**, into the adult cells and tissues with their specific functions.

These three layers of tissue include an outer, middle, and inner layer. The outer layer or **ectoderm** gives rise to the skin, nervous system, and sense organs. The middle layer or **mesoderm** gives rise to muscles, excretory organs, circulatory organs, gonads, and the skeleton. The innermost layer or **endoderm** lines the alimentary canal and become to the organs associated with digestion and breathing.

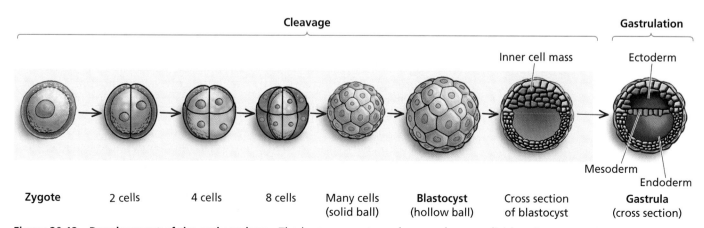

Figure 20.12 Development of the early embryo. The human zygote undergoes cleavage divisions to produce the blastocyst. Continued cell divisions along with specialization result in the formation of the three-layered gastrula.

This differentiation occurs in response to chemical signals that pass between the different cell layers and adjacent cells, turning on some genes in a given cell and leaving others off. In this manner, the specialized cells that comprise the adult body all arise from the single fertilized egg cell that undergoes mitotic cell division and cell differentiation.

Exposures to toxins of any kind very early in development tend to have an all-or-nothing impact. Exposures can be so damaging that they kill many cells, and the early embryo is spontaneously aborted—often without a women ever realizing that she was pregnant. Conversely, exposures can have very little effect since cells are dividing so rapidly that if a few cells die due to exposure, they are quickly replaced. Exposures of greatest concern occur while organs are developing. The development of reproductive organs in particular can be derailed by exposures to endocrine-disrupting chemicals.

Development of Human Reproductive Organs. Reproductive organs develop at around week 7 in humans. Prior to that time, male and female embryos are indistinguishable except at the level of their chromosomes.

In the early embryo, the structures that will become the gonads are located in the abdomen. The same starting material can be used to produce either the paired ovaries of a female or the paired testes of a male. Differentiation of the embryonic gonads into ovaries or testes is determined by the expression of sex-specific genes. Once differentiated, the male gonads produce androgens; but the female gonads do not start to produce estrogen until puberty.

The cells of the embryonic testes make the androgen testosterone, which directs the development of internal, gamete-carrying, ductal structures. Two different ductal structures exist side by side in the developing embryo. In each sex, one embryonic duct system persists and the other regresses, based in part on the presence or absence of androgens. In males, the ductal structure that persists develops into the vas deferens, epididymis, and urethra. In female embryos, the other ductal structure develops into the oviducts, uterus, cervix, and vagina.

Androgens also affect the development of external genitalia. In male and female embryos, one embryonic structure can be molded into either the male or female external genitalia. In fact, the penis and clitoris arise from the same starting tissue, as do the scrotum and vulva. Table 20.3 summarizes the different developmental pathways followed by embryonic males and females.

In this section, we have highlighted the important role that androgens play in driving the development of male and female reproductive structures. Endocrine disruptors that interfere with the action of androgens can alter this developmental pathway.

Endocrine Disruptors and the Development of Reproductive Organs. Earlier in the chapter, we related the story of DES—the synthetic estrogen taken by women in the 1960s and 1970s that caused some of their daughters to have deformed uteruses. The development of male reproductive structures also appears to be vulnerable to the effects of environmental contamination.

According to a report issued by the Tufts University School of Medicine, mice exposed to commonly used fungicides and herbicides have a higher-than-average rate of undescended testes, a condition called **cryptorchidism**. This anomaly decreases reproductive success because the sperm of males with this condition are held inside the abdomen, where the temperature is higher than optimum for the development of sperm. It is thought that these fungicides and herbicides function as endocrine disruptors by blocking androgen receptors.

Cryptorchidism occurs in 2% to 4% of human male newborns and is usually surgically corrected. According to a 2000 study performed by scientists in Germany, assessments of adipose tissues from boys undergoing such surgery shows increased pesticide residues when compared with adipose tissue from children undergoing surgery for other reasons.

Table 20.3 Development of reproductive structures. Development is color coded, using the same color for adult structures that develop from a common embryonic structure.

Embryonic structure	Structure in adult female	Structure in adult male
Gonads	**Ovaries**	**Testes**
Gamete-carrying ducts	Oviduct / Uterus / Cervix / Ovary / Vagina	Bladder / Seminal vesicle / Prostate gland / Penis / Epididymis / Urethra / Vas deferens / Testicles
External genitalia	Clitoris / Urethra / Labium minora / Labium majora / Vaginal opening / Anus	Glans of penis / Urethra / Shaft of penis / Scrotum / Anus

Deformities in reproductive or other structures in humans occur because certain chemicals have the ability to cross from the mother's bloodstream to the developing offspring during pregnancy. This is why pregnant women must be very careful to limit their exposure to toxins.

Pregnancy

Pregnancy or gestation involves carrying a developing baby within the female reproductive tract. Gestation encompasses the period of time between fertilization and childbirth that, in humans, lasts about 38 weeks.

To sustain a pregnancy, the placenta must form inside the uterus. The placenta is produced by the cooperative actions of the embryo and the mother's uterus. By the time the embryo has made its way to the uterus, it is composed of the inner cell mass, which will become the fetus, and an outer ring of cells called the **trophoblast**, which becomes part of the placenta. Cells of the trophoblast secrete enzymes that enable the embryo to implant in the wall of the uterus. At about the seventh day after fertilization, the implanted trophoblast begins to infiltrate the uterine lining, forming fingerlike projections that are able to carry blood. Maternal uterine blood vessels erode and spurt blood into the area surrounding the fetal blood vessels. This close positioning of fetal and maternal blood supplies allows the exchange of nutrients and wastes to occur. Blood cells and bacteria do not normally pass between fetal and maternal blood supplies. Substances that can be freely exchanged between the fetus and the mother include oxygen, carbon dioxide, water, salts, hormones, viruses, and many drugs.

The cells of the early blastocyst synthesize human chorionic gonadotropin (HCG), which serves to maintain the corpus luteum. Recall that the corpus luteum is the ruptured, empty follicle present on the ovary after the ovulation of its egg cell. Continued activity of the corpus luteum results in a steady increase in the levels of progesterone and estrogen during the first 5 weeks of pregnancy. Once the placenta has established its own hormone production, the corpus luteum is no longer necessary. Consequently, HCG is secreted only during the first months of pregnancy.

An additional hormone synthesized by the placenta is called **somatomammotropin**, which works with estrogen, progesterone, and another hormone, **prolactin**, to stimulate development of the mother's mammary glands. Near term, the mother's estrogen level drops, and prolactin is produced by the pituitary gland of the brain. Prolactin continues to stimulate the production and secretion of breast milk during breastfeeding or **lactation**.

From the time the embryo implants itself in the lining of the uterus until delivery, it undergoes remarkable growth and development. A month-old embryo is about 7 mm long and is beginning to develop a brain and spinal cord. It has buds that will give rise to limbs (Figure 20.13). A 9-week-old developing fetus has all of its organs and limbs in place (Figure 20.14). It can move its arms and legs and turn its head. For the remainder of the pregnancy, the baby increases in size; its features become more refined with developments such as eyebrows, eyelashes, fingernails, and toenails. During the last stages of development, the fetal circulatory and respiratory systems undergo changes that will allow the fetus to breathe air.

During pregnancy, a woman must be very careful about exposing a developing fetus to many different chemicals. It has been clearly established that environmental chemicals can have negative effects on fetal development. For example, exposure to secondhand cigarette smoke is known to cause decreased birth weight, exposure to mercury can affect brain development, and exposure to alcohol can cause severe developmental problems. Exposure to endocrine disruptors may also have negative effects.

Endocrine Disruptors and Pregnancy.   Polychlorinated biphenyls (PCBs) are pollutants that once were used in hundreds of products including electrical and hydraulic equipment, plastic and rubber products, and paints and dyes. Concern over the toxicity and persistence of PCBs in the environment led Congress in 1976 to prohibit their manufacture and distribution. Prior to this ban, billions of pounds of PCBs were manufactured in the United States.

Figure 20.13   Human embryo at 1 month.

Fetal exposure to PCBs is correlated with low birth weight and premature birth. A 2003 study of children born in New York state showed that children born to mothers who live near PCB-contaminated sites show a 6% increased likelihood of being born underweight. Low-birth-weight babies weigh less than 2500 grams (5 pounds, 8 ounces) at birth and are at increased risk of respiratory and heart problems at birth as well as of cerebral palsy and learning problems as they grow. Low birth weights are more likely when childbirth occurs prematurely—before development of the fetus is complete.

Childbirth

Giving birth involves both labor and delivery. **Labor** involves strong, rhythmic contractions of the uterus that result in the birth of the baby. Induction of labor is hormonally controlled. Maternal estrogen levels increase toward the end of pregnancy, triggering the formation of receptors on the uterus for a hormone called **oxytocin** that serves to stimulate the smooth muscles lining the wall of the uterus to contract. Oxytocin is produced by fetal cells and by the mother's pituitary gland. During the induction of labor, oxytocin and estrogen are regulated by positive feedback. Oxytocin causes uterine contractions. Uterine contractions stimulate the release of more oxytocin, resulting in progressively intense muscle contractions that ultimately expel the baby from the womb (Figure 20.15).

Uterine contractions normally signal the beginning of labor. Passage of the thick mucus plug blocking the cervix during pregnancy is another sign that labor is imminent. When the cervix begins to dilate, the plug loosens and is passed through the vagina. Beginning of labor can also be signaled by the rupture of a fluid-filled sac, called the **amnion**, that surrounds and protects the fetus. This is followed by the loss of fluid, occurring as either a gush of fluid or a slow trickle. It is often referred to as a woman's "water breaking" prior to delivery.

Figure 20.14 Fetus at 9 weeks.

Figure 20.15 Induction of labor.
Estrogen and oxytocin are involved in a feedback loop that stimulates labor.

Stage 1: Dilation   **Stage 2:** Expulsion   **Stage 3:** Delivery of placenta

Figure 20.16   The three stages of labor. During labor, (1) the cervix dilates, (2) the baby is expelled, and (3) the placenta is delivered.

Stages of Labor. Labor can be thought of as occurring in three stages (Figure 20.16). During the first stage of labor, the cervical opening enlarges or dilates, going from a tiny, pencil-lead-sized opening to 10 cm. Contractions of the uterine muscle tissues help to dilate the cervix because they apply pressure to the baby. The baby is forced against the cervical opening, which forces the opening to enlarge. The second stage of labor involves delivery of the baby through the narrow pelvic opening and birth canal, often aided by the mother's active "pushing" (contractions of the abdominal muscles); the third stage of labor is delivery of the placenta.

As any new parent knows, concern about exposure to toxic chemicals does not end when the baby is born. Newborns must be protected against toxic exposures as well.

Endocrine Disruptors and the Newborn. Concern about exposure to endocrine disruptors in newborns has again focused on plasticizers. Unease about materials leaching out of plastics has led the Consumer Product Safety Commission to recommend that parents dispose of any soft vinyl baby products such as pacifiers and teething rings that contain the plasticizer DEHP. Even though DEHP has not definitively been shown to act as an endocrine disruptor in humans, negative effects of DEHP exposure have not been ruled out.

Babies fed infant formula are also exposed to significant quantities of tap water. This brings us back to our original questions about endocrine disruptors in the water supply.

20.4   Is the Water Safe to Drink?

At first glance, the issue of endocrine-disrupting chemicals in the water supply seems quite disturbing, but how concerned do we really need to be? What is being done to monitor our tap water?

The U.S. Environmental Protection Agency (EPA) is the governmental organization charged with regulating the amount of various chemicals, including known endocrine disruptors, present in drinking water. City water suppliers collect water samples every 3 months and analyze contaminant levels. If the level of harmful chemicals is too high, steps are taken to lower the contaminant level, and the water supplier must immediately notify the public. Since there are regulations in place to protect the water supply, the more pertinent question becomes whether the EPA-allowed levels are low enough to prevent reproductive problems in humans. This we do not yet know.

What we do know is that wildlife is affected by endocrine disruptors found in aquatic living environments. We also know from laboratory studies of animals that exposure to endocrine-disrupting chemicals can have negative consequences for reproductive health and that animal studies can be predictive of what might happen in humans.

As the DES and TCE data make clear, the human body can mistake a synthetic compound for a naturally occurring hormone, and this is harmful to both female and male reproductive health. However, no study to date has been able to show that endocrine disruptors in the water supply have caused reproductive problems. All studies to date have been correlative. Abnormal sperm, decreased ability of sperm to fertilize eggs, and cryptorchidism are correlated with endocrine disrupter exposure, as is endometriosis. You learned in Chapter 1 that correlation does not equal causation. For instance, there is a high correlation between milk consumption and the number of teachers living in a particular geographic area. If we relied on correlation only, we would assume that teachers drink more milk than do people with other occupations. In reality, another factor is causing the increase in milk consumption. Where there are many teachers, there are also many students and consequently high levels of milk consumption. The correlation between endocrine-disrupting chemicals and reproductive health could also result from a shared factor rather than demonstrating a cause-and-effect relationship. For instance, men in agricultural areas may be exposed to more atrazine than urban men are, but they may also have jobs with a greater component of manual labor. It may be that the consistently higher body temperatures experienced by manual laborers interfere with sperm production and that atrazine exposure has little effect.

While there is no conclusive evidence that endocrine disruptors in the water supply are harming human reproductive health, neither is there any reason to disregard this hypothesis. Further research is necessary before the role of endocrine disruptors in human reproductive health can be firmly established.

CHAPTER REVIEW

Summary

20.1 Principles of Animal Reproduction

- Asexual reproduction is a method of reproduction whereby one parent produces offspring that are genetically identical to the parent (p. 516).

- Large numbers of offspring can be reproduced quickly during asexual reproduction, but the absence of any genetic diversity makes the offspring less able to adapt to changes in the environment (p. 517).

- Sexual reproduction requires a mating between two parents. Males and females produce gametes in structures called gonads. Gametes unite at fertilization to produce genetically distinct offspring (p. 517).

20.2 Human Reproduction

- The penis of the male reproductive system is involved in sperm delivery; the urethra delivers both sperm and urine, the scrotum houses the testes, and the testes produce sperm and androgens (pp. 520–521).

- Sperm are produced when cells in the seminiferous tubules undergo cell division. Sperm are stored in the epididymis

that rests atop the testes. At ejaculation, sperm pass through the vas deferens and out the urethra (p. 521).

- The seminal vesicles, prostate, and bulbourethral glands add secretions to sperm that help them develop and provide a source of energy. Semen is composed of the secretions from these glands combined with sperm (p. 521).

- The female reproductive system consists of the external vulva and clitoris, the internal vaginal passageway, the cervix at the base of the uterus that opens during childbirth, the uterus that houses the developing baby, oviducts for the passage of gametes, and ovaries that produce egg cells and hormones (p. 522).

- Spermatogenesis occurs in the seminiferous tubules of the testes and begins at puberty. The diploid cell that begins meiosis is called a primary spermatocyte. After the first meiotic division, the haploid cell is called a secondary spermatocyte. These cells then undergo meiosis II to produce spermatids. Removal of cytoplasm produces spermatozoa composed of a small head containing the DNA, a midpiece that has mitochondria to provide energy for the journey to the oviduct, and a flagellum. An acrosome at the sperm's head contains enzymes that help the sperm gain access to the egg (pp. 526–527).

- Oogenesis occurs in the ovaries and results in the production of egg cells. This process begins while the female is still in her mother's uterus, then pauses until puberty. At puberty, development of preexisting eggs continues each month until menopause. The diploid cells that begin meiosis are called oogonia. When paused at prophase I, they are called primary oocytes. When surrounded by nourishing follicle cells, the structure is called a primary follicle. At puberty, the follicle cells divide and secrete estrogen. Meiosis restarts, producing a smaller polar body and a secondary oocyte. The ovulated egg cell is called the secondary oocyte. If the secondary oocyte is fertilized, it finishes meiosis and produces the ovum (pp. 528–529).

- The hypothalamus secretes GnRH, which causes the pituitary gland to secrete FSH and LH. These two hormones regulate the release of an egg cell at ovulation. High levels of estrogen cause increased FSH and LH production. High levels of progesterone cause decreased FSH and LH secretion (p. 531).

- Estrogen, produced by the follicle, begins to rise during menstruation, eventually stimulating GnRH release and causing a spike in FSH and LH levels. LH causes ovulation to occur. The remnant of the follicle that stays in the ovary after ovulation is the corpus luteum, which secretes progesterone and estrogen (p. 531).

- If fertilization does not occur, the corpus luteum degenerates, and progesterone and estrogen levels fall. Menstruation occurs. Then LH and FSH levels rise, and the cycle starts over (p. 532).

- If fertilization has occurred, the embryo implants itself within the uterus and together they form the placenta, which secretes HCG to extend the life of the corpus luteum. Progesterone and estrogen levels remain high. Eventually the placenta begins to produce these hormones (p. 532).

Web Tutorial 20.1 **The Male Reproductive System**

Web Tutorial 20.2 **The Female Reproductive System**

20.3 Human Development

- To fertilize an egg cell, sperm slide through the follicular cells that surround the egg to reach the zona pellucida. Binding of the sperm head to the zona pellucida triggers the release of acrosomal enzymes, which break down the zona pellucida and allow the sperm to reach the egg cell's plasma membrane (p. 536).

- After fusion of the egg and sperm plasma membranes, the sperm nucleus enters the egg cell, and the zygote can begin to develop (p. 536).

- The zygote undergoes rapid cleavage divisions to produce the blastocyst. The inner cell mass of the blastocyst becomes the embryo and eventually the fetus. The blastocyst undergoes many divisions to produce a three-layered gastrula. Each layer of the gastrula (ectoderm, mesoderm, and endoderm) differentiates into specific tissues (p. 537).

- Embryonic gonads become either testes or ovaries. Ductal structures exist side by side in male and female embryos. In each sex, one structure regresses. Male and female external genitalia are fashioned from the same embryonic material. The expression of sex-specific genes and hormones determines whether a particular embryo will become male or female (pp. 538–539).

- Pregnancy hormones are produced by the placenta, which forms in the uterus. These hormones help maintain the uterus, prevent ovulation, and prepare the breasts for lactation (p. 540).

- The embryo grows in size and increases in complexity during gestation (p. 540).

- Childbirth involves both labor and delivery. Labor involves contractions that help expel the baby from the uterus. During labor, the cervix dilates to allow the baby's passage out of the uterus. After her child's birth, the mother delivers the placenta (pp. 541–542).

20.4 Is the Water Safe to Drink?

- The effects of endocrine disruptors in drinking water on human health and development are unclear (pp. 542–543).

Learning the Basics

1. In what ways do asexual and sexual reproduction differ?

2. Describe the male and female reproductive structures.

3. Describe gametogenesis in males and females.

4. A sperm cell follows which path?

   A. seminiferous tubules, epididymis, vas deferens, urethra; **B.** urethra, vas deferens, seminiferous tubules, epididymis; **C.** seminiferous tubules, vas deferens, epididymis, urethra; **D.** epididymis, seminiferous tubules, vas deferens, urethra; **E.** epididymis, vas deferens, seminiferous tubules, urethra

5. An egg cell that is not fertilized follows which path?

   A. ovary, oviduct, uterus, cervix, vagina; **B.** ovary, uterus, oviduct, cervix, vagina; **C.** oviduct, ovary, cervix, uterus, vagina; **D.** oviduct, ovary, uterus, cervix, vagina; **E.** ovary, oviduct, cervix, uterus, vagina

6. Gametogenesis _____.

   A. begins at puberty in males and females; **B.** requires that the Leydig cells of males produce semen; **C.** results in the production of diploid cells from haploid cells; **D.** begins at puberty in females; **E.** requires that meiosis halve the chromosome number so that sperm and eggs carry half the number of chromosomes.

7. Gametogenesis _____.

A. begins in the vas deferens of males; B. begins when the embryo implants in the uterus; C. begins in the epididymis of males; D. begins in the seminiferous tubules of males; E. begins in the oviduct of females

8. Menstruation is regulated such that _____.

A. increasing estrogen levels have a positive feedback effect on FSH and LH; B. increasing FSH levels lead to ovulation; C. as progesterone levels increase, so do FSH and LH levels; D. ovulation occurs on the fifth day of the cycle; E. the placenta produces FSH, which stimulates ovulation

9. Gastrulation _____.

A. consists of rapid cell divisions that form the blastocyst; B. allows a zygote to implant in the lining of the uterus; C. produces the placenta; D. results in the formation of three layers of cells that will give rise to specific adult tissues; E. causes sperm and egg release in frogs

10. The path of a sperm during fertilization is _____.

A. follicle cells, egg plasma membrane, zona pellucida; B. egg plasma membrane, zona pellucida, follicle cells; C. egg plasma membrane, follicle cells, zona pellucida; D. zona pellucida, follicle cells, egg plasma membrane; E. follicle cells, zona pellucida, egg plasma membrane;

Analyzing and Applying the Basics

1. Why does the continuous dose of estrogen provided by the birth-control pill prevent ovulation?

2. In what ways do spermatogenesis and oogenesis differ? In what ways are they similar?

3. How is sperm from one species prevented from fertilizing an egg from another species?

Connecting the Science

1. What kinds of experiments would let scientists know for certain that endocrine disruptors in the water are affecting humans?

2. Why should pregnant women be very careful about limiting their exposures to toxins?

Attention Deficit Disorder

Brain Structure and Function

Many schoolchildren
are taking the
prescription
drug . . .

Ritalin...

to control attention deficit disorder.

Some college students who have *not* been prescribed the drug use it illegally to pull all-night study sessions.

L unchtime at a typical United States elementary school includes more than crustless peanut butter sandwiches, juice boxes, potato chips, and harried lunchroom aides. For a growing number of children, the lunch period also includes a trip to the nurse's office for a dose of methylphenidate, more commonly known by its trade name, Ritalin®. At least 6% of all school-aged boys and 2% of school-aged girls line up to take the drug, aimed at treating a syndrome called attention deficit disorder (ADD).

In 1980, ADD first appeared in the third edition of psychiatry's most used reference text, *The Diagnostic and Statistical Manual of Mental Disorders,* or *DSM-III*, published by the American Psychiatric Association. This book guides thousands of mental health professionals and millions of their patients. The editors of the fourth edition, released in 1987, attempted to increase awareness of the hyperactivity component of ADD by renaming the syndrome as attention deficit hyperactivity disorder (ADHD). However, most people still call the syndrome ADD.

The ADD diagnosis is typically given to children who consistently display a suite of behaviors including forgetfulness, impulsivity, distractibility, fidgeting, impatience, and restlessness. Virtually all children will display one or a few of these traits; therefore, the diagnosis is usually given only if a child shows most or all of these behaviors and does so in a manner that causes him or her problems in at least two settings, such as at school and at home.

Ritalin belongs to a class of drugs known as stimulants. Stimulants temporarily increase activity in an organism. Giving a stimulant to someone who is having trouble concentrating seems illogical. However, Ritalin has a paradoxical effect in that it actually causes people, especially children, to slow down and pay closer attention. The fact that the drug helps people concentrate has led to its illegal use, sometimes by college students who need to cram for an exam.

Some people argue that ADD is caused by brain malfunction, and so it is best to treat ADD with medicines. Others argue that if ADD is not caused, then it is at least exacerbated, by our fast-paced society and is therefore better treated by changing the environment of the affected child. To comprehend the debate, an understanding of the biology of the human nervous system is required.

21.1 The Nervous System

Every second, millions of signals make their way to your brain and inform it about what your body is doing and feeling. Your **nervous system** receives and interprets these messages and decides how to respond. Receiving and responding to stimuli requires the actions of specialized cells called **neurons**. Neurons carry electrical and chemical messages back and forth between your brain and other parts of the body. These electrochemical message-carrying cells can be bundled together, producing structures called **nerves**.

Neurons form an extensive network throughout your body. They are highly concentrated in your brain and spinal cord, which together form the **central nervous system (CNS)**. Other neurons in sense organs such as your eyes, ears, or skin transmit information from the world around you to your CNS. There the information is processed, and the appropriate response relayed by yet other neurons out of the CNS back to your body (Figure 21.1).

To carry information between parts of the body, the cells of the nervous system pass electrical and chemical signals to each other. Signals are

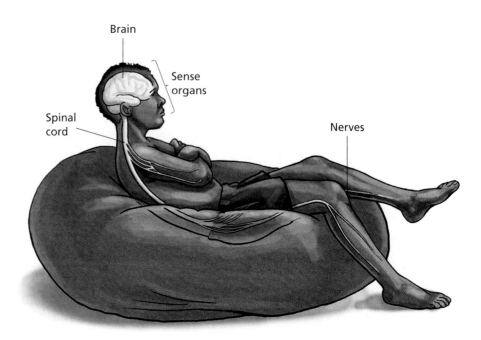

Figure 21.1 The nervous system. The nervous system includes the brain, spinal cord, sense organs, and the nerves that interconnect those organs.

transmitted from one end of a neuron to the other end, between neurons, and from neurons to the cells of tissues, organs, or glands that respond to nerve signals. These responsive tissues are called **effectors**. Information is typically carried along nerves by electrical changes called **nerve impulses** and transmitted between neurons and from neurons to the cells they act on by chemicals, called **neurotransmitters**, released from nerve cells.

Neurons can be grouped into three general categories: (1) **sensory neurons**, which carry information toward the CNS; (2) **motor neurons**, which carry information away from the CNS toward effector tissues; and (3) **interneurons**, which are located between sensory and motor neurons within the brain or spinal cord. Most actions involve input from all three sources; sensory input is followed by integration and motor output (Figure 21.2).

Sensory input is detected by **sensory receptors**. Receptors are usually neurons or other cells that communicate with sensory neurons. They detect changes in conditions inside or outside the body. When receptors are stimulated, signals are generated and carried to the brain.

The **general senses** are temperature, pain, touch, pressure, and body position, also called proprioception. The sensory receptors for the general senses are scattered throughout the body. The **special senses** are smell, taste, equilibrium, hearing, and vision. The sensory receptors for these five special senses are found in complex sense organs (see Table 21.1 on page 550).

As sensory information is passed to and from your brain, it travels through the main nerve pathway, the **spinal cord**. Your spine, which protects your spinal cord from injury, is made up of bones called **vertebrae**. Spinal nerves branch out between the vertebrae and go to every part of the body (Figure 21.3).

In addition to transmitting messages to and from the brain, the spinal cord also serves as a reflex center. **Reflexes** are automatic responses to a stimulus.

Figure 21.2   Neurons. The nervous system's three main functions are sensory input, integration, and motor output. Most actions, such as being tempted to eat cookies, involve input from sensory neurons, followed by integration via interneurons and motor output from motor neurons.

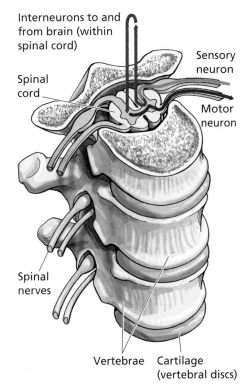

Figure 21.3   Spinal nerves and vertebrae. Spinal nerves branch out between the vertebrae and go to all parts of the body.

Table 21.1   **The senses.** There are two groups of senses: general and special. Both groups are illustrated here, along with their locations and functions.

Sense and its location	Sensory receptors	Description
Temperature, pain, pressure, touch		The skin's sensory nerve endings react to temperature, pain, pressure, and touch. These specialized receptors respond to different levels of sensation (e.g., slight cold versus freezing cold) and send information to the brain.
Proprioception		Specialized neurons sense joint position, tension in joints and ligaments, and muscle contractions. These neurons relay messages to the brain, along with balance information from the inner ear.
Smell		Paired olfactory (smell) organs with specialized receptors are located within the nasal cavity. Chemicals stimulate the receptors on hair-like projections called *cilia*. The receptors then send nerve impulses to the brain.
Taste		A tongue's surface contains taste receptors called taste buds. When food or drink contacts the taste buds, nerve impulses travel to the brain. Taste buds are specialized to respond to one of four primary taste sensations: sweet, salty, sour, or bitter.
Vision		Neuron receptors in the eye's retina allow for sight and relay sensory information to the brain.
Equilibrium and hearing		The inner ear is responsible for equilibrium and hearing. Equilibrium determines the body's position in space, monitoring gravity, acceleration, and rotation. Hearing allows us to detect and interpret sound waves. Receptors for both senses are located on the cilia, which move and generate nerve impulses to the brain.

Left margin labels:
General senses—throughout body
Special senses—found in complex sense organs

They are prewired in a circuit of neurons called a **reflex arc**, which often consists of a sensory neuron that receives information from a sensory receptor, an interneuron that passes the information along, and a motor neuron that sends a message to the muscle that needs to respond.

Reflexes allow a person to react quickly to dangerous stimuli; for instance, the withdrawal reflex occurs when you encounter a dangerous stimulus such as touching something hot. When you touch something hot, sensory neurons from touch receptors send the message to your spinal cord. Within the spinal cord, interneurons send the message to motor neurons to withdraw your hand from the hot surface (Figure 21.4).

While the spinal reflexes are removing your hand from the source of the heat, pain messages are also being sent through your spinal cord to your brain. This takes a little longer because the distance to the brain is longer than the distance to the spinal cord. Therefore, by the time the pain message reaches your brain, your hand has already been removed from the hot surface.

Even though these parts of the nervous system work together, they are separated into two different subdivisions. You have already learned that the CNS consists of your brain and spinal cord. The CNS is responsible for interpreting and acting on information received by the senses and is the seat of functions such as intelligence, learning, memory, and emotion. The second subdivision of the nervous system, the **peripheral nervous system (PNS)**, includes the network of nerves that radiates out from your brain and spinal cord (Figure 21.5). Not all organisms have such complex nervous systems. Essay 21.1 (pages 552–553) outlines some evolutionary trends in the development of nervous systems from more primitive to more complex organisms.

Many scientists who study ADD believe that the brains of those affected differ from the brains of those who are not affected. In the next section, we take a closer look at this incredible structure.

Figure 21.4   A reflex arc. A reflex arc can consist of a sensory receptor, a sensory neuron, an interneuron, a motor neuron, and an effector. Touching a hot baking sheet evokes the withdrawal reflex.

Figure 21.5   The central and peripheral nervous systems. The CNS (brain and spinal cord) integrates, processes, and coordinates information from the PNS. The nerves of the peripheral nervous system communicate sensory and motor information to the rest of the body.

Essay 21.1 Focus on Evolution: The Nervous System

The defining feature of the animal kingdom is the ability to move about the environment and to sense living prey. Animals perform these activities by using two tissues unique to their kingdom: muscles and nerves. Like muscle tissue, nervous tissue in all animals is remarkably similar in basic structure, but the arrangement of these tissues into systems differs in the various animal phyla due to their different lifestyles.

The beginnings of a nervous system are found in invertebrate organisms such as jellyfish and sea anemones, whose nerve cells are distributed throughout their bodies in what are called diffuse nervous systems. Diffuse nervous systems allow organisms some limited ability to respond to stimuli (Figure E21.1a).

Slightly more complex nervous systems called nerve nets arose in more complex animals such as hydra. Nerve nets are a meshlike network of individual nerve cells that, like diffuse nervous systems, are distributed throughout the body of the organism. However, the cells of the nerve net can communicate with each other. The ends of nerve nets form synapse-like structures that secrete substances similar to neurotransmitters (Figure E21.1b).

Communication among nerve cells can allow more complex movements to evolve. In some invertebrates, such as insects, the development of more complex nervous systems has permitted a wider range of behaviors. Concentration of nerves in the head region allows for greater complexity; insect brains, like human brains, are subdivided into different regions that control different functions. One region of the insect brain receives information from the eyes and other sense organs and controls complex behaviors. A second region controls antennae. The third region controls the mouthparts and digestive canal (Figure E21.1c). In many of these invertebrates, longitudinal nerve cords provide a conduit for information that is transmitted to and from the extremities.

Vertebrates are members of the phylum Chordata. All chordates have, at some point in their life cycle, a rod-shaped bar called a notochord. The notochord runs down the animal's dorsal, or back, surface between the digestive tube and nerve cord (Figure 21.1d). The cord develops in a unique manner in chordates, leading to a unique feature in vertebrates—a ridge of tissue called the neural crest, which eventually develops into part of the skull. The skull protects the sensitive nerve tissue of the brain, allowing for much larger brains to develop. As a result, vertebrates display the most complex behaviors among animals, including a significant capacity for learning.

Figure E21.1a   Jellyfish have a diffuse nervous system.

Figure E21.1b   The cells comprising the nerve net of hydra can communicate with each other.

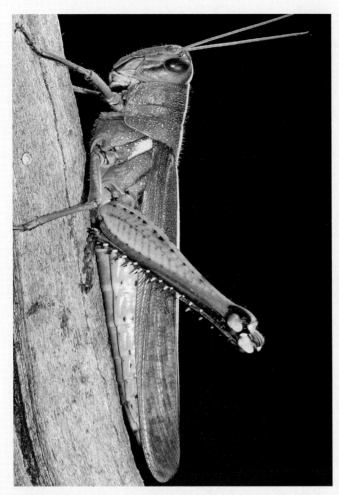

Figure E21.1c   The nervous systems of insects show a concentration of nerves in the brain.

Figure E21.1d   Vertebrates have skulls that protect their brains.

21.2   The Brain

The brain is the region of the body where decisions are reached and where bodily activities are directed and coordinated. The human brain is about the size of a small cantaloupe. Brains of the evolutionary ancestors to humans tended to be smaller and less complex (Figure 21.6). The brain is housed inside the skull, where it sits in a liquid bath, called **cerebrospinal fluid**, that protects and cushions it.

In addition to housing 100–200 billion neurons, the brain is composed of other cells called **glial cells**. There are 10 times as many glial cells in the brain as there are neurons. In contrast to neurons, glial cells do not carry messages. Instead, they support the neurons by supplying nutrients, helping to repair the brain after injury, and attacking invading bacteria. Structurally, the brain is subdivided into many important anatomical regions including the cerebrum, thalamus, hypothalamus, cerebellum, and brain stem (see Figure 21.7 on page 554).

Figure 21.6   Evolution of larger brains.   These skulls of early human ancestors show that the volume of the skull taken up by the brain has increased over the last 3–5 million years from about 400 ml to 1400 ml.

553

(a)

(b)

Cerebrum
controls language, memory, sensations, and decision making.

Cerebrospinal fluid

Skull

Thalamus
is the relay center for sensory information to cerebrum and motor response to cerebrum.

Hypothalamus
is the control center for sex drive, pleasure, pain, hunger, thirst, blood pressure, body temperature, and so on.

Cerebellum
is responsible for balance, muscle movement, and coordination.

Brain stem
governs reflexes, heartbeat, breathing, swallowing, and other automatic functions.

Figure 21.7   Anatomy of the brain. The photo (a) and illustration (b) show the cerebrum, thalamus, hypothalamus, cerebellum, and the brain stem.

Cerebrum

The **cerebrum** fills the whole upper part of the skull. This part of the brain controls language, memory, sensations, and decision making. The cerebrum has two hemispheres, each divided into four lobes (Figure 21.8):

1. The **temporal lobe** is involved in processing auditory information and some visual information, as well as memory and emotion.
2. The **occipital lobe** processes visual information from the eyes.
3. The **parietal lobe** processes information about touch and is involved in self-awareness.
4. The **frontal lobe** processes voluntary muscle movements and is involved in planning and organizing future expressive behavior.

The deeply wrinkled outer surface of the cerebrum is called the **cerebral cortex**. In humans, the cerebral cortex, if unfolded, would be the size of a 16-inch pizza. The folding of the cortex increases the surface area and allows this structure to fit inside the skull. The cortex contains areas for understanding and generating speech, areas that receive input from the eyes, and areas that receive other sensory information from the body. It also contains areas that allow planning.

The cerebrum and its cortex are divided from front to back into two halves—the right and left cerebral hemispheres—by a deep groove called a **fissure**. At the base of this fissure lies a thick bundle of nerve fibers, the **corpus callosum**, which functions as a communication link between the hemispheres. The **caudate nuclei** are paired structures found deep within each cerebral hemisphere. These structures function as part of the pathway that coordinates movement.

Thalamus and Hypothalamus

Deep inside the brain, lying between the two cerebral hemispheres, are the **thalamus** and the **hypothalamus**. The thalamus relays information between the spinal cord and the cerebrum. The thalamus is the first region of the

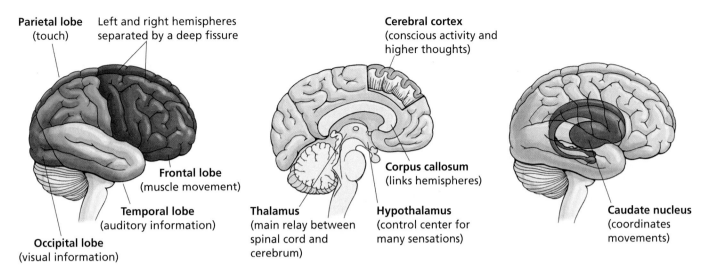

Figure 21.8   Structure of the cerebrum. The highly convoluted cerebral cortex is divided down the midline into left and right hemispheres, and each hemisphere is divided into four lobes: temporal, occipital, parietal, and frontal. The hemispheres are connected by the corpus callosum. Deep inside and between the two cerebral hemispheres are the thalamus and the hypothalamus. A caudate nucleus is located within each cerebral hemisphere.

brain to receive messages signaling such sensations as pain, pressure, and temperature. The thalamus suppresses some signals and enhances others, which are then relayed to the cerebrum. The cerebrum processes these messages and sends signals to the spinal cord and to neurons in muscles when action is necessary. The hypothalamus, located just under the thalamus and about the size of a kidney bean, is the control center for sex drive, pleasure, pain, hunger, thirst, blood pressure, and body temperature. The hypothalamus also releases hormones that regulate the production of sperm and egg cells as well as the menstrual cycle.

Cerebellum

The **cerebellum** controls balance, muscle movement, and coordination (*cerebellum* means "little brain" in Latin). Since this brain region ensures that muscles contract and relax smoothly, damage to the cerebellum can result in rigidity and in severe cases, jerky motions. The cerebellum looks like a smaller version of the cerebrum (see Figure 21.7). It is tucked beneath the cerebral hemispheres and, like the cerebrum, also has two hemispheres connected to each other by a thick band of nerves. Additional nerves connect the cerebellum to the rest of your brain.

Brain Stem

The **brain stem** lies below the thalamus and hypothalamus (see Figure 21.9 on page 556). It governs reflexes and some spontaneous functions such as heartbeat, respiration, swallowing, and coughing.

The brain stem is composed of the **midbrain, pons,** and **medulla oblongata.** Highest on the brain stem is the midbrain, which adjusts the sensitivity of your eyes to light and of your ears to sound. Below the midbrain is the pons (*pons* means "bridge"). The pons functions as a bridge, allowing messages to travel between the brain and the spinal cord. The medulla oblongata is a continuation of the spinal cord. It conveys information between the spinal cord and other parts of the brain.

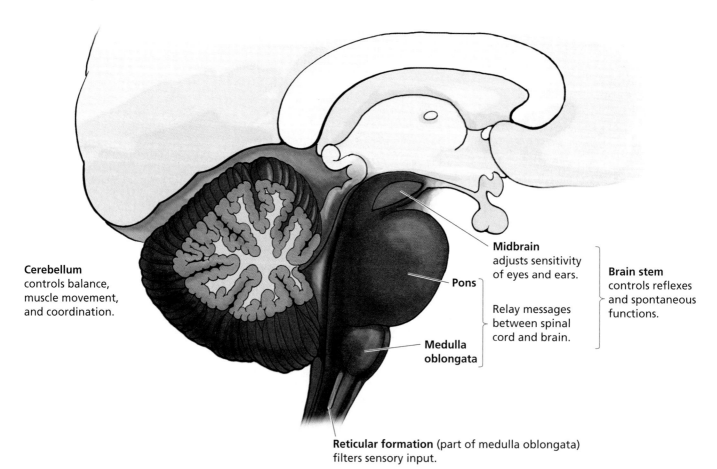

Cerebellum
controls balance,
muscle movement,
and coordination.

Midbrain
adjusts sensitivity
of eyes and ears.

Pons

Relay messages
between spinal
cord and brain.

**Medulla
oblongata**

Brain stem
controls reflexes
and spontaneous
functions.

Reticular formation (part of medulla oblongata)
filters sensory input.

Figure 21.9   The cerebellum and brain stem. The brain stem consists of the midbrain, pons, and medulla oblongata. The reticular formation lies within the medulla oblongata.

The functions of the brain are divided between the left and right hemispheres. Because many nerve fibers cross over each other to the opposite side, the brain's left hemisphere controls the right half of the body, and vice versa. The areas that control speech, reading, and the ability to solve mathematical problems are located in the left hemisphere, while areas that govern spatial perceptions (the ability to understand shape and form) and the centers of musical and artistic creation reside in the right hemisphere.

The **reticular formation** (Figure 21.9), found in the medulla oblongata, is an intricate network of neurons that radiates toward the cerebral cortex. The reticular formation functions as a filter for sensory input; it analyzes the constant onslaught of sensory information and filters out stimuli that require no response. This prevents the brain from having to react to repetitive, familiar stimuli such as the sound of automobile traffic outside your dorm room or the sound of your roommate's breathing while you are trying to sleep.

The reticular formation also functions as an activating center by keeping the cerebral cortex alert. While conscious activity originates in the cerebral cortex, it can do so only if the reticular formation is keeping the cortex alert.

ADD and Brain Structure and Function

In the same way that X-rays help physicians check for broken bones, scientists have access to brain-scanning technologies that enable them to view the brain. Some **neurobiologists**—biologists who study the nervous system—use these technologies to find physical evidence in the brain structures of people diagnosed with **attention deficit disorder (ADD)**. Some neurobiologists believe they have found subtle differences in the structure and function of the brains of people with ADD.

In one study comparing the brains of people with and without ADD, researchers found that the corpus callosum (the band of nerve fibers connecting the two hemispheres of the brain) was slightly smaller in people with ADD. A similar study discovered size differences in the caudate nuclei (involved in coordinating movement). The caudate nucleus in the right hemisphere of people with ADD was slightly smaller than it was in people who did not have the disorder. Other studies have shown that the cortex and cerebellum of affected persons tend to be smaller than those in unaffected people.

Additional studies have detected functional differences between the brains of people with and without ADD. Some researchers have gathered evidence indicating that the reticular formation is not very effective at filtering out signals; it allows too much information to be sent on to the cerebral cortex in people diagnosed as having ADD. One research team observed decreased blood flow through the right caudate nucleus in people with ADD. Another study showed that an area of the cortex metabolizes less glucose in adults with ADD. Glucose metabolism is an indicator of activity—the more glucose that is metabolized, the more active an area is. This finding could indicate that this area of the cortex, which is involved in predicting the consequences of various actions, might be less active in people who have ADD.

Brain differences between people with ADD and people who do not have the diagnosis may have arisen because of each person's unique experiences. The brains of people with ADD could reflect inborn biological differences. Or, brain differences may arise from some combination of these two factors, in much the same way that people develop different muscles in response to exercise.

Differences in brain structure and function, whether present at birth or developed later in life, are closely related to how neurons work. By investigating the ways in which messages are transmitted by neurons, researchers have been able to propose additional hypotheses about the causes and effects of ADD.

21.3   Neurons

Neurons are highly specialized cells that usually do not divide. Therefore, damage to a neuron cannot be repaired by cell division and often results in permanent impairment. For example, damage to spinal motor neurons results in lifelong paralysis because messages can no longer be transmitted from the CNS to muscles. Likewise, injury to the brain can result in permanent brain damage if neurons in the brain are harmed.

Neuron Structure

Neurons are composed of branching **dendrites** that radiate from a bulging **cell body**, which houses the nucleus and organelles, and a long, wirelike **axon**, terminating in knobby structures called the **terminal boutons** (Figure 21.10). A nerve impulse usually travels down the axon of one neuron and is transmitted to the dendrites of another neuron.

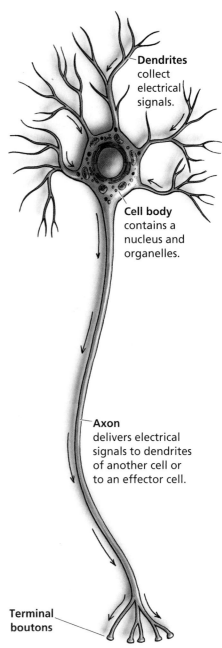

Figure 21.10   The structure of a generalized neuron. A neuron consists of the branching dendrites, the cell body, the axon, and terminal boutons. Nerve impulses are propagated in the direction of the arrows.

(a) Axons covered in myelin sheaths   **(b)** Nerve impulses travel more quickly on myelinated than on unmyelinated nerves.

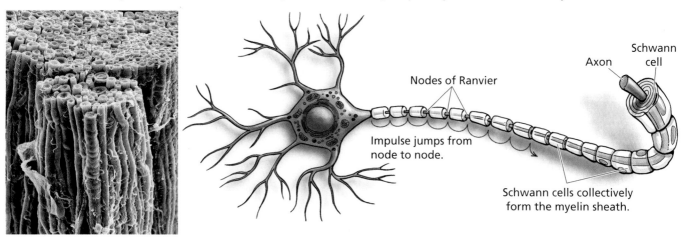

Figure 21.11   Myelination. (a) This photo shows a cut bundle of nerve fibers. Each separate nerve consists of a nerve-cell axon (greenish) covered in an insulating layer of myelin (whitish). (b) By jumping from one node of Ranvier to the next, nerve impulses travel more quickly on myelinated nerves than on unmyelinated nerves.

The axons of many neurons are coated with a protective layer called the **myelin sheath** (Figure 21.11a). The myelin sheath is formed by supporting cells such as **Schwann cells** (named for the German cell biologist, Theodore Schwann). The myelin sheath is composed mostly of lipids and has the same appearance as animal fat. Due to its white color, nervous tissue composed of myelinated cells is called **white matter**. The myelin sheath functions like insulation on a wire by preventing sideways message transmission, thus increasing the speed at which the electrochemical impulse travels down the axon.

Long stretches of the myelin sheath are separated by small interruptions in the Schwann cells, leaving tiny, unmyelinated patches called the **nodes of Ranvier**. Nerve impulses "jump" successively from one node of Ranvier to the next (Figure 21.11b). This jumping transmission of nerve impulses is up to 100 times faster than signal conduction would be on a completely unmyelinated axon—nerve impulses can travel along myelinated axons at a rate of 100 meters per second (more than 200 miles per hour).

Some neurons are not myelinated and thus transmit nerve impulses more slowly. Unmyelinated axons, combined with dendrites and the cell bodies of other neurons, look gray in cross sections and thus are called **gray matter**.

Neuron Function

Neurons transmit impulses from one part of the body to another. Many kinds of stimuli, including touch, sound, light, taste, temperature, and smell, cause neurons to fire off a response. When you touch something, signals from touch sensors travel along sensory nerves from your skin, through your spinal cord, and into your brain. Your brain then sends out messages through your spinal cord to the motor nerves, telling your muscles how to respond. To evoke this response, nerve cells must transmit signals along their length and from one cell to the next. Another name for nerve impulse is action potential.

Action Potential.   An **action potential** transmits a signal along the length of a neuron. An action potential is a brief reversal of the electrical charge of the membrane of a nerve cell. The action potential is propagated as a wave of electrical current down the length of the neuron at a speed of several hundred feet per second.

In the resting neuron, a charge difference exists between the inside and outside of the cell. The resting neuron is negatively charged on the inside compared to the outside of the cell. Since opposite charges tend to move toward each other, the membrane serves as a source of stored energy, like a battery, by keeping these opposite charges apart. This difference in the charge across the membrane is called **polarization**. This polarization exists because the concentration of potassium (K^+) inside the cell is greater than the amount of sodium (Na^+) on the outside of the cell. This is because potassium channels are "leaky" and allow potassium to leak out of the cell down its own concentration gradient, leaving a negative charge behind (other ions inside the cell contribute to its overall negative charge).

A protein pump in the membrane of neurons called the **sodium-potassium pump** uses the energy of ATP to move 3 sodium molecules out of the cell for every 2 potassium molecules returned back into the cell. Thus the sodium potassium pump returns potassium into the cell to maintain the concentration gradient. (see Figure 21.12a on page 560).

Stimulation of a neuron partially opens the gates on the sodium-channel proteins in the membrane. When these channels are open, sodium can enter the cell, shifting that particular region inside the cell toward a less negative state of polarization. If this local change reaches a critical level, then the sodium channels open completely and sodium floods the cell, quickly eliminating the charge difference across the cell and causing an action potential. This loss of charge difference or **depolarization** moves in a wave down the cell, activating sodium channels in adjacent parts of the membrane and thereby moving the impulse along the length of the neuron (see Figure 21.12b on page 560).

The entry of sodium into the neuron must be balanced by the exit of another positively charged ion, or the action potential could not decline and return the neuron to its resting state. The action potential declines as sodium channels close and potassium channels open, allowing potassium out of the cell. **Repolarization** occurs when potassium ions leach out of the cell, and the inside of the cell again becomes more negative than the outside.

Neurons can also be inhibitory; that is, rather than causing a depolarization in the neighboring cell, they cause it to become even more polarized or **hyperpolarized**.

Synaptic Transmission.   Once the signal has traveled along the length of the axon, it must be passed to the next neuron or to the effector organ, muscle, or gland. Since most neurons are not directly connected to each other, the signal must be transmitted to the next neuron across a gap between the two neurons, called the **synapse**. The synapse consists of the terminal boutons of the **presynaptic** neuron, the space between the two adjacent neurons, and the cell membrane of the **postsynaptic** neuron. Saclike structures called **vesicles** are found in the terminal boutons of the presynaptic neuron. Each vesicle in a particular neuron is filled with a specific chemical neurotransmitter. When an electrical impulse arrives at the terminal bouton of a nerve cell, neurotransmitters are released. Once released, they diffuse across the synapse and bind to specific receptors on the cell membrane of the postsynaptic neuron. The binding of a neurotransmitter to the postsynaptic cell membrane once again stimulates a rapid change in the uptake of sodium ions by the postsynaptic neuron. The sodium channels open, and sodium ions flow inward, causing another depolarization and generating another action potential. In this manner, the nerve impulse is propagated

(a) Resting nerve cell

Nodes of Ranvier

Nerve cell

Outside cell   Inside cell

Na+   K+   K+
Na+   Na+   K+   K+   Na+   Na+
Na+   Na+   K+
Na+   Na+   K+
Na+   K+
Na+   K+
Na+   K+ channel
Na+   K+
Na+   Na+ channel

Na+   Na+   Na+   Na+   Na+   Na+   Na+

All channels are closed. The inside of the cell has a more negative charge than the outside of the cell.

(b) Propagation of an action potential or nerve impulse

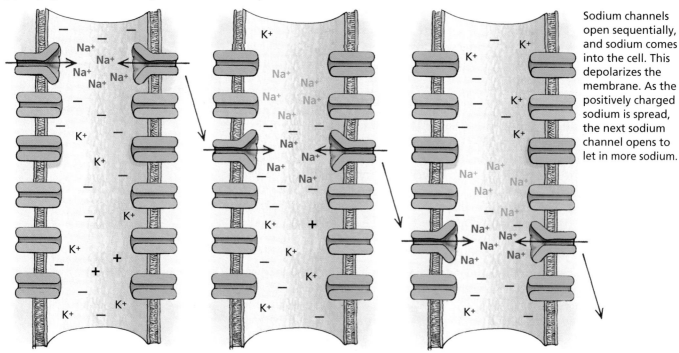

Sodium channels open sequentially, and sodium comes into the cell. This depolarizes the membrane. As the positively charged sodium is spread, the next sodium channel opens to let in more sodium.

Figure 21.12 Generating and propagating a nerve impulse. (a) In a resting nerve cell, the inside of a neuron is more negatively charged than the outside. This difference in charge is maintained by protein pumps that pump 3 sodium ions (Na+) out of cells for every 2 potassium ions (K+) pumped into the cell. Closed channels stopping these ions from moving back across the membrane prevent the ions from diffusing back toward their chemical equilibrium. (b) A nerve impulse starts with an influx of positively charged sodium ions, which depolarizes the nerve cell. A domino effect occurs as the sodium ions spread their charges toward each successive sodium channel, causing each to open and let in more sodium, thus propagating the depolarization along the length of the neuron.

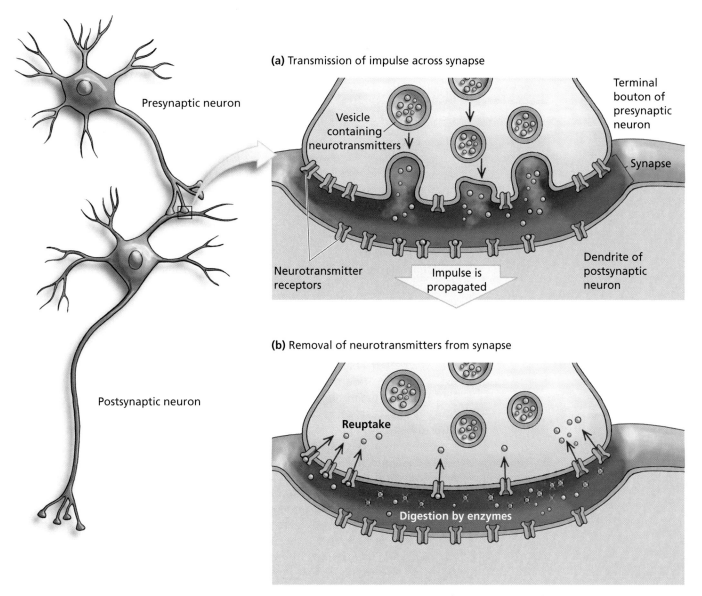

(a) Transmission of impulse across synapse

Presynaptic neuron

Vesicle containing neurotransmitters

Terminal bouton of presynaptic neuron

Synapse

Neurotransmitter receptors

Impulse is propagated

Dendrite of postsynaptic neuron

(b) Removal of neurotransmitters from synapse

Postsynaptic neuron

Reuptake

Digestion by enzymes

Figure 21.13   Propagating the nerve impulse between neurons. (a) The nerve impulse is transmitted from the terminal bouton of one neuron to the dendrite of the next. Neurotransmitters are released from the presynaptic neuron, travel across the synapse to the postsynaptic neuron, and bind to receptors on dendrites of the postsynaptic neuron. (b) After the neurotransmitter evokes a response, it is removed from the synapse by enzymes present in the synapse that break the neurotransmitter apart or by reuptake via the presynaptic cell.

from one neuron to the next until the signal reaches the tissue that it will affect (Figure 21.13a).

Over a dozen known chemicals function as neurotransmitters. Examples of neurotransmitters include glutamate, which is an important CNS neurotransmitter, and acetylcholine, which is an important neurotransmitter in the muscle. Each type of neuron secretes one type of neurotransmitter, and each cell responding to it has receptors specific to that neurotransmitter.

After the neurotransmitter evokes a response, it is removed from the synapse. Some neurotransmitters are broken down by enzymes in the synapse; others are reabsorbed by the neuron that secreted them via a process called **reuptake** (Figure 21.13b). Reuptake occurs when neurotransmitter receptors on the presynaptic cell permit the neurotransmitter to reenter the cell. This allows the neuron that released the neurotransmitter to use it again. Both breakdown by enzymes and reuptake by receptors prevent continued stimulation of the postsynaptic cell.

Neurotransmission, Alzheimer's, Depression, Parkinson's, and ADD

There is a great deal of evidence linking disturbances in neurotransmission to various diseases. For example, **Alzheimer's disease**, a progressive mental deterioration in which there is memory loss along with the loss of control of bodily functions that eventually results in death, is thought to involve impaired function of the neurotransmitter **acetylcholine** in some neurons. Drugs that inhibit the enzyme **acetylcholinesterase**, which breaks down acetylcholine, can temporarily improve mental function but cannot stop this progressive illness.

Depression is a disease that involves feelings of helplessness, despair, and thoughts of suicide. No one knows for sure what causes depression; but experts believe that a genetic susceptibility, coupled with environmental factors such as stress or illness, may trigger an imbalance in neurotransmitters and lead to depression. Three neurotransmitters—serotonin, dopamine, and norepinephrine—have been implicated in depression, but scientists do not know whether changes in these neurotransmitters are a cause or result of depression. Antidepressants blocking the actions of enzymes that degrade these neurotransmitters or inhibit its reuptake help alleviate many symptoms of this disease in some people. Prozac® and Zoloft®, for instance, fall into the class of specific serotonin reuptake inhibitors (SSRIs) that are prescribed to treat depression.

Parkinson's disease is thought to be caused by the malfunctioning of neurons that produce dopamine. The loss of dopamine causes nerve cells to fire without regulation; as a result, patients have difficulty in controlling their movements and experience tremors, rigidity, and slowed movements. There is no cure for this progressive disease, and scientists do not know whether there is a genetic component.

Abnormal levels of the neurotransmitter **dopamine** may also be involved in producing the symptoms of ADD. This neurotransmitter controls emotions as well as complex movements. Some researchers hypothesize that people with ADD symptoms may have lower dopamine levels than other people do. Decreased dopamine levels may cause ADD symptoms because dopamine suppresses the responsiveness of neurons to new inputs or stimuli. Therefore, someone with a low concentration of dopamine may respond impulsively to situations in which pausing to process the input would be more effective.

The cause of decreased dopamine levels in some people with ADD may actually involve an overabundance of dopamine receptors on the presynaptic cell. During reuptake, dopamine receptors on the presynaptic cell remove dopamine from the synapse to prevent continued stimulation of the postsynaptic neuron. Some studies have shown elevated numbers of these receptors in people with ADD. **Ritalin**, the drug usually prescribed to treat symptoms of ADD, may work to decrease the impact of these extra reuptake receptors.

Ritalin®

Ritalin is thought to increase dopamine's ability to stimulate the postsynaptic cell by blocking the actions of the dopamine reuptake receptor (Figure 21.14).

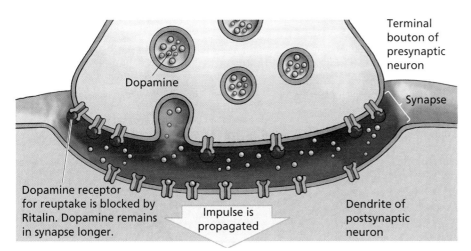

Dopamine

Terminal bouton of presynaptic neuron

Synapse

Dopamine receptor for reuptake is blocked by Ritalin. Dopamine remains in synapse longer.

Impulse is propagated

Dendrite of postsynaptic neuron

Figure 21.14   The mechanism of Ritalin action. Ritalin blocks dopamine receptors on the presynaptic neuron. This blocks reuptake and thereby allows dopamine access to the postsynaptic neuron for a longer time.

This leaves more dopamine around for a longer time, resulting in a decrease in the symptoms of ADD.

Other stimulants work through similar mechanisms. In addition to their direct effects on dopamine levels, all stimulants—whether legal like Ritalin and obtained by prescription or illegal like cocaine and amphetamines (collectively called speed)—have similar effects on the body. These drugs affect the heart, blood vessels, and lungs on their way to the brain by increasing heart rate, elevating blood pressure, and expanding airways in the lungs.

When stimulants are taken in high doses, the user feels euphoric and has more energy and endurance, a sense of power, and a feeling of mental sharpness. As the drug wears off, heightened fatigue, insomnia, poor concentration, irritability, tearfulness, and depression are common; other effects include personality changes, skin rashes, fever, nausea, and headaches. Abusing stimulants can lead to psychotic episodes, delusions, seizures, hallucinations, and sudden death.

Many elementary school students who grow up taking Ritalin to help them with schoolwork are still having their prescriptions filled when they head off to college. Once on campus, these students no longer stand in line for the school nurse to provide the prescribed dose. Instead, they self-administer the stimulant. Some take the prescribed dose, others use the drug only to pull an all-nighter, and still others give pills to their friends.

Due to its availability and status as a prescription drug—not a street drug, such as speed or other amphetamines—students may see Ritalin as a less-serious drug than other stimulants, and these students are often unaware of the dangers of abusing this drug. Recent studies indicate that about 15% of college students report having used the drug recreationally.

If you have not been prescribed Ritalin but use it to help focus for an exam, there is no evidence that it actually helps you learn. This is because you probably do not have low dopamine levels to begin with. For Ritalin abusers, the euphoria they experience, coupled with the fact that they are more likely to sit in a chair and focus on their schoolwork, may make them *believe* that Ritalin is helping them learn. As with any drug, the potential for dangerous side effects should be seriously considered prior to use. Table 21.2 (pages 564–565) describes many recreational drugs and their effects on the body.

The dopamine reuptake theory is not the only hypothesis about the biological cause of ADD symptoms, nor does it explain all cases of ADD.

Table 21.2 Recreational drugs and the nervous system.

Drug	Mechanism of action	Desired mental effect	Side effect
Alcohol is a byproduct of fermentation. When yeast cells ferment barley, beer is produced. Likewise, when yeast cells break down the sugars in grapes, wine is produced.	• A depressant that diffuses easily across cell membranes and does not require specific receptors • Inhibits neurotransmission in the reticular formation, interfering with the activity of many neurons in the brain	• Reduced anxiety and a sense of well-being, loss of concern for social constraints	• Impaired judgment, slurred speech, unsteady gait, slower reaction times, uncontrollable emotions • Chronic alcohol abuse leads to loss of intellectual ability by damaging nerve cells. Cells in the frontal lobes (center for judgment, thought, and reason) are the first to die • Liver damage
Amphetamines are used legally to treat obesity, asthma, and narcolepsy. Methamphetamine is an illegal amphetamine. A crystalline form of methamphetamine called *ice* is smoked to produce effects similar to crack cocaine.	• Stimulants that increase activity in the reticular formation • Mimic the actions of the neurotransmitter norepinephrine, a hormone produced in response to stress • Block reuptake and inhibit the enzyme that breaks down norepinephrine, resulting in prolonged stimulation of the postsynaptic cell	• Small doses make a person feel more energetic, alert, and confident	• Effects wear off quickly, causing sudden "crashes" from depleted neurotransmitter stores and resulting in depression and fatigue • Prolonged use results in aggressiveness, delusions, hallucinations, and violent behaviors. • Can cause blood vessels to spasm, clots to form, insufficient blood to enter the heart, fluid to build up in the lungs, and death
Caffeine is a naturally occurring chemical found in plants such as coffee, tea, and cocoa.	• A general stimulant that affects all cells, not just those of the central nervous system • Increases cellular metabolic activities	• Mental alertness, increased energy	• Insomnia, anxiety, irritability, and increased heart rate
Cocaine is extracted from the leaves of the coca plant of South America. It can be inhaled or injected. A more potent form, crack, is smoked.	• A stimulant that increases levels of dopamine and norepinephrine by decreasing reuptake	• A rush of intense pleasure, increased self-confidence, and increased physical vigor	• Increased heart rate and blood pressure, narrowing of blood vessels, dilation of pupils, a rise in body temperature, and reduction of appetite • Crash involves deep depression, anxiety, and fatigue. • Abuse may change neurons and prevents the abuser from experiencing positive feelings without the drug

Drug	Mechanism of action	Desired mental effect	Side effect
Ecstasy or MDMA is a white crystalline powder that is primarily ingested in pill or capsule form.	• A stimulant and hallucinogenic that acts to prevent serotonin reuptake • Also floods neurons with several other neurotransmitters	• Euphoria, enhanced emotional and mental clarity, increased energy, heightened sensitivity to touch, and enhanced sexual response	• Confusion, anxiety, paranoia, depression, and sleeplessness that may last for several weeks • May permanently damage neurons involved in utilizing serotonin and may also result in permanent memory damage • When combined with physical exercise, ecstasy use can lead to severe dehydration and the inability to regulate body temperature
Lysergic acid diethylamide (LSD) and Mushrooms LSD is a derivative of the fungus *Claviceps purpurea*, which grows on rye. It is related to the compound psilocybin, found in certain mushrooms.	• Hallucinogenic that binds to serotonin receptors in the brain, increasing the normal response to serotonin	• Heightened sensory perception and bizarre changes in thought and emotion, hallucinations	• Hallucinations can lead users to dangerous actions. • Heavy use leads to permanent brain damage, including losses of memory and attention span, and can lead to psychosis.
Marijuana is the leaves, flowers, and stems of the Indian hemp plant *Cannabis sativa*. Also contains a drug called delta-9-tetrahydrocannabinol (THC).	• Receptors for THC in the areas of the brain that influence mood, pleasure, memory, pain, and appetite. • THC is thought to work by increasing dopamine release	• Altered sense of time, enhanced feeling of closeness to others, increased sensitivity to stimuli • Large doses can cause hallucinations	• Slowed reaction time, decreased coordination, and inability to judge time, speed, and distance. Impaired short-term memory, slowed learning, decreased attention span, and lessened ability to store and acquire information • Damage to THC receptors on the hypothalamus, the area of the brain that regulates sex-hormone secretion, can decrease testosterone production and disrupt menstruation
Nicotine is found in tobacco plants. Just 1 of over 1000 chemicals in cigarette smoke. It is one of the most likely to affect the brain.	• A stimulant that triggers neuron receptors in the cerebral cortex to produce acetylcholine, epinephrine, and norepinephrine	• Increased alertness and awareness, appetite suppression, relaxation	• Cigarette smoking increases the odds of obtaining virtually every type of cancer • Also causes increased heart rate and blood pressure
Opiates are derived from the opium poppy. Heroin, morphine, and codeine are opiates. Morphine and codeine can be used legally to control pain.	• Binds to opiate receptors in neurons that control feelings of pleasure • Opiate receptors have evolved to bind to opiates produced by the brain in response to exercise; these opiates are called endorphins	• A quick, intense feeling of pleasure, followed by a sense of well-being and drowsiness	• Addiction, poor motor coordination, depression • High doses can cause coma and death • Thought to change the brain's ability to respond to normal pleasures

Not all drugs that alleviate ADD symptoms have any impact on dopamine levels; and many times, drugs must be combined with changes in environmental conditions for the symptoms of ADD to improve.

21.4 The Environment and ADD

The nervous system is a dynamic, changing system, and many different parts of the brain are responsive to environmental influences. There is some evidence of biological differences in the brains of people diagnosed with ADD; but there is also evidence that environmental factors may exacerbate the ADD situation. In the last decade, the number of people diagnosed with ADD has risen from about 1 million to 6–11 million. Such a dramatic increase in a very short time is not likely due to biology alone. It would take thousands of years for evolution to affect this magnitude of biological change within a population.

As is often the case in science, the data on increased diagnosis of ADD are difficult to interpret, and there are many different ways to explain the same findings. Doctors may simply be getting better at diagnosing ADD cases, or they may be misdiagnosing overactivity as ADD. It also may be that societal changes are having a negative impact on children. This is an area of active research and debate involving a wide variety of people including scientists, medical doctors, psychologists, teachers, and parents.

Some research suggests that one factor affecting the likelihood of ADD diagnosis is simply sleep deprivation in young children. This goes hand in hand with the belief of many experts that the increased number of people diagnosed with ADD reflects the increased pressures that children and adults face in our fast-paced, success-driven society. In the United States, most caretakers of elementary schoolchildren work full time—one parent in single-parent families, and both parents in two-parent families—according to the 2000 census data. Today's children are involved in more and more time-consuming after-school activities such as sports, dance, music, theater, and art lessons, leaving less time for sleep. Some opponents of Ritalin point out that, with so much to pay attention to, it is not surprising that some kids have a hard time staying focused.

Deciding whether ADD is caused by biology or environment influences treatment options. If this disease is solely one of neurobiology, then using medication alone to treat it makes sense, especially if the medication has few or no side effects. However, like other stimulants, Ritalin use has serious side effects including insomnia, loss of appetite, and nervousness; it even may be associated with decreased growth rates. It is unclear whether those who were prescribed Ritalin as children are at increased risk for stimulant use and addiction as adults. Given the serious side effects and potential for abuse of Ritalin, research on the cause of ADD and investigation of alternative treatments becomes more critical. It is easier to accept biological explanations for many problems and treat them with medicines than to work on changing the environment that is contributing to the problem. People take antacids to treat stomach disorders that, in some cases, would be more effectively treated through dietary changes. Likewise, heart disease is medicated when, for many people, dietary changes coupled with exercise would work as well or better. In some situations, modifying the environment may be the best way of dealing with the problem.

Environmental changes have been shown to help many people with ADD. Behavior modification therapies can improve ADD symptoms in children who are, and are not, taking Ritalin. For instance, parents of ADD children are counseled about effective parenting strategies, such as not overscheduling their ADD child and clearly defining and enforcing rules. Teachers of ADD students often try to minimize distractions by having these students sit close to the front of the room, away from hallways or windows, and by giving them detailed schedules and timelines for assignments.

The success of approaches that combine medical and environmental interventions shows that it may be more useful to think of ADD as an imbalance between the brain's inborn tendencies and the demands of a person's environment (Figure 21.15).

(a) Biological factors that may influence ADD

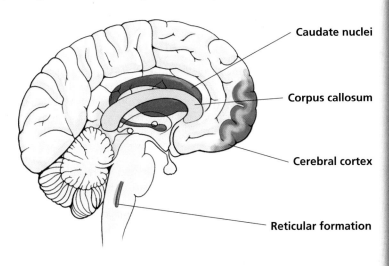

	Person without ADD	Person with ADD
Caudate nuclei	Right caudate nucleus larger than left caudate nucleus; normal blood flow.	Two caudate nuclei same size? Decreased blood flow?
Corpus callosum	Normal size	Slightly smaller?
Cerebral cortex	Normal activity	Parts of cortex less active?
Reticular formation	Normal traffic flow	Not as effective at limiting the number of traffic signals to cortex?

(b) Societal factors that may influence ADD

Sleep deprivation

Increase in number of families with two working parents

Increased participation in multiple activities

Increased number of children going to college

(c) Biological and societal factors may combine

Child with ADD brain + Stress ⟶ ADD behavior

Figure 21.15   Biological and societal factors that may influence ADD. Many factors influence ADD; these include (a) differences in brain structure and function as well as (b) stresses produced by modern society, or (c) both.

CHAPTER REVIEW

Summary

21.1 The Nervous System

- The nervous system allows an organism to respond to external and internal stimuli (p. 548).

- Neurons are specialized cells of the nervous system that carry chemical and electrical messages (p. 548).

- Sensory neurons carry information toward the brain; motor neurons carry information away from the brain. Interneurons are located between sensory and motor neurons (p. 549).

- Sensory receptors for general senses, found throughout the body, relay information to the brain. Receptors for the special senses are found in more complex sense organs (pp. 549–550).

- Reflexes are prewired responses that carry messages from sensory receptors to interneurons to a motor neuron for a response (pp. 549–551).

- The nervous system consists of the brain and spinal cord of the central nervous system as well as the nerves of the peripheral nervous system that carry information to and from the brain (p. 551).

21.2 The Brain

- The cerebrum of the brain is where most thinking occurs. The two hemispheres of the cerebrum consist of four lobes—temporal, occipital, parietal, and frontal. The outer surface of the cerebrum is the cortex (p. 554).

- The thalamus and hypothalamus are located between the two cerebral hemispheres. The thalamus relays information between the spinal cord and the cerebrum, and the hypothalamus regulates many vital bodily functions (pp. 554–555).

- The cerebellum is located at the base of the brain, beneath the cerebral hemispheres. It regulates balance and coordination (p. 555).

- The brain stem is located below the thalamus and hypothalamus. It controls many unconscious functions (p. 555).

- Many studies have shown structural and functional differences in the brains of people affected by ADD (p. 557).

21.3 Neurons

- Neurons are specialized cells whose structure consists of the branching dendrites, the cell body, the axon, and terminal boutons (p. 557).

- Axons that are covered in myelin conduct impulses faster and comprise the white matter (p. 558).

- Unmyelinated axons conduct impulses more slowly and comprise the gray matter (p. 558).

- Nerve impulses or action potentials are generated when depolarization of the cell membrane occurs. A resting neuron is polarized in the sense that it is negatively charged in comparison to the outside of the cell. This polarization is maintained by sodium-potassium pumps in the membrane that pump 3 sodium molecules out of the cell for every 2 potassium molecules it pumps inside the cell (pp. 558–559).

- Stimulation of a neuron partially opens the gates on the sodium-channel proteins in the membrane, allowing sodium to diffuse into the cell and depolarizing the cell. The depolarization moves in a wave down the cell, activating sodium channels in adjacent parts of the membrane and thereby moving the impulse along the length of the neuron (p. 559).

- The electrical impulse is propagated to the ends of the axon, which house chemical neurotransmitters (p. 559).

- Nerve impulses cause neurotransmitters to be released into the synapse, from where they can bind to receptors on the next neuron (p. 559).

- Binding of the neurotransmitter to its receptor on the next neuron or effector causes that cell to be depolarized, generating another impulse; in some cases hyperpolarization results, preventing other impulses from occurring (p. 559).

- Neurological disorders can be caused by structural abnormalities of the brain and/or by defects in neurotransmission. Some scientists believe that ADD is caused by differences in brain structures or by deficiencies in the neurotransmitter dopamine (p. 562).

- Ritalin helps people with ADD by blocking dopamine reuptake receptors on the presynaptic nerve cell, giving dopamine more time to access the receptors on the post-synaptic nerve cell (pp. 562–563).

Web Tutorial 21.1 **Communication Within Neurons: The Axon**
Web Tutorial 21.2 **Communication Between Neurons: The Synapse**

21.4 The Environment and ADD

- Some people involved in the Ritalin debate wonder whether our fast-paced society is increasing the incidence of people with ADD (p. 566).

Learning the Basics

1. How does the CNS differ from the PNS?

2. Where are the receptors for neurotransmitters located?

3. How is a nerve impulse transmitted from one neuron to the next?

4. Which brain structure controls balance and coordination?
   A. cerebrum; **B.** thalamus; **C.** hypothalamus; **D.** cerebellum; **E.** brain stem

5. Where in the brain is the thalamus located?
   A. in the pons; **B.** in the medulla; **C.** in the cerebral cortex; **D.** between the cerebral hemispheres; **E.** in the occipital lobe

6. Myelin _____.
   A. is a protective layer that coats some neurons; **B.** is formed by cells called Schwann cells; **C.** gives neurons a white appearance; **D.** prevents sideways transmission of nerve impulses, thereby increasing the rate of transmission; **E.** all of the above

7. Neurons maintain a negative charge in their cytoplasm relative to outside the cell via the _____.
   A. Krebs cycle; **B.** electron transport chain; **C.** dopamine receptor; **D.** sodium-potassium pump

8. An action potential _____.
   A. is a brief reversal of temperature in the neuronal membrane; **B.** is propagated from the terminal bouton towards the cell body; **C.** begins when sodium channels close in response to stimulation; **D.** is propagated as a wave of depolarization moves down the length of the neuron

9. Neurotransmitters _____.
   A. are electrical charges that move down myelinated axons; **B.** are released across the nodes of Ranvier to hasten nerve impulse transmission; **C.** are released when an electrical impulse arrives at the terminal bouton of the postsynaptic neuron; **D.** diffuse across a synapse and bind to receptors on the cell membrane of the postsynaptic neuron; **E.** are steroid hormones released from the thalamus

10. The effects of a neurotransmitter could be increased by _____.
    A. increasing the number of receptors on the post-synaptic cell; **B.** preventing reuptake; **C.** providing more enzymes involved in synthesizing the neurotransmitter; **D.** inhibiting enzymes involved in breakdown of the neurotransmitter from the synapse; **E.** all of the above

Analyzing and Applying the Basics

1. Why might decreased glucose consumption in a brain structure indicate an underlying problem?

2. How might blocking neurotransmitter breakdown in the synapse affect neural function?

3. Propose a mechanism for decreasing the amount of dopamine present in the synapse, aside from increased dopamine transport back into the presynaptic neuron.

Connecting the Science

1. Some forms of depression are thought to be caused by decreased amounts of the neurotransmitter serotonin. Does this mean that all depression has a biological cause? Why or why not?

2. Suppose that your friend is an engineering major. This friend finds it very difficult to study for the courses in his major because he finds them uninteresting. He talks to you about his idea that taking Ritalin might help him study. What advice would you have for your friend?

CHAPTER 22

Feeding the World

Plant Structure and Growth

Agricultural practices in the Midwest were partially responsible for enormous dust storms in the 1930s.

The damage to soil that occurred during this period caused many families to lose their farms and become homeless.

Similar dust storms have occurred more recently in China, carrying soil across the Pacific Ocean.

The Dust Bowl is now one of the most productive regions on Earth, thanks to modern technology. But is this production sustainable?

A*nd then the dispossessed were drawn west.... Car-loads, caravans, homeless and hungry; twenty thousand and fifty thousand and a hundred thousand and two hundred thousand. They streamed over the mountains, hungry and restless—restless as ants, scurrying to find work to do—to lift, to push, to pull, to pick, to cut—anything, any burden to bear, for food. The kids are hungry. We got no place to live.*

This passage from John Steinbeck's classic novel, *The Grapes of Wrath*, describes the movement of people from the southern Great Plains of the United States to California during the period known as the Dust Bowl. From 1931 to 1936, a combination of intense drought, severe storms, and inappropriate farming practices in the shortgrass prairies of the Great Plains led to crop failures and widespread soil loss. Impoverished residents of the panhandles of Oklahoma and Texas faced a desperate choice of waiting out the bad times or pulling up stakes and leaving their homes. Nearly 25% of the population in these regions left over a period of just 3 or 4 years.

The farming practices that helped create the Dust Bowl were initially heralded as great improvements. Slow horse-drawn plows and harvesters were replaced with gasoline-powered tractors and combines, allowing more and more of the native prairie to be turned into wheat fields—but plowing prairie grasses destroyed the root systems that held the soil in place. Once these grasses were gone, soil could be picked up by fierce midwestern winds and blown thousands of miles in tremendous dust storms. Aggressive efforts to reverse soil erosion, combined with the return of normal rainfall levels, finally ended the Dust Bowl era. Thanks to the advances of modern agriculture, including the use of fertilizers and irrigation, this former disaster area is now part of the most productive corn- and wheat-producing region on Earth. But for how long?

Fast forward to March 2001, when a giant dust cloud born on the northern plains of China was blown thousands of miles across the Pacific Ocean and obscured the view of the Rocky Mountains from downtown Denver, Colorado. What created this dust cloud? Drought, severe storms, and inappropriate agricultural practices were the causes—a familiar story. Hundreds of thousands of people are moving away from this Asian Dust Bowl to cities along China's Pacific coast. Similar migrations have occurred in west-central Africa, parts of the Middle East, southern Russia, and elsewhere around the world as agriculture expands to ever-larger areas of Earth's surface. The mistakes of the Dust Bowl era are being repeated again and again.

Is the inevitable cost of agriculture massive environmental damage? Can newly destroyed regions recover in the way that areas depleted in the American Dust Bowl did? Or is the Dust Bowl recovery an illusion maintained by huge inputs of nonrenewable resources that will eventually run out? To answer these questions, we must first understand the objectives and methods of agricultural practice—beginning by understanding the source of our crops and what these plants need to produce abundant food.

22.1   Plants as Food

For the first 240,000 years of modern humans' existence, our ancestors survived by hunting game and gathering fruits, seeds, roots, and other edible plant parts. Beginning about 11,000 years ago, human groups in what is now modern Iraq, Mexico, Peru, China, New Guinea, and elsewhere independently began to practice **agriculture**—that is, to plant seeds, cultivate the soil, and harvest crops. The development of agriculture may have been triggered by climate change, the needs of growing human populations, and advances in human knowledge about the requirements for plant growth.

The Evolution of Agriculture: Food Plant Diversity

Nearly all agricultural plants belong to a single phylum of plants—the **angiosperms**, or flowering plants. Certain angiosperms that were suitable for cultivation may have become more common around the time that agriculture developed due to favorable changes in climate and increased human impacts on natural ecosystems.

Clues about past climate conditions can be gathered from a variety of sources, including plant remains trapped in sediments, the size of annual rings in ancient trees and fossil wood, and air bubbles trapped in 400,000-year-old ice sheets. According to this record, Earth's climate changed dramatically beginning about 11,000 years ago from a cool and wet period to a warmer, drier condition. The new climate also included more dramatic seasonality—wet seasons alternating with dry seasons.

Angiosperms that can grow from seed to flower and then produce fruit and new seeds in the course of a single season apparently were very successful in these new conditions. These **annual plants** are well suited to environmental variability, whether in climate or resulting from disturbance of the soil. Annuals are also quite suitable for agriculture, providing a rapidly produced, abundant, and easily transported food source. Nearly all of the major crops developed during the origins of agriculture are annual plants.

The annuals that provided the majority of calories in the diets of most early agricultural societies were weedy; that is, they were well suited to growing on open and disturbed land. Increasing human populations led to more disturbed areas and thus more habitat for weedy species. In most human settlements, the hardiest weeds were from the same two plant families: Poaceae, grasses in the class known as **monocots**, and Fabaceae, peas from the class called eudicots, but more commonly known as **dicots**. Monocots and dicots get their names from the number of **cotyledons**, or embryonic leaves, they produce in their seeds (Figure 22.1). The seeds of grasses (known in agriculture as **cereal grains**) provide a rich source of carbohydrates and 80% of the calories in a typical diet. On the other hand, pea seeds (and other members of the Fabaceae, commonly called **legumes**) are rich in proteins, especially in certain amino acids that are essential to our survival but are produced only in low concentrations in cereal grains. Thus, these two groups of plants together provide a reasonably balanced diet. In the Middle East, the grain-legume combinations are wheat or barley with lentils or peas; in China, it is rice and soybeans; and in Mexico, corn and kidney beans.

Humans who realized that nutritious and highly productive wild grasses and peas were colonizing "waste areas" could deliberately have created additional waste areas or scattered leftover seed to encourage the plants' growth. Over time, the wild weeds became **domesticated** so that their growth and reproduction came under human control. This process of artificial selection produced crop plants that cannot survive without human intervention—and human societies that cannot survive without crop plants (see Figure 22.2 on page 574). Grains and peas are not the only domesticated plants, of course, but they are among the most important. The **staple crops**, or major sources of calories, in Europe, Asia, and Central America are wheat, rice, and corn, respectively. But in other places, people get most of their calories from different crops—primarily sweet potatoes, white potatoes, and

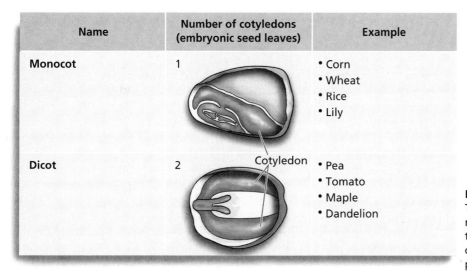

Name	Number of cotyledons (embryonic seed leaves)	Example
Monocot	1	• Corn • Wheat • Rice • Lily
Dicot	2	• Pea • Tomato • Maple • Dandelion

Figure 22.1 Monocot and dicot seeds. The two major classes of angiosperms are named for the number of cotyledons contained in the seed. Monocots like corn contain one cotyledon, while dicots like peas contain two.

(a) Teosinte, ancestor of modern corn   (b) Modern corn

Figure 22.2   The effect of domestication. Artificial selection of wild plants over thousands of generations has led to agricultural crop plants that are unable to reproduce without human intervention.
Teosinte, the ancestor of corn, produces a few small, loosely held kernels in a fragile "ear" that
disintegrates when the kernels are ripe. In contrast, corn's kernels are large, numerous, and do not
detach from the cob.

manioc (also known as cassava or tapioca). While these three crops are also
angiosperms, their food value comes not from their fruit or seeds but from
other plant organs.

Plant Structure

Although fruits and seeds are often packed with calories, all plant parts are
potentially edible. Angiosperm plants have a relatively simple structure, consisting of three nonreproductive or **vegetative organs** and two **reproductive
organs**. The three vegetative organs are the roots, stems, and leaves; the reproductive organs are flowers and fruit (Figure 22.3). These organs are made
up of only four types of tissue: epidermal, vascular, ground, and meristematic.
The relative amounts of these types of tissue in a vegetative organ often determine its nutritional value.

Roots.   The primary function of a plant's **root system** is to absorb water
and minerals from the substrate in which the plant is growing. The roots
of most monocots are fibrous and spread out broadly in the soil, giving
these plants the ability to exploit water as soon as it becomes available
in the soil. In contrast, the single large taproot of most dicots permits
these plants to access minerals and water found much deeper in the soil
(Figure 22.4). Most of the water and minerals obtained by a plant are
absorbed across the surface of **root hairs**, tiny projections on the cells
covering the surface of the root near the root tip (see Figure 22.5 on
page 576).

Root hairs are modifications of the plant's **epidermal tissue**, the tissue system that serves as an outer covering of the plant body. Water entering the root
hairs is pulled into cells in the **vascular tissue**, which is made up of **xylem**, the
water-conducting tissue; and **phloem**, the tissue that conducts food throughout the plant. Roots that are modified for food storage, such as sweet potatoes

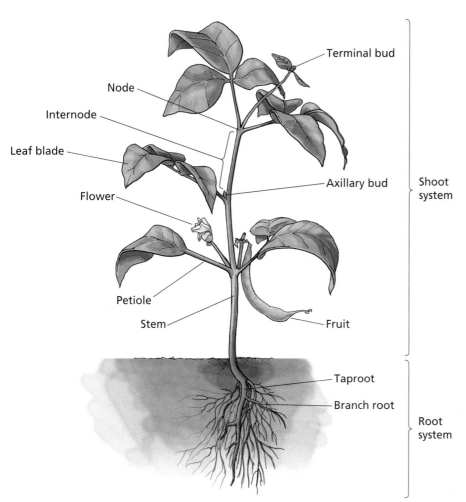

Figure 22.3 Plant structure. Plants consist of five organs: the vegetative structures—roots, stems, and leaves—and the reproductive structures—flowers and fruit.

and manioc, contain carbohydrates brought via the phloem and stored in the **ground tissue**, which makes up the remainder of the plant body. Ground tissue has a variety of functions—from storage in roots, to support in stems, to metabolism in leaves. Roots extend through the soil by growing at the root tip made up of **meristematic tissue**, which has cell division as its primary

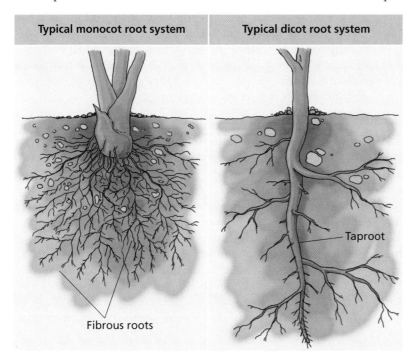

Figure 22.4 Monocot and dicot root systems. Most monocots produce a fibrous root system that competes well for water but is not a strong support structure; most dicots produce a much sturdier, deeper taproot.

Figure 22.5   Root hairs. The growing ends of a plant's roots are covered with tiny, hairlike extensions of the epidermal cells. These extensions greatly increase the surface area of the root, across which water and nutrients are absorbed.

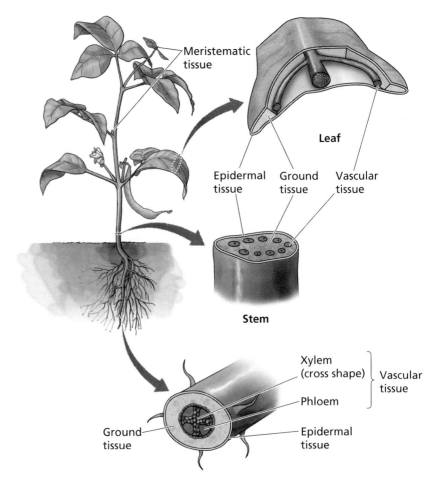

Figure 22.6   Plant tissue systems. Plants possess four tissue types—epidermal, ground, vascular, and meristematic.

Figure 22.7   Manioc. The storage roots of this plant, also known as cassava, are a staple in the diet of people throughout Brazil and elsewhere in the tropics. The roots contain large amounts of a cyanide-producing acid and must be extensively processed to make them edible.

function. Figure 22.6 provides an overview of the tissue systems in each of a plant's vegetative organs.

Many plants with storage roots are not annuals but **perennials** that live for many years. Perennials survive difficult environmental conditions by storing food in good times and using the stored food to produce new stems and leaves when the good times return. Perennials grown as crops, such as sweet potatoes and manioc, do not use their stored carbohydrates to produce a new season's worth of growth; instead, they are harvested when the roots contain sufficient sugar or starch for human use (Figure 22.7).

Stems. The principal role of the **stem** is to serve as a support structure for the other organs of the **shoot system**. At the apex or top of the stem is the **terminal bud**, which encloses the growing tip, or **apical meristem**, of the plant. The **nodes** of the stem, the point where leaves are attached, are separated from each other by **internodes**. At the junction between the leaf and stem at each node is an **axillary bud**, an embryonic shoot containing meristematic tissue. Most axillary buds have the ability to grow into a new lateral node-internode system (that is, a vegetative branch), although some will become flowering branches that cannot branch further. Most stems contain significant vascular tissue, which is fibrous and of limited food value. However, like roots, stems can be modified for food storage, and some storage stems have become staple crops. White potatoes and yams are **tubers**, enlarged structures found on underground stems called **rhizomes**. The "eyes" on a potato are axillary buds, each with the ability to sprout into a stem containing nodes and internodes (Figure 22.8).

Leaves.   The "food factories" of a plant are its **leaves**, the primary site for photosynthesis. While no staple crops are leaves, leafy greens including lettuce, spinach, chard, kale, and bok choy are rich in essential vitamins and minerals and are important components of diets in many parts of the world. The leaf part that is more commonly consumed is the flat, photosynthetic **blade**, rather than the stiff and fibrous **petiole**, which consists primarily of vascular tissue. Humans have also domesticated plants that store food in leaves. Onions and other similar plants produce **bulbs**, underground storage structures that consist of short stems and fleshy, starch-filled leaves.

Humans have domesticated crops based on all imaginable plant parts (Figure 22.9). However, the vast majority of our foods are based on reproductive organs, primarily fruits and seeds.

Plant Reproduction

We may not think of our breakfast toast and cornflakes as belonging to the same category of foods as our orange juice or bananas, but all of these foods are products of **fruit**, defined by botanists as mature flower ovaries containing seeds. Flowers and fruit are structures for sexual reproduction that are unique to angiosperms. In Essay 22.1 (on pages 578-579), we explore how plants that do not produce flowers reproduce.

Figure 22.8   A modified stem. A tuber is a modified stem, as the presence of sprouting buds on these potatoes indicates. These tubers were produced on another stem modified for underground life, a rhizome.

Seed   Fruit   Leaf   Tuber (modified stem)

Flower   Root   Endosperm (of seed)   Bulb (modified stem)

Figure 22.9   Plants as food. Our diet includes a diversity of plant parts, including fruits, seeds, roots, and flowers.

Essay 22.1 Focus on Evolution: Plant Reproduction

Land plants have a unique life cycle compared to that for animals. Instead of one multicellular "adult" form, plants have two, which alternate with each other; in fact, the life cycle of plants is commonly referred to as **alternation of generations**. The two life-forms result because the product of meiosis in plants is not a gamete; it is a haploid structure called a **spore**, which undergoes mitosis to produce a multicellular plant called the **gametophyte** (or gamete-producing plant). The gametophyte produces sperm or eggs by mitosis. When gametes fuse, they produce a diploid embryo that divides and eventually produces spores by meiosis; thus it is the **sporophyte**. The generalized life cycle is illustrated in Figure E22.1a. The trend in land plant evolution has been that the gametophyte generation has become more and more reduced, while the sporophyte has become more dominant.

In mosses and other nonvascular plants, the gametophyte generation is dominant. Sperm and eggs are produced in structures at the tips of the plant, and sperm swim in a film of water, such as a drop of rain, to reach the eggs. The embryo develops in the female reproductive organ, eventually producing a sporophyte that emerges from the tip of the gametophyte. Tiny spores produced in the sporophyte are dispersed on the wind and germinate if they land in moist conditions (Figure E22.1b).

Figure E22.1b   Moss gametophytes (leafy green structures) with attached sporophytes.

Ferns and other seedless vascular plants have a dominant sporophyte. Spores are produced in structures on modified leaves (Figure E22.1c) and are dispersed by wind or water to germinate in warm, wet conditions. The small independent gametophyte that emerges may produce both sperm and eggs. Sperm swim on a film of water to reach the egg, and the embryo grows out of the female reproductive structure.

Gymnosperms such as the pine in Figure E22.1d produce tiny, immature male gametophytes called pollen. Pollen is carried by the wind (or occasionally insects) to the female reproductive structures, which are often produced in persistent cones. Here the pollen undergoes further development and releases sperm to fertilize the egg. The egg is produced inside the female gametophyte, which is retained in the parent sporophyte plant. The embryo begins development inside the gametophyte, and the embryo, gametophyte, and surrounding sporophyte tissue become the seed. The seed is the primary dispersal unit; the adult sporophyte develops when the seed has landed in an appropriate site for growth.

As described in this chapter, flowering plants have continued the trend in plant evolution toward a much reduced, dependent gametophyte. In these plants, the female gametophyte consists of only seven cells, a tiny structure inside the female part of the flower.

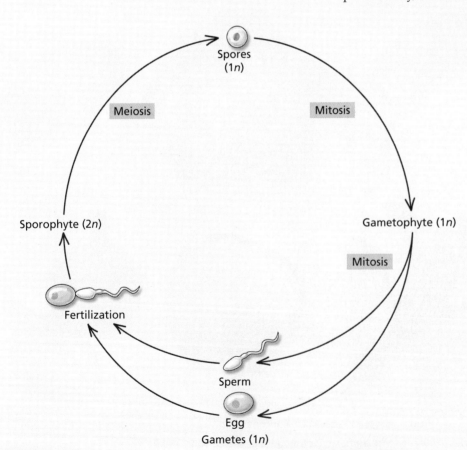

Figure E22.1a   Alternation of generations.

Figure E22.1c Fern sporophyte with spore-producing structures.

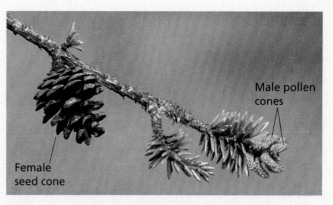

Figure E22.1d Gymnosperm sporophyte with cones containing male and female gametophytes.

Flowers are the sexual organs of angiosperms (Figure 22.10). The typical flower produces both male and female structures. The male gametes of flowering plants are contained in pollen. The pollen is produced in structures called **anthers** that are held up by the **filaments** of the male reproductive organ, called the **stamen**. Female gametes are contained in structures called **ovules** produced in a chamber, called the **ovary**, of a flower's female reproductive organ, the **carpel**. The transfer of pollen to the **stigma** of a carpel is called **pollination**; when pollen grains grow tubes down the length of the carpel's **style** to reach the ovules, sperm are released from the pollen, and fertilization occurs.

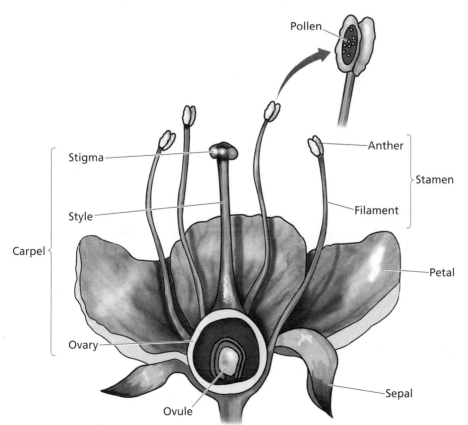

Figure 22.10 Flower structure. The typical flower contains both male and female reproductive organs, although many plant species produce two types of flowers: male-only and female-only.

Pollination may occur via wind, as in corn and wheat, or via insects or other animals, as in soybeans. A flower's **petals** are adapted to facilitate pollination; in the case of wind-pollinated grass, the petals are much reduced, whereas in insect-pollinated legumes, they are colorful and contain nectar to attract and reward their bee pollinators (Figure 22.11a–b). Many plants have very specific relationships with their pollinators, such that they are pollinated only by a single species or group of species. These relationships evolved as pollinators selected plants that gave them the best rewards, and plants developed traits that were highly attractive to particular pollinators (Figure 22.11c–e). Recall that Chapter 14 explores humans' dependence on these plant-pollinator relationships. A developing flower is enclosed in protective, modified leaves called **sepals** that in some plants function similarly to petals by facilitating pollination.

When a pollen tube reaches an ovule, it releases two sperm. In a process known as **double fertilization**, one sperm fertilizes the egg to become the embryo and the other fuses with two nuclei within the ovule to produce a cell containing 3 copies of each chromosome. This triploid cell divides and develops into **endosperm**, tissue that nourishes the developing embryo. The primary component of wheat flour is the starchy endosperm of wheat seeds;

(a) Wind-pollinated flower

(b) Bee-pollinated flower

(c) Bird-pollinated flower

(d) Bat-pollinated flower

(e) Fly-pollinated flower

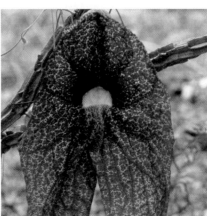

Figure 22.11   Petals and pollination. A flower's petals are adapted to aid in the process of pollination. (a) The most successful wind-pollinated plants have reduced petals that do not interfere with pollen transfer. (b) Bee-pollinated flowers often provide a nectar reward at the base of a petal tube, while (c) bird-pollinated flowers often have petals fused into a single deep, highly visible floral tube. (d) Flowers that are pollinated by bats usually open their petals only at night. (e) The petals of fly-pollinated flowers often have the appearance—and odor—of rotting flesh.

(a) Wind-dispersed fruit   **(b)** Water-dispersed fruit   **(c)** Animal-dispersed hitchhiker fruit   **(d)** Animal-dispersed edible fruit

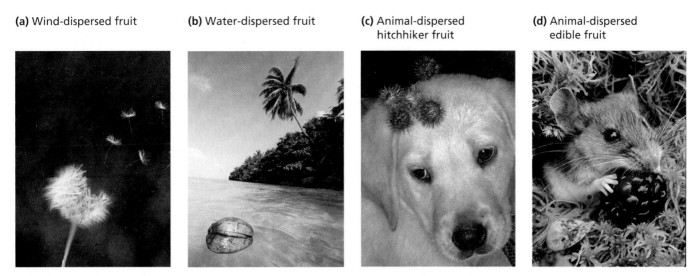

Figure 22.12   Seed dispersal units. Fruits are adapted to disperse seeds away from the parent plant. Many mechanisms are possible, including (a) wind, (b) water, (c) mammal fur, and (d) through an animal's digestive tract.

whole wheat flour also contains the embryo (the germ) and accessory tissues (the bran), whereas white flour is strictly the endosperm. In some species, such as in legumes, the embryo consumes the endosperm as it develops and the seed matures.

The embryo, endosperm, and remainder of the ovule tissues make up the **seed**. Triggered by successful fertilization, other parts of the flower—in most cases, the ovary—develop into a fruit. Most fruits are adapted to be vehicles for dispersing seeds far from the parent plant. Some fruits help seeds disperse on the wind or water; other fruits help seeds disperse by using animals to carry the fruit away and drop its seeds elsewhere (Figure 22.12). Many of the crop plants that we rely on were selected from fruits adapted to attract mammal dispersers; these plants include even grains, which are attractive to rodents. Figure 22.13 illustrates the relationship between flowers, seeds, and fruit in angiosperms.

① Flower petals facilitate pollination, which leads to fertilization.

② The products of fertilization are an embryo and endosperm, a food source. Both are contained in the seed.

③ Fruits package seed in a structure that aids their dispersal, such as a tasty fruit or a parachute.

Stamen produces pollen (containing sperm).

Carpel contains eggs.

Seed coat

Embryo

Endosperm

Seed

Figure 22.13   A summary of angiosperm reproduction.

Producing abundant fruit for human consumption requires more than simply selecting plants for optimal fruit production. The agricultural practices of people around the world are designed to maximize the growth of these plants.

22.2   Plant Growth Requirements

Agricultural practices such as turning over soil, irrigating soil with water, and adding nutrients to soil developed independently everywhere that plants were domesticated by people. To understand why these techniques enhance agricultural production, we must understand first how plants grow.

How Plants Grow

The first stage of plant growth is the emergence of the seedling, supported by its internal food stores. Later plant growth can be divided into two types—growth in length, called **primary growth**, and growth in girth or circumference, called **secondary growth**.

Germination.   The emergence of an embryo from a seed is called **germination**. The process begins when the seed begins to take up water, expands, and ruptures its confining outer capsule. This outer covering is called the **seed coat**; as long as the seed coat is impermeable to water, the seed remains **dormant** and will not germinate. Dormancy allows seeds to act as "time travelers," passing through poor environmental conditions in a state of suspended animation. Many seeds require exposure to specific environmental conditions before they can escape dormancy. For example, in temperate regions with cold winters, some seeds need to be exposed to low temperatures for a certain amount of time before they can germinate. This requirement ensures that the young plant will emerge only when temperatures are warming and not during a brief midwinter thaw.

The first structure to emerge from a germinating seed is the root, which enables the developing young plant, or **seedling**, to become anchored in the soil and absorb water. Growth of the shoot soon follows, as the seed's stored food is depleted and photosynthesis becomes necessary (Figure 22.14). The time between germination and when the seedling is established as an independent organism is the most vulnerable period in the life of a plant. Damage to the plant during this period is usually fatal. Once a seedling becomes established, however, damage may be overcome by continued growth.

(a)   (b)   (c)   (d)

Figure 22.14   Germination. (a) Seed germination begins when the seed takes up water. (b) The first organ to emerge is the root. (c) Water obtained by the root makes the cells in the embryonic shoot enlarge, allowing the shoot to emerge from the soil. (d) Germination is complete when the seedling is functioning independently of the stored food in the seed.

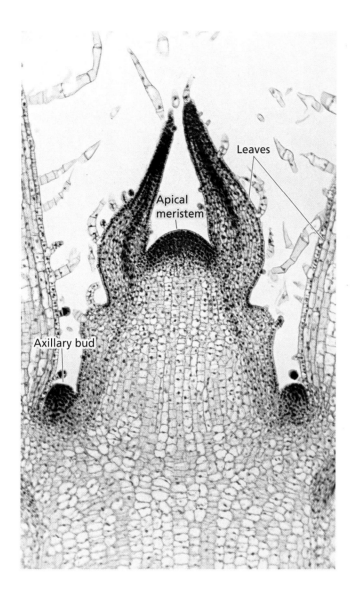

Figure 22.15   Primary growth in stems. In this microscope image of an apical meristem, the repeating node-internode system can be readily seen. Every shoot's apical meristem produces additional meristems at each node in the axillary bud.

Leaves

Apical meristem

Axillary bud

Primary Growth.   In plant stems, primary growth takes place through the production of additional nodes and internodes by growth at the shoot tips and by the growth of axillary buds. Growth in length occurs when cells at a meristem divide in a single plane, creating a column of daughter cells below them and pushing the apical meristem farther up into the air (Figure 22.15). A hormone produced by the terminal bud inhibits cell division in the meristems of the axillary buds. Once the terminal apical meristem has been pushed by growth away from a particular axillary bud, its meristem begins to divide and produces a lateral branch, also containing axillary buds. The node-internode system of the main stem and branches supports the **indeterminate growth** of a plant's shoot system—its ability to grow throughout its lifetime. This is in contrast to the **determinate growth** of most animals, which grow to a set size that they typically reach early in life.

In roots, primary growth at the tip occurs due to cell division in the apical meristem. Unlike the shoot apical meristem, cell division in the root occurs on two planes, resulting in a column of cells that elongate and push the root tip forward through the soil, as well as a loosely organized **root cap** that covers the tip of the root. As the root tip pushes through the soil, exterior cells of the root cap are continually worn off and replaced by younger cells more recently produced by the apical meristem. The primary meristematic tissue producing lateral roots is found near the center of the root, surrounding the vascular tissue. As a result, lateral roots can form anywhere along the length of a root (see Figure 22.16 on page 584).

Figure 22.16 **Lateral root formation.** Lateral roots form when meristematic tissue near the center of the root begins to divide. The origin from the central cylinder of vascular tissue ensures that water obtained by these lateral roots will be carried into the main body of the plant.

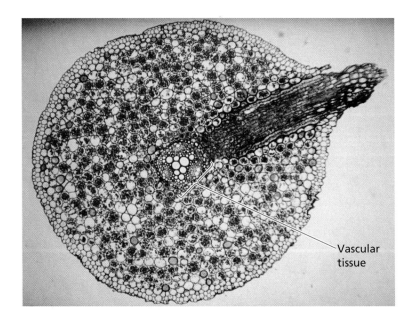

Vascular tissue

Primary growth results in plants that are taller, bushier, and have more extensive root sytems. In longer-lived plants, continued primary growth presents a dilemma—a tall plant cannot continue to support itself against gravity with a skinny stem. Many plants are able to increase their stability by making wider stems and roots via secondary growth.

Secondary Growth. All of the flowering plants that can undergo true secondary growth are dicots. In these plants, a ring of cells running through the bundles of vascular tissue in the stem becomes a meristem called the **vascular cambium** (Figure 22.17). This cambium produces phloem cells to the outside of the stem and xylem cells to the inside. The additional xylem cells make up the bulk of the material in these stems, and their stiffened cell walls produce the stem's **wood**. Epidermis on the surface of the stem breaks apart and is shed as the stem's surface area increases. Some of these cells become another meristem (called the **cork cambium**) before they are lost. Cell division in the cork cambium results in the production of **cork**, a water-

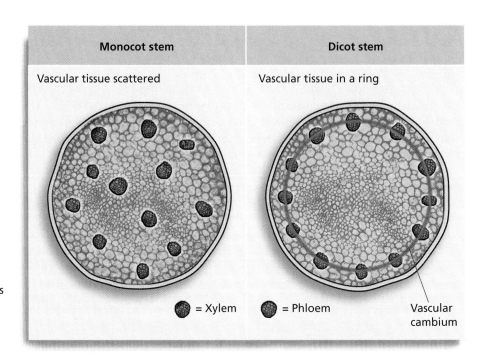

Monocot stem	Dicot stem
Vascular tissue scattered	Vascular tissue in a ring

= Xylem   = Phloem

Vascular cambium

Figure 22.17 **Monocot and dicot stems.** Unlike monocot stems, which have scattered bundles of vascular tissue, dicot stems have vascular bundles arranged in a ring. This organization allows a single ring of meristematic tissue, the vascular cambium, to form between the xylem and phloem.

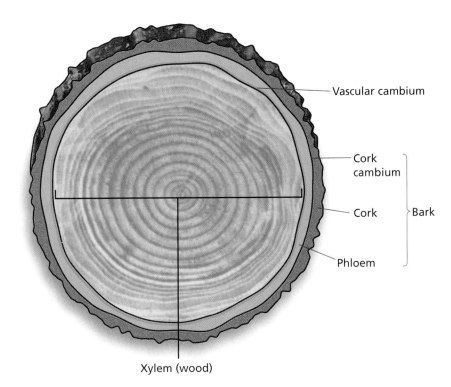

Vascular cambium

Cork cambium

Cork — Bark

Phloem

Xylem (wood)

Figure 22.18   Dicot wood. The wood of a dicot stem is made up of xylem cells, while the bark consists of phloem and cork.

resistant outer covering. Phloem, cork, and cork cambium make up the **bark** of a woody dicot stem. All the components of bark are continually shed and replenished by secondary phloem cells, which are produced by the vascular cambium as the tree increases in diameter (Figure 22.18). Interestingly, although dicot trees are much more common than monocot trees, the only tree-based staple food is the monocot coconut. The thick stem of coconut palms consists of additional strands of vascular tissue, produced by a **thickening meristem** that surrounds the apical meristem. The structure of this stem makes palm trees incredibly flexible (Figure 22.19).

Primary and secondary growth of plants can be maintained only by abundant energy. Plants obtain this energy by transforming light into chemical energy in the process of photosynthesis. Our agricultural practices are intended to maximize the photosynthesis of crop plants so that they will produce a surplus of carbohydrates that we can harvest.

Maximizing Plant Growth: Water, Nutrients, and Pest Control

Photosynthesis is described in detail in Chapter 4; but for the purposes of this chapter, it is sufficient to understand the summary equation for this reaction:

$$\text{carbon dioxide} + \text{water} + \text{light energy} \longrightarrow \text{carbohydrate} + \text{oxygen gas}$$

In other words, photosynthesis in plants converts energy from sunlight, carbon dioxide from the atmosphere, and water from the soil into energy-rich carbohydrate molecules. If we look only at this chemical reaction, it is clear that plants require carbon dioxide, water, and light to produce carbohydrates, but the process also requires a few more components. For instance, to survive, plants require much more water than that needed simply for the chemical reactions of photosynthesis. Plants also require other chemical inputs in smaller amounts. For example, photosynthesis requires the pigment **chlorophyll** to proceed, and magnesium is an essential element in chlorophyll. Inputs such as magnesium and nitrogen, an essential element in proteins, are obtained from the soil. Additionally, animals that consume plant organs as well as bacterial, fungal, and protistan pathogens,

(a)

(b)

Figure 22.19   Monocot "wood." Secondary growth in monocots, such as this palm, produces fibrous wood (a), which is not nearly as strong as dicot wood but is incredibly flexible (b).

Figure 22.20 Plant growth requirements. Plants require the basic ingredients for photosynthesis—water, carbon dioxide, and light—as well as nutrients from the soil and freedom from damage to produce excess carbohydrates for human consumption.

Plant growth requirements:

① Solar energy (sunlight)

② Carbon dioxide

③ Freedom from damage

④ Water

⑤ Inorganic nutrients

reduce the production of crop plants by damaging photosynthetic tissue or consuming fruits and seeds before harvest. Therefore, to increase crop production, modern farming practices attempt to maximize the amount of light, water, carbon dioxide, and nutrients available to crop plants as well as to minimize damage caused by plant predators (Figure 22.20). The primary concern of farmers in almost any environment is that their crops receive adequate water.

Water and Sunlight. To hold its leaves perpendicular to the sun's rays, a plant's cells must be full of water; that is, they must be **turgid**. Recall that the membrane of a plant cell is surrounded by a stiff cell wall. The cell wall is tough, but it is also elastic, meaning that it can stretch and still hold its basic shape. As a plant cell's central vacuole fills with water, the cell wall balloons slightly, becoming turgid. It is the pressure of adjoining turgid cells that holds the leaves upright (Figure 22.21a). When cells lose water, the pressure of cells pushing against each other decreases; the cells become flaccid, and the plant wilts (Figure 22.21b). Thus, part of maximizing a plant's exposure to light is ensuring that it has plenty of water.

To reduce the chance that their crops will wilt during hot, dry conditions, farmers try to provide the plants with an optimal amount of water. In many

Turgid cells   **(a)** Hydrated plant   **(b)** Wilted plant   **Flaccid cells**

Cell wall
Cell membrane
Vacuole filled with water

Cell wall
Cell membrane
Vacuole has lost water.

Figure 22.21   Water in plant cells. (a) When a plant is fully hydrated, the cell's vacuoles balloon with water, and the cells press against each other so that the plant remains "crisp." (b) As the plant dries out, the vacuoles lose water, and the cells no longer support each other, causing the plant to wilt.

regions, adding water through **irrigation** is necessary. The enormous production of corn, wheat, and soybeans in the former Dust Bowl region is a result of an equally enormous amount of irrigation. The water for irrigation in this dry region comes from a huge underground reservoir called the Ogallala aquifer (Figure 22.22).

Another way to increase the amount of water that reaches preferred plants is to reduce the amount of competition that the plants face. Farmers do this by trying to minimize the number of non-preferred plants, commonly called

(a) Irrigation

(b) Water source

South Dakota

OGALLALA AQUIFER

Nebraska

Kansas

Colorado

New Mexico

Oklahoma

Texas

Figure 22.22   Irrigation. (a) The bright green circles in this aerial photo are formed by lush plant growth. The circle pattern derives from center-pivot irrigation sprayers, which spray water in a complete arc from the center point. (b) The Ogallala aquifer is the largest single pool of freshwater in the world; it underlies and supplies the "bread basket" region of the United States.

weeds, in agricultural fields. Any plant can be a weed if it grows in the wrong place—just ask a gardener who maintains a turf lawn and must continually pull turf grass out of the flower bed. Minimizing the number of weeds growing with a crop has the additional benefit of minimizing the chance that a weed will shade the preferred plants.

Controlling Weeds.   Farmers have several techniques for controlling weeds. One is to remove competitors before the crop is planted. This process, called **tilling**, involves turning over the soil to kill weeds that have sprouted from seed. After a crop begins growing, it is possible to prevent the growth of competitors as well as reduce water loss from the soil by **mulching** the crop. To mulch, farmers spread a thick layer of straw, other dead plant material, sheets of newspaper, or dark plastic around the base of preferred plants. The mulch blocks sunlight from reaching the soil, preventing weed seeds from germinating.

Although tilling sets back the growth of weeds, it cannot kill all of them, and some spring back to life quickly. In these cases, farmers may use an **herbicide**, a chemical that kills plants. Herbicides that exploit chemical differences between dicot and monocot plants are the most commonly used sprays in agriculture. These herbicides are used to control dicot weeds in monocot crops, such as corn and wheat. Broad-spectrum herbicides, such as glyphosate (Roundup®), kill all plants; until recently, these herbicides have been useful only for controlling weeds before planting. Development of genetically modified organisms (GMOs) has increased the applicability of broad-spectrum herbicides to many more crop-and-weed combinations in the past decade, as described in Chapter 8.

Even when plants are fully hydrated and not wilting, lack of water can interfere with growth. This is because plant strategies for conserving water can impose a cost to photosynthetic production.

Water and Carbon Dioxide.   The primary adaptation that plants have to reduce water loss is a waxy covering, called the **cuticle**, on the epidermis of the shoot system. The cuticle reduces water loss, but it also restricts the flow of carbon dioxide into the plant. To enable plant cells to obtain adequate carbon dioxide for photosynthesis, the shoot epidermis is pocked with tiny pores called **stomata** (singular: stoma). However, this porosity comes with a price—stomata also provide portals through which water can escape.

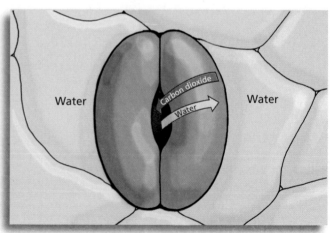

When water is abundant:

Water   Carbon dioxide   Water
                          Water

- Guard cells swell.
- Stoma (pore) is large.
- High carbon dioxide uptake
- High water loss

When water is scarce:

Water   Carbon dioxide   Water
                        Water

- Guard cells shrink.
- Stoma (pore) is small.
- Low carbon dioxide uptake
- Low water loss

Figure 22.23   Guard cells. Guard cells help to regulate the size of stomata on a plant's surface, thus minimizing water loss and maximizing carbon dioxide intake.

Table 22.1   Plant nutrients and their functions.

Macronutrient	Primary functions	Micronutrient	Primary functions
Nitrogen	Component of proteins, DNA, and RNA	Chlorine	Involved in moving water into and out of cells
Potassium	Involved in opening and closing of stomata	Iron	Required for chlorophyll synthesis and aerobic respiration
Calcium	Component of cell walls, involved in cell membrane permeability	Boron	Influences calcium use; required for DNA and RNA synthesis
Magnesium	Component of chlorophyll	Manganese	Required for integrity of chloroplast membrane
Phosphorus	Component of ATP, DNA and RNA, cell membranes	Zinc	Component of many enzymes
Sulfur	Component of proteins	Copper	Component of some enzymes
		Nickel	Essential for nitrogen metabolism
		Molybdenum	Required for nitrogen metabolism

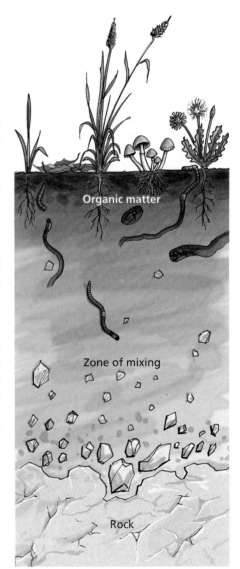

In most plants, each stoma is encircled by a pair of **guard cells** that regulate the size of the pore (Figure 22.23). When the guard cells have abundant water, they increase in size, and their unique shape causes the pore to open. When the guard cells are water deprived, they shrink and relax, causing the pore to be tiny. Because of their changing shape, guard cells help minimize water loss under dry conditions and maximize carbon dioxide uptake under wet conditions. Therefore, maximizing the carbon dioxide uptake of crop plants requires the same input as maximizing light exposure—abundant water.

Abundant water is crucial for adequate plant growth. However, where precipitation or irrigation is sufficient, the most important factor affecting plant growth is nutrient availability.

Nutrients and Soil.   Elements needed by plants in large amounts, such as nitrogen and magnesium, are called **macronutrients**. Nutrients needed in much smaller amounts—such as iron, which is required in the process of chlorophyll synthesis and aerobic respiration—are known as **micronutrients** (Table 22.1). Both categories of nutrients are available in the soil.

Soil is a unique material, composed of both living and nonliving components. Healthy soil consists of a base layer of recently eroded rock; a top layer made up of the dead and decaying remains of plants, animals, and their wastes along with the organisms that eat them; and a middle layer where these two components are mixed by earthworms and other soil organisms (Figure 22.24). One way to maximize a crop plant's access to soil nutrients is to reduce the access that weeds have to these nutrients. Thus, techniques that reduce competition for light, water, and carbon dioxide between crops and weeds also increase nutrient availability to crops.

Figure 22.24   Soil structure and development. Soil forms from the breakdown of rock and the accumulation of organic matter from dead and decaying plants, fungi, animals, and animal wastes. Organisms in the soil decompose the organic matter and "mix" the soil through their movements, making nutrients available for plant growth.

Plant protein

Fertilizer (usually inorganic) is added to soil to replace nitrogen.

Nitrogen in the form of plant protein is removed from system during harvest.

Nitrogen is removed from soil by growing plants and converted to protein.

Excess nitrogen in soil washes into waterways.

Figure 22.25   A broken nutrient cycle. The movement of nitrogen in an agricultural system is illustrated here. Notice that nutrients are not recycled to the soil in an agricultural system; they are lost because plants are removed from the system. Fertilizer is added to the soil to make up the loss.

In natural systems, nutrients become available to plants through nutrient cycling, as described in Chapter 15. In agricultural systems, the nutrient cycle is often broken. Instead of plants being eaten by animals, which are then eaten by other animals and decompose all in the same location, the plants from agricultural fields are removed and trucked to other locations where they will be consumed and their nutrients released. Most of the nutrients that are consumed end up in human or animal waste, which often flows into lakes or oceans. On croplands, nutrients are not recycled back into the soil but mined from it (Figure 22.25). Because the nutrient cycle is broken, farmers must add nutrients to agricultural soils in the form of **fertilizer**. Fertilizer that is applied to the soil can be organic—that is, carbon-containing molecules made up of the partially decomposed waste products of plants and animals. Fertilizer can also be inorganic—meaning simple molecules that lack carbon, such as ammonia and nitrate, produced by an industrial process. Because plants require nutrients in inorganic form, soil organisms must first decompose organic fertilizer before its nutrients are available for plant growth. Most farmers prefer inorganic fertilizer not only because the nutrients in it are more concentrated and much easier to transport, store, and apply than most organic fertilizers but also because plants can utilize these nutrients immediately. Although the vast majority of farmers in the United States fertilize with inorganic nitrogen, they also may apply large amounts of inorganic phosphorus and potassium, and occasionally other macronutrients such as calcium.

Farmers can also replace some nutrients in the soil through **crop rotation**—that is, by not putting the same crop in the same field year after year. Surprisingly, some plants actually increase the levels of nutrients in the soil. Plants that can add nitrogen to the soil do this by having a mutualistic relationship with **nitrogen-fixing bacteria** (Figure 22.26). Nitrogen-fixing bacteria have the ability, unique among living organisms, to "fix" nitrogen gas from the atmosphere into a solid form that can be taken up by plant roots. Legumes develop structures called **nodules** on their roots in response to chemical signals from certain species of nitrogen-fixing bacteria. The nodules

Nodule

Figure 22.26 Nitrogen-fixing crop. (a) Planting alfalfa can increase the nitrogen content of soil. (b) The nodules on the roots of alfalfa contain nitrogen-fixing bacteria. The plant provides a home and food source for the bacteria, and the bacteria convert nitrogen from the air into a form that can be taken up by the plant.

house the bacteria and supply them with carbohydrates. In return, the plants receive the excess nitrogen that the bacteria fix. When a legume is growing well, the bacteria fix so much nitrogen that the excess is released into the surrounding soil. If legumes are planted in alternating years with a crop that does not fix nitrogen (such as corn or wheat), the need for nitrogen fertilizer can be greatly reduced. The benefits of crop rotation may be one reason that so many early agricultural societies relied on staples of grains and legumes.

Once a farmer has maximized the growth of his or her plants by providing abundant water and nutrients, these plants become attractive targets to other consumers—including other mammals, birds, insects, fungi, and bacteria. To ensure that most of the food produced by an agricultural crop goes to feed people, a farmer must try to reduce the damage these competitors do to the crop.

Freedom from Pest Damage. Living organisms that cause damage to agricultural products are generally referred to as **pests**. Up to 40% of the potential food production in major crops is lost to pests every year. The primary tools farmers use to reduce pest impact are **pesticides**, chemicals designed to kill or reduce the growth of a target pest. (Herbicides are often considered to be a class of pesticides. In this book however, we use the word *pesticide* only when referring to chemicals that control insect, fungal, and bacterial pests.) Pesticide use became widespread in the years following World War II. Today, millions of tons of these chemicals are applied to crops all over the world.

Surprisingly, pesticide use has not considerably reduced the overall loss of crops to pests. Instead, these chemicals have dramatically influenced farming methods—primarily by allowing farmers to plant a single crop over a wide acreage. This practice is called **monoculture**. Monocultural production of crops is very efficient. Farmers have one planting, fertilizing, and pest-control schedule. They require only the planting and harvesting equipment specific to their crop, and they can use huge mechanical harvesters to quickly collect

(a) Monoculture

(b) Polyculture

Figure 22.27   Monoculture and polyculture. (a) Planting the same crop over many hundreds of acres (monoculture) allows for the efficient use of massive planting and harvesting equipment. (b) Planting many different crops together (polyculture) can reduce pest problems and buffer farmers from crop failures.

enormous amounts of product (Figure 22.27a). Monoculture and the accompanying mechanization of agriculture have greatly reduced the amount of labor required to produce crops. The percentage of the population working on farms in the United States has dropped from 25% to less than 2% in the past 50 years due to this change in farming practice.

On the other hand, monocultures are very susceptible to outbreaks of pests. If a small population of insects or disease infestation consumes or kills a host plant, another appropriate host for the pest is right next door. The easy availability of resources to pests in monocultural crops allows their populations to grow exponentially and can rapidly lead to the loss of an entire year's production. Thus, while monoculture increases the efficiency of agricultural production, it results in an increased need for pesticides just to keep pest damage at historical levels.

Farmers can also use **cultural control** to minimize pest populations. Because most pests have evolved to attack a single crop, they will not cause damage to other, unrelated crops. Crop rotation, in addition to conserving soil nutrients, is a method of cultural control that moves plants away from their pests. After the growing season, a pest population spends the winter in the soil or on plant waste left in the field. If the same crop is planted the following year, the pest population has easy access to more resources and can continue to increase in population. However, if a different crop is planted, the pests must disperse in the spring in an attempt to find their required crop. Only some of these dispersers will find their host, and the population of pests in the new site will be relatively small.

Another form of cultural control is **polyculture**—planting many different crop plants over a single farm's acreage (Figure 22.27b). Unlike monocultures, polycultural planting keeps pest populations relatively small because, after the pests have consumed a patch of plants, they must disperse long distances to find another source of food. Polyculture also minimizes risk to farmers by ensuring that a pest outbreak on a single crop does not destroy all of the farm's production for that year.

Over the last 60 years, the use of pesticides, fertilizer, and large farm machinery has dramatically changed the way food is produced around the world. The revolution in biology—particularly the new science of genetics—has also played a role in this change.

Designing Better Plants: Hybrids and Genetic Engineering

Traditionally, farmers used artificial selection to modify the characteristics of their crop plants and animals; that is, they selected for next year's seed stock or breeding animals those that performed best under the conditions on their farm. However, this process limits farmers to select from only the variation that is currently available. Two major developments in plant science during the twentieth century reduced and finally eliminated this limitation.

Beginning in the 1920s, agricultural scientists began using plant breeding to produce better crops. Plant breeding generally consists of creating **hybrids**, the offspring of two different varieties of an agricultural crop. One of the first hybrids produced on a large scale was dwarf spring wheat. One parent variety was a nitrogen-loving, high-producing wheat that was difficult to harvest because it tended to fall over when the wheat grains were ripe. The other parent variety was a lower-producing dwarf plant. The hybrid of these two varieties was high producing but short, so it responded well to fertilizer application and was easy to harvest with large machines. Because hybrids are the first-generation offspring of two distinct varieties, new hybrid seeds can be produced only by allowing one variety to pollinate the other, a time- and resource-consuming process.

In addition to the difficulty of producing hybrid seeds, plant breeding limits agricultural scientists to the genetic variation present within a single species. More recently, agricultural scientists have applied DNA technology to produce a host of novel agricultural organisms containing genes from other species. Herbicide-resistant plants are genetically modified organisms (GMOs) as well as the pest-resistant corn and ripen-on-demand tomatoes described in Chapter 8. In the United States, over 40% of farmland planted with corn, more than 70% of the cotton crop, and nearly 80% of soybean fields are planted with GM varieties. However, as we noted in Chapter 8, many questions remain about GMOs, including their safety as human food, the likelihood that their widespread use will cause the evolution of resistance in pests, their effects on non-target organisms, and the consequences of the "escape" of engineered genes into non-crop plants.

The concerns related to genetically modified organisms should not be downplayed. However, in several ways, these concerns pale in comparison to those created by conventional agricultural practices, including environmental degradation and loss of soil. We address these environmental problems, as well as some proposed solutions, in the final section of the chapter.

22.3 The Future of Agriculture

The use of pesticides, herbicides, fertilizer, and irrigation has increased food production dramatically. However, many biologists argue that current high levels of production are **unsustainable**—they compromise the ability of future generations to grow enough to feed the human population. The future of agricultural production depends on finding ways to maximize food production while minimizing environmental damage.

Modern Agriculture Causes Environmental Damage

What evidence supports the hypothesis that current agricultural practices are unsustainable? First, the production and application of agricultural chemicals such as pesticides and fertilizers requires large amounts of non-renewable fossil fuels; second, the water sources used to provide irrigation water are limited and under increasing demand; and third, the environmental costs of fertilizer and pesticide use may be greater than the human population is able to bear.

Precious Liquids: Oil and Water.   Petroleum is required not only to run large pieces of farm equipment; it is also a raw material for the production of fertilizer and pesticides. While scientists and oil industry experts disagree about if and when the world's oil supply will run out, it is clear that demand for oil continues to increase. Dependence on the finite and ever more costly resource of oil makes current agricultural practices unsustainable.

Some environmental scientists argue that freshwater is the "new oil." Several statistics support this view. In the United States, 65% of total water withdrawals not used in power generation are used to irrigate crops. In western states, the portion of water use devoted to irrigation exceeds 90%. During the 1990s, it became clear to geologists studying the Ogallala aquifer (see Figure 22.22b on page 587) that irrigation in many of the regions above the aquifer was using water much more quickly than it was being replaced by precipitation. Analysis of the rate of water loss indicated that some areas of the aquifer would dry up within decades. As demands increase on limited supplies, water will no longer be essentially free, and an industry that collects, transports, and charges high fees for water may develop.

In addition to its reliance on nonrenewable or limited resources, such as oil and water, many of the processes and inputs of modern agriculture can cause serious environmental damage.

Fertilizer Pollution.   The minerals in inorganic fertilizer that are not taken up quickly by growing plants are carried off by water moving through the soil. This **runoff** accumulates in water reservoirs in lakes, streams, and oceans or deep underground in aquifers. When the nutrients from inorganic fertilizer reach high levels in the water, it becomes unsafe for human consumption. Increasing levels of nitrate fertilizer in water wells in the American Midwest correlate with increasing rates of miscarriage and bladder cancer among women in the surrounding communities. Some forms of nitrogen in drinking water can cause brain damage or death among infants who drink it.

Fertilizer runoff from farms can also cause eutrophication (described in Chapter 14) and result in extensive fish kills. If the use of fertilizer increases as human populations grow, the loss of drinking water sources and fishing grounds will probably also increase.

The Problem of Pesticides.   Over the last 50 years, the amount and toxicity of pesticides applied to crops in the United States has increased tenfold due to a phenomenon known as the pesticide treadmill. Pesticides directed toward a specific crop pest may lead to an increase in another pest whose populations were held in check by competition with the target pest. These infestations of secondary pests then result in the use of more pesticides. Additionally, the use of pesticides can lead to the evolution of pest resistance in the target pest population. If a few individuals in a target pest population are genetically resistant to a pesticide, the resistant individuals will survive pesticide application and may initiate the next outbreak. To control populations of secondary pests or now-resistant target pests, higher levels of pesticide or new, more toxic pesticides must be applied. Thus the term "**pesticide treadmill**," because farmers must continually expend energy and resources simply to "stay in place" and keep pest populations under control.

One side effect of the increase in pesticide toxicity and applications has been a steady increase in the number of reported incidents of wildlife poisonings, especially of birds. Although data collection is poor and trends can only be inferred, it appears that the number of wildlife poisonings has tripled since 1980. In addition, at low doses many pesticides may have sublethal effects that do not result in death but may significantly weaken individuals and make them more susceptible to predation, less able to compete, or even affect their endocrine system, as described in Chapter 20.

An additional problem associated with some types of pesticides is **biomagnification**, the concentration of toxic chemicals within higher levels of a food web. Biomagnification of chemicals that are not easily degraded or eliminated from organisms is a product of the trophic pyramid, the bottom-heavy relationship between the biomass (total weight) of populations at each level of a food chain, as described in Chapter 14. Thanks to the activism of ecologist Rachel Carson (Figure 22.28), the problem of biomagnification was recognized in the late 1960s—particularly the negative effects of accumulations of high concentrations of the pesticide DDT in fish-eating birds such as bald eagles. This recognition led to a ban on DDT application. However, many persistent organic pesticides are still used in the United States and elsewhere, and these can accumulate in any organism that is high on a food chain—including humans. Long-term health effects of low-level exposure to these chemicals are still mostly unknown, but some evidence suggests that these effects include increased risks of cancer, birth defects, and permanent nerve damage. The pesticide treadmill demands ever more powerful and toxic chemicals, and high pesticide levels that result from biomagnification are likely to negatively affect more and more of the human population.

Wise use of fertilizers and pesticides may help minimize the environmental effects of modern agriculture; the negative consequences fade when their use is reduced. However, one consequence of modern agriculture results in environmental damage that is not easily recovered—the loss of soil via erosion.

Figure 22.28   Rachel Carson. *Silent Spring*, Rachel Carson's most influential book, made well known the hypothesis that pesticides were killing birds and negatively affecting people. Her work led directly to a ban on DDT use in the United States.

Soil Erosion.   Soil is held together both by the organisms living within it and by the roots of the plants growing throughout it. Modern agricultural techniques destroy soil by regularly removing the plants (through tilling or harvest) as well as by causing the death of soil organisms. Inorganic fertilizer in the form of salts is very harmful to soil organisms. When soil loses its structure, it is easily picked up by strong winds and blown away—the Dust Bowl and the dust storms in China are dramatic examples of this type of **soil erosion** caused by agricultural practices. People in the northern plains of the United States still experience localized dust storms that result from soil blowing and drifting off of agricultural fields.

Irrigation water running over the surface of disturbed soil can also result in erosion; but in addition, irrigation can cause **salinization**—the "salting" of soil. Water from aquifers contains tiny amounts of mineral salts. When the water is applied to the soil, some evaporates, leaving the salts behind. Eventually, so much salt accumulates that the soil becomes completely infertile. The development of soil takes many years; soil lost through erosion and salinization is not easily replaced. Agricultural practices that degrade soil are unsustainable.

If we also include the uncertainty about how future climate change will affect agriculture, it seems clear that current agricultural practices will not be sufficient to feed a growing human population indefinitely. Fortunately, some alternatives to current agricultural practices may help provide enough food without as much environmental damage.

How to Reduce the Damage

As farmers and consumers begin to recognize the unsustainable nature of current agricultural practice, efforts are growing to craft a more **sustainable** system—one that meets the needs of the current generation of humans without compromising the ability of future generations to meet their needs. The most obvious strategy for approaching sustainability is for farmers to change some of their methods of growing food. However, another strategy is for consumers to change their eating and food-buying habits.

Reducing Harmful Inputs. Many of the techniques that farmers can use to reduce the environmental costs of agriculture have already been mentioned. Minimizing monocultural plantings may be the single most effective strategy. Not only will planting polycultures reduce the need for pesticides, it will help minimize fertilizer inputs and preserve soil by increasing crop rotation. Planting polycultures also helps to insulate farmers from environmental factors that destroy a single crop. Moving away from monocultural production will not be an easy task; it would probably need to be encouraged by changes in government policies and agricultural support systems.

Even without a switch to polyculture, farmers have an incentive to reduce their use of fertilizer and pesticides—after all, these inputs cost money. Many farmers have begun employing technology to monitor the production of their agricultural fields and now apply fertilizer and pesticides in a more targeted manner. One technique to reduce inputs has been the use of GMOs that produce pesticide directly. In non-GMO crops that require high levels of pesticide, farmers have employed a technique called **integrated pest management (IPM)**, which utilizes releases of predator insects, introduction of pest competitors, and changes in planting techniques among other strategies to reduce pesticide use. However, the use of neither GMOs nor IPM has been a complete solution. Since their introduction in conventional agriculture, pesticide and fertilizer use has still continued to rise in the United States.

The Role of the Consumer. Farmers are only one-half of the food-production equation. Consumers also determine what food is produced and how. If consumers consistently select certain types of foods, farmers will try to supply them. In many cases, consumer demand is encouraging some of the more damaging environmental consequences of modern agriculture.

Most of the pesticides sprayed on vegetable and fruit crops are applied in order to reduce crop "losses" that are caused by consumers who do not purchase fruits and vegetables with superficial signs of pest damage. The only way to produce pristine products is to apply hundreds of pounds of pesticides to completely eliminate pest insects. If consumers were willing to accept a visible but small amount of pest damage to fresh vegetables, then the amount of pesticides used on these crops could drop tremendously (Figure 22.29).

(a)

(b)

Figure 22.29 Cosmetic pesticide use. (a) This apple was produced on a conventional farm, with high levels of pesticides to control superficial damage. (b) This apple was produced on an organic farm, with no pesticides. If consumers were willing to accept the level of superficial "damage" seen on the organic apple, then the amount of pesticides used on fruit and vegetable crops would greatly decline.

Consumer demand for meat also fuels unsustainable farming practices. Most of the beef, pork, and poultry produced in the United States comes from animals that are fed field-grown grain in enormous feed lots, called "factory farms" (Figure 22.30). The same production process holds for hogs, chickens, and turkeys. In fact, 66% of the cereal grain consumed in the United States, including 80% of the corn and 95% of the oats, is used to feed these animals. Because much of the energy consumed by an animal is used for maintenance rather than growth, the grain used as animal feed yields a much smaller number of calories in the form of meat. Put another way, a 10-acre field of corn can support 10 people if they eat the corn directly, but only 2 people if they feed the corn to cattle and then eat the beef. When individuals consume many of their calories as meat and dairy products, they are forcing the high-intensity farming that causes the environmental problems discussed in this chapter. Reducing meat consumption would also reduce the number of acres farmed; as a result, any soils that are highly susceptible to degradation through erosion or salinization could be left undisturbed. Less land needed for agricultural production also leaves more natural habitat for other species to survive.

Factory farming of cattle, hogs, and poultry carries environmental costs in addition to its prodigious use of crops. For example, the U.S. Department of Agriculture estimates that the meat industry produced 1.4 billion tons of animal waste in 1997; this is 130 times the amount of waste produced by the human population of the United States. Very large feedlots lead to large concentrations of animal waste and create a storage and disposal problem for the industry. Failures of manure storage ponds and inadequate disposal have led to spills, water contamination, and fish kills. In addition, the high-density populations of animals in factory farms provide prime conditions for the spread of disease. Many meat producers in the United States handle this risk by feeding large amounts of antibiotics to all of their animals, sick and healthy alike. These antibiotics help control disease and, surprisingly, increase the growth rate of healthy animals. The use of antibiotics in agricultural settings is a significant factor in the development of antibiotic resistance in several dangerous human pathogens, including *E. coli* O157:H7, a bacterial pathogen found in beef that sickens thousands of people—and kills over 100—every year. While the efficiencies of factory farming have allowed Americans to enjoy plentiful and inexpensive meat and dairy products, these efficiencies also result in a cost that is borne by the environment, rural communities, and future generations.

Organic Farming.   There is little doubt that consumer demand can change agricultural practices. In recent years, many consumers with concerns about the environmental and human health costs of modern agriculture have changed their buying habits. As a result, some farms have changed to **organic farming**, supported by practices that do not require industrially produced inputs or the use of GMOs. The U.S. Department of Agriculture reported 20% growth in the retail sales of organically grown food in every year from 1990 to 2002. Organic farm acreage in the United States doubled between 1992 and 1997 and is projected to represent 10% of total farm acreage by 2010. Organic farms are still very much the exception in American agriculture, but consumer interest in organically grown food is having a noticeable impact on farming practices.

The focus of organic farming is the natural maintenance of soil fertility (using crop rotation and organic fertilizers) and the development of biological diversity on the farm (using polycultures) as a method to control pests and provide income stability. Organically grown food has lower levels of pesticide residues and is produced in a way that minimizes the environmental costs of agriculture (see Table 22.2 on page 598). Organic food does often cost more for

Figure 22.30   Modern meat production. Most of the meat and dairy products available in modern supermarkets are produced in high-density "factory farms," such as the one pictured here.

Table 22.2   Reducing the environmental impact of modern agriculture. This table summarizes the environmental costs of modern agriculture along with the individual and collective actions we can employ to reduce these costs.

Environmental problem	Solution	Actions we can take
Bioaccumulation of persistent pesticides	• Minimize monocultural plantings. • Reduce use of pesticides. • Increase use of integrated pest management.	• Accept produce that has superficial pest damage. • Purchase organic foods. • Add diversity of foods to our diets to promote polyculture. • Reduce meat consumption, and thus overall crop production. • Support policies that reduce use of monocultural production (for example, elimination of price supports for crops). • Support integrated pest management research.
Eutrophication of surface water and nitrate pollution of groundwater	• Reduce use of inorganic fertilizer.	• Purchase organic foods. • Purchase foods grown in polyculture, where crop rotation is employed. • Reduce meat consumption.
Salinization of soil	• Reduce number of acres irrigated. • Increase efficiency of irrigation.	• Reduce meat consumption. • Support research and national agricultural policies to reduce agricultural water use.
Soil erosion and desertification	• Reduce use of tilling on marginal lands. • Reduce amount of land used for crop production.	• Reduce meat consumption. • Support research on effective nonchemical means of weed control, and see the following suggestions.
Loss of habitat for nonhuman species	• Minimize the amount of land converted to agricultural production.	• Reduce meat consumption. • Support policies that effectively lead to decreased human population growth rates. • Support policies that provide developing countries with the tools to increase crop yield sustainably.

consumers; grocery bills average 60% higher according to several analyses, although these cost differences are smaller when food is purchased direct from the grower at farm markets and when seasonal foods are purchased. And because food is relatively inexpensive in the United States, a 60% increase in food cost for a family of four totals about $3500 a year—about what an average family spends annually on meals at restaurants.

Feeding the human population in a way that preserves environmental quality and leaves room for the diversity of other life forms that share the planet will require thoughtful decision making about agricultural policy. These decisions include those we make as individuals, such as how much meat to consume and what sort of produce we buy, and those we make collectively, such as how federal subsidies for crops are distributed. Better knowledge of plant and agricultural science and an understanding of the power and limitations of scientific knowledge can help you become an effective participant in this decision-making process.

CHAPTER REVIEW

Summary

22.1 Plants as Food

- Nearly all major crops belong to the angiosperm phylum, flower- and fruit-producing plants (p. 572).

- Roots absorb water and minerals from the soil but also may be modified to store carbohydrates (pp. 574–575).

- Stems serve as support for leaves and reproductive organs but may also be modified for carbohydrate storage (p. 576).

- Leaves are the photosynthetic organs of the plant and contain some essential nutrients for humans (p. 577).

- Flowers contain both male and female reproductive organs. When pollen is transferred from the male to the female, sperm fertilize the egg and two other nuclei inside the ovule, leading to the production of a seed. The seed remains inside the flower structure, which becomes modified into a fruit that functions in seed dispersal (pp. 579–581).

22.2 Plant Growth Requirements

- Plants undergo primary growth, in the length of stems and roots, and secondary growth, in the width of these organs (p. 582).

- Seeds may survive difficult environmental conditions by becoming dormant. The emergence of a plant embryo from a seed is called germination (p. 582).

- Primary growth occurs as a result of cell division in the apical meristems. This meristem can produce additional meristems, leading to indeterminate growth patterns in plants (p. 583).

- Secondary growth in dicot stems occurs when the vascular cambium produces additional xylem, which becomes wood, and phloem, which becomes the inner part of the bark (p. 584).

- Plants require abundant water for photosynthesis, for effective interception of light, and to maximize the uptake of carbon dioxide. Farmers not only provide water through irrigation but also remove non-crop plants that compete for water (pp. 585–588).

- Competing plants can be removed via tilling, mulching, or the application of herbicides (p. 588).

- Plants regulate water loss but also cut off the uptake of carbon dioxide by closing the stomata on their leaf and stem surfaces (pp. 588–589).

- Soil is a matrix composed of minerals from eroded rock, dead and decaying material, and the living organisms that decompose and mix the components. Soil is the major source of nutrients for plant growth (p. 589).

- In agricultural systems, nutrients are mined from the soil. Farmers make up for the loss of nutrients by adding fertilizer and growing nitrogen-fixing crops (pp. 590–591).

- Farmers minimize the loss of plants to pests by using cultural control and pesticides (pp. 591–592).

- Widespread use of pesticides has led to monocultural production of crops. Before pesticides, polycultural plantings kept pest populations down and insulated farmers from pest outbreaks (pp. 591–592).

- The production of hybrid strains and the advent of genetic engineering have led to improvements in crop production (p. 593).

Web Tutorial 22.1 **Primary and Secondary Growth**

22.3 The Future of Agriculture

- Petroleum and water are essential to modern agriculture and are in limited or finite supply (p. 594).

- Fertilizer runs off into waterways, causing human health problems and damage to fish and other animal populations (p. 594).

- Pesticide use continually increases as pests become resistant to them. Persistent pesticides accumulate in organisms at higher levels of the food web, including humans (p. 594).

- In many areas, soil is being lost through erosion and salinization faster than it is being replaced (p. 595).

- Fertilizer and pesticide use can be minimized through the use of polyculture, by crop rotation, and by careful monitoring of the agricultural environment (pp. 595–596).

- Consumers can help reduce the environmental costs of modern agriculture by accepting a small amount of pest damage on produce, eating less meat, and purchasing organically grown products (pp. 596–598).

- Organic agriculture may provide a sustainable model for future food production (pp. 597–599).

Learning the Basics

1. Sketch and label the parts of a plant and flower.

2. Describe angiosperm reproduction, beginning with pollination and ending with the production of a seed.

3. All of the following plant organs have been modified in certain plants to store large amounts of carbohydrate, *except* _____.

   A. fruit; **B.** stems; **C.** roots; **D.** leaves; **E.** apical meristems

4. During the process of fertilization in flowering plants, _____.

   A. pollen is picked up at the stigma by an animal for transfer to an anther; **B.** a single sperm is released by a pollen grain; **C.** a seed begins to absorb water and break free of its coat; **D.** a triploid tissue called endosperm is formed; **E.** a carpel is produced that later becomes a seed

5. In dicot woody plants, secondary growth results in _____.

   A. production of new xylem and phloem; **B.** increase in the girth or width of the stem; **C.** the production of additional nodes and internodes; **D.** a and b are correct; **E.** a, b, and c are correct

6. The function of stomata is _____.

   A. to prevent pests from damaging the photosynthetic surfaces of the plant; **B.** to allow carbon dioxide to enter the plant body; **C.** to sense water availability in the air; **D.** to provide stability for a growing plant; **E.** to increase nutrient availability in soil

7. The function of essential nutrients in plants includes _____.

   A. components of proteins; **B.** components of amino acids; **C.** components of membranes; **D.** components required for essential metabolic processes; **E.** all of the above

8. Soils in agricultural systems require fertilizer because _____.

   A. farming mines nutrients from the soil; **B.** weeds compete with crop plants for soil nutrients; **C.** most crop plants have nitrogen-fixing bacteria in their roots and need lots of nitrogen; **D.** most nutrients run off soils into waterways, causing eutrophication; **E.** pests remove most nutrients from crop plants

9. When inorganic fertilizer ends up in waterways, _____.

   A. the waterways become eutrophic; **B.** algae populations explode in surface waters; **C.** fish kills result; **D.** drinking water may become contaminated; **E.** all of the above

10. Consumers can help reduce the environmental effects of modern agriculture by _____.

   A. accepting small levels of superficial damage from pests when buying fresh fruit; **B.** purchasing foods that require low levels of fertilizer and pesticide; **C.** eating less grain-fed meat; **D.** b and c are correct; **E.** a, b and c are correct

Analyzing and Applying the Basics

1. Agricultural scientists have been successful at creating only a few types of seedless fruits through traditional breeding—notably bananas, apples, grapes, pineapples, navel oranges, and a few melons. Why might it be difficult to produce seedless fruits?

2. A simple way to kill trees is called "girdling," in which a strip of bark is removed from the tree's trunk. This process starves the roots of the plant. Use your understanding of plant secondary growth to explain how girdling a tree can cause its roots to die.

3. Your friend has been planting the same variety of tomatoes in the same spot in his garden for the past 8 years. In recent years, he has noticed that his plants are less healthy and that pests damage more of the tomato fruits. Apply what you have learned in this chapter to explain why his tomato crop is declining, and suggest several ways that he can improve his tomato yield.

Connecting the Science

1. The human population relies on approximately 6 staple crops and about 24 additional crops for most of our calories. In the United States, nearly 80% of the cropland is planted with one of three species: corn, wheat, and soybeans. What are the risks and benefits of relying on so few agricultural species? Are you concerned about the lack of diversity in agricultural production? Why or why not?

2. Have you ever considered the environmental impact of your diet? Are you willing to change your diet to reduce its environmental impact? Why or why not?

23

Growing a Green Thumb

Plant Physiology

What does it take to
have a garden this
beautiful?

How can you make a backyard vegetable plot yield abundant produce?

Or fill your home with thriving houseplants?

Many of us know people who are described as "having a green thumb"—men and women whose homes are a jumble of thriving houseplants, who produce enough vegetables in their backyard gardens to saturate friends with their excess tomatoes and zucchini, or who have flower gardens that make the neighbors, well, green with envy. How do they do it? Is it simply a gift, a green thumb that some are born with and some without?

A significant contributor to a gardener's success is time. Years of trial and error, weeks of preparation for a growing season, and hours of work tending to gardens and indoor plants will yield an experienced grower who can get the most out of his or her plants. Part of what a gardener gains through experience is a more thorough understanding of how plants work. This knowledge helps a plant lover choose the right plants for the environment, design a landscape that is always in bloom, and even shape the form of plants to fit their needs.

Fortunately for those of us who are eager but inexperienced gardeners, it is possible to start developing our green thumb by learning the basics of plant structure and function as described by biologists. In the previous chapter, we introduced you to the physical characteristics of plants as well as their basic needs for growth. Here we go deeper into the dynamic life of plants to

With knowledge of plants, gained through experience and understanding.

603

provide a better understanding of how they interact with and respond to their environment. Our discussion of plant physiology fits within the context of planning a plant-filled space, such as a garden or houseplant collection. Let's get growing!

23.1 The Right Plant for the Place: Water Relations

One of the first considerations of gardening, whether indoors or out, is choosing plants that can thrive in the available environment. Many factors influence plant growth in a particular space: the soil's chemical nature, light availability, and the length of the **growing season**, which is the time from the last frost in the spring to the first frost in the fall. As discussed in Chapter 22, one of the primary factors influencing plant production is the amount of water available to the plant. How do plants obtain and manage water, and how can we use our understanding of this process to help plants thrive in our gardens?

Transpiration

Water and dissolved minerals obtained from the soil, together called **xylem sap**, travel up the water-conducting cells inside a plant. This occurs primarily due to the loss of water vapor from the leaves. Through this process, called **transpiration**, the plant essentially pulls a column of sap from the soil, through the roots, up the xylem tubes in the stem, and into the leaves. Xylem sap can be pulled via transpiration from the roots to the leaves of even the tallest trees (over 100 meters, or 330 feet) because of the strong hydrogen bonds formed by water molecules (described in Chapter 2). Transpiration evolved in plants after they colonized land 400 million years ago, and it has allowed plants to colonize nearly all areas of Earth's surface (see Essay 23.1 on pages 606–607).

The tendency for molecules to stick together via hydrogen bonding is referred to as **cohesion**, while the tendency for polar molecules to form hydrogen bonds with different types of polar molecules is called **adhesion**. Water is very cohesive and it adheres to cellulose in plant-cell walls. These two forces together maintain the continuity of a sap column as it is pulled up xylem tubes.

The pulling force on a column of xylem sap is generated by the evaporation of water from the leaves. Recall from Chapter 22 that pores on leaf surfaces control the rate of gas exchange and water loss in land plants. When water evaporates out of these stomata, the forces of cohesion and adhesion on the remaining water molecules inside the leaves create **tension**, or negative water pressure. Tension is the force that allows a column of water to be pulled into a syringe when the plunger is pulled up. In plants, tension causes water molecules from the xylem to be drawn out into the leaf to replace the water that has evaporated. This in turn increases tension on the water molecules immediately below them in the xylem, causing them to move toward the leaves, and so on, all the way down to the roots and soil (Figure 23.1). When the continuity of a stream of water is disrupted by injuring the stem, the tension is released, and water retreats into the xylem from the damaged point. Gardeners who understand this know to cut flowers in the early morning, when the plants are maximally hydrated and tension is lowest, and to reestablish a continuous stream of water by cutting the flower stem again once it is placed in water.

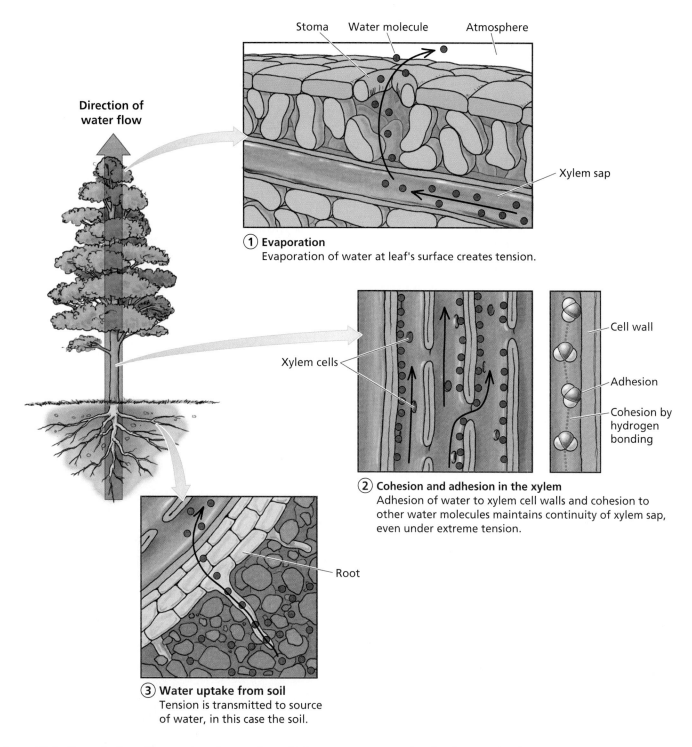

Figure 23.1 Transpiration. Evaporation of water from the leaves creates a tension that is transmitted to a column of water in the xylem. Water flows upward against the pull of gravity and maintains its continuity thanks to the forces of cohesion and adhesion.

While transpiration is a process that requires only the energy of the sun, causing evaporation, and the strong hydrogen bonds in liquid water, plants can actively modify the *rate* of transpiration by regulating the size of stomata, as discussed in Chapter 22 and illustrated in Figure 23.2 (page 607). In addition to these active movements, certain plants also possess adaptations in photosynthesis or leaf shape that can affect the rate of transpiration. The presence or absence of these adaptations help determine which plants are best suited for particular climates and environments.

Essay 23.1   Focus on Evolution: Plants on Land

The ancestors of plants were green algae, photosynthetic organisms that spent their lives immersed in water. A significant challenge faced by plants as they moved onto land was the physical separation from the resources they require for survival—light above ground, water and other nutrients below ground. The earliest land plants were small and close to the ground, with no water- or food-conducting tissues. Some groups of mosses evolved primitive water-conducting tissues, but true xylem and phloem evolved in the ancestors of modern ferns and seed plants. In some early seed plants, a pattern of secondary growth from a vascular cambium evolved that allowed continued production of food- and water-conducting tissues in expanding stems. Flowering plants possess additional adaptations that greatly increase the speed of food and water transport. These adaptations have allowed seed plants to dominate Earth's surface.

Thalloid Liverworts

The simplest land plants, the hornworts and liverworts, consist primarily of thin, platelike leaves with no roots or stems (Figure E23.1a). These plants obtain water via diffusion from the wet soil, and carbon dioxide enters the plant through stomata-like pores on the leaf surface. Hornworts and liverworts have no internal water-conducting tissues; therefore, they must remain small and are limited to land areas that are consistently damp.

Figure E23.1a   Thalloid liverwort.

Mosses

The center of the "stem" of many mosses consists of water-conducting cells called hydroids (Figure E23.1b). Like xylem cells, hydroids are dead at maturity; but they have thin cell walls and are not functional as support structures. Some groups of moss also possess food-conducting cells called leptoids, which are similar in structure to true phloem and may illustrate an intermediate step in the development of true vascular tissue.

Figure E23.1b   Moss.

Ferns

Ferns belong to the oldest group of plants with true xylem and phloem—the seedless vascular plants (Figure E23.1c). Tracheids, xylem tube cells with stiff secondary walls that provide support for stems, first appeared in the ancestors of modern ferns. These allow true vascular plants to grow much taller than nonvascular plants and to grow in much drier places. The food-conducting cells of ferns and other seedless vascular plants are also identical to those in all other seed plants and probably evolved at the same time. The vascular tissue in ferns forms a solid ring of mixed xylem and phloem tubes just under the stem's surface.

Figure E23.1c   Fern.

Conifers

Gymnosperms, like the pine tree in Figure E23.1d, are plants that produce seeds but no flowers. Gymnosperms organize their vascular tissue in discrete strands within the stem. Production of new vascular tissue during secondary growth (growth in girth) occurs at a "bifacial" vascular cambium—a ring of tissue that

Figure E23.1d   Conifer.

Figure E23.1e   Flowering plant.

produces new phloem to the outside and new xylem to the inside. This growth pattern provides a solid core of stiffened xylem stems to support the increasing height of the tree. This adaptation permits the immense height of some species of seed plants, which can grow up to 110 meters (360 feet) tall.

Flowering Plants

The most successful group of plants on Earth is the angiosperms, or flowering plants (Figure E23.1e). A key adaptation that appears to have contributed to their success is more efficient vascular tissue. Vessel elements, which result in large-diameter xylem tubes, are highly efficient water-conducting cells that can move water from soil to plant tissues much more quickly than can the narrower tracheids found in nonflowering vascular plants. Sieve-tube members in phloem produce wider and more efficient food-conducting tubes, possibly allowing for the faster growth of organs in response to environmental opportunities as compared to other vascular plants.

Adaptations That Affect Transpiration

Plants have evolved several strategies for reducing transpiration water loss. These strategies fit into two general categories: (1) modifications to photosynthesis to make carbon dioxide acquisition more efficient, and (2) modifications to leaf shape and stomata placement to increase water conservation.

(a)

(b)

 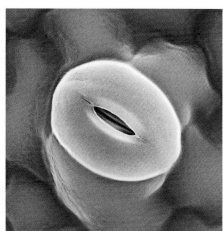

Figure 23.2   Regulating water loss.
(a) Open stomata allow carbon dioxide to enter the plant leaf for photosynthesis but also allow water to escape. (b) Plants can close their stomata to reduce transpiration when water is limited.

Figure 23.3   A C4 plant. Crabgrass has the C4 photosynthetic pathway, which gives it an advantage over Kentucky bluegrass in hot, dry mid summer and late summer. This advantage makes crabgrass a very successful weed in urban lawns throughout the United States.

Photosynthetic Adaptations.   Chapter 4 introduced two adaptations in photosynthetic pathways that reduce water loss: C4 and crassulacean acid metabolism (CAM) photosynthesis. Plants with C4 photosynthesis have an additional chemical process that precedes the Calvin cycle, the process that converts carbon dioxide into sugar. This additional chemical cycle essentially pumps carbon dioxide from the air into the center of the leaf, allowing photosynthesis to proceed even when the stomata are just barely open. Many C4 plants, including corn, sugarcane, and crabgrass, belong to the grass family. These plants have an advantage over non-C4 plants in hot, sunny, and dry conditions because they will continue photosynthesizing even when carbon dioxide levels within the leaf are low due to closed stomata. Homeowners across the United States can see the relative advantage of C4 crabgrass over lawn grasses like Kentucky bluegrass during the warmest parts of summer. At these times the yellowish, broad leaves of crabgrass will become dominant as growth of the less-tolerant bluegrass slows (Figure 23.3). If the crabgrass spreads, patches of bluegrass may die off, leading to a patchy lawn by the end of summer when the crabgrass completes its life cycle and dies off.

CAM plants accumulate carbon dioxide from the air during the cooler night, when less water will evaporate through the stomata. The carbon dioxide is converted to a simple carbohydrate and is stored in the vacuoles of leaf cells in the form of malic acid. At sunrise, the stomata close, and the accumulated acid is degraded, releasing the carbon dioxide for photosynthesis. The growth of CAM plants is greatly limited by the relatively small amount of carbon dioxide they can use in a day—only what can be chemically stored. These plants are most common in extremely dry environments, although they can be successful in other environments. One of the driest environments a plant can experience is a house in winter, and many popular houseplants use the CAM process (Figure 23.4).

So, what about the plants without either C4 or CAM photosynthesis? Their strategies for reducing the rate of water loss through stomata are limited to the structural adaptations of their leaves.

Leaf Adaptations.   The amount of water lost by a plant via transpiration is generally correlated to the photosynthetic surface area of the plant. In particular, a plant with many large leaves tends to transpire more water than does a plant with fewer small leaves. Gardeners use their understanding of this relationship between leaf surface area and transpiration to help transplanted plants survive. Because removing plants from soil (no matter how carefully) inevitably results in root damage, newly transplanted plants have a reduced capacity to acquire water. To compensate for this, gardeners remove leaves to reduce the rate of water loss.

Of course, smaller leaf area also means less photosynthesis, so plants with smaller leaves tend to grow more slowly than plants with broader leaves in the same environment. Plants that thrive in shady sites tend to have broad leaves because evaporation is lower in shade and intercepting light is more crucial, while narrow-leaved plants tend to be more successful in bright sun. Leaf size is reduced to its extreme in cacti, which have photosynthetic stems and leaves modified into spines (Figure 23.5). Cacti grow quite slowly, but they make for another ideal houseplant that can survive even the most inattentive gardener.

Some broad-leaved plants have other adaptations to reduce water loss, such as the placement of stomata on the cooler, shaded undersurface of leaves or in hair-lined pits, which keep humidity high near them and thus reduce water loss even when the stomata are open (Figure 23.6). Where periods of drought occur reliably on an annual basis, many broad-leaved plants have evolved to become **deciduous**, that is, to actively drop their leaves in order to reduce the rate of transpiration. In Section 23.2, we describe this process in greater detail.

In environments where water is abundant, plants compete with each other for access to light. In these environments, faster-growing plants often have an

Figure 23.4   Jade, a CAM houseplant. Plants with CAM photosynthesis are adapted to very dry conditions and thus are ideal houseplants to survive the dry indoor air of northern winters.

Figure 23.5 Small leaves limit transpiration loss. Cacti show the most extreme modification of leaves to reduce transpiration—they are modified to non-photosynthetic spines. The stems are the major photosynthetic surface. Note that the stems are oriented vertically, so they do not intercept the maximum amount of light and thus have limited rates of photosynthesis.

advantage. Plants that have evolved strategies for maximizing transpiration, including changes in water-conducting tissues, can grow very rapidly and quickly tower over their neighbors.

Xylem Adaptations.   When a column of xylem sap is under extreme tension inside a plant, the forces of cohesion and adhesion may not be strong enough to prevent the water column from breaking. Breakage of the water column results in the formation of a bubble, called an embolism, that inhibits water flow within the xylem. In general, larger-diameter xylem tubes are more likely to suffer embolisms than are smaller-diameter tubes, so many drought-adapted species have only very narrow, tapering **tracheids** as the water-carrying cells in their xylem. However, the narrow passageway produced by a column of tracheids creates increased friction, reducing the rate of water flow so that tracheid-rich plants are at a disadvantage in moist environments. In these environments, plants with wider, perforated **vessel elements** have an advantage; their xylem cells create less friction on the sap column and can more rapidly move water to

Epidermal hairs   Stomata

Figure 23.6 Shaded stomata help conserve leaf water. The stomata of this leaf are on the shaded underside, where temperatures are lower and evaporation slower. In addition, the stomata are in hair-lined pits, which trap evaporating water molecules and keep humidity high, reducing the rate of water loss.

(a) Cross section of pine wood

Narrow diameter = low risk of embolism, but more friction on water flow.

Tracheids

Water flows through pits in cell wall = greater friction.

(b) Cross section of oak wood

Vessel elements = less friction on water flow, but greater risk of embolism.

Vessel elements

Water flows through open ends of cells = less friction.

Figure 23.7   Xylem cells. Water is moved in plants through long, narrow tracheids or short, wide vessel elements. (a) Plants in dry environments have a high proportion of tracheids to prevent against embolism. These narrow tubes carry water relatively slowly. (b) In moist environments, vessel elements, which transport water at about 10 times the flow rate of an equivalent cross section of tracheids, are the most dominant xylem cell type.

their leaves (Figure 23.7). Trees with vessel elements are able to supply water to their leaves at rates that are 10 times faster than those for trees with only tracheids, resulting in a proportionally faster growth rate. The diameter of the xylem tubes in any given plant represents a trade-off between the risk of embolism and maximum growth rate. This pattern is especially visible among trees. For example, trees in cold climates, where liquid water may be unavailable for a large part of the year, are at high risk of embolism and have much less porous-looking wood than do trees in warmer, wetter climates where embolism is rare but competition for light is intense.

The growth of a plant in a natural landscape is strongly influenced by the balance between water availability and transpiration rate. However, in gardens and houses, water is rarely limited due to the availability of sprinklers and watering cans. In fact, a common problem of houseplants in the hands of inexperienced gardeners is too much water, not excessive dryness.

Overwatering

Excess water, by filling the air spaces in soil, interferes with root activity and actually decreases the transpiration rate, causing the plant to wilt. Inexperienced gardeners respond to an overwatered, wilting plant by watering it, further destroying the roots and eventually killing it. As a rule of thumb, potting soil should be allowed to dry to the touch down to 2 centimeters ($^3/_4$ in.) below the surface between waterings in order to avoid this problem.

Overwatering is a typical problem for houseplants, but garden and agricultural plants can also suffer from excess moisture. Soils that are high in clay, the tiniest mineral particles, are prone to becoming waterlogged because each clay particle can trap a shell of water around it, eventually filling up all of the soil's air spaces and leading to the same symptoms as those in overwatered houseplants. Gardeners with clay soils often need to amend, or improve, the soil before planting to avoid this problem. The most common soil amendment is organic matter, such as compost or peat moss. Additions of organic matter help separate clay particles, provide air holes in the soil, and have the additional benefit of adding nutrients when the organic matter decomposes. Sandy soil, which has large particles and cannot hold as much water, can be amended by adding water-holding organic matter as well.

Even in the best soils, overwatering can become a problem when water remaining on leaves and the soil surface encourages the growth of bacteria and fungi that damage plants. In addition, because plants produce roots where water is available, consistently wet soils promote shallow rooting and thus less stable plants. In most areas of the country, plants require about 2.5 centimeters (1 inch) of water per week, through rainfall or supplemental watering, for optimal growth. Experienced gardeners know to look for signs of water stress (limp or wilting plants, a yellow or bluish cast to leaves) and to check how dry the soil is a few centimeters below the surface before watering their plants.

A gardener's ability to supplement natural rainfall allows the growth of water-needy plants even in dry landscapes. However, it is far more difficult for gardeners to control the temperatures that their plants are exposed to, so most garden plant choices are based on temperature requirements. Ability to tolerate freezing is one factor that determines whether a plant can survive in a particular temperature range. Cold tolerance is primarily a function of a plant's ability to control the freezing of water within its cells.

Water Inside Plant Cells

Ice formation in plants damages cells in three ways: (1) When water outside of plant cell walls turns into ice, liquid water from inside the cells will begin flowing out by osmosis, that is, down a concentration gradient across the cell membrane. As the cell loses more and more water to the ice crystal, normal cell processes become disrupted and may cease to function. (2) When water freezes inside a plant cell, it produces an ice crystal. The jagged edges of this crystal can pierce cell membranes, allowing cytosol (the fluid inside the cell) to leak out, thereby killing the cell (Figure 23.8). (3) When ice freezes within

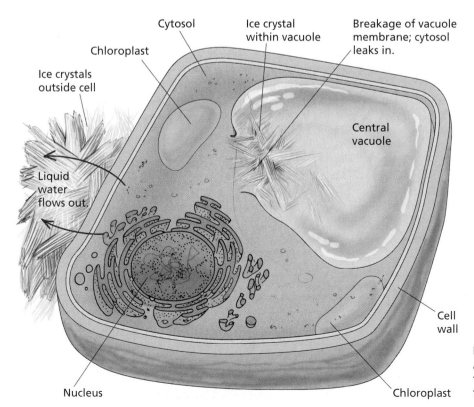

Figure 23.8 Freezing inside plant cells. An ice crystal inside a plant cell can pierce the cell membrane, allowing the cytosol to leak out.

xylem tubes, air comes out of the water solution, forming embolisms and blocking future water uptake. All three of these processes can severely damage or kill plants.

A plant's tolerance of cold temperatures, known as its **hardiness**, depends on its ability to either prevent or manage freezing inside of cells. Water freezes at temperatures lower than 0°C (32°F) when it has solutes dissolved in it, in the same way that added antifreeze prevents water in a car's radiator from freezing. Tomato plants can survive temperatures down to −1°C (31°F), apple fruits −4°C (24°F), and cabbage as low as −9°C (16°F) due to their sugar-based cell antifreeze. At lower temperatures, antifreeze is ineffective, and adaptations rely on freezing tolerance. The ability to survive freezing includes controlling how ice forms within cells to prevent membrane-damaging sharp edges and the presence of internal structures to trap air bubbles and prevent xylem embolisms.

Many cold-tolerant plants require a gradual cooling period to become maximally hardy and uninterrupted cold temperatures to remain hardy. Cold snaps that occur in late summer and early fall may damage plants that can normally thrive in even colder temperatures because they have not become **hardened**, or fully able to resist the cold. Woody plants can become damaged by cold even after they are fully hardened when bright winter sunshine warms up their stems, allowing water to transpire when it cannot be replaced by frozen water in the soil. This is a common occurrence in evergreens in landscaping, which can suffer from winter "burn." Freezing temperatures are most likely to cause damage during the spring, however, when intolerant leaves and flowers are beginning to emerge from formerly cold-tolerant buds.

Because of the extreme damage that freezing temperatures can cause, gardeners must know two things before planning an outdoor garden: (1) the hardiness zone of their landscape, that is, how cold it gets (Figure 23.9), and (2) the rated hardiness—that is, the cold tolerance—of the plants they wish to grow. Experienced gardeners know, however, that with special care, less cold-tolerant plants can survive freezing conditions. Plants that are grown outside of their hardiness zone can be placed on the north sides of buildings, where they are shaded from the sun (preventing winter burn) and where cooler local temperatures retard budding and leafing until the threat of frost

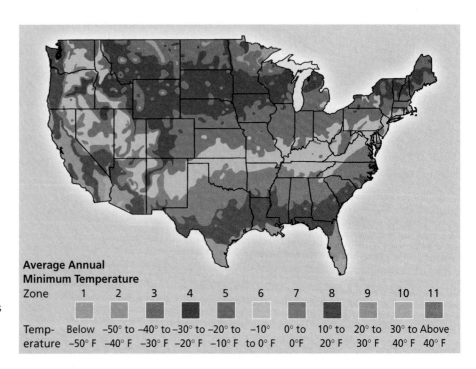

Figure 23.9   Hardiness zones. The U.S. Department of Agriculture generated this hardiness zone map based on average lowest temperatures experienced in thousands of sites across the country.

Average Annual Minimum Temperature

Zone	1	2	3	4	5	6	7	8	9	10	11
Temperature	Below −50° F	−50° to −40° F	−40° to −30° F	−30° to −20° F	−20° to −10° F	−10° to 0° F	0° to 10° F	10° to 20° F	20° to 30° F	30° to 40° F	Above 40° F

has decreased. Plants can also be covered with mulch such as straw or snow, and they can be protected from unusually early or late frosts by being covered with plastic tents. By insulating the plants from cold air, both of these strategies help to reduce the temperature extremes that plants experience.

Less-tolerant plants are also better able to survive cold winters if they have sufficient water and food reserves. These materials allow the plants to produce adequate antifreeze at the appropriate times during the hardening process. Gardeners can ensure that their sensitive plants are ready for winter by giving them plenty of water during the fall and by preventing them from using up their stored starch and sugar reserves. To best manage a plant's sugar supply, a gardener needs to understand how plants store and move their food reserves. This knowledge helps gardeners not only to ensure that their favorite plants survive harsh conditions but also to maximize the beauty and productivity of these plants.

23.2 A Beautiful Garden: Sap Translocation, Photoperiodism, and Flower and Fruit Production

The first step to planning a garden, either indoors or out, consists of choosing plants that will thrive in the environmental conditions. Just as important to most gardeners is ensuring that the garden provides what they want—from a summer-long supply of fresh vegetables, to a continually changing palette of showy flowers, to an aesthetically pleasing but low-maintenance decor. Because plants exhibit indeterminate growth, growing throughout their lives rather than to a set size, each plant can allocate its resources in many possible directions at any time—to greater leaf production, flower production, fruit size, and so on. By understanding how plants allocate resources, gardeners can encourage plants to transfer these resources to the gardeners' preferred plant organ at appropriate times.

Translocation of Sugars and Nutrients

Sugars and other dissolved nutrients are moved around a plant's body in phloem tissue. The solution in the phloem, or **phloem sap**, always flows from a sugar **source**, where it is in high concentration, to a sugar **sink**, where sugar is being used or accumulated. As a result, the movement of phloem sap, called **translocation**, can be in any direction within a plant. Different organs can act as sources and sinks throughout the year; in fact, separate phloem tubes in the same branch can even be translocating sugars in opposite directions at the same time.

Translocation of phloem sap results from relative pressure differences along a phloem tube between a source and a sink. Figure 23.10 illustrates

① Sugar is actively loaded into phloem tubes at the source.

② Water flows into phloem passively, down its concentration gradient.

Contents of phloem move from source to sink due to the water-pressure difference between them.

③ Sugar moves into sink cell (actively if it is a storage cell, passively if it is an active cell).

④ Water flows out of the phloem tube passively, down its concentration gradient.

Figure 23.10 The pressure flow mechanism of phloem translocation. Phloem sap moves from source to sink because of pressure differentials in the phloem tube.

the hypothesized **pressure flow mechanism** that causes phloem sap movement. The first step in this mechanism is for sugar to be actively loaded into a phloem tube at the source. As sugar accumulates, water flows in passively via osmosis from nearby tissues and xylem tubes. At the same time, respiring cells at the sink are acquiring sugar from the phloem tube, either through active transport or passively, down a concentration gradient from the sugary phloem sap into a sugar-consuming cell. When this happens, the sugar concentration at the sink declines, and water exits the phloem tube passively. Due to sugar loading at the source and unloading at the sink as well as subsequent osmosis of water molecules across the cell membranes of the phloem tube, water pressure is high at the source and low at the sink. A pressure differential causes the bulk movement of water in any tube. Just as in a garden hose, where high water pressure at the faucet and low pressure in the open air causes water to flow out, phloem sap moves in bulk from the high-pressure source end to the lower-pressure sink end in the phloem tubes of a plant.

Managing Translocation

The key feature of translocation in plants is that materials always move from source to sink. Gardeners who recognize which plant organs are functioning as sources or sinks at any given time can modify a plant's structure in order to promote flowering, increase the quantity and quality of produce, and prepare plants for winter.

Promoting Flowering. How a gardener manages translocation in order to increase flower production varies depending on the plant type. Plants that grow from seed and then flower, produce fruit and seeds, and die all within a single year are called annuals. Many common garden flowers, including petunia, pansy, morning glory, and cosmos, are annuals (Figure 23.11a). The

(a) Annual plants

(b) Biennial plants

(c) Perennial plants

Figure 23.11 Plant seasonal growth cycles. (a) Annual plants live only a single season or year, growing from a seed and producing a new seed crop at the end of their lives. Their evolutionary strategy is to put a significant amount of energy into flower production. This makes them ideal for garden displays. (b) Biennial plants typically make a low-profile set of leaves in the first year and send sugar into a storage root. In the second year, sugar from storage supplies a large floral display, after which the plant dies. (c) Perennials live for many years, putting energy into flower production and storage roots every year. Gardeners maintain their perennial plantings by reducing crowding among individual plants.

evolutionary strategy of annual plants is to produce as many seeds as possible in their single year of life; to do this, annuals produce many flowers to increase their chances of pollination and fertilization. The larger the sugar source available for flower production, the greater the number of flowers (sinks) can be produced. To manage flower production in annuals, gardeners remove, or **prune**, many of the flowers on the young transplants they purchase in order to encourage the plant to send sugars to different sinks—meristems that are developing more leaves. Once the annual has reached sufficient size, the plant can be encouraged to produce flowers instead of more leaves; this can often be accomplished by restricting both water and fertilizer. If the plant is grown just for its flowers, individual blooms are removed once they fade (a process called deadheading) to eliminate the competitive sink of developing fruits and seeds.

Plants that live for 2 years—producing leaves the first year, surviving dormancy by storing nutrients in roots or underground stems, and producing flowers the second year—are called **biennials**. Some of the best-loved garden flowers, including hollyhock, forget-me-not, and delphinium, are biennial plants (Figure 23.11b). Gardeners who plant biennials from seed know that the role of the plant in its first year is to send sugars into the storage roots, so that the floral display the following year is as large as possible. The most effective way to do this is to encourage leaf growth early in summer, thus maximizing the amount of sugar sent into storage in late summer. Typically, this means ensuring that the biennial leaves are not overly shaded by taller neighbors.

Plants that live for many years by storing nutrients in roots or shoots, often stored underground, are called perennials. Favorite garden perennials produce flowers every year and include phlox, peony, day lilies, and tulips (Figure 23.11c). Perennials with stiffened above-ground stems, such as trees and shrubs, are called **woody plants**; most garden perennials do not have woody stems and instead die back to ground level every year. Because of their long lives, the evolutionary strategy for perennials is more balanced; sinks are evenly divided between flower and fruit production and storage organ production. As with annuals, if perennials are grown for flowers, they are deadheaded to minimize the fruit and seed sink. Many gardeners will remove perennial flowering stalks well before the last flower has bloomed, so that adequate sugars go into the storage sink before **dormancy**, the plant's yearly rest period. In addition, because the underground storage organs continue to enlarge and become crowded over time, older perennials may have difficulty accessing water and light. To deal with this problem, gardeners must regularly divide perennials by lifting the storage organs from the soil, removing old tissue that no longer produces buds, and replanting smaller, more vigorous portions.

Increasing Produce Quality and Quantity.   When annuals, biennials, and perennials are grown for fruit or vegetable production, the goal is to increase the number and quality of these sinks. Many of the same techniques for flower production apply. For example, for annual crops, making sure that the plant puts energy into producing leaves and stems early ensures better production of fruit later. Other gardening strategies focus on different sinks; for instance, plants grown for leaves or stems instead of fruit (such as lettuce and asparagus) may be actively discouraged from flowering or setting seed since these activities will rob their edible parts of quality. In many crop plants grown for fruit, gardeners must strike a balance between the number of fruits produced and their quality. If plants are allowed to continue making new fruit sinks, each fruit will be smaller and less flavorful than if fewer sinks are produced. Experienced gardeners know to prune additional tomato blossoms after a plant has a significant load of fruit for best quality and to pick off a

38 cm

15 in

Fruits to remove   Fruits to remove

Midsummer

Autumn

Figure 23.12 Maximizing fruit quality over quantity. Thinning fruits in midsummer ensures that each apple is adequately supplied by enough source leaves. A flavorful apple requires about 40 leaves to support it. A good rule of thumb is to thin fruit until they are spaced about the width of your hand apart.

proportion of the developing apples on a tree in order to produce larger, sweeter fruit (Figure 23.12).

The same source-sink principle can be applied when harvesting fruits in order to maximize flavor. Once a fruit is harvested, living cells within the fruit continue to act as a sink; however, because the fruit no longer has an external source, it begins to use its own stored sugars as an energy source. As these sugars are used up, the flavor of the fruit suffers. Plant cells use sugar at higher rates when they are warm, so keeping fruit as cool as possible during harvesting helps retain maximum sweetness. To keep vegetable quality high, gardeners try to harvest during the coolest parts of the day and load their vegetables in wicker baskets that allow for air movement.

Preparing Plants for Winter. As discussed in the previous section, one of the most difficult challenges many plants face is surviving freezing temperatures. Because many plants produce carbohydrate antifreeze throughout the late fall, perennials with inadequate carbohydrate stores are susceptible to freezing damage. To minimize damage, gardeners must not encourage the growth of new leaves late in the year because this causes the plant to use previously stored carbohydrates to produce new stems. Avoiding pruning and fertilization in late summer and fall keeps stem growth down and promotes carbohydrate storage. Managing fruit production so that moderate numbers are produced and harvesting fruit early rather than allowing it to continue to act as a sink are two ways of increasing the amount of sugar that goes into storage.

Because plants develop through indeterminate growth, gardeners can manage plants and their environments to promote the qualities they wish in their gardens. However, much of a plant's response to its environment is a result of innate traits. In particular, the timing of plant flowering and the beginnings of dormancy are often controlled by genetic factors. One of these factors, photoperiodism, ties plant growth to day length.

Photoperiodism

In the early part of the twentieth century, scientists from the U.S. Department of Agriculture began studying a unique tobacco plant that had appeared in a Maryland farmer's field. This plant grew to twice the height of normal tobacco plants and continued producing leaves until the end of the growing season. Obviously, farmers were thrilled with this variety, except for one small problem—it rarely flowered in the field and thus produced little seed with which to sow an entire crop. The scientists who worked on the problem suspected that the "Maryland Mammoth" was not flowering because it was not responding to day length in the same way that other tobacco plants did. Through a process of trial and error, researchers discovered that these tobacco plants could be triggered to flower if the day length (that is, total sunlight) they experienced was shorter than a critical number of hours. Later investigations of other plants illustrated that many, like tobacco, exhibit **photoperiodism**—a biological response that is tied to a change in the proportion of light and dark within a 24-hour cycle.

Tobacco, like chrysanthemums, strawberries, primroses, and poinsettias, is considered a short-day plant because it will not flower if day length surpasses a critical value. Short-day plants tend to be spring and fall blooming. Other plants will flower only if day length is longer than a critical value. These long-day plants, like irises, spinach, and lettuce, are typically summer flowers. Not all plants exhibit flowering photoperiodism; those that do not are called day-neutral plants.

Long-day and short-day plants are actually responding to the relative concentrations of two forms of a photosensitive chemical called **phytochrome**. Phytochrome is produced by plants in one form and changes chemical shape, or conformation, when exposed to red light, a wavelength that is abundant in sunlight (Figure 23.13). In its light-activated conformation, phytochrome acts to trigger flowering in long-day plants and to delay flowering in short-day plants. Phytochrome is a complex of two proteins that is produced continuously within a photoperiodic plant. The switch between the inactive and active form of phytochrome is nearly instantaneous upon exposure to light, but

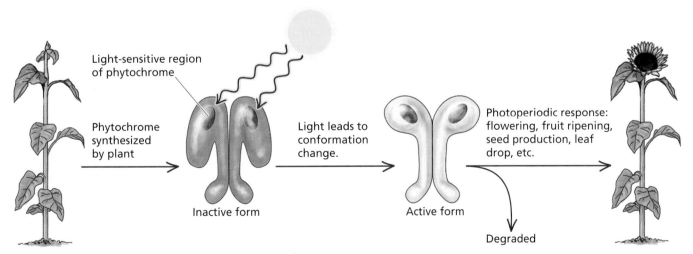

Light-sensitive region of phytochrome

Phytochrome synthesized by plant

Inactive form

Light leads to conformation change.

Active form

Degraded

Photoperiodic response: flowering, fruit ripening, seed production, leaf drop, etc.

Figure 23.13 How plants tell time. The ratio between the active and inactive form of phytochrome allows plants to "sense" day length. A plant synthesizes phytochrome in an inactive form, while exposure to sunlight turns it into its active form. In the active form, phytochrome can have a number of effects, but it is slowly degraded over time.

**Active phytochrome triggers flowering
in long-day plants such as lettuce.**

24
hours

Active form of
phytochrome is
abundant for
longer than the
critical night length.

Active form of
phytochrome is
abundant for
shorter than the
critical night length.

Darkness

Flash of light —— Active form of
phytochrome
produced—time
that it is present
in the plant is
artificially
increased.

*Critical night
length*

Light

**Active phytochrome prevents flowering
in short-day plants such as tobacco.**

Figure 23.14 Phytochrome, short-day plants, and long-day plants. Plants that will flower only if night
is longer than a certain critical length are called short-day plants, while those that flower only if night is
shorter than this length are called long-day plants. The presence of the active form of phytochrome has
the opposite effect on the two types of plant. Plants can be "tricked" by using light during the night to
modify phytochrome levels.

the active form slowly degrades over time; thus, the amount of time that active phytochrome is present provides a measure of day length. In reality, photoperiodic plants measure the length of uninterrupted night; brief exposure to light during the night causes phytochrome to switch to the active form, causing the plant to react as if the day length is much longer. All photoperiodic plants thus react to a **critical night length**; if night is shorter than this critical length, the active form of phytochrome can exert its effect (Figure 23.14). Indoor gardeners can manipulate a photoperiodic plant's exposure to artificial light during the night, or shade it during the day, in order to promote or prevent flowering.

Flowering is not the only activity that is triggered by light cues. The opening of leaf buds in spring and the dropping of leaves in fall are also photoperiodic in many woody plants. **Abscission**, or leaf drop, is a complex process that is first triggered by increasing night length as winter approaches. In preparation for abscission, leaves cease to produce chlorophyll for photosynthesis. As chlorophyll levels decline, other pigments in the leaf become visible. Many of these pigments are minor players in the process of photosynthesis, but some scientists hypothesize that they may actually serve as "warning signals" to potential parasites. For example, a tree that has a very colorful fall display is likely to be healthy and well defended and thus a poor host for overwintering insects. As the leaf turns, cells at the boundary between leaf and stem break down, allowing the leaf to separate from the stem and leaving a protective cap over the junction (Figure 23.15).

Experienced gardeners know that they should match the photoperiod of a plant's native environment to their own environment. The same species of

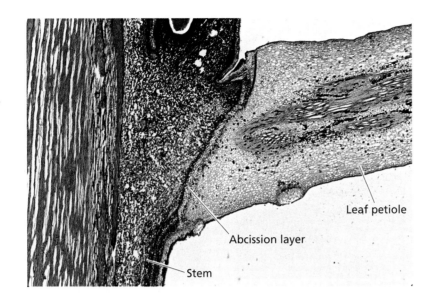

Figure 23.15 Abscission. Leaf drop occurs because cells at the junction between a leaf and a stem break down, forming a weak layer that separates and leaving a layer of scar tissue.

Leaf petiole

Abcission layer

Stem

plant growing in populations at different latitudes may display different critical night lengths and will thus respond differently to the same light conditions. Spinach, strawberries, and onions are all especially susceptible to being planted "out of place" and thus flower, fruit, and produce bulbs at inappropriate times, either too early or too late in the growing season. Southern varieties of widespread deciduous tree species can also suffer from damage if they are planted so far north that they freeze before dropping their leaves.

In addition to finding the right plant for the site and ensuring that their plants produce when and how they wish them to, gardeners are also concerned with the aesthetic form of their gardens. For many, this means actively managing the way that plants fill and share space with each other. Modifying the form of a plant requires understanding how plants respond via growth to environmental cues.

23.3 Pleasing Forms: Tropisms and Hormones

On the surface, plants seem to be captives to their environment. They cannot run away when something tries to eat them. They cannot seek shade on hot days or find a sheltered spot from a freezing wind. They do not even seem to have any way of knowing whether they are in a good spot for growth or a bad one. However, plants are not as inert as they appear. We learned in the previous section that many plants are able to measure day length. Here, we will see that plants can respond to many different environmental signals, sometimes in surprising ways.

Tropisms

Although a rooted plant cannot physically move itself from one environment to another, as a result of indeterminate growth, plants can grow into the best possible environment. Directional growth in plants is called **tropism**. Tropic responses are visible in various plant organs in response to several environmental cues. Roots in germinating seeds show positive **gravitropism** (growing downward toward the source of gravity); stems

(a) Gravitropism

(b) Phototropism

(c) Thigmotropism

Figure 23.16   Plant tropisms. (a) This plant stem is exhibiting positive gravitropism, orienting itself opposite the pull of gravity. The plant was allowed to reorient in the dark, indicating that the response is not to light. (b) This houseplant is growing toward a light source, exhibiting phototropism. (c) The tendril on this vine is curled around the support as a result of thigmotropism.

show negative gravitropism (Figure 23.16a), ensuring that emerging plant roots maximize exposure to water in the soil and that young shoots maximize exposure to light. Houseplants often demonstrate **phototropism**, directional growth in response to light, when placed on a windowsill (Figure 23.16b). Other plants grow in response to touch, exhibiting **thigmotropism**. Vining plants show thigmotropism by twining around support stakes (Figure 23.16c).

Indoor gardeners can use their understanding of plant tropisms to maintain the beauty of their plants. Rotating plants regularly ensures that the phototropic growth of a plant is balanced around the whole pot. Handling plants occasionally, by touching their leaves or stems, can promote a thigmotropic response of shorter, sturdier stems. It may be that the widespread belief that talking to plants is good for them comes from this response—not to the kind words, but to the increased amount of touch and movement that a well-tended plant experiences.

Plants bend toward light, or away from gravity, when cells on one side of a stem increase in length and cells on the other side do not, forcing the top of the plant to turn away from the expanding cells (Figure 23.17). This differential cell expansion is caused by different amounts of growth hormone in these cells.

Hormones

Plant hormones regulate a plant's internal environment and control its responses to environmental conditions. Plant hormones are similar to animal hormones in several ways: (1) They are produced in tiny amounts but have large effects; (2) they are mobile so that they can be produced in one organ and have an effect on another; and (3) they have multiple, at times complex, effects and interactions. Table 23.1 lists the five major plant hormones, their effects, and their commercial applications. We will focus on the first plant hormone identified, and the one most commonly manipulated, auxin.

Auxin is the hormone that triggers differential expansion of cells in tropic responses, as illustrated in Figure 23.17. This important regulator affects not only the amount of plant growth, but its pattern as well. Like all hormones, auxin has sometimes counterintuitive effects on different plant

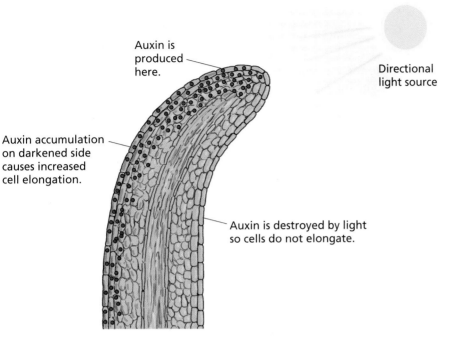

Auxin is produced here.

Auxin accumulation on darkened side causes increased cell elongation.

Auxin is destroyed by light so cells do not elongate.

Directional light source

Figure 23.17   Directional growth.
Tropisms are caused when one side of a plant organ (in this case, a stem) grows more rapidly than the opposite side, causing the tip to bend, in this case, toward a light source.

organs. While auxin causes cells to expand in a main stem, it inhibits the growth of axillary buds along the sides of the stem. Thus auxin is responsible for **apical dominance**—the tendency for the plant apex, or uppermost bud, to grow more rapidly than the lower buds. Auxin produced at the

Table 23.1   Plant hormones.

Hormone type	Effects	Commercial applications
Auxin	Promotes cell expansion, apical dominance, tropic responses, differentiation of vascular tissue; promotes roots on plant cuttings; inhibits fruit and leaf abscission; stimulates fruit development; stimulates ethylene production (in high doses)	Rooting powder for plant cuttings; prevents fruit and flower drop; promotes seedless fruit production; used as weed killer
Gibberellin	Stimulates leaf and stem cell elongation and division; hastens seed germination, breaks seed dormancy; stimulates fruit development	Used to cause uniform germination of barley used in brewing; prevents flower initiation; promotes seedless fruit production
Cytokinin	Promotes cell division; triggers lateral bud growth even in presence of auxin; with auxin, regulates root and shoot production; delays leaf aging	Used to trigger growth of whole plants from genetically engineered plant cells
Ethylene	Inhibits cell expansion; promotes fruit ripening; promotes abscission	Hastens ripening of tomatoes, walnuts, and grapes after harvest; used as fruit-thinning agent in commercial orchards; used to defoliate plants
Abscisic Acid	Prevents seed germination; promotes dormancy	Inhibits growth and promotes flowering of potted plants for florists

shoot tip strongly inhibits branching right below the tip. Further from the shoot tip, auxin concentrations are lower, and its inhibitory effects are less dramatic. Auxin-mediated apical dominance is responsible, for instance, for the "Christmas tree" shape of young conifers.

Apical dominance increases the height and leaf area of a plant at the expense of bushiness as well as flower and fruit production. For this reason, gardeners are often interested in reducing the amount of apical dominance that a plant exhibits. The primary method for reducing apical dominance is light pruning, also called heading back or pinching back. When the terminal bud is pruned, the plant's auxin levels are reduced; inhibition on lateral buds is lifted, and branches form. Eventually, these branches begin to exert apical dominance and must also be headed back to maintain a plant's bushy nature (Figure 23.18).

Auxin also interacts with other growth regulators. Abscission of leaves, as well as of fruits and flowers, is promoted by the hormone **ethylene**. Recall that during abscission, cells at the boundary between the stem and a leaf, flower, or fruit break down so that the abscised organ can separate from the plant. Auxin reduces the sensitivity of cells at this boundary layer to the effects of ethylene and thus can delay abscission. Levels of auxin production are positively correlated to a plant's overall health; as a result, stressed plants tend to abscise fruits and leaves earlier than do healthy plants. The anti-abscission role of auxin makes it useful in commercial operations; for instance, citrus growers spray their orchards with auxin to prevent early fruit drop.

All of the plant hormones identified by scientists have been adopted for some commercial uses, as described in Table 23.1 on page 621. Knowledge of the effects of these hormones, along with knowledge of transpiration, translocation, and photoperiodism, can contribute to the development of a gardener's "green thumb." But this knowledge has a deeper meaning as well. The application of our understanding of plant hormones to commercial use reflects the way that the evolution of life has resulted in both highly complex biological processes and highly complex human brains, which in turn has enabled us to modify organisms to meet human needs. As a result, our species has the incredible opportunity to improve lives and ensure that our descendants and other organisms benefit from our knowledge.

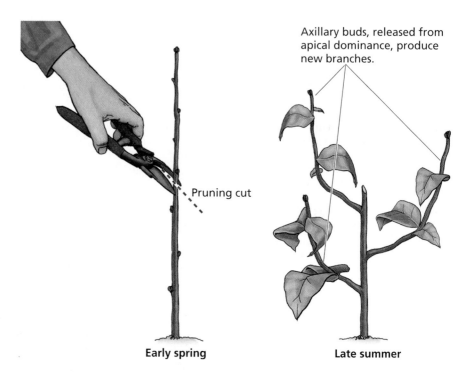

Figure 23.18 Overcoming apical dominance. Apical dominance can be countered by light pruning, removing the terminal bud. Loss of the terminal bud reduces auxin concentration, lifting growth suppression on the lateral buds, which sprout. These branches may be pruned back to induce more branching, if desired.

Axillary buds, released from apical dominance, produce new branches.

Pruning cut

Early spring   **Late summer**

Figure 23.19 Earth as a garden. The future of humanity depends on the wise use of our biological heritage. Learning about the natural world ideally inspires people to live within natural processes rather than trying to control them.

An experienced gardener understands that he or she gets the best results by working with natural processes, not by ignoring them. This basic gardening principle might be a reasonable guide for all of our interactions with the biological world (Figure 23.19).

CHAPTER REVIEW

Summary

23.1 The Right Plant for the Place: Water Relations

- Water is drawn from the soil and into the stems and leaves of a plant because of evaporation at the leaf's surface (p. 604).

- Loss of water results in tension, which pulls on the water column. The column maintains its continuity via cohesion among water molecules and adhesion between water and the xylem walls (p. 604).

- Certain plants have modifications to photosynthesis that reduce water loss. These include C4 photosynthesis, which pumps carbon dioxide to photosynthetic cells, and CAM photosynthesis, which allows carbon dioxide to be harvested from the air at night (p. 608).

- Smaller size, shaded stomata, and deciduousness are leaf adaptations that reduce the rate of transpiration (p. 608).

- Narrow tracheids in xylem prevent the occurrence of embolisms, which destroy a plant's capacity to uptake water. However, narrow vessels deliver water more slowly, making this trait a disadvantage in moist environments (pp. 609–610).

- Overwatered plants wilt because their roots die from lack of oxygen. Without roots, water uptake is too slow to keep up with the plant's needs (pp. 610–611).

- When water inside cells freezes, the cell membrane can be destroyed, and cell contents can leak out. Freezing can be inhibited by the presence of carbohydrate-based antifreeze. Cold-resistant plants control the rate of freezing inside cells in order to prevent cell damage (pp. 611–612).

- When water freezes inside xylem cells, embolisms can form, blocking water flow. If ice in stem xylem cells melts while water in root xylem is still frozen, transpiration of the melted water can lead to dehydration of the plant (p. 612).

23.2 A Beautiful Garden: Sap Translocation, Photoperiodism, and Flower and Fruit Production

- Sap moves in phloem from source to sink. Sugar is loaded into phloem tubes at the source, causing water to flow in as well. Sugar is unloaded at the sink, and water flows out. The difference in water pressure between the tube at the source and the tube at the sink causes sap to travel through the phloem tubes (pp. 613–614).

- Gardeners can modify relationships between sources and sinks to ensure that energy is directed to the gardener's preferred organ. These modifications include pruning flowers and fruit and encouraging or discouraging vegetative growth at different times of the year (pp. 614–616).

- Many plants exhibit photoperiodism—sensing light and responding to it. One system for sensing light relies on a chemical called phytochrome (p. 617).

- Light-mediated responses in plants include the initiation of flower or fruit production, bud break, and leaf abscission in deciduous trees (pp. 618–619).

23.3 Pleasing Forms: Tropisms and Hormones

- Plants grow in response to environmental cues such as the direction of a light source, gravity, and contact with other objects (pp. 619–620).

- Directional growth is caused by differential cell expansion on opposite sides of an organ (p. 620).

- Plant growth regulators, called hormones, tune a plant to its environment (p. 620).

- Auxin causes cell expansion in stem cells but inhibits the growth of lateral buds, leading to apical dominance. Light pruning can reduce apical dominance (pp. 620–621).

- Auxin interacts with another hormone, ethylene, to control leaf abscission (p. 622).

Web Tutorial 23.1 Solute Transport in Plants
Web Tutorial 23.2 Tropisms and Hormones

Learning the Basics

1. Why does a column of water in a xylem tube remain intact despite being under tremendous tension?

2. How do the C4 and CAM adaptations to photosynthesis reduce water loss from plants?

3. What light conditions promote flowering in a long-day plant?

4. Water moves up a plant's stem as a result of _____.

   A. the xylem pump; **B.** diffusion of water into roots; **C.** translocation of phloem sap; **D.** evaporation of water from the leaves; **E.** photosynthesis in xylem cells

5. All of the following are adaptations to reduce the rate of transpiration *except* _____.

   A. closing stomata in response to decreased water availability; **B.** large-diameter xylem vessels; **C.** CAM photosynthesis; **D.** small leaf size; **E.** stomata placed on the undersurface of leaves

6. Sap travels in phloem as a result of _____.

   A. evaporation of sugar water from the leaves; **B.** the phloem pump; **C.** a countercurrent to xylem flow; **D.** differences in water pressure in a phloem tube between a sugar source and a sugar sink; **E.** cohesion and adhesion of water to the phloem walls

7. Each of the following pairs represents a likely sugar source and its sink *except* _____.

   A. an apple seed and the fruit in which it is contained; **B.** mature leaves of a tomato plant and a developing tomato; **C.** newly emerged leaves on a bean seedling and the seedling's developing roots; **D.** a sugar beet in spring and its developing flower stalk; **E.** the roots of an oak tree and its leaf buds in spring

8. A short-day plant can be prevented from flowering by _____.

   A. keeping it constantly in the light; **B.** preventing phytochrome from converting into the active form; **C.** ensuring that day length does not exceed the plant's critical length; **D.** growing it in the same environment as a long-day plant; **E.** more than one of the above is correct

9. Apical dominance is caused by _____.

   A. phototropism causing branches to grow toward the sun; **B.** auxin suppression of lateral bud growth; **C.** auxin stimulation of cell elongation in the main stem; **D.** b and c are correct; **E.** a, b, and c are correct

10. Plant hormones are unlike animal hormones because _____.

    A. they are produced only in small amounts; **B.** they have different effects on different organs; **C.** they are not produced in specialized glands; **D.** they are mobile and can move throughout the plant body; **E.** interactions among different plant hormones can determine their ultimate effect on an organ

Analyzing and Applying the Basics

1. You have decided to divide an overgrown day lily and move a portion of it to a new spot in your yard. Thinking about water relations in plants, how should you prepare the site, and the lily, for transplanting?

2. Poinsettias are short-day plants with a critical night length of about 14 hours. Describe how you could induce a poinsettia to flower during the long days of summer.

3. Roses are among the most sensitive cut flowers. They drop their petals more rapidly if they are stored in a cooler with ripening fruit. Given your understanding of plant hormones, why might this be the case?

Connecting the Science

1. In even the smallest apartments, there is usually enough window space and light to grow a few herbs or small lettuce plants, and people with larger yards can grow a significant amount of the fresh produce they consume during the growing season. Do you grow food to eat? Why or why not? Do you see a value in growing some of your own food?

2. In the last paragraph of the chapter, we suggest taking a gardener's approach—working with natural processes and having a desire to understand—in all of our interactions with the natural world. Describe how this approach could be applied to a current issue concerning human modification of nature. Do you agree that it might be helpful to "think like a gardener," or do you see disadvantages of this attitude as well?

Appendix
Metric System Conversions

To Convert Metric Units:	Multiply by:	To Get English Equivalent:
Length		
Centimeters (cm)	0.3937	Inches (in)
Meters (m)	3.2808	Feet (ft)
Meters (m)	1.0936	Yards (yd)
Kilometers (km)	0.6214	Miles (mi)
Area		
Square centimeters (cm^2)	0.155	Square inches (in^2)
Square meters (m^2)	10.7639	Square feet (ft^2)
Square meters (m^2)	1.1960	Square yards (yd^2)
Square kilometers (km^2)	0.3831	Square miles (mi^2)
Hectare (ha) (10,000 m^2)	2.4710	Acres (a)
Volume		
Cubic centimeters (cm^3)	0.06	Cubic inches (in^3)
Cubic meters (m^3)	35.30	Cubic feet (ft^3)
Cubic meters (m^3)	1.3079	Cubic yards (yd^3)
Cubic kilometers (km^3)	0.24	Cubic miles (mi^3)
Liters (L)	1.0567	Quarts (qt), U.S.
Liters (L)	0.26	Gallons (gal), U.S.
Mass		
Grams (g)	0.03527	Ounces (oz)
Kilograms (kg)	2.2046	Pounds (lb)
Metric ton (tonne) (t)	1.10	Ton (tn), U.S.
Speed		
Meters/second (mps)	2.24	Miles/hour (mph)
Kilometers/hour (kmph)	0.62	Miles/hour (mph)

To Convert English Units:	Multiply by:	To Get Metric Equivalent:
Length		
Inches (in)	2.54	Centimeters (cm)
Feet (ft)	0.3048	Meters (m)
Yards (yd)	0.9144	Meters (m)
Miles (mi)	1.6094	Kilometers (km)
Area		
Square inches (in^2)	6.45	Square centimeters (cm^2)
Square feet (ft^2)	0.0929	Square meters (m^2)
Square yards (yd^2)	0.8361	Square meters (m^2)
Square miles (mi^2)	2.5900	Square kilometers (km^2)
Acres (a)	0.4047	Hectare (ha) (10,000 m^2)
Volume		
Cubic inches (in^3)	16.39	Cubic centimeters (cm^3)
Cubic feet (ft^3)	0.028	Cubic meters (m^3)
Cubic yards (yd^3)	0.765	Cubic meters (m^3)
Cubic miles (mi^3)	4.17	Cubic kilometers (km^3)
Quarts (qt), U.S.	0.9463	Liters (L)
Gallons (gal), U.S.	3.8	Liters (L)
Mass		
Ounces (oz)	28.3495	Grams (g)
Pounds (lb)	0.4536	Kilograms (kg)
Ton (tn), U.S.	0.91	Metric ton (tonne) (t)
Speed		
Miles/hour (mph)	0.448	Meters/second (mps)
Miles/hour (mph)	1.6094	Kilometers/hour (kmph)

Metric Prefixes

Prefix		Meaning
giga-	G	10^9 = 1,000,000,000
mega-	M	10^6 = 1,000,000
kilo-	k	10^3 = 1,000
hecto-	h	10^2 = 100
deka-	da	10^1 = 10
		10^0 = 1
deci-	d	10^{-1} = 0.1
centi-	c	10^{-2} = 0.01
milli-	m	10^{-3} = 0.001
micro-	μ	10^{-6} = 0.000001

$$°C = \frac{°F - 32}{1.8}$$

$$°F = (1.8 \times °C) + 32$$

Answers to Learning the Basics

CHAPTER 1

1. A double blind experiment ensures that participants do not influence the results by reporting what they think is expected based on the hypothesis, and it ensures that data collectors do not err in the measurement of results by emphasizing supportive data and minimizing nonsupportive data.
2. A statistically significant result is one that shows a difference between the experimental and control groups larger than the difference expected due to chance variations between the groups.
3. A correlation allows researchers to test hypotheses that are difficult to test via controlled experiments by looking for a relationship between two factors. They have the disadvantage of lacking controls, meaning that not all alternative hypotheses for the relationship can be excluded.
4. b, **5.** a, **6.** e, **7.** b, **8.** c, **9.** a, **10.** c

CHAPTER 2

1. Sugars; amino acids; glycerol and fatty acids.
2. See Table 2.1
3. The cell is composed of phospholipids arranged in a bilayer, proteins, and cholesterol
4. b, **5.** a, **6.** a, **7.** d, **8.** d, **9.** d, **10.** d

CHAPTER 3

1. All of the building up and breaking down chemical reactions that occur in a cell. Human metabolism is affecetd by sex, age, genetics, exercise level, and weight.
2. Substances that can pass by diffusion include nonpolar molecules and small polar molecules. Carbon dioxide, oxygen, and water can pass. Ions and larger molecules require the help of proteins to pass through membranes.
3. d, **4.** c, **5.** c, **6.** b, **7.** d, **8.** d, **9.** e, **10.** b

CHAPTER 4

1. Cellular respiration reactants are glucose and oxygen, products are carbon dioxide and water. Photosynthesis reactants are carbon dioxide and water, products are glucose and oxygen.
2. c, **3.** d, **4.** c, **5.** e, **6.** a, **7.** c, **8.** e, **9.** d

CHAPTER 5

1. Proto-oncogenes and tumor suppressor genes.
2. Cell division.
3. d, **4.** a, **5.** e, **6.** d, **7.** b, **8.** c, **9.** d **10.** a diploid; b diploid; c haploid; d haploid

CHAPTER 6

Answers to Learning the Basics:

1. A gene is a segment of DNA containing information about making a protein. A protein is a chemical that either makes up part of the structure of a cell or helps perform an essential function of the cell. A trait is the physical outcome of the activity of proteins.
2. Quantitative variation can occur when many genes, each with more than one allele, influence a single trait and/or when the environment influences how a trait develops.
3. We cannot exactly predict the phenotypes of offspring from looking at the phenotypes of their parents, but in many cases we can determine the probability that a particular offspring will have a particular phenotype.
4. a, **5.** b, **6.** b, **7.** b, **8.** c, **9.** e, **10.** d

Answers to Genetics Problems:

1. All have genotype Tt and are tall.
2. 25% have genotype TT and are tall, 50% have genotype Tt and are tall, 25% have genotype tt and are short.
3. Both must be heterozygote (Aa).
4. 50% of their offspring are expected to develop Huntington's disease.
5. (a) 0%; (b) 25%.
6. (a) yellow is dominant; (b) YY (yellow) and yy (green).
7. Yy (yellow) and yy (green).
8. (a) 50%; (b) 50%; (c) 25%; (d) yes.
9. Male parent is Rrhh; female parent is RR Hh. Possible offspring genotypes and phenotypes: RRHh (Red coat, horns), RRhh (Red coat, hornless), RrHh (Roan coat, horns), Rrhh (Roan coat, hornless).
10. (a) *BRCA2* is dominant; (b) It is unlike the typical pattern because not all individuals with *BRCA2* have the trait of early breast or ovarian cancer.

CHAPTER 7

1. See Table 7.1
2. DNA is isolated and cut with restriction enzymes. The fragments are loaded on a gel and an electrical current is applied to separate them; DNA fragments are transferred to a filter, and then probed. The probe binds in complementary regions. X-ray film is placed over filter to generate a fingerprint.
3. Females inherit one X chromosome from their mom in the egg cell and one from their dad in the sperm cell. Males inherit an X chromosome from their mom in the egg cell and a Y chromosome from their dad in the sperm cell.
4. $I^a i$ (dad) $I^B i$ (mom)
5. $1/4$ A+; $1/4$ A−; 1/8 AB−; 1/8 AB+; 1/8 B+; 1/8 B−
6. c, **7.** b, **8.** b, **9.** e **10.** $1/4$ $X^C X^C$ (normal female); $1/4$ $X^C Y$ (normal male); $1/4$ $X^C X^c$ (normal female); $1/4$ $X^c Y$ (colorblind male)

CHAPTER 8

1. GCUAAUGAAU.
2. gln, arg, ile, leu.
3. mRNA, ribosome, amino acids, tRNAs.
4. b, **5.** c, **6.** b, **7.** c, **8.** c, **9.** b, **10.** d

CHAPTER 9

1. The theory of common descent describes that the similarities among living species can be explained as a result of their descent from a common ancestor.
2. Darwin observed that organisms on different islands in the archipelago were similar, but not identical to each other, and were clearly related to each other and to similar organisms on the South American mainland.
3. A vestigial structure is one that has little or no function in an organism but appears to be similar to a more useful structure in other organisms. These structures provide support for the theory of common descent because they are best explained as having evolved from a functional structure in a common ancestor.
4. d, **5.** d, **6.** b, **7.** d, **8.** e, **9.** e, **10.** d

CHAPTER 10

1. Fitness is a term that describes the survival and reproductive output of an individual in a population relative to other members of the population.
2. Artificial selection is a process of selection of plants and animals by humans who control the survival and reproduction of members of a population in order to increase the frequency of human-preferred traits. Artificial selection is like natural selection in that it causes evolution; however, it differs because humans are directly choosing which organisms reproduce. In natural selection, environmental conditions cause one variant to have higher fitness than other variants.
3. Within a single patient, an HIV population can consist of up to 1 billion different virus variants. These variants differ in a number of traits, including their susceptibility to a particular anti-HIV drug. When the patient takes an anti-HIV drug, the individual viruses that are more resistant to the drug have higher fitness, and natural selection leads to the evolution of a virus population that is resistant to the drug.

4. c, **5.** d, **6.** e, **7.** e, **8.** c, **9.** d, **10.** a

CHAPTER 11

1. The gene pools of populations must become isolated from each other, they must diverge (as a result of natural selection or genetic drift) while they are isolated, and reproductive incompatibility must evolve.
2. Yes, as long as the gene pools of the populations are isolated. An example is when the timing of reproduction in two populations of a species is different.
3. Genetic drift can occur as a result of founder's effect, the bottleneck effect, or by the chance loss of alleles from small, isolated gene pools. By changing allele frequencies, all of these events result in the evolution of a population.

4. d, **5.** b, **6.** a, **7.** b, **8.** d, **9.** c, **10.** b

CHAPTER 12

1. Double fertilization, which triggers increased food supply to the embryo; the development of fruit, aiding in dispersal; development of the flower, increasing the probability of fertilization; the development of defensive compounds, deterring herbivores.
2. Competition between organisms and the risk of predation favors the evolution of defensive chemicals, some of which may be valuable to humans. Organisms that live on or in other living organisms must have some way to evade their host's immune system, another potentially valuable characteristic.
3. By comparing the DNA sequences of organisms to see if the pattern of similarity and difference matches what is predicted; and by examination of the fossil record, looking for the record of evolutionary change in the group of organisms.

4. b, **5.** e, **6.** a, **7.** d, **8.** d, **9.** c, **10.** b

CHAPTER 13

1. A decrease in death rate, especially a decrease in infant mortality, has led to increasing growth rates. In addition, the process of exponential growth lends itself to these population explosions, as the number of people added in any year is a function of the population of the previous year.
2. In most populations, growth rate declines because death rate increases (or birth rate decreases) as resources are "used up" by the population.
3. Demographic momentum occurs because a population continues to grow in number even if birth rates drop, as large numbers of individuals who have not yet reached reproductive maturity add offspring to the population in subsequent years.

4. b, **5.** c, **6.** b, **7.** b, **8.** d, **9.** d, **10.** e

CHAPTER 14

1. By examining the fossil record and determining the average life span of various species. The most serious problem with these estimates is that the fossil record over-represents long-lived species and thus overestimates the life span of these species.
2. Habitat fragmentation endangers species because it interferes with their ability to disperse from unsuitable to suitable habitat and because it increases their exposure to humans and human-modified environments. Species that have very specialized habitat requirements, those that cannot move rapidly, and those that are very susceptible to human disturbance are most negatively affected by fragmentation.
3. Mutualism is a relationship among species where all partners benefit. Predation is a relationship among species where one benefits and others are consumed. Competition is a relationship among species where all partners are harmed by the presence of the others.

4. d, **5.** c, **6.** d, **7.** b, **8.** a, **9.** d, **10.** a

CHAPTER 15

1. The tilt of Earth's axis means that as the planet revolves around the sun, the northern hemisphere is tilted toward the sun during part of the year and away during part of the year. The temperature at a given point on Earth's surface is determined in large part by the solar irradiance (i.e. the strength of sunlight) striking the surface. Because the northern hemisphere is tilted away in the winter, solar irradiance is low, as are temperatures. The opposite occurs in the summer.
2. Water is slow to gain and lose heat, so when surrounding land areas cool as a result of declining solar irradiance, warmer air from over the water keeps the temperature high. The opposite is true as the surrounding land warms as a result of increasing solar irradiance.
3. All forests require significant precipitation to support the growth of trees. The major difference between the types is in average temperature. Tropical forests are in climate regions that rarely or never reach freezing. The forests are dominated by broad-leaf evergreens. Temperate forests experience hot/cold seasonality, and are dominated by broad-leafed deciduous trees, which drop their leaves during the freezing (and thus, dry) winters. Boreal forests form where winters are very long and growing seasons short. The evergreen coniferous trees there are able to quickly take advantage of warm weather for photosynthesis, although they can survive long periods of freezing without needing to drop their leaves.

4. c, **5.** e, **6.** e, **7.** e, **8.** c, **9.** a, **10.** d

CHAPTER 16

1. Epithelia are tightly packed sheets of cells that cover and line organs, vessels, and body cavities. Connective tissues are loosely organized tissues that bind organs and tissues to each other. Muscle tissue consists of bundles of long, thin cylindrical muscle fibers that are able to contract. Nervous tissue is composed mainly of neurons that can conduct and transmit electrical impulses.
2. Different connective tissues are composed of different cell types each embedded in a matrix with a unique composition.
3. Muscle tissues differ based on the presence or absence of striations and whether the muscle is voluntary or involuntary. Cardiac muscle is striated

and involuntary; skeletal is striated and voluntary; smooth muscle is not striated and involuntary.

4. c, **5.** e, **6.** a, **7.** a, **8.** d, **9.** b, **10.** c

CHAPTER 17

1. Contraction of the diaphragm flattens out this dome-shaped muscle at the base of the chest cavity. This has the effect of increasing the size of the lung capacity, lowing air pressure and allowing outside, oxygen-rich air to flow through the upper respiratory system and into the lungs.
2. Newly oxygenated blood from the lungs flows into the left atrium of the heart during the relaxation (diastole) phase of a heart beat. Contraction of the heart forces this blood from the left atrium into the left ventricle, where it is pumped into the systemic circulation. The blood flows through arteries to capillaries in the body tissues, where its oxygen is unloaded, nutrients and wastes are picked up and/or unloaded, and carbon dioxide is picked up. The blood then enters the veins, which carry it to the right atrium. A contraction of the heart sends this blood to the right ventricle and then into the lungs, where carbon dioxide is unloaded and a new load of oxygen is picked up.
3. The first step, filtration, occurs when blood plasma is forced out of tiny holes in the capillaries surrounding the nephron into the nephron. The second stage is reuptake, when water and other important constituents are actively recovered from the filtrate and returned to the blood stream. Secretion occurs near the end of the nephron, when toxins and other wastes in low concentration in the blood stream are actively removed.

4. e, **5.** e, **6.** d, **7.** c, **8.** b, **9.** d, **10.** e

CHAPTER 18

1. Bacteria are single celled organisms that have no nucleus or membrane bounded organelles. They house their DNA in the nucleiod region and are surrounded by a cell wall. Viruses are composed of genetic material surrounded by a protein coat called the capsid. Some viruses are also surrounded by a membranous envelope.
2. Bacteria reproduce by binary fission which involves copying the DNA and separating the cell into two. Viruses reproduce by copying their genetic material and synthesizing new coat components. To reproduce viruses must be inside a host cell.
3. B cells produce receptors that are found on cell surfaces or secreted as antibodies and destroy invaders. T cells also produce receptors but not secretable ones.

4. d, **5.** b, **6.** c, **7.** d, **8.** b, **9.** c, **10.** c

CHAPTER 19

1. The hypothalamus produces GnRH. GnRH acts on the pituitary gland, which secretes FSH and LH. FSH and LH act on the testes and ovaries. The testes produce testosterone; the ovaries produce estrogen and progesterone. The adrenal glands produce sex hormones in small amounts.
2. Bone tissue can be tightly or loosely packed. Compact bone forms the hard outer shell of bones, and spongy bone is the porous honeycomb-like inner bone. It has blood vessels throughout. Bone marrow inside bones helps to produce blood cells.
3. A muscle contains bundles of parallel muscle fibers. Within each muscle fiber is a parallel array of thread-like filaments called myofibrils. Each myofibril is a linear arrangement of sarcomeres. A sarcomere is composed of thin filaments of the protein actin and thick filaments of the protein myosin.

4. c, **5.** d, **6.** a, **7.** b, **8.** b, **9.** b, **10.** a

CHAPTER 20

1. Asexual reproduction occurs when one parent produces offspring that are genetically identical to the parent and to each other. Sexual reproduction requires genetic input from two parents.
2. See Figures 20.5 and 20.6
3. See Figures 20.7 and 20.8

4. a, **5.** a, **6.** e, **7.** d, **8.** a, **9.** d, **10.** e

CHAPTER 21

1. The CNS is composed of the brain and spinal cord and is responsible for integrating, processing, and coordinating information taken in by the senses. It is also the seat of functions such as intelligence, learning, memory, and emotion. The PNS includes the network of nerves outside the brain and spinal cord and functions to link the CNS with senses.
2. Receptors for neurotransmitters are located on the pre- and post-synaptic neuron of many brain structures.
3. The binding of neurotransmitter to the postsynaptic membrane stimulates a rapid change in the uptake of sodium ions by the postsynaptic cell. Sodium channels open and sodium ions flow inward, causing another depolarization and generating another action potential.

4. d, **5.** d, **6.** e, **7.** d, **8.** d, **9.** d, **10.** e

CHAPTER 22

1. See Figures 22.3 and 22.10
2. Pollen is carried by an animal or the wind from the anther of one flower to the carpel of another. The pollen lands on the stigma of the carpel, where it germinates into a long tube that grows down the style into an ovary. Once the pollen tube reaches an ovule, two sperm nuclei are released; one fuses with the egg and another fuses with nuclei in accessory tissue, leading the formation of a triploid tissue called endosperm. The resulting embryo and endosperm begin to divide, forming the embryonic plant and its food source. The accessory tissues of the ovule become the seed coat, while the remains of the flower become a fruit.

3. e., **4.** d, **5.** d, **6.** b, **7.** e, **8.** a, **9.** e, **10.** e

CHAPTER 23

1. Hydrogen bonding among water molecules (cohesion) and between water molecules and xylem cell walls (adhesion) provides a force that counters the strong negative pressure.
2. The C4 adaptation pumps carbon dioxide towards actively photosynthetic cells, allowing stomata to remain only partially open. The CAM adaptation allows plants to accumulate carbon dioxide during the cooler night when less water will evaporate so that they may keep their stomata closed during the hottest parts of the day.
3. A dark period that is shorter than the critical night length. Alternatively, if the dark period is longer than the critical night length, a brief flash of light during this period should promote flowering.

4. d, **5.** b, **6.** d, **7.** a, **8.** a, **9.** d, **10.** c

Glossary

ABO blood system A system for categorizing human blood based on the presence or absence of carbohydrates on the surface of red blood cells. (Chapter 7)

abscisic acid A hormone in plants associated with dormancy in seeds and buds. (Chapter 23)

abscission The dropping off of leaves, flowers, fruits, or other plant organs. (Chapter 23)

accessory organ Organs including the pancreas, liver, and gallbladder, which aid the digestive system. (Chapter 16)

acetylcholine A neurotransmitter that functions by binding to receptors and altering the permeability of the post-synaptic neuron. Impaired transmission of this neurotransmitter is thought to contribute to Alzheimer's disease. (Chapter 21)

acetylcholinesterase An enzyme in nerve cells that breaks down acetylcholine. (Chapter 21)

acid A substance that increases the concentration of hydrogen ions in a solution. (Chapter 2)

Acquired Immune Deficiency Syndrome (AIDS) Syndrome characterized by severely reduced immune system function and numerous opportunistic infections. Results from infection with HIV. (Chapter 10)

acrosome An organelle at the tip of the sperm cell containing enzymes that help the sperm penetrate the egg cell. (Chapter 20)

actin A protein found in muscle tissue that, together with myosin, facilitates contraction. (Chapters 16, 19)

action potential Wave of depolarization in a neuron propagated to the end of the axon—also called a nerve impulse. (Chapter 21)

activation energy The amount of energy that reactants in a chemical reaction must absorb before the reaction can start. (Chapter 3)

activator A protein that enhances the transcription of a gene. (Chapter 8)

active site Substrate-binding region of an enzyme. (Chapter 3)

active smoker An individual who smokes tobacco. (Chapter 17)

active transport The ATP-requiring movement of substances across a membrane against their concentration gradient. (Chapter 3)

ADD See Attention deficit disorder. (Chapter 21)

adaptation Trait that is favored by natural selection and increases an individual's fitness in a particular environment. (Chapters 9, 10)

adaptive radiation Diversification of one or a few species into large and very diverse groups of descendant species. (Chapter 12)

adenine Nitrogenous base in DNA, a purine. (Chapters 2, 4, 5, 8)

adenosine diphosphate (ADP) A molecule composed of adenine, a sugar, and two phosphate groups. Produced by the hydrolysis of the terminal phosphate bond of ATP. (Chapter 4)

adenosine triphosphate (ATP) A molecule composed of adenine, a sugar, and three phosphate groups that can be hydrolyzed to release energy. Form of energy that cells can use. (Chapters 2, 3, 4)

adhesion The sticking together of unlike materials—often water and a particular surface. (Chapter 23)

adipose tissue Fat-storing connective tissue. (Chapter 16)

adrenal glands Endocrine glands located adjacent to the kidney. Secrete adrenaline in response to stress or excitement, maintains water and salt balance, and secretes small amounts of sex hormones. (Chapter 19)

adrenaline A hormone secreted by the adrenal glands in response to stress or excitement. (Chapter 19)

aerobic An organism, environment, or cellular process that requires oxygen. (Chapter 4)

aerobic respiration Cellular respiration that uses oxygen as the electron acceptor. (Chapter 4)

agarose gel A jelly-like slab used to separate molecules on the basis of molecular weight. (Chapter 7)

agribusiness Farming as a large-scale business operation, including the production, processing, and distribution of food as well as manufacture and distribution of agricultural equipment and chemicals. (Chapter 8)

agriculture Cultivation of crops, raising of livestock; farming. (Chapter 22)

algae Photosynthetic protists. (Chapter 12)

alimentary canal Part of the digestive system that forms a tube extending from the mouth to the anus. Also called the digestive tract. (Chapter 16)

allele frequency The percentage of the gene copies in a population that are of a particular form, or allele. (Chapters 11, 14)

alleles Alternate versions of the same gene, produced by mutations. (Chapters 5, 6, 7, 8, 10, 14)

allopatric Geographic separation of a population of organisms from others of the same species. Usually in reference to speciation. (Chapter 11)

alternation of generations A reproductive cycle in which a haploid phase, called a gametophyte, alternates with a diploid phase, called a sporophyte. (Chapter 22)

alternative hypotheses Factors other than the tested hypothesis that may explain observations. (Chapter 1)

alveoli Sacs inside lungs, making up the respiratory surface in land vertebrates and some fish. (Chapter 17)

Alzheimer's disease Progressive mental deterioration in which there is memory loss along with the loss of control of bodily functions, ultimately resulting in death. (Chapter 21)

amenorrhea Cessation of menstruation. (Chapters 3, 19)

amino acid Monomer subunit of a protein. Contains an amino, a carboxyl, and a unique side group. (Chapters 2, 3, 8)

amnion The fluid-filled sac in which a developing embryo is suspended. (Chapter 20)

anabolic steroids See steroids. (Chapter 19)

anaerobic An organism, environment, or cellular process that does not require oxygen. (Chapter 4)

anaerobic respiration A process of energy generation that uses molecules other than oxygen as an electron acceptor. (Chapter 4)

anaphase Stage of mitosis during which microtubules contract and separate sister chromatids. (Chapter 5)

anchorage dependence Phenomenon that holds normal cells in place. Cancer cells can lose anchorage dependence and migrate into other tissues or metastasize. (Chapter 5)

androgens Masculinizing hormones, such as testosterone, secreted by the adrenal glands. (Chapters 19, 20)

anecdotal evidence Information based on one person's personal experience. (Chapter 1)

anemia Illness resulting from the loss of red blood cells. (Chapter 18)

angiogenesis Formation of new blood vessels. (Chapter 5)

angiosperms Plants in the phyla Angiospermae, which produce seeds borne within fruit. (Chapters 12, 22, 23)

Animalia Kingdom of Eukarya containing organisms that ingest others and are typically motile for at least part of their life cycle. (Chapters 9, 12)

annual growth rate Proportional change in population size over a single year. Growth rate is a function of the birth rate minus the death rate of the population. (Chapter 13)

annuals Plants that complete their life cycle in a single growing season. (Chapters 15, 22, 23)

anorexia Self-starvation. (Chapter 3)

antagonistic muscle pairs A set of muscles whose actions oppose each other. (Chapter 19)

anthers The pollen-containing structure on the stamen of a flower. (Chapter 22)

antibiotics Chemicals that kill or disable bacteria. (Chapter 12,18)

antibody Protein made by the immune system in response to the presence of foreign substances or antigens. Can serve as a receptor on a B cell or be secreted by plasma cells. (Chapters 5, 10, 18)

anticodon Region of tRNA that binds to a mRNA codon. (Chapter 8)

antigen Short for antibody-generating substances, an antigen is a molecule that is foreign to the host and stimulates the immune system to react. (Chapters 10, 18)

antigen receptors Proteins in B- and T-cell membranes that bind to specific antigens. (Chapter 18)

antioxidants Substances present in some foods that are thought to protect the body from the damaging effects of oxygen-free radicals. (Chapter 3)

antiparallel Feature of DNA double helix in which nucleotides face "up" on one side of the helix and "down" on the other. (Chapters 2, 5)

antisense nucleotide sequence Nucleotide sequence that, when transcribed, produces a mRNA complementary to the normally transcribed mRNA of another gene. Binding of antisense and sense mRNA decreases gene expression. (Chapter 8)

apical dominance The influence exerted by the terminal bud in suppressing the growth of axillary or lateral buds. Mediated by the hormone, auxin. (Chapter 23)

apical meristem The actively dividing cells at the tip of stems and roots in plants. (Chapter 22)

appendicular skeleton The part of the skeleton composed of the bones of the hip, shoulder, and limbs. (Chapter 19)

applied research Research that has an immediate and potentially profitable application. (Chapter 8)

aquaporins A transport protein in the membrane of a plant or animal cell that facilitates the diffusion of water across the membrane (osmosis). (Chapter 3)

aquatic Of, or relating to, water. (Chapter 15)

Archaea Domain of prokaryotic organisms made up of species known from extreme environments. (Chapters 2, 12)

aromatase The enzyme that converts testosterone into estrogen. (Chapter 20)

arrector pili Muscles at the base of hairs that raise them above the level of the skin. In humans, stimulation of these muscles results in "goose bumps." (Chapter 9)

arteries Blood vessels that carry oxygenated blood from the heart to body tissues. (Chapter 17)

artificial insemination The practice of collecting semen from a male and manually injecting it into a female's reproductive tract. (Chapter 6)

artificial selection Selective breeding of domesticated animals and plants to increase the frequency of desirable traits. (Chapters 6, 10)

asexual reproduction A type of reproduction in which one parent gives rise to genetically identical offspring. (Chapters 5, 11, 20)

assortative mating Tendency for individuals to mate with someone who is like themselves. (Chapter 11)

asthma A respiratory disease characterized by spasmodic inflammation of the air passages in the lungs and overproduction of mucus. Often triggered by air contaminants. (Chapter 17)

asymptomatic Stage in an infection that is characterized by relatively unnoticeable, or absent symptoms of illness. (Chapter 10)

atherosclerosis Accumulation of fatty deposits within blood vessels; hardening of the arteries. (Chapter 17)

atom The smallest unit of matter that retains the properties of an element. (Chapter 2)

atomic number The number of protons in the nucleus of an atom. Unique to each element, this number is designated by a subscript to the left of the symbol for the element. (Chapter 2)

ATP synthase Enzyme found in the mitochondrial membrane that helps synthesize ATP. (Chapter 4)

atrium An upper chamber of the heart which receives blood from the body or lungs and pumps it to a ventricle. (Chapter 17)

Attention Deficit Disorder (ADD) Syndrome characterized by forgetfulness, distractibility, fidgeting, restlessness, impatience, difficulty sustaining attention in work, play, or conversation, or difficulty following instructions and completing tasks in more than one setting. (Chapter 21)

autoimmune diseases Referring to diseases resulting from an attack by the immune system on normal body cells. (Chapter 18)

autosomes Non-sex chromosomes, of which there are 22 pairs in humans. (Chapters 5, 6, 7)

auxin A class of plant hormones that control cell elongation, among other effects. (Chapter 23)

AV valves Heart valves between the atria and the ventricles. (Chapter 17)

axial skeleton Part of the skeleton that supports the trunk of the body and consists largely of the bones making up the vertebral column or spine and much of the skull. (Chapter 19)

axillary bud A bud located at the junction between a plant stem and leaf. (Chapters 22, 23)

axon Long, wire-like portion of the neuron that ends in a terminal bouton. (Chapter 21)

AZT Drug that inhibits replication of HIV's genetic material while having relatively little effect on the normal replication and function of a patient's cells. (Chapter 10)

B lymphocytes (B cells) White blood cells that develop in bone marrow and recognize and react to small, free-living microorganisms such as bacteria and the toxins they produce. B lymphocytes provide an immune response called humoral immunity. (Chapters 10,18)

background extinction rate The rate of extinction resulting from the normal process of species turnover. (Chapter 14)

Bacteria Domain of prokaryotic organisms. (Chapters 2, 4, 8, 12, 14, 18)

ball and socket joints Joints in the hips and shoulders that enable arms and legs to move in three dimensions (Chapter 19)

bark The outer layer of a woody stem, consisting of cork, cork cambium, and phloem. (Chapter 22)

Barr body Inactivated X chromosome visible in female mammalian cells as a dark body in the nucleus. (Chapter 7)

basal metabolic rate Resting energy use of an awake, alert person. (Chapter 3)

base A substance that reduces the concentration of hydrogen ions in a solution. (Chapter 2)

base-pairing rules In DNA, A pairs with T, and C pairs with G. (Chapters 2, 5)

basic research Research for which there is not necessarily a commercial application. (Chapter 8)

behavioral isolation Prevention of mating between individuals in two different populations based on differences in behavior. (Chapter 11)

benign Tumor that stays in one place and does not affect surrounding tissues. (Chapter 5)

bias Influence of research participants' opinions on experimental results. (Chapter 1)

biennials Plants that complete their life cycle over the course of two growing seasons. (Chapter 23)

bile Mixture of substances produced in the liver that aids in digestion. (Chapter 16)

binary fission Asexual form of bacterial reproduction. (Chapters 18, 20)

binomial Name composed of two parts. (Chapter 9)

biodiversity Variety within and among living organisms. (Chapters 12, 14)

biogeography The study of the geographic distribution of organisms. (Chapter 9)

biological classification Field of science attempting to organize biodiversity into discrete, logical categories. (Chapters 9, 12)

biological diversity Entire variety of living organisms. (Chapter 12)

biological evolution See evolution. (Chapter 9)

biological population Individuals of the same species that live and breed in the same geographic area. (Chapters 9, 11)

biological race Populations of a single species that have diverged from each other. Biologists do not agree on a definition of "race." See also subspecies. (Chapter 11)

biological species concept Definition of a species as a group of individuals that can interbreed and produce fertile offspring but typically cannot breed with members of another species. (Chapter 11)

biomagnification Concentration of toxic chemicals in the higher levels of a food web. (Chapter 22)

biomass The mass of all individuals of a species, or of all individuals on a level of a food web, within an ecosystem. (Chapter 14)

biome A broad ecological community defined by a particular vegetation type (e.g. temperate forest, prairie), which is typically determined by climate factors. (Chapter 15)

biophilia Humans' innate desire to be surrounded by natural landscapes and objects. (Chapter 14)

biopiracy Using the knowledge of the native people in developing countries to discover compounds for use in developed countries. (Chapter 12)

bioprospecting Hunting for new organisms and new uses of old organisms. (Chapter 12)

biopsy Surgical removal of some cells, tissue, or fluid to determine if cells are cancerous. (Chapter 5)

bipedal Walking upright on two limbs. (Chapter 9)

birth rate Number of births averaged over the population as a whole. (Chapter 13)

bladder An organ of the excretory system, which stores urine after it is excreted from the kidneys. (Chapter 17)

blade The broad part of a leaf. (Chapter 22)

blastocyst An embryonic stage consisting of a hollow ball of cells. (Chapter 20)

blind experiments Tests in which subjects are not aware of exactly what they are predicted to experience. (Chapter 1)

blood The combination of cells and liquid that flow through blood vessels in the circulatory system; made up of red blood cells, plasma, white blood cells, and platelets. (Chapters 16, 17, 18)

blood clotting The process by which a blood clot, a mass of the protein fibrin and dead blood cells, forms in the region of blood vessel damage. (Chapter 17)

blood pressure The force of the blood as it travels through the arteries; partially determined by artery diameter and elasticity. (Chapters 3, 17)

blood vessels The structures that carry blood via the circulatory system throughout the body; arteries, capillaries, and veins. (Chapter 17)

body mass index (BMI) Calculation using height and weight to determine a number that correlates to an estimate of a person's amount of body fat with health risks. (Chapter 3)

bolus A ball of chewed food. (Chapter 16)

bone A type of connective tissue consisting of living cells in a matrix rich in collagen and calcium. (Chapters 16, 17, 19)

bone marrow See marrow. (Chapters 17, 18, 19)

boreal forests A biome type found in regions with long, cold winters and short, cool summers. Characterized by coniferous trees. (Chapter 15)

botanist Plant biologist. (Chapter 12)

brain Region of the body where most neural activity takes place, where decisions are reached, and where bodily activities are directed and coordinated. (Chapters 16, 18, 21)

brain stem Region of the brain that lies below the thalamus and hypothalamus that governs reflexes and some involuntary functions such as breathing and swallowing. (Chapter 21)

branch root A plant root that arises from another older root. (Chapter 22)

bronchi The large air passageways from the trachea into the lungs. (Chapter 17)

bronchioles The branching air passageways inside the lungs. (Chapter 17)

bronchitis Inflammation of the bronchi and bronchioles in the lungs. (Chapter 17)

budding An asexual mechanism of propagation in which outgrowths of the parent form and pinch off, producing new individuals. (Chapter 20)

bulbourethral glands A pair of glands at the base of the penis that secrete acid-neutralizing fluids (Chapter 20)

bulbs Plant organs consisting of short stems and multiple leaves modified for food storage. (Chapter 22).

bulimia Binge eating followed by purging. (Chapter 3)

C$_3$ plants Plants that use the Calvin cycle of photosynthesis to incorporate carbon dioxide into a 3-carbon compound. (Chapter 4)

C$_4$ plants Plants that perform reactions incorporating carbon dioxide into a 4-carbon compound that ultimately provides carbon dioxide for the Calvin cycle. (Chapters 4, 23)

calcitonin A hormone produced by the thyroid that helps lower blood calcium levels. (Chapter 19)

calcium Nutrient required in plant cells for the production of cell walls and for bone strength and blood clotting in humans. (Chapters 3, 19)

Calorie A kilocalorie or 1000 calories. (Chapter 3)

calorie Amount of energy required to raise the temperature of one gram of water by 1 °C. (Chapter 3)

Calvin cycle A series of reactions that occur in the stroma of plants during photosynthesis that utilize NADPH and ATP to reduce carbon dioxide and produce sugars. (Chapter 4)

CAM plants Plants that use Crassulacean acid metabolism, a variant of photosynthesis during which carbon dioxide is stored in sugars at night (when stomata are open) and released during the day (when stomata are closed) to prevent water loss. (Chapters 4, 23)

Cambrian explosion Relatively rapid evolution of the modern forms of multicellular life that occurred approximately 550 million years ago. (Chapter 12)

cancer A disease that occurs when cell division escapes regulatory controls. (Chapter 5)

capillaries The smallest blood vessels of the circulatory system, connecting arteries to veins and allowing material exchange across their thin walls. (Chapters 5, 17)

capillary bed A branching network of capillaries supplying a particular organ or region of the body. (Chapter 17)

capsid Protein coat that surrounds a virus. (Chapter 18)

capsule Gelatinous outer covering of bacterial cells that aids in attachment to host cells during an infection. (Chapter 18)

carbohydrate Energy-rich molecule that is the major source of energy for the cell. (Chapters 2, 3, 4)

carbon dioxide Abundant molecule in the atmosphere (CO_2). (Chapters 3, 4, 17, 22)

carcinogens Substances that damage DNA or chromosomes. (Chapters 5, 17)

cardiac cycle The cycle of contraction and relaxation that a normally functioning heart undergoes over the course of a single heart beat. (Chapter 17)

cardiac muscle Muscle that forms the contractile wall of the heart. (Chapter 16)

cardiovascular disease Malfunction of the cardiovascular system, including, but not limited to, heart attack, stroke, and hypertension. (Chapter 17)

cardiovascular system The organ system made up of the heart and circulatory system, including arteries, capillaries, and veins. (Chapter 17)

carpel The female structure of a flower, containing the ovary, style, and stigma. (Chapter 22)

carrier Individual who is heterozygous for a recessive disease allele. (Chapters 6, 7)

carrying capacity Maximum population that the environment can support. (Chapter 13)

cartilage Connective tissue found in the skeletal system that is rich in collagen fibers. (Chapter 16)

catalyst A substance that lowers the activation energy of a chemical reaction, thereby speeding up the reaction. (Chapter 3)

caudate nucleus (*plural:* caudate nuclei) Structure within the cerebral hemispheres that functions as part of the pathway that coordinates movement patterns. (Chapter 21)

CD4 + cell Class of immune-system cells that are susceptible to HIV infection. Most are T4 lymphocytes; they are named for the CD4 receptor on the cell surface. (Chapter 10)

cell Basic unit of life, an organism's fundamental building-block units. (Chapters 2, 3, 9)

cell body Portion of the neuron that houses the nucleus and organelles. (Chapter 21)

cell cycle An ordered sequence of events in the life cycle of a eukaryotic cell from its origin until its division to produce daughter cells. Consists of M, G$_1$, S, and G$_2$ phases. (Chapter 5)

cell division Process a cell undergoes when it makes copies of itself. Production of daughter cells from an original parent cell. (Chapters 5, 6)

cell plate A double layer of new cell membrane that appears in the middle of a dividing plant cell and divides the cytoplasm of the dividing cell. (Chapter 5).

cell wall Tough but elastic structure surrounding plant and bacterial cell membranes. (Chapters 2, 4, 5, 18, 22, 23)

cell-mediated immunity Immunity conferred by T-cell response. (Chapter 18)

cellular respiration Process requiring oxygen by which cells use food to make ATP. (Chapter 4)

cellulose A structural polysaccharide found in cell walls and composed of glucose molecules. (Chapters 2, 4, 5)

central nervous system (CNS) Includes brain and spinal cord and is responsible for integrating, processing, and coordinating information taken in by the senses. It is the seat of functions such as intelligence, learning, memory, and emotion. (Chapter 21)

centriole A structure in animal cells that helps anchor for microtubules during cell division. (Chapters 2, 5)

centromere Region of a chromosome where sister chromatids are attached and to which microtubules bind. (Chapter 5)

cereal grains Plants in the grass family that have been domesticated for human consumption, including rice, wheat, and barley. (Chapter 22)

cerebellum Region of the brain that controls balance, muscle movement, and coordination. (Chapter 21)

cerebral cortex Deeply wrinkled outer surface of the cerebrum where conscious activity and higher thought originate. (Chapter 21)

cerebrospinal fluid Protective liquid bath that surrounds the brain within the skull. (Chapter 21)

cerebrum Portion of the brain in which language, memory, sensations, and decision making are controlled. The cerebrum has two hemispheres—each of which has four lobes. (Chapter 21)

cervix Located at the bottom of the uterus and the top of the vagina, enlarges during childbirth. (Chapter 20)

chaparral A biome characteristic of climates with hot, dry summers and mild, wet winters and a dominant vegetation of aromatic shrubs. (Chapter 15)

checkpoint Stoppage during cell division that occurs to verify that division is proceeding correctly. (Chapter 5)

chemotherapy Using chemicals to try to kill rapidly dividing (cancerous) cells. (Chapter 5)

chlorophyll Green pigment found in the chloroplast of plant cells. (Chapter 4)

chloroplast An organelle found in plant cells that absorbs sunlight and uses the energy derived to produce sugars. (Chapters 2, 4, 12, 23)

cholesterol A steroid found in animal cell membranes that affects membrane fluidity. Serves as the precursor to estrogen and testosterone. (Chapters 2, 3, 19)

chondrocytes Cartilage cells. (Chapter 16)

chromosomes Housed inside the nucleus, subcellular structures composed of a long single molecule of DNA and associated proteins. (Chapters 5, 6, 7, 8, 11)

chyme Partially digested food and enzymes that is passed from the stomach to the intestine. (Chapter 16)

circulatory system The blood vessels which transport blood, nutrients, and waste around the body. (Chapters 5, 16, 17, 18)

cladistic analysis A technique for determining the evolutionary relationships among organisms that relies on identification and comparison of newly evolved traits. (Chapter 12)

classification systems Methods for organizing biological diversity. (Chapters 9, 12)

cleavage Rapid cell division that occurs during animal development. (Chapters 5, 20)

climate The average temperature and precipitation as well as seasonality. (Chapter 15)

climax community The group of species that is stable over time in a particular set of environmental conditions. (Chapter 15)

clinical trial Controlled scientific experiment to determine the effectiveness of novel treatments. (Chapter 5)

clitoris Erectile tissue found in the external genitalia of females that functions in sexual arousal. (Chapter 20)

clonal population Population of identical cells copied from the immune cell that first encounters an antigen. The entire clonal population has the same DNA arrangement, and all cells in a clonal population carry the same receptor on their membrane. (Chapter 18)

cloning Making copies of a gene or an organism that are genetically identical. (Chapter 8)

closed circulatory system A circulatory system in which vessels are continuous and retain the circulating fluid. (Chapter 17)

clumped distribution A spatial arrangement of individuals in a population where large numbers are concentrated in patches with intervening, sparsely populated areas separating them. (Chapter 13)

coagulation When blood turns from a liquid into a solid. (Chapter 18)

codominant Alleles that result in a new protein with a different, but not dominant, activity compared to the normal protein. (Chapters 6, 7)

codons Three-nucleotide sequences of mRNA that tRNA binds with to add an amino acid to the growing protein. The genetic code is read from the mRNA codon. (Chapter 8)

coenzyme (or cofactor) Substances such as vitamins that help enzymes catalyze chemical reactions. (Chapter 3)

cohesion The tendency for molecules of the same material to stick together. (Chapters 2, 23)

colon Tube between the small intestine and anus of the digestive system. (Chapter 16)

combination drug therapy Treatment with at least three different anti-HIV drugs, from two different classes of drugs. The therapy of choice for HIV patients. (Chapter 10)

common descent The theory that all living organisms on Earth descended from a single common ancestor that appeared in the distant past. (Chapter 9)

community A group of interacting species in the same geographic area. (Chapter 14)

compact bone The hard, outer shell of bones. (Chapter 19)

competition Interaction that occurs when two species of organisms both require the same resources within a habitat; competition tends to limit the size of populations. (Chapter 14)

competitive exclusion Process of establishing harmless organisms in an ecosystem that serves to reduce levels or harmful organisms. (Chapter 14)

competitors Species that survive on the same food source or otherwise compete for the same resources. (Chapter 12)

complement proteins Proteins in the blood with which an antibody–antigen complex can combine in order to kill bacterial cells. Enhances the immune response on many levels. (Chapter 18)

complementary base pair Nitrogenous bases that hydrogen bond to each other. In DNA, adenine is complementary to thymine, and cytosine is complementary to guanine. In RNA, adenine is complementary to uracil and guanine to cytosine. (Chapters 2, 5, 7, 8)

complete digestive tract A digestive tube running from the mouth to anus. (Chapter 16)

complete proteins Dietary proteins that contain all the essential amino acids. (Chapter 3)

complex carbohydrates Highly branched polysaccharides. (Chapter 3)

compound A substance consisting of two or more elements in a fixed ratio. (Chapter 2)

coniferous Pertaining to trees and shrubs that produce cones for reproduction. (Chapter 15)

connective tissue Animal tissue that functions to bind and support other tissues. Composed of a small number of cells embedded in a matrix. (Chapter 16)

consilience The unity of knowledge. Used to describe a scientific theory that has multiple lines of evidence to support it. (Chapter 9)

contact inhibition Property of cells that prevents them from invading surrounding tissues. Cancer cells may lose this property. (Chapter 5)

contagious Spreading from one organism to another. (Chapter 18)

continuous variation A range of slightly different values for a trait in a population. (Chapter 6)

control Subject for an experiment who is similar to experimental subject except is not exposed to the experimental treatment. Used as baseline values for comparison. (Chapter 1)

convergent evolution Evolution of same trait or set of traits in different populations as a result of shared environmental conditions rather than shared ancestry. (Chapters 9, 11)

copulation Sexual intercourse (Chapter 20)

coral reef Highly diverse biome found in warm, shallow salt water, dominated by the limestone structures created by coral animals. (Chapters 12, 15)

cork A tissue produced by the cork cambium that provides protection to a woody stem by forming outer bark. (Chapter 22)

cork cambium A meristematic tissue in bark that arises from phloem cells and creates cork cells before being shed. (Chapter 22)

corpus callosum Bundle of nerve fibers at the base of the cerebral fissure that provides a communication link between the cerebral hemispheres. (Chapter 21)

corpus luteum Hormone-producing tissue (the ovarian follicle after ovulation) that makes progesterone and estrogen and degenerates about 12 days after ovulation if fertilization does not occur. (Chapter 20)

correlation Describes a relationship between two factors. (Chapters 1, 6, 11)

cotyledons The first leaves produced by an embryonic seed plant. (Chapter 22)

countercurrent exchange The process by which streams of fluid move in parallel but opposite directions, maximizing material, gas, or heat exchange between them. (Chapter 17)

covalent bond A type of strong chemical bond in which two atoms share electrons. (Chapter 2)

critical habitat Defined by the Endangered Species Act as a habitat designated as crucial to the survival of an endangered species. (Chapter 14)

critical night length The night length that a plant must experience in order to trigger a photoperiodic response. (Chapter 23)

crop rotation The practice of growing different crops at different times on the same field in order to maintain soil fertility and decrease pest damage. (Chapter 22)

cross Mating of two organisms. (Chapter 6)

crossing over Exchange of genetic information between members of a homologous pair of chromosomes. (Chapters 5, 6)

cryptorchidism A developmental defect in which the testes fail to descend to the scrotum. (Chapter 20)

cultural carrying capacity Maximum human population of Earth that provides not only adequate food for all but an adequate quality of life as well. (Chapter 13)

cultural control Control of agricultural pests through environmental techniques, such as crop rotation and maintenance of native vegetation. (Chapter 22)

cuticle The waxy layer on the outer surface of plant epidermal cells. (Chapter 22)

cyst Noncancerous, fluid-filled growth. (Chapter 5)

cytokinesis Part of the cell cycle during which two daughter cells are formed by the cytoplasm splitting. (Chapter 5)

cytoplasm The entire contents of the cell (except the nucleus) surrounded by the plasma membrane. (Chapters 2, 8)

cytosine Nitrogenous base, a pyrimidine. (Chapters 2, 4, 5, 8)

cytoskeleton A network of tubules and fibers that branch throughout the cytoplasm. (Chapter 2)

cytosol The semifluid portion of the cytoplasm. (Chapters 2, 4, 23)

cytotoxic T cells Immune-system cells that attack and kill body cells that have become infected with a virus before the virus has had time to replicate. Release a chemical that causes the plasma membranes of the target cells to leak. (Chapter 18)

data Information collected by scientists during hypothesis testing. (Chapter 1)

daughter cells Offspring cells that are produced by the process of cell division. (Chapter 5)

death rate Number of deaths averaged over the population as a whole. (Chapter 13)

deciduous Pertaining to woody plants that drop their leaves at the end of a growing season. (Chapters 15, 23)

decomposers Organisms, typically bacteria and fungi in the soil, whose action breaks down complex molecules into simpler ones. (Chapter 14)

decomposition The breakdown of organic material into smaller molecules. (Chapter 15)

deductive reasoning Making a prediction about the outcome of a test; "if . . . then" statements. (Chapter 1)

deforestation The removal of forest lands, often to enable the development of agriculture. (Chapters 4, 14)

degenerative disease Disease characterized by progressive deterioration. (Chapter 8)

dehydration A decrease in an organism's required water level. (Chapters 3, 23)

deleterious Causing a negative outcome; especially pertaining to particular alleles. (Chapter 14)

demographic momentum Lag between the time that humans reduce birth rates and the time that population numbers respond. (Chapter 13)

demographic transition The period of time between when death rates in a human population fall (as a result of improved tech-

nology) and when birth rates fall (as a result of voluntary limitation of pregnancy). (Chapter 13)

denaturation In proteins, the process where proteins unravel and change their native shape, thus losing their biological activity. For DNA, the breaking of hydrogen bonds between the two strands of the double-stranded DNA helix, resulting in single-stranded DNA (Chapters 4, 7)

dendrite Portion of the neuron radiating from the cell body. (Chapter 21)

density-dependent factors Factors related to a population's size that influence the current growth rate of a population—for example, communicable disease or starvation. (Chapter 13)

density-independent factors Factors unrelated to a population's size that influence the current growth rate of a population—for example, natural disasters or poor weather conditions. (Chapter 13)

deoxyribonucleic acid (DNA) Molecule of heredity that stores the information required for making all of the proteins required by the cell. (Chapters 2, 5, 6, 7, 8, 10, 12, 18)

deoxyribose The five-carbon sugar in DNA. (Chapters 2, 5, 6, 8)

depolarization Reduction in the charge difference across the neuronal membrane. (Chapter 21)

depression Disease that involves feelings of helplessness and despair, and sometimes thoughts of suicide. (Chapter 21)

desert The biome found in areas of minimal rainfall. Characterized by sparse vegetation. (Chapter 15)

determinate growth Growth of limited duration. (Chapter 22)

developed countries Countries that have completed the process of industrial development and have a high per capita income level. (Chapter 13)

developing world Countries beginning the process of industrial development. (Chapter 13)

development All of the progressive changes that produce an organism's body. (Chapters 12, 20)

diabetes Disorder of carbohydrate metabolism characterized by impaired ability to produce or respond to the hormone insulin. (Chapter 3)

diaphragm Dome-shaped muscle at the base of the chest cavity. Contraction of this muscle helps draw air into the lungs. (Chapter 17)

diastole The stage of the cardiac cycle when the heart relaxes and fills with blood. (Chapter 17)

diastolic blood pressure The lowest blood pressure in the arteries, occurring during diastole of the cardiac cycle. (Chapters 3, 17)

dicots The class of angiosperms characterized by having two cotyledons. (Chapter 22)

differentiation Structural and functional divergence of cells as they become specialized. (Chapter 20)

diffusion The spontaneous movement of substances from a region of their own high concentration to a region of their own low concentration. (Chapters 3, 17)

digestive system or tract See alimentary canal. (Chapters 16, 17)

dihybrid cross A genetic cross involving two different genes. For example *AABB* × *aabb*. (Chapter 6)

diploid cell A cell containing homologous pairs of chromosomes (2n). (Chapters 5, 6, 20)

directional selection Natural selection for individuals at one end of a range of phenotypes. (Chapter 10)

disaccharide A double sugar consisting of two monosaccharides joined together by a glycosidic linkage. (Chapter 2)

diverge In evolution, divergence occurs when gene flow is eliminated between two populations. Over time, traits found in one population begin to differ from traits found in the other population. (Chapters 9, 11, 12)

diversifying selection Natural selection for individuals at both ends of a range of phenotypes but against the "average" phenotype. (Chapter 10)

dizygotic twins Fraternal twins (nonidentical) that develop from separate zygotes. (Chapter 6)

DNA See deoxyribonucleic acid. (Chapters 2, 5, 6)

DNA fingerprinting Powerful identification technique that takes advantage of differences in DNA sequence by utilizing electrohporesis and single-stranded probes. (Chapter 7)

DNA polymerase Enzyme that facilitates base pairing during DNA synthesis. (Chapter 5)

DNA replication The synthesis of two daughter DNA molecules from one original parent molecule. Takes place during the S phase of interphase. (Chapters 5, 7)

DNA sequence The linear order of nucleotides in a DNA molecule. (Chapters 7, 8, 9, 12)

domain Most inclusive biological category. Life is grouped by many biologists into three major domains. (Chapters 9, 12)

domesticated Referring to animals or plants that are now largely dependent on humans for reproduction and survival as a result of artificial selection. (Chapter 22)

dominant Applies to an allele with an effect that is visible in a heterozygote. (Chapter 6)

dopamine Neurotransmitter that controls emotions and complex movements. (Chapter 21)

dormancy A condition of arrested growth and development in plants. (Chapter 23)

dormant In a state of dormancy; i.e. alive, but not growing or developing. (Chapter 22)

double blind Experimental design protocol when both research subjects and scientists performing the measurements are unaware of either the experimental hypothesis or who is in the control or experimental group. (Chapter 1)

double circulation The flow of blood through two circuits in an animal body, from heart to lungs, back to heart, and to the body. (Chapter 17)

double fertilization The fusion of egg and sperm and the simultaneous fusion of a sperm with nuclei in the ovule of angiosperms. (Chapter 22)

drug resistance In pathogens, it occurs when the pathogen is no longer susceptible to the effects of a drug,; thus, infections are no longer controlled by drug treatment. (Chapters 10, 18)

ecological footprint A measure of the natural resources used by a human population or society. (Chapter 15)

ecology Field of biology that focuses on the interactions between organisms and their environment. (Chapters 12, 13)

ecosystem All of the organisms and natural features in a given area. (Chapter 14)

ecosystem services Proper functioning of the natural world's ecosystems. (Chapter 14)

ecotourism The visitation of specific geographical sites by tourists interested in natural attractions, especially animals and plants. (Chapter 14)

ectoderm The outermost of the three germ layers that arise during animal development. (Chapter 20)

ectotherms An animal that must use energy from the sun and other environmental sources to regulate its body temperature. (Chapter 16)

effector Muscle, gland, or organ stimulated by a nerve. (Chapter 21)

egg cell Gamete produced by a female organism. (Chapters 5, 6, 7, 11, 12, 19, 20)

electron A negatively charged subatomic particle. (Chapters 2, 4)

electron transport chain A series of proteins in the mitochondrial and chloroplast membranes that move electrons during the redox reactions that release energy to produce ATP. (Chapter 4)

element A substance that cannot be broken down into any other substance. (Chapter 2)

embryo Developing individual. (Chapters 6, 7, 8, 12, 20)

emerging viruses Viruses whose incidence is on the rise. (Chapter 18)

emphysema A lung disease caused by the breakdown of alveoli walls; characterized by shortness of breath and an expanded chest cavity. (Chapter 17)

encephalopathy Pathology, or disease, of the brain. (Chapter 18)

Endangered Species Act (ESA) U.S. law intended to protect and encourage the population growth of threatened and endangered species enacted in 1973. (Chapter 14)

endocrine disrupters Chemicals that disrupt the functions of the hormone-producing endocrine system. (Chapter 20)

endocrine glands Glands that secrete hormones into the bloodstream. (Chapter 19)

endocrine system The internal system of chemical signals involving hormones, the organs and glands that produce and secrete them, and the target cells that respond to them. (Chapter 19)

endocytosis The uptake of substances into cells by a pinching inward of the plasma membrane. (Chapter 3)

endoderm The innermost of the three germ layers that arise during animal development. (Chapter 20)

endometriosis The abnormal occurrence of functional endometrial tissue outside the uterus resulting in painful menstrual cycles. (Chapter 20)

endometrium Lining of the uterus, shed during menstruation. (Chapter 20)

endoplasmic reticulum (ER) A network of membranes in eukaryotic cells. When rough, or studded with ribosomes, functions as a workbench for protein synthesis. When devoid of ribosomes, or smooth, it functions in phospholipid and steroid synthesis and detoxification. (Chapter 2)

endoskeleton A hard skeleton buried within the soft tissues of an animal. (Chapter 19)

endosperm The triploid structure formed by the fusion of a sperm and two nuclei in the ovule of an angiosperm plant. (Chapter 22)

endosymbiotic theory Theory that organelles such as mitochondria and chloroplasts in eukaryotic cells evolved from prokaryotic cells that took up residence inside ancestral eukaryotes. (Chapter 12)

endotherms An animal that uses its own metabolic energy to maintain a constant body temperature. (Chapter 16)

energy shell Different states of energy for electrons in an atom. (Chapter 2)

environmental tobacco smoke (ETS) The tobacco smoke in the air that results from smoldering tobacco on the lit ends of cigarettes and pipes as well as the smoke exhaled by active smokers. (Chapter 17)

enzyme Protein that catalyzes and regulates the rate of metabolic reactions. (Chapters 2, 3, 4, 10)

epidemic Contagious disease that spreads rapidly and extensively among many individuals. (Chapter 18)

epidemiologist Scientist who attempts to determine who is prone to a particular disease, where risk of the disease is highest, and when the disease is most likely to occur. (Chapter 18)

epidemiology The study of the cause and spread of disease. (Chapter 17)

epidermal tissue The outermost layer of cells of the leaf and of young stems and roots. (Chapter 22)

epididymis A coiled tube located adjacent to the testes where sperm are stored. (Chapter 20)

epiglottis Flap that blocks the windpipe so food goes down the pharynx, not into the lungs. (Chapter 16)

epithelial tissue (epithelia) Tightly packed sheets of cells that line organs and body cavities. (Chapter 16)

equator The circle around Earth that is equidistant to both poles. (Chapter 15)

esophagus Tube that conducts food from the pharynx to the stomach. (Chapter 16)

essential amino acids The 8 amino acids that humans cannot synthesize and thus must be obtained from the diet. (Chapter 3)

essential fatty acids Fatty acids that animals cannot synthesize and must be obtained from the diet. (Chapter 3)

estrogens Feminizing hormones secreted by the ovary in females and adrenal glands in both sexes. (Chapters 19, 20)

estuary An aquatic biome that forms at the outlet of a river into a larger body of water such as a lake or ocean. (Chapter 15)

ethylene A plant hormone involved in fruit ripening and abscission. (Chapter 23)

eugenics Science of "improving" the human species through selective breeding. (Chapter 6)

eukaryotes Cells that have a nucleus and membrane-bounded organelles. (Chapters 2, 8, 12)

eutrophication Process resulting in periods of dangerously low oxygen levels in water, sometimes caused by high levels of nitrogen and phosphorus from fertilizer runoff that result in increased growth of algae in waterways. (Chapters 14, 22)

evolution Changes in the features (traits) of individuals in a biological population that occur over the course of generations. See also theory of evolution. (Chapters 1, 2, 9, 10)

evolutionary classification System of organizing biodiversity according to the evolutionary relationships among living organisms. (Chapter 12)

excretion The release of urine from the kidneys into the storage organ of the bladder. (Chapter 17)

excretory system The organ system that filters blood and removes water soluble wastes from the blood. (Chapter 17)

exocytosis The secretion of molecules from a cell via fusion of membrane-bounded vesicles with the plasma membrane. (Chapter 3)

exoskeleton A hard outer encasement on the surface of an animal. (Chapter 19)

experiments Contrived situations designed to test specific hypotheses. (Chapter 1)

exponential growth Growth that occurs in proportion to the current total. (Chapter 13)

external fertilization Fertilization that does not require that the parents ever have physical contact with each other. Seen in many aquatic invertebrates and most fish and amphibians. (Chapter 20)

extinction Complete loss of a species. (Chapter 14)

facilitated diffusion The spontaneous passage of molecules, through membrane proteins, down their concentration gradient. (Chapter 3)

falsifiable Able to be proved false. (Chapter 1)

fat Hydrophobic molecule composed of a 3-carbon glycerol skeleton bonded to three fatty acids. Energy source that contains more calories than an equal weight of carbohydrates or proteins. (Chapters 2, 3, 4)

fatty acid A long acidic chain of hydrocarbons bonded to glycerol. Fatty acids vary on the basis of their length and on the number and placement of double bonds. (Chapters 2, 3)

feces Waste material excreted from the rectum. (Chapter 16)

femurs Bones extending from the pelvis to the knee, also called thighbones. (Chapter 19)

fermentation A process that makes a small amount of ATP from glucose without using an electron transport chain. Ethyl alcohol and lactic acid are produced by this process. (Chapter 4)

fertility Ability to produce healthy sperm and egg cells. (Chapter 20)

fertilization The fusion of haploid gametes (egg and sperm) to produce a diploid zygote. (Chapters 5, 6, 7, 20)

fertilizer Any of a variety of growth medium amendments that increase the level of plant nutrients. (Chapter 22)

fetus The term used to describe a developing human from the ninth week of development until birth. (Chapter 20)

fibrin A protein produced during the process of blood clotting that makes up a net to trap and block blood flow from a damaged blood vessel. (Chapter 17)

fibroblasts A protein-secreting cell found in loose connective tissue. (Chapter 16)

fibrous connective tissue A dense tissue found in tendons and ligaments composed largely of collagen fibers. (Chapter 16)

fibrous roots A root system, typically found in monocots, dominated by many thin roots rather than a single tap root. (Chapter 22)

filament The stalk of a stamen. (Chapter 22)

filtration In the kidneys, the removal of plasma from the blood stream through capillaries surrounding a nephron. (Chapter 17)

fissure Groove in the brain that divides the cerebrum and its cortex from front to back, into the right and left cerebral hemispheres. (Chapter 21)

fitness Relative survival and reproduction of one variant compared to others in the same population. (Chapters 10, 14)

flagellum (*plural:* flagella) Whip-like structure found on some bacterial cells, which aids in motility. (Chapter 18)

flower Reproductive structure of a flowering plant. (Chapters 22, 23)

flowering plants Division of the kingdom Plantae containing members that produce flowers and fruit. (Chapter 12)

fluid mosaic model The accepted model for how membranes are structured with proteins bobbing in a sea of phospholipids. (Chapter 2)

follicle Structure in the ovary that contains the developing ovum and secretes estrogen. (Chapters 5, 19, 20)

follicle cells A single layer of flattened cells that surrounds each primary oocyte. (Chapter 20)

follicle-stimulating hormone (FSH) Hormone secreted by the pituitary gland involved in sperm production, regulation of ovulation, and regulation of menstruation. (Chapters 19, 20)

food chain The linear relationship between trophic levels from producers to primary consumers, and so on. (Chapter 14)

food web The feeding connections between and among organisms in an environment. (Chapter 14)

foramen magnum Hole in the skull that allows passage of the spinal cord. (Chapter 9)

forests Terrestrial communities characterized by the presence of trees. (Chapter 15)

fossil fuels Nonrenewable resources consisting of the buried remains of ancient plants that have been transformed by heat and pressure into coal and oil. (Chapters 4, 14)

fossil record Physical evidence left by organisms that existed in the past. (Chapter 9)

fossils Remains of plants or animals that once existed, left in soil or rock. (Chapters 9, 12, 14)

founder effect Type of sampling error that occurs when a small subset of individuals emigrates from the main population to found a new population. Results in differences in the gene pools of source population and the new population. (Chapter 11)

founder hypothesis The hypothesis that the diversity of unique forms in isolated habitats results from divergence of different species from a single founding population. (Chapter 11)

frameshift mutation A mutation that occurs when the number of nucleotides inserted or deleted from a DNA sequence is not a multiple of three. (Chapter 8)

free radical A substance containing an unpaired electron that is therefore unstable and highly reactive causing damage to cells. (Chapter 3)

frontal bone Upper front portion of the cranium, or the forehead. (Chapter 19)

frontal lobe The largest and most anterior portion of each cerebral hemisphere. (Chapter 21)

fruit A mature ovary of an angiosperm, containing the seeds. (Chapters 22, 23)

Fungi Kingdom of eukaryotes made up of members that are immobile, rely on other organisms as their food source, and are made up of hyphae that secrete digestive enzymes into the environment and that absorb the digested materials. (Chapters 9, 12, 14)

gall bladder Organ that stores bile and empties into small intestine. (Chapter 16)

galls Tumor growths on a plant. (Chapter 8)

gamete Specialized sex cells (sperm and egg in humans) that contain half as many chromosomes as other body cells and are therefore haploid. (Chapters 5, 6, 20)

gamete incompatibility An isolating mechanism between similar species in which sperm from one species cannot fertilize eggs from another. (Chapter 11)

gametogenesis The production of gametes. (Chapter 20)

gametophyte In terrestrial plants, the haploid, gamete-producing generation or phase. (Chapter 22)

gas exchange The passage of gases as a result of diffusion, from one compartment to another. For example, in the lungs, oxygen from air spaces in the lungs diffuses into the capillaries, while carbon dioxide in capillaries diffuses into the lung's air spaces. (Chapter 17)

gastrula The 2-layered, cup-shaped stage of embryonic development. (Chapter 20)

gel electrophoresis The separation of biological molecules on the basis of their size and charge by measuring their rate of movement through an electric field. (Chapter 7)

gene Discrete unit of hereditary information consisting of a sequence of DNA that contains information about genetic traits, thus they code for specific proteins. (Chapters 5, 6)

gene expression Turning a gene on or off. A gene is expressed when the protein it encodes is synthesized. (Chapter 8)

gene flow Spread of an allele throughout a species' gene pool. (Chapter 11)

gene gun Device used to shoot DNA-coated pellets into plant cells. (Chapter 8)

gene pool All of the alleles found in the individuals of a species. (Chapter 11)

gene therapy Replacing defective genes (or their protein products) with functional ones. (Chapter 8)

genealogical species concept A scheme that identifies as separate species all populations with a unique lineage. (Chapter 11)

general senses Senses including temperature, pain, touch, pressure, and body position, or proprioception with sensory receptors scattered throughout the body. (Chapter 21)

Generally Recognized As Safe (GRAS) A modified food that does not need to undergo FDA testing because it contains substances that have already been tested. (Chapter 8)

genetic code Table showing which mRNA codons code for which amino acids. (Chapters 8, 9)

genetic drift Change in allele frequency that occurs as a result of chance. (Chapters 11, 14)

genetic engineering Using technology to change one or more genes in an organism. (Chapters 8, 22)

genetic variability All of the forms of genes, and the distribution of these forms, found within a species. (Chapter 14)

Genetically Modified Organisms (GMOs) Transgenic organisms or organisms that have been genetically engineered. (Chapter 8)

genome Entire suite of genes present in an organism. (Chapters 8, 18)

genotype Genetic composition of an individual. (Chapters 6, 7)

genus Broader biological category to which several similar species may belong. (Chapters 9, 11)

germ line gene therapy Gene therapy that changes genes in a zygote or early embryo, thus the embryo will pass on the engineered genes to their offspring. (Chapter 8)

germination The beginning of growth by a seed. (Chapter 22)

germ theory The scientific theory that all infectious diseases are caused by microorganisms. (Chapter 1)

gibberellin A plant harmone that causes the elongation of stem cells. (Chapter 23)

gills The respiratory exchange surface in many aquatic animals, consisting of thin sheets of tissue containing many capillaries. (Chapter 17)

glans penis The head of the penis. (Chapter 20)

glial cells Cells within the brain that do not carry messages but rather support neurons by supplying nutrients, repairing the brain after injury, and attacking invading bacteria. (Chapter 21)

global warming Increases in average temperatures as a result of the release of increased amounts of carbon dioxide and other greenhouse gases into the atmosphere. (Chapters 4, 14)

glycolysis The splitting of glucose into pyruvate which helps drive the synthesis of a small amount of ATP. (Chapter 4)

Golgi apparatus An organelle in eukaryotic cells consisting of flattened membranous sacs that modify and sort proteins and other substances. (Chapter 2)

gonadotropin-releasing hormone (GnRH) Hormone produced by the hypothalamus that stimulates the pituitary gland to release FSH and LH, thereby stimulating the activities of the gonads. (Chapters 19, 20)

gonads The male and female sex organs; testicles in humans males or ovaries in human females. (Chapters 5, 19, 20)

gradualism The hypothesis that evolutionary change occurs in tiny increments over long periods of time. (Chapter 11)

grana Stacks of thylakoids in the chloroplast. (Chapters 2, 4, 22)

grasslands Biomes characterized by the dominance of grasses, usually found in regions of lower precipitation. (Chapter 15)

gravitropism Directional plant growth in response to gravity. (Chapter 23)

gray matter Unmyelinated axons, combined with dendrites and cell bodies of other neurons that appear gray in cross section. (Chapter 21)

greenhouse effect The retention of heat by carbon dioxide and other greenhouse gases. (Chapter 4)

ground tissue The plant tissues other than the vascular tissue, epidermis, and meristem. (Chapter 22)

growing season The length of time from last freeze in spring to first freeze in autumn. (Chapter 23)

growth factor Protein that stimulates cell division. (Chapter 5)

growth rate Annual death rate in a population subtracted from the annual birth rate. A species' growth rate is influenced by how long the species takes to reproduce, how often it reproduces, the number of offspring produced each time, and the death rate of individuals under ideal conditions. (Chapter 13)

guanine Nitrogenous base in DNA, a purine. (Chapters 2, 4, 5, 6, 8)

guard cells Paired cells encircling stomata that serve to regulate the size of the stomatal pore, helping to minimize water loss under dry conditions and maximize carbon dioxide uptake under wet conditions. (Chapters 4, 22)

gymnosperms A group of plants made up of several phyla that produce seeds, but not true flowers. (Chapter 22, 23)

habitat Place where an organism lives. (Chapter 14)

habitat destruction Modification and degradation of natural forests, grasslands, wetlands, and waterways by people; primary cause of species loss. (Chapter 14)

habitat fragmentation Threat to biodiversity caused by humans that occurs when large areas of intact natural habitat are subdivided by human activities. (Chapter 14)

half-life Amount of time required for one-half the amount of a radioactive element that is originally present to decay into the daughter product. (Chapter 9)

haploid Cells containing only one member of each homologous pair of chromosomes (n); e.g. sex cells. (Chapters 5, 20)

hardened Relating to plants that have undergone the appropriate physiological changes that allow them to survive freezing temperatures. (Chapter 23)

hardiness In plants, the ability to survive freezing temperatures. (Chapter 23)

Hardy-Weinberg theorem Theorem that holds that allele frequencies remain stable in populations that are large in size, randomly mating, and experiencing no migration or natural selection. Used as a baseline to predict how allele frequencies would change if any of its assumptions were violated. (Chapter 11)

heart attack An acute condition, during which blood flow is blocked to a portion of the heart muscle, causing part of the muscle to be damaged or die. (Chapters 3, 17)

heart The muscular organ that pumps blood via the circulatory system to the lungs and body. (Chapters 16, 17)

heat The total amount of energy associated with the movement of atoms and molecules in a substance. (Chapter 4)

helper T cells Immune-system cells that enhance cell-mediated immunity and humoral immunity by secreting a substance that increases the strength of the immune response. Also see T4 cell. (Chapters 10, 18)

hemoglobin An iron containing protein that carries oxygen in red blood cells. (Chapter 17)

hemophilia Rare genetic disorder caused by a sex-linked recessive allele that prevents normal blood clotting. (Chapter 7)

hepatocytes Liver cells. (Chapter 16)

herbicide A chemical that kills plants. Herbicides may be specific to particular groups of plants or may be broader spectrum. (Chapter 22)

heritability The amount of variation for a trait in a population that can be explained by differences in genes among individuals. (Chapter 6)

hermaphrodite An individual with both male and female reproductive organs. (Chapter 20)

heterozygote Individual carrying two different alleles for a particular gene. (Chapters 11, 14)

heterozygous Genotype containing two different alleles for a gene. (Chapter 6)

high-density lipoproteins (HDL) A cholesterol carrying particle in the blood that is high in protein and low in cholesterol. (Chapters 3, 17)

hinge joint A joint that allows back and forth movement. (Chapter 19)

HIV See Human Immunodeficiency Virus. (Chapter 10)

homeostasis The steady state condition an organism works to maintain. (Chapters 2, 16)

hominins Humans and human ancestors. (Chapters 9, 11)

homologous pairs Sets of two chromosomes of the same size and shape with centromeres in the same position. Homologous pairs of chromosomes carry the same genes in the same locations but may carry different alleles. (Chapters 5, 6)

homology Similarity in characteristics as a result of common ancestry. (Chapter 9)

homozygous Genotype containing identical alleles for a gene. (Chapters 6, 14)

hormones Substances that travel through the circulatory system and act as signals that turn on genes in their target cells. (Chapter 19)

host Organism infected by a pathogen or parasite. (Chapters 10, 18)

Human Genome Project Effort to determine the nucleotide base sequences and chromosomal locations of all human genes. (Chapter 8)

Human Immunodeficiency Virus (HIV) Agent identified as causing the transmission and symptoms of AIDS. (Chapter 10)

humoral immunity B-cell-mediated immunity that occurs when a B-cell receptor binds to an antigen. The B cell divides to produce a clonal population of memory cells; the cell also produces plasma cells. (Chapter 18)

hybrid Offspring of two different strains of an agricultural crop (see also interspecies hybrid). (Chapters 11, 22)

hybridization and assimilation hypothesis Hypothesis about the origin of modern humans stating that *Homo sapiens* evolved in Africa and spread around the world, interbreeding with *H. erectus* populations already present in Asia and Europe. (Chapter 11)

hydrocarbon A molecule consisting of carbons and hydrogens. (Chapters 2, 3)

hydrogen atom One negatively charged electron and one positively charged proton. (Chapters 2, 4)

hydrogen bond A type of weak chemical bond. In DNA this type of bond forms between nitrogenous bases across the width of the helix. (Chapters 2, 4)

hydrogen ion A single proton with a charge of +1. (Chapter 2)

hydrogenation Adding hydrogen gas under pressure to make liquid oils more solid. (Chapter 3)

hydrophilic Water-loving molecule. (Chapter 2)

hydrophobic Water-hating molecule. (Chapter 2)

hydrostatic skeleton A skeletal system composed of fluid in a closed body compartment. (Chapter 19)

hyperpolarized An electrical state in which the inside of the cell is made more negative than the outside of the cell. (Chapter 21)

hypertension High blood pressure. (Chapters 3, 17)

hyphae Thin, stringy fungous material that grows over and within a food source. (Chapter 12)

hypothalamus Gland that helps regulate body temperature; influences behaviors such as hunger, thirst, and reproduction; and secretes a hormone (GnRH) that stimulates the activities of the gonads. (Chapters 19, 20, 21)

hypothesis Tentative explanation for an observation that requires testing to validate. (Chapters 1, 9, 11)

IDDM (insulin-dependent diabetes mellitus) Type of diabetes that requires insulin injections, Type I. (Chapter 3)

immortal Property of cancer cells that allows them to divide more times than normal cells. (Chapter 5)

immune deficiency Poor immune-system function, usually resulting in increased opportunistic infections. (Chapters 8, 10)

immune response Ability of the immune system to respond to an infection resulting from increased production of B cells and T cells. (Chapters 6, 10, 18)

immune system The organ system that produces cells and cell products, such as antibodies, that help remove pathogenic organisms. (Chapters 8, 10, 18)

insulin A hormone secreted by the pancreas that lowers blood glucose levels by promoting the uptake of glucose by cells and the storage of glucose as glycogen in the liver. (Chapter 3)

insulin-dependent diabetes mellitus (IDDM) See IDDM. (Chapter 3)

In vitro fertilization Fertilization that takes place when sperm and egg are combined in glass or a test tube. (Chapter 8)

inbreeding Mating between related individuals. (Chapter 14)

inbreeding depression Negative effect of homozygosity on the fitness of members of a population. (Chapter 14)

incomplete digestive system A digestive system in which there is only one opening that serves as mouth and anus. (Chapter 16)

incomplete dominance A type of inheritance where the heterozygote has a phenotype intermediate between both homozygotes. (Chapters 6, 7)

independent assortment See Law of independent assortment. (Chapter 6)

indeterminate growth Unrestricted or unlimited growth. (Chapter 22)

induced fit A change in shape of the active site of an enzyme so that in binds tightly to a substrate. (Chapter 3)

inductive reasoning A logical process that argues from specific instances to a general conclusion. (Chapter 1)

infant mortality Death rate of infants and children under the age of 5. (Chapter 13)

infectious When a pathogen finds a tissue inside the body that will support its growth. (Chapter 18)

infertility The inability to conceive after one year of unprotected intercourse. (Chapter 20)

inflammatory response A line of defense triggered by a pathogen penetrating the skin or mucus membranes. (Chapter 18)

inner cell mass A cluster of cells in the blastocyst that eventually develops into the embryo. (Chapter 20)

inorganic Chemical compounds that do not contain carbon. Inorganic fertilizers consist of simple molecules produced by an industrial process. (Chapter 22)

integrated pest management (IPM) A strategy for pest control that relies on a mix of pesticides, cultural control, and better knowledge of pest populations. (Chapter 22)

interferon A chemical messenger produced by virus infected cells that helps other cells resist infection. (Chapter 18)

intermembrane space Space between two membranes. (Chapters 2, 4)

internal fertilization Fertilization that occurs as a result of copulation, or sexual intercourse. (Chapter 20)

interneurons Neurons located between sensory and motor neurons that function to integrate sensory input and motor output. (Chapter 21)

internodes The regions of a plant stem between nodes. (Chapter 22)

interphase Part of the cell cycle when a cell is preparing for division and the DNA is duplicated. Consists of G_1, S and G_2. (Chapter 5)

interspecies hybrid Organism with parents from two different species. (Chapter 11)

intertidal zone The biome that forms on ocean shorelines between the high tide elevation and the low tide elevation. (Chapter 15)

introduced species A nonnative species that was intentionally or unintentionally brought to a new environment by humans. (Chapter 14)

invertebrates Animals without backbones. (Chapter 12)

involuntary muscle Muscle tissue whose action requires no conscious thought. (Chapter 16)

ionic bond A chemical bond resulting from the attraction of oppositely charged ions. (Chapter 2)

IQ test Tool for measuring intelligence that compares an individual's performance with that of peers. (Chapter 6)

irrigation The technique of supplying additional water to crop plants, typically via flooding or spraying. (Chapter 22)

karyotype Picture of chromosomes prepared from blood cells that organizes the chromosomes in homologous pairs. (Chapters 5, 7)

keystone species A species that has an unusually strong effect on the structure of the community it inhabits. (Chapter 14)

kidneys Major organ of the excretory system, responsible for filtering liquid waste from the blood. (Chapters 16, 17)

kingdom In some classifications, the most inclusive group of organisms; usually five or six. In other classification systems, the level below domain on the hierarchy. (Chapters 9, 12)

Krebs cycle A chemical cycle occurring in the matrix of the mitochondria that breaks the remains of sugars down to produce carbon dioxide. (Chapter 4)

labia majora Paired thick folds of skin that enclose and protect the labia minor. (Chapter 20)

labia minora Paired thin folds of skin that enclose the urinary and vaginal openings and clitoris. (Chapter 20)

labor Strong rhythmic contractions that force a developing baby from the uterus through the vagina during childbirth. (Chapter 20)

lactase The enzyme that cleaves the disaccharide lactose into glucose and galactose. Missing or deficient in people with lactose intolerance. (Chapter 3)

lactation Production of milk to nurse offspring. (Chapter 20)

lactose intolerance Inability to digest lactose resulting in bloating, gas, and diarrhea. (Chapter 3)

lakes An aquatic biome that is completely land locked. (Chapter 15)

laparoscope A thin tubular instrument inserted through an abdominal incision and used to view organs in the pelvic cavity and abdomen. (Chapter 5)

large intestine Colon, portion of the digestive system located between the small intestine and the anus that absorbs water and forms feces. (Chapter 16)

larynx A portion of the upper respiratory tract made up primarily of stiff cartilage. Also known as the "voice box." (Chapter 17)

latent Inactive or dormant phase of a disease. (Chapter 18)

law of independent assortment The pattern, in genetic inheritance, in which each chromosomal member of a homologous pair is inherited independently of the other member of the homologous pair. (Chapter 6)

law of segregation The pattern, in genetic inheritance, in which alleles for the same gene are separated before being passed on to the next generation. (Chapter 6)

leaves The primary photosynthetic organs in plants. (Chapters 22, 23)

legumes A family of angiosperms that produces seeds in pods and which establish mutualistic relationships with nitrogen-fixing bacteria. (Chapter 22)

leptin A substance produced by fat cells that may be involved in the regulation of appetite. (Chapter 3)

Leydig cells Cells scattered between the seminiferous tubules of the testicles that produce testosterone and other androgens. (Chapter 20)

life cycle Description of the growth and reproduction of an individual. (Chapter 6)

light reactions A series of reactions that occur on thylakoid membranes during photosynthesis and serve to convert energy from the sun into the energy stored in the bonds of ATP and evolve oxygen. (Chapter 4)

linked genes Genes located on the same chromosome. (Chapter 5)

lipids Hydrophobic molecules including fats, phospholipids, and steroids. (Chapters 2, 3)

lipid bilayer The plasma membrane that surrounds cells is composed two layers of lipids. (Chapter 3)

lipoproteins Cholesterol-carrying proteins. (Chapter 3)

liver Organ with many functions including the production of bile to aid in the absorption of fats. (Chapter 16)

lobules Subdivisions of the lobes of the liver. (Chapter 16)

logistic growth Pattern of growth seen in populations that are limited by resources available in the environment. A graph of logistic growth over time typically takes the form of an S-shaped curve. (Chapter 13)

loose connective tissue Connective tissue that serves to bind epithelia to underlying tissues and to hold organs in place. (Chapter 16)

low-density lipoproteins (LDLs) Cholesterol carrying subsance in the blood that is high in cholesterol and low in protein. (Chapters 3, 17)

lungs The primary organ of the respiratory system; the site where gas exchange occurs. (Chapter 17)

luteinizing hormone (LH) Hormone involved in sperm production, regulation of ovulation, and regulation of menstruation. (Chapters 19, 20)

lymph nodes Organs located along lymph vessels that filter lymph and help defend against bacteria and viruses. (Chapters 5, 18)

lymphatic system A system of vessels and nodes that return fluid and protein to the blood. (Chapters 5, 18)

lymphocyte White blood cells that make up part of the immune system. (Chapters 10, 18)

lysosome A membrane-bounded sac of hydrolytic enzymes found in the cytoplasm of many cells. (Chapter 2)

macromolecules Large molecules including polysaccharides, proteins and nucleic acids, composed of subunits joined by dehydration synthesis. (Chapters 2, 4)

macronutrients Nutrients required in large quantities. (Chapters 3, 22)

macrophage Phagocytic white blood cell that swells and releases toxins to kill bacteria. (Chapter 18)

malaria An infectious disease by a protozoan which is transmitted to humans by the bite of an infected female mosquito. (Chapter 18)

malignant Tumor that invades surrounding tissues. (Chapter 5)

mandible Bone of the lower jaw. (Chapter 19)

marine Of, or pertaining to, salt water. (Chapter 15)

mark-recapture method A technique for estimating population size, consisting of capturing and marking a number of individuals, releasing them, and recapturing more individuals to determine what proportion are marked. (Chapter 13)

marrow Network of connective tissue that fills bones and is involved in the production of red blood cells. (Chapters 17, 18, 19)

mass extinctions Losses of species that are rapid, global in scale, and affect a wide variety of organisms. (Chapter 14)

mass number The sum of the number of protons and neutrons in an atom's nucleus. (Chapter 2)

matrix In a mitochondrion is the semifluid substance inside the inner mitochondrial membrane, which houses the enzymes of the Kreb's cycle (Chapters 2, 4); in connective tissue, a nonliving fibrous substance embedded in the ground substance. (Chapter 16)

mechanical isolation A form of reproductive isolation between species that depends on the incompatibility of the genitalia of individuals of different species. (Chapter 11)

medulla oblongata Region of the brain stem that is a continuation of the spinal cord and conveys information between the spinal cord and other parts of the brain. (Chapter 21)

meiosis Process that diploid sex cells undergo in order to produce haploid daughter cells. Occurs during gametogenesis. (Chapters 5, 7, 11)

memory cells Cells of a clonal population, programmed to respond to a specific antigen, that help the body respond quickly if the infectious agent is encountered again. (Chapter 18)

menopause Cessation of menstruation. (Chapters 19, 20)

menstrual cycle Changes that occur in the uterus and depend on intricate interrelationships among the brain, ovaries, and lining of the uterus. (Chapter 20)

menstruation The shedding of the lining of the uterus during the menstrual cycle. (Chapter 20)

meristematic tissue Undifferentiated plant tissue from which new plant cells arise. (Chapter 22)

mesoderm The middle of three germ layers that arise during animal development. (Chapter 20)

messenger RNA (mRNA) Complementary RNA copy of a DNA gene, produced during transcription. The mRNA undergoes translation to synthesize a protein. (Chapters 8, 10)

metabolic rate Measure of an individual's energy use. (Chapter 3)

metabolism All chemical reactions occurring in the body. (Chapters 2, 3, 4)

metaphase Stage of mitosis during which duplicated chromosomes align across the middle of the cell. (Chapters 5, 11)

metastasis When cells from a tumor break away and start new cancers at distant locations. (Chapter 5)

microbe Microscopic organism, especially Bacteria and Archaea. (Chapters 12, 18)

microbiologists Scientists who study microscopic organisms, especially referring to those who study prokaryotes. (Chapter 12)

microevolution Changes that occur in the characteristics of a population. (Chapter 9)

micronutrients Nutrients needed in small quantities. (Chapters 3, 22)

microorganism See microbe. (Chapter 12)

microtubules Protein structures that move chromosomes around during mitosis and meiosis. (Chapters 2, 5, 6)

microvilli Fine finger-like projections of epithelial cells of the small intestine that function in absorption. (Chapter 16)

micturation Release of urine from the bladder. Also known as urination. (Chapter 17)

midbrain Uppermost region of the brain stem, which adjusts the sensitivity of the eyes to light and the ears to sound. (Chapter 21)

mineral Inorganic nutrient essential to many cell functions. (Chapter 3)

mitochondria Organelles in which products of the digestive system are converted to ATP. (Chapters 2, 4, 12)

mitosis The division of the nucleus that produces daughter cells that are genetically identical to the parent cell. Also, portion of the cell cycle in which DNA is apportioned into two daughter cells. (Chapter 5)

model organisms Nonhuman organisms used in the Human Genome Project that are easy to manipulate in genetic studies and help scientists understand human genes because they share genes with humans. (Chapters 1, 8)

mold A fungal form characterized by rapid, asexual reproduction. (Chapters 9, 12)

molecular clock Principle that DNA mutations accumulate in the genome of a species at a constant rate, permitting estimates of when the common ancestor of two species existed. (Chapter 9)

molecule Two or more atoms held together by covalent bonds. (Chapter 2)

monocot One of the two classes of flowering plants, also called narrow-leaved plants. Monocot seeds contain one leaf (cotyledon). (Chapter 22)

monoculture Practice of planting a single crop over a wide acreage. (Chapter 22)

monomer Individual subunit. (Chapter 2)

monosaccharide Simple sugar. (Chapter 2)

monosomy A chromosomal condition in which only one member of a homologous pair is present. (Chapter 7)

monozygotic twins Identical twins that developed from one zygote. (Chapter 6)

morphological species concept Definition of species that relies on differences in physical characteristics among them. (Chapter 11)

morphology Appearance or outward physical characteristics. (Chapter 11)

motor neurons Neurons that carry information away from the brain or spinal cord to muscles or glands. (Chapter 21)

mulching Covering the ground around favored plants with light-blocking material to prevent weed growth. (Chapter 22)

multicellular The condition of being composed of many coordinated cells. (Chapter 12)

multiple allelism A gene for which there are more than 2 alleles segregating in the population. (Chapter 7)

multiple hit model The notion that many different genetic mutations are required for a cancer to develop. (Chapter 5)

multiregional hypothesis Hypothesis about the origin of modern humans that states that *Homo sapiens* evolved from *H. erectus* separately in Africa, Asia, and Europe. (Chapter 11)

muscle fibers Single cells that align in parallel bundles to form muscles. (Chapter 19)

muscle tissue Tissue composed of muscle cells that contract when stimulated. (Chapter 16)

mutagens Substances that increase the likelihood of mutation occurring; increases the likelihood of cancer. (Chapter 5)

mutations Changes to DNA sequences that may result in the production of altered proteins. (Chapters 5, 6, 8, 10)

mutualism Interaction between two species that provides benefits to both species. (Chapter 14)

mycologists Scientists who specialize in the study of fungi. (Chapter 12)

myelin sheath Protective layer that coats many axons, formed by supporting cells such as Schwann cells. The myelin sheath increases the speed at which the electrochemical impulse travels down the axon. (Chapter 21)

myofibrils Fibrils found in muscle cells and composed of thin filaments of actin and thick filaments of myosin. (Chapter 19)

myosin A type of protein filament that, along with actin, causes muscle cells to contract. (Chapters 16, 19)

NAD (nicotinamide adenine dinucleotide) Molecule that helps transfer enzymes during oxidation reduction reactions. (Chapter 4)

natural experiments Situations where unique circumstances allow a hypothesis test without prior intervention by researchers. (Chapter 6)

natural killer cells Nonspecific cells of the immune system. (Chapter 18)

natural landscape Landscape that is not strongly modified by humans. (Chapter 14)

natural selection Process by which individuals with certain traits have greater survival and reproduction than individuals who lack these traits, resulting in an increase in the frequency of successful alleles and a decrease in the frequency of unsuccessful ones. (Chapters 9, 10, 11, 20)

negative feedback A mechanism of maintaining homeostasis in which the product of the process inhibits the process. (Chapters 16, 20)

nephrons The functional structures within kidneys, where waste filtration and urine concentration occurs. (Chapter 17)

nerve Bundle of neurons; nerves branch out from the brain and spinal cord to eyes, ears, internal organs, skin, and bones. (Chapter 21)

nerve impulses Electrochemical signals that control the activities of muscles, glands, organs, and organ systems. (Chapter 21)

nerve net A weblike system of neurons. (Chapter 21)

nervous system Brain, spinal cord, sense organs, and nerves that connect organs and link this system with other organ systems. (Chapter 21)

nervous tissue Tissue composed of neurons and associated cells. (Chapters 16, 21)

net primary production (NPP) Amount of solar energy converted to chemical energy by plants, minus the amount of this chemical energy plants need to support themselves. A measure of plant growth, typically over the course of a single year. (Chapter 13)

neurobiologist Biologist who studies the nervous system. (Chapter 21)

neurons Specialized message-carrying cells of the nervous system. (Chapters 16, 21)

neurotransmitters Chemicals released by the presynaptic neuron into the synapse, which then diffuse across the synapse and bind to receptors on the membrane of the postsynaptic neuron. Each type of neuron secretes one type of neurotransmitter. Each cell responding to the neurotransmitter has receptors specific to that neurotransmitter. (Chapter 21)

neutral mutation A genetic mutation that confers no selective advantage or disadvantage. (Chapters 8, 21)

neutrons An electrically neutral particle found in the nucleus of an atom. (Chapter 2)

nicotine The active drug in tobacco that stimulates dopamine receptors in the brain. (Chapter 17)

NIDDM (non-insulin-dependent diabetes mellitus) Type of diabetes that does not require insulin injections, Type II. (Chapter 3)

nitrogen Important nutrient for plants in terrestrial systems (N). The fourth most abundant chemical in plant cells. (Chapter 22)

nitrogen-fixing bacteria Organisms that convert nitrogen gas from the atmosphere into a form that can be taken up by plant roots; some species live in the root nodules of legumes. (Chapters 14, 22)

nitrogenous bases Nitrogen-containing bases found in DNA: A, C, G, and T and RNA: U. (Chapters 2, 4, 5, 6, 8)

nodes of Ranvier Small indentations separating segments of the myelin sheath. Nerve impulses "jump" successively from one node of Ranvier to the next. (Chapter 21)

nodes Points on a plant stem where a leaf and axillary bud arise. (Chapter 22)

nodules Compartments housing nitrogen-fixing bacteria, produced on the roots of legume plants. (Chapter 22)

nondisjunction The failure of members of a homologous pair of chromosomes to separate from each other during meiosis. (Chapter 7)

non-insulin-dependent diabetes mellitus See NIDDM. (Chapter 3)

nonrenewable resources Resources that are a one-time supply and cannot be easily replaced. (Chapter 13)

nonspecific defenses Defense system against infection that do not distinguish one pathogen from another. Includes the skin, secretions, and mucous membranes. (Chapter 18)

normal distribution Bell-shaped curve, as for the distribution of quantitative traits in a population. (Chapters 6, 7)

notochord A long flexible rod that runs through the axis of the vertebral body in the future position of the spinal cord. (Chapter 21)

nuclear envelope The double membrane enclosing the nucleus in eukaryotes. (Chapters 2, 5)

nuclear transfer Transfer of a nucleus from one cell to another cell that has had its nucleus removed. (Chapter 8)

nucleases Enzymes that cleave DNA and RNA into their component nucleotides. (Chapter 8)

nucleic acids Polymers of nucleotides that comprise DNA and RNA. (Chapters 2, 4, 5)

nucleoid region The region of a prokaryotic cell where the DNA is located. (Chapter 18)

nucleotides Building blocks of nucleic acids that include a sugar, a phosphate, and a nitrogenous base. (Chapters 2, 4, 5, 6, 7, 8, 10)

nucleus Cell structure that houses DNA; found in eukaryotes. (Chapters 2, 5, 8, 10, 22, 23)

nutrient cycling Process by which nutrients become available to plants. Nutrient cycling in a natural environment relies upon a healthy community of decomposers within the soil. (Chapter 14)

nutrients Atoms other than carbon, hydrogen, and oxygen that must be obtained from a plant's environment for photosynthesis to occur. (Chapters 3, 22, 23)

obesity Condition of having a BMI of 30 or greater. (Chapters 3, 5)

objective Without bias. (Chapter 1)

observations Measurements of nature. (Chapters 1, 10)

observer bias Systematic errors in measurement and evaluation of results made by researchers. (Chapter 1)

occipital lobe The posterior lobe of each cerebral hemisphere, containing the visual center of the brain. (Chapter 21)

ocean A biome consisting of open stretches of salt water. (Chapter 15)

oncogenes Mutant versions of proto-oncogenes. (Chapter 5)

oogenesis Formation and development of female gametes, which occurs in the ovaries and results in the production of egg cells. (Chapter 20)

oogonia Developing egg cells that begin meiosis but pause at prophase I. (Chapter 20)

open circulatory system A circulatory system in which the circulating liquid is released into the body cavity and is returned to the system via pores in the vessels. (Chapter 17)

opportunistic infection Diseases that only occur when a weakened immune system allows access. (Chapter 10)

organ systems Suites of organs working together to perform a function or functions. (Chapter 16)

organelle Subcellular structure found in the cytoplasm of eukaryotic cells that performs a specific job. (Chapters 2, 4)

organic chemistry The chemistry of carbon-containing substances. (Chapter 2)

organic farming Agricultural technique that requires no manufactured chemical inputs (such as inorganic fertilizer or chemical pesticides), and relies on developing and maintaining the health of soil. (Chapter 22)

organic Carbon-containing compound. Alternatively, a fertilizer consisting of complex molecules made up of the partially decomposed waste products of plants and animals. (Chapters 2, 8, 14, 22)

organs A specialized structure composed of several different types of tissues. (Chapter 16)

osmosis The diffusion of water across a selectively permeable membrane. (Chapter 3)

ossa coxae Paired bones that form the bony pelvis. (Chapter 19)

osteoblasts Bone-forming cells responsible for the deposition of collagen. (Chapter 19)

osteoclasts Bone-reabsorbing cells that liberate calcium. (Chapter 19)

osteocytes Highly branched cells found in bone. (Chapter 16)

osteoporosis A condition resulting in an elevated risk of bone breakage from weakened bones. (Chapter 3)

out-of-Africa hypothesis Hypothesis about the origin of modern humans that states that modern *Homo sapiens* evolved in Africa and replaced *H. erectus* populations. (Chapter 11)

ovary (1) In plants, a chamber of the carpel, containing the ovules. (Chapter 22); (2) In animals, the paired abdominal structures that produce egg cells and secrete female hormones. (Chapters 5, 19, 20)

overexploitation Threat to biodiversity caused by humans that encompasses over-hunting and overharvesting. Overexploitation occurs when the rate of human destruction or use of a species outpaces the ability of the species to reproduce. (Chapter 14)

overshoot Occurs when a population exceeds the carrying capacity of the environment. Typically followed by a population crash. (Chapter 13)

oviduct Egg-carrying duct that brings egg cells from ovaries to uterus. (Chapters 5, 20)

ovulation Release of an egg cell from the ovary. (Chapters 5, 20)

ovules Structures in angiosperm plants consisting of eggs and accessory tissues. After fertilization, will develop into a seed. (Chapter 22)

ovum Larger of the two cells produced during oogenesis, and the cell that receives the majority of the cytoplasmic nutrients and organelles. (Chapter 20)

oxytocin Pituitary hormone that stimulates the contraction of smooth muscle of the uterus during labor and facilitates secretion of milk from the breast during nursing. (Chapter 20)

pacemaker A patch of heart tissue or an implanted device that produces a regular electrical signal, setting the rhythm for the cardiac cycle. (Chapter 17)

paleontologist Scientist who searches for, describes, and studies ancient organisms. (Chapter 9)

pancreas Gland that secretes digestive enzymes and insulin. (Chapters 16, 19)

parasites Organisms that feed on other living organisms. (Chapters 14, 18)

parathyroid glands Four endocrine glands located in the thyroid that secrete parathyroid hormone in order to regulate blood calcium levels. (Chapter 19)

parietal lobe Part of the brain that which processes information about touch and is involved in self-awareness. (Chapter 21)

Parkinson's disease Disease that results in tremors, rigidity and slowed movements. May be due to faulty dopamine production. (Chapter 21)

particulates Tiny airborne particles found in smoke and other pollutants. (Chapter 17)

passive smoker An individual who inhales environmental tobacco smoke, but who does not actively consume tobacco. (Chapter 17)

passive transport The diffusion of substances across a membrane with their concentration gradient and not requiring an input of ATP. (Chapter 3)

pathogens Disease-causing organisms. (Chapter 18)

pectinase An enzyme that breaks down cell wall constituents, resulting in ripening of fruit. (Chapter 8)

pedigree Family tree that follows the inheritance of a genetic trait for many generations. (Chapter 7)

peer review The process by which reports of scientific research are examined and critiqued by other researchers before they are published in scholarly journals. (Chapter 1)

penis The copulatory structure in males. (Chapter 20)

peptide bond Chemical bond that joins adjacent amino acids. (Chapter 2)

perennials Plants that live for many years. (Chapters 15, 22, 23)

peripheral nervous system (PNS) Network of nerves outside the brain and spinal cord that links the CNS with sense organs. (Chapter 21)

peristalsis Rhythmic muscle contractions that move food through the digestive system. (Chapter 16)

permafrost Permanently frozen soil. (Chapter 15)

pest Any organism that competes with humans for agricultural production or other resources. (Chapter 22)

pesticide treadmill The tendency for pesticide toxicity and total applications to increase once a farmer begins to employ pesticides. (Chapter 22)

pesticides Chemicals that kill or disable agricultural pests. (Chapter 22)

petal A flower part which typically functions to attract pollinators. (Chapter 22)

petiole The stalk of a leaf. (Chapter 22)

pH A measure of the hydrogen ion concentration ranging from 0-14 with lower numbers equaling higher hydrogen ion concentrations. (Chapter 2)

phagocytosis Ingestion of food or pathogens by cells. (Chapter 18)

pharynx Tube and muscles connecting the mouth to the esophagus; throat. (Chapter 16)

phenotype Physical and physiological traits of an individual. (Chapters 6, 7)

phloem The portion of a plant's vascular tissue modified for multidirectional transport of nutrients throughout the body of the plant. (Chapter 22)

phloem sap The liquid, often containing dissolved carbohydrates, carried in the phloem tubes in a plant. (Chapter 23)

phosphate group A functional group important in energy transferring reactions. (Chapters 2, 4, 5, 8)

phosphodiester bond Chemical bond that joins nucleotides in DNA. (Chapter 6)

phospholipid bilayer The membrane that surrounds cells and organelles and is composed of phospholipids (along with proteins and sometimes cholesterol). (Chapter 2)

phospholipid Molecule that makes up the plasma membrane, with a hydrophilic head and a hydrophobic tail. (Chapter 2)

phosphorus Element required for plant growth that is an important component of proteins and DNA (P). (Chapter 22)

phosphorylation Addition of a phosphate group, thereby energizing some other substance. (Chapter 4)

photoperiodism Response to the duration and timing of day and night length in plants. (Chapter 23)

photorespiration A series of reactions triggered by the closing of stomatal openings to prevent water loss. (Chapter 4)

photosynthesis Process by which plants, along with algae and some bacteria, transform light energy to chemical energy. (Chapters 4, 23)

phototropism Growth in response to directional light. (Chapter 23)

phyla (singular: phylum) The taxonomic category below kingdom and above class. (Chapters 9, 12)

phylogeny Evolutionary history of a group of organisms. (Chapter 12)

phytochrome The light sensitive chemical in plants responsible for photoperiodic responses. (Chapter 23)

pili Surface appendages in some bacteria that facilitate the exchange of genetic information. (Chapter 18)

pituitary dwarfism Lack of normal growth due to a malfunction of the growth hormone producing pituitary gland. (Chapter 8)

pituitary gland Gland located at the base of the skull that secretes growth hormone in addition to other hormones. (Chapters 8, 19)

pivot joint A type of joint that allows freedom of movement. (Chapter 19)

placebo Sham treatments in experiments. (Chapter 1)

placenta Membrane produced by a developing fetus that releases a hormone to extend the life of the corpus luteum. (Chapter 20)

plant hormones Substances in plants that help tune the organisms' response to the environment. Produced in small amounts but are mobile and have large effects on growth and development. (Chapter 23)

Plantae Multicellular photosynthetic eukaryotes, excluding algae. (Chapter 12)

plasma The liquid portion of blood. (Chapter 17)

plasma cells Cells produced by a clonal population that secrete antibodies specific to an antigen. (Chapter 18)

plasma membrane Structure that encloses a cell, defining the cell's outer boundary. (Chapters 2, 3, 18)

plasmid Circular piece of bacterial DNA that normally exists separate from the bacterial chromosome and can make copies of itself. (Chapters 8, 18)

platelets Cells in the blood that carry constituents required for the clotting response. (Chapter 17)

pleiotropy The ability of one gene to affect many different functions. (Chapter 7)

polar molecule A molecule that carries opposite charges on opposite sides. Water is a polar molecule. (Chapter 2)

polar body Smaller of the two cells produced during oogenesis, and therefore does not have enough nutrients to undergo further development. (Chapter 20)

polarization The difference in charge between the inside and outside of the resting neuron cell. (Chapter 21)

poles Opposite ends of a sphere, such as of a cell (Chapter 5) or of a planet such as Earth (Chapter 15).

pollen The male gametophyte of seed plants. (Chapters 11, 12, 14, 22)

pollination The process by which pollen is transferred to the female structures of a plant. (Chapter 22)

pollinators Organisms, such as bees, that transfer sperm (pollen grains) from one flower to the female reproductive structures of another flower. (Chapter 14)

pollution Human-caused threat to biodiversity involving the release of poisons, excess nutrients, and other wastes into the environment. (Chapter 14)

polyculture Practice of planting many different crop plants over a single farm's acreage. (Chapter 22)

polygenic traits Traits influenced by many genes. (Chapters 6, 7)

polymer Combination of monomers. (Chapters 2, 8)

polymerase chain reaction (PCR) A laboratory technique that allows the production of many identical DNA molecules. (Chapter 7)

polyploidy A chromosomal condition involving more than two sets of chromosomes. (Chapter 11)

polysaccharide Complex carbohydrate. (Chapter 2)

polyunsaturated A property of fatty acids resulting from carbon carbon double bonding of fatty acid tails. (Chapter 3)

pond An aquatic biome that is completely landlocked. (Chapter 15)

pons Region of the brain stem that relays messages between the brain and spinal cord. (Chapter 21)

populations Subgroup of a species that is somewhat independent from other groups. (Chapters 11, 13)

population bottleneck Dramatic but short-lived reduction in population size followed by an increase in population. (Chapter 11)

population crash Steep decline in number that may occur when a population grows larger than the carrying capacity of its environment. (Chapter 13)

population cycle In some populations, the tendency to increase in number above the environment's carrying capacity, resulting in a crash, following by an overshoot of the carrying capacity and another crash, continuing indefinitely. (Chapter 13)

population goal Defined by the Endangered Species Act to be the population of an endangered species that would allow it to be removed from the endangered species list. (Chapter 14)

population pyramid A visual representation of the number of individuals in different age categories in a population. (Chapter 13)

positive feedback A relatively uncommon homeostatic mechanism in which the product of a process intensifies the process, thereby promoting change. (Chapters 16, 20)

post-fertilization (postzygotic) Barrier to reproduction that occurs when fertilization occurs as a result of mating between two members of different species, but the resulting offspring does not survive or is sterile. (Chapter 11)

postsynaptic After the synapse. (Chapter 21)

potassium Element required for plant growth that is essential in regulating water balance within cells (K). (Chapter 22)

prairie A grassland biome. (Chapter 15)

precipitation When water vapor in the atmosphere turns to liquid or solid form and falls to Earth's surface. (Chapter 15)

predation Act of capturing and consuming an individual of another species. (Chapter 14)

predator Organism that eats other organisms. (Chapter 14)

prediction Result expected from a particular test of a hypothesis in the hypothesis were true. (Chapter 1)

pre-fertilization (prezygotic) Barrier to reproduction that occurs when individuals from different species either do not attempt to mate,

or if they do mate, fail to produce a fertilized egg. (Chapter 11)

pressure flow mechanism The process by which phloem sap moves from a sugar source to a sugar sink within a plant. (Chapter 23)

presynaptic Before the synapse. (Chapter 21)

primary consumers Organism that eats plants. (Chapter 14)

primary follicle Primary oocyte plus surrounding follicle cells. (Chapter 20)

primary growth Growth occurring at the tips of a plant, originating in the apical meristems. (Chapter 22)

primary oocyte Developing egg cell paused at prophase I of meiosis. (Chapter 20)

primary sources Articles written by researchers and reviewed by the scientific community. (Chapter 1)

primary spermatocyte Diploid cell that begins meiosis in the production of sperm. (Chapter 20)

prion Normally occurring protein produced by brain cells that, when misfolded, causes spongiform encephalopathy; prion is the shortened form of proteinaceous infectious particle. (Chapter 18)

probe Single-stranded nucleic acid that has been radioactively labeled. (Chapter 7)

producers Organisms that produce carbohydrates from inorganic carbon; typically via photosynthesis. (Chapter 14)

products The results of a chemical reaction. (Chapter 2)

progesterone Hormone produced by the ovary, high levels of which have a negative feedback effect on the hypothalamus, causing GnRH secretion to decrease. (Chapter 20)

prokaryotes Cells that do not have a nucleus or membrane-bounded organelles. (Chapters 2, 8, 12)

prolactin A hormone produced by the pituitary gland that stimulates the development of mammary glands. (Chapter 20)

promoter Sequence of nucleotides to which the polymerase binds to start transcription. (Chapter 8)

prophase Stage of mitosis during which duplicated chromosomes condense. (Chapter 5)

prostate A gland in human males that secretes an acid neutralizing fluid into semen. (Chapter 20)

protein Cellular constituents made of amino acids coded for by genes. Proteins can have structural, transport, or enzymatic roles. (Chapters 2, 3, 4, 8, 10, 14)

protein synthesis Joining amino acids together, in an order dictated by a gene, to produce a protein. (Chapter 8)

Protista Kingdom in the domain Eukarya containing a diversity of eukaryotic organisms, most of which are unicellular. (Chapter 12)

proton A positively charged subatomic particle. (Chapters 2, 4)

proto-oncogenes Genes that encode proteins that regulate the cell cycle. Mutated proto-oncogenes (oncogenes) can lead to cancer. (Chapter 5)

prune To trim back the growing tips of a plant. (Chapter 23)

pseudopodia A cellular extension of amoebas used for feeding and in motility. (Chapter 18)

puberty The point in human development when male and female hormones are triggered in the body. Males produce sperm at this time, and females begin the menstrual cycle. (Chapter 19)

pulmonary circuit The path of blood through vessels from the heart, through the lungs, and back to the heart; one half of the double circulation. (Chapter 17)

pulse The volume of blood that passes into the arteries as a result of the heart's contraction. (Chapter 17)

punctuated equilibrium The hypothesis that evolutionary changes occurs rapidly, and in short bursts followed by long periods of little change. (Chapter 11)

Punnett square Table that lists the different kinds of sperm or eggs parents can produce relative to the gene or genes in question and predicts the possible outcomes of a cross between these parents. (Chapter 6)

purine Nitrogenous base (A or G) with a two-ring structure. (Chapter 2)

pyrimidine Nitrogenous base (C, T or U) with a single-ring structure. (Chapter 2)

pyruvic acid The 3-carbon molecule produced by glycolysis. (Chapter 4)

Q angle Angle of the femur in relation to a horizontal line drawn through the kneecap. (Chapter 19)

qualitative traits Traits that come in distinct categories. (Chapter 6)

quantitative traits Traits that have many possible values. (Chapter 6)

race See biological race. (Chapter 11)

racism Idea that some groups of people are naturally superior to others. (Chapter 11)

radiation therapy Focusing beams of reactive particles at a tumor to kill the dividing cells. (Chapter 5)

radioactive decay Natural, spontaneous breakdown of radioactive elements into different elements, or "daughter products." (Chapter 9)

radioimmunotherapy Experimental cancer treatment with the goal of delivering radioactive substances directly to tumors without affecting other tissues. (Chapter 5)

radiometric dating Technique that relies on radioactive decay to estimate a fossil's age (Chapters 9, 14)

random alignment When members of a homologous pair line up randomly with respect to maternal or paternal origin during metaphase I of meiosis, thus increasing the genetic diversity of offspring. (Chapters 5, 7)

random assignment Placing individuals into experimental and control groups randomly to eliminate systematic differences between the groups. (Chapter 1)

random distribution The dispersion of individuals in a population without pattern. (Chapter 13)

random fertilization The unpredictability of exactly which gametes will fuse during the process of sexual reproduction. (Chapter 6)

reabsorption Reuptake of water and other essential substances by a nephron from the filtrate initially squeezed into the kidney. (Chapter 17)

reactants In a chemical reaction, the starting materials. (Chapter 2)

reading frame The grouping of mRNAs into 3 base codons for translation. (Chapter 8)

receptor Protein on the surface of a cell that recognizes and binds to a specific chemical signal. (Chapters 5, 9)

recessive Applies to an allele with an effect that is not visible in a heterozygote. (Chapter 6)

recombinant Produced by manipulating a DNA sequence. (Chapter 8)

recombinant bovine growth hormone (rBGH) Growth hormone produced in a laboratory and injected into cows to increase their size and ability to produce milk. (Chapter 8)

recovery plan Defined by the Endangered Species Act as the plan of action put in place designed to remove a species from the endangered species list. (Chapter 14)

red blood cells Primary cellular component of blood, responsible for ferrying oxygen throughout the body. (Chapters 5, 7, 17)

reflex Automatic response to a stimulus. (Chapter 21)

reflex arc Nerve pathway followed during a reflex consisting of a sensory receptor, a sensory neuron, an interneuron, a motor neuron, and an effector. (Chapter 21)

remission The period during which the symptoms of a disease subside. (Chapter 5)

repolarization In a nerve cell, when potassium ions leave the cell and the inside of the cell again becomes more negative than the outside. (Chapter 21)

repressors Proteins that suppress the expression or transcription of a gene. (Chapter 8)

reproductive isolation Prevention of gene flow between different biological species due to failure to produce fertile offspring; can include premating barriers and postmating barriers. (Chapter 11)

reproductive organs Internal and external genitalia involved in production and delivery of gametes. (Chapters 20, 22)

resources Food, water, shelter, and area required for the survival of a population. (Chapter 13)

respiratory surface Body surface across which gas exchange occurs. (Chapter 17)

respiratory system The organ system involved in gas exchange between an animal and its environment. In humans, the lungs and air passages. (Chapter 17)

restriction enzymes Enzymes that cleave DNA at specific nucleotide sequences. (Chapters 7, 8)

restriction fragment length polymorphisms (RFLP) Differences among members of a population in the number and size of DNA fragments generated by cutting DNA with restriction enzymes. (Chapter 7)

reticular formation Extensive network of neurons that runs through the medulla of the brain and projects toward the cerebral cortex. It functions as a filter for sensory input and activates the cerebral cortex. (Chapter 21)

reuptake Process by which neurotransmitters are reabsorbed by the neuron that secreted them. (Chapter 21)

reverse transcriptase Enzyme in RNA viruses that participates in copying the viral DNA. It performs the reverse of transcription by producing DNA from RNA. (Chapters 10, 18)

Rh factor Surface molecule found on some red blood cells. (Chapter 7)

rhizomes Underground stems in plants. (Chapter 22)

ribose The five-carbon sugar in RNA. (Chapters 5, 8)

ribosomal RNA (rRNA) RNA that makes up part of the structure of ribosomes. (Chapters 8, 12)

ribosomes Cellular structures that help translate genetic material into proteins by anchoring and exposing small sequences of mRNA. (Chapters 2, 8, 12)

risk factors Exposures or behaviors that increase the likelihood of disease. (Chapter 5)

Ritalin Stimulant used to treat ADD. (Chapter 21)

river Aquatic biome characterized by flowing water. (Chapter 15)

RNA (ribonucleic acid) Information-carrying molecule composed of nucleotides. (Chapters 8, 10, 18)

RNA polymerase Enzyme that synthesizes mRNA from a DNA template during transcription. (Chapter 8)

root cap Loosely organized cells that cover the growing tip of a root and are continually shed as the root pushes through the soil. (Chapter 22)

root hairs Extensions of the epidermal cells on young roots, which maximize the surface area for water and nutrient uptake. (Chapter 22)

root system The plant organ system responsible for water and nutrient uptake and anchorage. (Chapter 22)

rubisco Ribulose bisphosphate carboxylase oxygenase, the enzyme that catalyzes the first step in the Calvin cycle of photosynthesis. (Chapter 4)

runoff Water moving across a land surface, picking up contaminants and eventually flowing into lakes, streams, or groundwater. (Chapter 22)

safer sex Practice of minimizing contact with a partner's bodily fluids during sexual activity as prevention against the transmission of HIV and other sexually transmitted diseases. (Chapter 10)

salinization Degradation of soil by mineral salts deposited as a result of irrigation. (Chapter 22)

salivary amylase An enzyme secreted by salivary glands to break down starch. (Chapter 16)

sample Small subgroup of a population used in an experimental test. (Chapter 1)

sample size Number of individuals in both the experimental and control groups. (Chapter 1)

sampling error Effect of chance on experimental results. (Chapter 1)

sarcomere The fundamental repeating unit of striated muscle, bounded by Z-lines. (Chapter 19)

saturated fatty acid Fatty acid in which carbons are bound to as many hydrogens as possible and therefore no carbon carbon double bonds occur. (Chapters 2, 3)

savanna Grassland biome containing scattered trees. (Chapter 15)

Schistosomiasis Tropical disease caused by infestation with certain parasitic worms. (Chapter 18)

Schwann cell Cells that form the myelin sheath along the axons of nerve cells. (Chapter 21)

scientific theory Body of scientifically accepted general principles that explain natural phenomena. (Chapters 1, 9)

scrotum Sac of skin that houses the testes. (Chapter 20)

secondary consumers Animals that eat primary consumers; predators. (Chapter 14)

secondary growth Growth in girth of plants, as a result of cell division in lateral meristems. (Chapter 22)

secondary oocyte Egg cell that is ovulated, having entered metaphase II of meiosis. If the secondary oocyte is fertilized, meiosis begins again from metaphase II. (Chapter 20)

secondary sources Books, news media, and advertisements as sources of scientific information. (Chapter 1)

secondary spermatocyte Haploid cell produced by the first meiotic division in the production of sperm. (Chapter 20)

secretion A step in waste removal in the kidney, in which contaminants in low concentration in the blood are actively absorbed into the urine. (Chapter 17)

seed A plant embryo packaged with a food source and surrounded by a seed coat. (Chapters 12, 22)

seed coat Layer of tissue surrounding a plant embryo and endosperm, arising from accessory tissue in the ovule. (Chapter 22)

seedling A young plant that develops from a germinating seed. (Chapter 22)

segregation See Law of segregation. (Chapter 6)

selective breeding Controlling the breeding of individual organisms to influence the phenotype of the next generation. (Chapter 8)

semen Sperm and energy-rich associated fluids. (Chapter 20)

semilunar valves Heart valves controlling blood flow from the ventricles into blood vessels leading away from the heart. (Chapter 17)

seminal vesicle Gland of the male reproductive system that produces a fructose-rich fluid that helps supply sperm with energy. (Chapter 20)

seminiferous tubules Highly coiled tubes in the testicles where sperm are formed. (Chapter 20)

semipermeable membrane A biological membrane that allows some substances to pass but prohibits the passage of others. (Chapter 2)

sense strand DNA strand of the double helix that codes for a protein. (Chapter 8)

sensory neurons Neurons that carry information toward the brain. (Chapter 21)

sensory receptors Cellular systems that collect information the environment inside or outside the body and transmit that information to the brain. (Chapter 21)

sepals The outermost floral structures, usually enclosing the other flower parts in a bud. (Chapter 22)

separate types hypothesis Hypothesis that numerous types of organisms (e.g., birds, cats, ferns) appeared on Earth separately and each type diversified into many species via evolutionary processes. (Chapter 9)

Sertoli cells Cells that secrete substances that remove some of the excess cytoplasm from spermatids, converting them to spermatozoa (sperm). (Chapter 20)

Severe Combined Immunodeficiency Disorder (SCID) Illness caused by a genetic mutation that results in the absence of an enzyme, and a severely weakened immune system. (Chapter 8)

sex chromosomes The X and Y chromosomes in humans. (Chapters 5, 7)

sex determination Determining the biological sex of an offspring. Humans have a chromosomal mechanism of sex determination in which two X chromosomes produce a female and an X and a Y chromosome produce a male. (Chapter 7)

sex hormones Hormones that affect reproducitve functions inlcuding estrogen and testosterone. (Chapter 19)

sex-linked genes Genes linked to and inherited along with the X and Y chromosomes. (Chapter 7)

sexual reproduction Reproduction involving two parents that give rise to offspring that have unique combinations of genes. (Chapters 5, 12, 20)

sexual selection Form of natural selection that occurs when a trait influences the likelihood of mating. (Chapter 11)

shaman Indigenous healer. (Chapter 12)

shoot system The organs of a plant that are modified for photosynthesis; stem and leaf. (Chapter 22)

side group Of an amino acid, varies from one amino acid to the next and gives an amino acid its particular chemistry. (Chapters 2, 3)

signal transduction When a change in a cell or its environment is relayed through various molecules and results in a cellular response. (Chapter 19)

sink A plant organ that is using carbohydrate, either for growth and metabolism or for conversion into starch. (Chapter 23)

sinoatrial (SA) node Region of the heart muscle that generates an electrical signal which controls heart rate. (Chapter 17)

sister chromatids Duplicated identical copies of a chromosome. (Chapter 5)

skeletal muscle Striated muscle involved with voluntary movements. (Chapter 16)

skeletal system Bones and cartilage that provide support and protection, aid in movement, and store minerals. (Chapter 19)

sliding filament model A theory for how muscles contract hypothesizing that actin filaments slide across myosin filaments shortening the sarcomere. (Chapter 19)

small intestine Large tubular structure in the digestive system that helps break down food and absorbs nutrients into the blood. (Chapter 16)

smog Products of fossil fuel combustion in combination with sunlight, producing a brownish haze in still air. (Chapter 15)

smooth muscle Nonstriated muscle. Lack of striations due to uniform distribution of myosin filaments. (Chapter 16)

social construct Product of history and learned attitudes. (Chapter 11)

sodium potassium pump Protein in the membranes of some animal cells that moves sodium out of the cell and potassium into the cell against their concentration gradients. (Chapter 21)

soil Medium for plant growth made up of mineral particles, partially decayed organic matter, and living organisms. (Chapter 22)

soil erosion Loss of topsoil. (Chapter 22)

solar irradiance The amount of solar energy hitting Earth's surface at any given point. (Chapter 15)

solid waste Garbage. (Chapter 15)

solstice When the sun reaches its maximum and minimum elevation in the sky. (Chapter 15)

solute A substance that is dissolved in a solution. (Chapter 2)

solvent A liquid, such as water, that a solute is dissolved in. (Chapter 2)

somatic cell gene therapy Changes to malfunctioning genes in somatic or body cells. These changes will not be passed to offspring. (Chapter 8)

somatic cells The body cells in a an organism. All cells that are not gametes. (Chapter 5)

somatomammotropin A hormone produced by the placenta that plays a role in stimulating milk production. (Chapter 20)

source Plant organs that are actively generating sugar. (Chapter 23)

spatial isolation A mechanism for reproductive isolation that depends on the geographic separation of populations. (Chapter 11)

special creation The hypothesis that all organisms on Earth arose as a result of the actions of a supernatural creator. (Chapter 9)

special senses Smell, taste, equilibrium, hearing, and vision. The sensory receptors for these five special senses are found in complex sense organs. (Chapter 21)

speciation Evolution of one or more species from an ancestral form; macroevolution. (Chapter 11)

species A group of individuals that regularly breed together and are generally distinct from other species in appearance or behavior. In Linnaeus' classification system, a group in which members have the greatest resemblance. (Chapters 2, 9, 11, 12, 14)

species–area curve Graph describing the relationship between the size of a natural landscape and the relative number of species it contains. (Chapter 14)

specific defense Defense against pathogens that utilizes white blood cells of the immune system. (Chapter 18)

specificity Phenomenon of enzyme shape determining the reaction the enzyme catalyzes. (Chapters 3, 18)

sperm Gametes produced by males. (Chapters 5, 6, 7, 12, 19, 20)

spermatids Cells produced when secondary spermatocytes undergo meiosis II to produce haploid cells that no longer have duplicated chromosomes. (Chapter 20)

spermatogenesis Production of sperm. (Chapter 20)

spermatozoa Mature sperm composed of a small head containing DNA, a midpiece that contains mitochondria, and a tail (flagellum). (Chapter 20)

spinal cord Main nerve pathway between the brain and the rest of the body. (Chapters 16, 21)

spleen An abdominal organ that functions to filter foreign substances from the blood and produce lymphocytes. (Chapter 21)

spongy bone The porous, honeycomb-like material of the inner bone. (Chapter 19)

spores Reproductive cells in plants and fungi that are capable of developing into an adult without fusing with another cell. (Chapters 12, 18, 22)

sporophyte Diploid stage in the alternation of generations in plants, which produces spores via meiosis. (Chapter 22)

SRY Sex determining region of Y chromosome. (Chapter 7)

stabilizing selection   Natural selection that favors the average phenotype and selects against the extremes in the population. (Chapter 10)

stamen   Male flower part, producing the pollen. (Chapter 22)

staple crops   Agricultural crops that make up the majority of calories in human societies. (Chapter 22)

static model   Discarded hypothesis about the origin of living organisms that states that they appeared in the recent past and have not changed over time. (Chapter 9)

statistical tests   Tests that help scientists evaluate whether the results of a single experiment demonstrate the effect of treatment. (Chapter 1)

statistically significant   Low probability that experimental groups differ simply by chance. (Chapter 1)

statistics   Specialized branch of mathematics used in the evaluation of experimental data. (Chapter 1)

stem   The part of vascular plants that is aboveground, as well as similar structures found underground. (Chapter 22)

stem cell   Cells that can divide indefinitely and can differentiate into other cell types. (Chapters 8, 18)

steppe   Biome characterized by short grasses, found in regions with relatively little annual precipitation.(Chapter 15)

steroids   Fat-soluble hormones that can cross cell membranes readily. (Chapters 2, 19)

stigma   Sticky pad on the tip of a flower's carpel where pollen is deposited. (Chapter 22)

stimulant   Drugs that increase CNS activity. (Chapter 21)

stomach   A large organ that plays a major role in digesting food. (Chapter 16)

stomata   Pores on the photosynthetic surfaces of plants that allow air into the internal structure of leaves and green stems. Stomata also provide portals through which water can escape. (Chapters 4, 22, 23)

stop codon   An mRNA codon that does not code for an amino acid and causes the amino acid chain to be released into the cytoplasm. (Chapter 8)

stream   Biome characterized by flowing water, sometimes seasonal. Typically smaller than rivers. (Chapter 15)

striated muscle   Skeletal and cardiac muscle, that is distinguished from smooth muscle by the in register banding patterns of myosin fibers. (Chapter 16)

stroke   Acute condition caused by a blood clot that blocks blood flow to an organ or other region of the body. (Chapters 3, 17)

stroma   The fluid inside a chloroplast. (Chapters 2, 4)

strong inference   A strong statement about the truth of a given hypothesis possible when an experimental protocol greatly minimizes the number of alternative hypotheses that can explain a result. (Chapters 1, 10)

style   Stalk connecting stigma to ovary in a flower's carpel. (Chapter 22)

subject expectation   Conscious or unconscious modeling of behavior the subject thinks a researcher expects. (Chapter 1)

subspecies   Subdivision of a species that is not reproductively isolated but represents a population or set of populations with a unique evolutionary history. See also biological race, variety. (Chapter 11)

substrate   The chemicals metabolized by an enzyme-catalyzed reaction. (Chapter 3)

succession   Replacement of ecological communities over time since a disturbance, until finally reaching a stable state. (Chapter 15)

sugar-phosphate backbone   Series of alternating sugars and phosphates along the length of the DNA helix. (Chapters 2, 5, 6)

superfamily   Taxonomic category between family and order. (Chapter 9)

supernatural   Not constrained by the laws of nature. (Chapter 1)

sustainable   Referring to human activities that are able to continue indefinitely with no degradation of the resources on which they depend. (Chapter 22)

symbiosis   A relationship between two species. (Chapter 12)

sympatric   In the same geographic region. (Chapter 11)

synapse   Gap between neurons consisting of the terminal boutons of the presynaptic neuron, the space between the two adjacent neurons, and the membrane of the postsynaptic neuron. (Chapter 21)

systematist   Biologist who specializes in describing and categorizing a particular group of organisms. (Chapter 12)

systemic circuit   Flow of blood from the heart to body capillaries and back to the heart. One half of double circulation. (Chapter 17)

systole   Portion of the cardiac cycle when the heart is contracting, forcing blood into arteries. (Chapter 17)

systolic blood pressure   Force of blood on artery walls when heart is contracting. Highest blood pressure in arteries. (Chapters 3, 17)

T cell receptor   Antigen receptors found in T cells. Unlike antibodies, these receptors are not secreted. (Chapter 18)

T lymphocytes (T cells)   Immune-system cells that develop in the thymus gland and recognize and respond to body cells that have gone awry, such as cancer cells or cells that have been invaded by viruses, as well as transplanted tissues and organisms such as fungi and parasitic worms. T cells provide an immune response called cell-mediated immunity. (Chapter 18)

T4 cell   Immune-system cell that helps coordinate the body's specific response to a pathogen; also called a helper T cell. (Chapters 10, 18)

tap root   A single deep primary root, common among the root systems in eudicots. (Chapter 22)

Taq **polymerase**   A polymerase enzyme that can withstand high temperatures and is used in PCR reactions. (Chapters 7, 12)

target cells   Cells that respond to regulatory signals such as hormones. (Chapter 19)

telomerase   An enzyme that helps prevent the degradation of the tips of chromosomes, active during development and sometimes reactivated during cancer. (Chapter 5)

telophase   Stage of mitosis during which the nuclear envelope forms around the newly produced daughter nucleus, and chromosomes decondense. (Chapter 5)

temperate forest   Biome dominated by deciduous trees. (Chapter 15)

temperature   A measure of the intensity of heat or kinetic energy. (Chapters 4, 15)

temporal bones   Paired bones found near the temple. (Chapter 19)

temporal isolation   Reproductive isolation between populations maintained by differences in the timing of mating or emergence. (Chapter 11)

temporal lobe   Part of the cerebral hemisphere that processes auditory and visual information, memory and emotion. (Chapter 21)

tension   Negative water pressure. (Chapter 23)

terminal bouton   Knoblike structure at the end of an axon. (Chapter 21)

terminal bud   Apical meristem on the tip of a stem or branch. (Chapter 22)

testable   Possible to evaluate through observations of the measurable universe. (Chapter 1)

testes   The male gonad, involved in gametogenesis and secretion of reproductive hormones. (Chapters 19, 20)

testicles   See testes. (Chapters 19, 20)

testimonial   Statement made by an individual asserting the truth of a particular hypothesis because of personal experience. (Chapter 1)

testosterone   Masculinizing hormone secreted by the testes. (Chapters 19, 20)

thalamus   Main relay center between the spinal cord and the cerebrum. (Chapter 21)

theory of evolution   Theory that all organisms on earth today are descendants of a single ancestor that arose in the distant past. See also evolution. (Chapters 1, 2, 9)

theory   See scientific theory. (Chapters 1, 9)

therapeutic cloning   Using early embryos as donors of stem cells for the replacement of damaged tissues and organs in another individual. (Chapter 8)

thermoregulation   The ability to maintain a body temperatures within a narrow range. (Chapter 16)

thickening meristem   Dividing tissue near the apical meristem in monocots that increases stem girth. (Chapter 22)

thigmotropism   Growth in response to contact with a solid object. (Chapter 23)

thylakoids   Flattened membranous sacs located in the chloroplast stroma that function in photosynthesis. (Chapters 2, 4)

thymine   Nitrogenous base in DNA, a pyrimidine. (Chapters 2, 4, 5, 8)

thymus   An endocrine gland located in the neck region that helps establish the immune system. (Chapters 18, 19)

thyroid gland   An endocrine gland in the neck region that stimulates metabolism and regulates blood calcium levels. (Chapter 19)

Ti plasmid   Tumor-inducing plasmid used to genetically modify crop plants. (Chapter 8)

tilling   Turning over the soil to kill weed seedlings, with the goal of removing competitors before a crop is planted. (Chapter 22)

tissue   A group of cells with a common function. (Chapter 16)

tongue   A muscular structure in the mouth that has taste buds that help you taste food. It aids in breaking down food for the digestive process. (Chapter 16)

totipotent   The ability of a cell to specialize into any cell type. (Chapter 8)

toxin   Causes disease when produced by bacterial cells. (Chapters 17, 18)

trachea   Air passage from upper respiratory system into lower respiratory system. Also called "windpipe." (Chapters 16, 17)

tracheids   Narrow xylem vessels with pitted walls. (Chapter 23)

transcription   Production of an RNA copy of the protein coding DNA gene sequence. (Chapters 8, 10, 18)

transfer RNA (tRNA)   Amino-acid-carrying RNA structure with an anticodon that binds to a mRNA codon. (Chapter 8)

transformation hypothesis   Hypothesis about the origin of living organisms stating that each arose separately in the past and has changed over time, but not into new species. (Chapter 9)

transgenic organism   Organism whose genome incorporates genes from another organism; also called genetically modified organism (GMO). (Chapter 8)

translation   Process by which an mRNA sequence is used to produce a protein. (Chapters 8, 10, 18)

translocation   Movement of phloem sap around the body of a plant. (Chapter 23)

transpiration   Movement of water from the roots to the leaves of a plant, powered by evaporation of water at the leaves and the cohesive and adhesive properties of water. (Chapters 4, 23)

trisomy   A chromosomal condition in which three copies of a chromosome exist instead of the two copies of a chromosome normally present in a diploid organism. (Chapter 7)

trophic level   Feeding level or position on a food chain; e.g. producers, primary consumers, etc. (Chapter 14)

trophic pyramid   Relationship among the mass of populations at each level of a food web. (Chapter 14)

trophoblast   The outer layer of a developing blastocyst that supplies nutrition to the embryo. (Chapter 20)

tropical forest   Biome dominated by broad-leaved, evergreen trees; found in areas where temperatures never drop below the freezing point of water. (Chapter 15)

tropism   In plants, directional growth. (Chapter 23)

tuber   Underground storage structure in plants, made up of modified stem. (Chapter 22)

tumor   Mass of tissue that has no apparent function in the body. (Chapter 5)

tumor suppressors   Proteins that stop tumor formation by suppressing cell division, but when mutated lead to increased likelihood of cancer. (Chapter 5)

tundra   Biome that forms under very low temperature conditions. Characterized by low-growing plants. (Chapter 15)

turgid   In plants, cells that are filled with enough water so that the cell walls are deformed. (Chapter 22)

undifferentiated   A cell that is not specialized. (Chapter 8)

unicellular   Made up of a single cell. (Chapter 12)

uniform distribution   Occurs when individuals in a population are disbursed in a uniform manner across a habitat. (Chapter 13)

uniformitarianism   The principle that processes occurred in the past in the same way and at the same rate as they occur currently. (Chapter 9)

unsaturated fatty acid   Fatty acid with many carbon–carbon double bonds. (Chapter 3)

unsustainable   Relating to practices that may compromise the ability of future generations to support a large human population with an adequate quality of life. (Chapter 22)

uracil   Nitrogenous base in RNA, a pyrimidine. (Chapters 2, 6, 8)

urban sprawl   The tendency for the boundaries of urban areas to grow over time as people build housing and commercial districts farther and farther from an urban core. (Chapter 15)

ureter   Tube that delivers urine from a kidney to the bladder. (Chapter 17)

urethra   Urine-carrying duct that also carries sperm in males. (Chapters 17, 20)

urine   Liquid expressed by the kidneys and expelled from the bladder in mammals, containing the soluble waste products of metabolism. (Chapter 17)

uterus   Pear-shaped muscular organ in females that can support pregnancy and that undergoes menstruation when its lining is shed. (Chapter 20)

vaccine   A preparation of a weakened or killed pathogen, or portion a pathogen that will stimulate an immune response in a recipient thus facilitating the mounting of a longer term defense against the pathogen. (Chapter 18)

vacuole   A membrane enclosed sac in a plant cell that functions to store many different substances. (Chapters 2, 16, 22, 23)

vagina   Tubular structure in females located between the cervix and vulva. (Chapter 20)

valence shell   The outermost energy shell of an atom containing the valence electrons which are most involved in the chemical reactions of the atom. (Chapter 2)

variable number tandem repeats (VNTRs)   DNA sequences that vary in number to make the restriction fragments in a variety of lengths. They are nucleotide sequences that all organisms carry but in different numbers. These VNTRs are used during the process of DNA fingerprinting. (Chapter 7)

variance   Mathematical term for the amount of variation in a population. (Chapter 6)

variant   An individual in a population that differs genetically from other individuals in the population. (Chapter 10)

variety   Subgroup of a species with unique traits relative to other subgroups of the species. Equivalence of this term to biological race or subspecies is disputed by biologists. (Chapter 11)

vas deferens   Duct in males that carries sperm from the epididymis to the urethra. (Chapter 20)

vascular cambium   Meristematic tissue that forms in vascular bundles of dicot roots and stems and permits secondary growth. (Chapter 22)

vascular system   Plant tissue system responsible for the delivery of water and dissolved solutes throughout the plant body. (Chapter 17)

vascular tissue   Cells that transport water and other materials within a plant. (Chapters 12, 22)

vector   A shuttle that facilitates the movement of genes into cells during genetic engineering. (Chapter 18)

vegetative organs   Plant organs that are not involved in sexual reproduction. (Chapter 22)

vegetative reproduction   Asexual cloning of plants. (Chapter 20)

veins   Vessels that carry blood from the body tissues back to the heart. (Chapter 17)

ventricle   Chamber of the heart that pumps blood from the heart to the lungs or systemic circulation. (Chapter 17)

vernalization   Physiological changes that occur in a plant or seed that prepare it to begin growth. (Chapter 22)

vertebrae   Bones of the spine. (Chapter 21)

vertebrates   Animals with backbones. (Chapters 7, 12)

vesicles   Membrane-bounded sac-like structures. In neurons, these structures are found in the terminal bouton and store neurotransmitters. (Chapter 21)

vessel elements   Wide xylem cells with perforated ends, found only in angiosperms. (Chapter 23)

vestigial traits Modified with no, or relatively minor function compared to the function in other descendants of the same ancestor. (Chapter 9)

villi Small fingerlike projections on the in side of the small intestine. (Chapter 16)

viral envelope Layer formed around some virus protein coats (capsids) that is derived from the cell membrane of the host cell and may also contain some proteins encoded by the viral genome. (Chapter 18)

virus Infectious intracellular parasite composed of a strand of genetic material and a protein or fatty coating that can only reproduce by forcing its host to make copies of it. (Chapters 1, 8, 10, 18)

vitamin Organic nutrient needed in small amounts. Most vitamins function as coenzymes. (Chapter 3)

voluntary muscle Muscle that requires conscious thought. (Chapter 16)

vulva The outer portion of the female external genitalia including the labia majora, labia minora, clitoris, vaginal and urethral openings. (Chapter 20)

wastewater Liquid wastes produced by humans. (Chapter 15)

water One molecule of water consists of one oxygen and two hydrogen atoms. (Chapters 2, 3, 4, 15, 22, 23)

weather Current temperature and precipitation conditions. (Chapter 15)

weeds Common term for non-preferred plants. (Chapter 22)

wetlands Biome characterized by standing water, shallow enough to permit plant rooting. (Chapter 15)

white blood cells General term for cells of the immune system. (Chapter 17)

white matter Nervous system tissue made of myelinated cells. (Chapter 21)

whole foods Foods that have not undergone processing. (Chapter 3)

wood Xylem cells produced by secondary growth of stems and roots. (Chapter 22)

X inactivation The inactivation of one of two chromosomes in the XX female. (Chapter 7)

X-linked genes Genes located on the X chromosome. (Chapter 7)

xylem Plant vascular tissue, dead at maturity, that carries water and dissolved minerals in a one-way flow from the roots to the shoots of a plant. (Chapters 22, 23)

xylem sap Water and dissolved minerals flowing in the xylem vessels. (Chapter 23)

yeast Single-celled eukaryotic organisms found in bread dough. Often used as model organisms and in genetic engineering. (Chapters 4, 9, 12)

Y-linked genes Genes located on the Y chromosome. (Chapter 7)

Z line (Z discs) The borders of a sarcomere in muscle. (Chapter 19)

zona pellucida The substance surrounding an egg cell that a sperm must penetrate for fertilization to occur. (Chapter 20)

zoologists Scientists who specialize in the study of animals. (Chapter 12)

zygote Single cell resulting from the fusion of gametes (egg and sperm). (Chapters 5, 7, 20)

Credits

12.2 D. Cavagnaro/DRK Photo 12.3a Michael Viard/Peter Arnold, Inc. 12.3b Mauro Fermariello/SPL/Science Source/Photo Researchers, Inc. 12.6 J. William Schopf, University of California at Los Angeles 12.7a Dr. Dennis Kunkel/Visuals Unlimited 12.7b Carolina Biological Supply/Visuals Unlimited 12.7c Marli Miller/Visuals Unlimited Table 12.2(1) M. I. Walker/Photo Researchers, Inc. Table 12.2(2) Dr. Dennis Kunkel/Visuals Unlimited Table 12.2(3) Wim van Egmund/Visuals Unlimited Table 12.2(4) Bill Beatty/Visuals Unlimited Table 12.2(5) Dee Breger/ Photo Researchers, Inc. Table 12.2(6) Hal Beral/Corbis Bettmann Table 12.2(7) Wim van Egmund/Visuals Unlimited 12.9a Dr. Jeffrey Lord 12.9b Professor J.L. Kerwin 12.9c David M. Phillips/Visuals Unlimited 12.9d Walter H. Hodge/Peter Arnold, Inc. 12.10 Chip Clark/National Museum of Natural History © 2003 Smithsonian Institution 12.11 Wayne Lynch/DRK Photo Table 12.3(1) Hal Beral/Visuals Unlimited Table 12.3(2) Hiroya Minakuchi/Minden Pictures Table 12.3(3) Robert Calentine/Visuals Unlimited Table 12.3(4) (c) Dr. Richard Kessel & Gene Shih/Visuals Unlimited Table 12.3(5) David B. Fleetham/Visuals Unlimited Table 12.3(6) E.R. Degginger/Photo Researchers, Inc. Table 12.3(7) Tom McHugh/Photo Researchers, Inc. Table 12.3(8) Randy Morse/Animals Animals/Earth Scenes Table 12.3(9) Tom McHugh/Photo Researchers, Inc. 12.12a Michael Fogden/DRK Photo 12.12b David Wrobel/Visuals Unlimited 12.12c David Barron/Animals Animals/Earth Scenes 12.13 Gregory Ochocki/Photo Researchers, Inc. Table 12.4(1) Dr. M.F. Brown/Visuals Unlimited Table 12.4(2) Corbis RF Table 12.4(3) H. Zetti/Zefa/Corbis/Bettmann 12.15 Photo Researchers, Inc. Table 12.5(1) John Lemker/Animals Animals/Earth Scenes Table 12.5(2) Clay Perry/Corbis, NY Table 12.5(3) (c) Prenzel Photo /Animals Animals/Earth Scenes Table 12.5 (4) Getty Images/Digital Vision 12.16a Ed Reschke/Peter Arnold, Inc. 12.16b Frans Lanting/Minden Pictures 12.16c Murray, Patti/ Animals Animals/Earth Scenes 12.18a George D. Lepp/ Photo Researchers, Inc. 12.18b Joe McDonald/Visuals Unlimited 12.18c Bill Dyer/Photo Researchers, Inc. 12.20 Alison Wright/CORBIS CHAPTER 13 Opener (L) Reto Stockli/NASA Goddard Laboratory for Atmospheres Opener (R) Top AP/Wide World Photos Middle Pat Hamilton/Reuters New Media, Inc./CORBIS 13.1a Peter Haigh/Almany Images 13.1b Tom Payne/Almany Images 13.1c Carr Clifton/Minden Pictures 13.5a Dung Vo Trung/Corbis/Symga 13.5b Robert Pickett/CORBIS 13.5c Michael Gadomski/Animals Animals/Earth Scenes 13.5d Steve Wolper/DRK Photo 13.8 Wolfgang Kaehler/CORBIS CHAPTER 14 Opener (L) Tupper Ansell Blake/WSFWS Opener (R) Top AP/Wide World Photos Middle © Stephan Anderson/The Nature Conservancy Bottom AP/Wide World Photos 14.1 Prentice Hall School Division 14.5 Konrad Wothe/Minden Pictures 14.7a NASA Space Center/Visuals Unlimited 14.7b Buddy Mays/CORBIS 14.7c David Dennis/Animals Animals/Earth Scenes 14.7d Belinda Wright/DRK Photo 14.7e Allen Blake Sheldon/Animals Animals/Earth Scenes 14.7f Martha Holmes/Nature Picture Library 14.9a David Muench/Corbis, NY 14.9b Professor John Doebley/University of Wisonsin 14.9c Scott Bauer/USDA ARS 14.11 Courtesy of www.plantworksuk.co.uk and www.friendlyfungi.co.uk 14.12 M. Harvey/DRK Photo 14.13a Ron Austin/Photo Researchers, Inc. 14.13b Jerry Gildemeister 14.15a Siri Stafford/Getty Images, Inc.—Stone Allstock 14.15b Peter Wieman/Animals Animals/Earth Scenes 14.17a,b Dave Hansen, University of Minnesota Agricultural Experiment Station 14.18 Nik Wheeler/Corbis, NY 14.20a Photo by Rollie White 14.20b Kenneth W. Fink/Photo Researchers, Inc. 14.21 © Steven Holt/VIREO 14.24a Andrew Syred/Science Photo Library/Photo Researchers, Inc. 14.24b Astrid & Hanns-Frieder Michler/Science Photo Library/Photo Researchers, Inc. CHAPTER 15 Opener (L) Michael Newman/PhotoEdit

Opener (R) Top Tom & Pat Leeson/Photo Researchers, Inc. Middle Wes Thomson/Corbis, NY Bottom Don Farrall/Getty Images-Photodisc 15.3 V. Rumyantsev, Crimean Astrophysical Observatory 15.6 (c) Frances Roberts/Alamy 15.9a Getty Images, Inc.—Botanica 15.9b Ross M. Horowitz/Getty Images, Inc. 15.9c Bryan and Cherry Alexander/Almany Images 15.10 Geoff Bryant/Photo Researchers, Inc. 15.11 Jesse Allan/NASA/Goddard Space Flight Center 15.12 Hans Dieter Brand/Frank Lane Picture Agency/Corbis, NY 15.13 B. Runk/S. Schoenberger/Grant Heilman Photography, Inc. 15.14 George Reszeter 15.15 Walt Anderson/Visuals Unlimited E15.1a (c) Jan Baks/Alamy E15.1b Raymond Gehman/National Geographic/Getty Images, Inc. E15.1c Getty—National Geographic Society 15.16 Kennan Ward/Corbis, NY 15.17 Getty—National Geographic Society 15.18 Adam Jones/Photo Researchers, Inc. 15.19 Getty Images Inc—Botanica 15.20 Stephen Krasemann/Photo Researchers, Inc. 15.21a Getty—National Geographic Society 15.22c Jerome Wexler/Photo Researchers, Inc. 15.23 Lowell Georgia/Corbis, NY 15.23b Index Stock Imagery, Inc. 15.24 Cisca Castelijns/Foto Naturals /Minden Pictures 15.25 David Muench/Getty Images, Inc.-Stone Allstock 15.26 Jeffrey Rottman/Corbis, NY 15.27 Millard H. Sharp/Photo Researchers, Inc. 15.28 © 1990. American Association for the Advancement of Science 15.29 Steven Wayne Rotsch 15.30 David R. Frazier Photolibrary, Inc. 15.31 James Leynse/Corbis, NY 15.32 Steven Chiang/Getty Images 15.33 Courtesy of Virginia Borden CHAPTER 16 Opener (L) Jeff Forster Opener (R) Top James Holmes/Science Photo Library/Photo Researchers, Inc. Middle Tony Cordoza/Index Stock Imagery, Inc. Bottom Age Fotostock/SuperStock, Inc. 16.6a © Children's Hospital & Medical Center/CORBIS All Rights Reserved 16.6b-c Dr. Fred Hossler/Visuals Unlimited CHAPTER 17 Opener (L) Christian Zachariasen/PhotoAlto/Getty Images Opener (R) Top William Campbell/Corbis Sygma Middle David Young-Wolff/Getty Images, Inc./Stone-Allstock Bottom Mark Gibson/Corbis, NY E17.1a Tom Adams/Visuals Unlimited E17.1b Jef Meul/Foto Natura /Minden Pictures E17.1c Robert Myers/Visuals Unlimited E17.1d Patrice Ceisel/Visuals Unlimited E17.e Rod Planck/Photo Researchers, Inc. E17.f Pete Oxford/Minden Pictures 17.3a Custom Medical Stock Photo, Inc. 17.3b James A. Sugar/National Geographic Society/Getty 17.4b Eye of Science/Photo Researchers, Inc. 17.7 Science Photo Library/Photo Researchers, Inc. 17.9 Arthur Glauberman/Photo Researchers, Inc. 17.10a St. Bartholomew's Hospital/Science Photo Library/Photo Researchers, Inc. 17.10b Simko/Visuals Unlimited 17.10c Dr. Richard Kessel & Dr. Randy Kardon/Visuals Unlimited 17.10d NIBSC/Photo Researchers, Inc. 17.11 CNRI/Photo Researchers, Inc. 17.17a Photo Researchers, Inc. 17.17b Alfred Pasieka/Science Photo Library/Photo Researchers, Inc. CHAPTER 18 Opener (L) TEK/SPL/Photo Researchers, Inc. Opener (R) Top Dr. Robert J. Higgins/UC Davis Middle AP/Wide World Photos Bottom SIPA Press Table 18.1(1) Gary Gaugler/Visuals Unlimited Table 18.1(2) Oliver Meckes/Photo Researchers, Inc. Table 18.1(3) Tina Carvalho/Visuals Unlimited E18-1 Michael Newman/PhotoEdit 18.4(1) CNRI/Science Photo Library/Photo Researchers, Inc. 18.4(2) Dr. Linda Stannard, UCT/SPL/Photo Researchers, Inc. 18.6 Dr. Linda Stannard, UCT/Photo Researchers, Inc. 18.7a Arthur Siegelman/Visuals Unlimited 18.7b Dr. Richard Kessel & Dr. Gene Shih/Visuals Unlimited 18.10 AP/Wide World Photos 18.11a Scott Camazine/Photo Researchers, Inc. 18.11b David Spears/Corbis, NY CHAPTER 19 Opener (L) Steve Burmeister/Steve Burmeister Opener (R) Top Josh Meltzer/Duluth News Tribune Middle Robert Laberge/Getty Images Bottom Courtesy IMS Photo by Ron McQueeney/Indy Racing League E19.1a Duncan McEwan/Nature Picture Library E19.1b Hal Beral/Corbis, NY 19.11b Dr. Don W. Fawcett/Visuals Unlimited 19.13 Pete Soloutos/ZEFA/SuperStock CHAPTER 20

Opener (L) Carl & Ann Purcell/Corbis, NY Opener (R) Top Natalie Fobes/Corbis, NY Middle David R. Frazier Photolibrary, Inc./Almany Images Bottom Comstock Royalty Free Division 20.1a Copyright © Azure Computer & Photo Services/Animals Animals Earth Scenes—All Rights Reserved 20.1b Tom Branch/Photo Researchers, Inc. 20.1c © Constantinos Petrinos/nature-pl.com 20.2 Allen Blake Sheldon/Animals Animals/Earth Scenes 20.3 Copyright Sea Studios Foundation/WGBH Film & Video Resource Center 20.7b Eye of Science/Photo Researchers, Inc. 20.8b Getty Images/Stone-Allstock Table 20.2(1) Don Farrell/Getty Images/Photodisc Table 20.2(2) Andrew McClenaghan/Science Photo Library/Photo Researchers, Inc. Table 20.2(3) Scott Camazine/Photo Researchers, Inc. Table 20.2(4) SuperStock, Inc. Table 20.2(5) Conceptus Incorporated Table 20.2 (6) Frank LaBua/Pearson Education Inc./PH College Table 20.2(7) Garo/Photo Researchers, Inc. Table 20.2(8) Scott Camazine/Sue Trainor/Science Photo Library/Photo Researchers, Inc. Table 20.2(9) Scott Camazine and Sue Trainor/Photo Researchers, Inc. Table 20.2(10) Gary Watson/Science Photo Library/Photo Researchers, Inc. Table 20.2(11-13) Joel Gordon Photography Table 20.2(14) Don Farrell/Getty Images/Photodisc 20.13 Steve Allen/Science Photo Library/Photo Researchers, Inc. 20.14 G. Moscoso/Photo Researchers. Inc. CHAPTER 21 Opener (L) Michael Newman/PhotoEdit Opener (R) Top From Newsweek 3/10/96 © 1996 Newsweek, Inc. All rights reserved. Reprinted by permission. Middle David Young-Wolff/PhotoEdit Bottom Dex Images/CORBIS E21.a Pascal Goetgheluck/Photo Researchers, Inc. E21.b Science Pictures Limited/Photo Researchers, Inc. E21.c David M. Dennis/Animals Animals/Earth Scenes E21.d © G.T. Andrewartha/Frank Lane Picture Agency/CORBIS. All Rights Reserved 21.6 Pascal Goetgheluck/Photo Researchers, Inc. 21.7a Ralph Hutchings/Visuals Unlimited 21.11a SPL/Photo Researchers, Inc. Table 21.2(1) Don Farrell/Getty Images, Inc./Photodisc Table 21.2(2) David Hoffman/Almany Images Table 21.2(3) Corbis Table 21.2(4) Corbis Table 21.2(5) Scott Houston/Corbis, NY Table 21.2(6) George Post/Science Photo Library/Photo Researchers, Inc. Table 21.2(7) David Buffington/Getty Images—Photodisc Table 21.2(8) Age Fotostock/Superstock, Inc. Table 21.2(9) ImageState/Almany Images CHAPTER 22 Opener (L) USDA/NRCS/Natural Resources Conservation Service Opener (R) Top Corbis/Bettmann Middle Seawifs/NASA John F. Kennedy Space Center Bottom Getty Images/Digital Vision 22.2a Professor John Doebley/University of Wisconsin 22.2b Ross Durant/Picture Arts Corporation 22.5 Dr. Richard Kessell & Dr. Gene Shih/Visuals Unlimited 22.7 Photo by Dr. Steve O'Hair. Courtesy of Professor Stephen M. Olson, North Florida Research and Education Center 22.8 Barry Runk/Grant Heilman Photography 22.9(4) Corbis 22.9(6) C. Squared Studios/Photodisc/Getty Images E22.1b B. Runk/S. Schoenbreger/Grant Heilman Photography, Inc. E22.1c David M. Dennis/Animals Animals/Earth Scenes E22.1d Ed Reschke/Peter Arnold, Inc. 22.11a Jerome Wexler/Photo Researchers, Inc. 22.11b Carson Baldwin, Jr./Animals Animals/Earth Scenes 22.11c Anthony Mercieca/Photo Researchers, Inc. 22.11d Merlin Tuttle/Photo Researchers, Inc. 22.11e Heather Angel/Natural Visions/Almany Images 22.12a Steve Bloom Images/Almany Images 22.12b Krimmer/ PictureQuest 22.12c Scott Camazine/Almany Images 22.12d Bruce Coleman, Inc./Almany Images 22.14a Jim Strawser/Grant Heilman Photography 22.14b Dr. J. Burgess/Photo Researchers, Inc. 22.14c B. Runk/S. Schoenberger /Grant Heilman Photography 22.14d Adam Hart-Davis/Photo Researchers, Inc. 22.15 Ed Reschke/Peter Arnold, Inc. 22.16 Patrick Lynch/Science Photo Library/Photo Researchers, Inc. 22.19a Wayne P. Armstrong, Palomar College 22.19b AP/Wide World Photos 22.22a James L. Amos/Corbis/Bettmann 22.26a Arthur C. Smith, III/Grant Heilman Photography 22.26b Science Photo Library/Photo Researchers, Inc.

Index

Page references that refer to figures are in **bold**. Page references that refer to tables are in *italics*.

Lysergic acid diethylamide (LSD), 332
   and nervous system, *565*
Lysine, **48**
Lysosomes, *37, 39*

M

Macromolecules
   metabolism of, **86**
   on other planets, 34
   structure and function of, 29–30
Macronutrients, 46–50, 589, *589*
Macrophages, 473, **482**, 483, 489
Mad cow disease, 469, 491
Magnesium, 52, **52**, **589**
Malaria, 476
   resistance to, 150
Male condom, *534*
Male infertility, 528
Male reproductive anatomy, 520–521, **521**
Male sterilization, *533*
Malignant, 103
Malignant ovarian tumor, **121**
Mammal forelimbs
   homology of, **238**
Mandible, 505
Manganese, **589**
Manioc, **576**
Marijuana
   and nervous system, *565*
Marine biomes, 411–412
Mark-recapture method, 346–347
Marrow, 456, 484
Mass extinction, **361**, 361–362
Mass number, 27
Matrix, 423
Mattesia oryzaephili, **327**
MDMA
   and nervous system, *565*
Mean, 154, **154**
Meat, 425
   production, **597**
Mechanical isolation, 285
Mechanical work, 78
Medulla oblongata, 555, **556**
Meiosis, 123–133, **172**, 174, 186, **292**, 523
   and DNA fingerprinting, 185–186
   interphase, **127**
   vs. mitosis, **132**
Meiosis I, 127
Meiosis II, 127–128
Memory cells, 488, **489**
Mendel, Johann Gregor, 147, **147**
Mendelian genetics, 146–153
   extensions of, 168–172, *170*
Menopause, 499, 528
Menstrual cycle, 530, **530**, 531–532
   and drinking water, 528
   feedback during, **531**
Menstruation, 530, 532
Meristematic tissue, 575
Mesoderm, 537

Mesozoic era, *335*
Messenger RNA (mRNA), 197
Metabolic rate, 56–57
   and age, 56
   basal, 56
Metabolism, 24, 54–57
Meta female, *175*
Metaphase, 109
Metastasis, 103, **104**
Methylphenidate
   Ritalin, 547
Microbes, 324, 470
Microbiologists, 324
Micronutrients, 46, 50–52, 589, *589*
Microtubules, 109
Microvilli, 435
Micturation, 463
Midbrain, 555, **556**
Migration, **409**
   leading to speciation, **289**
Milk production in cows
   artificial selection increasing, **156**
Millennium man, 247
Minerals, 52, **52**
Minipill, *533*
Miss America, **61**
Missing link, **248**, 248–249
Mitochondria, 39, 81, **81**
Mitochondrion, *36*
Mitosis, 108–112
Model organisms, 10, **10**, 215
Modern agriculture
   causing environmental damage,
      593–595
   reducing environmental impact of, *598*
Modern humans
   history, 297–298
   origins of, **297**
Modern meat production, **597**
Modern organisms
   origin hypotheses, **233**, **250**
Modified stem, **577**
Mold, 332
Molecular clock, 242
Molecule, 25
Mollusca, *329*
Molybdenum, **589**
Monera, *321*
Monocot, 573
   root system, **575**
   seeds, **573**
   stems, **584**
   wood, **585**
Monoculture, 591, **592**
Monomers, **47**, **57**
Monosaccharides, **61**
Monosomy, 174
Monozygotic twins, 145, **146**
Montagnier, Luc, 259
Morphological species
   comparison of, *296*, 296–297
Moss, 606, **606**
   gametophyte, **578**

Motor neurons, 549, *551*
mRNA. *See* Messenger RNA (mRNA)
Mucous membranes, **482**, 482–483
Mulching, 588
Mules, 287, **287**
Multiple allelism, 169, *170*
Multiple drug resistance, 275
Multiple drug resistant HIV, 275–276
Multiple hit model, 117
Multiple sclerosis, 487
Multiplicity, **193**
Murray, Richard, 160, 161
Muscle
   cell, **82**
   contraction, **508**
      sliding-filament model of, 508–509
   contraction of, 506–509
   sex differences in, 509
   structure of, 506–509
   tissues, **425**, 425–426
Muscular dystrophy, *177*
Muscular system, *430*, 506–509
Mussels
   circulatory system of, **447**
Mutant allele, **141**
Mutated Her2 receptor, **115**
Mutation, 113, 141, **141**, 141–142, **269**
Mutualism, 371–372, **372**
Mycologists, 331
Mycteria americana, **338**
Myelination, **558**
Myelin sheath, 558
Myofibrils, 507
Myosin, 425, 508

N

NAD. *See* Nicotinamide adenine
   dinucleotide (NAD)
National Cancer Institute (NCI), 337
Natural experiments, 158
Natural killer cells, 483
Natural resources, 414–415, 415
Natural selection, 227, 256–279, **269**,
   306–308
   acting on inherited traits, 268–269
   does not imply progression, **270**
   fruit flies, 267, **267**
   HIV, 272–273
   in lab, 266–267
   modern understanding of, 267–268
   not resulting in perfection, 269–270
   resulting from current environmental
      conditions, 270–271
   subtleties of, 268–269
   testing, 266–267
   in wild populations, 267
Nature
   innate appreciation of, **378**
   psychological effects of, 378
NCI. *See* National Cancer Institute (NCI)
Neanderthal man, 246

Salt water, **26**
Saltwater, 411–412
Sample, 12
Sampling error, 13
Sarcomeres, 507
SARS. *See* Severe acute respiratory syndrome (SARS)
Saturated fat, 49, **49**
Savannas, 406, **406**
    maintenance, **406**
Saving species, 379–386
Schistosoma, 476
Schistosomiasis, 476
Schwann cells, 558
SCID. *See* Severe combined immunodeficiency (SCID)
Science
    in news, 16–17
        evaluation guide for, *18*
    process of, 2–3
Scientific information
    evaluating, 15–16
Scientific method, 1–19, **4**, *189*
Scientific theory, 6–7, 228
Scrotum, 521
Sea levels, **71**, 76
Sea squirt
    asexual reproduction in, **517**
Secondary consumers, 366
Secondary growth, 582, 584–585
Secondary oocytes, 528
Secondary sources, 16, 17–19
Secondary spermatocyte, 526
Secretion, 463, **464**
Seed, 333, **336**, **577**, 581
    coat, 582
    dispersal units, **581**
Seedling, 582
Selection, 265
Selenium, *53*
Self *versus* nonself, 486–487
Semen, 521
Semilunar valves, 457
Seminal vesicles, 521
Seminiferous tubules, 521
Semipermeable, 38
Sense RNA, 208
Senses, *550*
Sensory neurons, 549, *551*
Sensory receptors, 549
Sepals, 580
Separate types, **233**
Sertoli cells, 526, **527**
Severe acute respiratory syndrome (SARS), 327, 475, **475**
Severe combined immunodeficiency (SCID), 216–217
    gene therapy in, **217**
Sex cells, 517
Sex chromosomes, 124, **124**, 172, 174
Sex determination, **172**, 172–173, *173*
Sex hormones, 496

Sex linkage, 176
Sex-linked genes, 176
Sexually transmitted diseases (STD), 480, *524*
    causing infertility, 523
Sexual reproduction, 105, 517–519
    and environmental contaminants, 519–520
Sexual selection, 312
    effects of, **312**
Shamans, 340–341, **341**
Shared characteristics
    among humans and apes, **237**
Shoot system, 576
Shortnose suckers, 360, 386
Shrublands, 400–401
Sickle-cell allele, 150, **150**, **301**
    in malarial environments, **307**
Sickle-cell anemia
    Punnett square, **152**
Signal transduction, 496
Simon, Julian, 345
Simple diffusion, 446
Sink, 613
Sinoatrial (SA) node, 457
Sister chromatids, 105
Sixth extinction, 360–368
Skeletal muscle, **425**, 426
    structure of, **507**
Skeletal system, *430*, 500–506
Skeleton, 500–502, **501**
    sex differences in, **505**
Skin, **482**, 482–483
    cancer
        risk factors of, *119*
    color
        genetic and environmental influence, 155, **155**
        UV exposure, **308**
Skulls
    vertebrates, **553**
Slowed rate of ripening
    tomatoes, **209**
Sludge, 416
Small intestines, **57**, 434
Small populations
    chance events in, **310**
Small population size, **381**
Smell, *550*
Smog, 417
Smoke
    and cardiovascular disease, 460–462
    effects on cardiovascular system, 455–462
    effects on respiratory system, 444–454
    particles
    and lung function, 452–454
Smoking
    and excretory system, 463–465
Smooth muscles, 425, **425**, 426

Sodium, *52*, 559
Sodium-potassium pump, 559
Soil, 589
    erosion of, 595, *598*
    fertility, 597
    and nutrients, 589
    salinization of, *598*
    structure and development, **589**
Solar irradiance, 393
    on earth's surface, **393**
Solid waste, 416
Solstice, 395
Solute, 25
Solvent, 25
Somatic cell gene therapy, 216
Somatic cells, 123
Somatomammotropin, 540
Sorenstam, Annika, 495
Source, 613
Soybeans
    genetically modified, 211
Spatial isolation, 285
Special creation, **225**, 231
Special senses, 549, *550*
Speciation, 288–289
Species, 40, 226, **235**, 335–341
    definition of, 284, 294
    differences between, **293**
Species and race, 282–314
Species-area curve, 363
Species concepts
    comparison of, *296*
Species interactions, *375*
Species overexploitation, **365**, 367–368
Specific defense, 483
Specificity, 54, 484
Sperm, **136**, **138**, 139, 143, **143**, **146**
    developing, 526
    with egg, **529**
    production of, 526
Spermatids, 526
Spermatocyte, 526
Spermatogenesis, 526–527, **527**
Spermatozoa, 526
Sperm counts
    and drinking water, 528
Spermicides, *534*
Spinal cord, 549
Spinal nerves, **549**
Spleen, 484
Sponges
    asexual reproduction in, **517**
Spongiform encephalopathies, 490
Spongy bone, 503
Spores, 331, 472, 578
Sporophytes, 578, **579**
Spruce bark beetle, 87
SRY gene, 177
St. Anthony's Fire, 332
Stabilizing selection, 270, **271**
Stamen, 579, **581**
Staphylococcus aureus, 332